9th International Conference on Inertial Fusion Sciences and Applications (IFSA 2015)

Journal of Physics: Conference Series
Volume 717

Seattle, Washington, USA
20 – 25 September 2015

Editors:

William H. Goldstein
Bruce Hammel

ISBN: 978-1-5108-2455-3
ISSN: 1742-6588

Printed from e-media with permission by:

Curran Associates, Inc.
57 Morehouse Lane
Red Hook, NY 12571

Some format issues inherent in the e-media version may also appear in this print version.

Printed by Curran Associates, Inc. (2016)

For permission requests, please contact the Institute of Physics
at the address below.

Institute of Physics
Dirac House, Temple Back
Bristol BS1 6BE UK

Phone: 44 1 17 929 7481
Fax: 44 1 17 920 0979

techtracking@iop.org

Additional copies of this publication are available from:

Curran Associates, Inc.
57 Morehouse Lane
Red Hook, NY 12571 USA
Phone: 845-758-0400
Fax: 845-758-2633
Email: curran@proceedings.com
Web: www.proceedings.com

Table of contents

Volume 717

9th International Conference on Inertial Fusion Sciences and Applications (IFSA 2015)20–25 September 2015, Seattle, Washington, USA

Accepted papers received: 29 April 2016 Published online: 26 May 2016

Preface

Papers

Keynotes

Teller Lecture

Physics of Inertial Fusion

High Energy Density Physics & Applications

Lasers, Particle Beams, Fusion Technology

Teller Lecture 2

Preface

The Ninth International Conference on Inertial Fusion Science and Applications (IFSA) was held on September 20-25, 2015 at the Hyatt Regency Bellevue on Seattle's Eastside, Washington, U.S.A. The event was hosted by the University of California and was organized by the Lawrence Livermore National Laboratory. It brought together more than 370 participants from 16 countries.

The goal, as for all previous IFSA Conferences, was to bring together scientists in the fields of inertial fusion science and high-energy-density physics, and their applications. Three hundred twenty seven papers were presented emphasizing the science of high-energy and high-intensity laser, pulsed-power, and particle-beam interactions with matter, the associated high-energy-density physics, and their application to fusion concepts. Results presented included theory, modeling, and experimental results from facilities worldwide.

In recent years, significant advances have been made in high-energy-density science using lasers, Z-pinches, and particle beam systems with dramatic technical achievements in areas such as central-hot-spot ignition, fast and impulse ignition, material properties at extreme conditions, warm dense matter, particle acceleration and laser-plasma interactions. For the first time in the laboratory, x-ray driven ignition experiments, performed at the National Ignition Facility (NIF) in the United States, have exhibited self-heating. In the month following the Conference, the first plasma experiments were performed at Laser Mégajoule (LMJ) in France, and ignition scale projects are under way in China and Russia. Other approaches, such as magnetic compression on the Z-machine at Sandia National Laboratories and direct drive experiments at the University of Rochester, have produced exciting new results which were reported on at the Conference.

Second-generation petawatt short-pulse laser systems such as the highest-energy petawatt laser systems LFEX (FIREX) in Japan, OMEGA-EP in the United States, PETAL in France, and Vulcan PW in the United Kingdom, have allowed the study of novel approaches to producing and probing high-energy-density matter and exploring alternative ignition schemes. These scientific studies contribute to the physics understanding needed to develop inertial fusion energy as a commercial energy source and were also reported on at the Conference.

Third-generation rep-rated and high-energy petawatt lasers are coming on line for strong field science studies and have opened up a new field of relativistic plasma physics. The extreme states achieved in these laser-plasma interactions are relevant to high-intensity x-ray applications, material processing, novel accelerators, and laboratory astrophysics.

Recent activities and research in all these fields were discussed in 327 presentations, including three Keynote lectures, two Teller lectures, 21 plenary talks, 135 orals, and 166 posters. The proceedings collected in this volume underwent a rigorous peer-review process. The Editor is thankful to all the authors who agreed to act as referees on papers written by their fellow authors, and who worked both hard and diligently to complete this task in record time. We also thank the Technical Program Committee members who helped make this Ninth IFSA Conference possible.

The American Nuclear Society Teller Award was bestowed during IFSA2015 on two key contributors to ICF science, Hiroshi Azechi and Jie Zhang. In addition, prizes were awarded to four students for their excellent presentations.

On behalf of the various Committees, we extend our appreciation to everyone who helped make IFSA2015 a great success.

William H. Goldstein

Bruce Hammel

Peer review statement

All papers published in this volume of *Journal of Physics: Conference Series* have been peer reviewed through processes administered by the proceedings Editors. Reviews were conducted by expert referees to the professional and scientific standards expected of a proceedings journal published by IOP Publishing.

Journal of Physics: Conference Series **717** (2016) 011003 doi:10.1088/1742-6596/717/1/011003

Members of IFSA 2015 Governing Committee

General Co-Chairs

William H. Goldstein, Lawrence Livermore National Laboratory (LLNL), USA

Patrick Mora, Institut Laser et Plasmas (ILP), France

Hiroshi Azechi, Institute of Laser Engineering (ILE), Osaka University, Japan

Organizing Co-Chairs

John Edwards, LLNL, USA

Sylvie Jacquemot, Laboratoire pour l'Utilisation des Lasers Intenses (LULI), France

Ryosuke Kodama, ILE, Osaka University, Japan

Teller Award Chair

Robert Kauffman, LLNL, USA

Local Organizing Committee (LOC)

LOC Chair

Wayne Meier, LLNL, USA

LOC Co-Chair

Trina Voelker, LLNL, USA

LOC Members

Don Haynes, Los Alamos National Laboratory (LANL), USA

Mila Shapovalov, LLNL, USA

Dan Sinars, Sandia National Laboratories (SNL), USA

Steve Obenschain, Naval Research Laboratory (NRL), USA

Tom Schenkel, Lawrence Berkeley National Laboratory (LBNL), USA

Abbas Nikroo, LLNL, USA

David Meyerhofer, University of Rochester/Laboratory for Laser Energetics (UR/LLE), USA

Journal of Physics: Conference Series **717** (2016) 011003 doi:10.1088/1742-6596/717/1/011003

Technical Program Committee (TPC)

TPC Co-Chairs

Bruce Hammel, LLNL, USA

Dimitri Batani, Centre Lasers Intenses et Applications (CELIA), France

Hiroyuki Shiraga, ILE, Osaka University, Japan

TPC Members

Stefano Atzeni, Sapienza University of Rome, Italy

John Barnard, LLNL, USA

Farhat Beg, University of California, San Diego (UC San Diego), USA

Hongbo Cai, Institute of Applied Physics and Computational Mathematics (IAPCM), China

Debbie Callahan, LLNL, USA

Benoit Canaud, Commissariat a l`Energie Atomique (CEA), France

Jean Christophe Chanteloup, LULI, France

Jerry Chittenden, Imperial College, UK

Dan Clark, LLNL, USA

Gilbert Collins, LLNL, USA

Tom Cowan, Forschungszentrum Dresden-Rossendorf (FZD), Germany

Sylvie Depierreux, CEA, France

Jean-Michel Di Nicola, LLNL, USA

Juan Fernandez, LANL, USA

David Fittinghoff, LLNL, USA

Shinsuke Fujioka, ILE, Osaka University, Japan

Pascal Gauthier, CEA, France

Leonida Gizzi, Laboratoro Laser Intensi (ILIL), Italy

Gianluca Gregori, University of Oxford, UK

Gary Grim, LLNL, USA

Cristina Hernandez-Gomez, Central Laser Facility, STFC Rutherford Appleton Laboratory, UK

Denise Hinkel, LLNL, USA

Junji Kawanaka, ILE, Osaka University, Japan

Joe Kilkenny, General Atomics (GA), USA

Ondrej Klimo, Institute of Physics, Academy of Sciences, Czech Republic

Jim Knauer, UR/LLE, USA

Bruno Le Garrec, ELI Beamlines, Czech Republic

Dick Lee, Lawrence Berkeley National Laboratory, USA

Patrick Mckenty, UR/LLE, USA

Masakatsu Murakami, ILE, Osaka University, Japan

Mitsuo Nakai, ILE, Osaka University, Japan

Abbas Nikroo, LLNL, USA

Hiroaki Nishimura, ILE, Osaka University, Japan

Takayoshi Norimatsu, ILE, Osaka University, Japan

Peter Norreys, University of Oxford, UK

Jose Manuel Perlado, Universidad Politécnica de Madrid, Spain

Kyle Peterson, SNL, USA

Tomasz Plewa, Florida State University (FSU), USA

Sean Regan, UR/LLE, USA

David Riley, Queen's University Belfast, UK

Harry Robey, LLNL, USA

Nina Rohringer, Center for Free-Electron Laser Science (CFEL), Germany

Frank Rosmej, Université Pierre et Marie Curie (UPMC), France

Hitoshi Sakagami, National Institute for Fusion Science (NIFS), Japan

Youichi Sakawa, ILE, Osaka University, Japan

Andy Schmitt, NRL, USA

Gonzalez de Vicente Sehila Maria, International Atomic Energy Agency (IAEA), Austria

Boris Sharkov, Facility for Antiproton and Ion Research (FAIR), Germany

Zheng-Ming Sheng, Shanghai Jiao Tong University, China

Keisuke Shigemori, ILE, Osaka University, Japan

Max Tabak, LLNL, USA

Toshihiro Taguchi, Setsunan University, Japan

Martin Tolley, Central Laser Facility, STFC Rutherford Appleton Laboratory, UK

Roger Vesey, SNL, USA

Akifumi Yogo, ILE, Osaka University, Japan

Proceedings Editor

Erik Storm, LLNL, USA

Journal of Physics: Conference Series **717** (2016) 011003 doi:10.1088/1742-6596/717/1/011003

Conference Series Sponsoring Organizations

UNIVERSITY OF CALIFORNIA

INSTITUT LASERS ET PLASMAS

Institute of Laser Engineering

Sponsoring Organizations

Lawrence Livermore National Laboratory

LCLS

Los Alamos NATIONAL LABORATORY EST. 1943

NIF&PS

Physical and Life SCIENCES

Sandia National Laboratories

UR LLE

Journal of Physics: Conference Series **717** (2016) 011003 doi:10.1088/1742-6596/717/1/011003

Endorsements

American Nuclear Society

APS physics

EPS

Fusion Power Associates

OSA® The Optical Society

Exhibitors

CHINESE LASER PRESS

GENERAL ATOMICS

LUXEL

NORTHROP GRUMMAN

Journal of Physics: Conference Series **717** (2016) 012001 doi:10.1088/1742-6596/717/1/012001

The quest for laboratory inertial fusion burn in the United States

D H Crandall

Independent

crandalldh@gmail.com

Abstract. Ignition and significant fusion yield from inertial confinement fusion (ICF) remains a grand scientific challenge with significant near-term and long-term applications. The ICF community in the U.S. is executing a coordinated effort to explore three viable approaches: laser x-ray drive, laser direct drive, and magnetic direct drive. Cooperative efforts from multiple institutions are directed at the physics basis of each of the three approaches with advancing diagnostics, precision targets, and improved simulations being the basis for the quest. X-ray drive experiments between 2010 and 2012 at the National Ignition Facility (NIF) gave yields much lower than expected because of both challenging hydrodynamics associated with high capsule convergence (~35×) and laser–plasma instabilities (LPI's) in the hohlraum. Recent experiments employing a variation of the laser pulse and resulting in lower convergence and lower hydrodynamic instability growth gave higher yields approaching 10^{16} neutrons (for the first time with significant fusion heating of the fuel), roughly in agreement with predictions for that approach. At the Omega Laser Facility the direct laser drive of the capsule is being developed to determine what could be expected if the NIF were reconfigured for spherical direct drive. Recent experiments on OMEGA, hydrodynamically scaled to the NIF, project to yields similar in nature to those of the best x-ray drive cases. Mitigation of cross-beam energy transfer (CBET) is required for improvement in direct-drive implosions. At the Z pulsed-power facility, a new approach of magnetically pinching a cylinder containing magnetized and laser-heated plasma shows promise for attaining significant fusion yield. Improvements in this technique are being addressed at a number of facilities. In addition to the fusion-yield experiments, a number of basic science studies use the advanced facilities to study plasma physics, materials science, and astrophysical phenomena at extreme parameters not previously available in a laboratory.

1. The U.S. ICF community seeks understanding of physics leading toward greater fusion yield

Through coordinated research, three approaches to laboratory fusion burn are being developed, following the community-developed plan for inertial confinement fusion (ICF) [1]. Highly capable facilities, including the National Ignition Facility (NIF; Livermore, CA), the Omega Laser Facility (Rochester, NY), and the Z pulsed-power facility (Albuquerque, NM) are central to this research.

Advanced diagnostics, precision targets, and detailed simulations are critical to all three approaches. Greater resolution in time and space of the fusion capsule along with emissions from it is a particularly important objective of diagnostics development. Experiments that appropriately agree with detailed simulations are expected to serve as touch points for developing physics understanding and seeking increased fusion yields. In all three cases, advances have been achieved over the past two years that provide the basis for further research and development.

2. Laser x-ray drive has achieved fusion heating for the first time

At the NIF, depicted in the first panel of figure 1, laser beams are directed to the inside walls of a hohlraum, where they generate x rays to fill the cavity and drive the ablation of the spherical capsule. The capsule has an outer ablator material (CH, Be, or high-density carbon) with an inner frozen shell of deuterium and tritium (DT) ice and central DT gas. Experiments at the NIF from 2010 through 2012 used a target and laser pulse designed to achieve ignition and modest fusion gain. The laser pulse had four steps in energy selected to drive the capsule on a low adiabat (~1.5) to a greater-than-35× convergence. The experiments achieved a fusion yield significantly lower than expected because of the growth of hydrodynamic instabilities, high convergence, and time variation of the x-ray drive. The hohlraums employed in these experiments were dominated by laser–plasma instabilities (LPI's), limiting the accuracy and quantitative understanding of the time- and space-varying radiation field needed for the demanding implosions required for ignition.

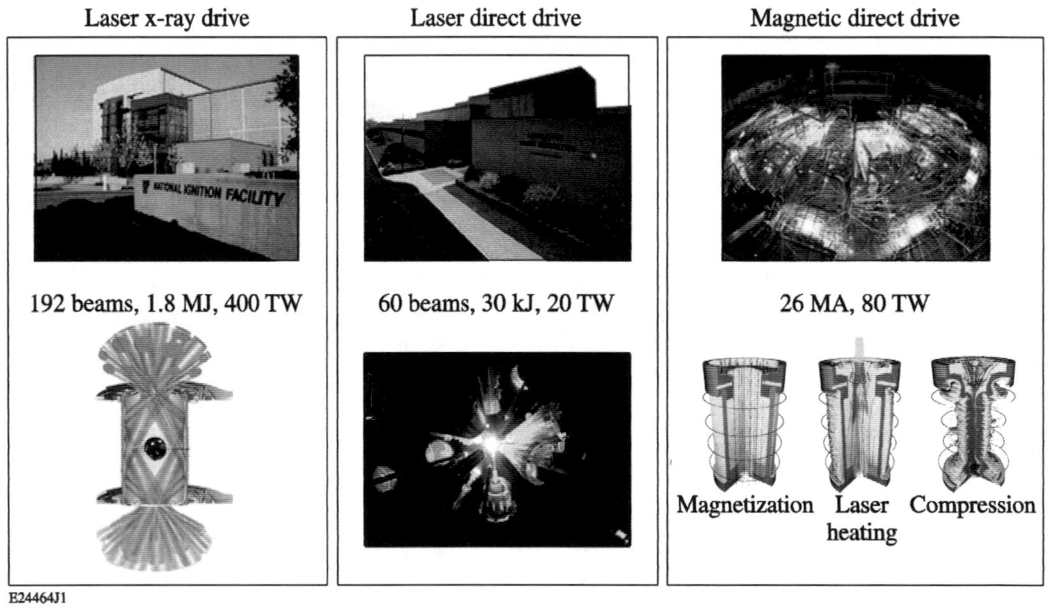

Figure 1. The three approaches to fusion burn currently in development in the U.S. ICF program and the major facilties where research is being conducted.

To drive the capsule to lower convergence with lower hydrodynamic instability growth, a laser pulse with higher initial power (foot) and fewer energy steps was employed: the "high-foot" implosions [2]. These experiments are on a higher adiabat (~2.5) and are not predicted to result in ignition (more fusion yield than laser energy input to the target). In these experiments, yield increased by about a factor of 10 to within about a factor of 2 of prediction, and for the first time significant heating of the fuel by fusion alphas resulted in an increased fusion yield (see figure 2). These experiments are within about a factor of 2 of the hot-spot density, temperature, and pressure required for ignition. Under these conditions the yield is slightly more than doubled as a result of alpha heating [2,3].

Understanding and improving the hohlraum energetics and x-ray drive at the NIF are critical to achieving the ignition goal. Hohlraums with different shapes and different gas fills are being investigated [4] along with different ablators for the capsule that have either improved hydrodynamic behavior or are better matched to a specific hohlraum and radiation pulse. A number of papers within this volume discuss these investigations; in particular see Hurricane *et al.* [2].

3. Laser direct-drive experiments seek an understanding of what might be achieved by reconfiguration of the NIF for spherical direct drive

Most experiments on laser direct drive are conducted at the Omega Laser Facility, depicted on the center panel of figure 1. Experiments are also conducted at the Nike KrF laser at the Naval Research Laboratory and at the NIF [with polar direct drive (PDD) [5], which approximates symmetric drive without spherical laser beam input]; these experiments address specific direct-drive physics issues. Experiments on OMEGA investigate spherical drive of layered DT-filled capsules with an extensive suite of diagnostics. Hydrodynamic scaling of parameters used on OMEGA to NIF laser energies allows one to infer what may be accomplished with spherical drive at the NIF. The NIF can be reconfigured for spherical direct drive but the time and cost of reconfiguration lead to a requirement for high confidence as to the fusion yield that might be obtained. The physics of spherical direct drive presents a compelling opportunity for laser-driven fusion. Relative to x-ray laser drive, the direct coupling requires smoother laser beams but, for a given laser energy, will allow ~7× more fuel to be compressed at the same implosion velocity, in turn requiring 3×- to 4×-lower stagnation pressure (~120 Gbar versus ~350 Gbar for NIF indirect drive designs) and lower convergence as a result.

Current DT implosions on OMEGA, scaled to NIF conditions, would give alpha heating with about the same results as x-ray drive (see figure 3), with somewhat greater yield (~100 kJ) because of the increased amount of fuel compressed. These results are presented by Goncharov *et al.* [6] in this volume. The best OMEGA direct-drive result to date produced 56 Gbar in the hot spot of the converging capsule. A pressure of about 120 Gbar is required for ignition at the NIF with spherical direct drive.

Observations on OMEGA provide evidence that laser–plasma interactions in the plasma ablating from the capsule surface limit laser coupling to the capsule through cross-beam energy transfer (CBET). Some of the incident laser energy is coupled to the outgoing portion of laser beams missing the converging target, reducing ablation pressure by up to 50%. Techniques to mitigate CBET by zooming [7] the laser beams to a smaller spot size as the capsule implodes, together with other improvements, are being implemented on OMEGA with the goal of demonstrating $P_{fuel} > 100$ Gbar by the end of the decade. Realizing this goal would provide reasonable confidence that ignition might be achieved at the NIF through reconfiguration for spherical direct drive.

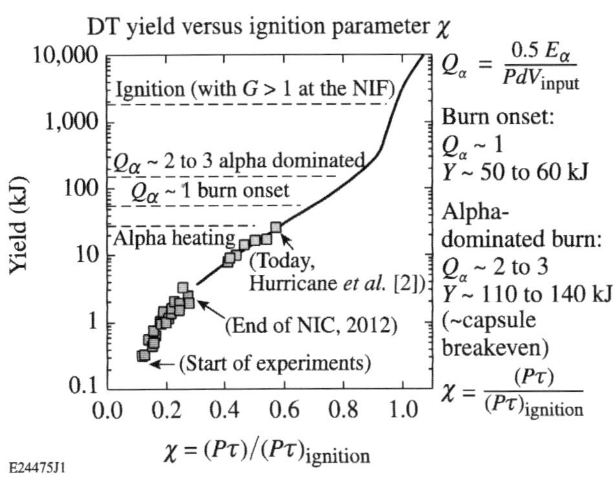

Figure 2. Fusion yield with laser x-ray drive at the NIF as a function of the Lawson scaling parameter [2,3].

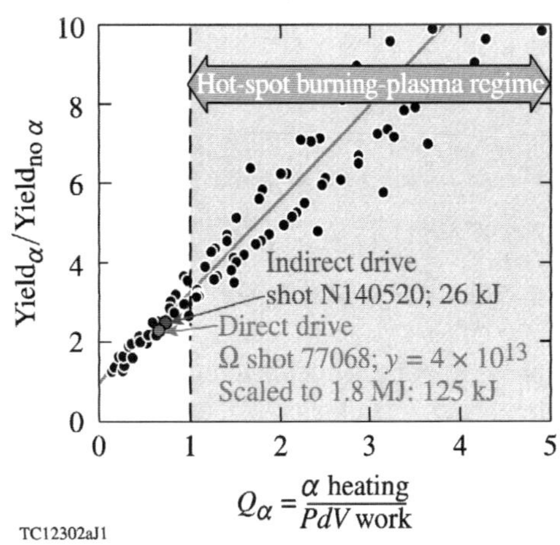

Figure 3. Fusion yield showing the best result with laser direct drive on OMEGA, scaled to NIF energy [3,6].

4. Magnetic direct drive of a magnetized and laser-heated plasma can provide fusion yield

At the Z pulsed-power facility, a new approach is being developed for fusion yield as depicted in the third panel of figure 1 and in figure 4. A cylindrical beryllium liner, 1 cm in length and containing a magnetized plasma heated by a laser to ~200 eV, is pinched by a strong, fast electrical pulse directly driving the plasma to fusion conditions [8]. The Z operates at a current of up to 26 MA and power up to 77 TW. Recent experiments [8] obtained 2×10^{12} DD fusion neutrons but, as expected, increases in both the applied magnetic field (B_{z0}) and laser heating are required to obtain more-significant yield. Evidence has been developed that the Z Beamlet laser (ZBL) with an unconditioned beam and a thick laser entrance window (~3 μm) coupled only several hundred joules rather than the ~1 kJ of expected energy into preheating the plasma, so this is an area for further research. Experiments on OMEGA and the NIF that can help address this issue are currently underway. The use of multiple facilities and research teams can rapidly advance this concept. A D–D fusion yield equivalent to 100 kJ of D–T yield is predicted for experiments at Z if 6 to 8 kJ of preheat, and ~30-T B field, and ~25 MA of current can be successfully applied to the target, provided that instability and mix can be adequately controlled.

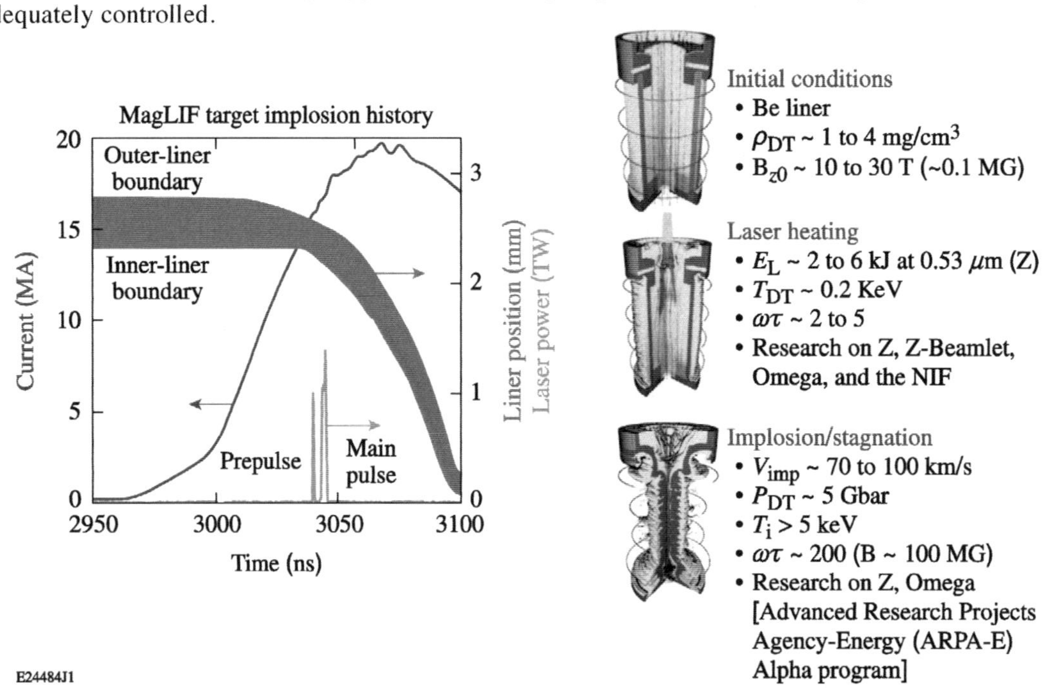

Figure 4. Magnetic liner implosion fusion (MagLIF) with a beryllium cylinder filled with a magnetized deuterium plasma and preheated by the Z Beamlet Laser (ZBL) directly pinched in the Z.

5. The ICF facilities provide extreme conditions that can advance other physics areas

The NIF, OMEGA, and Z all provide an opportunity to study plasma physics, atomic and nuclear physics, material behavior, and astrophysical phenomena at conditions of temperature and pressure not previously available in the laboratory. As an example, material properties at pressures relevant to planet formation are being studied on the NIF [9], Z, and OMEGA. Experiments on OMEGA and Z show distinct plasma-flow–generated instabilities [10] and important opacity of iron [11], respectively (see figure 5 for configuration and data examples). Significant contributions in the future are expected in a wide range of high-energy physics topics using these state-of-the-art facilities. Contributions to planetary science, material science, fundamental plasma physics, stellar, and astrophysics are expected over the next decade.

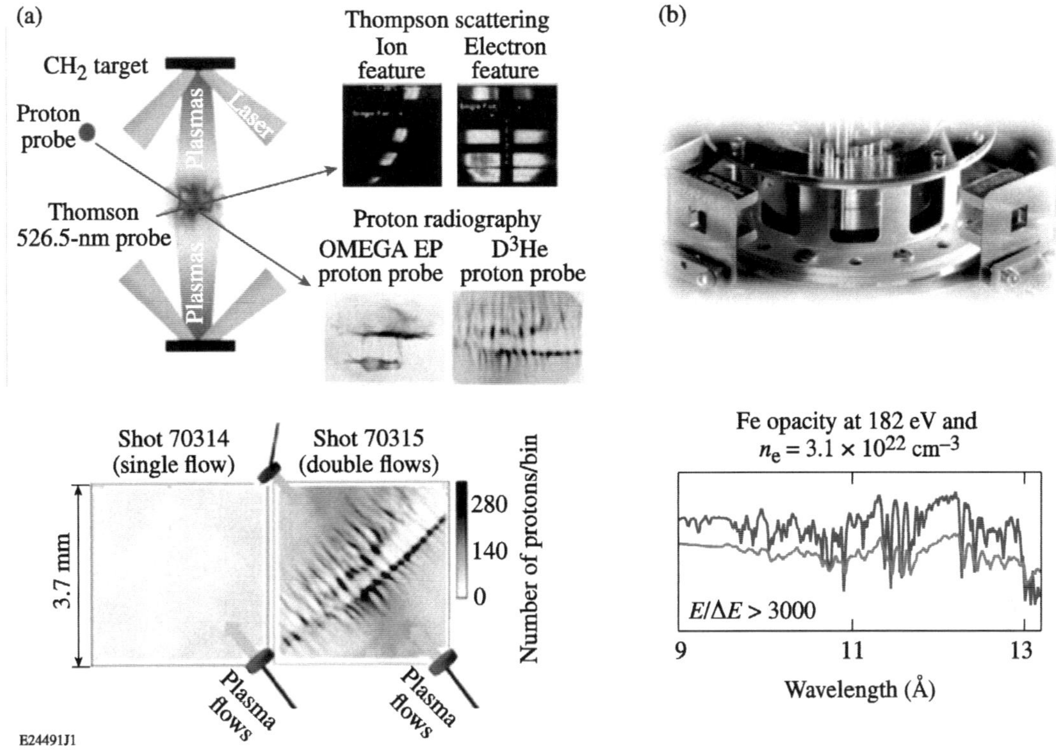

Figure 5. (a) Experiments on OMEGA [10] and (b) Z [11] provide advanced physics results.

6. Conclusion

The U.S. Inertial Confinement Fusion (ICF) Program continues the quest for fusion yield in the laboratory through the cooperation of multiple institutions and research teams addressing the physics constraints. The three approaches may provide different advantages in the various application of ICF. Over the next few years, research will focus on the science and understanding needed for ignition and either achieving it or determining what is needed and in advancing high-energy-density physics.

Acknowledgment

A number of people specifically participated in preparation of the talk at the IFSA Conference, which was the basis for this paper: T. C. Sangster, E. M. Campbell, and the Publications and Design Department of the University of Rochester, Laboratory for Laser Energetics; M. J. Edwards, R. P. J. Town, and Rip Collins of Lawrence Livermore National Laboratory; D. B. Sinars of Sandia National Laboratory; S. H. Batha of Los Alamos National Laboratory; S. P. Obenschain of Naval Research Laboratory; R. D. Petrasso of Massachusetts Institute Technology; C. A. Back and J. D. Kilkenny of General Atomics; and N. Petta of Schafer Corporation.

This material is based upon work supported by the Department of Energy National Nuclear Security Administration under Award Number DE-NA0001944, the University of Rochester, and the New York State Energy Research and Development Authority. The support of DOE does not constitute an endorsement by DOE of the views expressed in this article.

References

[1] December 2012 National Nuclear Security Administration's Path Forward to Achieving Ignition in the Inertial Confinement Fusion Program: Report to Congress Washington, DC: U.S. Department of Energy

[2] Hurricane O A *J. Phys.:Conf. Ser.* this volume

Hurricane O A, Callahan D A, Casey D T, Celliers P M, Cerjan C, Dewald E L, Dittrich T R, Döppner T, Hinkel D E, Berzak Hopkins L F *et al.* 2014 *Nature* **506** 343

[3] Betti R, Christopherson A R, Spears B K, Bose A, Nora R, Woo K M, Howard J, Edwards M J and Sanz J *J. Phys.:Conf. Ser.* this volume

Betti R, Christopherson A R, Spears B K, Nora R, Bose A, Howard J, Woo K M, Edwards M J and Sanz J 2015 *Phys. Rev. Lett.* **114** 255003

[4] Callahan D A *J. Phys.:Conf. Ser.* this volume

[5] Hohenberger M, Radha P B, Myatt J F, LePape S, Marozas J A, Marshall F J, Michel D T, Regan S P, Seka W, Shvydky A *et al.* 2015 *Phys. Plasmas* **22** 056308

[6] Goncharov V N, Regan S P, Sangster T C, Betti R, Boehly T R, Bonino M J, Davis A, Edgell D H, Epstein R, Forrest C J *et al. J. Phys.:Conf. Ser.* this volume

Goncharov V N, Sangster T C, Betti R, Boehly T R, Bonino M J, Collins T J B, Craxton R S, Delettrez J A, Edgell D H, Epstein R *et al.* 2014 *Phys. Plasmas* **21** 056315

[7] Froula D H, Goncharov V N, Kessler T J, Igumenshchev I V, Huang H, Kelly J H, Kosc T Z, Meyerhofer D D, Regan S P, Shvydky A *et al. J. Phys.:Conf. Ser.* this volume

[8] D. B. Sinars, *J. Phys.:Conf. Ser.* this volume

Sefkow A B, Slutz S A, Koning J M, Marinak M M, Peterson K J, Sinars D B and Vesey R A 2014 *Phys. Plasmas* **21** 072711

[9] Eggert J H *J. Phys.:Conf. Ser.* this volume

Smith R F, Eggert J H, Jeanloz R, Duffy T S, Braun D G, Patterson J R, Rudd R E, Biener J, Lazicki A E, Hamza A V *et al.* 2014 *Nature* **511** 330

Lazicki A, Rygg J R, Coppari F, Smith R, Fratanduono D, Kraus R G, Collins G W, Briggs R, Braun D G, Swift D C *et al.* 2015 *Phys. Rev. Lett.* **115** 075502

[10] Huntington C M, Fiuza F, Ross J S, Zylstra A B, Drake R P, Froula D H, Gregori G, Kugland N L, Kuranz C C, Levy M C *et al.* 2015 *Nat. Phys.* **11** 173

[11] Bailey J E, Nagayama T, Loisel G P, Rochau G A, Blancard C, Colgan J, Cosse P, Faussurier G, Fontes C J, Gilleron F *et al. J. Phys.:Conf. Ser.* this volume

Bailey J E, Nagayama T, Loisel G P, Rochau G A, Blancard C, Colgan J, Cosse P, Faussurier G, Fontes C J, Gilleron F *et al.* 2015 *Nature* **517** 56

Journal of Physics: Conference Series **717** (2016) 012002 doi:10.1088/1742-6596/717/1/012002

Will fusion be ready to meet the energy challenge for the 21st century?

Yves Bréchet – Haut-Commissaire à l'Energie Atomique
CEA Saclay 91191 Gif-sur-Yvette, France
Thierry Massard
CEA DAM-Ile de France, Bruyères le Chatel, 91297 Arpajon, France

Abstract. Finite amount of fossil fuel, global warming, increasing demand of energies in emerging countries tend to promote new sources of energies to meet the needs of the coming centuries. Despite their attractiveness, renewable energies will not be sufficient both because of intermittency but also because of the pressure they would put on conventional materials. Thus nuclear energy with both fission and fusion reactors remain the main potential source of clean energy for the coming centuries. France has made a strong commitment to fusion reactor through ITER program. But following and sharing Euratom vision on fusion, France supports the academic program on Inertial Fusion Confinement with direct drive and especially the shock ignition scheme which is heavily studied among the French academic community. LMJ a defense facility for nuclear deterrence is also open to academic community along with a unique PW class laser PETAL. Research on fusion at LMJ-PETAL is one of the designated topics for experiments on the facility. Pairing with other smaller European facilities such as Orion, PALS or LULI2000, LMJ-PETAL will bring new and exciting results and contribution in fusion science in the coming years.

1. Introduction

The energy challenge is critical for todays' world. This is motivated on one hand by the finite amount of fossil fuel and by the emerging of new economies and new countries with growing populations and energy demand. At the same time there is a strong appeal towards the reduction of all means of energy productions which produce large amounts of CO_2. Global warming appears to be the threat of the century. This is why the European Union has taken a strong commitment to reduce its CO_2 production in the coming years and to develop energy production greenhouse gas free. [Fig. 1]

2. The flaw in renewable energies: not enough material

Renewable energies will probably play an important role in this new energy balance but their intrinsic efficiency in terms of space occupation, and their intermittency pose serious problems for intensive production of energy coupled to stable distribution network, at least as long as mass storage is limited. A simple common sense idea, often forgotten: In renewable energies one always insists on the renewable «source of energy» (wind, water, sun) without ever addressing the question of the means to make this energy useful: dams for hydroelectricity, wind-turbines for the wind, and solar cells of thermal devices for the sun.

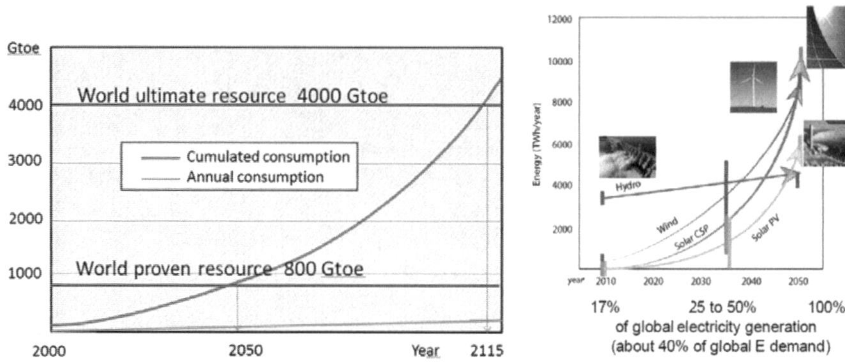

Figure 1 – World energy consumption - The scenarios of IEA and Ecofys rely on a strong increase in the share of solar and wind energy in the total production of energy

These devices consume space and matter. This space and matter consumption, in a finite world have a substantial impact on the available reserves of space and matter.

It has been shown that this limitation may be a key issue, and not only for so called « critical materials» such as Rare Earths which can be replaced or recycle, but also for non-specific materials such as steel, copper or aluminum. A similar analysis for space can be carried out.

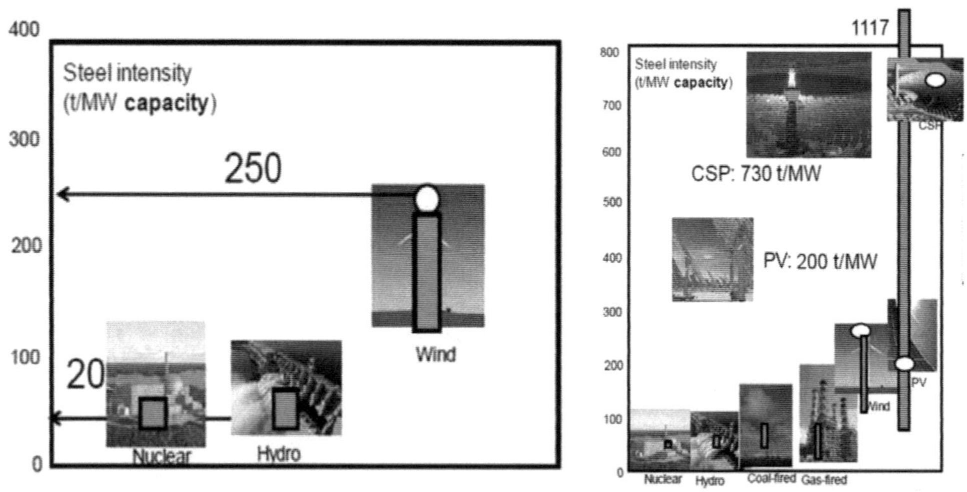

Figure 2 – the amount of steel per MW of produced energy is more than ten times larger for wind turbines than for nuclear or hydro plants and reaches 50 times for concentrated solar production

Despite all the common idea that new renewable energy sources are "free" and "practically unlimited" in a sense that no one owns the sun nor the wind, the fact remains that in order to capture these energies, one needs an immense construction effort. This, unfortunately, is neither free nor unrestricted in the material sense. As shown by Vidal, Goffé & Arndt Ref. 1] projected renewable energy deployments would very soon outstrip the current global production of several key materials. [Fig. 2]

3. Developing nuclear energy fission and fusion : a priority for the energy mix

In this context, the development nuclear energy appears as a necessary task to be addressed in order to satisfy the demand of energy of tomorrow's world, at least in a transient stage. Fission nuclear energy has a long history and especially France has developed a strong industrial sector. Promising vision of nuclear fission include the generation 4 (GENIV) type of nuclear reactors which will be able to burn depleted uranium and to minimize the amount of ultimate long live waste. Plutonium becomes a resource instead of being a waste.

Journal of Physics: Conference Series **717** (2016) 012002 doi:10.1088/1742-6596/717/1/012002

In the longer term, nuclear fusion seems to be even more attractive promising a production of energy with no limit on fuel materials and limited long term nuclear waste – only fission products - in the production cycle. However this attractive goal faces scientific and technical challenges which will not be easily overcome and require advanced investigations. In parallel, the common points between different possible routes and the possibility of joining forces appear as strong assets.
In recent years several important results have been achieved in the field of nuclear fusion. In 1997 the European tokamak machine JET, in Culham, succeeded in generating 16 megawatts of fusion power, i.e. 70 percent of input power. Last year, at NIF, the high-foot implosions were able to demonstrate a significant alpha-heating of the thermonuclear fuel, thereby really entering a new physical domain in the physics of imploded plasmas [Fig 3]. These achievements are really significant and show how substantial progress understanding nuclear fusion has really been made.

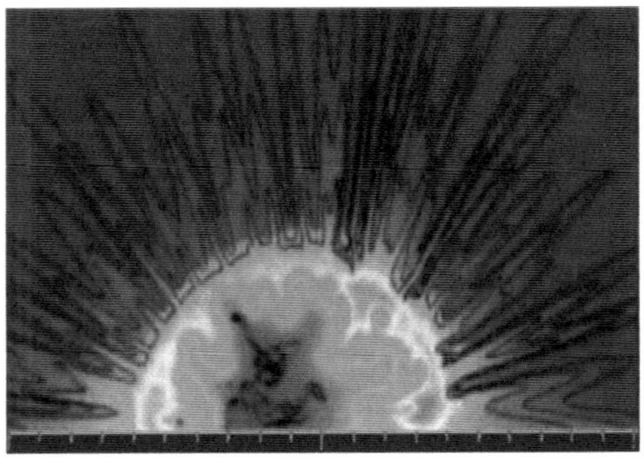

Figure 3 - The experimental results have matched computer simulations much better than previous experiments, providing an important benchmark for the models used to predict the behavior of matter under conditions similar to those generated during a nuclear explosion, a primary goal for NIF. (source LLNL)

At the same time, these results show that the path to energy production by nuclear fusion is still not a straight road, and the development of fusion-based reactors is not at hand. Europe and France in particular are in the front-line of research on nuclear fusion. France has the honor of hosting two leading research facilities for the study of nuclear fusion.

4. ITER the priority path for fusion in Europe

The first one is the ITER facility currently under construction in Cadarache in the south of France. ITER will be an advanced tokamak machine which is designed to achieve the plasma energy breakeven point, and aims at producing more power than it consumes: for 50 MW of input power, 500 MW of output power is the target. ITER will also allow studying the conditions of a burning plasma, the radiation flux from the plasma and the effects on the materials composing the machine. Therefore it will pose the basis for the development of future reactor demonstrators. ITER is an international enterprise, including Europe, USA, Russia, China, Japan and India, but France plays a particular role being the host country for the reactor.
At Cadarache, CEA is modifying its Tore Supra plasma facility which, once transformed, will become a test platform open to all ITER partners : the WEST project (acronym derived from W Environment in Steady-state Tokamak, where W is the chemical symbol for tungsten). The goal is to equip the tokamak with an actively cooled tungsten divertor, benefitting from its unique long pulse capabilities, its high level of additional power and the unique experience of operation with actively cooled component [Fig. 4]. The divertor is a key component which faces the largest part of the heat and particle fluxes coming from the core plasma during experiments. The current option, in order to avoid erosion is Tungstene but the potential pollution of the plasma and the need for huge heat extractor remains critical issues on the path to success

Figure 4 - First WEST Stabilization Coil inside Tore Supra Tokamak

5. The Megajoule Laser in France : a key facility for the Nation's security

At the same time, CEA is running the Megajoule Laser facility on its site near Bordeaux. LMJ, a brother laser to NIF, is already operational for physics experiments although with a limited number of beams, and more beams are currently added to reach the final configuration of 176 beams and over 20 permanent diagnostics. The LMJ has been built with the main goal of contributing to assuring the French nuclear deterrence. [Fig 5]. CEA, in agreement with the Region Aquitaine, has decided to open the facility to academic civilian research for a quota up to 30% of the laser time. In particular a Petawatt laser, companion to the LMJ, PETAL has been developed explicitly as a research instrument for academic research and will be open through the same process as LMJ to the international academic community. [Fig 6].

The first call for proposals for civilian experiments was issued end of 2013 under the management of ILP, Institute for Lasers and Plasmas, which federated the research groups which work, in France, on the physics of laser-produced plasmas. Finally, 4 experiments out of 6 proposals have been selected by an independent International Scientific Committee nominated by ILP. Experiments are planned to start in 2017. These experiments include:
1) The study of strong shock generation by laser plasma interaction in the context of shock ignition studies proposed by Sophie Baton (LULI), Xavier Ribeyre (CELIA) and al.
2) The study of radiative shocks, of astrophysical relevance, in the laboratory proposed by Michel Koenig and al.
3) The study of primordial magnetogenesis proposed by Gianluca Gregori and al.
and finally
4) The study of the interplay between B-field and heat transport in ICF conditions proposed by Roch Smets and al.

The next call for experiments will appear early next year for a deadline for submission mid-2016
Academic civilian experiments are an important point in the research policy of CEA. At the same time they offer the possibility of starting a research program dedicated to the study of nuclear fusion for the production of energy. Such research program conducted jointly by the European scientific community and by CEA researchers will need to grow and aggregate new teams in the next years in order to be able to cope with the challenges of nuclear fusion. Such "consolidation" implies a large scientific community working in Europe on the subject, larger funds and more technical means, increased dedicated laser time.

One could wonder what can be the role of the European scientific community working in ICF compared to much larger communities such as the US or the Asian communities. What will be in particular the contribution of experiments performed on LMJ-PETAL. Certainly the advancements done by our US colleagues during the National Ignition Campaign are enormous and there are many things, which we need to learn by collaborating with them.

Figure 5 – The LMJ facility near Bordeaux is in operation since 2014 supporting the Simulation Program. The LMJ is also open part time for academic access.

At the same time, the results of NIF experiments have clearly shown that there are still many difficulties to be overcome to reach ignition. On LMJ, scientists will put most of their efforts to perform direct-drive experiments. Using PETAL as an advanced radiographic diagnostic they will be able to check the uniformity of implosion needed to create the conditions of extreme densities and temperatures required at the center of the fuel pellet.

Petal
PETAWATT AQUITAINE LASER

Figure 6 - The PETAL laser is on the right of the picture in one of the laser bay of LMJ. The architecture of the amplifier section is similar to a LMJ chain, 4 passes architecture with 16 amplifier slabs. The main difference is the longitudinal chromatism corrector installed in the U-turn which allows to correct 1,5 ps in two passes. PETAL has reached 1.2 PW and will be capable of 7PW , 500 fs, 3kJ when proper technology for transport mirrors will be at hand.

6. European studies on direct drive ignition

Indeed the difficulties of indirect-drive experiments on NIF seem to re-open the chances for direct drive and, in this field, promising new schemes are appearing, in particular the new approach called "Shock Ignition" introduced by the colleagues of the University of Rochester, which decouple the ignition process from the compression process allowing much more flexibility and potentially higher efficiency.
This seems to be a very interesting path for academic experiments on LMJ, and for the development of a European research program on inertial fusion for energy production.

Such effort, which in Europe has to be created, should not to be seen as an alternative to the main recognized approach of Magnetic Confinement Fusion. Interesting in the last few years, research programs involving both the ICF and the MCF communities are starting and need to be extended. After many years of "confrontation" and cold war between MCF and ICF we are now in a new phase of beginning of collaborations around some points of common interests: the behavior of material irradiated by high radiation and neutron fluxes, the need to develop new diagnostics tools adapted to burning plasmas, the problems of the interaction chamber and the breeding of nuclear fuel. This is certainly a very positive change and CEA will work to further favor it.

Another important element of the European panorama is certainly the presence of several smaller laser facilities. Such "intermediate" laser facilities are indeed essential for the development of a strong European program on inertial fusion for energy (IFE). Of course one cannot imagine realizing experiments on LMJ-PETAL without a proper preparation on a dedicated facility.
Therefore all these laser systems are essential: LULI in France, Vulcan and Orion in the UK [Fig. 7], Phelix in Germany, Pals in the Czech Republic. Here European scientists will be able to test the ideas for experiments on LMJ-PETAL, to develop and calibrate diagnostics and, even more important, such facilities will continue to assure the formation of a new generation of scientists working in the field.

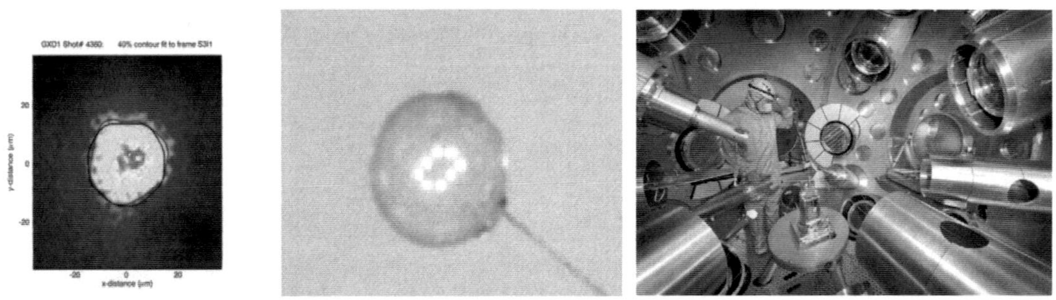

Figure 7 – ORION progresses in direct drive implosion experiments. Initial experiments have achieved >10^9 DD neutron yield

At the same time, we need to begin addressing other issues related to drivers for future fusion facilities. We need to develop new laser technologies which are able to simultaneously offer large delivered energy, high overall efficiency, high repletion rates. Without such laser systems, inertial fusion energy will probably remain a dream even after it has been scientifically demonstrated. Research programs addressing the required technology developments have already started in France, the LEAP project again in Aquitaine within collaboration between CEA, the University of Bordeaux and the industrial world, but also in other countries, in the UK (Dipole project), in Germany (Polaris program at Jena), in the Czech Republic (at ELI-beamlines in Prague).

7. Conclusion and perspectives for fusion studies

It is clear that the development of such new generations of powerful and efficient laser systems will benefit not only the world of scientific research but also the economic and industrial world via the generated activity, and finally every aspects of society, lasers finally being a pervasive technology in the 21st century. Finally, this is the final point, which I would like to stress: by pursuing the research on Inertial Confinement Fusion, we are not only advancing science but we are providing top-level scientific and technological research. We are contributing to maintaining high technological standard and preserve the interest of our economies and of our industry. We are contributing to the training of new scientists and a new technological leadership for tomorrow's society.

[Ref. 1] Vidal, Goffé, Arndt, Nature Geoscience 6, 894–896 (2013).

Fusion Studies in Japan

Yuichi OGAWA

Graduate School of Frontier Sciences, the University of Tokyo
5-1-5 Kashiwanoha, Kashiwa, Chiba, 277-8568, JAPAN
ogawa@ppl.k.u-tokyo.ac.jp

Abstract. A new strategic energy plan decided by the Japanese Cabinet in 2014 strongly supports the steady promotion of nuclear fusion development activities, including the ITER project and the Broader Approach activities from the long-term viewpoint. Atomic Energy Commission (AEC) in Japan formulated the Third Phase Basic Program so as to promote an experimental fusion reactor project. In 2005 AEC has reviewed this Program, and discussed on selection and concentration among many projects of fusion reactor development. In addition to the promotion of ITER project, advanced tokamak research by JT-60SA, helical plasma experiment by LHD, FIREX project in laser fusion research and fusion engineering by IFMIF were highly prioritized. Although the basic concept is quite different between tokamak, helical and laser fusion researches, there exist a lot of common features such as plasma physics on 3-D magnetic geometry, high power heat load on plasma facing component and so on. Therefore, a synergetic scenario on fusion reactor development among various plasma confinement concepts would be important.

1. Introduction

Japanese government is promoting fusion research and development based on so-called Phase Basic Program formulated by Atomic Energy Commission (AEC). Currently the third phase basic program started in 1992 has been on progress. This program sets forth key elements of research and development for a DEMO reactor, and is aiming at achieving necessary conditions for the self-ignition and at realizing the long pulse burning plasmas in a tokamak experimental reactor.

Based on this principal policy, the ITER has been assigned as a core device in the third phase basic program. While, just before the start of ITER project, this third phase basic program has been reviewed, and report entitled by Future Fusion Research and Development Strategy in the Third Phase Basic Program of Fusion Research and Development was issued in 2005 [1]. In this report action items of recent research and development including ITER project have been described.

In 2011 we had a Great East Japan Earthquake, and quite serious nuclear accidents have taken place in Fukushima nuclear power plants. Therefore, in Japan the government has reconsidered energy strategy, and the Cabinet has decided a new strategic energy plan in 2014. In this new strategic energy plan, the Government declares that although nuclear power is expected to be a baseload power supply, Japan will minimize its dependency on nuclear power. While, concerning to nuclear fusion, this report declares that the Japanese government steadily promotes nuclear fusion development activities, including the ITER project, which is being implemented through international cooperation, and the Broader Approach Activities from the long-term viewpoint [2].

2. Japanese Policy on Fusion Energy Development

The report compiled by AEC in 2005 describes the role of fusion energy as an important option in solving energy and environmental problems in addition to energy security viewpoints, and following key points are listed up.

(1) Reduction of CO_2
If highly economic fusion energy is soon introduced in addition to energy-saving and use of renewable energy, possible significant reduction of carbon dioxide concentration in the end of this century might be feasible. Early realization of fusion energy is important from the viewpoint of solving the global environmental problem. Tokimatsu *et al.* have analyzed the feasibility on the introduction of fusion energy in this 21st century, and concluded that fusion energy might have a potentiality to supply electric power demand of 10 ~ 20 % in the world at 2100, if the first fusion reactor could be introduced around 2050 with a reasonable cost [3].

(2) Necessity of introduction to developing countries
The increase of energy consumption is coming from developing counties in the near future. Active introduction to developing countries is also expected, since fusion energy has less resource localization and less restriction in introduction to society. Japan should take a leadership in early realization of fusion energy toward resolution of energy and environmental problems, because Japan is a responsible country to actively contribute to the world prosperity and promotion of international cooperation as a nation based on the creativity of science and technology.

(3) Energy security
It is also important to ensure the energy security of Japan by increasing its self-sufficiency presently in low level in energy resources.

This report expects that if a switchover to the DEMO phase is done in the early 2020s followed by prompt construction of DEMO, it will be foreseeable to start test program and improvement using DEMO from 2030s with the objective of demonstrating continuous power generation, safety, economical viability and operation reliability. If so, it is foreseeable to expect commercialization by the mid-21st century.

A schematic drawing of strategy for fusion reactor development was compiled in the report, as shown in Fig. 1. Now third phase basic program is strongly promoted, and ITER is a core device. In the previous 2nd phase a large tokamak JT-60 device was constructed, and in the 4th phase program construction of DEMO reactor is expected. In addition, along with the strong promotion of ITER project, necessity of complemental projects is described in this report, as well.

ITER will advance burning physics, while deep understanding and further improvement of core plasma might be necessary. For that purpose promotion of tokamak plasma experiments is proposed, by modifying JT-60 device to JT-60SA. Of course, the promotion of fusion engineering is indispensable toward DEMO. For example, material development and blanket/divertor technologies in addition to safety research. Now advanced tokamak research and fusion engineering development are promoted in the framework of Broader Approach (BA) activities.

A first candidate of DEMO is a tokamak at present, because plasma performance is better than other approaches. While, there are other attractive approaches for fusion reactors, as well. In Japan helical and laser fusion have been prioritized as other attractive approaches, compared with other basic researches. Therefore, LHD project in helical and FIREX project in laser have been strongly supported by the Government.

3. Research Activities on magnetic fusion

The Government has been strongly supporting ITER project from the start in 1988, and Japan is playing an important role in ITER project with other Parties. The main contribution is an in-kind procurement in ITER project, and Japan is contributing to many components such as (i) Toroidal Field Coil, (ii) Center Solenoid Coil, (iii) Outer Target of Divertor, (iv) Neutral Beam H&CD, (v) Electron Cyclotron H&CD, (vi) Blanket Remote Handling System, (vii) Diagnostics and (viii) Detritiation System [4].

As described in the previous section, for the DEMO reactor development, promotion of further development of key technologies for core plasmas and fusion engineering field is indispensable in order to establish a technological basis for DEMO as well as to play supporting and complementary roles for the ITER Project. For this purpose Broader Approach activities have been initiated as a joint project between Japan and EU, and BA agreement was effectuated in 2007. Main sites are located in Japan, i.e., Rokkasho in Aomori prefecture and Naka in Ibaraki prefecture. The total cost is evenly shared by JA and EU. This is a 10-years project, and remains in force thereafter unless terminated by either Party [5].

Rokkasho is a new site established for this BA activities. Two projects are promoted in Rokkasho site; one is a so-called IFMIF/EVEDA project and another is IFERC project. For the development of fusion reactor materials the strong 14 MeV neutron source such as the International Fusion Materials Irradiation Facility; i.e., IFMIMF, might be indispensable. In the framework of BA activities, therefore, R&D for IFMIF has been promoted, as Engineering Validation and Engineering Design Activities; i.e., EVEDA. In IFMIF/EVEDA project, the prototype accelerator with 9 MeV and 125mA beam is under development. At present, as the first step of the validation test of the prototype accelerator, the injector and its peripheral systems have been introduced from EU, and installed in Rokkasho site. The beam commissioning was started in November 2014. Another project in Rokkasho site is International Fusion Energy Research Center, so-called IFERC, project. The purposes of this project are to coordinate DEMO design and R&D, and to manage computational simulation center and ITER Remote Experimentation Center.

In Naka site a satellite tokamak program is ongoing; i.e., the JT-60 tokamak is upgraded to JT-60SA, which is a superconductor device largest in the world except ITER. The plasma experiment will start in 2019. Nine vacuum vessel sectors of 10 have been installed sequentially on the cryostat base in 2015 and they have been welded with each other, remaining an opening of 20 degree sector. Since the toroidal field coil is linked with the vacuum vessel in the tokamak device, the next step is an installation of the toroidal field coil through this opening sector region. The toroidal field coil is just under fabrication in Europe, and the first TF coil will be delivered to Japan in 2016.

Along with tokamak research activities, helical plasma research is strongly promoted in Japan, because the helical plasma has an attractive feature; i.e., steady-state operation is feasible. One of the largest helical device LHD was constructed in NIFS. Since 1998 LHD has conducted 18th experimental campaign, and remarkable progress has been demonstrated. That is, about 10 keV temperature plasma has been achieved, and long pulse operation during about 50 minutes has been demonstrated [6].

Since helical plasma should be considered 3-D geometrical effects into plasma physics, quite comprehensive 3-D theory and simulation codes have been developed in helical plasma society. Recently, even in tokamak plasmas 3-D effect seems to play an important role. For example, in ITER in order to control ELM burst, the in-vessel coil will be installed inside the vacuum vessel. Since this in-vessel coil produces the magnetic island and/or stochastic field, 3-D simulation code is indispensable. Therefore, collaboration between tokamak and helical is strongly enhanced in the magnetic fusion community. This is a good example of synergetic effect between tokamak and helical plasmas on the scenario of fusion reactor development.

4. Research Activities on fast ignition laser fusion and reactor designs

In 1972 Central ignition scenario was proposed by Nuckolls, and major facilities such as NIF and LMJ have been constructed to explore this central ignition scenario. While in 1983 a new idea so-called fast ignition was proposed by Prof. Tatsuhiko Yamanaka and other colleagues. Sometimes, the central ignition is likened to diesel engine, and the fast ignition is to gasoline engine. The idea of the fast ignition

Journal of Physics: Conference Series **717** (2016) 012003 doi:10.1088/1742-6596/717/1/012003

seems to be quite elegant, and direct-drive fast-ignition could provide potentially high energy gain with small driver, but the development of PW-level high power laser is indispensable.

Osaka University is strongly promoting a fast ignition campaign [7]. Design and construction of LFEX laser has started in 2003, and in 2008 target irradiation with high-power beam started. The LFEX laser system consist of 4 beams, and the first one beam was constructed in 2009, and the preliminary experiments for irradiation of fast ignition has started. In 2010 the second beam was fabricated, and the fast ignition experiment with two beams was conducted in 2010. However, due to the great earthquake in 2011, the facility of the company for fabricating LFEX laser components has been damaged. For that reason, the construction of LFEX laser system was delayed by 2 years. In 2014 four beams have been completed, and plasma mirror has been introduced, as well. In 2015 the fast ignition campaign with 4 full beam system has been conducted, especially paying attention to high pulse contrast. The maximum beam energy is, at present, 2 kJ with a pulse length of 1 – 1.5 psec [8]. Preliminary experiments on fast ignition have been conducted, and it has been found that efficient heating is achievable by controlling generation and transport of laser-produced electron beam. For this purpose, the guide of the electron beam by introducing strong magnetic field has been proposed in addition to the elegant tailoring of the laser pulse form.

Concerning to laser fusion reactor, KOYO-F was designed in Osaka University, based on fast ignition concept [9]. The cooled ytterbium:YAG ceramic laser of 16 Hz is employed, where 1.1 MJ of the laser energy is for compression and 100kJ for heating. An energy gain of 160 and fusion yield of 200 MJ are expected. One of the critical issues in laser fusion reactor is the chamber wall, where a quite large thermal load should be protected. In this design a flowing liquid LiPb is employed. This is a quite elegant design, but technical barrier seems to be quite high.

To overcome this difficulty, a dry wall laser fusion reactor FALCON-D has been designed in CRIEPI and the University of Tokyo [10]. The design concept of FALCON-D reactor is based on fast ignition scenario, dry wall chamber and high repetition laser. In fast ignition scenario it might be available to achieve a relatively high gain with low laser energy. For example, about 0.3-0.4 MJ might be sufficient for fast ignition. This is an advantage of fast ignition scenario. Therefore, in FALCON-D reactor design, the fusion yield in one pulse is decreased and chamber radius is increased, so as to survive the first wall heat load. While the repetition frequency is increased up to 20 - 40 Hz.

Nevertheless, the gap between present experimental results and future DEMO reactor seems to be quite large. Therefore, to mitigate technical gap between single shot present experiments and future commercial plant, the experimental plant LIFT was proposed in 2015 [11]. In LIFT laser-diode pumped Yb:YAG ceramic laser operated at 225K is adopted. The laser energy is 500kJ for compression and 150kJ for heating, and the repetition frequency is 4 Hz with the efficiency of 12%. Since the fusion gain of around 100 could be expected, the fusion yield of 40 – 80 MJ per shot can be achieved.

The interesting feature of this LIFT project is that the project will be promoted step by step according to the progress of the experimental research, by utilizing the same laser facility. The mission of the Phase I is the repeated fusion burn, and blanket is not installed. In addition, repetition around 100 shots is planned in one experiment series. The Phase II is a power generation, and the solid blanket will be installed in this Phase. From the viewpoint of power generation, the slightly long time operation is indispensable. Therefore, in this Phase II, operation during one week is expected. Finally, in the Phase III continuous operation during one month or more is expected. In addition, the solid blanket will be replaced with liquid one, so as to explore the future commercial reactor. Since the self-supply of tritium is required, the tritium breeding function is indispensable. In these three phases the common laser system will be utilized, and replacement of chamber room is sufficient for advancing to the next phase . Since replacement of the chamber core region seems to be relatively easy compared with magnetic fusion systems such as tokamak and helical, the stepwise development might be feasible for laser fusion reactor research.

Next let me briefly review the necessity of the technical development of laser facility for laser fusion reactor. In general, a high repetition high power laser around 30 Hz is indispensable in a laser fusion reactor, while LFEX laser facility in Osaka university can be operated in 3 shots per day. That means the repetition frequency of $9x10^{-5}$ Hz. This is quite low level, because LFEX laser facility consists of Nd-doped glass laser with flash lamp pumped, and air cooling operation at room temperature. To achieve the repetition frequency of 30 Hz from $9x10^{-5}$ Hz, several advanced technologies should be introduced such as laser diode, ceramic window, Yb-dopant, 200 K operation and liquid cooling [12]. By taking these advanced technologies into account, we can expect the remarkable increase of the laser repetition frequency from the present low level to 30 Hz level.

5. Recent activities toward DEMO

In order to reinforce strategy to DEMO reactor development, a special team called Joint-Core Team was organized in the fusion community under the Government, and intensively discussed on critical issues for DEMO reactor. Concerning to the strategy to DEMO development, the Joint-Core team has summarized a report on the basic concept of DEMO and structure of technological issues in 2014 [13]. This Joint-Core team has sorted out tasks regarding the development of the design of DEMO, and research and development programs to resolve the key issues as follows; (i) Superconducting coils, (ii) Blanket, (iii) Divertor, (iv) Heating and current drive systems, (v) Theory and numerical simulation research, (vi) Reactor plasma research, (vii) Fuel systems, (viii) Material development and establishment of codes and standard, (ix) Safety of DEMO and safety research, (x) Availability and maintainability, (xi) Diagnostics and control systems and (xii) Newly required facilities and platforms.

The Joint-Core team has compiled this report based on tokamak DEMO reactor. However, some of these critical issues are common or similar between magnetic and inertial fusion reactors. For example, blanket might be common. Therefore, the technology of the blanket developed by tokamak DEMO might be applicable to inertial fusion reactor. In addition, fuel systems; i.e., tritium circulation systems might be common, except for the fabrication of pellet itself. Of course, material development and establishment of codes and standards might be common issue. Safety is also common issue. While, in laser fusion reactor it is not necessary to develop superconducting coils and heating and current drive systems. Instead, there are several specific R&D issues for laser fusion reactor. That is, high repetition laser, final optics and chamber wall should be developed. Of course, Ignition physics should be developed in addition to pellet fabrication and tracking technologies.

By the way, in the inertial fusion reactor the divertor is not necessary. However, the divertor heat load problem might be quite similar to the chamber wall issue in the inertial fusion reactor. F. Najmabadi has compared heat load conditions between tokamak divertor with IFE armor, and pointed out that although the pulse duration is quite different between IFE and Type-I ELM, the deposition energy and frequency are quite similar in both cases [14].

Although the basic concept is quite different between different confinement concepts, we can see a lot of common features among tokamak, helical and laser fusion reactors. Therefore, it should be emphasized that we have to consider a synergetic scenario on fusion reactor development among various plasma confinement concepts. It is quite important so as to promote fusion reactor development efficiently and smartly.

Acknowledgments

An author acknowledges to Profs. H. Azechi, H. Shiraga, T. Norimatsu and S. Fujioka in Osaka University for their support and information on laser fusion research activities.

References

[1] http://www.aec.go.jp/jicst/NC/senmon/kakuyugo2/siryo/kettei/houkoku051026/index.htm (in Japanese)
[2] http://www.enecho.meti.go.jp/en/category/others/basic_plan/pdf/4th_strategic_energy_plan.pdf
[3] K. Tokimatsu, et al., 2002 *Nuclear Fusion* **42** 1289.

[4] http://naka-www.jaea.go.jp/english/iter-e/iter.html
[5] http://www.naka.jaea.go.jp/english/ba/index.html
[6] K. Ida, et al., 2015 *Nuclear Fusion* **55** 104018(10pp).
[7] Azechi H. et al 2009 *Nucl. Fusion* **49** 104024
[8] http://phys.org/news/2015-08-world-powerful-laser-trillion-watts.html
[9] T. Norimatsu, et al., 2009 *Fusion Science and Technology* **56**, No1, 361-368.
[10] Y. Ogawa, T. Goto, K. Okano, Y. Asaoka, R. Hiwatari and Y. Someya, 2008 The fifth International Conference on Inertial Fusion Sciences and Applications (IFSA2007), Journal of Physics: Conference Series. **112** 032033.
[11] IFE Forum Report (March 2015, in Japanese).
[12] H. Azechi, private communication.
[13] http://www.mext.go.jp/b_menu/shingi/gijyutu/gijyutu2/056/shiryo/__icsFiles/afieldfile/2015/01/26/1354643_5.pdf
[14] F. Najmabadi, http://aries.ucsd.edu/LIB/MEETINGS/0310-USJ-PPS/

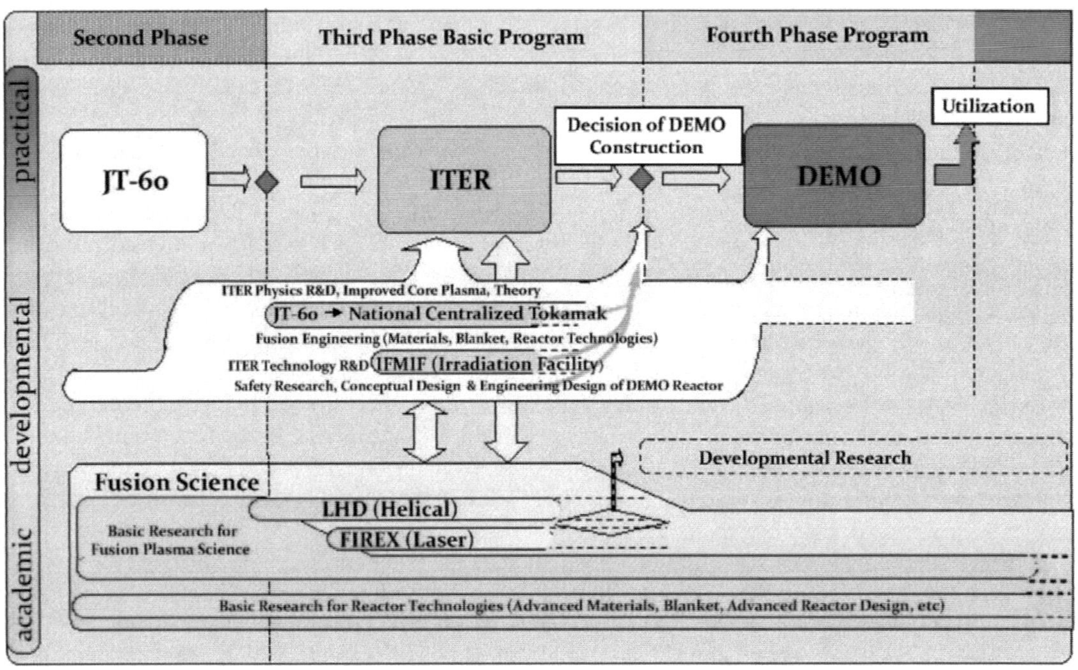

Fig. 1 A schematic drawing of strategy for fusion reactor development decided by AEC in Japan.

Journal of Physics: Conference Series **717** (2016) 012004 doi:10.1088/1742-6596/717/1/012004

Studies of high energy density physics and laboratory astrophysics driven by intense lasers

J Zhang[1,2], Y T Li[2,3], L M Chen[2,3], Q L Dong[2,4], J Y Zhong[2,5], W M Wang[2,3], Z M Sheng[1,2,6] and G Zhao[7]

[1]Key Laboratory for Laser Plasmas of Ministry of Education and Department of Physics and Astronomy, Shanghai Jiao Tong University, Shanghai, China

[2]Collaborative Innovation Centre of Inertial Fusion Science and Applications, Shanghai Jiao Tong University, Shanghai, China

[3] Beijing National Laboratory for Condensed Matter Physics, Institute of Physics, Chinese Academy of Sciences, Beijing 100190, China

[4]School of Physics and Optoelectronics, Ludong University, Yantai 264025, China

[5]Department of Astronomy, Beijing Normal University, Beijing 100875, China

[6]SUPA, Department of Physics, University of Strathclyde, Glasgow, United Kingdom

[7]Key Laboratory of Optical Astronomy, National Astronomical Observatories, Chinese Academy of Sciences, Beijing 100012, China

E-mail: jzhang1@sjtu.edu.cn

Abstract. Laser plasmas are capable of creating unique physical conditions with extreme high energy density, which are not only closely relevant to inertial fusion energy studies, but also to laboratory simulation of some astrophysical processes. In this paper, we highlight some recent progress made by our research teams. The first part is about directional hot electron beam generation and transport for fast ignition of inertial confinement fusion, as well as a new scheme of fast ignition by use of a strong external DC magnetic field. The second part concerns laboratory modeling of some astrophysical phenomena, including 1) studies of the topological structure of magnetic reconnection/annihilation that relates closely to geomagnetic substorms, loop-top X-ray source and mass ejection in solar flares, and 2) magnetic field generation and evolution in collisionless shock formation.

1. Generation and transport of laser produced hot electrons

1.1. Surface emission of fast electrons

Fast electron beams produced in ultra-intense laser-interactions with plasmas have attracted broad interest due to their potential applications in the fast ignition (FI) of inertial confinement fusion (ICF) [1], ultrafast X-ray and terahertz radiation, and so on [2]. About 20 - 50% of laser energy can be converted into hot electrons at laser intensity higher than 10^{18}Wcm^{-2}. While some of the hot electrons will be transported into overdense plasma regions, the rest of the electrons may be emitted along the target surface (or transport laterally) under certain conditions [3]. The latter can considerably enhance the hot electron density in the cone guided scheme for FI [4,5].

We have investigated the lateral transport of hot electrons comprehensively. When an intense laser pulse is incident upon a solid target with a large incident angle such as $\theta\sim70^{\circ}$, as shown in figure 1(a)

a collimated hot electron beam with a divergence angle of 15°(FWHM) is emitted along the front target surface [6]. It is found that this surface electron jet is dominated for large laser incidence angles and steep electron density gradients. If there are pre-pulses with an energy of several mJ before the main laser pulse, the surface electron beam emission will be affected significantly. Another evidence for the lateral hot electron transport is obtained by K_α imaging. For $\theta \sim 70°$, a horizontally asymmetric K_α halo structure has also been observed at the front surface of a Cu target, as shown in figure 1(b). The left wing of the halo extends to ~ 310 μm.

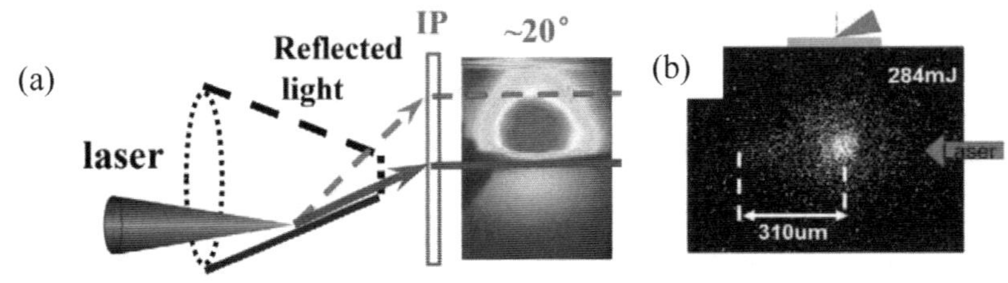

Figure 1.Measured angular distribution of > 300 keV hot electron beam (a) and K_α image (b) for θ=70°.

Hot electron beams with tight collimation, narrow energy spread and high charge are required for FI. In our recent work, highly collimated surface electron beams with MeV quasi-monoenergetic spectra are generated both in femtosecond and picosecond laser-solid interactions. In the case with fs lasers irradiating on a Cu target with a large incident angle, MeV-level monoenergetic electron beams with a divergence angle of only 1° and an average beam charge of tens of pC were obtained by optimizing the pre-pulse intensity [7]. By using the PHELIX laser in GSI and the TITAN laser at LLNL, surface electron beams with divergence angle as small as 2°, several nC charge with MeV peak energy were also generated [8].

1.2. Hot electron temperature scaling in a wide range of laser intensity

When the incident laser is at high intensity, most of the absorbed laser energy converts to the kinetic energy of hot electrons. The hot electrons follow a bi-Maxwellian distribution, i.e., $dN_e(E)/dE \sim A\exp(-E/T) + B\exp(-E/T_{hot})$, where T is the background electron temperature and T_{hot} is the hot electron temperature. Generally T_{hot} depends upon the plasma conditions (the plasma density scalelength) and the laser parameters (laser intensity, pulse duration, and laser incident angle). It scales with the laser intensity in the form of $T_{hot} \sim f(\tau, L)(I\lambda^2)^\alpha$, where $f(\tau, L)$ is a function of the laser pulse duration and the plasma density scalelength, I and λ are respectively the laser intensity and wavelength. The parameter α is a constant, which has been considered in many publications. Typically $1/4 \le \alpha \le 1$ under different conditions [9,10,11]. Based upon PIC simulations, we have investigated the hot electron temperature in a large range of parameters [12]. The results reveal that the value of α has a strong dependence on the dominant absorption mechanism. That is, $\alpha \sim 1/2$ when the laser intensity is high and the pondermotive acceleration dominates the absorption, which qualitatively agrees with the pondermotive potential scaling [10]. However, the value of T_{hot} can be several times higher than the ponderomotive potential, which can be attributed to the factor $f(\tau, L)$. The scaling with $\alpha \sim 1/3$ occurs at the transitional regime between resonance absorption (at lower intensities) and ponderomotive acceleration (at higher intensities).

Even though the hot electron temperatures from laser-solid interaction are confined to the scaling discussed above, there are ways to enhance the hot electron temperatures by the use of proper target designs. These introduce new absorption and/or electron acceleration mechanisms. One of the ideas is the use of a solid target with sub-wavelength structures at the front [13]. Both numerical simulations and experimental studies suggest this enhances laser energy coupling to the target, therefore both the hot electron and X-ray temperatures are enhanced [14,15]. Another idea is to adopt a target with multiple thin-slices along the laser propagation direction [16]. In this case, the inverse-free-electron acceleration mechanism works at the solid surface of the thin-slices [6,17,18], which can effectively enhance the hot electron temperatures. The produced hot electrons are also well-collimated in the forward direction, which is beneficial both for fast ignition and laser-driven ion acceleration.

1.3. Proposal of a magnetically assisted fast ignition scheme
A key issue with the FI scheme is to enhance the heating efficiency of high intensity lasers. The first integrated FI experiment in 2001 [4] demonstrated a high energy coupling of 20%, while some subsequent experiments performed in 2008-2011 achieved a much lower coupling efficiency of only 0.7-5% [19,20,21]. Explaining the discrepancy and then further improve the efficiency is crucial for the success of FI. Integrated simulations including both generation and transport of fast electrons to simulate the plasma with densities spanning 5 orders of magnitude will be required. Recently we have realized such integrated simulations via the so-called "two-systems" PIC approach. We find that the pre-plasma formed in the cone significantly affects the coupling, which can explain the contradictory experimental observations mentioned above. To overcome such pre-plasma effect, we propose a new FI scheme [22], imposing a multi-MG magnetic field across a cone-free target, which can also avoid the asymmetry in target compression. Our integrated simulation results show that the laser-to-core coupling efficiency is enhanced 7 fold up to 14% with an external magnetic field of 20 MG, where we take two counter-propagating1 PW laser pulses of 6 ps and achieve a compressed target with a peak density of 300 g cm^{-3} and areal density of 0.49 g cm^{-2} at the core. This is attributed to the constrained fast-electron motion along the magnetic field. This simulation suggests that the magnetically assisted scheme is promising.

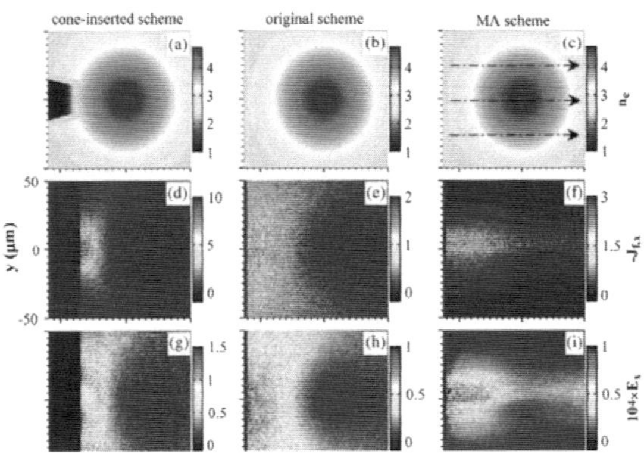

Figure 2. Snapshots of electron densities lg(n_e/n_c) initially (first row), fast-electron currents (unit of $en_c c$) at 2 ps (second row), and resistive electric fields (unit of $m_e\omega$/c) at 2 ps (third row). The three columns correspond to the cone-inserted, original, and magnetically assisted (MA) schemes, respectively. (f) and (i) shows that the magnetic field well confines the fast electron transport.

1.4. Controllable way for generating magnetic field larger than 100 teslas
As shown above, the magnetically assisted scheme for FI can significantly enhance the laser energy coupling to the target. Strong magnetic field of hundreds of Teslas or several MG was demonstrated on the experimental platform of ShenGuang(SG)-II laser facility by irradiating 2 kJ in 1 ns on a planar plate attached on the end of one open-ended coil [23]. The magnetic field is generated by cold background electrons to neutralize the positively charged laser foci, which has a different physical mechanism from the previous study of magnetic fields driven by hot electron current in a capacitor-coil target [24]. The strong magnetic fields generated is interesting not only for FI research, but also for other research areas including astrophysics, material science, and atomic and molecular physics, *etc.*

2. Laboratory studies of astrophysical and astronomical phenomena
Using high-power laser systems, unprecedented extreme physical conditions can be achieved at laboratories nowadays, allowing one to explore actively and controllably some important astrophysical problems. This is now called high energy density laboratory astrophysics (HEDLA) [25]. On the experimental platform of the SG II laser facility, we made some progress in HEDLA as presented below.

2.1. Modelling solar and geo-magnetic activities in laboratory

Clear evidence for magnetic reconnection in solar flares was captured [26] in X-ray and optical regions. For the slowly-developing/long-lived two-ribbon flares, a cusp-shaped bright loop formed as the reconnected separatrices are heated up by the accelerated electrons from magnetic reconnection (MR). For the short compact flare, a loop-top hard x-ray source could be produced when MR occurs above the magnetic loop as shown in figure 3(a). Solar activities also affect space environment around the Earth. Substorms of Earth's magnetosphere are produced when the solar wind arrives, compressing the geomagnetic field at magnetopause and stretching it at magnetotail. Then reconnection sites are formed locally, producing high energy particles which arrive two poles of Earth along geomagnetic field lines and excite the aurora.

Laser plasmas (LP) have strong spontaneous magnetic fields through Biermann battery effects. Therefore, experiments at the SG II laser facility can be performed by configuring a MR topology to help investigate solar eruptions and geomagnetic substorms [27,28]. In the experiments, we simply put a solid Cu target in front of the expected MR outflow. A bright spot with an area comparable to the cross section of the ion diffusion region is found when the MR outflow is ejected with enough kinetic energy. Figure 3(b) shows the comparison between the experimental result and the solar compact flare (figure 3(a)). The cartoon in figure 3(a) invokes the MR process to explain the formation of hard X-ray sources.

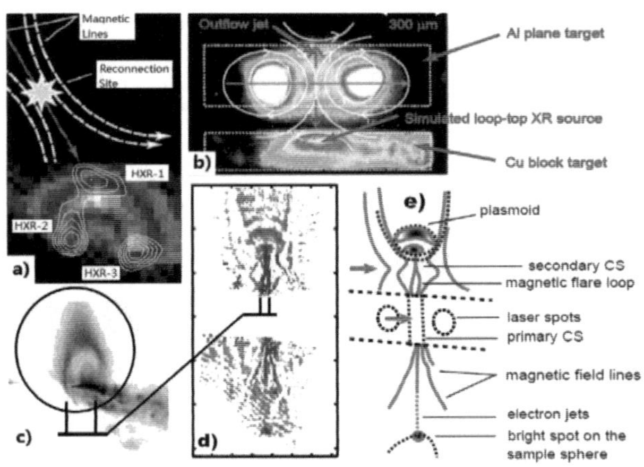

Figure 3. X-ray images of compact /short-period solar flare (a) and large/long-period solar flares (c), by Yohkoh telescope; Laser -plasma experiment results for loop-top hard x-ray emission (b); Detailed structure of MR region; (e) is the schematic cartoon of (d).

During the same series of laser plasma experiments, the moving plasmoid was also observed to stretch surrounding magnetic fields forming a flare with bright magnetic loops . Figure 3(d) gives the raw polarized image taken at 532 nm in the same series. For the clear signatures to be emphasized, a schematic picture of the experimental image is given in figure 3(e), with solid lines representing the bright lines. It is clear that the graph has two kinds of features above and below the line connecting the two laser focal spots, respectively. The upper half part of the image presents complicated features. As a whole, the fan-like region is confined by the two separatrices (or side-EDRs) as in the lower part, partially determined by the symmetrical configuration of the whole system. The detailed structure includes a bright current sheet (the 2nd CS), on the top of which exists a plasmoid with a bright spot on the bottom of it, a dark cavity, and then a bright edge located at the front. The bright spot on the bottom of the plasmoid is also connected to the 2nd CS. Attached to the other end of the 2nd CS are two bright ridges with foot-points located on the ending point of the primary/1st CS between the two mutually approaching laser-plasma bubbles, which mimics solar observations of long period flares as shown in figure 3(c). Our experimental observation confirms the theoretical prediction of possible generation of anomalous plasmoids from a current sheet. In solar CMEs, one also observed a stringy, concave-outward U-loop half-circling the ejected plasmoid from the coronal bottom. It is suspected that the U-loop in these cases might represent a detached magnetic field formed by reconnected open field lines, or some kinds of wave fronts with unclear generation mechanisms. The observation in present experiments is a plasmoid that is not yet completely detached from the secondary current sheet within the observation duration, confirming the astronomically observed U-loop or V-loop profiles of coronal mass ejection as the reconnected magnetic field lines.

The lower part of the image presents three bright jets from the central MR site. It argues that those jets are actually three electron diffusion regions, two of which are along the separatrices. The angle between the two side-EDRs is $2\theta_{rec}{\sim}40^{o}$, giving the reconnection rate of 0.25-0.35. The two side-EDRs can find their counterparts in Hall-MHD simulations and astronomic observations. Mozer et al. in 2005 summarized 19 observations of EDRs along separatrices in the magnetosheath. The third/ center-EDR is rather surprising in that, besides its appearance in PIC simulations, only one event was recorded by Cluster spacecrafts monitoring the magnetotail as reported by Phan et al. in 2007, and that the astronomic observations of side-EDRs and center-EDR were made separately in different missions. In the laser-plasma experiment, however, the three EDRs were recorded in one shot. The center-EDR appeared much later than the two side-EDRs. Comparisons between astronomic observations and experimental results indicate that the lack of simultaneous observation of the three EDRs is because MR has different characteristics in different evolution stages, or the center-EDR is not within the same plane of the two side-ones, i.e., MR is truly a 3D physical process.

2.2. Laboratory study of collisionless shocks mediated by magnetic fields

Magnetic field generation, amplification and self-organization play important or even key roles in many cosmic phenomena including collisionless shocks around supernova remnants and ultrahigh energy cosmic rays that are expected to originate from the former through the so-called Fermi acceleration mechanism. For this hypothesis to be verified in laboratory, three stages of magnetic field evolution have to be confirmed, with further requirements of observations of magnetic mediated collisionless shock and its acceleration of charged particles [29].

Laser plasmas produced by high power laser pulses usually have tremendous kinetic energy. In experiments designed for the above purpose, observations of endogenous electromagnetic (EM) field transformed from kinetic energy through various instabilities within counter-streaming collisionless laser plasma is rather preferred over the Biermann magnetic field. The re-organization behaviors of the latter usually blur the identification of magnetic generation mechanisms. To realize this, two plasmas need to penetrate without collision through each other and magnetic field diagnosis should be avoided at the interface where Biermann battery effect usually dominates [30,31,32]. With such criteria, experiments were conducted in 2011 with the SG II laser facility, and the ion-plasma Weibel filamentation instability was observed with interferometry and shadowgraphy techniques [33,34]. The plasma temperature could be calculated through the measured soft x-ray emission spectra.

Figure 4(a) shows a typical shadowgraph recorded at 4.5ns after the first bunch of 4 laser beams launched to produce the left-side plasma. The filaments can be divided into two groups according to their lengths. The longer filaments of 400-600 μm are supposed to be located in or parallel to face-on planes containing two laser spots, while shorter filaments are supposed to be the projections of those filaments with larger angles to face-on planes. The dashed line gives the position where the filament distribution pattern is analyzed in figure 4(b) by using different band FFT filters. The red line presents the filtered profile involving all (short and long) filaments spaced by 90-250 μm (9.3μm/pixel), while the blue line presents only long filaments spaced by 400-600 μm as indicated by the solid triangles. Both FFT-filtered distributions reproduce the corresponding peaks /filaments in the experimental profile. It can be expected that the space among real filaments in 3D is in the range of 400-600 μm, and the smaller value of 90-250 μm is due to the "forest-effect" that the projection process usually has. Calculation of the ion-plasma Weibel instability growth rate indicated that the Weibel instability grew at $\gamma{\sim}4\times10^{-5}\omega_{pe}$ and $k\,c/\omega_{pe}{\sim}0.17$, which means, for plasmas with electron density of 5×10^{18}cm^{-3}, the space between filaments is around 480 μm, well in agreement with the measurements. The larger filament length in the range of 400-600 μm is another evidence for the ion-plasma Weibel instability as the inertial length is around 450 μm for the ions propagating at 500-1000km/s in plasma of 50 eV.

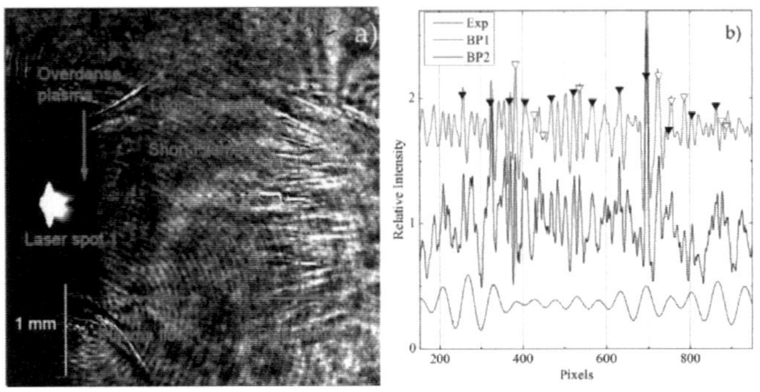

Figure 4. (a) Shadowgraph of counter-streaming plasmas, showing Weibel filamentation; (b) Filament profile analysis with BP1 for band passing of 90-250um (red line), and BP2 for 400-600um (blue line).

Acknowledgments

The authors wish to acknowledge all domestic and international colleagues contributing to the work. This work is supported by the National Basic Research Program of China (Grant Nos. 2013CBA01500), the National Nature Science Foundation of China, and Chinese Academy of Sciences.

References

[1] Tabak M *et al* 1994 *Phys. Plasmas* **1** 1626
[2] Zhang J *et al* 2005 *Appl. Phys. B* **80** 957; Zhang J *et al* 2005 *High Energy Density Physics* **1** 61
[3] Nakamura T *et al* 2004 *Phys. Rev. Lett.***93** 265002
[4] Kodama R *et al* 2001 *Nature* **412** 798
[5] Sentoku Y *et al* 2004 *Phys. Plasmas* **11** 3083
[6] Li Y T *et al* 2006 *Phys. Rev. Lett.* **96** 165003
[7] Mao J Y *et al* 2015 *Appl. Phys. Lett.***106** 131105
[8] Ma Y *et al* 2015 *Sci Sin-Phys Mech Astron* **45** 025202 (in Chinese)
[9] Sheng Z M *et al* 2015 *Chin. Phys. B* **24** 015201
[10] Wilks S C *et al* 1992 *Phys.Rev. Lett.* **69** 1383
[11] Beg F N *et al* 1997 *Phys. Plasmas* **4** 447
[12] Cui Y Q *et al* 2013 *Plasma Phys. Control. Fusion* **55** 085008 and references therein
[13] Wang W M, Sheng Z M and Zhang J 2008 *Phys. Plasmas* **15** 030702
[14] Kahaly S *et al* 2008 *Phys. Rev. Lett.* **101** 145001
[15] Mondal S *et al* 2011*Phys. Rev. B* **83** 035408
[16] Zheng J *et al* 2011 *Phys. Plasmas* **18** 113103
[17] Chen M *et al* 2006 *Opt. Exp.***14** 3093
[18] Pukhov A, Sheng Z M and Meyer-ter-Vehn J 1999 *Phys. Plasmas***6**2847
[19] Key M H *et al* 2008 *Phys. Plasmas* **15**, 022701
[20] Theobald W *et al* 2011 *Phys. Plasmas* **18**, 056305
[21] Shiraga H *et al* 2011 *Plasma Phys. Controlled Fusion* **53** 124029
[22] Wang W M *et al* 2015 *Phys. Rev. Lett.***114** 015001; *Phys. Rev. E* **91** 013101
[23] Zhu B J *et al* 2015 *Appl. Phys. Lett.* **107**, 261903
[24] Fujioka S *et al* 2013 *Rep. Sci.***3** 1170
[25] Remington B A *et al* 2006 *Rev. Mod. Phys* **78** 755
[26] Masuda S *et al* 1994 *Nature* **371** 495; Tsuneta S *et al.* 1992 *Publ. Astron. Soc. Japan* **44** L63-L69
[27] Zhong JY *et al* 2010 *Nat. Phys.* **6** 984
[28] Dong QL *et al* 2012 *Phys. Rev. Lett.* **108** 215001
[29] Cui YQ *et al* 2015 *Sci. China-Phys. Mech. Astron.* **58** 105201
[30] Fox W *et al* 2013 *Phys. Rev. Lett.* **111** 225002
[31] Huntington C M *et al* 2015 *Nat. Phys.* **11** 173
[32] Kugland N L *et al* 2012 *Nat. Phys.* **8** 809
[33] Liu X *et al* 2011 *New J. Phys.* **13** 093001
[34] Yuan D W *et al* 2013 *High Energy Density Physics* **9** 239

9th International Conference on Inertial Fusion Sciences and Applications (IFSA 2015) IOP Publishing
Journal of Physics: Conference Series **717** (2016) 012005 doi:10.1088/1742-6596/717/1/012005

Overview of Progress and Future Prospects in Indirect Drive Implosions on the National Ignition Facility

O A Hurricane on behalf of the ICF Program
Lawrence Livermore National Laboratory, PO Box 808, L-472, Livermore, CA 94550

E-mail: hurricane1@llnl.gov

Abstract. Alpha-particle self-heating, the process of deuterium-tritium (DT) fusion reaction products depositing their kinetic energy locally within the fusion reaction region and thus increasing the temperature in the reacting region with a concomitant exponential increase in the fusion reaction-rate, is the essential process needed for a fusion plasma to ignite. For the first time in the laboratory, significant alpha-heating in a fusion plasma was inferred in experiments and fusion fuel gain was demonstrated on the U.S. National Ignition Facility (NIF). Experiments on the NIF have achieved the highest yet recorded stagnation pressures ($P_{stagnation}$ > 150–230 Gigabar) of any facility based inertial confinement fusion (ICF) experiments, albeit they are still short of the pressures required for ignition on the NIF (i.e. ~ 300 – 400 Gbar), and have exhibited undesirable shape distortions that waste kinetic energy. We review the issues that have been uncovered and discuss the program strategy and plan that we are following to systematically address the known issues as we press on.

1. Introduction

The physics basis for indirect drive ignition on the NIF has been documented previously [1, 2, 3] as was the subsequent outcome of the National Ignition Campaign (NIC) [4] - the three-year effort to achieve ignition from 2009 to 2012. The NIC implosion focused upon a design that emphasized maximizing fuel ρR for the purpose of minimizing the energy required for ignition while achieving high gain. In this paper, we concentrate on the status and prospects based upon the last few years of post-NIC progress and scientific exploration on the NIF that leverage the remarkable diagnostic suite developed under the NIC aegis.

2. Progress since 2012

Since 2012, progress has been made towards ignition conditions (see Fig. 1) by using pulse shapes that are resistant to hydrodynamic instabilities [such as the ablative Rayleigh-Taylor (A-RT) instability] over a range of ablator materials, pulse-shapes, and hohlraum gas-fills. In a strategy of performing implosions that are less stressing of the physics by using laser pulse-shapes that mitigate the ablative Rayleigh-Taylor instability, reducing capsule convergence ratios, and making small systematic steps in a complex parameter space of initial conditions, notable accomplishments have been: (1) Fusion fuel-gain $G_{fuel} > 1$ with significant alpha-particle self-heating [5] (not to be confused with 'ignition' or 'target gain'); (2) Demonstrating stagnation pressures in excess of 200 Gbar [6, 7] at temperatures of about 5 keV; (3) Achieving more than a

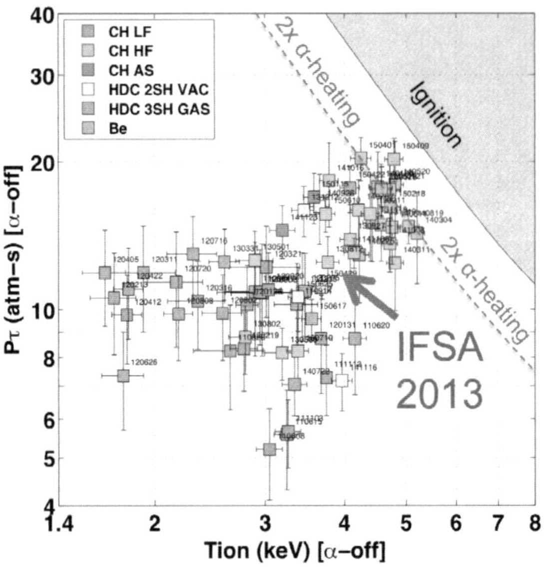

Figure 1. The classical Lawson Confinement parameter, $P\tau$, is plotted against ion temperature, T_{ion} for the various implosion campaigns on the NIF. The NIC low-foot implosion shots are shown in blue, the high-foot implosion shots are shown in green, the adiabat-shaping shots (both 3-shock and 4-shock) are shown in magenta, the near-vacuum hohlraum high-density-carbon (HDC) shots are shown in yellow, the high-gas-fill hohlraum HDC shots are shown in orange, and the one beryllium ablator shot to date is shown in grey.

doubling of the fusion yield due to alpha-particle self-heating with fusion yields of 26 kJ[6, 7, 8, 9]; (4) Demonstrated mitigation of the ablative Rayleigh-Taylor instability via laser pulse-shape modifications [10, 11, 12, 13, 14] in a way that was consistent with theoretical expectations [15, 16, 17]; (5) Demonstration of DT implosions that range in levels of fuel compression from ρR =0.6 g/cm^2 to ρR =1.1 g/cm^2 [18, 19, 20, 21, 22, 23, 24, 25] with fuel velocities in excess of 380 km/s [6, 7, 26] with no indications of mixing of ablator material into the hot-spot.

However, in the process of obtaining the above results a significant body of the data imply that the higher convergence and higher velocity indirectly-driven implosions of the NIF are not behaving like the ideal "text-book" [1, 27] implosion in-spite of mitigating the growth of hydrodynamic instabilities. This deviation from ideal behavior was not entirely unexpected, since numerous issues were suspected in degrading the performance of NIC implosions, and mitigating the A-RT only addressed one (dominant) issue. These newer A-RT mitigated and less stressful implosions on NIF serve as integrated diagnostics of the remaining difficulties.

2.1. High-foot Implosions

As was documented in a series of experiments that incrementally increased laser drive (Fig. 2) on high-foot implosions [18, 19], symmetry of the implosions' central hot-spot degraded as laser-power was increased in hohlraums with 1.6 mg/cm^3 of helium gas fill, resulting in oblate and sometimes even toroidal hot-spots. Increasing laser energy, instead of laser power, appeared to be preferable in terms of hot-spot symmetry control and net fusion performance, in contradiction to the conventionally expected behavior. A key element of managing implosion symmetry in the high gas-fill hohlraum was a reliance upon manipulating cross-beam energy transfer (CBET). This laser-plasma instability was used to force more energy onto the middle circumference of

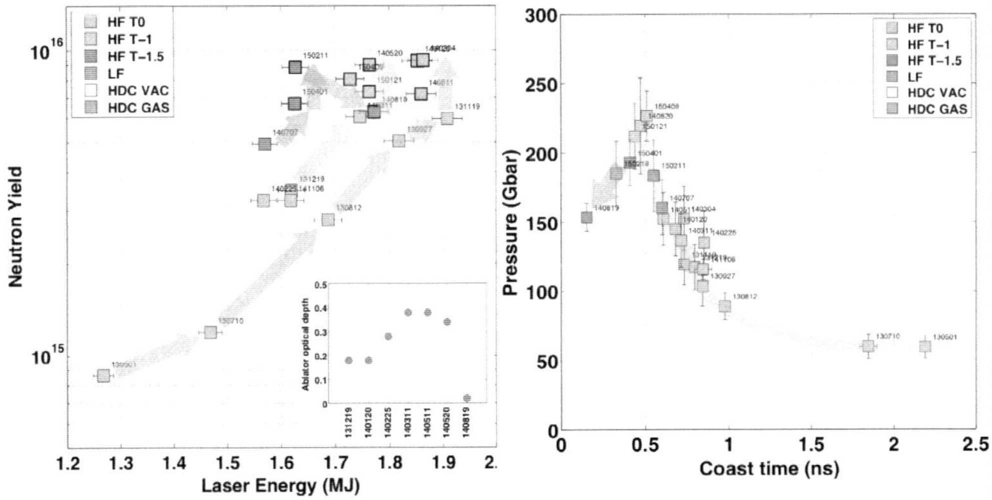

Figure 2. (Left Frame) High-foot DT implosion shots are plotted for three different ablator thicknesses: 195 μm ("T0," green), 175 μm ("T-1," cyan), and 165 μm ("T-1.5," magenta). Inset is the measured ablator optical depth inferred from the south-pole-bang-time (SPBT) x-ray diagnostic. The 165 μm shot that appeared to break the observed scaling of yield vs. laser energy (which is proportional to capsule absorbed kinetic energy) has little optical depth along the southern line-of-sight. The increase in yield observed in the plot at fixed laser energy is a result of switch the hohlraum material from pure gold (Au) to Au-lined depleted uranium (U). (Right Frame) Inferred stagnation pressure for the high-foot series vs. coast-time (the time difference between bang-time and the time the laser is turned off). Interestingly, small changes in coast-time ($\sim$ 100's of ps) result in large changes in stagnation pressure (and yield). At very short coast-times ($<$ 500 ps), the stagnation pressure appears to drop (albeit the number of data in that region is small). Very little ablator material remains for implosions with very small coast-times, placing the implosion at risk for burn-through or loss-of-confinement through thin-spots in the ablator.

the hohlraum [28, 29, 30, 31], but there are limits to this technique as there are limits to the maximum amount of energy the laser itself can supply. Moreover, even after years of using CBET, to successfully manage gross aspects of implosion symmetry, considerable uncertainty exists surrounding the detailed time-dependence of the energy transfer.

Further improvements in hot-spot symmetry control and net fusion performance were obtained by transitioning from a gold (Au) to depleted uranium (DU) hohlraum [7]. For a fixed amount of laser drive, DU effectively increases the peak hohlraum radiation temperature, T_r, and therefore ablation pressure. DU atomic physics differences also increase the emissivity at the hohlraum waist thereby driving the waist of the implosion slightly harder than in Au, and reducing the tendency towards oblate implosions. To a limit, reducing ablator thickness was also effective at improved symmetry control and net fusion performance [26]. Thinner ablators, while risking more instability because of increased in-flight-aspect-ratio, obtain higher implosion speed for a given peak T_r, and also inject less ablator plasma into the hohlraum thereby facilitating the propagation of the laser beams. While fusion yields and stagnation pressures increased as the implosion velocity was increased and reproducibility of nuclear performance was demonstrated [26], eventually a performance maximum and cliff were identified [6]. The observed performance cliff showed no evidence of hot-spot mix, but low measured optical depth along at least one diagnostic line-of-sight (Fig. 2), suggested thin-spots in the shell of the implosion that resulted in a loss of confinement [8].

Integrated two-dimensional (2D) post-shot simulations of the high-foot target series [32] and three-dimensional (3D) simulations [33] are largely self-consistent with data and imply that implosion asymmetry is a key mechanism responsible for the measured fusion performance of the

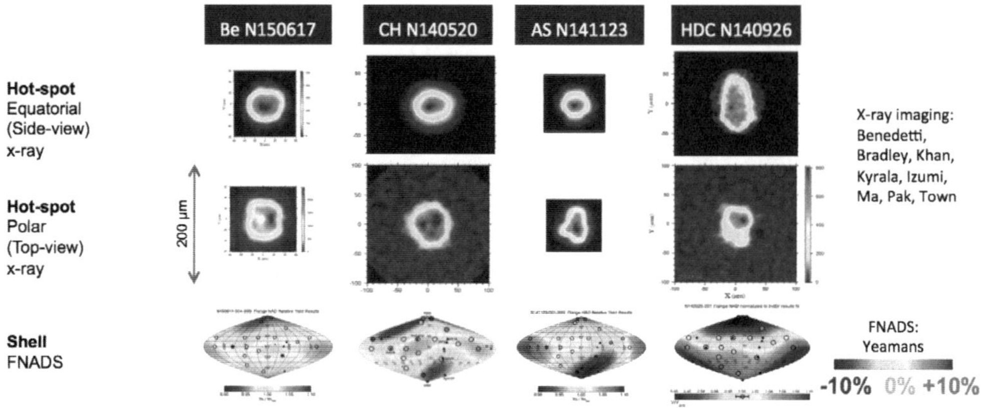

Figure 3. Time-integrated x-ray emission imaging data shows a variety of hot-spot morphologies for a sample of different NIF implosions (from left-to-right the beryllium DT implosion, one of the highest yield and stagnation pressure high-foot implosions, the highest fuel ρR adiabat-shaped implosion, and the highest yielding near-vacuum HDC implosion. The top row is the equatorial view of the implosion hot-spot and the middle row is the polar view of the implosion hot-spot. Similar shapes are seen in time-resolved x-ray emission data and in neutron imaging data (neither of which are shown here). Indirectly, the flange neutron activation data (FNAD) gives some reflection of the shell of DT fuel ρR distribution. Even in cases where the hot-spot emission appears round (and it usually isn't), the FNAD's indicates large deviations from a 1D implosion in the confining shell – this is a problem.

high-foot targets, but not the only mechanism (the tent and 3D features seem to be of increasing importance for the high-velocity high-foot experiments). Additionally, while mitigating A-RT reduces the damage to the implosion coming from the tent mounting membrane, data indicate the hot-spot [34] and shell [35] of the high-foot implosions are still impacted. Hot-electrons, generated in the plastic window plasma at the "laser-entrance-holes" (LEH's) [36, 37] are another potential source of asymmetry. Regardless of their origin, asymmetries degrade the conversion of kinetic energy of the imploding fuel shell into internal energy upon stagnation [32, 38] and, if severe enough, can lead to a loss of confinement through thin-spots that develop in the fuel and remaining ablator. Data and post-shot simulations indicate that, presently, many indirectly driven NIF implosions have poor shell symmetry [39, 40, 41] even when the hot-spot appears fairly round regardless of ablator or pulse-shape (Fig. 3).

2.2. Alternate Ablator Materials and Adiabat-Shaping

Remarkably, experiments using alternate and more efficient ablator materials such as beryllium (Be) [42, 43, 44] or high-density-carbon (HDC, i.e. diamond) [22] had the same essential fusion performance as CH ablators when driven by the same gas-filled hohlraum and high-foot pulse-shape (Fig. 4), even having the same observed hot-spot shape in the case of HDC [25]. Results which imply that the gas-filled hohlraum is dominating the performance of these implosions.

Attempts at retaining the good hydrodynamic stability properties of the high-foot targets, while returning to the higher levels of compression of the NIC implosion via adiabat-shaping [17, 45, 46], were successful in increasing the fuel ρR by $\sim 20\%$ from where the high-foot fuel ρR saturated in a high gas-fill hohlraum [20, 21, 47] and in greatly reducing the signatures of hot-electrons effusing from the hohlraum (at 350 TW and 1.6 MJ of laser power and energy). However, the fusion yield performance and inferred stagnation pressure were essentially not improved as compared to high-foot implosions with equivalent target geometry and laser drive (Fig. 4). These results suggest that the dense fuel-shell and central hot-spot properties are not as coupled as one would expect for a uniform 1D implosion. At higher laser drive (388 TW

Journal of Physics: Conference Series **717** (2016) 012005 doi:10.1088/1742-6596/717/1/012005

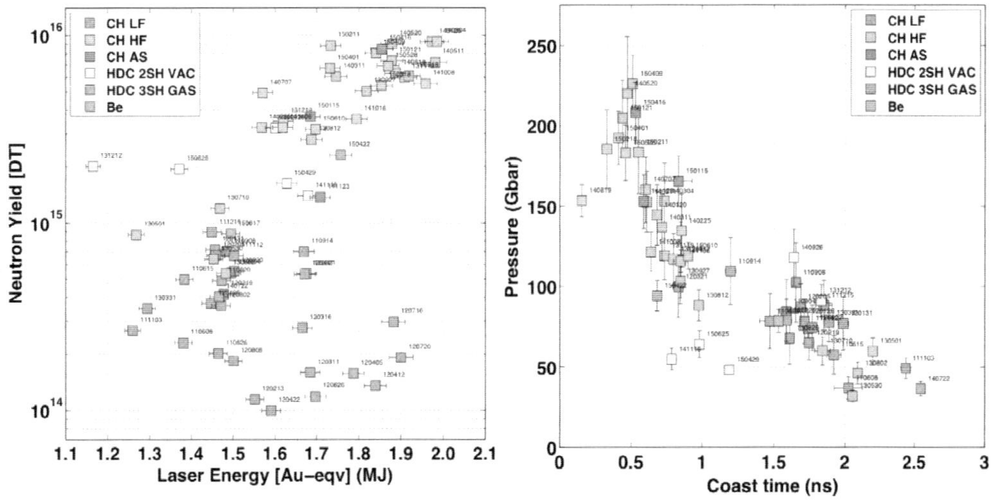

Figure 4. The left and right frames are similar to Fig. 2 in what is being shown, except now all DT implosion shots on NIF are plotted. The color key scheme of the shots is the same as is in Fig. 1. Some interesting patterns are observed. For example, for laser energies below 1.5 MJ, the near-vacuum HDC implosions have the highest yields, likely reflecting the superior ablation efficiency of HDC. The hot-spot performance as reflected in yield and stagnation pressure for high-foot and adiabat-shaped shots are essentially the same, even though the adiabat-shaped shot generally had $\sim 20\%$ larger total measured DT fuel ρR.

and 1.74 MJ) and with a thinner (175 μm) ablator, the total neutron yield and fuel ρR were comparable, to within error bars, for adiabat-shaped and equivalently driven high-foot shots [48], perhaps as a result of late-time hot-electron preheat at high drive.

2.3. Improved Hohlraums and Engineering Features

Larger case-to-capsule ratio (the ratio of inner hohlraum wall and outer capsule radii) near-vacuum hohlraums, which are well suited to the very short laser-pulse durations of the HDC ablator, have shown $\sim 98\%$ levels of laser-to-hohlraum energy coupling, the ability to control symmetry directly through laser cone-fraction without needing CBET, and levels of hot-electrons $100-1000\times$ less (over an energy range of 50-300 keV) [23, 24]. The addition of a small amount of helium gas (0.3 - 0.6 mg/cc), for a low gas-fill hohlraum configuration, makes it possible to utilize less efficient ablators such as CH yet retain much of the benefit of the near-vacuum hohlraum [49] – a direction which indirect-drive work on NIF is just now developing to address the obstacle of implosion shape control. Work on alternate geometry hohlraums (e.g. the "rugby"), which have also demonstrated low hot-electron levels is also ongoing in tight collaboration with CEA in France [50].

3. Conclusions

Work to clearly assess and mitigate the negative impact of engineering features in contact with the capsule is also just developing. Once the major challenges of implosion time-dependent shape control and capsule mounting features are addressed, it is anticipated that other challenges will emerge from the data. Progress will be made towards ignition by systematically addressing each challenge uncovered in a disciplined series of steps, getting into the burning plasma regime next, in order to better master implosion finesse. Experiments that back away from highly stressing (e.g. high convergence ratio) implosions will be pursued to assess where numerical simulations depart from observation in order to determine what improvements are needed in our predictive capability. In parallel, efforts focusing upon delivering more energy to the imploding DT fuel, which lessens the need for high finesse, will also be developed.

4. Acknowledgments

This work was performed under the auspices of the U.S. Department of Energy by Lawrence Livermore National Laboratory under Contract No. DE-AC52-07NA27344.

5. References

[1] J. D. Lindl, Inertial Confinement Fusion, (Springer, New York, 1998).
[2] Lindl, J.D., *et al. Phys. of Plasmas*, **11**, 2, 339-491, (2003).
[3] Haan, S.W. *et al. Phys. Plasmas*, **18**, 051001 (2011).
[4] Lindl, J., *et al.*, *Phys. Plasmas*, **21**, 020501 (2014). Op cit. Lindl, J. et al., *Phys. Plasmas*, **21**, 129902 (2014).
[5] Hurricane, O.A., *et al.* Nature, **506**, 343-348, (2014). doi:10.1038/nature13008. Op cit. doi:10.1038/nature13533.
[6] Callahan, D., *et al. Phys. Plasmas*, **22**, 056314 (2015).
[7] Doeppner, T., *et al.* Phys. Rev. Lett., **115**, 055001 (2015).
[8] Hurricane, O.A., *et al. Nature Physics*, doi: NPHYS3720, (2016).
[9] Betti, R., et al. *Phys. Rev. Lett.*, **114**, 255003 (2015).
[10] Raman, KS., *et al. Phys. Plasmas*, **21**, 072710 (2014).
[11] Peterson, J.L., *et al. Phys. Plasmas*, **22**, 056309 (2015).
[12] Smalyuk, V.A., *et al. Phys. Plasmas*, **22**, 072704 (2015).
[13] Casey, DT., *et al. Phys. Rev. E*, **90**, 011102 (2014).
[14] MacPhee, A., *et al. Phys. Plasmas*, **22**, 080702 (2015).
[15] Clark, D.S., *et al. Phys. Plasmas*, **22**, 022703 (2015).
[16] Dittrich, T. R., *et al. Phys. Rev. Lett.* 055002 (2013).
[17] Peterson, J.L., *et al. Phys. Rev. E*, *91*, 31101, (2015).
[18] Park, H.-S., *et al. Phys. Rev. Lett.*, **112**, 055001 (2014).
[19] Hurricane, O.A., *et al. Phys. Plasmas*, **21**, 056314 (2014).
[20] Casey, D. *et al. Phys. Rev. Lett.*, **115**, 105001 (2015).
[21] Smalyuk, V., *et al. Phys. Plasmas*, **22**, 080703 (2015).
[22] Mackinnon, A.J., *et al. Phys. Plasmas*, **21**, 056318-1 (2014).
[23] Berzak Hopkins, L.F., *et al. Phys. Rev. Lett.*, **114**, 175001 (2015).
[24] Meezan, N.B., *et al. Phys. Plasmas*, **22**, 062703-1 (2015).
[25] Ross, J.S., *et al. Phys. Rev. E*, **91**, 021101, (2015).
[26] Ma, T., *et al. Phys. Rev. Lett.*, **114**, 145004 (2015).
[27] Atzeni, S. and J. Meyer-ter-vehn, The Physics of Inertial Fusion, (Clarendon, Oxford, 2004)
[28] Michel, P. *et al. Phys. Rev. Lett.*, **102**, 025004 (2009).
[29] Michel, P. *et al. Phys. Plasmas*, **17**, 056305 (2010).
[30] Moody, J. D. *et al. Nature Physics*, **8**, 344-349 (2012).
[31] Callahan, D. A. *et al. Phys. Plasmas*, **19**, 056305 (2012).
[32] Kritcher, A., *et al. Phys. Plasmas*, in press, (2016)
[33] Clark, D.S. *et al. Phys. Plasmas*, **23**, 056302 (2016).
[34] Nagel, S.R., *et al. Phys. Plasmas*, **22**, 022704 (2015).
[35] Tommasini, R. *et al. Phys. Plasmas*, **22**, 32309 (2015).
[36] Regan S.P. *et al.*, *Phys. Plasmas*, **17**, 020703 (2010)
[37] Dewald, E.L., *et al. Phys. Rev. Lett.*, accepted, (2016).
[38] Kritcher, AL., *et al. Phys. Plasmas*, **21**, 042708 (2014).
[39] Spears, B.K., *et al. Phys. Plasmas*, **22**, 056317 (2015).
[40] Zylstra, A.B., *et al. Phys. Plasmas*, **22**, 056301 (2015).
[41] Pak, A., *et al. Phys. Plasmas*, **22**, 122701 (2015).
[42] Simakov, A. *et al.*, *Phys. Plasmas*, **21**, 022701 (2014).
[43] S. Yi, *et al. Phys. Plasmas*, **21**, 092701 (2014).
[44] Kline, J. *et al. Phys. Plasmas*, *accepted*, (2016).
[45] Clark, DS., *et al. Phys. Plasmas*, **21**, 112705 (2014).
[46] Baker, K., *et al. Phys. Plasmas*, **22**, 052702, (2015).
[47] Milovich, J., *et al. Phys. Plasmas*, **22**, 122702 (2015).
[48] Robey, H., *et al. Phys. Plasmas*, *accepted*, (2016).
[49] Hinkel, D.E., *et al.* Bull. of the American Physical Society 57th Annual Meeting of the Division of Plasma Physics, **60**(19), BO4.00004 (2015).
[50] Leidinger J.-P., *et al.*, NIF Rugby High Foot Campaign from the design side, Proc. of the 9th International Conference on Inertial Fusion Science and Applications, J. of Phys. Conf. Ser., submitted (2015).

The status of Fast Ignition Realization EXperiment (FIREX) and prospects for inertial fusion energy

H. Azechi and FIREX Project Team
Institute of Laser Engineering, Osaka University, Osaka Japan.
Email: azechi@ile.osaka-u.ac.jp

Abstract. Here we report recent progress for the fast ignition inertial confinement fusion demonstration. The fraction of low energy (< 1 MeV) component of the relativistic electron beam (REB), which efficiently heats the fuel core, increases by a factor of 4 by enhancing pulse contrast of heating laser and removing preformed plasma sources. Kilo-tesla magnetic field is studied to guide the diverging REB to the fuel core. The transport simulation of the REB accelerated by the heating laser in the externally applied and compressed magnetic field indicates that the REB can be guided efficiently to the fuel core. The integrated simulation shows > 4% of the heating efficiency and > 4 keV of ion temperature are achievable by using GEKKO-XII and LFEX, properly designed cone-fuel and an external magnetic field.

Under the auspices of Fast Ignition Realization EXperiment (FIREX) project [1], demonstration of the fast ignition inertial confinement fusion is being performed at Osaka University. Figure 1 shows a schematic of the FIREX experiment. A world-largest 2-petawatt LFEX laser system [2], which is currently capable of delivering 3 kJ in a 1.5 ps pulse using 4 laser beams with a pulse contrast of 10^{-10}, has been constructed next to the GEKKO-XII laser facility as shown in figure 1. The LFEX laser pulses are focused to generate a relativistic electron beam (REB), which heats a dense fuel core up to the ignition temperature (5 keV).

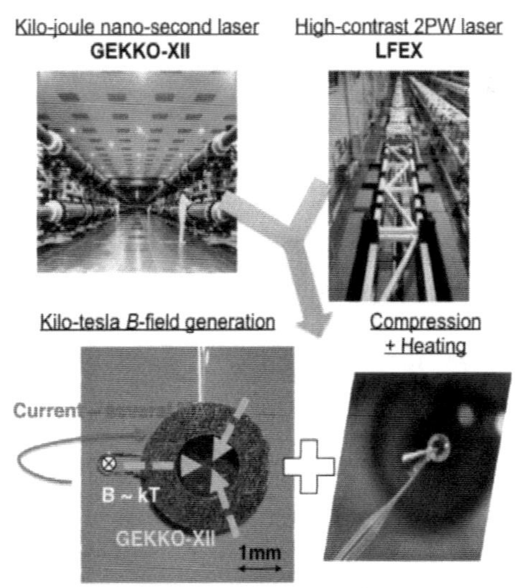

Figure 1. A spherical fuel target is compressed by the GEKKO-XII laser beams and the compressed core is heated by the high-contrast 2PW laser. Kilo-tesla magnetic field is generated by a capacitor-coil target driven by part of the GEKKO laser.

One of the critical problems of the FI scheme is the generation of high energy electrons. A pico-second LFEX laser pulse is inevitably preceded by several types of pulses. The initial pulse produces a plasma before the main pulse arrives, then laser-plasma interactions in a long-scale preformed plasma generate electrons that are too energetic (> 10 MeV) to efficiently heat the fuel core. The intensity contrast between the main pulse and the preceding pulses has already been improved in the LFEX laser system to 10^{-10} by introducing saturable absorbers in its front-end system. The good pulse contrast results in the enhancement of the high energy conversion efficiency (12%) from the LFEX laser to the low energy component of the REB, whose mean energy is close to the optimum value (1 MeV) [3].

The large divergence angle of the laser accelerated REB is the other critical problem of the FI scheme. Guiding the REB using an externally applied kilo-tesla magnetic field has been studied by simulations [4], and seems to be able to solve the REB divergence problem. A kilo-tesla magnetic field has been generated by using the capacitor-coil target [5], and kilo-tesla magnetic field strength was measured using a B-dot probe and proton radiography [6]. Guiding the REB using an externally applied magnetic field was previously demonstrated experimentally in a planar geometry.

In an actual FI experiment, a magnetic mirror structure associated with the fuel compression is generated. Compression of the fuel and magnetic field, and REB transport in the compressed magnetic field were studied numerically as shown in figure 2. Figure 2 (b) shows compressed magnetic field strength, a moderate mirror ratio ($R_m \sim 3$) between the cone tip (300 T) and compressed core (1 kT) can be generated using a spherically converging shock wave due to fast diffusion of the embedded magnetic field in the shock compressed region [7]. This moderate mirror structure is adequate to efficiently guide the REB to the fuel core without significant REB reflection induced by the mirror magnetic field. The integrated experiment designed by the integrated simulation is under preparation for being conducted in FY2016.

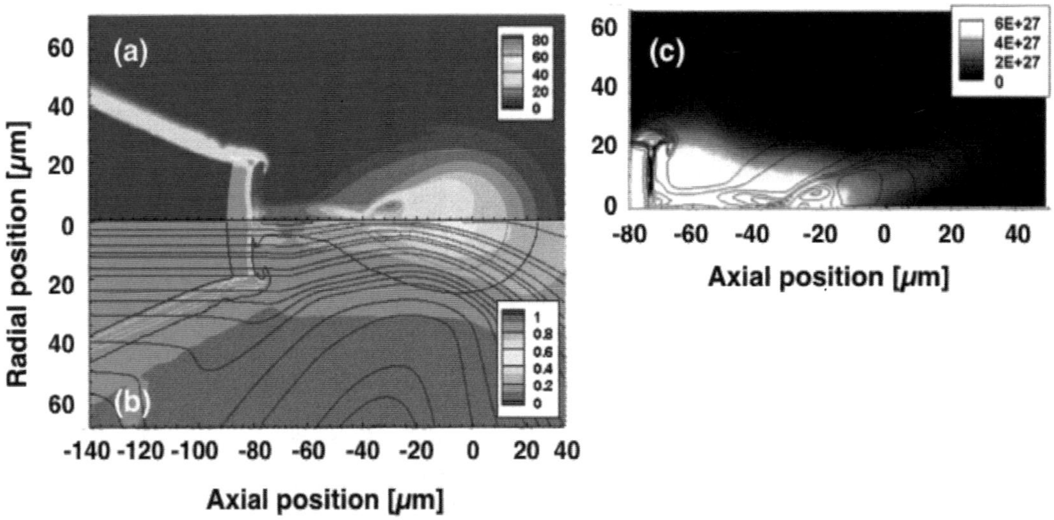

Figure 2. (a) Mass density profile of compressed plasma core under the magnetic field. (b) Lines are examples of the field lines, and color scale shows field strength in kT unit. Field strength at the cone tip is 300 T and the maximum compressed field strength at the maximum compressed area is 1 kT. Moderate mirror ratio ($R_m \sim 3$) is generated. (c) Energy density of the REB guided by the compressed external magnetic field.

After the achievement of the FIREX-I project, we will take the next step toward development of the inertial fusion energy. It will consist of ignition demonstration with the FI scheme at the FIREX-II project and development of an integrated engineering test. International collaboration and cooperation are essential to fast-track the FIREX-II project. We will launch the international planning committees of the FIREX-II project. On the other hand, the integrated engineering test will be focused on clarifying the engineering feasibility of the IFE reactor technologies such as the integration of high-repetition high-power laser, target injection-tracking, laser-beam steering, and fusion chamber-and-blanket technologies. This facility has a potential to generate ten to hundred trillion fusion neutrons per second.

References

[1] H. Azechi *et al.*, Nucl. Fusion **49**, 104024 (2009).
[2] N. Miyanaga *et al.*, J. Phys. IV **133**, 81 (2006).
[3] S. Fujioka *et al.*, submitted to Phys. Plasmas.
[4] T. Johzaki *et al.*, Nucl. Fusion **55**, 053022 (2015).
[5] S. Fujioka *et al.*, Sci. Rep. **3**, 1170 (2013).
[6] K. F. F. Law *et al.*, submitted to Appl. Phys. Lett.
[7] H. Nagatomo *et al.*, Nucl. Fusion **55**, 093028 (2015).

Journal of Physics: Conference Series **717** (2016) 012007 doi:10.1088/1742-6596/717/1/012007

Alpha Heating and Burning Plasmas in Inertial Confinement Fusion

R. Betti, A.R. Christopherson, A. Bose, K.M. Woo
University of Rochester, Rochester NY 14534, USA
E-mail: betti@lle.rochester.edu

Abstract.
Assessing the degree to which fusion alpha particles contribute to the fusion yield is essential to understanding the onset of the thermal runaway process of thermonuclear ignition. It is shown that in inertial confinement fusion, the yield enhancement due to alpha particle heating (before ignition occurs) depends on the generalized Lawson parameter that can be inferred from experimental observables. A universal curve valid for arbitrary laser-fusion targets shows the yield amplification due to alpha heating for a given value of the Lawson parameter. The same theory is used to determine the onset of the burning plasma regime when the alpha heating exceeds the compression work. This result can be used to assess the performance of current ignition experiments at the National Ignition Facility.

1. Introduction

In inertial confinement fusion (ICF) [1, 2] a shell of cryogenic deuterium and tritium (DT) ice is imploded at high velocities (300 - 400 km/s) and low entropy to achieve high central temperatures and high areal densities. The final fuel assembly consists of a relatively low-density (30 -100 g/cc), high-temperature (3 - 7 keV) core -the hot spot- surrounded by a dense (300 -1000 g/cc), cold (200 - 400 eV) fuel layer - the compressed shell. Fusion alphas are produced in the D+T fusion reactions with an energy of 3.5 MeV and slow down primarily through collisions with the plasma electrons. The alpha-heated electrons transfer part of their energy to the DT ions, thereby increasing the fusion reaction rate. The process of alpha-energy deposition to the hot spot of a compressed ICF capsule is called alpha heating. Ignition is a direct consequence of alpha heating and of its feedback on the thermal energy and fusion reaction rate. When this feedback process becomes unstable, it leads to a thermal runaway within the central hot spot. An ignited hot spot drives a burn wave in the surrounding dense shell, leading to fusion-energy outputs in the megajoule range that greatly exceed the thermal and kinetic energy supplied to the DT fuel by the implosion alone ($\sim$ tens of kilojoules).

To make progress toward ignition, it is important to assess the performance of current implosions with respect to the level of alpha particle heating. Heating by the alphas enhances the fusion yield to varying degrees depending on the fraction of deposited alpha particle energy to the total hot spot energy. Here we consider the yield amplification occurring before ignition takes place and therefore limit the amplification to within a factor of about 10 folds. Yield amplification $\leq$ 10 are of most interest to the current experiments at the National Ignition Facility (NIF) [3]. We show that the fusion yield enhancement due to alpha heating depends only on the Lawson parameter $P\tau$ [4] through a universal curve valid for any ICF ignition target from small to large, both directly and indirectly driven [5, 6]. The generalized Lawson parameter [7, 8] can be easily estimated from experimental observables and the yield enhancement due to alpha heating can be inferred from the universal curves [5, 6].

Recent experiments on NIF (High-Foot targets [9, 10]) have demonstrated significant alpha heating using indirect drive. To make progress toward ignition on NIF [3], it is crucial to be able to measure the level of alpha heating and to identify intermediate plasma states where the alpha heating is the leading source of input energy (alpha dominated or burning plasmas). In magnetic confinement fusion (MCF) [11], the burning plasma regime is identified through the thermonuclear Q= *fusion power output/external power input.* Since the alpha energy is about 1/5 of the total fusion energy, a Q=5 denotes the state where the alpha power equal the input power. For convenience, in this paper we use Q_α= *alpha power /input power*=Q/5 and define the onset of a burning plasma at Q_α=1.

While determining Q_α for a steady state MCF device is straightforward, the definition for ICF is complicated by the transient nature of an ICF implosion and by the fact that the vast majority of the input energy does not reach the DT plasma. Since this paper is only concerned with the physics of burning plasmas and not with the prospects for fusion energy, the relevant input energy is the one reaching the DT plasma where the fusion reactions occur. Therefore, the parameters Q, Q_α used here refer to the DT fuel and should not be confused with the engineering-Q used for fusion reactors [11].

2. A simple alpha-heating model

To illustrate this concept, we first consider the simplest ignition model valid for a static plasma at rest with a given pressure $P(0)$. For an ICF plasma, this represents the conditions of the central hot spot at stagnation before the alpha heating occurs. We will refer to the initial conditions as $no\alpha$ conditions. We assume that the alpha particles are locally deposited at times $t > 0$, the plasma temperature is large enough that the fusion reaction rate can be approximated with $< \sigma v > \sim T^2$ and the radiation losses are negligible compared to the alpha heating. In this case, the plasma pressure P is obtained from the energy conservation

$$\frac{dP}{dt} = \frac{P}{\tau}\left[\frac{P\tau}{S_\alpha(T)} - 1\right], \tag{1}$$

where τ is the energy confinement time (assumed constant), $S_\alpha \equiv 24T^2/\varepsilon_\alpha < \sigma v >$ and ε_α =3.5 MeV. Note that $S_\alpha \sim$ constant and Eq. (1) can be rewritten in the dimensionless form

$$\frac{d\hat{P}}{d\hat{t}} = \hat{P}[\chi_{no\alpha}\hat{P} - 1], \tag{2}$$

where $\chi_{no\alpha} \equiv P(0)\tau/S_\alpha$, $\hat{P} = P/P(0)$, and $\hat{t} = t/\tau$. Note that the subscript $no - \alpha$ in χ is used because $P(0)$ is the pressure without alpha heating ($P_{no\alpha}$) and $\chi_{no\alpha}$ is the Lawson parameter $P_{no\alpha}\tau$ normalized with the term S_α (e.g. χ is the normalized Lawson parameter). The solution of Eq. (2) can be written in the following simple form

$$\hat{P} = \frac{1}{\chi_{no\alpha} - (\chi_{no\alpha} - 1)e^{\hat{t}}} \tag{3}$$

and exhibit an explosive singularity for $\chi_{no\alpha} = 1$. The singularity defines the onset of ignition and therefore S_α represents the minimum value of $P_{no\alpha}\tau$ required for ignition. Following Ref. [7, 8], the ignition parameter $\chi_{no\alpha}$ sets the ignition condition for the $no\alpha$ parameters (the hydrodynamic pressure in this case) and can be rewritten as

$$\chi_{no\alpha} = \frac{P_{no\alpha}\tau}{[P_{no\alpha}\tau]_{ign}^{min}}, \tag{4}$$

Journal of Physics: Conference Series **717** (2016) 012007 doi:10.1088/1742-6596/717/1/012007

where the denominator ($= S_\alpha$) is the minimum requirement for ignition in terms of purely hydrodynamic parameters without the α-heating contribution. If the plasma pressure follows Eq. (1), one can easily calculate the total neutron yield $Y \sim V_p \tau P_{no\alpha}^2 \int_0^\infty \hat{P}^2 d\hat{t}$ where V_p is the plasma volume. Note that without α's, the pressure decays according to Eq. (3) with $\chi_{no\alpha} = 0$ leading to $\hat{P}_{no\alpha} = e^{-\hat{t}}$. Therefore, one can compute a neutron yield with or without alphas and measure the effect of alpha heating through the amplification of the yield due to the alphas

$$\hat{Y}_{amp} \equiv \frac{Y_\alpha}{Y_{no\alpha}} = \frac{2}{\chi_{no\alpha}^2} \left[ln \left(\frac{1}{1 - \chi_{no\alpha}} \right) - \chi_{no\alpha} \right]. \tag{5}$$

Note that the yield amplification is only a function of the $no\alpha$ Lawson parameter $\chi_{no\alpha}$ and it can be inferred from Eq. (5) if $\chi_{no\alpha}$ is known in the experiments. To relate $\chi_{no\alpha}$ to the stagnation properties of an ICF implosion, we follow Ref. [8] and estimate the confinement time from Newton's law $M_{sh} R_{hs}'' = 4\pi P R_{hs}$ applied to the dense shell of mass M_{sh} surrounding the hot spot of radius R_{hs}. By ordering $R_{hs}'' \sim R_{hs}/\tau^2$, $M_{sh} \sim (\rho_{sh}\Delta_{sh}) R_{hs}^2$ (where Δ_{sh} and ρ_{sh} are the shell thickness and density), and the yield as $Y \sim P^2 V_{hs} \tau$, we can rewrite χ as

$$\chi^3 \sim (\rho_{sh}\Delta_{sh})^2 \frac{Y}{M_{sh}} \sim (\rho R)^2 \frac{Y}{M_{sh}}. \tag{6}$$

Assuming that most of the areal density comes from the dense shell, the scaling $\rho_{sh}\Delta_{sh} \sim \rho R$ has been used in Eq. (6) and ρR is the measured total areal density. Except for the mass dependence, Eq. (6) has the same form of the experimental Ignition Threshold Factor ITFx [12]. The absence of the mass in the ITFx is due to the fact that the ITFx was derived by fitting the ignition condition of gain unity for targets with the same DT mass. Since the areal density is weakly dependent on alpha heating for yield amplifications ≤ 10, Eq. (6) leads to two χ parameters depending on whether the yield with or without alphas is used. The neutron yield measured in the experiments includes the contribution from the alphas and therefore the only measurable χ is χ_α that includes the effects of alpha heating on the yield. However the χ that determines the yield amplification in Eq. (5) and the ignition conditions is $\chi_{no\alpha}$. The constant of proportionality for χ in Eq. (6) can be obtained from Ref. [5] where the ignition parameter without alphas has been fit to a large database of hydrodynamic simulations leading to

$$\chi \simeq (\rho R)^{0.61} \left(\frac{0.24 Y_{16}}{M_{DT}^{unab}} \right)^{0.34} \tag{7}$$

where ρR is in g/cm^2, Y_{16} is the yield in 10^{16}, and M_{DT}^{unab} is the unablated DT mass in mg. Equation (7) provides a way to infer the parameter χ_α from the experimental observables (neutron yield, areal density and DT mass). To find the yield amplification from Eq. (5) requires $\chi_{no\alpha}$ that is related to the measurable parameter χ_α using the relation from Eq. (6) leading to $\chi_{no\alpha} \simeq \chi_\alpha / \hat{Y}_{amp}^{1/3}$. Substituting the latter into Eq. (5) leads to the following unique relation between the yield amplification $\hat{Y}_{amp}$ and the measurable parameter χ_α

$$\hat{Y}_{amp}^{1/3} = \frac{2}{\chi_\alpha^2} \left[ln \left(\frac{\hat{Y}_{amp}^{1/3}}{\hat{Y}_{amp}^{1/3} - \chi_\alpha} \right) - \frac{\chi_\alpha}{\hat{Y}_{amp}^{1/3}} \right]. \tag{8}$$

The yield in the presence of alpha heating can be obtained from Eq. (3) leading to

$$Y_\alpha = \frac{3}{2} \frac{V_p \tau P_{no\alpha}^2}{S_\alpha \varepsilon_\alpha} \frac{1}{\chi_{no\alpha}^2} \left[ln \left(\frac{1}{1 - \chi_{no\alpha}} \right) - \chi_{no\alpha} \right]. \tag{9}$$

The definition of χ given in (4) and the yield amplification in (5) are used to rewrite the yield in the following form,

$$Y_\alpha = \frac{3}{4}\frac{P_{no\alpha}V_p}{\varepsilon_\alpha}\chi_{no\alpha}\hat{Y}_{amp}. \tag{10}$$

Notice that $3/2P_{no\alpha}V_p$ is the hot spot energy from hydrodynamic compression which, in the limit of negligible radiation losses, comes from the pdV work of the imploding shell on the hot spot. It follows that the ratio of the alpha heating to the pdV work (Q_α) can be determined from Eq. (10) leading to,

$$Q_\alpha = \frac{1}{2}\chi_{no\alpha}\hat{Y}_{amp}(\chi_{no\alpha}) = \frac{1}{2}\chi_{no\alpha}(\hat{Y}_{amp})\hat{Y}_{amp}, \tag{11}$$

indicating that Q_α is a unique function of the Lawson parameter $\chi_{no\alpha}$ or alternatively a unique function of the yield amplification. A comprehensive analysis of the simplified alpha heating model is given in Ref. [6] where a more realistic expansion model is used for the hot spot.

3. Simulations and comparison with the comprehensive model

A more accurate model for the hot spot dynamics and the alpha heating process is given in Ref. [6, 5] where the temporal evolution of the hydrodynamic quantities is determined from the beginning of the shell deceleration phase up to peak compression and the following expansion phase. The fusion rate is approximated with $< \sigma v >\sim T^3$ which is fairly accurate in the interesting range 3–7 keV. The hot spot model consists of the equations for conservation of mass, momentum and energy with an ideal gas equation of state. The shell is described as a compressible plasma divided into two regions by the return (or rebound) shock. The inner portion of the shell is called the "shocked shell" while the outer portion is the "free-fall shell." The free-fall shell implodes inward with the implosion velocity while the shocked shell stagnates and converts its kinetic energy into internal energy. A fraction of the alphas produced within the hot spot escapes into the surrounding cold shell and deposits its energy near the inner shell surface thereby ablating shell material into the hot spot. This process is similar to the shell ablation driven by the heat flux leaving the hot spot. Details of the model are provided in Ref. [6] and the model is referred here as the "analytic model". Similarly to Eq. (5), the solution of the comprehensive analytic model exhibits a singularity (ignition) when $\chi_{no\alpha} \simeq 1$. After solving the analytic model with alphas ($0 < \chi_{no\alpha} < 1$) and without alphas ($\chi_{no\alpha} = 0$), the corresponding neutron yields with and without alphas are computed, and their ratio determines the yield amplification due to alpha heating. The solution of the analytic model is compared with the results of one- and two-dimensional hydrodynamic simulations using the codes LILAC [13] and DRACO [14], respectively. The properties with and without alphas are obtained from hydrodynamic simulations where the alpha deposition is turned on and off, respectively.

Figure 1 shows the yield amplification due to alpha heating as a function of the Lawson parameter without ($\chi_{no\alpha}$) and with (χ_α) alpha heating. The Lawson parameter is computed using the yield, areal densities and unablated DT mass from the simulations into Eq. (7). One- (black dots) and two-dimensional (red dots) simulations of ICF implosions are carried out with different target sizes ranging from 860 μm to 3.5 mm in diameter, laser energies from 30 kJ to 2 MJ, adiabats from 1 to 4 and velocities from 300 to 500 km/s. In the 2-D simulations, the level of implosion nonuniformities is varied by changing the initial DT ice roughness. Each simulation is carried out with and without alpha deposition, and the results are compared with the analytic model of Ref. [5] (solid curve). Note that the yield amplification is a strong function of the Lawson parameter χ with or without alpha heating and it can be fitted with the simple analytic fitting formulas

$$\hat{Y}_{amp} \approx \frac{1}{(1-1.04\chi_{no\alpha})^{0.75}}, \tag{12}$$

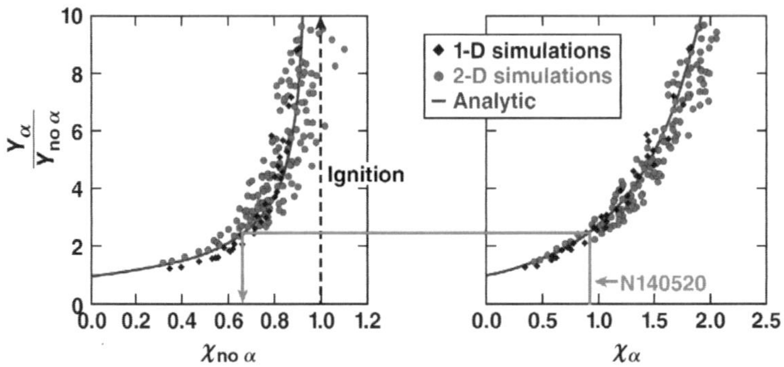

Figure 1. Yield amplification from alpha heating versus the Lawson parameter χ [Eq. (7)] without (left) and with (right) alpha-particle energy deposition. The black and red dots are from 1D and 2D simulations respectively. The simulations use directly driven targets with laser energies from 30 kJ to 2 MJ. The solid curves are from the analytic alpha-heating model of Ref. [5]. NIF high-foot implosion N140520 [10] exhibits $\chi_\alpha \simeq 0.93$, $\chi_{no\alpha} \simeq 0.66$ and a yield amplification of about 2.5×.

$$\hat{Y}_{amp} \approx \exp(\chi_\alpha^{1.2}). \tag{13}$$

When compared to the results of Spears and Lindl [15] for the NIF indirect-drive-ignition target ($M_{DT} \approx 0.18$ mg), the yield amplification curves are in good agreement with the data points from the simulation database of that specific target. In this paper, the analysis is extended to all targets, large or small, directly or indirectly driven as long as the ignition parameter χ_α is calculated using Eq. (7). For the high-foot shot N140520 [10] that achieved a yield of about 9×10^{15} neutrons, areal density of ≈ 0.8 g/cm^2, with $M_{DT} \approx 0.18$ mg, we find that $\chi_\alpha \approx 0.93$ and the yield amplification is ~ 2.5 (close to the value of 2.2 quoted in Ref. [10]). The corresponding $\chi_{no\alpha} \simeq 0.66$ is inferred from the left plot and the yield amplification.

Following the discussion in section 2 [Eq. (11)], the implosion performance with respect to the burning plasma parameter Q_α is assessed through Fig. 2 showing the yield amplification as a function Q_α. Here $Q_\alpha^{hs} = 1/2E_\alpha/W_{hs}$ where W_{hs} is the pdV work done by the shell on the hot spot and E_α is the total alpha particle energy. The factor 1/2 is due to the transient character of the implosion. Since the pdV work on the hot spot is positive only up to stagnation, we exclude the contribution from the alphas that are produced after stagnation (about 1/2 of the total). In addition to the hot-spot burning-plasma parameter Q_α^{hs}, Fig. 2 also shows the value of the total Q_α defined as the ratio of the alpha heating to the total compression work, $Q_\alpha^{tot} = 1/2E_\alpha/W_{tot}$ (where W_{tot} = pdV work to the hot spot and to the shell). The black dots are the results of 1D simulations using the code LILAC [13], the blue solid curve is the analytic model and the red curve is for a steady-state plasma. The red dot represents the High-Foot implosions N140520 [10] that achieved $Q_\alpha \simeq 0.6$ and $Q_{tot} \simeq 0.3$. The yellow dots in Fig. 2 show possible improvements of the High-Foot implosion achieving burning plasma conditions within the hot spot ($Q_\alpha^{hs} \simeq 1$) with about 50 kJ of total fusion-energy yield, as well as the more

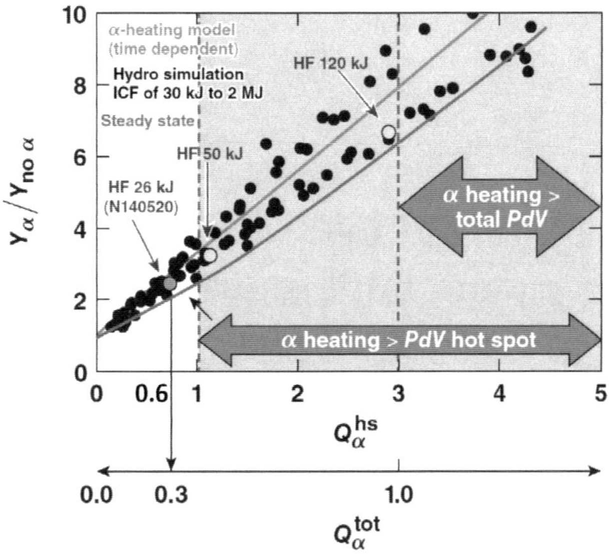

Figure 2. Yield amplification from alpha heating versus the burning plasma parameters Q_α^{hs} and Q_α^{tot}. The black dots are from 1D simulations. The solid blue curve is from the analytic model of Ref. [5]. The red curve is the steady state plasma result. The red dot represents the HF implosion 140520 [10]. The yellow dots show hypothetical improvements of the HF implosion to achieve the hot-spot burning plasma regime ($Q_\alpha^{hs} = 1$) and the state where the alpha heating equal the total input work to the hot spot and shell ($Q_\alpha^{tot} = 1$).

advanced burning plasma state where the alpha heating exceeds the total pdV work ($Q_\alpha^{tot} = 1$) with about 120 kJ of fusion yield. This material is based upon work supported by the DOE under Cooperative Agreements DE-FC02-04ER54789 (OFES) and under Award Number DE-NA0001944 (NNSA), the University of Rochester, and the New York State Energy Research and Development Authority. The support of DOE does not constitute an endorsement by DOE of the views expressed in this article.

References

[1] J.H. Nuckolls, L. Wood, A. Thiessen, G.B. Zimmermann, Nature **239**, 139(1972)
[2] S. Atzeni and J. Meyer-ter-vehn, *The Physics of Inertial Fusion* (Clarendon, Oxford, 2004); J. D. Lindl, *Inertial Confinement Fusion* (Springer, New York, 1998)
[3] M. J. Edwards et al., Phys. Plasmas **20**, 070501 (2013)
[4] J.D. Lawson, *Proc. Phys. Soc. London* B **70**, 6 (1957)
[5] R. Betti, et al., Phys. Rev. Lett., **114**, 255003 (2015)
[6] A. R. Christopherson, et al, submitted to Physics of Plasmas
[7] P. Y. Chang et al, Phys. Rev. Lett. **104**, 135002 (2010)
[8] R. Betti et al, Phys. Plasmas **17**, 058102 (2010)
[9] O. A. Hurricane et al, Nature **506**, 343 (2014)
[10] T. Doppner et al, Phys. Rev. Lett., **115**, 055001 (2015)
[11] J.P. Freidberg, *Fusion Energy and Plasma Physics* (Cambridge, Cambridge, 2007)
[12] B. K. Spears et al., Phys. Plasmas **19**, 056316 (2012)
[13] J. Delettrez et al, Phys. Rev. A **36**, 3926 (1987)
[14] P. B. Radha et al., Phys. Plasmas. **12**, 056307 (2005)
[15] J.D. Lindl, O. Landen, J. Edwards, E. Moses and the NIC Team, Phys. Plasmas 21, 020501 (2014) and Phys. Plasmas 21, 129902 (2014)

Demonstrating ignition hydrodynamic equivalence in direct-drive cryogenic implosions on OMEGA

V N Goncharov[1,2], S P Regan[1], T C Sangster[1], R Betti[1,2,3], T R Boehly[1], E M Campbell[1], J A Delettrez[1], D H Edgell[1], R Epstein[1], C J Forrest[1], D H Froula[1,3], V Yu Glebov[1], D R Harding[1], S X Hu[1], I V Igumenshchev[1], F J Marshall[1], R L McCrory[1,2,3], D T Michel[1], J F Myatt[1,2], P B Radha[1], W Seka[1], A Shvydky[1], C Stoeckl[1], W Theobald[1,2], B Yaakobi[1], and M Gatu-Johnson[4]

[1]Laboratory for Laser Energetics, University of Rochester, Rochester, New York 14623, USA
[2]Also at Department of Mechanical Engineering, University of Rochester, Rochester, New York 14623, USA
[3]Also at Department of Physics and Astronomy, University of Rochester, Rochester, New York 14623, USA
[4]Massachussets Institute of Technology, *Plasma Science and Fusion Center*, Cambridge, Massachusetts 02139, USA

E-mail: vgon@lle.rochester.edu

Abstract. Achieving ignition in a direct-drive cryogenic implosion at the National Ignition Facility (NIF) requires reaching central stagnation pressures in excess of 100 Gbar, which is a factor of 3 to 4 less than what is required for indirect-drive designs. The OMEGA Laser System is used to study the physics of cryogenic implosions that are hydrodynamically equivalent to the spherical ignition designs of the NIF. Current cryogenic implosions on OMEGA have reached 56 Gbar, and implosions with shell convergence CR$<$ 17 and fuel adiabat $\alpha > 3.5$ proceed close to 1-D predictions. Demonstrating hydrodynamic equivalence on OMEGA will require reducing coupling losses caused by cross-beam energy transfer (CBET), minimizing long-wavelength nonuniformity seeded by power imbalance and target offset, and removing target debris occumulated during cryogenic target production.

1. INTRODUCTION

The main approach to ignition by means of laser-driven inertial confinement fusion [1] (ICF) currently pursued by the National Ignition Facility [2] (NIF) is x-ray (or indirect) drive. In the other ICF laser approach, direct drive, the target is driven by laser irradiation directly coupled to the plasma blowing off the imploding capsule. The main advantage of the indirect-drive approach is reduced sensitivity of drive uniformity to short-scale beam nonuniformities. The main advantage of direct drive is higher coupling efficiency (by a factor of 3 to 5) of the laser energy into kinetic energy of the shell (hydrodynamic efficiency) compared to that of x-ray drive. The OMEGA Laser System [3] and the KrF laser NIKE at the Naval Research laboratory [4] have been the principal facilities for direct-drive experiments in the United States. Early challenges in achieving beam uniformity required for ignition have been resolved over the last several decades by introducing several beam smoothing techniques. These include distributed

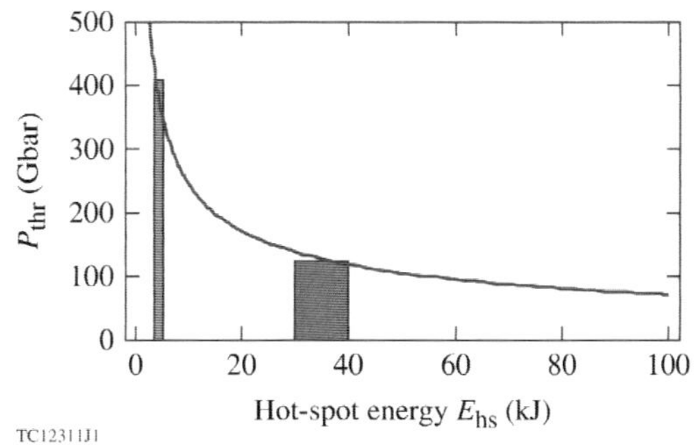

Figure 1. Threshold hot-spot pressure p_{thr} as a function of hot-spot internal energy. Typical hot-spot energies in indirect- and direct-drive implosions for a NIF-scale laser energy are shown with blue and red boxes, respectively.

phase plates [5] (DPP's), polarization smoothing with birefringent wedges [6], and smoothing by spectral dispersion [7] (SSD). In addition, implementing adiabat-shaping techniques [8, 9] significantly reduced the impact of Rayleigh–Taylor (RT) instability growth [1] during shell acceleration. Also, imprint reduction was demonstrated by using mid-Z–doped ablators [10] and high-Z target overcoats [11]. These developments have made the direct-drive approach very attractive. Such progress and the challenges in achieving ignition on the NIF using x-ray drive suggest considering direct drive as a viable alternative for developing a burning plasma platform in a laboratory.

Compared to the x-ray drive, direct-drive targets couple a larger fraction of laser energy into shell kinetic energy and internal energy of the neutron-producing central region of the target (hot spot) at peak fuel compression. This relaxes the requirement on shell convergence and hot-spot pressure in an igniting target. This can be shown with the help of a commonly used ignition condition (which can also be derived from generalized Lawson criterion [12]) according to which plasma self-heating is initiated by pdV work of the shell and alpha heating if the product of areal density and ion temperature inside the shot spot satisfies [1, 13]

$$(\rho R)_{\rm hs} \times T \gtrsim 0.3 \text{ g/cm}^2 \times 5 \text{ keV}, \quad (1)$$

where ρ, $R_{\rm hs}$, and T are the hot-spot density, radius, and temperature. Substituting expressions for the pressure $p_{\rm hs} = (1+Z)\rho T/m_{\rm i}$ (Z is the average ion charge and $m_{\rm i}$ is the average ion mass) and internal energy $E_{\rm hs} = 3/2\ p_{\rm hs} V_{\rm hs}$ ($V_{\rm hs}$ is the neutron-averaged hot-spot volume) into Eq. (1) gives a minimum pressure requirement (threshold) for ignition,

$$p_{\rm hs} > p_{\rm thr} = 250 \text{ Gbar}\left(\frac{E_{\rm hs}}{10 \text{ kJ}}\right)^{-1/2}, \text{ or } \bar{P} = \frac{p_{\rm hs}}{p_{\rm thr}} = \left(\frac{p_{\rm hs}}{250 \text{ Gbar}}\right)\sqrt{\frac{E_{\rm hs}}{10 \text{ kJ}}} > 1, \quad (2)$$

where $\bar{P}$ is the ignition pressure parameter.

The dependence of the threshold pressure $p_{\rm thr}$ on the hot-spot internal energy is shown in Fig. 1. Spherically symmetric direct-drive cryogenic designs on OMEGA couple up to 0.44 kJ (out of 26-kJ incident laser energy) into the hot-spot internal energy. When hydrodynamically scaled to the NIF-size laser energy (1.5 MJ to 1.8 MJ), these designs are predicted to couple 5× to 10× more energy into the hot spot [25 kJ to 40 kJ, depending on laser coupling efficiency (see red-shaded region in Fig. 1)] compared to that of indirect drive [4 kJ to 5 kJ (see blue-shaded region in Fig. 1)], resulting in 2.5× to 3× lower hot-spot pressures required for ignition. The hot-spot size also gets larger with $E_{\rm hs}$, leading to smaller shell convergence (Cr$\sim$ 22 compared to 35 to 40 in present x-ray–drive ignition designs) and less-demanding long-wavelength drive uniformity requirements.

2. OMEGA cryogenic implosions

To separate one-dimensional (1-D) factors that limit the target performance (drive efficiency, adiabat, etc.) from three-dimensional (3-D) effects, dedicated experiments are performed on OMEGA with the purpose of improving physics understanding and accuracy of 1-D code predictions. To identify critical implosion parameters, the 1-D scaling laws for peak pressure, hot-spot energy, and the ignition pressure parameter are written in terms of implosion velocity $v_{\rm imp}$ (defined as the peak mass-averaged shell velocity), the drive (ablation) pressure p_a, and the in-flight shell adiabat α (ratio of the shell pressure to Fermi pressure at shell density) [14]:

$$p_{\rm hs}^{\rm 1-D} \sim \frac{p_a^{1/3} v_{\rm imp}^{10/3}}{\alpha}, \; E_{\rm hs}^{\rm 1-D} \sim E_{\rm kin} \frac{v_{\rm imp}^{4/3}}{\alpha^{2/5} p_a^{4/15}}, \; \bar{P}_{\rm 1-D} \sim \frac{\sqrt{E_{\rm kin}} v_{\rm imp}^4 p_a^{1/5}}{\alpha^{6/5}}. \tag{3}$$

The implosion velocity and shell kinetic energy are inferred in an experiment by measuring ablation-front trajectory and mass ablation rate using self-emission imaging [15]. The ablation pressure is inferred from simulations that match the measured ablation-front trajectory, mass ablation rate, bang time [16], and scattered-light power and spectrum [17]. Finally, the shock-induced adiabat is inferred by measuring shock velocities early in the pulse using VISAR [18]. An additional fuel-adiabat increase caused by hot-electron preheat is estimated by measuring the hard x-ray signal [19] and areal density [20, 21] in mid- to high-adiabat implosions (the areal density in 1-D, for a given laser energy, depends mainly on shell adiabat [22], $\rho R \sim \alpha^{-0.5}$). A detailed comparison of 1-D simulation results using the hydrocode *LILAC* [23] with the data [14] shows good agreement between the two for a variety of target designs and drive conditions. One-dimensional simulations include nonlocal thermal transport model [24], a ray-based cross beam energy transfer (CBET) model [25], and first-principle EOS (FPEOS) models [26] for both the DT ice and CD ablator.

An analysis of direct-drive implosions on OMEGA has shown that coupling losses related to CBET [25] significantly limit the ablation pressure (as much as 40% on OMEGA and up to 60% on the NIF-scale targets), implosion velocity, and shell kinetic energy. Considering such losses, demonstrating the hydrodynamic equivalence of implosions on OMEGA to ignition designs on the NIF requires the shell IFAR to exceed the current stability threshold level (~ 22) [14]. One of the CBET mitigation strategies [27] involves using laser illumination with a laser beam diameter that is smaller than the initial shell diameter. This strategy, as demonstrated both theoretically and experimentally, recovers some coupling losses and increases the ablation pressure. Since the effect of CBET is small early in the implosion when the density scalelength and laser intensity are small, beam zooming schemes [28] consider the beam focal spot at an early time to be at the initial target radius (to maximize the illumination uniformity), reducing it to 0.6× to 0.7× of that size at the beginning of the main drive.

While zooming implementation on OMGEA is still a few years away, a test of the CBET reduction strategy was performed using "static" DPP's that produce the focal spots smaller than the initial target size through the entire drive pulse. The focal-spot radius (defined as the radius of a 95% beam energy contour) in these experiments was fixed at $R_{\rm b} = 410\ \mu$m. The ratio of $R_{\rm b}$ to target size ($R_{\rm target}$) was changed by varying the $R_{\rm target}$ from 400 μm to 500 μm. SSD was off during the main pulse for larger targets with $R_{\rm target} = 450\ \mu$m, 480 μm, and 500 μm (to increase on-target laser energy), while the targets with $R_{\rm target} = 400\ \mu$m, 430 μm, and 450 μm used SSD for the entire pulse. Hence, the ratio $R_{\rm b}/R_{\rm target}$ changed from 1.025 to 0.78. According to simulation results (which matched the observables), the smallest target ($R_{\rm target} = 400\ \mu$m) had $v_{\rm imp} = 3.5$ to 3.6×10^7 cm/s and a hydroefficiency of $f_{\rm hydro} = 3.5\%$, while the largest target had similar implosion velocity $v_{\rm imp} = 3.6$ to 3.7×10^7 cm/s, but more than twice larger hydroefficiency $f_{\rm hydro} = 7.2\%$. Such an increase in hydroefficiency was partially caused by smaller refraction losses experienced by the larger target (smaller $R_{\rm b}/R_{\rm target}$ and

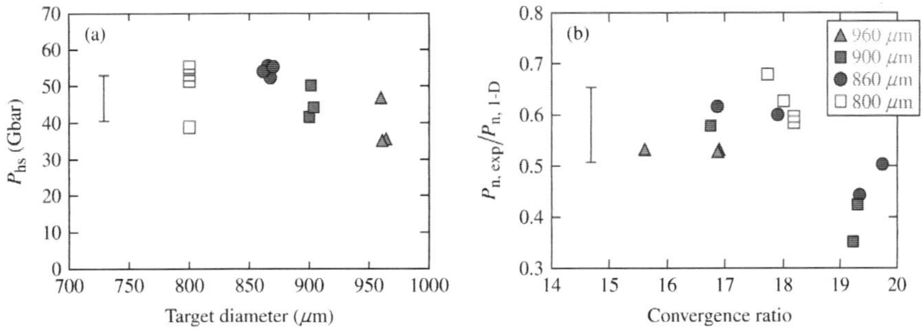

Figure 2. (a) Hot-spot pressure, inferred from experimental observables, as a function of target size. (b) Inferred hot-spot pressure normalized to 1-D code predictions versus the predicted shell convergence at 1-D bang time. The vertical lines on both plots show a typical experimental error bar.

larger density scale length) and partially by reduced CBET losses. To quantify each effect, a simulation was performed with $R_{\text{target}} = 500\ \mu\text{m}$, where the R_{b} was increased to match R_{target} and the laser energy was increased from 27.5 kJ (used in an experiment) to 32 kJ to avoid coasting chase. In such a simulation, implosion velocity was dropped by 11% to $v_{\text{imp}} = 3.2 \times 10^7$ cm/s and the shell hydroefficiency was reduced by 26% to $f_{\text{hydro}} = 5.3\%$.

Figure 2 shows the target performance (p_{hs}) for different target diameters. The hot-spot pressure is inferred [29] by using the measured neutron yield, burn duration Δt_{burn} [using both the neutron temporal diagnostics (NTD) [16] and framing-camera measurement of the x-ray-burn duration], neutron-average ion temperature $\langle T_{\text{i}} \rangle_{\text{n}}$, and hot-spot size R_{17} (defined as the radius of 17% of peak emission contour for x rays in the 4-keV to 7-keV energy range) at bang time using high-resolution ($\sim 6\mu$m in space and $\sim$ 30 ps in time) time-resolved Kirkpatrick–Baez framing camera [30]. Assuming an isobaric hot spot and fitting the burn history with a Gaussian with FWHM= Δt_{burn}, the maximum burn rate N_{max} is related to neutron yield Y as $N_{\text{max}} = 2Y\sqrt{\ln 2/\pi}/\Delta t_{\text{burn}}$, where $N_{\text{max}} = n_T n_D T^2 \int_{V_{\text{hs}}} dV \langle \sigma v \rangle / T^2$. Therefore, pressure at bang time can be determined using $p_{\text{hs}} \simeq \left[8Y\sqrt{\ln 2/\pi}/(f_D f_T \Delta t_{\text{burn}} \int_{V_{\text{hs}}} dV \langle \sigma v \rangle / T^2)\right]^{1/2}$, where $\langle \sigma v \rangle$ is the cross section for D-T reactions, and f_D and f_T are the fractions of D and T in the fuel, respectively. In evaluating the spatial integral in the latter equation, the following spatial profile for the ion temperature (obtained using simulation results) is assumed: $T(r) = T_{\text{c}}[1 - (r/R_{\text{hs}})^2(1 - 0.15^{3/2})]^{2/3}$, where T_{c} is the maximum hot-spot temperature, determined by matching $(\int_{V_{\text{hs}}} dV \langle \sigma v \rangle / T)/(\int_{V_{\text{hs}}} dV \langle \sigma v \rangle / T^2)$ with the measured $\langle T_{\text{i}} \rangle_{\text{n}}$], and, as follows from code predictions, R_{hs} and R_{17} are related using $R_{\text{hs}} = 1.06\ R_{17}$.

To understanding trends shown in Fig. 2, the effect of shell nonuniformity must be considered. The evolution of long-wavelength nonuniformities seeded by target offset, beam geometry, beam-power imbalance, and mispointing is studied using the 3-D hydrocode *ASTER* [31]. These simulations show that the bubbles (region of low-density material that protrudes from the central region into the higher-density shell), which develop because of the RT growth of long-wavelength perturbations ($l \lesssim 5$) during shell deceleration, increase the volume of central region V_{cntr}, reducing the hot-spot pressure ($p_{\text{hs}} \sim 1/V_{\text{cntr}}^{5/3}$) and neutron yield. As the shell continues to converge, the bubbles eventually break out of the shell, quenching hot-spot confinement and neutron yield (see Fig. 3). Earlier peak burn leads to sampling implosion conditions when shell convergence and central pressure have not yet reached the peak values. The 3-D effects also increase the central region volume, reducing the hot-spot pressure and preventing fuel material from reaching stagnation. This limits conversion efficiency of shell kinetic energy into the internal

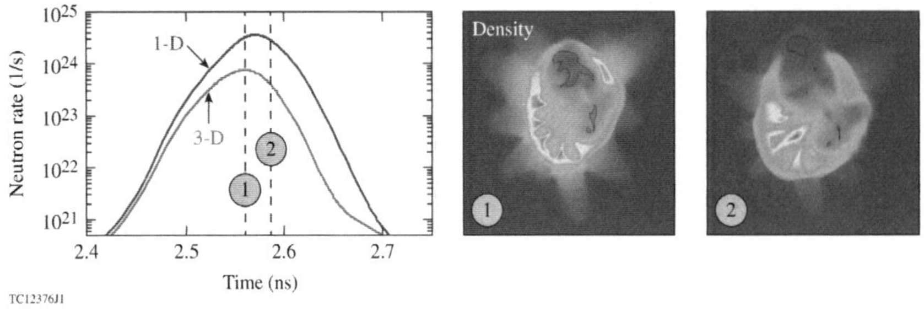

Figure 3. Neutron-production rate calculated using the code *ASTER* without (labeled "1-D") and with (labeled "3-D") effects of long-wavelength nonuniformity growth. Also shown are simulated shell-density maps at times labeled by (1) and (2).

energy of the hot spot. To account for the first effect (early pressure sampling), Fig. 4 plots the inferred hot-spot pressure normalized to the predicted pressure at the measured (earlier) bang time as a function of 1-D shell convergence calculated at the experimental bang time. Figure 4 shows that implosions with a fuel adiabat of $\alpha > 3.5$ proceed close to 1-D predictions up to a shell convergence of $C_r \sim 17$. An additional shell convergence does not lead to an increased *pdV* work on the hot-spot because of the RT growth of low-l modes. Further reduction in the target performance at lower fuel adiabat is caused by compromised shell integrity due to short-wavelength nonuniformity growth during shell acceleration seeded by shell imperfections (mainly target debris).

In summary, the cryogenic campaign with reduced beam size relative to the target size ($R_b/R_{target} < 1$), performed on OMEGA to reduce CBET losses, demonstrated increased laser coupling and hydrodynamic efficiency. This coupling enhancement, however, did not improve the target performance. Numerical simulations indicate that long-wavelength nonuniformities caused by target offset and power imbalance lead to an increased target central volume and early burn truncation. This effect is exacerbated by a reduction in beam overlap when the target size is increased relative to beam size.

Acknowledgments
This material is based upon work supported by the Department of Energy National Nuclear Security Administration under Award Number DE-NA0001944, the University of Rochester, and the New York State Energy Research and Development Authority. The support of DOE does not constitute an endorsement by DOE of the views expressed in this article.

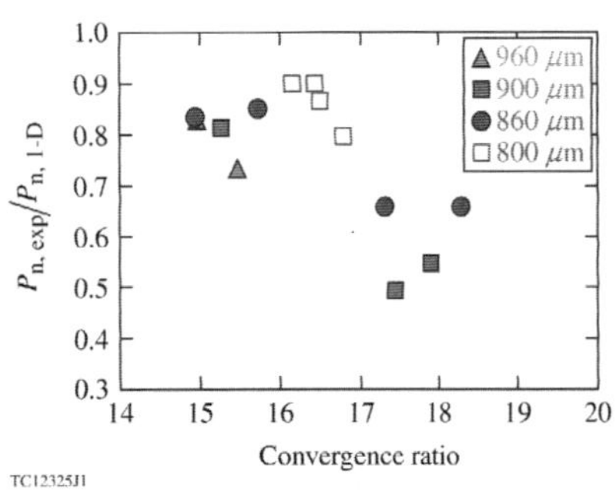

Figure 4. Inferred hot-spot pressure normalized to the 1-D predictions calculated at the experimental bang time versus 1-D shell convergence at the experimental bang time.

Journal of Physics: Conference Series **717** (2016) 012008 doi:10.1088/1742-6596/717/1/012008

References

[1] Atzeni S and Meyer-ter-Vehn J 2004 *The Physics of Inertial Fusion: Beam Plasma Interaction, Hydrodynamics, Hot Dense Matter* International Series of Monographs on Physics (Oxford: Clarendon Press)

[2] Paisner J, Boyes J D, Kumpan S A, Lowdermilk W H and Sorem M S 1994 *Laser Focus World* 30 75

[3] Boehly T R, Brown D L, Craxton R S, Keck R L, Knauer J P, Kelly J H, Kessler T J, Kumpan S A, Loucks S J, Letzring S A et al. 1997 *Opt. Commun.* 133 495

[4] Obenschain S P, Bodner S E, Colombant D, Gerber K, Lehmberg R H, McLean E A, Mostovych A N, Pronko M S, Pawley C J, Schmitt A J et al. 1996 *Phys. Plasmas* **3** 2098

[5] Kessler T J, Lin Y, Armstrong J J and Velazquez B 1993 in *Laser Coherence Control: Technology and Applications* Powell H T and Kessler T J eds (Bellingham, WA: SPIE) vol 1870 p 95

[6] Boehly T R, Smalyuk V A, Meyerhofer D D, Knauer J P, Bradley D K, Craxton R S, Guardalben M J, Skupsky S and Kessler T J 1999 *J. Appl. Phys.* **85** 3444

[7] Skupsky S, Short R W, Kessler T, Craxton R S, Letzring S and Soures J M 1989 *J. Appl. Phys.* **66** 3456

[8] Goncharov V N, Knauer J P, McKenty P W, Radha P B, Sangster T C, Skupsky S, Betti R, McCrory R L and Meyerhofer D D 2003 *Phys. Plasmas* **10** 1906

[9] Goncharov V N, Sangster T C, Boehly T R, Hu S X, Igumenshchev I V, Marshall F J, McCrory R L, Meyerhofer D D, Radha P B, Seka W et al. 2010 *Phys. Rev. Lett.* **104** 165001

[10] Hu S X, Fiksel G, Goncharov V N, Skupsky S, Meyerhofer D D and Smalyuk V A 2012 *Phys. Rev. Lett.* **108** 195003

[11] Obenschain S P, Colombant D G, Karasik M, Pawley C J, Serlin V, Schmitt A J, Weaver J L, Gardner J H, Phillips L, Aglitskiy Y et al. 2002 *Phys. Plasmas* **9** 2234

[12] Betti R, Chang P Y, Spears B K, Anderson K S, Edwards J, Fatenejad M, Lindl J D, McCrory R L, Nora R and Shvarts D 2010 *Phys. Plasmas* **17** 058102

[13] Betti R, Anderson K, Goncharov V N, McCrory R L, Meyerhofer D D, Skupsky S and Town R P J 2002 *Phys. Plasmas* **9** 2277

[14] Goncharov V N, Sangster T C, Betti R, Boehly T R, Bonino M J, Collins T J B, Craxton R S, Delettrez J A, Edgell D H, Epstein R et al. 2014 *Phys. Plasmas* **21** 056315

[15] Michel D T, Sorce C, Epstein R, Whiting N, Igumenshchev I V, Jungquist R and Froula D H 2012 *Rev. Sci. Instrum.* **83** 10E530

[16] Stoeckl C, Glebov V Yu, Roberts S, Sangster T C, Lerche R A, Griffith R L and Sorce C 2003 *Rev. Sci. Instrum.* **74** 1713

[17] Seka W, Edgell D H, Knauer J P, Myatt J F, Maximov A V, Short R W, Sangster T C, Stoeckl C, Bahr R E, Craxton R S et al. 2008 *Phys. Plasmas* **15** 056312

[18] Barker L M and Hollenbach R E 1972 *J. Appl. Phys.* **43** 4669

[19] Stoeckl C, Glebov V Yu, Meyerhofer D D, Seka W, Yaakobi B, Town R P J and Zuegel J D 2001 *Rev. Sci. Instrum.* **72** 1197

[20] Frenje J A, Li C K, Séguin F H, Casey D T, Petrasso R D, Sangster T C, Betti R, Glebov V Yu and Meyerhofer D D 2009 *Phys. Plasmas* **16** 042704

[21] Forrest C J, Radha P B, Glebov V Yu, Goncharov V N, Knauer J P, Pruyne A, Romanofsky M, Sangster T C, Shoup M J, III, Stoeckl C et al. 2012 *Rev. Sci. Instrum.* **83** 10D919

[22] Betti R and Zhou C 2005 *Phys. Plasmas* **12** 110702

[23] Delettrez J, Epstein R, Richardson M C, Jaanimagi P A and Henke B L 1987 *Phys. Rev. A* **36** 3926

[24] Goncharov V N, Sangster T C, Radha P B, Betti R, Boehly T R, Collins T J B, Craxton R S, Delettrez J A, Epstein R, Glebov V Yu et al. 2008 *Phys. Plasmas* **15** 056310

[25] Igumenshchev I V, Edgell D H, Goncharov V N, Delettrez J A, Maximov A V, Myatt J F, Seka W, Shvydky A, Skupsky S and Stoeckl C 2010 *Phys. Plasmas* **17** 122708

[26] Hu S X, Militzer B, Goncharov V N and Skupsky S 2010 *Phys. Rev. Lett.* **104** 235003; Hu S X, Collins L A, Goncharov V N, Kress J D, McCrory R L, and Skupsky S 2015 *Phys. Rev. E* **92** 043104

[27] Igumenshchev I V, Froula D H, Edgell D H, Goncharov V N, Kessler T J, Marshall F J, McCrory R L, McKenty P W, Meyerhofer D D, Michel D T et al. 2013 *Phys. Rev. Lett.* **110** 145001

[28] Froula D H, Kessler T J, Igumenshchev I V, Betti R, Goncharov V N, Huang H, Hu S X, Hill E, Kelly J H, Meyerhofer D D et al. 2013 *Phys. Plasmas* **20** 082704

[29] Cerjan C, Springer P T and Sepke S M 2013 *Phys. Plasmas* **20** 056319

[30] Marshall F J, Goncharov V N, Glebov V Yu, Regan S P, Sangster T C and Stoeckl C *Framed X-Ray Imaging of Cryogenic Target Implosion Cores on OMEGA* presented at the Ninth International Conference on Inertial Fusion Sciences and Applications Seattle WA 2025 September 2015 Paper Th.Po.17

[31] Igumenshchev I V, Goncharov V N, Marshall F J, Knauer J P, Froula D H and Regan S P *Three-Dimensional Modeling of Direct-Drive Cryogenic Implosions* submitted to Physics of Plasmas

Polar-direct-drive experiments at the National Ignition Facility

P B Radha[1,6], M Hohenberger[1], F J Marshall[1], D T Michel[1], J Bates[2], T R Boehly[1], T J B Collins[1], R S Craxton[1], J A Delettrez[1], S N Dixit[3], D H Edgell[1], J A Frenje[4], D H Froula[1], V N Goncharov[1], S X Hu[1], M Karasik[2], J P Knauer[1], S LePape[3], J A Marozas[1], R L McCrory[1], P W McKenty[1], D. D. Meyerhofer[5], J F Myatt[1], S Obenschein[2], R D Petrasso[4], S P Regan[1], M J Rosenberg[1], T C Sangster[1], W Seka[1], A Shvydky[1], H Sio[4], S Skupsky[1] and A Zylstra[4]

[1]Laboratory for Laser Energetics, University of Rochester,
250 East River Road, Rochester, NY 14623-1299, USA
[2]Naval Research Laboratory, SW Washington, DC 20375, USA
[3]Lawrence Livermore Laboratory, Livermore, CA 94550, USA
[4]Plasma Fusion Science Center, Massachusetts Institute of Technology, Cambridge, MA 02139, USA
[5]Los Alamos National Laboratory, NM 87545, USA

rbah@lle.rochester.edu

Abstract. Polar-direct-drive experiments at the National Ignition Facility (NIF) are being used to validate direct-drive–implosion models. Energy coupling and fast-electron preheat are the primary issues being studied in planar and imploding geometries on the NIF. Results from backlit images from implosions indicate that the overall drive is well modeled although some differences remain in the thickness of the imploding shell. Implosion experiments to mitigate cross-beam energy transfer and preheat from two-plasmon decay are planned for the next year.

1. Introduction

Direct-drive couples 3× to 5× more laser energy into the kinetic energy of imploding shell than x-ray drive, making it possible to create ignition-relevant designs in a less-challenging regime. Experiments on the National Ignition Facility (NIF) [1] are necessary to study the interaction of the laser with the coronal plasma in an imploding geometry at coronal-density scale relevant to ignition. Validation of laser deposition and heat conduction models at ignition-relevant scales and demonstrating mitigation strategies for various laser plasma interactions are important to gaining confidence in direct-drive–ignition predictions.

Implosion experiments on the NIF, as described in this paper, are performed in the "polar-direct-drive" (PDD) [2] configuration. Since the NIF is configured for x-ray drive, beam ports are missing at the equator. To compensate for the reduced equatorial drive, beams located closer to the poles are displaced toward the equator. Additionally, to improve equatorial drive, beam energies of some of the

[6] Author to whom any correspondence should be addressed.

beams are adjusted. PDD offers a platform where the physics of direct-drive implosions can be studied. These implosions use existing hardware on the NIF including indirect-drive–ignition phase plates and indirect-drive smoothing.

2. Physics issues in ablatively driven NIF implosions

In direct drive, the implosion velocity and the ablation pressure are primarily determined by the coupling of the laser into the coronal plasma and the conduction of heat to the ablation surface. The equation of state has also been shown to influence these quantities, although to a smaller extent [3]. Modifications to collisional absorption caused by cross-beam energy transfer (CBET) have been used in OMEGA [4] experiments to explain observables including capsule trajectory, scattered-light spectra and time histories, bang times, etc. [5]. The CBET gain scales with the volume of the resonance region. This volume in the NIF implosions is larger than those in OMEGA implosions, necessitating experiments to validate models.

Nonlocal heat conduction has also been identified in OMEGA experiments as being critical to modeling the imploding shell [4]. The coronal plasma is treated as a thermal source and the energetics tails in the Maxwellian-distributed electrons in the corona conduct heat to the ablation surface, providing additional drive. Ongoing NIF implosions have higher coronal temperatures (~3.2 keV) compared to OMEGA-scale coronal temperatures (~2.75 keV). Again, the nonlocal model requires validation in this regime.

Plasma waves excited by the laser can accelerate electrons to energies ($\gtrsim$50 keV) that can be detrimental to compression. On OMEGA and NIF implosions, the magnitude of the hot-electron source, as measured by the x rays generated by bremsstrahlung radiation from these electrons, has been shown to scale with the threshold parameter η [6], given by

$$\eta = I_{n_c/4}\left(\times 10^{14}\ \mathrm{W/cm^2}\right) L_{n_c/4}\left(\mu\mathrm{m}\right) \Big/ \left[233\, T_e\left(\mathrm{keV}\right)\right], \tag{1}$$

where $I_{n_c/4}$ is the laser intensity, $L_{n_c/4}$ is the density scale length, and T_e is the electron temperature, all at the quarter-critical surface. The threshold parameter is larger for NIF implosions compared to OMEGA implosions. For the same ignition-relevant on-target intensity at the initial radius of ~8 × 10^{14} W/cm^2, OMEGA-scale targets have density scale lengths of ~150 μm compared to the NIF-scale density scale lengths of ~360 μm. Therefore, NIF experiments are important to understanding the two-plasmon–decay (TPD) source and its effect on compression.

3. Results

Room-temperature plastic (CH) shells of ~1200-μm radius filled with 20 atm of deuterium gas are imploded on the NIF in PDD geometry. Existing indirect-drive–ignition phase plates, defocused to 1 cm, are used to irradiate these capsules. Trajectories are inferred from framing camera images located at the pole and the equator. The equatorial camera captures images obtained by backlighting the implosion with an iron backlighter. The radius of peak absorption as a function of polar angle is identified [7], providing information about the nonuniformity in the polar angle. The average of this radius provides the location of the peak absorption as a function of time, i.e., a trajectory of the peak absorption location. The polar camera captures images of the emitting target and a similar procedure is used to extract the radius as a function of polar angle and average radius as a function of time [8].

Simulations are performed with the two-dimensional (2-D) axisymmetric code *DRACO* [9]. Laser deposition is modeled using a fully three-dimensional (3-D) ray trace that models collisional absorption. The deposition from CBET using the formulation of Randall *et al.* [10] modifies the laser deposition. Nonlocal heat conduction is modeled diffusively [11]. The far-field defocused spot shapes that are used in *DRACO* are calculated using *Zhizhoo'* [12] based on the near-field spot shapes [13]. The first-principles equation of state [3] is used with opacities from the LANL astrophysical tables [14]. Simulations are then post-processed with the ray-tracing package *Spect3D* [15] to generate synthetic images. Both the measured and simulated images are post-processed similarly to extract radius as a function of polar angle and time.

Figure 1 plots the average locations of the radius versus time for the simulated and measured images. The backlit trajectory is reproduced very well with the simulations indicating that the global energetics is also modeled well with simulations. Simulations indicate that CBET reduces the implosion velocity by ~15% and the ablation pressure by ~45% in these implosions. As figure 1 also indicates, the trajectories from the self-emission images are overdriven in the simulation relative to the experiment. This indicates that the shell is thicker in the experiment than the simulation. Although the exact reasons for this decompression are unknown, potential reasons include fast-electron and/or radiative preheat and instability growth from laser imprint. Experimentally, the magnitude of the hot-electron source has been inferred using the filter-fluorescer x-ray (FFLEX) diagnostic [16], which measures the x-ray emission from bremsstrahlung of the fast electrons in ten channels in the 20- to 500-keV range. Using the measured radiation and the stopping power of electrons in CH, a total energy of ~2.5±0.3 kJ with a hot-electron temperature of ~46±2 keV is inferred [17]. This extent of preheat has a small influence on the trajectories in simulations, suggesting that fast-electron preheat is not the source of this decompression. *DRACO* simulations with imprint with modes $\ell \leq 200$ show significant distortion of the ablation surface without affecting the shell. Images from these simulations indicate that imprint could provide a plausible mechanism for the apparent decompression of the shell. Low-intensity implosions, where preheat is reduced, will be studied. Improved smoothing with multi-FM [18] is required to further disentangle the effect of preheat and laser imprint on shell decompression.

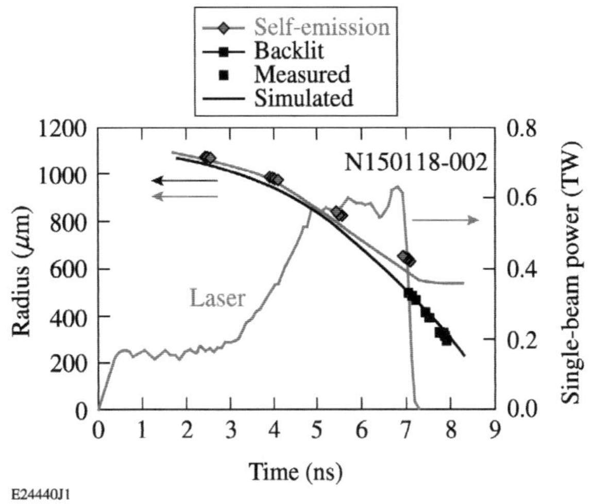

Figure 1. Trajectories of the peak self-emission surface (red) and peak absorption surface (black) inferred from framing camera images in the experiment (symbols) and simulated framing camera images (lines).

4. Future work

Reducing the extent of energy transfer between the rays in NIF implosions will be studied by exploiting the NIF capability of irradiating the target with somewhat different wavelengths for the various cones [19]. This will move the stimulated Brillouin scattering resonance to regions that are less detrimental to laser-energy deposition. These experiments will begin in FY16. CBET, however, is expected to influence TPD. With only collisional absorption in the models, the intensity at the quarter-critical surface is higher by ~60%, while not significantly influencing the scale length or the electron temperature. Therefore, mitigating CBET in NIF implosions will increase η by the same factor of 60%, increasing the hot-electron source. Mid-*Z* ablators have been demonstrated to decrease the hot-electron source by primarily increasing the temperature at the quarter-critical surface. Mid-*Z* ablators will be imploded to study the extent of TPD in NIF-scale implosions with the wavelength-detuning mechanism for CBET mitigation in FY16.

5. Conclusions

Results from PDD NIF implosions have been presented. The goal of these experiments is to validate modeling relating to implosion energetics on coronal density scale lengths larger than those available on the OMEGA laser, which has been traditionally used for direct-drive–implosion studies. Cross-beam energy transfer and nonlocal heat conduction are primary physics issues being investigated on the NIF. Comparison of the simulated and measured images and trajectories indicate that energetics is well modeled. Mitigation of CBET via wavelength detuning of the resonance will be studied in the following year. It is expected that mitigating CBET will increase the hot-electron source from the corona that can potentially preheat the imploding shell, thereby compromising compression. Mid-*Z* ablators have been demonstrated on OMEGA and the NIF to reduce this source. Implosions with CBET mitigation through wavelength detuning will be studied with mid-*Z* ablators to estimate the hot-electron preheat in the future.

Acknowledgments
This material is based upon work supported by the Department of Energy National Nuclear Security Administration under Award Number DE-NA0001944, the University of Rochester, and the New York State Energy Research and Development Authority. The support of DOE does not constitute an endorsement by DOE of the views expressed in this article.

References

[1] Lindl J D and Moses E I 2011 *Phys. Plasmas* **18** 050901
[2] Skupsky S, Marozas J A, Craxton R S, Betti R, Collins T J B, Delettrez J A, Goncharov V N, McKenty P W, Radha P B, Boehly T R *et al.* 2004 *Phys. Plasmas* **11** 2763; Kyrala G, Delamater N, Wilson D, Guzik J, Haynes D, Gunderson M, Klare K, Watt R W, Wood W M, and Varnum, W 2004 Los Alamos, NM: Los Alamos National Laboratory Report LA-UR-04-6178
[3] Hu S X, Collins L A, Goncharov V N, Kress J D, McCrory R L and Skupsky S 2015 *Phys. Rev. E* **92** 043104
[4] Boehly T R, Brown D L, Craxton R S, Keck R L, Knauer J P, Kelly J H, Kessler T J, Kumpan S A, Loucks S J, Letzring S A *et al.* 1997 *Opt. Commun.* **133** 495
[5] Igumenshchev I V, Seka W, Edgell D H, Michel D T, Froula D H, Goncharov V N, Craxton R S, Divol L, Epstein R, Follett R *et al.* 2012 *Phys. Plasmas* **19** 056314
[6] Simon A, Short R W, Williams E A and Dewandre T 1983 *Phys. Fluids* **26** 3107
[7] Marshall F J, McKenty P W, Delettrez J A, Epstein R, Knauer J P, Smalyuk V A, Frenje J A, Li C K, Petrasso R D, Séguin F H *et al.* 2009 *Phys. Rev. Lett.* **102** 185004
[8] Michel D T, Sorce C, Epstein R, Whiting N, Igumenshchev I V, Jungquist R and Froula D H 2012 *Rev. Sci. Instrum.* **83** 10E530
[9] Radha P B, Goncharov V N, Collins T J B, Delettrez J A, Elbaz Y, Glebov V Y, Keck R L, Keller D E, Knauer J P, Marozas J A *et al.* 2005 *Phys. Plasmas* **12** 032702
[10] Randall C J, Albritton J R and Thomson J J 1981 *Phys. Fluids* **24** 1474
[11] Cao D, Moses G and Delettrez J 2015 *Phys. Plasmas* **22** 082308
[12] Marozas J A, Collins T J B, Zuegel J D, Radha P B and McKenty P W 20–25 September 2015 presented at the Ninth International Conference on Inertial Fusion Sciences and Applications IFSA 2015 Seattle WA paper Mo.Po.65
[13] Montgomery D S 2009 *Los Alamos National Laboratory* private communication
[14] MacFarlane J J, Golovkin I E, Wang P, Woodruff P R and Pereyra N A 2007 *High Energy Density Phys.* **3** 181
[15] Huebner W F, Merts A L, Magee N H, Jr. and Argo M F 1977 Los Alamos, NM: Los Alamos National Laboratory Report LA-6760-M
[16] Hohenberger M, Albert F, Palmer N E, Lee J J, Döppner T, Divol L, Dewald E L, Bachmann B, MacPhee A G, LaCaille G *et al.* 2014 *Rev. Sci. Instrum.* **85** 11D501
[17] Hohenberger M *et al.* 2015 *Phys. Plasmas* **22** 056308
[18] 2008 *LLE Review Quarterly Report* **114** 86 Laboratory for Laser Energetics University of Rochester Rochester NY LLE Document No. DOE/NA/28302-826 NTIS Order No. DE2008935224
[19] Marozas J A, Collins T J B, McKenty P W and Zuegel J D 16–20 November 2015 presented at the 57th Annual Meeting of the APS Division of Plasma Physics Savannah GA paper JO5.00005

Journal of Physics: Conference Series **717** (2016) 012010 doi:10.1088/1742-6596/717/1/012010

Experimental evidence of a bubble-merger regime for the Rayleigh-Taylor Instability at the ablation front

A Casner, S Liberatore and L Masse
CEA, DAM, DIF, F-91297 Arpajon, France

D Martinez, S W Haan, J Kane, A S Moore, R Seugling, M Farrell, E Giraldez, A Nikroo, V A Smalyuk and B A Remington
Lawrence Livermore National Laboratory, Livermore, California 94550, USA

E-mail: alexis.casner@cea.fr

Abstract. Under the Discovery Science program, the longer pulses and higher laser energies provided by the National Ignition Facility (NIF) have been harnessed to study, first time in indirect-drive, the highly nonlinear stage of the Rayleigh-Taylor Instability (RTI) at the ablation front. A planar plastic package with pre-imposed two-dimensional broadband modulations is accelerated for up to 12 ns by the x-ray drive of a gas-filled gold radiation cavity with a radiative temperature plateau at 175 eV. This extended tailored drive allows a distance traveled in excess of 1 mm for a 130 μm thick foil, a factor 3x larger than previously achieved on other laser facilities. As a consequence, we have measured the ablative RTI in transition from the weakly nonlinear stage up to the deep nonlinear stage for various initial conditions. A bubble merger regime has been observed and the ablative stabilization strength varied by changing the plastic dopant from iodine to germanium.

1. Introduction

The recent experimental results, acquired either in indirect drive (ID) or direct drive (DD), have reminded us that hydrodynamics instabilities are one of the major difficulties to be overcome to achieve thermonuclear ignition in Inertial Confinement Fusion (ICF) [1]. Cryogenic ID implosions at the National Ignition Facility (NIF) have underlined that the ablative Rayleigh-Taylor Instability (RTI) is an important driver of hot spot mix [2]. Any tiny perturbations of the capsule surface, such as those induced by the supporting tent [3], or even modulations of the oxygen contamination in the bulk of the capsule [4] are amplified during the convergence and could break through the shell. This increased sensitivity has motivated the switch to a more forgiving higher adiabat implosion design [5]. Adiabat shaping has also been used to mitigate the ablation front growth, providing a new path to improved performance at the high compression necessary for ignition. In DD ICF shell instability is also the main candidate for performance degradation in low-adiabat cryogenic implosions on OMEGA [6]. It seems therefore that ablative RTI is more aggressive than previously thought and deserves a deeper investigation. To investigate the uncharted regime of ablative RTI at the highly nonlinear stage, a novel long-duration indirect-drive has been commissioned on the NIF [7]. Here we present some

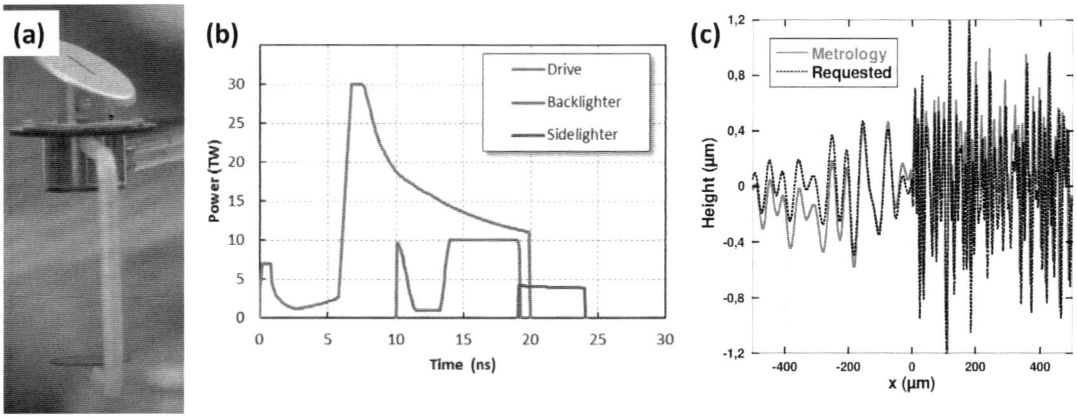

Figure 1. (a) Photograph of the indirect-drive halfraum. The face-on backlighter source is (b) Laser concatenation in time of the different beams used. (c) Illustration of the excellent reproductibility and control of our initial experimental conditions: comparison of the requested and machined pre-imposed modulations pattern.

of the results obtained for the growth of two-dimensional (2D) broadband perturbations pre-imposed on planar plastic samples. By changing the doping material from iodine to germanium, the strength of the ablative stabilization has been varied and diagnosed through face-on time-resolved x-ray radiographs.

2. Experimental set-up

The experimental configuration has been described in details elsewhere [8]. A gold (Au) half-hohlraum is driven at one end by 60 NIF UV beams (at 44.5° and 50°) with a shaped laser pulse (cf Fig. 1a) and b)). The accelerated sample was a 130 μm thick plastic (CH) foil doped with a 0.6 % iodine dopant by atomic number (or a 3 % germanium dopant) attached to the opposite end of the hohlraum. The doping fraction is experimentally constrained by the optimization of radiograph contrast. A crucial point for the success of these experiments is the perfect control and knowledge of the initial modulations pre-imposed on the package, as illustrated in Fig. 1c). The hohlraum is gas-filled (room temperature neopentane at 400 torr) to delay wall plasma expansion in order to keep a clear radiography line-of-sight at late time. The x-ray drive was measured through the laser entrance hole with Dante, a time-resolved, low resolution x-ray spectrometer consisting of 18 channels of filtered x-ray diodes. A noticeable result is the creation of a flat temperature plateau at 175 eV lasting 12 ns [9]. Two x-ray radiographs were performed simultaneously on each shot. The concatenation in time of the drive and radiography laser beams is shown in Fig. 1b)). Trajectory measurements performed with a streaked or gated imager inserted in equatorial plane in allow to infer a traveled distance $\approx$ 1.5 mm at 21 ns, whereas NOVA and OMEGA experiments were limited to $\approx$ 200 μm [10, 11]. The primary diagnostic is a four strip gated X-ray detector (GXD) [12]inserted in the polar diagnostic insertor and coupled with an charge-coupled device (CCD) camera. Depending on the dopant, a vanadium (for iodyne doped samples) or scandium (for germanium doped samples) face-on backlighter was used.

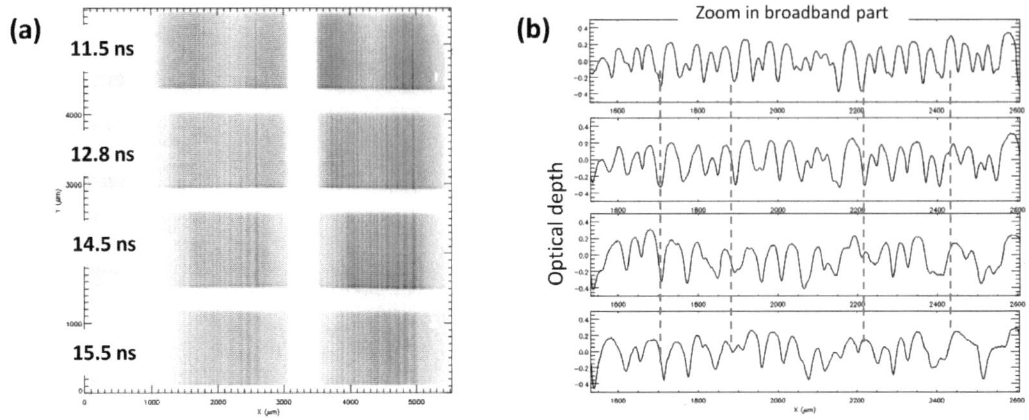

Figure 2. (a) Experimental face-on radiograph acquired on shot N141109-002 with a germanium-doped sample CHGe (3 % atomic fraction). The acquisition time of each strip is shown on the left. The denser RTI spikes correspond to the thinner white vertical lines. (b) Corresponding lineouts showing the development of large bubbles in the broadband part of the spectrum as a consequence of the bubble-competition, bubble-merger process. Vertical dotted are guide to the eyes. 29 bubbles merged into 21 between 11.5 ns and 15.5 ns.

3. Broadband modulations RTI growth under different ablative stabilization strengths

A typical face-on radiograph acquired in the case of a germanium-doped plastic foil CHGe (3 % at. fraction) is shown in Fig.2(a) (NIF Shot N141109-001). The pre-imposed pattern was constituted by a broadband modulated central part, whose root-mean-square (RMS) amplitude is $0.5\,\mu m$. The succession of black and white vertical lines in the face-on radiograph corresponds to 2D bubbles and spikes. Optical-depth (OD) variations (used in the analysis below) were obtained by taking the natural logarithm of the framing-camera image transmission intensities after x-ray backgrounds were subtracted. Lineouts across the image shown in Fig.2(b) is evidence that bubbles merging is observed [13], despite the stabilizing effect of ablation. 29 bubbles are present at t = 11.5 ns leading to 21 principal bubbles at the latest time (15.5 ns). It seems that, with our ID RTI platform, we have entered into the uncharted physics regime of ablative destabilization, thanks to the very-long acceleration drive provided by NIF. In fact, whereas ablation is stabilizing in the linear phase, ablation off the unstable surface transfers vorticity from the ablation front to the ablated plasma, in contrary to classical RTI for which vorticity remains localized at the interface. The vorticity fills the bubble void leading to vorticity accumulation inside the bubble, thus accelerating the bubble growth [14].

The lineouts in optical depth of Fig.2(a) were Fourier analyzed and the spectra from 11.5 ns and 14.5 ns are shown in Fig.3 after correction by the Modulated Transfer Function of the polar GXD. The peak of the spectrum shifts in time to the left (towards the lower spatial frequencies), as a consequence of the bubble merging process [13]. The spectra for CH Ge (3 % at.) clearly shows that the frequencies above $0.04\ \mu m^{-1}$ are attenuated, in accordance with the anticipated cut-off wavelength close to $25\ \mu m$ [7]. A iodine doped plastic sample machined with exactly the same broadband modulations was accelerated by the same x-ray drive conditions on the companion shot (NIF Shot N141109-001). The comparison of the spectra in Fourier space shown in Fig.3 clearly demonstrate that the ablative stabilization is larger for all spatial frequencies with the iodine dopant: the Fourier amplitude are always smaller than those for germanium. Nevertheless bubble merging was also observed with this dopant [13].

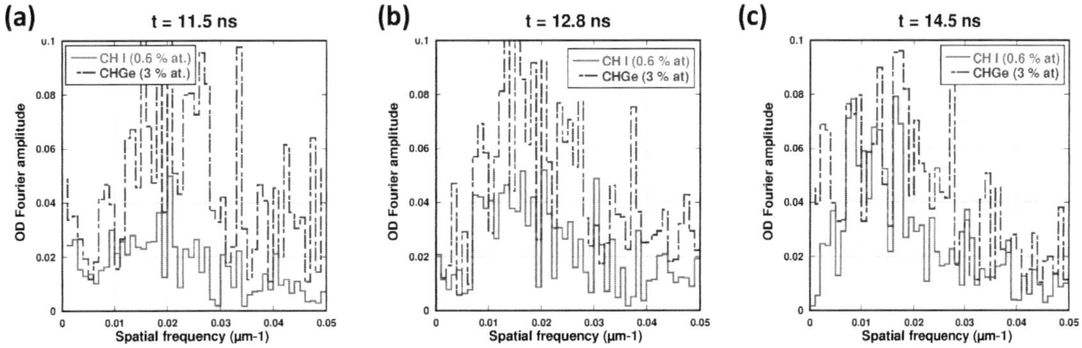

Figure 3. (a) to (c). The Fourier transform of the experimental radiographs lineouts for CH I (0.6 % at fraction) and CH Ge samples (3 % at fraction). The decreased ablative stabilization for the germanium doped samples is evidenced at all times. The data at t = 15.5 ns are not shown because the CHGe foil is broken at at that time due to the larger RTI growth rate.

4. Conclusion

In contrast to previous DD RTI experiments [15],the ablation velocity and density gradient scale length are significantly increased with indirect drive leading to cutoff wavelengths large enough, (about 25 microns) to be observable. For the first time, we have measured the ablation front RTI growth in a highly non-linear regime for the full unstable spectrum. A careful choice of the broadband modulation pattern has allowed us to evidence without doubt a bubble-merger regime for the RTI at the ablation front.These results are critical for ICF, and will be benchmarked against radiative hydrodynamic simulations performed with FCI2 and other theories predicting increased bubble growth due to ablation in nonlinear regime [17].

Acknowledgments

The authors sincerely thank the NIF operations staff who supported this work. We acknowledge also the support and help provided by P. Sharp of the NIF User Group Office.

References

[1] Atzeni S and Meyer-ter-Vehn J 2004 *The Physics of Inertial Fusion* (Oxford Science Publications)
[2] Regan S P, Epstein R, Hammel B A et al. 2013 *Phys. Rev. Lett.* **111**(4) 045001
[3] Nagel S R, Haan S W, Rygg J R et al. 2015 *Phys. Plasmas* **22** 022704
[4] Smalyuk V A, Weber S V, Casey D T et al. 2015 *Phys. Plasmas* **22** 072704
[5] Dittrich T R, Hurricane O A, Callahan D A et al. 2014 *Phys. Rev. Lett.* **112**(5) 055002
[6] Goncharov V N, Sangster T C, Betti R et al. 2014 *Phys. Plasmas* **21** 056315
[7] Casner A, Smalyuk V A, Masse L et al. 2012 *Phys. Plasmas* **19** 082708
[8] Casner A, Martinez D, Smalyuk V et al. 2015 *High Energy Density Physics* **17** 146
[9] Casner A, Masse L, Liberatore S et al. 2015 *Phys. Plasmas* **22** 056302
[10] Bradley D K, Braun D G, Glendinning S G et al. 2007 *Phys. Plasmas* **14** 056313
[11] Remington B A, Weber S V, Marinak M M et al. 1995 *Phys. Plasmas* **2** 241–255
[12] Kyrala G A, Dixit S, Glenzer S et al. 2010 *Rev. Sci. Instrum.* **81** 10E316
[13] Martinez D A, Smalyuk V A, Kane J O, et al. 2015 *Phys. Rev. Lett.* **114**(21) 215004
[14] Betti R and Sanz J 2006 *Phys. Rev. Lett.* **97** 205002
[15] Smalyuk V A, Sadot O, Delettrez J A et al. 2005 *Phys. Rev. Lett.* **95** 215001
[16] Oron D, Alon U and Shvarts D 1998 *Phys. Plasmas* **5** 1467–1476
[17] Sanz J, Ramirez J, Ramis R, Betti R and Town R P J 2002 *Phys. Rev. Lett.* **89**

Progress in detailed modelling of low foot and high foot implosion experiments on the National Ignition Facility

D S Clark, C R Weber, D C Eder, S W Haan, B A Hammel, D E Hinkel, O S Jones, A L Kritcher, M M Marinak, J L Milovich, P K Patel, H F Robey, J D Salmonson and S M Sepke

Lawrence Livermore National Laboratory, Livermore, CA, USA

clark90@llnl.gov

Abstract. Several dozen high convergence inertial confinement fusion ignition experiments have now been completed on the National Ignition Facility (NIF). These include both "low foot" experiments from the National Ignition Campaign (NIC) and more recent "high foot" experiments. At the time of the NIC, there were large discrepancies between simulated implosion performance and experimental data. In particular, simulations over predicted neutron yields by up to an order of magnitude, and some experiments showed clear evidence of mixing of ablator material deep into the hot spot that could not be explained at the time. While the agreement between data and simulation improved for high foot implosion experiments, discrepancies nevertheless remain. This paper describes the state of detailed modelling of both low foot and high foot implosions using 1-D, 2-D, and 3-D radiation hydrodynamics simulations with HYDRA. The simulations include a range of effects, in particular, the impact of the plastic membrane used to support the capsule in the hohlraum, as well as low-mode radiation asymmetries tuned to match radiography measurements. The same simulation methodology is applied to low foot NIC implosion experiments and high foot implosions, and shows a qualitatively similar level of agreement for both types of implosions. While comparison with the experimental data remains imperfect, a reasonable level of agreement is emerging and shows a growing understanding of the high-convergence implosions being performed on NIF.

1. Introduction

Substantial progress has been made in modeling and understanding ignition implosions on the National Ignition Facility (NIF) since the 2013 IFSA Conference. Since that time, hydrodynamic instability growth spectra have been measured for a range of perturbation mode numbers and for high foot as well as low foot implosion types [1]. These measurements have so far largely validated the simulation technique used to model NIF implosions, although measurements at higher convergence [2] and with better resolution are needed. In addition, the high foot implosion platform has been pushed to higher implosion velocities using higher laser powers and energies, and also thinner ablators, and appears to show evidence of a ceiling in performance. From the simulation perspective, detailed 3-D capsule-only simulations are showing increasing levels of agreement with NIF implosion measurements. In particular, these detailed simulations have now been compared for representative low foot [3] and high foot [4] implosion types, and show similar levels of agreement with the data for these quite different implosion types. This paper summarizes recent progress in 3-D simulations of

NIF implosions using the radiation hydrodynamics code HYDRA [5]. The different implosion characteristics, as revealed in simulations of low as compared to high foot implosions, and their different failure modes are particularly important as these indicate where each implosion type may be improved and how a route to ignition on NIF might finally be achieved.

2. 3-D simulation results

3-D capsule-only simulations were run following the methodology described in Ref. [6]. Since the simulations in Ref. [6] were completed, however, further 2-D simulation work suggested that the perturbation seeded by the capsule support tent was even larger than assumed in those simulations. The understanding of the low-mode radiation flux asymmetries imprinted on the capsule from the hohlraum has also improved. For these reasons, the NIF implosion simulated in Ref. [6] (N120321, the highest compression implosion yet fired on NIF) was rerun with these updated inputs. Two additional implosion experiments have also been simulated: N120405, a higher power and energy companion to N120321 that "mixed" heavily with ~1 μg of ablator material believed to have entered the hot spot; and N130927, a high foot implosion that showed the first evidence of "fuel gain," that is, a fusion neutron yield equal to or greater than the peak fuel kinetic energy. It is important to understand the behavior of N120405 as it clearly crossed a performance "cliff" compared to N120321, and understanding the origin of this cliff—the source of the hot spot mix mass in particular—is essential in avoiding this outcome in future implosions. It is equally important to contrast the results of the low foot implosions (N120321 and N120405) against the high foot to understand the different strengths, and weaknesses, of these quite different implosion types.

Figure 1 contrasts the results from 3-D simulations of each of these shots at their respective bang times (time of peak neutron production). In each case, the outer surface is the ablation front colored by the electron temperature, the left cutaway shows the ion temperature, and the right cutaway shows

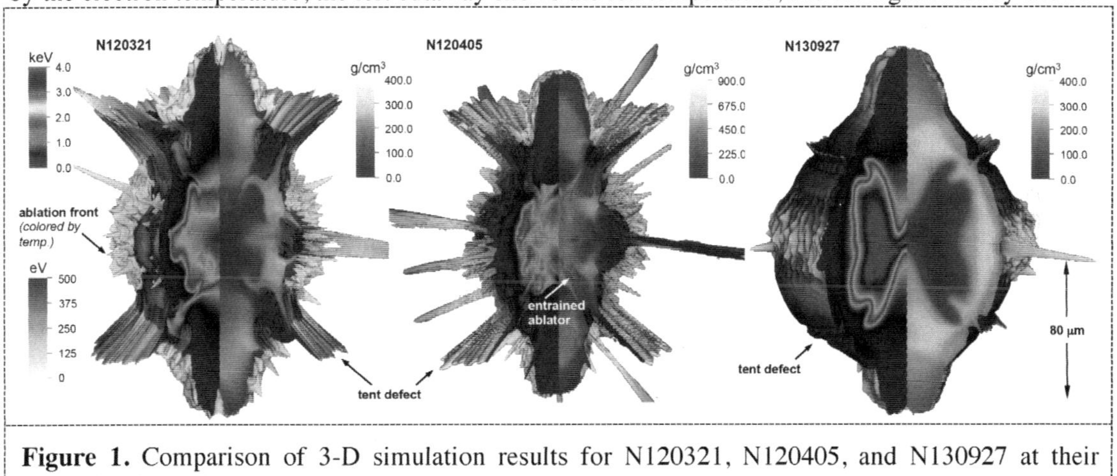

Figure 1. Comparison of 3-D simulation results for N120321, N120405, and N130927 at their respective bang times.

the density. The large "sombrero hat" features in both of the low foot implosions result from ablation front instability growth seeded by the capsule support tent. As is clear from the figure, the tent is the dominant perturbation in these low foot implosions. In fact, in the case of N120405, it can be seen that the tent perturbation has grown so extreme that it has entrained plastic ablator material into the center of the hot spot. This appears to explain the source of the hot spot mix mass observed for this shot: the stronger acceleration and hence ablation front instability growth of N120405 amplified the tent perturbation to such an extent that the already large perturbation on N120321 grew to the point of penetrating the hot spot on N120405. Note that the tent perturbation encircles the full azimuth of the capsule at each pole and hence gives a large area for ablator material to enter the hot spot suddenly once an amplitude threshold is passed. In contrast, the high foot implosion N130927 is clearly much less perturbed at the ablation front. While a small tent defect is evident in this simulation, it is much reduced relative to the low foot, and the dominant perturbation source is now the hohlraum radiation asymmetry. This asymmetry results in the large spikes or jets entering the north and south poles of the hot spot in this simulation. Even so, a much larger and hotter hot spot results compared to the low foot, and hence this implosion gives a much higher yield.

To quantify the relative importance of the various perturbation sources in the low foot and high foot implosion types, Figures 2 and 3 show the results of 2-D simulations run with each perturbation source included separately, namely, the hohlraum flux asymmetries alone, the tent perturbation alone,

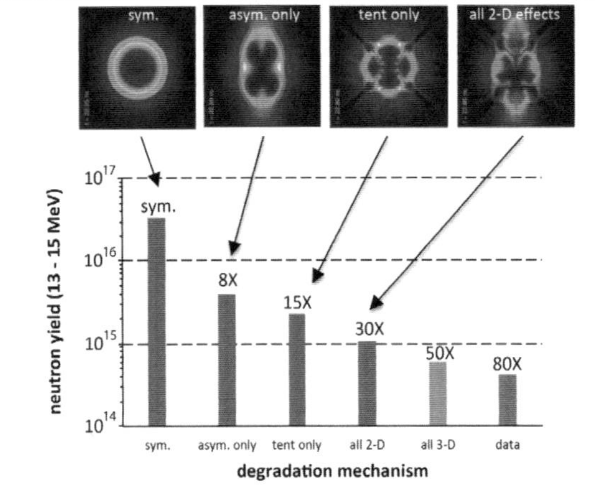

Figure 2. Relative effect of each perturbation source in 2-D simulations of N120321.

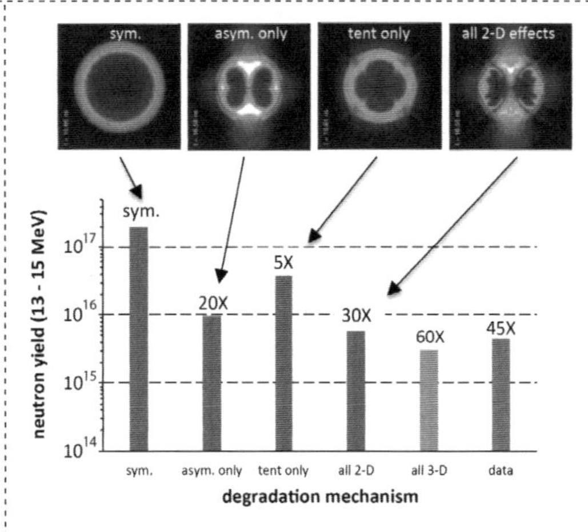

Figure 3. Relative effect of each perturbation source in 2-D simulations of N130927.

all 2-D effects in combination, and finally the 3-D result from the simulations described above. The histogram shows the impact on the neutron yield with each effect, and the insets show the 2-D simulation results at bang time. Note that α-particle deposition is included in all simulations.

For the low foot implosion N120321, the tent is quantitatively the largest impact resulting in a 15× yield degradation relative to 1-D. This is nearly twice the impact of the flux asymmetries that result in an 8× degradation. As shown in Figure 2, these roles reverse for the high foot. The better ablation front stability of the high foot reduces the impact of the tent to a 5× reduction in yield relative to 1-D, while the hohlraum asymmetries result in a 20× degradation. Thus, while the tent is the largest source of yield degradation for the low foot implosions, the hohlraum asymmetries are the largest degradation in this high foot implosion. Note that the high foot implosion achieves a higher symmetric yield than the low foot cases, despite its higher adiabat, due to its higher laser power and energy and hence higher implosion velocity. Interestingly, for the high foot, the 2-D simulation with all effects included is fairly close to the measured yield data and the 3-D simulation actually under-predicts the yield. By contrast, for the low foot, the 2-D simulation including all effects over-predicts the yield by more than a factor of two, while only the 3-D simulation is fairly close to the measured yield. This is indicative of the generally larger perturbation levels in low foot implosions that can only be accurately captured in a fully 3-D simulation.

Quantitative comparisons of the 3-D simulation results against the data for the three shots simulated are summarized in Table 1. The first column lists a number of the principle experimental observables and pairs of subsequent columns compare the simulation results against the data for each successive shot. The agreement is generally good for all three shots, although many quantities are not matched within the experimental error bars. In these cases, the simulation results are generally within two error bars of the data, however. In comparing the primary neutron image size (PNI P_0), down scattered neutron image size (DSNI P_0), burn-averaged ion temperature (T_{ion}), and neutron down scattered ratio (DSR), an important caveat should be pointed out. Given the scale of these 3-D simulations, current computing capabilities do not allow running the simulations with inline Monte Carlo neutronics, as is routinely done in 2-D. The simulation values listed in the table are hence taken from instantaneous post-processing of the simulations at bang time. As such, they represent snapshots of the state of the simulation at bang time and omit the time averaging over the duration of the burn that is inherent in the measurement and would be included if these quantities could be computed inline. This limitation in the current simulations may account for the noticeable discrepancies in the simulated DSR for N120321 and also in the ion temperatures for N120321 and N130927. It is, of course, possible that these discrepancies point to inadequacies in the physical models used in the simulations or are the result of imperfect knowledge of the initial and boundary conditions for these shots. At this time, it is not possible to resolve which of these possibilities is responsible. Nevertheless, the overall agreement between the simulations and the data is reasonably good. This is notable given that three quite different shots have been simulated, and each appears to agree equally well with the experimental results.

Table 1. Comparison of simulation and experimental results for N120321, N120405, and N130927.

	N120321		N120405		N130927	
	sim.	expt.	sim.	expt.	sim.	expt.
bang time (ns)	22.85	22.91±0.04	22.53	22.70±0.08	16.53	16.59±0.03
burn width (ps)	167	158±40	130	161±40	143.5	188±30
x-ray P_0 (μm)	21.9	20.1±1.4	23.9	23.4±0.85	31.4	35.3±3.0
x-ray M_0 (μm)	19.8	22.7±2.7	24.1	26.5±4.0	45.7	49.8±1.5
PNI P_0 (μm)	24.4	26±3	25.4	27±3	27.7	32±4
DSNI P_0 (μm)	38.4	35±3	31.7	43±6	51.1	55±4
T_{ion} (keV)	2.6	3.1±0.4	1.7	1.69±0.13	3.9	4.43±0.15
DSR (%)	5.0	6.2±0.6	5.5	5.14±0.29	3.5	3.48±0.17
$Y_{13-15\,MeV}$	6.0×10^{14}	$4.2 \pm 0.1 \times 10^{14}$	1.4×10^{14}	$1.3 \pm 0.1 \times 10^{14}$	3.1×10^{15}	$4.5 \pm 0.1 \times 10^{15}$

3. Conclusions

Three ignition implosion experiments from the NIF database have been simulated following the most up-to-date 3-D simulation methodology. All three show reasonably good, though not perfect, agreement with the experimental data. Given that these three shots explored quite different regions of implosion parameter space, the agreement suggests that a fairly robust simulation capability is developing for modeling NIF implosions. This validated simulation capability is clearly essential for assessing what design modifications can lead to further gains in implosion performance.

Acknowledgements

This work was performed under the auspices of the U.S. Department of Energy by Lawrence Livermore National Laboratory under Contract DE-AC52-07NA27344.

References

[1] Smalyuk V A, *et al.* 2014 *Phys. Plasmas* **21** 056301
[2] Hammel B A, *et al.* 2015, Bull. Am. Phys. Soc. **60** 164
[3] Edwards M J, *et al.* 2013 *Phys. Plasmas* **20** 070501
[4] Hurricane O A, *et al.* 2014 *Phys. Plasmas* **21** 056314
[5] Marinak M M, *et al.* 2001 *Phys. Plasmas* **8** 2275
[6] Clark D S, *et al.* 2015 *Phys. Plasmas* **22** 022703

Design options for polar-direct-drive targets from alpha heating to ignition

T J B Collins[1,3], J A Marozas[1], S Skupsky[1], D Cao[1], P W McKenty[1], J A Delettrez[1] and G Moses[2]

[2]University of Wisconsin,
Madison, WI 53706-1140 USA

[1]Laboratory for Laser Energetics, University of Rochester,
250 East River Road, Rochester, NY 14623-1299 USA

tcol@lle.rochester.edu

Abstract. Polar direct drive (PDD) makes it possible to perform direct-drive–ignition experiments at the National Ignition Facility while the facility is configured for x-ray drive. We present the first PDD ignition-relevant target designs to include the physical effects of cross-beam energy transfer (CBET) and nonlocal heat transport, both of which substantially affect the target drive. In the PDD configuration, a multiwavelength detuning strategy was found to be effective in mitigating the loss of coupling caused by CBET, allowing for implosion speeds comparable to those of previous designs. Two designs are described: a high-adiabat alpha-burning design and a lower-adiabat ignition design.

1. Introduction

In direct-drive inertial confinement fusion, laser ablation is used to implode a spherical shell composed largely of fuel (DT), producing a central volume of high density and ion temperature. The ions, briefly confined, undergo fusion reactions, producing alpha particles, some of which are stopped in the surrounding colder, denser shell. Ignition occurs when these deposit enough energy to launch a self-sustaining thermonuclear burn wave, which consumes a fraction of the fuel before the high pressure generated by the burn wave causes the target to disassemble. Ignition occurs only if the hot-spot ion temperatures are higher than ~10 keV and hot-spot areal density exceeds ~300 mg/cm^2. This process is prevented if either the imploding kinetic energy is insufficient or hydrodynamic instability prevents the required hot-spot temperature and areal density.

The hot-spot volume is reduced by perturbations on the inner edge of the shell that grow as a result of the Rayleigh–Taylor instability as the shell is being decelerated by the pressure of the gas interior to it. In its current configuration, the laser beam ports of the National Ignition Facility (NIF) [1] are preferentially located away from the equator of the target sphere, for use with x-ray–driven targets enclosed in a hohlraum. Polar-direct-drive (PDD) [2,3] implosions using this configuration must contend with low-mode perturbations caused by the lack of equatorial beams. Furthermore, direct-drive experiments on the NIF [4] have demonstrated that it is necessary to model both cross-beam

[3] Author to whom any correspondence should be addressed.

energy transfer (CBET) and nonlocal electron heat transport [5]. (CBET is a laser–plasma instability in which an incoming laser ray interacts with an outgoing ray by means of an intermediate ion-acoustic wave, depleting the energy of the incoming ray.) While nonlocal electron transport can increase the hydrodynamic efficiency of the implosion, CBET causes a sizable fraction of the incident laser energy to be scattered away, reducing the ablation pressure.

In this paper, we present a PDD-ignition design that addresses the two requirements given above, of shell kinetic energy and implosion uniformity, and is the first ignition design to include the effects of nonlocal heat transport and CBET. Previous ignition designs incorporated these processes in only an approximate way by using an *ad hoc* flux limiter applied to the classical expression for heat conduction. In this design, the loss of drive related to CBET is mitigated by the use of different laser wavelengths by different beams [6]. As with previous PDD designs, the drive asymmetries caused by the asymmetric disposition of the beams are controlled through a combination of individual cone pulse shapes and beam repointing. The resulting design achieves high gain for a moderate in-flight aspect ratio (IFAR). The second design we present does not ignite but has a lower IFAR for greater stability and is predicted to generate bootstrap heating producing over 10^{17} neutrons in a simulation incorporating only drive perturbations caused by beam geometry.

Section 2 describes the detuning configuration strategies. Section 3 presents the two target designs (ignition and alpha burning), followed by the conclusions in section 4.

2. Cross-Beam Energy Transfer and Wavelength Detuning

For direct-drive targets of sufficient density scale length and laser intensity, the stimulated Brillouin scattering (SBS) process is dynamically important. This process entails the parametric coupling of incident light with an ion-acoustic wave and a backscattered electromagnetic wave. The efficiency of energy transfer is determined by a resonance function of the parameter $\eta = \left[(\omega_i - \omega_o) - \mathbf{k}_a \cdot \mathbf{v}\right] / (c_a k_a)$, where ω_o and ω_i are the outgoing and incoming ray frequencies, respectively; c_a is the outflow sound speed; $\mathbf{k}_a = \mathbf{k}_o - \mathbf{k}_i$ is the ion-acoustic wave number; $\mathbf{v}$ is the outflow velocity; and $\eta > 0$ corresponds to energy transfer from the incoming ray to the outgoing ray [7]. CBET is well known in indirect-drive inertial confinement fusion (ICF), where it is used to transfer energy between cones of beams to affect low-mode capsule symmetry. In direct-drive ICF, it is more typical for an incoming low-energy ray from the edge of a beam to rob energy from an incoming ray in the central, higher-energy portion of another beam. The energy exchange caused by CBET is determined by an attenuation factor $\mathrm{d}\tau$, which is proportional to the resonance function $P(\eta) = \eta \cdot \nu_a \left[(\eta \nu_a)^2 + (1-\eta^2)^2\right]^{-1}$, where ν_a is the dimensionless ion-wave damping coefficient. SBS is particularly effective at robbing energy from the incoming rays because the matching condition is met over a potentially large volume near the surface where the expanding Mach number is unity. In PDD, this resonance region occurs preferentially over the equator where equatorial beams from each hemisphere overlap and where laser intensities, increased to offset the increased equatorial deposition radii, are the greatest.

CBET is mitigated in the present work by the use of wavelength detuning [6]. In this approach, the laser cavities are detuned slightly for different collections of beams to increase the frequency separation, which in turn alters the region over which the CBET efficiency is greatest. If the outgoing rays are blue shifted relative to the incoming rays, the resonance region moves to greater Mach numbers and correspondingly larger radii, where the beam overlap and corresponding energy transfer are reduced. If these rays are red shifted, the resonance region moves radially inward, and if the shift is large enough, the resonance region may, for a portion of the pulse, be hidden within the parabolic locus of turning points within which the laser rays cannot penetrate. Over time, for red-shifted outgoing rays, this resonance region moves radially outward, reducing or eliminating the CBET mitigation.

Several detuning configurations were investigated. The laser absorption efficiency for four of these configurations is shown in figure 1. The hemispherical scheme detunes the beams by hemisphere. This scheme greatly reduces the energy loss caused by the beams interacting across the equator, which is where the greatest scattering occurs. This scheme does not reduce, however, losses caused by interactions between beams from the same hemisphere. The beams on the NIF may be divided into four cones for each hemisphere: two inner cones nearer the pole and two outer cones nearer the equator. The banded scheme reverses the sign of the detuning for the two inner cones of beams in each hemisphere, increasing the coupling. The tricolor configuration improves on both of these by not detuning the inner cones, recovering over half of the drive lost to CBET. The tricolor scheme is more effective than banded because the inner cones interact with both the equatorial beams in the same hemisphere and with the equatorial beams in the opposite hemisphere; by not detuning the inner cones, more energy is retuned from the interaction across the equator than is lost to the intrahemispherical interaction. Figure 1(a) also shows the primary effect of nonlocal heat transport, an increase of hydrodynamic efficiency, especially near the equator, where the radial thermal gradient is greater. A comparison between the red and purple curves, representing implosions with and without nonlocal heat transport, shows an increase of ~30% in the absorbed laser energy, resulting in a much higher implosion speed. The effects of nonlocal heat transport on shock speeds have been measured experimentally [8].

All three of these schemes introduce a north–south asymmetry, as described above. This asymmetry is greatly reduced by use of a fourth configuration, balanced tri-color, in which the tricolor scheme is north to south in alternating quadrants. Figure 2(b) shows a projection in which color indicates the components of the three wavelengths. This projection plot shows a large purple region around the equator where red- and blue-shifted light overlaps, corresponding to effective CBET mitigation.

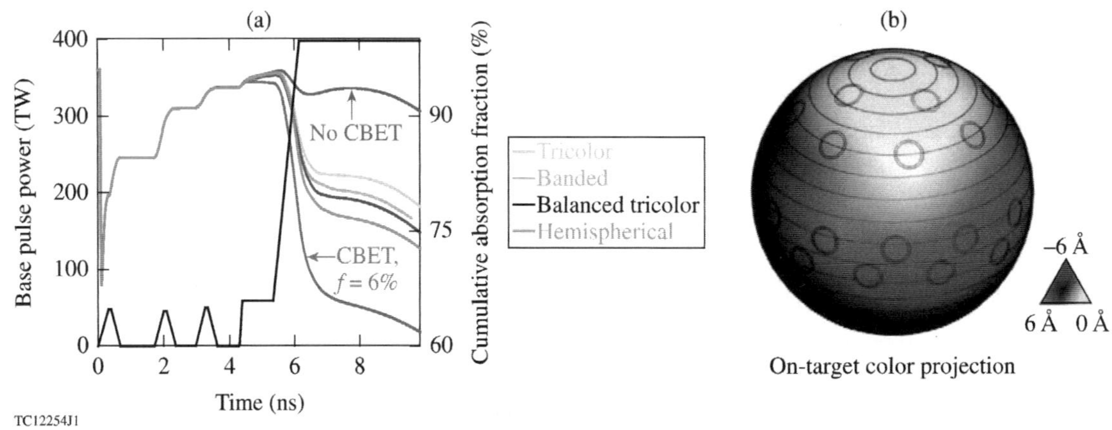

Figure 1. (a) Absorption efficiency is shown as functions of time for four detuning configurations for a PDD design tailored to achieve high gain and minimal hot-spot asymmetry when modeling nonlocal heat transport but without cross-beam energy transfer (CBET). The no-CBET case and the flux-limited CBET case are shown for comparison. (b) The projection on a surrogate target surface of the red-,blue-, and unshifted light.

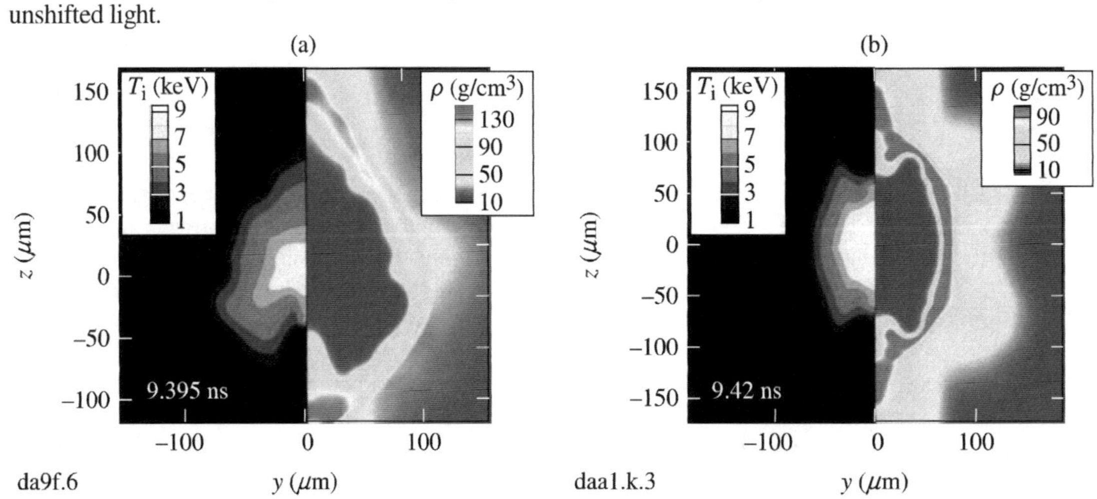

Figure 2. (a) The alpha-burning and (b) ignition designs are shown near peak compression. On the left of each contour plot is the ion temperature and on the right is the mass density.

3. Ignition and Alpha-Burning Designs Incorporating CBET and Nonlocal Heat Transport

The two designs shown here use a DT shell with a CH ablator. The alpha-burning (or "alpha-burner") design has 202 μm of DT, 41 μm of CH, and an outer radius of 1487 μm. The ignition design has 191 μm of DT, 41 μm of CH, and an outer radius of 1482 μm. Each design uses a triple-picket pulse shape to shape the adiabat, and both designs achieve a high implosion speed of ~400 μm/ns, sufficient to generate high burn-averaged hot-spot pressures of ~170 Gbar. Both designs have moderately low IFARs: the ignition design has an IFAR of 23 and a minimum end-of-pulse, density-weighted adiabat of 2.8, and the alpha burner has a somewhat lower IFAR of 21 with a larger ablator adiabat, resulting in an end-of-pulse, density-weighted adiabat of 4.6. Both of these IFARs are lower than that of a previous PDD-ignition design [9], which was shown to withstand the effects of laser imprint. (The simulations presented here include only perturbations related to port geometry, repointing, CBET, and nonlocal heat transport.) The peak areal density for the alpha burner is 1.4 g/cm^2, compared to 1.7 g/cm^2 for the ignition design.

The primary difference between these two designs is shown in the peak-convergence ion temperature and density plots of figure 2; the alpha-burning design, which uses the tricolor configuration, suffers from a larger north–south asymmetry, resulting in a south polar spike penetrating the hot spot. This spike does not preclude ignition; in fact, by stopping alpha particles near the center of the hot spot, it aids in ignition for a low-adiabat–igniting version of this design. However, the spike has a deleterious effect on the target margin and the integrity of the hot spot. The balanced tricolor design has much-greater north–south symmetry and, as a result, a more-uniform hot spot, making it a more-robust design. The alpha-burning design achieves a total neutron yield of 1.2×10^{17}, and the igniting design achieves a gain of 41. While the alpha-burning design does not ignite, it operates at a high adiabat for acceleration-phase stability and the neutrons generated by bootstrap heating are over 3× that generated by compression alone.

4. Conclusions

Ample experimental and theoretical evidence exists to suggest that both CBET and nonlocal electron transport are necessary for modeling PDD on the NIF; whereas for spherical direct drive, a time-varying flux limiter might suffice to mimic these effects. In PDD, the deposition and heat transport have a strong dependence on the polar angle. We have presented here models of 1.8-MJ PDD designs that incorporate these effects. The first design operates at a high adiabat to ensure acceleration-phase stability; it is predicted to generate ~10^{17} neutrons. The second design generates high gain (41), while maintaining an IFAR of 23, suggesting good stability with respect to short-wavelength modes.

Acknowledgment

This material is based upon work supported by the Department of Energy National Nuclear Security Administration under Award Number DE-NA0001944, the University of Rochester, and the New York State Energy Research and Development Authority. The support of DOE does not constitute an endorsement by DOE of the views expressed in this abstract.

References

[1] Paisner J, Boyes J D, Kumpan S A, *et al.*1994 *Laser Focus World* **30** 75
[2] Skupsky S, Marozas J A, Craxton R S, *et al.* 2004 *Phys. Plasmas* **11** 2763
[3] Marozas J A, Marshall F J, Craxton R S, *et al.* 2006 *Phys. Plasmas* **13** 056311
[4] Murphy T J, Krasheninnikova N S, Kyrala G A, S *et al.* 2015 *Phys. Plasmas* **22** 092707
[5] Cao D, Moses G, Delettrez J 2015, *Phys. Plasmas* **22** 082308
[6] Marozas J A, Collins T J B, Zuegel J D, Radha P B, Marshall F J and Seka W 8–13 June 2014 presented at the 44th Annual Anomalous Absorption Conference Estes Park CO
[7] Randall C J, Albritton J R, Thomson J J 1981 *Phys. Fluids* **24** 1474
[8] Goncharov V N, Gotchev O V, Vianello E, *et al.* 2006 *Phys. Plasmas* **13** 012702
[9] Collins T J B, Marozas J A, Anderson K S, *et al.* 2012 *Phys. Plasmas* **19** 05630

Journal of Physics: Conference Series **717** (2016) 012013 doi:10.1088/1742-6596/717/1/012013

Simulations of fill tube effects on the implosion of high-foot NIF ignition capsules

T R Dittrich[1], O A Hurricane[1], L F Berzak-Hopkins[1], D A Callahan[1], D T Casey[1], D Clark[1], E L Dewald[1], T Doeppner[1], S W Haan[1], B A Hammel[1], J A Harte[1], D E Hinkel[1], B J Kozioziemski[1], A L Kritcher[1], T Ma[1], A Nikroo[1], A E Pak[1], T G Parham[1], H-S Park[1], P K Patel[1], B A Remington[1], J D Salmonson[1], P T Springer[1], C R Weber[1], G B Zimmerman[1] and J L Kline[2]

[1]Lawrence Livermore National Laboratory, Livermore, CA 94551, USA
[2]Los Alamos National Laboratory, Los Alamos, NM 87545, USA

E-mail: dittrich1@llnl.gov

Abstract. Encouraging results have been obtained using a strong first shock during the implosion of carbon-based ablator ignition capsules. These "high-foot" implosion results show that capsule performance deviates from 1D expectations as laser power and energy are increased. A possible cause of this deviation is the disruption of the hot spot by jets originating in the capsule fill tube. Nominally, a 10 μm outside diameter glass (SiO_2) fill tube is used in these implosions. Simulations indicate that a thin coating of Au on this glass tube may lessen the hotspot disruption. These results and other mitigation strategies will be presented.

1. Introduction

The high-foot series of ignition capsule experiments [1,2,3] on NIF has shown encouraging results and indicate the achievement of alpha-particle heating of the fusion fuel. A superior performer in this series, shot N140520 (May 20, 2014), imploded the capsule shown schematically in Fig. 1 by means of a 297 eV peak x-ray radiation drive pulse, also shown in Fig. 1. This x-ray pulse was achieved in a depleted uranium hohlraum driven by a 388 TW, 1.76 MJ laser pulse. The capsule had a 178 μm thick graded doped shell enclosing a 69 μm thick layer of solid DT. Silicon doping of the capsule [4,5] varied radially in five steps, (0.00, 0.01, 0.02, 0.01 and 0.00, atomic fraction) as shown in Fig. 1. The structure of the x-ray drive pulse causes a three-shock implosion of the capsule. Shot N140520 produced a neutron yield of 7.6×10^{15} neutrons (13-15 MeV) with a peak neutron emission at 15.96 ns, a burn-weighted ion temperature of 5.5 keV and a DSR (down-scatter-ratio, an indicator of peak fuel ρr) value of 0.041. All quoted times are referenced to 0.0 ns in the laser pulse. This neutron production value is significantly lower than the simulated performance value of 1.2×10^{18} neutrons (1D, ignites) or 7.8×10^{17} neutrons (2D, no fill tube, ignites). The non-linear process of ignition can be removed from these simulations by turning off alpha-particle deposition; this results in a 1D yield of 1.0×10^{16} neutrons and a 2D no-tube yield of 8.9×10^{15} neutrons.

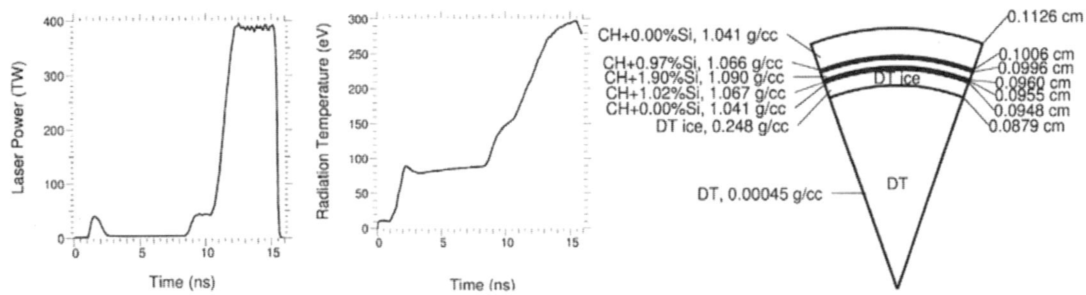

Figure 1. Laser pulse, x-ray drive pulse and schematic diagram and for NIF shot N140520 high-foot ignition capsule. All quoted times are referenced to 0.0 ns in the laser pulse.

This capsule had a fill tube attached in order to load the DT fuel prior to implosion. In this case, the tube was SiO_2 with an outside diameter of 10 μm and an inside diameter of 6 μm. The tube was inserted 44 μm radially into the CH shell. A tapered fill hole through the rest of the shell had a diameter of 3.6 μm at the inner shell surface next to the DT ice. Polar self-emission snapshot x-ray images from N140520 are shown in Fig. 2 before, at and after peak emission. The middle image occurs slightly before peak neutron production. Each image in Fig. 2 has its own emission-level to color-scale correspondence, with dark red corresponding to high emission and blue corresponding to low emission. In these images, the fill-tube orientation is at approximately 2:30 o'clock. The observed structure in the images appears to roughly align with the tube orientation.

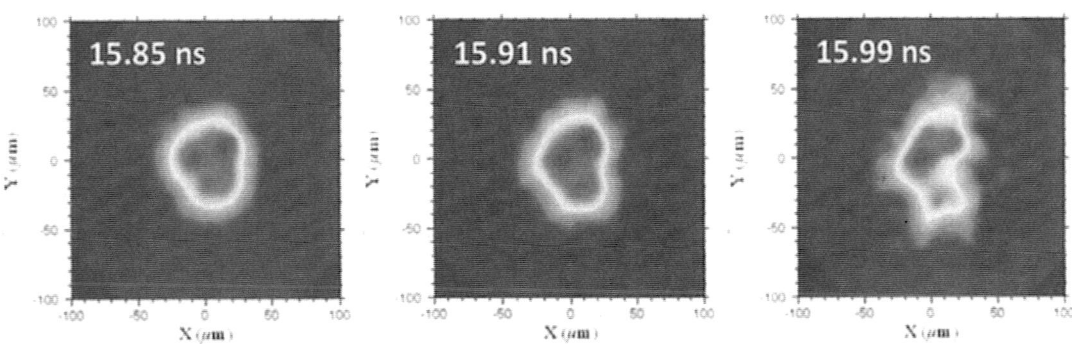

Figure 2. Polar self-emission x-ray images at three times for NIF shot N140520.

2. Simulation Technique

Implosion simulations of N140520 used a radiation-hydrodynamic computer code, LASNEX [6], to model the implosion of the capsule only, i. e., an x-ray source was applied to the outside of the ablator surface. An alternate approach would be to simulate the conversion of laser light to x-rays in the DU hohlraum that would then implode the capsule. In this two-dimensional, axially symmetric simulation a technique of "wedge expansion" was used that starts with an initial cone, or wedge angle of 5.625°; this was expanded to 11.25° at 3.4 ns, 22.5° at 8 ns, 45° at 13 ns, 90° at 14 ns and 180° at 14.7 ns. Since the number of angular zones in the simulation was held fixed the zonal angular resolution decreased at each wedge expansion. This technique allows high angular resolution at the start of the simulation when the fill tube and its perturbation are relatively small. Radial mesh management uses a technique called "donor line" remapping. The upper boundary of the simulation wedge is assumed (and verified) to be unperturbed by the tube evolution located on the symmetry axis; this upper boundary is called the "donor line". Wedge expansions are timed to maintain a one-dimensional (radial only)at the donor line. Radial node positions from this donor line are periodically broadcast over the entire simulation wedge using a uniform distribution of angular zones. This technique maintains concentric radial zoning. The wedge expansions are timed so that nodes on the donor line behave similarly to those in a 1-D implosion simulation. At late times, no-tube 2D simulations exhibit some non-symmetric behavior much reduced from that seen in 2D tube simulations. Efforts are continuing to achieve a more symmetric null simulation.

Results from simulating the implosion of N140520 are shown in Fig. 3, which depicts only "region-of-interest" density maps at eight times during the implosion, from initial configuration to bang time. Position coordinates and density-to-color-scale correspondence are not shown. The cylindrical axis of symmetry is at the bottom edge of each individual picture. The density map at 0.0 ns shows the initial configuration around the joint between the tube and capsule shell. The tube and capsule fill hole are contain solid DT; an epoxy glue fillet is shown at the tube-shell joint. Shell and tube ablation proceed at 1.8 ns. At 3.6 ns, a shaped-charge effect can be seen where the tube and shell meet; a non-radial shock is formed which reflects from the axis of symmetry, moving material azimuthally outward and creating a jet of material ahead of the main shock as seen clearly at 5.4 ns. By 15.6 ns this azimuthal movement of material has created a hole in the imploding shell. Further implosion causes this hole to collapse and create a large jet that penetrates the hotspot, as seen at 15.9 ns.

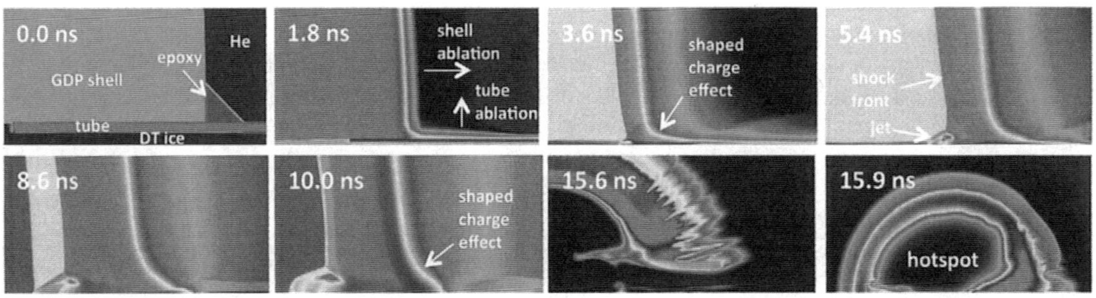

Figure 3. Density snapshots at eight times during the simulated implosion of N140520. No density-to-color-scale correspondence or position coordinates are shown. Dark red corresponds to high density; dark blue to low density.

3. Comparison of Data and Simulation

Figure 4 compares the observed polar x-ray emission of N140520 (Fig. 4a) with a simulation-generated image of the x-ray emission (Fig 4b) at the time of peak emission. The snapshot in Fig. 4a is at time 15.91 ns in the implosion and has the fill-tube orientation at 2:30 o'clock; the snapshot in Fig. 4b is at time 15.85 ns and has fill tube orientation at 3:00 o'clock. The two images are approximately the same size and show similar intensity fall-off from the image center to the outside of the emitting region. The non-symmetric structures seen in the two images are not inconsistent; each image has a flattening on the right-hand side and a dark region off the axis of the fill tube perhaps originating in the tube jet. Here dark red corresponds to areas of high x-ray emission and dark blue and gray correspond to areas of low emission.

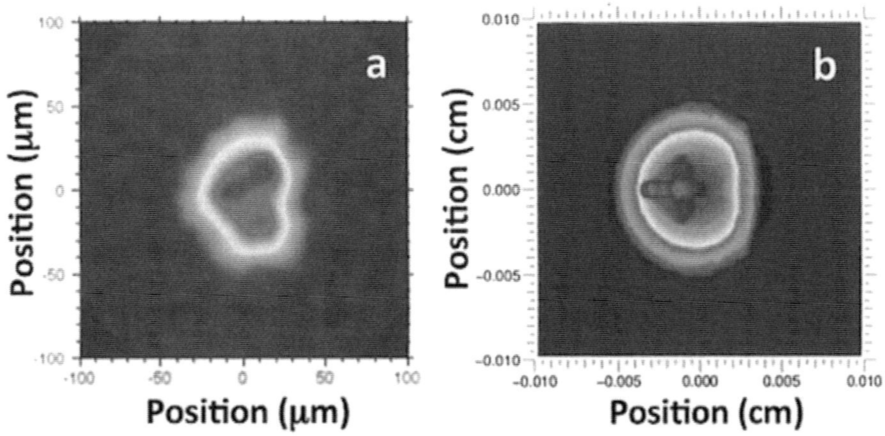

Figure 4. Observed (a) and simulated (b) x-ray images at peak emission for NIF shot N140520.

The simulation shown in Figs. 3 and 4 predicts a neutron output of 8.0×10^{16} (13-15 MeV). This is significantly lower than the no-tube igniting 2D simulation result of 7.8×10^{17} neutrons but more than the no-tube, alpha-deposition-off simulated result of 8.9×10^{15} neutrons. Removing alpha deposition from a simulated 2D implosion with the as-shot fill tube predicts a yield of 6.3×10^{15} neutrons.

4. Mitigation Strategies

Alternate fill-tube configurations may help mitigate the effects of the fill tube on capsule performance. Smaller diameter tubes may produce a smaller perturbation but would present additional fabrication and fielding difficulties. Larger diameter tubes would likely increase the size of the perturbation. Implosion simulations of a capsule with a gold-coated glass tube as well as simulations configured with the tube truncated at the capsule surface have predicted 30% and 100% improvement in neutron yield, respectively, compared with the nominal SiO_2 tube implosion.

5. Summary

Fill-tube hydrodynamic effects in high-foot ignition implosion N140520 may be responsible for degraded performance as compared with unperturbed implosion simulations. Both wedge-expansion and donor-line mesh-management techniques offer advantages for the simulation of this implosion.

The neutron output observed in high-foot shot N140520 is about 10% of the predicted output from a simulation of the as-built tube and capsule configuration. A polar x-ray image of this high-foot implosion shows hot-spot perturbations consistent with simulated hydro effects originating in the DT fill tube and fill hole.

Acknowledgments

This work was performed under the auspices of the Department of Energy, by Lawrence Livermore National Laboratory under Contract DE-AC52-07NA27344. Lawrence Livermore National Security, LLC.

References

[1] Hurricane O A, et al. 2014 *Nature* **506**, 343
[2] Dittrich T R, et al. 2014 *Phys. Rev. Lett.* **112** 055002
[3] Park H S, et al. 2014 *Phys. Rev. Lett.* **112** 055001
[4] Haan S W, et al. 2011 Phys. of Plas. **18**, 051001
[5] Chen K C, et al. 2011 Fusion Sci. Technol. **59** 8
[6] Zimmerman, G B, et al. 1975 Comments Plas. Phys. **2** 2

Journal of Physics: Conference Series **717** (2016) 012014 doi:10.1088/1742-6596/717/1/012014

Simulations of symcap and layered NIF experiments with top/bottom laser asymmetry to impose P1 drive on capsules

D Eder, B Spears, D Casey, A Pak, T Ma, N Izumi, B Pollock, C Weber, A Kritcher, O Jones, J Milovich, R Town, H Robey, D Hinkel, D Callahan, S Hatchett, J Knauer, C Yeamans, D Bleuel, S Nagel, R Hatarik, S Khan, D Sayre, J Caggiano, G Grim, M Eckart, D Fittinghoff, F Merrill and D Bradley

Lawrence Livermore National Laboratory, Livermore, CA USA

E-mail: deder@llnl.gov

Abstract. Integrated hohlraum/capsule post-shot simulation of the first full-scale cryogenic layered-DT experiment with top/bottom laser asymmetries of 8% is discussed. The imposed P1 Legendre mode drive on the capsule results in downward velocity of 85 ± 15 km/s as measured by neutron time of flight (NTOF) diagnostics and x-ray imagers, which is in excellent agreement with the calculated velocity of 87 km/s. The measured DT yield is approximately 30% less than the average of two comparable shots using the same 4 shock HiFoot pulse shape without drive asymmetry. The calculated DT yield of 5.0e15 is very close to the measured value of 4.86e15 for the shot with drive asymmetry, which implies that P1 effects dominate yield reduction. The neutron activation diagnostics (NADs) give clear indication of higher areal density in the direction of the north pole in excellent agreement with calculations. Integrated post-shot simulation of an earlier symcap (capsule with appropriate ablator thicknesses to act as a surrogate for an ignition capsule) experiment with laser asymmetries show that calculated neutron-wighted velocity is a strong function of capsule shape.

1. Introduction

In approximately 50% of NIF layered shots the neutron time of flight (NTOF) diagnostics show that the capsule has a bulk velocity of 50 km/s or greater for reasons that are not clearly understood[1]. Capsule only simulations have shown that radiation drive asymmetries and associated bulk capsule motions can be a source of yield reduction[2]. In order to understand the effects of bulk capsule motion a number of experiments have been conducted with top/bottom laser asymmetries using various scale targets and different capsule types. In this paper we focus on the latest one (N150318), which is a full-scale cryogenic layered-DT experiment with top/bottom laser asymmetries of 8%. We also discussed results from post-shot simulations of an earlier (N130625) CH symcap (capsule with appropriate ablator thicknesses to act as a surrogate for an ignition capsule) experiment with similar laser asymmetries.

Our integrated hohlraum/capsule post-shot simulations determine the drive asymmetry from laser asymmetries and hohlraum dynamics. We use measured laser powers but backscatter data is only available on a subset of inner and outer beams in the bottom half of the NIF chamber. We

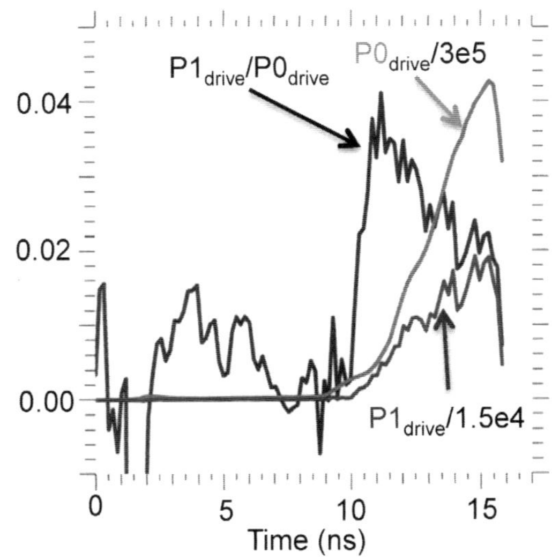

Figure 1. The calculated P0 and P1 components of the radiation drive on the capsule divided by 3.e5 and 1.5e4, respectively. We also give the ratio, which rises rapidly to approximately 0.4 at start of final rise and then drops to approximately 0.02 at peak drive.

assume that the backscatter power scales with laser power to estimate the backscatter for other beams. Separate plasma conditions at the upper and lower hohlraum laser entrance holes (LEHs) are used to calculate the crossbeam energy transfer between the outer and inner beams[3]. We use the same crossbeam saturation model for upper and lower beams.

2. Results

The NIF laser drive for shot N150318 is 4% higher/lower than nominal for the upper/lower beams, which produces a nonuniform radiation drive on the capsule giving an initial drive P1/P0 Legendre mode ratio of approximately 4%. However, radiation of x-rays within the hohlraum smooths the nonuniformity so that the drive P1/P0 ratio is approximately 2% at time of peak drive. This is shown in Figure 1 along with time dependence of the drive. This imposed P1 drive on the capsule results in downward velocity of 85 $\pm$ 15 km/s as measured by neutron time of flight (NTOF) diagnostics and x-ray imagers, which is in excellent agreement with the calculated neutron-weighted velocity of 87 km/s. The total laser energy was 1.78 MJ, a measured DSR (down scatter ratio, 10-12/13-15 MeV neutrons) of 2.6%, and a implosion velocity of 371 $\pm$20 km/s. The measured DT yield is approximately 30% less than the average of two comparable shots (N140520 and N150121) using the same 4 shock HiFoot pulse shape but no drive asymmetry. The calculated yield for shot N140520 is approximately 2 times larger than what was measured[4]. Such a difference is common when the simulation does not include detailed calculations of the capsule support tent and 3D effects. In contrast, for the shot being studied here with drive asymmetry the calculated DT yield of 5.0e15 is very close to the measured value of 4.86e15 implying P1 effects dominate yield reduction.

The neutron activation diagnostics (NADs) give clear indication of higher areal density in the direction of the north pole as seen in the right image of Figure 2 where the angular dependent DT yields relative to the average yield are shown. In order to compare with the 2D simulation, we give the relative yield as a function of angle from south pole to north pole in the left image. The calculated curve is in excellent agreement with the data. The reduction of yield in the direction of the north pole is due to neutron scattering by a high density region at the upper portion of the capsule as seen in Figure 3. The density is given at the time of peak of the x-ray emission (16 ns). The center of capsule has been shifted downward by approximately 25 microns at this time.

The post-shot simulation for N150318 uses a crossbeam transfer model that results in an

Journal of Physics: Conference Series **717** (2016) 012014 doi:10.1088/1742-6596/717/1/012014

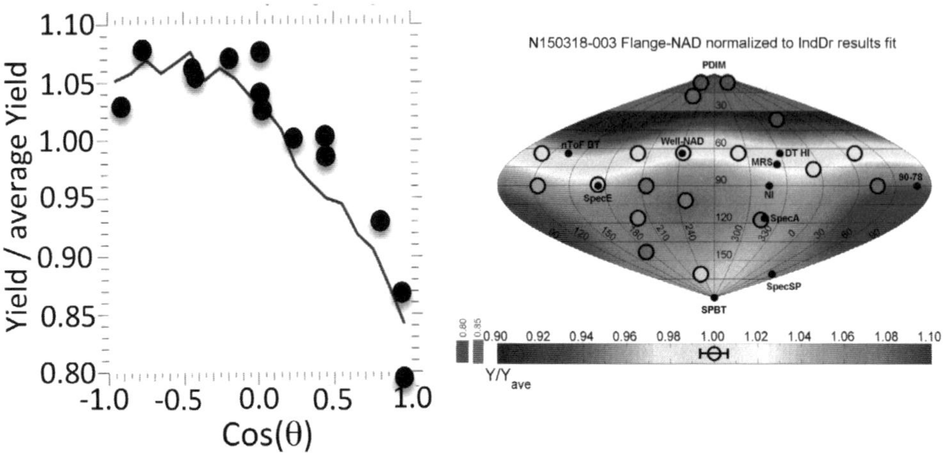

Figure 2. Left image is the measured (dots) change in yield going from south to north pole from the neutron activation diagnostics (NADs) and the calculated angular dependence (curve). Error in data is slightly less than size of dots. Right image shows the data for all locations.

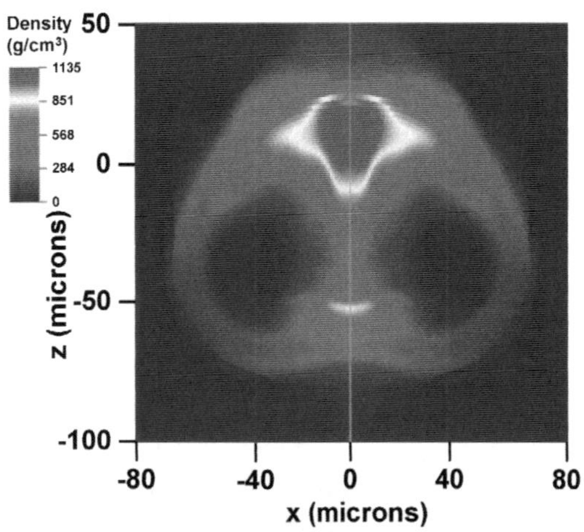

Figure 3. The calculated density at time of peak x-ray emission showing higher areal density in the direction of the north pole.

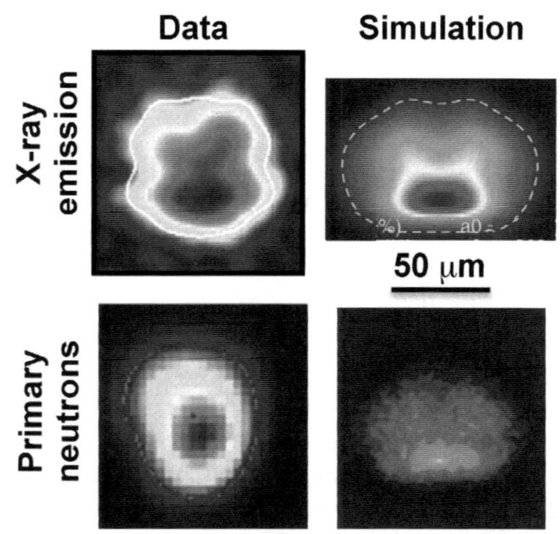

Figure 4. Left images are x-ray emission and primary neutrons and simulation results on right. The 17% contour shown on 3 images.

inner/outer laser-beam cone fraction that is slightly greater than 0.4 during the rise to peak power dropping to approximately 0.25 during peak as transfer is shutoff. The measured x-ray emission and primary neutron images and corresponding images generated from the simulation are given in Figure 4. The calculated images are only slightly more pancake than data. The post-shot simulation for N140520 that matches measured x-ray images uses a smaller saturation parameter that gives an inner/outer cone fraction of 0.3 during the rise and the same 0.25 during peak[3]. Using this amount of transfer for N150318 gives severely pancaked images that are not consistent with the data. Ideally the same saturation model for crossbeam transfer and laser-drive multipliers that give good agreement with data on x-ray emission shape and bang times on

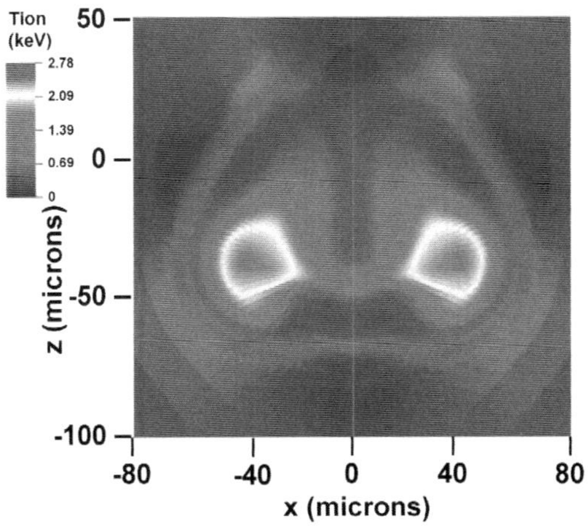

Figure 5. The calculated Tion at time of peak x-ray emission.

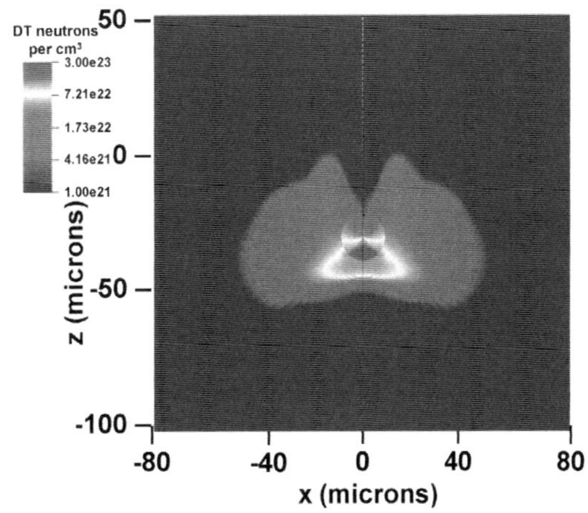

Figure 6. Calculated DT neutron production at time of peak x-ray emission.

the control shots are used for the corresponding shots with drive asymmetry. However, our post-shot simulations of an earlier symcap experiment (N130625) with top/bottom laser asymmetries showed that simulations with incorrect shapes can produce neutron-weighted that are an order of magnitude lower than measured[5]. Thus it can be necessary to adjust crossbeam parameters to match observed x-ray images, which is what is done for N150318. The same laser-drive multipliers that were used for N150520 are used for N150318.

The calculated ion temperature at time of peak x-ray emission is shown in Figure 5. The peak temperature at this time is less than 1/2 the peak value of 7 keV, which occurred approximately 200 ps earlier. The calculated DT neutron production per unit volume at 16 ns (log scale) is shown in Figure 6. The location of peak production is a function of density and Tion. The time of peak neutron production is approximately 50 ps later.

3. Summary

Results from an integrated hohlraum/capsule post-shot simulation of the first full-scale cryogenic layered-DT experiment with top/bottom laser asymmetries are in excellent agreement with the data. In particular, NADs give clear indication of higher areal density in the direction of the north pole in excellent agreement with calculations. In order to match x-ray emission and primary neutron images, a cross beam transfer model with more transfer during the rise to peak power is needed for a comparable shot without drive asymmetry. The agreement between calculated and measured yield implies that P1 effects dominate yield reduction.

4. Acknowledgements

Work was performed under the auspices of the U.S. Dept. of Energy by LLNL under Contract DE-AC52-07NA27344. Lawrence Livermore National Security, LLC. LLNL-JRNL-679579

References

[1] Spears B K, et al., this issue.
[2] Spears B K, et al. 2014 *Phys. of Plasma* **21** 042702
[3] Michel P, et al. 2010 *Phys. of Plasma* **17** 056305
[4] Kritcher A, et al., in prep for *Phys. of Plasma.*
[5] Eder D C, et al., 2014 *Bulletin of the American Physical Society* **59**

StarDriver: Recent results on beam smoothing and LPI mitigation

D Eimerl[1], S Skupsky[2] and E M Campbell[2]

(a) EIMEX Software and Consulting,107 Siebe Dr, Fairfield CA 94534
(b) University of Rochester LLE, 250 River Road, Rochester NY 14623

Email: deimerl@eimexinc.com

Abstract. StarDriver was recently proposed as a highly flexible laser driver for inertial confinement fusion and high energy density physics. It envisions a laser drive consisting of very many beams at an aperture and energy where the optical technology is well-developed, used in concert to create a large scale laser driver system. In this paper we describe a StarDriver–class laser with 5120 physical beamlets disposed about the target chamber in 80 evenly spaced ports, each port containing 64 beamlets, each beamlet having about ~1.5THz of 2D SSD bandwidth and suitable phase plates, an aperture of ~65mm, an energy of 80J, and frequency-converted to ~351nm.many beamlets at an aperture where optical technology is well-developed, and each beamlet has energy ~100J in a several times diffraction limited beam. The ensemble of beamlets has frequency bandwidth 2%-10%, thereby providing significant control of both hydrodynamic and laser-plasma instabilities The drive at the target is ~400kJ, has a well-behaved low L-mode spectrum, and smooths very rapidly, reaching an asymptotic smoothness of <1% in less than 1ns.

Background. To achieve the plasma conditions required for Inertial Confinement Fusion (ICF) and High Energy Density Physics (HEDP) drivers that effectively compress energy in space and time and whose energy can be effectively delivered to an appropriate target are required. The top-level requirements for ICF are currently under investigation at NIF (Lawrence Livermore National Laboratory) [1,2] and Omega (University of Rochester, NY) [3,4]. One of the challenges for ICF is obtaining the necessary uniformity to drive the fusion fuel capsule. For laser driven ICF with direct drive target concepts, modeling and the world-wide data base suggest the required pulse energy (1–3 MJ), and peak power (~400–500 TW) in a temporally shaped pulse for fusion ignition and gain. For civilian energy applications, namely Inertial Fusion Energy (IFE), lasers with average power of nominally 10–20 MW at repetition rates of ~10 Hz with driver efficiencies >10% are required. It is expected over the next decade that experiments on NIF will explore the physics of a variety of target concepts with MJ-sized plasmas representative of attractive "high gain" target configurations, further develop quantitative predictive simulation tools and hopefully demonstrate fusion ignition and modest fusion gain (G>10). These experiments and the development of quantitative and predictive modeling tools will be essential for the development of laser driven IFE.

In addition to these top-level laser energy and power requirements, experiments, theory, and computer simulations have conclusively demonstrated the importance of laser wavelength and "beam smoothing" [5,6] for ICF/IFE. To achieve the required convergence ratios (20-35) the ablative pressure on the capsule must be better than 1–2% *rms* for the low spatial frequencies (L mode ~10-20), and in addition to fabrication imperfection, hydrodynamic instabilities during target implosion are seeded by high spatial frequency non-uniformities in the laser drive. Laser-plasma instabilities in the plasma around the target can reduce the laser-target energy coupling efficiency, generate "suprathermal" electrons which "preheat" the target preventing efficient compression, and redistribute the light both reducing the coupling efficiency and degrading the symmetry in the ablative pressure.

To obtain the required ablation pressures and hydrodynamic efficiency while minimizing laser plasma instabilities, UV lasers have become the baseline for all ICF concepts. Experimental data have shown that beam smoothing technologies, or more accurately technologies that reduce spatial and temporal coherence, are also essential for controlling and minimizing laser-plasma instabilities and preventing the seeding of hydrodynamic instabilities for direct drive. Techniques for beam smoothing have been developed by NRL [5](Induced Spatial Incoherence (ISI)) and LLE[6] (Smoothing by Spectral Dispersion (SSD)) To date the former approach has only been applied to KrF lasers while the latter technique is applicable to solid state lasers including those up-converted to higher harmonics using non-linear frequency conversion techniques. Variations of SSD are found in essentially every solid state laser employed in fusion and high energy density physics research today.

Beam smoothing technologies incorporate bandwidth in a laser beam that is multi-spatial mode at the target. Regardless of the optical technology involved, it is essential that the coherence time of the radiation field at the target be short enough for the laser drive to control hydrodynamic and laser-plasma instabilities. In presently envisioned IFE laser drivers (KrF or up-converted solid state lasers) all of the apertures utilize the same gain media and therefore operate only over the limited bandwidth of that medium (~0.2% for KrF at 0.248 μm and 0.02% at 0.35 μm for tripled Nd:glass). Theory and modeling have long suggested that a much larger bandwidth would be most effective in minimizing or eliminating the deleterious effects described above [7]. This limitation motivates a key attribute of the StarDriver approach. StarDriver employs a frequency-diverse collection of many apertures that enables much larger system bandwidth and combines this with a large angular spread of the beamlets irradiating any region of the corona.

StarDriver [8,9] is motivated by the need to irradiate mm-scale targets at intensities of 10^{15} Watts/cm^2 at a standoff of >10 meters with a source that is essentially incoherent, in order to control hydrodynamic and laser-plasma instabilities. StarDriver envisions very many beams (~10,000), at an aperture where the optical technology is well-developed, used in concert to create a large scale laser driver system. StarDriver deploys a frequency-diverse collection of many apertures that enables much larger system bandwidth and combines this with a large angular spread of the beamlets irradiating any region of the corona. A large number of the frequency-diverse StarDriver beamlets overlap in the target corona, thus providing both a much shorter total laser coherence time for the laser field in the target (individual beam quality), as well as a much denser **k**-spectrum of the drive in the target corona, than any competing laser driver concept. The large number of beams also relaxes the tolerances on individual beamlet parameters compared to legacy ICF drivers, and enhances independent control of the mid-spatial frequencies and the high spatial frequencies in the drive. As mentioned above, StarDriver also offers unprecedented opportunity to optimize the laser drive including time-dependent focusing at the target in almost any parameter a laser beam may have. Such focal "zooming" is one strategy important to reducing cross-beam energy transfer in the corona and to maximize driver target coupling. The small laser aperture also reduces transverse ASE enabling the near full use of the gain media bandwidth.

We have simulated [9] the on-target time-dependent intensity profile of a StarDriver-class IFE driver with 5120 physical beamlets disposed about the target chamber in 80 evenly spaced ports, each port containing 64 beamlets, each beamlet having about ~1.5THz of 2D SSD bandwidth and suitable phase plates, an aperture of ~65mm, an energy of 80J, and frequency-converted to ~351nm. With less than 10,000 beamlets, each with 2D SSD, spanning 2% bandwidth, the laser drive on target is asymptotically less than 1% and is reached in less than 1ns. This performance is a very significant improvement over legacy ICF laser drivers and will ease, significantly, the constraints on target concepts arising from hydrodynamic instabilities. Figure 1 summarizes the smoothing performance.

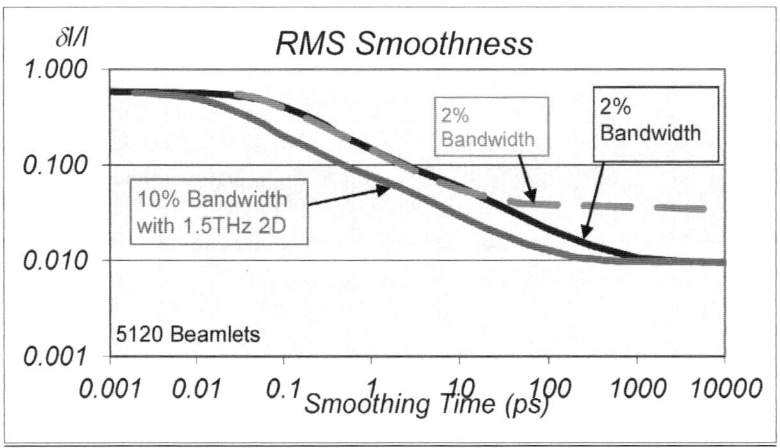

Figure 1: RMS Smoothness of a StarDriver-Class driver

We have also estimated the growth rates for the LPI process of CBET and two-plasmon decay (TPD) using a dispersion relation approach. We found that 2% bandwidth is adequate to suppress CBET. In the larger IFE-scale targets, it may also be adequate to suppress TPD, but in general, this may require as much as 5% bandwidth. Our LPI results must be taken as highly preliminary; much further work remains.

The most significant conclusions from this study are: (1) It appears that to reach 1% smoothness about 150,000 monochromatic beamlets are required. (2) The time development of the drive asymmetry is as expected and is very similar to ISI. (3) With SSD, significant improvement in the asymptotic smoothness is achieved. With 2D SSD, the number of beamlets is reduced to about 5000 beamlets.

A StarDriver-class IFE driver with 5120 beamlets in a ported configuration with 80 ports of 64 beamlets, each delivering 80J in a 65mm beam, and having 2% bandwidth would deliver 400kJ to a direct drive target and provide both extreme flexibility in the laser drive and a high level of control of all known instabilities. The energy on target could be increased to ~1.6MJ without compromising the smoothing performance by increasing the number of ports to 180, increasing the number of beamlets per port to 100, and/or increasing the energy of each beamlet. Although significant research remains, 5-10% bandwidth would likely control all the known instabilities in direct drive ICF.

Acknowledgements: One of us (DE) was supported in this work by the University of Rochester LLE, 250 River Road, Rochester NY 14623, under Contract U943038-15. We are grateful for productive discussions with John Marozas, Bob McCrory, Jason Myatt, Bill Krupke, Bob Lehmberg, Jason Zweiback, Steve Bodner, and Steve Obenschain.

References

1. C.A. Haynam etal, "National Ignition Facility laser performance status", Applied Optics, Vol. 46, No. 16 (1 June 2007), 3276.

2. J.A. Paisner, E.M. Campbell, W.F. Hogan, "The National Ignition Facility Project", ANS 11th Annual Conference on Fusion Energy, New Orleans, June 16th,1994.

3. "LLE Review", Scientific reports of the Laboratory for Laser Energetics, University of Rochester,, 250 E. River Rd, Rochester, Vols 1-126,

4. R. L. McCrory, "Laser-Driven Inertial Fusion Energy; Direct-Drive Targets Overview", NAS/NAE Committee on the Prospects for IFE Systems San Ramon, CA 29 January 2011

5. R.H. Lehmberg and S.P.Obenschain, "The Use of Induced Spatial Coherence for Uniform Illumination of Laser Fusion Targets" Optics Communications, 46, 27(1983)

6. S.Skupsky etal, "Improved laser-beam uniformity using the angular dispersion of frequency modulated light" J App. Phys. 66,3456(1989)

7. D.Eimerl and W.Kruer, E.M.Campbell, "Ultrabroad Bandwidth for ICF Applications", Comments in Plasma Physics 15, 85 (1993)

8 David Eimerl, E.Michael Campbell, William F. Krupke, Jason Zweiback, W.L.Kruer, J.Zuegel, J.Myatt, J.Kelly, D.Froula, and R.L.McCrory, "StarDriver: A Flexible Laser Driver for Inertial Confinement Fusion and High Energy Density Physics", Journal of Fusion Energy (2014) **33**, 476-488

9 David Eimerl, S Skupsky, J Myatt and EM Campbell, "A StarDriver-class laser achieving 1% beam uniformity in 1ns", Journal of Fusion Energy, (2016) 35(2), 459-469

Journal of Physics: Conference Series **717** (2016) 012016 doi:10.1088/1742-6596/717/1/012016

The first capsule implosion experiments on Orion*

W J Garbett, C J Horsfield, S G Gales, A E Leatherland, M S Rubery, J E Coltman, A E Meadowcroft, S J Rice, A J Simons, V E Woolhead

AWE plc, Aldermaston, Reading, Berkshire, RG7 4PR, UK

E-mail: warren.j.garbett@awe.co.uk

Abstract. Direct drive capsule implosions are being developed on the Orion laser at AWE as a platform for ICF and HED physics experiments. The Orion facility combines both long pulse and short-pulse beams, making it well suited for studying the physics of alternative ignition approaches. Orion implosions also provide the opportunity to study aspects of polar direct drive. Limitations on drive symmetry from the relatively small number of laser beams makes predictive modelling of the implosions challenging, resulting in some uncertainty in the expected capsule performance. Initial experiments have been fielded to evaluate baseline capsule performance and inform future design optimization. Highly promising DD fusion neutron yields in excess of 10^9 have been recorded. Results from the experiments are presented alongside radiation-hydrocode modelling.

1. Introduction

Direct drive capsule implosions are being developed on the Orion laser [1] at AWE as a platform for ICF and HED physics experiments. Although the energy scale of the facility (~5kJ) precludes access to very high yield implosions, there is significant potential to study relevant aspects of the implosion physics as well as more fundamental physics of ICF plasmas; an area of current interest being the use of implosions as a source of energetic charged particles for plasma stopping power experiments. The Orion facility is configured with ten long pulse (ns) beams and two additional PW (ps) beams, which make it well suited for studying the physics of alternative ignition approaches, including both fast ignition and shock ignition. Orion implosions also provide the opportunity to study aspects of polar direct drive [2], with particular relevance to configurations being explored for direct drive shock ignition at LMJ [3].

Limitations on drive symmetry from the relatively small number of laser beams, together with the complexity of polar drive configurations, makes predictive modeling of Orion implosions challenging, resulting in some uncertainty in the expected capsule performance. Accordingly initial experiments have been fielded to evaluate baseline capsule performance and inform future design and optimization.

Implosions were performed using thin-shell, deuterium-filled, silica glass capsules. These were intended to produce exploding pusher-like dynamics, in which the fusion yield is mainly produced during the initial shock phase, in order to minimize the impact of mix and asymmetry. A substantial suite of diagnostics was fielded to characterize the implosions, which included the first use of Orion's neutron total yield (NTYD) and time-of-flight (NTOF) diagnostics, and time-gated x-ray imaging (GXD) on multiple axes.

2. Capsule design

Suitable capsule designs were identified by a 1D study of capsule phase space using the Nym radiation-hydrocode [4]. As expected, clean yield was generally found to decrease rapidly with increasing shell thickness (figure 1), as shell inertia is increased and the drive shock is both delayed and weakened. The variation of yield with capsule radius is not monotonic, such that there is an optimal radius for a given shell thickness. Based on this study it does not appear possible to simultaneously achieve low convergence, high fusion yields and exploding pusher-like dynamics at the Orion energy scale without recourse to very high aspect ratio shells.

Consideration of drive uniformity requires a smaller capsule radius than the 1D optimum. The existing Orion phase plates are relatively small, being intended primarily for indirect drive, and at best focus result in a very low ratio of spot size to capsule radius, providing far from optimal illumination [5]. Improved illumination can be achieved by operating at ~2mm defocus, although this does compromise the quality of the beam spot. Polar drive is also necessary to optimize equatorial symmetry (figure 2).

The resulting nominal capsule design was a 250μm radius, 2.5μm thick silica glass shell with 10atm D_2 gas fill. These were imploded using a 600ps FWHM square pulse with ~2.5kJ laser energy.

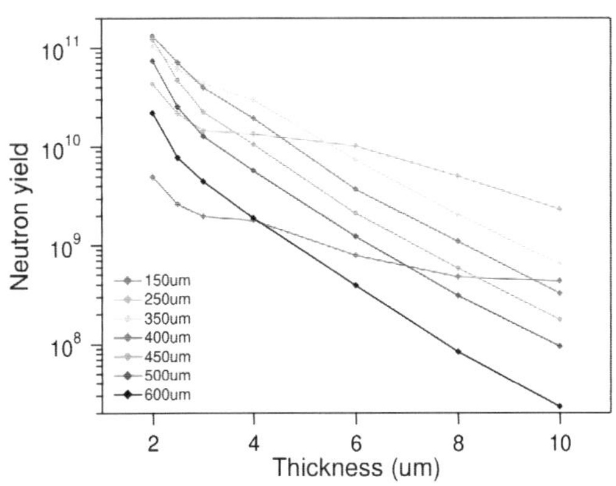

Figure 1. Clean neutron yield as a function of shell thickness and radius, as determined from 1D Nym simulation study. Only a subset of these capsules produce exploding pusher-like dynamics.

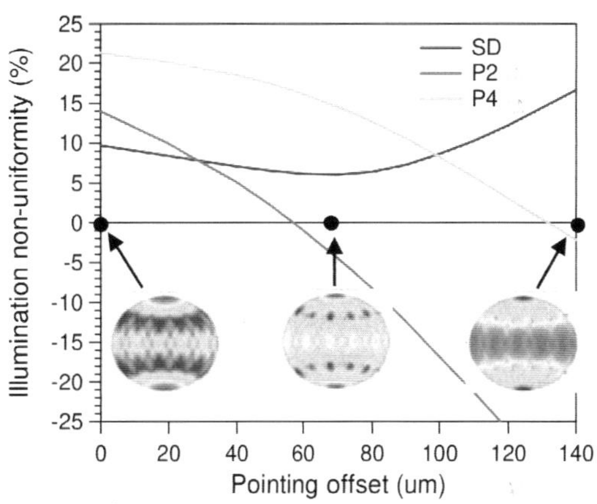

Figure 2. Static view-factor calculations show improved drive uniformity with polar drive pointing offset. Curves show standard deviation of incident laser power over the capsule surface (blue), with Legendre P2 (red) and P4 (green) modes. Inset images show incident power map at selected offsets.

3. Experimental results

Experimental fusion yields were measured using three different diagnostic methods: CR39 nuclear track detection for absolute proton yield; NTYD using activation of an indium sample for absolute neutron yield; and an NTOF spectrometer, which provided relative neutron yields. Excellent correlation between the different diagnostic yields (figure 3) provides confidence in the measured performance. This is particularly important for the highest yield shot, a single data point significantly separated from the others, which might otherwise be interpreted as anomalous. The highest recorded yield was 1.7 x 10^9 neutrons and corresponds to the thinnest capsule. A large increase in performance near this shell thickness was anticipated from design simulations (figure 1). However, it is difficult to definitively attribute this behaviour to the thinner shell; in general the yield data shows large variability and is not well-correlated with any single input parameter.

The experimental yields have been compared to post-shot modeling of the experiments using Nym (figure 4), which takes account of simultaneous variations in the 1D experimental parameters. The simulations were calibrated to the time of peak emission, as measured by the Dante diagnostic, by tuning the thermal conduction flux limiter and input laser energy, compensating for limitations in laser absorption modeling [6]. The use of existing GXD self-emission images to further constrain the modeling is also underway. Experimental neutron yields are typically ~15-35% of post-shot mix simulations. This difference indicates some variability or degradation mechanism not adequately captured in the 1D simulation model.

Ion temperatures ⟨Ti⟩ derived from the NTOF measurements of the neutron spectra range from ~3-5keV (figure 5). Although the line of best fit shows an upward trend, there is no strong correlation between ⟨Ti⟩ and neutron yield, which suggests that thermal broadening at the neutron source does not dominate the measured spectrum. Synthetic neutron spectra from post-processed simulations produce inferred temperatures significantly lower than the experiment, even when the effect of 1D bulk plasma velocity is included. This discrepancy could suggest additional motion due to asymmetry or turbulence that is not captured by the simulations.

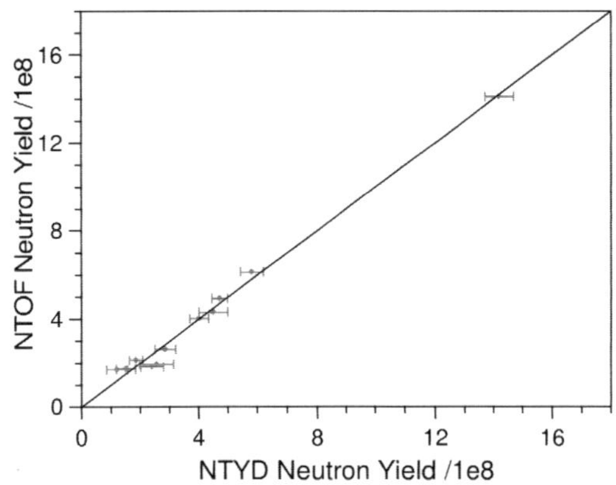

Figure 3. Comparison of NTOF and NTYD neutron yields, showing excellent correlation. The best-fit line has R^2=0.99.

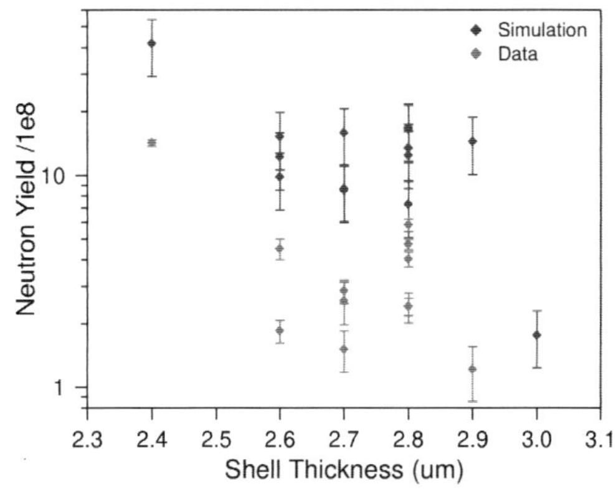

Figure 4. Measured (red) and simulated (blue) neutron yields as a function of capsule shell thickness. Although a large increase in performance is seen at 2.4μm, there is significant variability in the data.

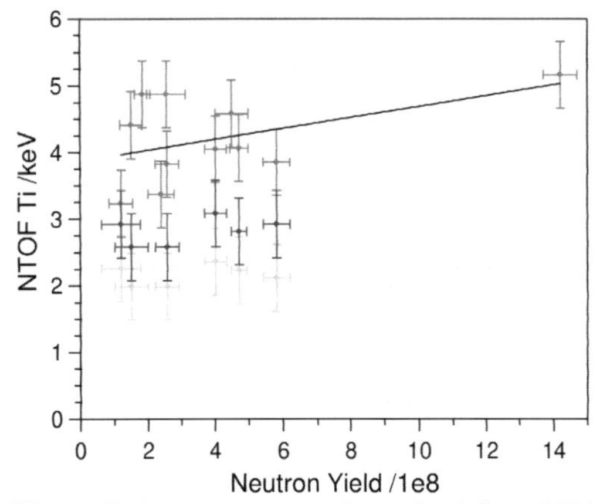

Figure 5. Ion temperature determined from NTOF (red) shows no significant correlation with neutron yield. Simulated ⟨Ti⟩ with (blue) and without (green) motion Doppler effects are significantly lower.

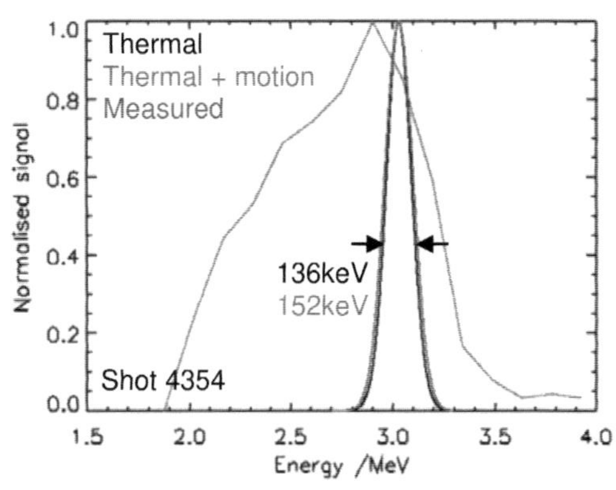

Figure 6. Preliminary proton spectral data (blue) shows significant broadening which is consistent with downshift through the residual shell material. Calculated source spectra with (red) and without (black) motion Doppler broadening are shown for reference.

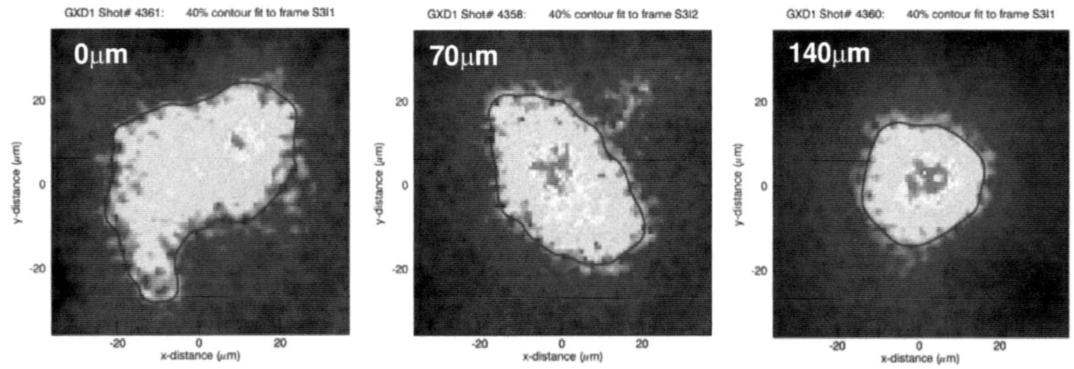

Figure 7. Gated x-ray images of the implosion self-emission taken from the equatorial view at bang time for three different polar drive pointing offsets (0, 70 and 140μm). Although implosion symmetry appears to improve with increased pointing offset, as expected, the experiment shows significant shot-to-shot variation.

Proton spectra from the DD reaction were obtained using the LLNL IPS magnetic spectrometer [7]. Preliminary analysis shows a small downshift in the peak energy ~100-200keV, but significant broadening which gives rise to a low energy tail (figure 6). Spectral data was obtained only for the thickest shells (2.7-2.8μm) for which simulations predict significant residual shell areal density at fusion bang time. The observed broadening is consistent with a simple energy loss calculation through this residual shell.

GXD images of the x-ray self-emission from the implosion core show obvious signs of asymmetry (figure 7). The data shows varying degrees of implosion asymmetry, with some shots appearing quite spherical, although this could also reflect cooling of perturbed peripheral regions due to mixed shell material. There is no significant correlation between neutron yield or convergence and symmetry. Gated equatorial images may hint at improved symmetry with an increase in the polar drive laser pointing offset, which is consistent with expectations from view-factor calculations. However, the data is inconclusive since these images exists for relatively few shots and there is significant shot-to-shot variation.

4. Conclusion

Direct drive capsule implosions are being developed on the Orion laser at AWE as a platform for ICF and HED physics experiments. Initial experiments have been fielded to evaluate the baseline capsule performance and inform future design and optimization. These experiments have successfully demonstrated the first use of Orion's neutron total yield and neutron time-of-flight fusion diagnostics. Data from multiple yield diagnostics confirms highly promising DD neutron yields up to $\sim 10^9$. Differences between 1D post-shot modeling and the experimental data are expected to be largely due to asymmetries in the implosions, as GXD images show asymmetric implosion cores on most shots. 2D post-shot modeling is now being undertaken. Future experiments will benefit from additional diagnostics and from new larger Orion phase plates, which will improve on-target beam quality.

References

[1] Hopps N *et al.* 2013 *Appl. Opt.* **52** 3597-3607
[2] Skupsky S and Lee K 1983 *J. Appl. Phys.* **54** 3662-71
[3] Temporal M, Canaud B, Garbett W J and Ramis R 2014 *Phys. Plasmas* **21** 012710
[4] Roberts P D, Rose S J, Thompson P C and Wright R J 1980 *J. Phys.* D **13** 1957
[5] Temporal M, Canaud B, Garbett W J and Ramis R 2013 *Eur. Phys. J.* D **67** 205
[6] Dodd E S, Benage J F, Kyrala G A, Wilson D C and Wysocki F J 2012 *Phys. Plasmas* **19** 042703
[7] Chen H *et al.* 2010 *Rev. Sci. Inst.* **81** 10D314

Configuration interaction effect on open M shell Fe and Ni LTE spectral opacities, Rosseland and Planck means

D Gilles[1], M Busquet[2], F Gilleron[3], M Klapisch[4] and J-C Pain[3]

[1]CEA, DSM, IRFU, F-91191 Gif-sur-Yvette, Cedex, France
[2]Research Support Instruments, Lanham, MD 20706, USA
[3]CEA, DAM, DIF, F-91297 Arpajon, France
[4]Berkeley Research Associates, Beltsville, MD 21042, USA

E-mail: dominique.gilles@cea.fr

Abstract. We have recently shown that iron and nickel open M-shell opacity spectra, up to $\Delta n = 2$ are very sensitive to Configuration Interaction (CI) treatments at temperature around 15 eV and for various densities. To do so we had compared extensive CI calculations obtained with two opacity codes HULLAC-v9 and SCO-RCG. In this work we extend these comparisons to a first evaluation of CI effects on Rosseland and Planck means.

1. Introduction

HULLAC-v9 [1, 2] and SCO-RCG [3, 4] detailed opacity codes can now provide precise spectral opacities for many applications like comparisons with experimental spectra and astrophysical studies [4, 5]. We shall discuss in this work an example of low temperature (kT = 15.3 eV) - low density ($n_e = 3\ 10^{-17}$ cm-3) plasma condition typical of stellar envelopes [5]. In such example a precise evaluation of the spectrum, and of the associated Rosseland (K_R) and Planck (K_P) means, supposes a careful evaluation of the billion of open M-shell transitions together with full Configuration Interaction treatment [6]. This is impossible in practise and only partial CI is included in recent improved opacity tables [7]. In a previous work we explored CI effect using HULLAC-v9 until convergence over the set of configurations [6]. We have shown that full CI treatment is really important for the first open M-shell transitions (n = 3, Δn = 0, 1, 2) by comparing "full CI" and "CI in one Non Relativistic Configuration (NRC)". But it is not practically possible to include higher excited levels (n > 5) using full CI treatment. A complete detailed description of the spectrum and of CI effect on the associated K_R and K_P requires mixed CI treatments and validation of the results are ongoing.

In this work we have taken advantage of interplaying comparisons [6] between the large possibilities and complementarities of HULLAC-v9 and SCO-RCG codes to discuss full CI influence on K_R and K_P means. The latter code is able to produce complete and precise LTE spectra, Rosseland and Planck means assuming CI in one NRC and the former can evaluate main full CI effects on the first M-shell transitions. On the example chosen above, we shall discuss and validate a method to provide corrections to K_R (and K_P) by using comparisons between HULLAC-v9 full CI and CIinNRC results and SCO-RCG ones. The Stark broadening of ion lines is now implemented using simple Dimitrijevic

& al. estimates [8] in both codes, but this effect is very small for such low density and will not be discussed in this paper.

2. Main features of the des opacity codes

2.1. *The SCO-RCG code.* The detailed opacity code SCO-RCG [3, 4, 6] is devoted to the diagnostics and interpretation of spectroscopy experiments in LTE plasmas. The data required for the calculation of the detailed transition arrays (Slater, spin-orbit and dipolar integrals) are obtained from the super-configuration code SCO, providing in this way a consistent description of the plasma screening effects on the wave functions. Then, the level energies and the lines (position and strength) are computed by an adapted RCG routine of Cowan. The extended list of configurations or super-configurations is generated automatically according to several criterions (on Boltzmann probability, number of successive excitations,...). DLA calculations are performed only for pairs of configurations giving rise to less than 800 000 lines. In other cases, transition arrays are represented statistically by Gaussian profiles in the UTA or SOSA formalisms. The strength of the hybrid approach is that we take into account many satellite lines and highly excited states, using an extension of the PRTA (Partially Resolved Transition Array) model, which enables us to replace many statistical transition arrays by small-scale DLA calculations. The SCO-RCG code, which includes many satellite lines and highly excited states, provides precise opacities, required for astrophysics, inertial confinement fusion, and for the interpretation of laser and Z-pinch experimental spectra. SCO-RCG iron spectral opacity is illustrated in Fig. 1, corresponding to the stellar envelope plasma condition of interest (15.3 eV, 3.2 10^{17} cm-3), showing the numerous open M-shell transitions contributing to Rosseland and Planck means for $1 < u = h\nu / kT < 20$ (see definitions on Fig. 1). To compute $1/K_R$ we split the spectral range in 3 domains. (a): below $u < 0.7$ (mainly f-f which gives a small contribution of 5 10^{-5}), (b): $0.7 < u < 10$, where K_{bb} is dominant and can be solely used as it contributes to 96% of the integral, (c): above u=10, where line and b-f contributions are negligible (3 10^{-6}). Same decomposition is applied to K_P.

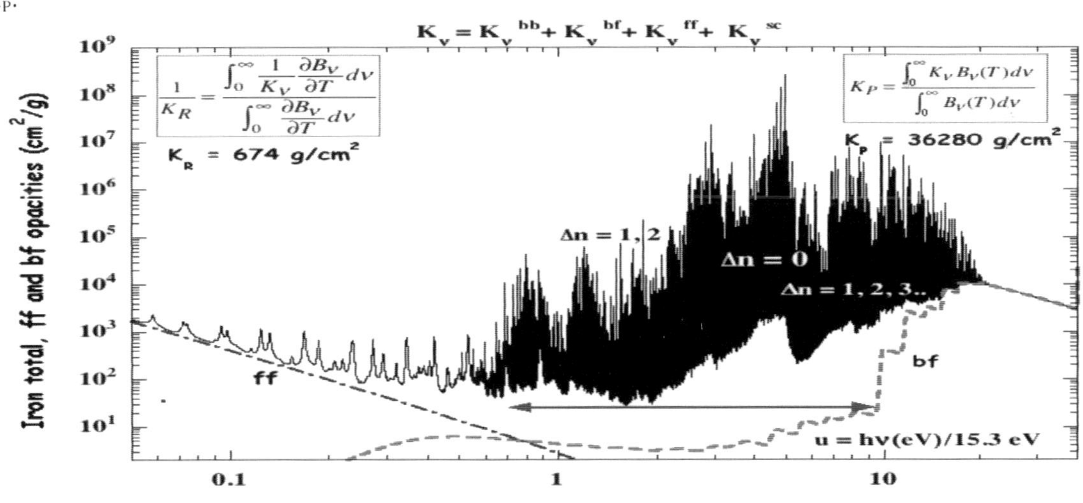

Fig 1: Iron SCO-RCG spectral f-f (blue dotted-dashed), b-f (red long dashed) and total opacities (full black line) at 15.3 eV and 3.2 10 ^{17}cm-3 ($\langle Z \rangle = 8.5$). K_R and K_P means definitions and values are reported on the figure.

2.2. *The HULLAC-v9 code.* This code is an integrated code for calculating atomic structure and cross sections for collisional and radiative atomic processes, in a jj coupling scheme [1, 2, 6] and references inside). In the Detailed Level description mode it is possible to take into account CI effects inside user defined groups of configurations (GroC). The diagonalization of eigenvectors for all levels of same J (in one charge state) is performed in each GroC. To be named "full or exact", CI treatment requires the inclusion of the whole set of fine structure levels in the same GroC, what means there will be one GroC per charge state. The keyword ClinNRC in the code generates coefficients of the Hamiltonian only between Relativistic Configurations pertaining to the same parent Non Relativistic Configuration (RCM), whereas CI lifts this restriction. The size of the matrices to be diagonalized obviously and dramatically increases with Δn. Following previous study conclusions our CI comparisons are performed over most intense and most sensitive to CI $\Delta n \leq 2$ transitions. More details on the validation are given in [2, 6]. Stark broadening of ion lines is now implemented using [8], together with Doppler and instrumental gaussian. Fig. 2 shows a comparison and a good agreement between SCO-RCG (all transitions) and HULLAC-v9 (assuming "CI in one NRC" and only $n = 3$, $\Delta n \leq 2$ transitions) over all the energy range of interest. This is true also for nickel (not illustrated) for same conditions.

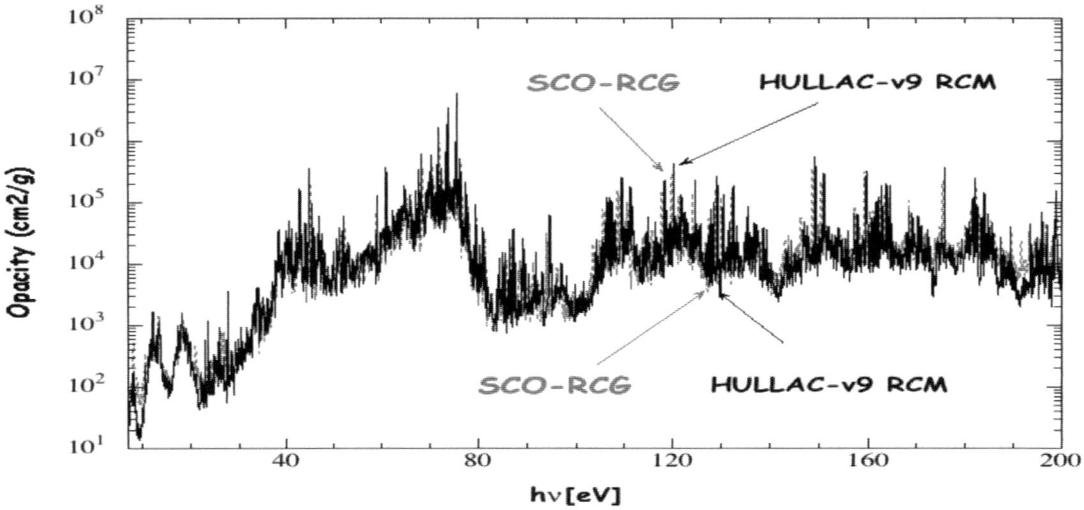

Fig. 2: Good agreement between SCO-RCG (all transitions) and HULLAC-v9 (assuming "CI in one NRC" $n = 3$, $\Delta n \leq 2$ transitions) spectral opacity for $0.7 < u < 13$ and same conditions as Fig. 1. A Gaussian broadening of 0.01 eV is applied to make the figure readable.

3. Evaluation of CI effect using HULLAC-v9 code.

In preceding sections we have shown that $\Delta n \leq 2$ b-b contributions are sensitive to CI and mainly contribute to K_R (and K_P) changes. Thus we can now continue and restrict our comparisons to HULLAC full CI and HULLAC ClinNRC b-b $\Delta n \leq 2$ contributions (Fig. 3 a, b). As we assume LTE, total opacity is just the opacity of different ionic species contributions, weighted by the populations. For our case we get: Fe VIII (0.03), Fe IX (0.41), Fe X (0.52), Fe XI (0.04). On Fig. 3 we have compared the two calculations for Fe VIII, spectral integrand (3a) and integrated over $\Delta h\nu = 0.1$ eV (3b). Fig. 3b reveals at which energies the integrals are most sensitive to CI. K_R is reduced by 38% because of CI. To obain K_R correction for any other density ones just needs to use the appropriated ionic populations. K_P linear means is less sensitive to CI (19.5%) and present a quite constant reduction for each individual charge states, 17% (Fe VIII), 19 % (Fe IX), 20 % (Fe X) and 18 % (Fe XI). Full CI and ClinNRC atomic data (energies, cross sections,...) have been stored for the different Fe (and Nickel not illustrated here) ions. We shall repeat this work for other temperatures in the range 10-30 eV to obtain quantitative corrections (ion, T, ρ) for all stellar envelopes conditions of interest [5]. For larger T we expect inclusion of more excited transitions to be required. These extensive studies are ongoing to calculate full CI and ClinNRC spectra using the atomic HULLAC-v9 data already calculated [1] and to characterize (T, ρ) regions of importance.

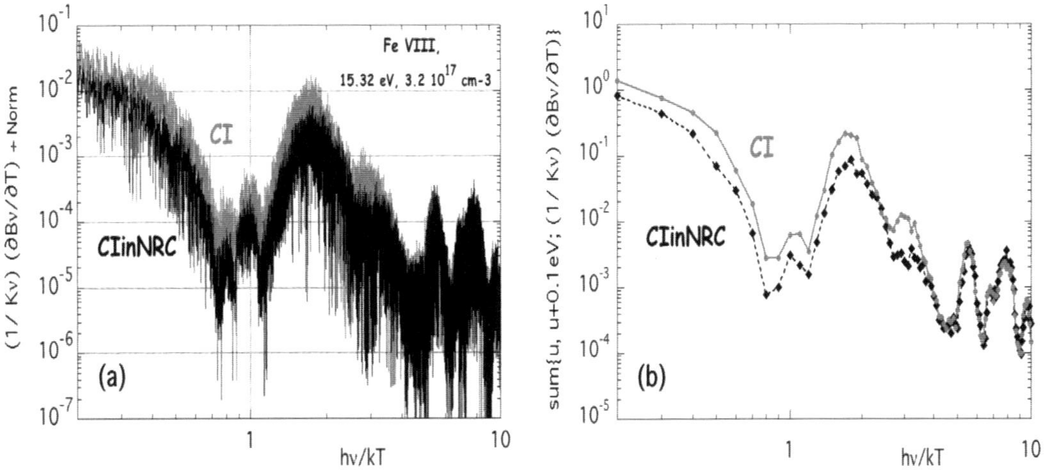

Fig. 3 a, b: Comparisons between iron HULLAC full CI and HULLAC ClinNRC b-b ($\Delta n \leq 2$) 1/ Kr contributions: (a) spectral integrand, (b) same after integration over $\Delta h\nu = 0.1$ eV for clarity.

4. Conclusion.

In a CI treatment both the CI effects and the number of configurations play a role, but the difficulty increases with the value of the principal quantum number of the upper shell of the transitions. Thus extensive detailed comparisons between SCO-RCG and HULLAC-v9 (including nickel opacities not illustrated in this paper) have been very instructive and useful for understanding the sensitivity of the opacity to CI, to the inclusion of highly excited transitions and to determine the spectral region of importance in the method proposed here. In this example full CI treatment up to $\Delta n \leq 2$ transitions is enough and make change in the b-b contribution to Rosseland means of about 38% (and around 20% for K_P). HULLAC-v9 and SCO-RCG have been upgraded to take into account Stark effects. As expected Stark broadening effect is small compared to Doppler effect for low density - low temperature application. After this first comparison on a typical stellar envelope condition, we shall use HULLAC-v9 CI and ClinNRC atomic data [6] to investigate more conditions (ρ-T) in the future and we shall also include Stark broadening in our corrections.

References

[1] Busquet M, Bar-Shalom A, Klapisch, Oreg J 2006 *J. Phys. IV* **133** 973.
[2] Busquet M, Klapisch and M Gilles 2013 D *EPJ Web of Conferences* **59** 14004;
Gilles D, Turck-Chièze S, Busquet M & al 2013 *EPJ Web of Conferences* **59** 14003.
[3] Pain J-C and Gilleron F 2015 *High Energy Density Phys.* **15** 30.
[4] Pain J-C and Gilleron *to be published in* ***9th*** *IFSA 2015 conference proceedings.*
[5] Gilles D, Tuck-Chièze S, G Loisel & al 2011 *High Energy Density Phys.* **7** 312;
Gilles D, Tuck-Chièze S, Busquet M, Thais F & al 2013 EAS publication series **58** 51;
Turck-Chièze S, Gilles D, Le Pennec M, Blenski T & al 2013 *High Energy Density Phys.* **9** 473.
[6] Gilles D, Busquet M, Klapisch M, Gilleron F, and Pain J-C 2015 *High Energy Density Phys.* **16** 1.
[7] New OPLIB (http://aphysics2.lanl.gov/opacity/lanl/); OPAL (https:// http://opalopacity.llnl.gov/).
[8] Dimitrijevic M S and Konjevic N 1980 J. Quant. Spectrosc. Radiat. Transfer **24** 451.

Speed of sound in solid molecular hydrogen-deuterium: Quantum Molecular Dynamics Approximation

Carlo Luis Guerrero and Jose Manuel Perlado

Nuclear Fusion Institute, ETSII, Polytechnic University of Madrid – Madrid 28006, Spain

guerrerocarlo@yahoo.es

Abstract. Uniformity of the solid layer is one of the critical points for an efficient ignition of the Deuterium-Tritium (DT) target. During the compression process this layer, perturbations grow as the Rayleigh-Taylor instability. Knowing the mechanical properties of this layer and its thermo-mechanical limits is necessary if we want to control or to minimize these instabilities. In this work we have used a simplified approach, replacing the DT ice system with a mixture of hydrogen-deuterium (HD) because beta decay of tritium complicates the analysis in the former case. Through simulation with ab initio methods we have calculated the elastic constants, the bulk modulus and sound velocity for hydrogen isotopes in solid molecular state. In this work we present the results for hydrogen-deuterium mixtures 50%-50%, at 15 K and with a compression which covers the range of 1 to 15 GPa. This system is interesting for study the early stages of the dynamic compression and provides conditions that are close to the manufacture of DT target in inertial confinement fusion. Discontinuities in the curve that have been observed on pure hydrogen, which are associated with phase transitions and the phase hysteresis.

1. Introduction

The study of the mechanical properties of solid molecular hydrogen (SMH) is of great importance for the development of inertial confinement fusion (ICF) system [1]. In the ICF system, fusion is achieved by compressing a sphere of Deuterium Tritium (DT) ice using high-power lasers [1]. The compression process is modeled to increase ignition efficiency of the target. Our purpose is to model systems composed of H, D and T and mixtures of them. In recent studies we have compared our results with experimental data, obtaining a good agreement, both for the structural changes undergone by the hydrogen and for the properties mechanical bulk modulus [2]. In this way we have demonstrated the reliability of the simulation in simple systems composed of hydrogen, deuterium and tritium. This work allows us to understand the changes in mechanical properties such as bulk modulus and sound velocity for systems with mixtures of H and D, a preliminary step for more detailed study for combinations of DT and HDT. Different studies have been conducted to analyze the mixture of isotopes HD. Phase diagram of HD is very different from having the H_2 and D_2 separately, Cui et al [3] show that the phase diagram has a different behavior in the pure H curve representing the transition from phase I and II for temperatures below 100 K. Several authors demonstrate the "broken symmetry" associated with the restriction of rotation of the rotor of the molecule HD [4,5]. The HD has been insufficiently studied, it becomes important to analyze the temperature and pressure region near to manufacture and early compression DT ice in ICF, for this we choose 15 K as initial temperature and pressure range of 1 to 15 GPa, to thereby representing dynamic compression system.

The method used in this study was applied in the study of various properties of SMH, describing the three phases of solid H pure [6], calculating the bulk modulus tritium [2] and relating changes in the sound velocity curve vs. pressure changes in the structural configuration of the solid [7]. The interest of these simulations is focused on generating parameters describing the behavior of the solid in conditions seldom studied experimentally. The temperature of interest in this article is near to that used in the storage and handling of DT ice spheres in ICF systems [8], analyzing factors that can induce problems in the solid layer of the target in early stage compression. For commercial applications targets can be produced in situ [9] or stored for later use; hence, storing thousands of targets is very relevant for systems of magnetic or inertial. The results obtained in this work also apply to models used to study geophysics and formation of exoplanets and giants planets, such as Jupiter and Saturn [10,11]. Finally, the paper presents the changes that occur in the bulk modulus and sound velocity of the HD solid. We show the methodology in section 2, the results of bulk and sound velocity in the range of 1-15 GPa with 15 K in section 3, and our conclusions and hypotheses for future work in Section 4.

2. Methodology

For HD simulation molecular solid we use a box with 576 atoms, 288 HD molecules, each molecule is composed of one hydrogen and one deuterium. The concentration of isotopes is variable in our code but the proposed concentration 1:1 for a sample basis and compared with pure H. The SIESTA code [12] is used, based localized orbitals on Density Functional Theory, the complete method has been described in previous papers [2,6,7]. To get the lowest-energy configuration of the HD system we use the method of conjugate gradients, whereby an initial structure that minimize the total energy of the system. Then we apply the Nose-Hoover thermostat and the Parrinello-Rahman barostat the crystal structure is obtained for a determinate pressure and temperature. For an understanding of the dynamic compression experiments, we applied the pressure as ramps. The initial pressure is 1 GPa maintained for 1 ps, with this structure we proceed to calculate the bulk modulus (B) using Hooke's law for this pressure applying deformation of 1% and observing stress in the structure. Then without deformation structure is compressed again applying 2 GPa pressure during the same period 1 ps, creating a new structure for each step of the pressure ramp, 1 GPa as increased variation of P, until the 15 GPa (see figure 1). The simulation is made with SIESTA standard basis double zeta polarized (DZP) and LDA functional, with MeshCutoff 100 Ry. The pseudo-potential used is Troullier-Martin type.

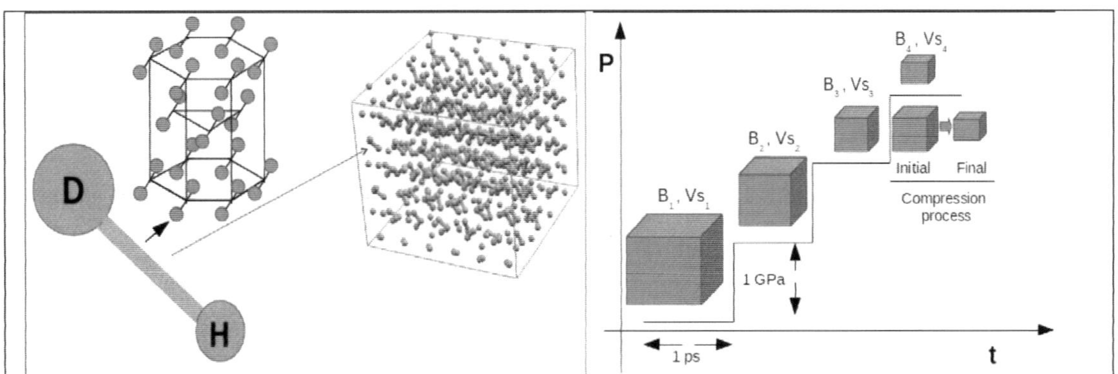

Figure 1. HCP configuration for the solid molecular HD 576 atoms (288 HD molecules). Ramp method to simulate the dynamic compression. B_i and Vs_i is obtained for each pressure level studied.

As a first approximation use the model of isotropic transverse material. HD is a molecular solid for low pressure in the Phase I rotor associated to molecule is free, the molecular rotation decreases with increasing pressure and rotor is restricted. The C_{ij} allow us to calculate each of the directions of deformation getting information about the speed of sound propagation in different aspects of the material [7].

$$\sigma = C_{ij}\,\epsilon \qquad (1)$$

where σ is the stress calculated by simulation with the deformation ϵ. Calculating the bulk modulus is achieved through equation of the Voigt average [13]:

$$9B = (C_{11}+C_{22}+C_{33}) + 2(C_{12}+C_{13}+C_{23}) \qquad (2)$$

For us $C_{22}=C_{11}$, $C_{12}=C_{13}$ and the sound velocity is, as a first approximation: $Vs = \sqrt{(B/\rho)}$, where ρ is solid density and Vs is the speed of sound in solid.

3. Results

The figure 2 shows a change in the magnitude of the bulk modulus and sound velocity. HD bulk modulus is greater than that of H (see figure 2a) and the difference increases with pressure. In the case of the HD sound velocity (see figure 2b) it is lower than that of H in the analyzed range. Variations in pressure below 5 GPa in the graph for bulk modulus and the sound velocity for HD, would be related to the phase transition from solid-solid [7]. There are two discontinuities in red and blue HD curves, one for 2 to 5 GPa and for P> 9 GPa, this last is larger than the first. This can be explained due to variations in the phase diagram of HD, and the same points are not present in the phase change of H [3]. For 1 and 2 ps the bulk modulus is similar between 5 GPa <P <9 GPa, while the rest of pressures the bulk modulus is different, this means that the range between 5 to 9 GPa change in time of application of pressure ramp (here 1ps and 2 ps), does not alter the system configuration and therefore the parameters calculated from this.

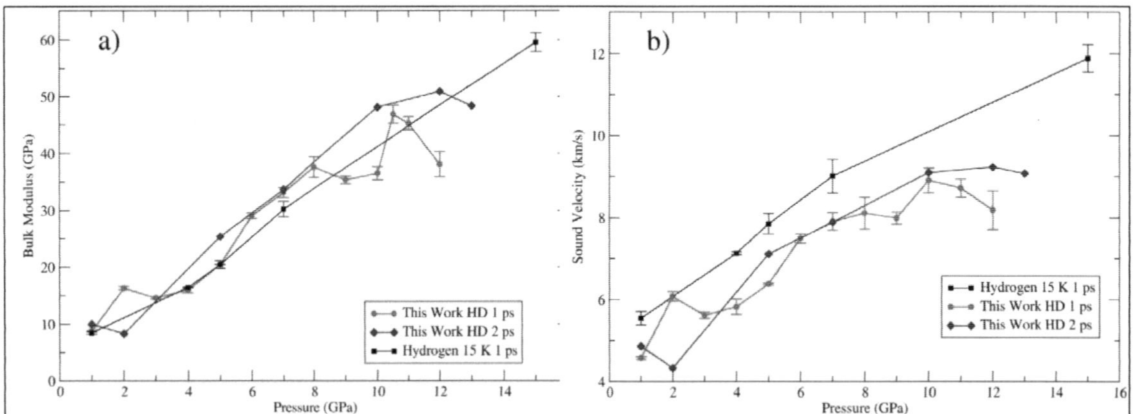

Figure 2. This shows the results a) bulk modulus and b) sound velocity for H and HD. These results are to 15 K and 1 to 15 GPa. The results of 1 ps and 2 ps are shown, there are differences in the bulk modulus and sound velocity for specific pressure ranges.

In the figure 2b the complexity of the curve for HD remains but also appears some differences in the magnitude of the sound velocity compared to that of H, this effect arises because the velocity depends on the solid density and this quantity is larger for HD than for H (50% larger than the original H). This density increase according to the ratio of deuterium, causing the HD phase diagram depends on the isotope ratio in the solid, changing its mechanical properties for a given P and T compared to H.

The discontinuities in the sound velocity curve (see figure 2b) have been associated with changes in the restriction of the molecules rotation in the solid H [7], changes being related to solid-solid phase [14]. For P < 2 GPa H_2 molecule center move outside the lattice point and the rotor is free, these effects have been studied in previous works [7], for 2 < P < 10 GPa H_2 molecule center coincides with lattice point and rotor is free, to P> 12 GPa the H_2 rotor is restricted [6,7]. The difference between 1ps and 2 ps is only appreciable in a range of pressure 5 GPa <P <10 GPa, for the rest of the curve the difference between H and HD does not exist. This shows that the application time of pressure ramp intervenes in the ability of the solid to respond to the change in the phase transition, this can be explained by the concept of hysteresis phase. This phenomenon is not present in experiments static compression [15] but is very important in dynamic compression experiments, especially compression laser ICF systems such as NIF experiments [16], where the pressure profile varies with time. The hysteresis phase has important implications for the process in which the target of DT ice is compressed, and has already been observed in experiments iron dynamic compression [17]. Remains to be studied more in depth this hysteresis of phase phenomenon in compression of SMH isotope, and see the effects on the generation of inhomogeneities in the front of the shock wave [18] that can alter conditions plasma heating.

4. Conclusions

The method presented in this work allows control of the H-D ratio in the volume, so we can get different relative concentrations and associate them with changes in their mechanical properties. We have shown that our methodology represents very well the experimental results for a wide range of temperature, 300 K [2] and 15 K (see figure 2) and a broad pressure range from 1 GPa to 200 GPa [6,7] (in this work between 1 and 15 GPa). The H:D ratio in the solid changes the magnitude of the sound velocity, which in turn changes the shape of the sound velocity and bulk modulus curves, producing variations in mechanical properties compared with that of pure H. The difference between 1 to 2 ps allows a deeper understanding of the effect of hysteresis in phase and pressure regions near the phase transition temperature. Getting to know the response of the solid-solid phase transition as a function of time of application P-ramps gives us a map of delays in the response of solid-liquid or solid-plasma transition in dynamic compression experiments.

Acknowledgments

The author acknowledges the computer resources and technical assistance provided by the Centro de Supercomputación y Visualización de Madrid (CeSViMa). To EUROfusion Consortium Task Agreement, WPENR: Enabling Research IFE, Project: AWP15-ENR-01/CEA-02. We thank O. Peña.

References

[1] Sangster TC et al 2007 *Phys. Plasmas* **14** 058101
[2] Guerrero C and Perlado J 2013 *Proc 25th Symposium on Fusion Engineering (SOFE)*, San Francisco, CA, USA, DOI:10.1109/SOFE.2013.6635457
[3] Cui L, Chen N and Silvera I 1995 *Phys. Rev.* B **51** 21 14987-97
[4] Crespo Y, Laio A, Santoro A and Tossatti E 2011 *Phys. Rev.* B **84** 144119
[5] Goncharenko I and Loubeyre P 2005 *Nature* **435** 1206-09
[6] Guerrero C, Cuesta-Lopez S and Perlado J 2013 *EPJ Web of Conferences*, **59** 16004
[7] Guerrero C, Cuesta-Lopez S and Perlado J 2014 *Europhysics Letters* **108** 26001
[8] Kozioziemski B et al 2007 *Nucl. Fusion* **47** 1-8
[9] Aleksandrova I et al 2013 *EPJ Web of Conferences* **59** 11003
[10] Duffy T, Vos W, Zha Ch, Hemley R and Mao H 1994 *Science* **263** 1590
[11] Alavi A, Parrinello M and Frenkel D 1995 *Science* **269** 1252
[12] Soler J et al 2002 *J. Phys. Condens. Mat.* **14** 2745
[13] Watt J and Peselnick L 1980 *J. Appl. Phys.* **51** 1525.
[14] Toledano P, Katzke H, Goncharov A and Hemley R 2009 *Phys. Rev. Lett.* **103** 105301
[15] Zha C, Duffy T, Mao H and Hemley R 1993 *Phys. Rev.* B **48** 13
[16] Hurricane O et al. 2014 *Nature* **506** 343-348
[17] Nguyen J and Holmes N 2004 *Nature* **427** 339-342
[18] Holst B, Redmer R, Gryaznov V, Fortov V and Iosilevskiy I 2012 *Eur. Phys. J.* D **66** 104

Propagation of intense short-pulse laser in homogeneous near-critical density plasmas

H Habara[1], S Nakaguchi[1], Y Uematsu[1], S D Baton[2], S N Chen[2], J Fuchs[2], T Iwawaki[1], M MacDonald[3,4], W Nazarov[5], C Rousseaux[6] and K A Tanaka[1]

[1]Osaka University, Suita, Osaka, Japan
[2]LULI, Ecole Polytechnique, Palaiseau, France
[3]University of Michigan, Ann Arbor, MI, USA
[4]SLAC National Accelerator Laboratory, Menlo Park, CA, USA
[5]University of St. Andrews, St. Andrews, Scotland, United Kingdom
[6]CEA, DAM, DIF, Arpajon, France

habara@eei.eng.osaka-u.ac.jp

Abstract. Ultra intense laser light propagation in a homogeneous overdense plasma was investigated using a plastic foam target filling a polyimide tube. Laser propagation into overdense plasma was measured via Doppler red shift of the reflected laser light from the moving plasma at 0.3-0.4 of speed of light. We also observed strongly collimated electron beam possibly caused by the magnetic field surrounding the plasma channel, and high energy X-rays emitted via synchrotron radiation by the oscillating electrons inside the channel. These features imply that UIL propagates inside the overdense plasma as predicted in PIC calculation, and are very important for direct irradiation scheme of fast ignition.

1. Introduction

Ultra intense laser (UIL) light can propagate in overdense plasma due to relativistic effects. For example, relativistic transparency enables the UIL propagation by effective reduction of the electron plasma frequency through the apparent electron mass enhancement. Increase of laser intensity by relativistic self-focusing of Gaussian beam also reinforces the transparency effect. We adapted these effects into Fast Ignition in order to make UIL accessible to the core plasma as close as possible [1]. For several years, we have investigated the UIL propagation into an implosion or a blow-off plasma having not only overdense but also long underdense regions. However, the investigation using these plasmas was somehow difficult because the critical density (N_c) position and/or the density scale length are not clear in such plasma. In addition, because UIL becomes unstable before reaching N_c by suffering several laser-plasma instabilities such as filamentation or bifurcation [2], quantitative discussions regarding the overdense region are difficult for reproducibility issues.

In order to overcome these difficulties, we performed an experiment to create characterized homogeneous plasmas (electron density of 0.8 and $1.3N_c$) over 300μm at ELFIE facility in LULI, Ecole Polytechnique, France [3]. In this experiment, we observed a significantly collimated electron beam (~11° FWHM) along the laser direction, possibly caused by magnetic collimation along the plasma channel. This result strongly suggests laser propagation into the overdense plasma [4]. In order to confirm these results, we recently performed an experiment for UIL propagation in this plasma

Journal of Physics: Conference Series **717** (2016) 012019 doi:10.1088/1742-6596/717/1/012019

where Doppler reflectometry of the backscattered light was measured in order to understand the propagation speed in the plasma.

2. Experimental setup

Figure 1 shows the experimental setup. In order to create a homogeneous plasma, a low-density plastic ($C_{15}H_{20}O_6$) foam target (ρ~5 and 10mg/cc) was inserted in a polyimide tube having 300μm diameter and 300μm length. A thin Cu foil (700nm, 5x5mm^2) was attached at the one side of the tube entrance hole. A sub-ns chirped pulse irradiated this foil with 500μm spot diameter to create X-rays which ionized the foam uniformly. Finally, an ultra intense laser light (1ω(=1054nm), 10J/300fs - 3×10^{19}W/cm^2) was focused into the foam plasma from the opposite entrance hole. The backscattered light was collected using a beam splitter located before the focusing off-axis parabola, and was then split to a photodiode and a spectrometer outside the interaction chamber. The electron energy spectra and electron emission distribution were observed in the laser propagation direction. Angular distribution of X-ray intensity was also obtained with the help of an imaging plate.

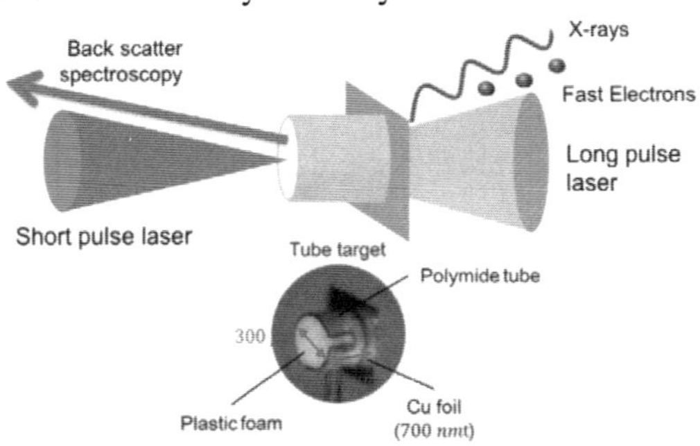

Fig.1 Experimental setup.

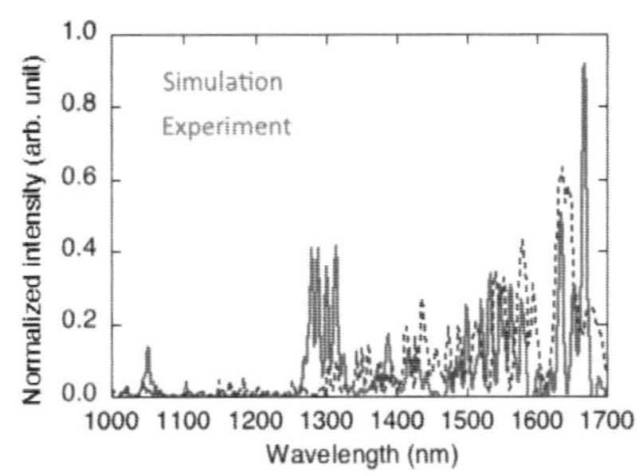

Fig.2 Backscattered light spectra from experiment (dashed line) and simulation (solid line).

3. Doppler reflectometry

Reflected spectra show a significant red shift for both foam densities. The solid curve in Fig. 2 indicates a typical experimental spectrum from 5mg/cc foam plasma irradiated by UIL. Typically, the peak of the reflected light appeared at 0.6ω (1.7μm) for 5mg/cc foam and 0.7ω (1.55μm) for 10mg/cc foam target, respectively. Here, the reflected light energies were about 5.3% and 7.2% of the incident laser energy for 5 and 10mg/cc plasmas respectively. These values are somehow similar to the observation in the relativistic laser region [5,6]. However, our plasma density is much higher than those experiments and is not favorable for Raman development even in relativistic regime. In addition, we only had a very short underdense plasma (could be less than a few microns) whereas SRS is very sensitive to the laser propagation length. In fact, we observed very similar reflected laser energies when the pulse width was stretched to 900fs and 1.5ps, whereas longer laser pulses were expected to drive more SRS in underdense plasma [5]. From these considerations we believe that the red shift was caused by Doppler shift at the moving piled up plasma at the laser front, rather than SRS. In this model, the plasma electrons inside the laser volume oscillate in the laser field at relativistic energies, with the result that the Lorentz factor of these electrons becomes larger than those at outside the laser region [7-9]. The local enhancement of the Lorentz factor results in a significant gap of the plasma frequency at the laser front, which acts as a mirror for the laser light. This surface moves with a velocity determined by the energy balance between the radiation pressure and the kinetic pressure around the surface, which is much slower than the relativistic group velocity given at its density.

Adapting this model, the experimental Doppler shifts (0.6ω and 0.7ω) give the mirror speed of 0.46c and 0.36c for 5 and 10mg/cc foams respectively. We also calculated the analytical recession velocity of the piled up plasma for various laser intensities and plasma density conditions, which shows that the experimental mirror speeds correspond to $1N_c$ and $2N_c$ plasmas with 2-3×10^{19}W/cm^2 [8,9] in 5 and 10mg/cc foams, in agreement with the experimental laser and plasma conditions. This model was evaluated with 1D and 2D PIC calculations [8,9]. The simulated reflected light spectrum for $1N_c$ plasma with I_L=3×10^{19}W/cm^2, as shown in dashed curve in Fig.2, also well reproduced the experimental one.

4. Electron measurement

We also observed the electron emission distribution by using a large image plate at the back side of the target with respect to the UIL direction. Figure 3 indicates the emission pattern obtained for (a) 5mg/cc foam target and (b) Al 50μm foil target. In order to assess the signal produced by electrons only, PET (Polyethylene terephthalate) plate 1mm and Al 12μm filters were inserted in front of imaging plate. These filters reject less than 1-keV X-rays and several MeV protons, so that the IP only records electrons having over 100-keV kinetic energies. Note that the detection probability of the image plate over 100-keV electrons is almost constant [9]. The horizontal black line in the middle of Fig. 3(a) represents the slit of the multi-channel electron spectrometer.

As shown in Fig.3, the electron emission pattern of the foam target (Fig. 3a) exhibits a significant reduction of beam divergence (10°-20° in FWHM), compared to that obtained from the aluminum foil (> 40°) (Fig. 3(b)), as observed in a previous paper [4]. With foam targets, the multi-channel spectrometer ESM (0°-20°) also shows the increase of the electron number observed near the peak position of the emission signal, compared to the side position, whereas no difference was observed with foil targets (not presented here).

2D PIC calculations also reproduce the beam collimation in the homogeneous plasma as already shown in a previous paper [8]. The calculation also indicates that the UIL creates a plasma channel inside the overdense plasma and a strong magnetic field along the plasma channel. This strong magnetic field collimates the fast electrons toward the laser direction. In order to check this collimation, we calculated trajectories of electrons having an initially isotropic emission distribution at the entrance of the field. We observed that the lower energy electrons are trapped inside the magnetic field, whereas the higher energy electrons are delivered toward the laser propagation direction, resulting in the collimation of the electron beam. The result of electron measurements also indicates laser propagation inside N_c plasma.

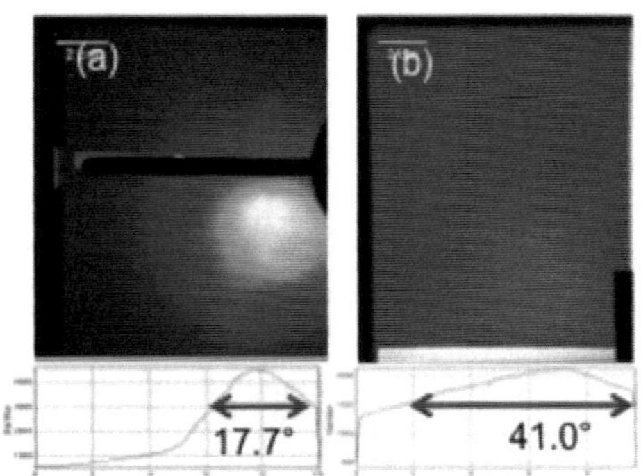

Fig.3 Emission distribution of fast electrons from (a) foam target and (b) Al 50μm foil target.

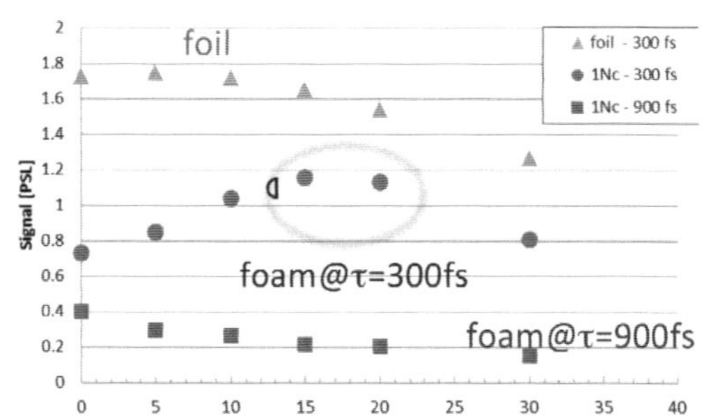

Fig.4 Angular distributions of X-ray emission for various conditions; Foam 5mg/cc, I=3×10^{19}W/cm^2 (circles), I=1×10^{19}W/cm^2 (squares), Al foil (triangles).

5. X-ray measurement

Angular distribution of X-ray intensity was separately measured from the electron measurement. An array of image plates was set in the laser direction from 0 to 30 degrees, every 5°, each one having an aperture of 4 msr. The electrons were removed from the path to the image plate by using strong magnets. We also inserted the PET and the Al filters, in order to record only X-rays over 1-keV. Figure 4 represents the angular distribution of X-rays for different conditions: foam 5mg/cc, I=3×10^{19}W/cm^2 / τ=300fs (circles) and I=1×10^{19}W/cm^2 / τ=900fs (squares), Al foil, I=3×10^{19}W/cm^2 /

τ=300fs (triangles). A side peak (at 15°-20° from the laser direction) is only observed with the foam plasma in the highest intensity case. If the X-rays were created via Bremsstrahlung emission, they should have a peak emission corresponding to the electron emission direction (0°-10°) [11], which is observed in the cases of Al foil and lower intensity foam targets. On the other hand, for synchrotron radiation, the X-ray peak emission is expected around $\theta\sim16°$ for our experimental conditions (laser intensity and plasma density) [12], which well agrees with the observation.

In order to better characterize the radiation mechanism of the side peak emission, we additionally inserted a 1 mm Cu filter in front of the image plate. Thus, the intensity around the side peak emission was reduced to 13% of the one without Cu filter. From this attenuation rate, the X-ray energy is estimated around 53keV. Analytically, the X-ray energy of synchrotron radiation given our experimental conditions is calculated to be 51.2keV [11], which well agrees with the experimental value. In addition, assuming that most of the X-rays are created through synchrotron radiation in a symmetry perpendicular to the polarization plane, the total X-ray photon number and energy are estimated to 5×10^{10} and 0.4mJ, respectively. According to Ref. [13], 7×10^{12} electrons are required to generate these photons; therefore the total energy of electrons is about 10% of the laser energy. From these estimations, we believe that the synchrotron radiation predominates the X-ray emission in the foam target case at the highest laser intensity. The observation of synchrotron radiation suggests electron oscillation inside the plasma channel, and therefore that the laser propagates in the dense plasma.

6. Summary

We have investigated ultra intense laser light propagation in a long, homogeneous, overdense plasma using a plastic foam target filling a polyimide tube. Laser propagation into overdense plasma was investigated via Doppler red shift of the reflected laser light from the moving plasma at 0.3-0.4 speed of light. The observed electron beam was strongly collimated probably by the magnetic field surrounding the plasma channel. High energy X-rays were emitted via synchrotron radiation by the oscillated electrons inside the channel. These features imply that UIL propagates inside the overdense plasma as predicted by PIC calculations. These results are very important for direct irradiation scheme of fast ignition.

Acknowledgement

The authors gratefully acknowledge the support of the staff of the ELFIE in the execution of this work (14TWF1). We also thank Professor H. Sakagami at National Institute for Fusion Science for his permission of use of 2D-PIC FISCOF code. A part of this work is supported by Grants-in-Aid for Scientific Research, type S (Grant No. 15H05751).

References

[1] A.L. Lei *et al.*, Phys. Plasmas **16** (2009) 056307.
[2] G. Li *et al.*, Phys. Rev. Lett. **100** (2008) 125002.
[3] S.N. Chen *et al.*, Sci. Report **6** (2016) 21495.
[4] T. Iwawaki *et al.*, Phys. Plasmas **21** (2014) 113103.
[5] C. Rousseaux *et al*, Phys. Plasmas. **9** (2002) 4261.
[6] T. Miyakoshi *et al*, Phys. Plasmas **9** (2002) 3552.
[7] S. Guerin, P. Mora, J. C. Adam, A. H´eron, and G. Laval, Phys. Plasmas **3** (1996) 2693.
[8] T. Iwawaki *et al.*, Phys. Rev. E **92** (2015) 013106.
[9] T. Iwawaki, Osaka Unverisity, 2015, Ph.D. thesis (in Japanese).
[10] K.A. Tanaka *et al.*, Rev. Sci. Instrum. **76** (2005) 013507.
[11] J. D. Jackson, *Classical Electrodynamics* (Wiley, New York, 1975).
[12] K.T. Phuoc et al., Phys. Plasmas **12** (2005) 023101.
[13] I. Kostyukov *et al.*, Phys. Plasmas **10** (2003) 4818.

9th International Conference on Inertial Fusion Sciences and Applications (IFSA 2015) IOP Publishing
Journal of Physics: Conference Series 717 (2016) 012020 doi:10.1088/1742-6596/717/1/012020

Fusion-neutron measurements for magnetized liner inertial fusion experiments on the Z accelerator

K D Hahn[1], G A Chandler[1], C L Ruiz[1], G W Cooper[2], M R Gomez[1], S Slutz[1], A B Sefkow[1], D B Sinars[1], S B Hansen[1], P F Knapp[1], P F Schmit[1], E Harding[1], C A Jennings[1], T J Awe[1], M Geissel[1], D C Rovang[1], J A Torres[1], J A Bur[1], M E Cuneo[1], V Yu Glebov[3], A J Harvey-Thompson[1], M C Herrman[1], M H Hess[1], O Johns[1], B Jones[1], D C Lamppa[1], J S Lash[1], M R Martin[1], R D McBride[1], K J Peterson[1], J L Porter[1], J Reneker[1], G K Robertson[1], G A Rochau[1], M E Savage[1], I C Smith[1], J D Styron[2] and R A Vesey[1]

[1] Sandia National Laboratories, Albuquerque, NM, USA
[2] University of New Mexico, Dept. of Nuclear Engineering, Albuquerque, NM, USA
[3] University of Rochester, Laboratory of Laser Energetics, Rochestor, NY, USA

Email: kdhahn@sandia.gov

Abstract. Several magnetized liner inertial fusion (MagLIF) experiments have been conducted on the Z accelerator at Sandia National Laboratories since late 2013. Measurements of the primary DD (2.45 MeV) neutrons for these experiments suggest that the neutron production is thermonuclear. Primary DD yields up to 3e12 with ion temperatures ~2-3 keV have been achieved. Measurements of the secondary DT (14 MeV) neutrons indicate that the fuel is significantly magnetized. Measurements of down-scattered neutrons from the beryllium liner suggest $\rho R_{liner} \sim 1$ g/cm^2. Neutron bang times, estimated from neutron time-of-flight (nTOF) measurements, coincide with peak x-ray production. Plans to improve and expand the Z neutron diagnostic suite include neutron burn-history diagnostics, increased sensitivity and higher precision nTOF detectors, and neutron recoil-based yield and spectral measurements.

1. Introduction

Measurements of the neutrons produced during magnetized liner inertial fusion (MagLIF) experiments [1]-[2] conducted at the Z pulsed-power accelerator [3] at Sandia National Laboratories suggest that the neutron production is predominantly thermonuclear [4]. Beginning in 2013, these experiments have been performed using deuterium fuel and produced primary DD (2.45 MeV) fusion neutron yields up to 3e12 with electron and ion stagnation temperatures in the 2-3 keV range. In section 2, we report several neutron measurements from MagLIF experiments. Plans for improvements to the nuclear diagnostic suite are presented in section 3.

MagLIF is an inertial confinement fusion (ICF) concept that also makes use of magnetic confinement via an externally applied magnetic field. In contrast with standard ICF concepts, pre-magnetizing the fuel is advantageous because this significantly relaxes the required stagnation pressures [1]-[2]. The MagLIF concept utilizes the implosion of a laser-preheated (~ 2.5 kJ), pre-magnetized (10 T) fuel by a solid cylindrical beryllium liner that is compressed by the 17-19 MA delivered by the Z pulsed-power accelerator [1]-[2]. Results from computer simulations using the magneto-hydrodynamics code GORGON are shown in figure 1(a) and illustrate the three main stages of the MagLIF concept. Figure 1(b) shows a time-integrated, self-emission x-ray image of the hot stagnated plasma column from shot 2613 which has a ~ 100 μm diameter and ~ 6 mm length.

Explanations for the helical structure in the column as well as variations in brightness are presently under investigation.

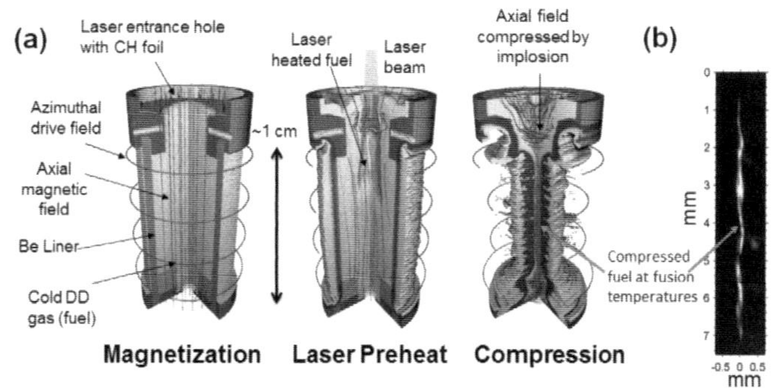

Figure 1. (a) Simulation results depict the three main stages of MagLIF: magnetization, laser preheat, and compression. (b) Time-integrated x-ray self-emission from plasma column.

2. Measurements of fusion neutrons on MagLIF experiments

There are several neutron diagnostics on Z used to characterize the primary DD and secondary DT neutron yields and spectra [5]. Neutron yields are principally measured using neutron activation of select materials such as indium [via ^{115}In(n,n')^{115m}In with threshold = 336 keV] for DD yields which are dominated by neutrons with energies between 2.3-2.6 MeV, but includes the significant contribution from down-scattered neutrons in the ~ 1.5-2.4 MeV range. Copper activation [via ^{63}Cu(n,2n)^{62}Cu with threshold = 11 MeV] is used to measure secondary DT yields which range between 11.8-17.2 MeV. The contribution to the indium activation from higher-energy secondary DT neutrons is << 1%, thus these materials are appropriate for isolating the DD and DT yields. Corrections for neutron attenuation and scattering from Z hardware are preformed using the Monte Carlo MCNP code [6]. Neutron spectra are measured using neutron time-of-flight (nTOF) detectors. Presently, there are two axial nTOF detectors at approximately 7 m and 8 m below the target. Three radial detectors are located at approximately 9.5 m, 11.5 m, and 25 m from the target. All the detectors are composed of a 2.54-cm thick, 7.62-cm diameter, BC-422Q (1%) scintillator coupled to a photomultiplier tube (PMT such as Hamamatsu 5946-mod4). Most of the nTOF detectors utilize extensive collimation and shielding both near the target and around the detectors. This is necessary to limit the amount of neutron down-scatter from hardware and the intense (> 1 MeV) bremsstrahlung-induced signals produced during the experiments that can saturate the detectors. In most cases, the activation-measured yields are compared with the yields inferred from nTOF measurements.

2.1. Neutron yield and ion temperature measurement results

As one of the most essential metrics to support the notion of thermonuclear neutron production, the DD neutron yields appear to be isotropic, and the radially and axially measured DD neutron spectral shapes appear to be very similar and have Gaussian shapes from which realistic fusion ion temperatures can be inferred. Figure 2 shows a plot of the primary and secondary neutron yields for several MagLIF shots as well as the ion and electron temperatures. The average primary yield shown in figure 2 is based on several indium activation measurements performed axially (above and below) and radially. The primary DD yields range from 1e11-3e12 and all measurements indicate asymmetries less than the measurement uncertainties which are ~ 20%. Secondary DT yields range from 1e9-5e10 and typical uncertainties are also ~ 20%. (It is important to note that the DT yields below 1e10 could not be verified with nTOF measurements.) Ion temperatures are also plotted and range from 1-3 keV with 30% uncertainties.

Figure 3 shows the axially and radially measured nTOF spectra from shot 2591 (2e12 DD yield) for data corrected for PMT response, light-output, and direct line-of-sight neutron attenuation and scattering. The ion temperature (T_{ion}) [7] inferred from these data is determined by fitting a Gaussian to the corrected, energy-transformed data. Significant down-scattered neutrons broaden the nTOF spectra on the lower energy side, hence the Gaussian is fitted to data in the ~ 2.4-3 MeV range in most cases. In figure 3, a Gaussian fit with corresponding T_{ion} of 2.5 keV, determined from the average of the spectra shown, is over-plotted and shows good agreement with the data. (We are presently pursuing a forward analysis similar to [8] to infer T_{ion} that also incorporates more realistic neutron scattering contributions to the nTOF signals.) Overall, the ion and electron temperatures (from x-ray spectroscopy measurements for some experiments) agree within the measurement uncertainties.

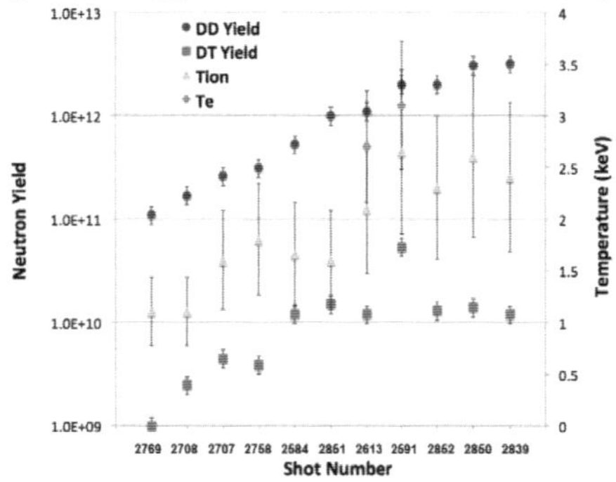

Figure 2. Plot of DD and DT neutron yields and ion and electron temperatures for several MagLIF shots.

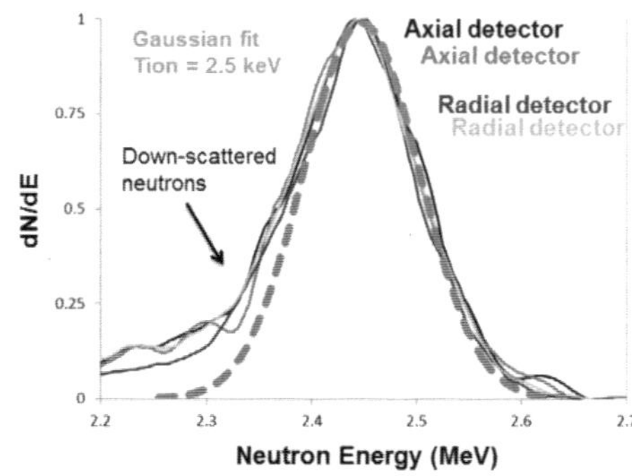

Figure 3. Axial and radial neutron spectra measured for shot 2591 are over-plotted with Gaussian fit.

Based on several filtered, absolutely calibrated x-ray measurements using diamond photo-conduction diodes, the neutron burn duration appears to be ≤ 2 ns FWHM [4]. Estimates of the neutron bang time from several neutron time-of-flight detectors indicate that the peak neutron production occurs near the time of peak x-ray emission, though uncertainties using the present far-away detectors are ~ 1 ns. Initial bang-time measurements using CVD diamond detectors show promise for measuring *actual* neutron burn widths and bang times with sub-ns uncertainties.

2.2. *Inference of fuel magnetization from secondary DT neutrons*

Production of secondary DT neutrons indicates that the fuel is significantly magnetized given the rather low fuel areal density in the radial direction (~ 2 mg/cm^2) [9]-[10]. Asymmetry in the DT spectral shapes measured radially and axially, shown in figure 4 for shot 2591, are yet another indicator of the degree of magnetization. The radially measured spectra appear narrower than the axially measured spectra. The asymmetry arises from the high aspect ratio cylindrical geometry where the 1 MeV tritons born in the fuel have increased probability of interacting along the axial direction. The tritons traveling in the axial direction are Doppler-shifted, and this leads to the double peaked structure observed. The inferred magnetization parameter BR, where B is the magnetic field and R is the inferred fuel radius, is 0.34 MG-cm. BR, rather than fuel areal density, is the fundamental confinement parameter relevant for MagLIF.

2.3. *Inference of liner areal density from down-scattered neutrons*

Down-scattered primary neutrons from the highly compressed beryllium liner reveal information about the *liner* areal density (ρR_{liner}). Figure 5 presents axially and radially measured neutron spectra from shot 2591 between 1-3 MeV, both of which reveal small peaks at ~ 1.6 MeV which are associated with neutrons down-scattered from beryllium. Results from MCNP simulations are over-plotted with the measured data. These simulations were performed for a 6-mm length, deuterium-fuel line source with a 2.5 keV ion temperature inside a cylindrical beryllium liner with ρR_{liner} of 1 g/cm^2. The simulated and measured peaks agree reasonably in terms of the ~ 1.6 MeV down-scatter amplitude and the peak location. Additional neutron down-scatter (from hardware, notably between 2-

2.4 MeV) may obscure the down-scatter peak amplitude since this background contribution is not fully understood. What is evident for the measurements and simulations is the subtle energy difference between the ~ 1.6 MeV peak amplitudes for the axial and radial spectra with the axial spectra located at slightly lower energy. It is encouraging that measurements of ρR_{liner} based on x-ray spectroscopy reveal the liner areal density to be ~ 0.9 g/cm^2, which is very close to the neutron measurement despite the background issues.

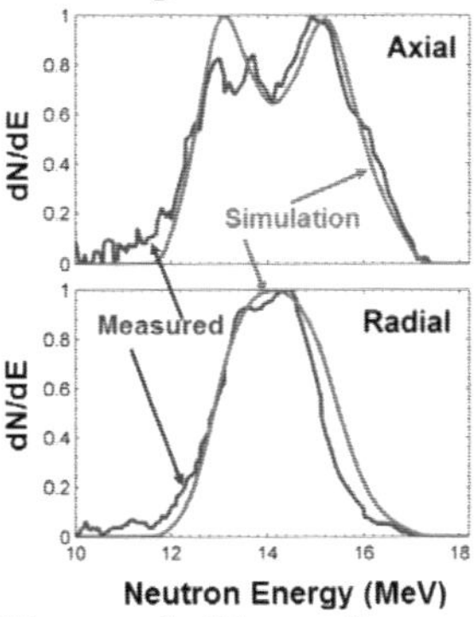

Figure 4. Plots of measured and simulated axial (top) and radial (bottom) DT neutron spectra.

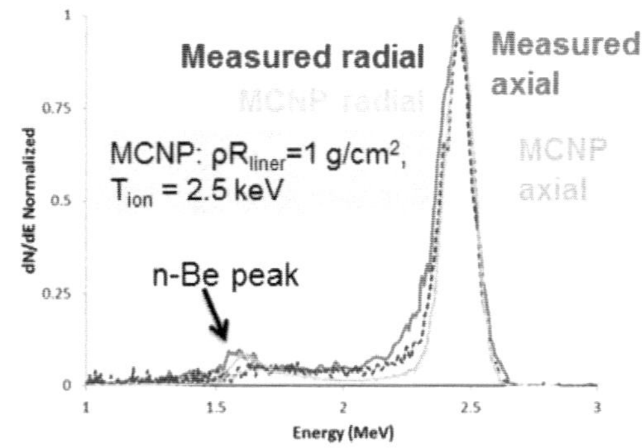

Figure 5. Down-scattered primary neutrons from the dense beryllium liner are indicative of ρR_{line} ~ 1 g/cm^2.

3. Future work and summary

One major area of improvement for neutron measurements on Z is to increase the sensitivity and precision for the nTOF measurements. We are presently testing and developing several gated nTOF detectors to gate out the large bremsstrahlung signals to better measure the smaller DT and DD signals. Also, we intend to improve the detector shielding and collimation to enable fielding of closer-in detectors and limit the down-scatter affecting all the detectors. Over the next few years, we will be exploring the incorporation of tritium fuel on Z experiments. In 2-3 years, our goal is to use ~ 1% tritium fuel mixed with deuterium. At this concentration, we expect primary DT yields to be of the same order of magnitude as the primary DD yields. With further improvements to MagLIF experiments (i.e., improved laser preheat coupling, higher magnetic fields, higher accelerator currents), we expect to produce yields in the ~1e13 range (DD and DT). At these higher DT yields, we intend to expand the Z neutron diagnostic suite to include neutron burn-history diagnostics, neutron imaging, and neutron recoil-based spectral measurements.

Sandia National Laboratories is a multi-program laboratory managed and operated by Sandia Corporation, a wholly owned subsidiary of Lockheed Martin Corporation, for the U.S. Department of Energy's National Nuclear Security Administration under contract DE-AC04-94AL85000.

References

[1] S. Slutz *et al.* Phys. Rev. Lett., **108**, 025003 (2012).
[2] S. Slutz *et al.* Phys. Plasmas, **17**, 056303 (2010).
[3] K. Matzen *et al.* Phys. Plasmas, **4**, 1519 (1997).
[4] M. Gomez *et al.* Phys. Rev. Lett. **113**, 155003 (2014).
[5] R. J. Leeper *et al.* J. Phys.: Conf. Ser. **112**, 032976 (2008).
[6] J. F. Briesmeister, MCNP - A General Monte Carlo N-particle Transport code, Version 4C, Los Alamos National Laboratory, 2000.
[7] V. Yu. Glebov *et al.* Rev. Sci. Instrum., **81**, 10D325 (2010).
[8] Hatarik *et al.* J. Appl. Phys. **118**, 184502 (2015).
[9] P. Schmit *et al.* Phys. Rev. Lett. **113**, 155004 (2014).
[10] P. F. Knapp *et al.* Phys. Plasmas, **22**, 056312 (2015).

Simulations and experiments of the growth of the "tent" perturbation in NIF ignition implosions

B A Hammel, R Tommasini, D S Clark, J Field, M Stadermann, C Weber

Lawrence Livermore National Laboratory, CA, USA

hammel1@llnl.gov

Abstract. NIF capsules are supported in the hohlraum by two thin (~15-110 nm) Formvar films ("tent"). Highly resolved HYDRA simulations indicate that a large (~40% peak-average) areal density (ρR) perturbation develops on the capsule during acceleration as a consequence of this support geometry. This perturbation results in a jet of dense DT and, in some cases, CH that penetrates and cools the hot spot, significantly degrading the neutron yield (~10-20% of 1D yield). We examine "low-foot" and "high-foot" pulse shapes, tent thicknesses, and geometries. Simulations indicate that thinner tents result in a smaller ρR perturbation, however, the departure angle of the tent from the capsule surface is important, with steeper angles resulting in larger perturbations.

1. Tent geometry

NIF capsules are supported in the hohlraum by two thin (~15-110 nm) Formvar films ("tent") that are stretched over the two poles of the capsule and supported at the joint between the upper and lower halves of the hohlraum [1][2][3]. Metrology of the assembled targets (See insert in Figure 1) indicates that a 50-nm-thick tent follows the curvature of the capsule for ~ 40° (measured from the pole), and then departs from the capsule at an angle that is ~15° steeper than tangential (i.e. 15° "off-tangent" (OT) angle). The angle continues to increase as the distance from the capsule surface increases. The tent angle likely varies azimuthally around the capsule since the tent support is affected in areas near diagnostic holes in the hohlraum. Recent measurements indicate that the departure angle depends on tent thickness, with thinner tents having steeper departure angles. These measurements are made for capsules at room temperature. Measurements of targets at cryogenic temperatures, where the tent angle could be affected, are in progress.

2. Simulation Method

Direct numerical simulations of the impact of the tent on NIF ignition target designs are difficult due to the challenge of resolving the tent thickness. Typical high-resolution simulations without a tent have a radial zone spacing of ~40 nm in the outer ablator region (~ 1 g/cm^3), and ~400 nm in the surrounding gas (~1 mg/cm^3). Fully resolving the tent (~ 5 zones across a 50 nm thick tent), which is embedded in the surrounding gas on a polar mesh, implies ~10 nm zones for the gas, and therefore ~1 nm zones in the denser ablator. Such fine radial zoning sets a limit on the angular zoning due to zone aspect ratio limits and run stability. Tent simulations reported here have ~ 5000 radial zones and 5120 angular zones (over 90°). The small zone size limits the Courant time step to impractically long run times. The strategy taken here is to rezone the ablated region after the tent has exploded, at ~ 1 ns after the start of the drive. Lower resolution simulations where the tent is under-resolved (e.g. 1 zone across the tent), or modeled at lower density and increased thickness (ρR constant), have shown qualitatively similar results, but with differences in the details of the resulting perturbation. In the simulations reported here we approximate the tent geometry and assume the tent departs the capsule at a fixed OT angle (i.e. the tent follows the curvature of the capsule and then departs in a straight line). These simulations are of the capsule and tent only, using an x-ray drive taken from integrated capsule-hohlraum simulations. The impact of the tent with other sources of perturbation included (e.g. surface roughness, drive asymmetry, fill-tube, etc.) has been assessed in lower resolution simulations that use an "equivalent" surface perturbation that match the ρR perturbation from the high-resolution simulations reported here.[4]

3. Tent seeded perturbation

The interaction of the tent with the capsule is complex. The departure of the tent from the capsule surface leads to a small jump in areal density $\rho R/\langle\rho R\rangle \sim 10^{-5}$, which increases with increasing OT angles. This ρR perturbation is small (0.1x) compared to typical capsule surface perturbations, and should be negligible. However, as the tent explodes and collides with the ablating capsules material, a pressure wave is created that propagates in the polar direction away from the contact point towards the equator. Resulting Kelvin-Helmholtz growth may create a seed for further hydrodynamic growth. In addition, as the tent explodes the gradient in the density of the ablated capsule material is misaligned with the gradient in the pressure from the exploding tent, creating vorticity in the ablated region $\omega \sim -\nabla P \times \nabla\rho \sim |\nabla P| \cdot |\nabla\rho| \cdot \sin(\theta)$. The induced vorticity increases with increasing OT angle. The perturbation seeded by these processes is then amplified by Rayleigh-Taylor (R-T) when the capsule accelerates inward.

4. Simulations and Experiments

The simulated fractional areal density perturbation $\rho R/\langle\rho R\rangle$, near the time of peak implosion velocity, resulting from a 50-nm-thick tent on a CH Symcap driven with a "low-foot" (LF) drive, for a range of OT angles, is shown in Figure 1. [5][6] The tent perturbation increases markedly with increasing OT angle, reaching a level of +35% to -40% peak-to-valley. Results for a "high-foot" (HF) drive [7], for the 0° OT angle case is also shown.

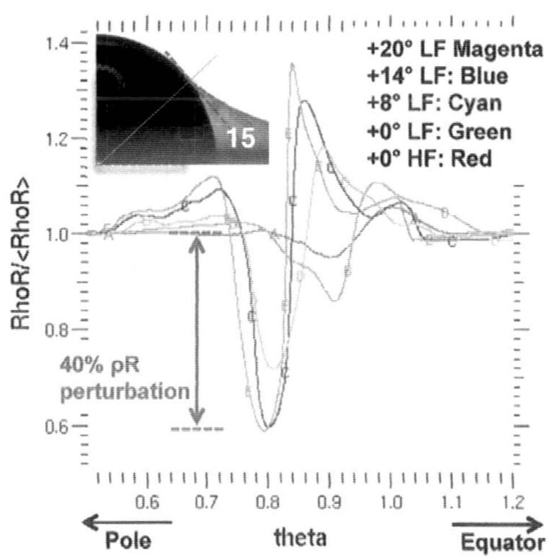

Figure 1. $\rho R/\langle\rho R\rangle$ *vs.* theta (rad) for different OT angle. Insert shows a pre-shot image of capsule and tent. Tent departs with ~15° OT angle. Simulations: Symcap targets.

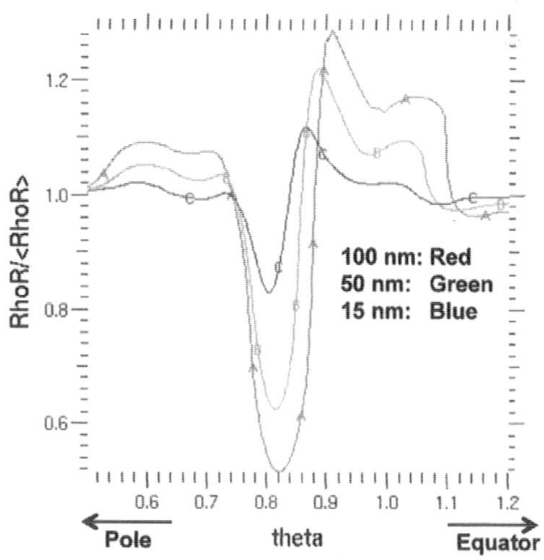

Figure 2. $\rho R/\langle\rho R\rangle$ *vs.* theta (rad) for different tent thickness. All with 14° OT angle and LF drive. Simulations: DT targets.

Simulations show that the perturbation for the HF case is slightly larger than LF after shock transit and prior to capsule acceleration, but the larger R-T growth for the LF case leads to a perturbation that is ~ 2.5x larger than HF. The simulated $\rho R/\langle\rho R\rangle$ for different tent thicknesses on CH DT targets (N120321), all with 14° OT angle, is shown in Figure 2. Thinner tents result in a smaller perturbation.

The ρR perturbation (Peak-to-Average) of implosions, near peak velocity (200 μm radius), are shown in Figures 3 [8] and 4. In Figure 3, pluses are the *actual* ρR perturbation from the simulations shown in Figure 1 (Symcap implosions). In Figure 4, pluses are the *actual* ρR perturbation from the simulations of DT implosions shown in Figure 2, and the open circle (for 50 nm thickness case) is for the equivalent Symcap target. The close agreement indicates that the ρR perturbation at peak velocity resulting from a given tent thickness is nearly identical for Symcap and DT implosions. In Figures 3 and 4, solid circles are *inferred* values of ρR perturbation from simulated x-ray radiographs at perfect resolution (see insert).[8] The excellent agreement with *actual* ρR perturbation from simulations (pluses) validates the unfold method used to infer ρR from radiographic transmission through the capsule $e^{-\kappa\rho L}$.[8] Squares are inferred ρR perturbation from simulated radiographs, blurred to match the experimental resolution (20 μm). In Figure 3, inferred values from experimental measurements of LF implosions fall in the range marked in blue, implying OT angles of 8-16°, generally consistent with pre-shot metrology. Note: the inferred values of 15%, at the 20 μm experimental resolution, imply *actual* ρR perturbations of ~30-40%. For HF implosions, the inferred perturbation is consistent with an actual ρR perturbation of ~5%. Figure 4 shows the inferred ρR perturbation from simulations for different tent thicknesses, all with 14° OT angle, and experimental measurements for LF implosions. The values inferred from simulations (20 μm resolution) and those inferred from experimental radiographic measurements, for LF implosions, are in good agreement except at the largest tent thicknesses. This may be due to a decrease in the tent's OT angle with increasing tent thicknesses. Inferred ρR perturbation from measurements on HF implosions is also shown. Simulations are in progress for this case.

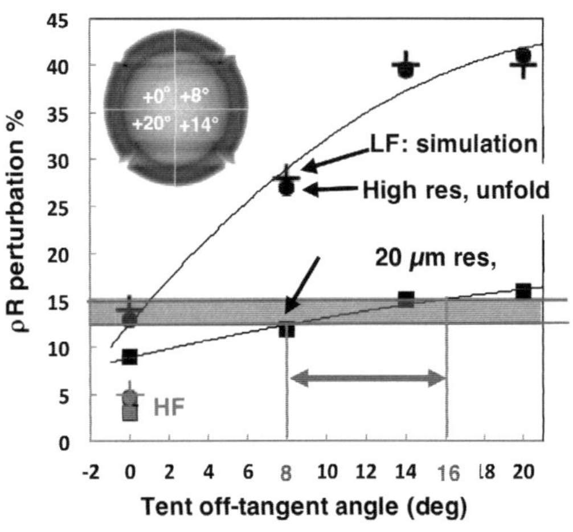

Figure 3. Areal density (ρR) perturbation (Peak-to-Average) *vs.* tent departure angle. Black: LF drive, Red: HF drive. Insert: Simulated radiographs for different OT angle from LF drive, Symcap implosions.

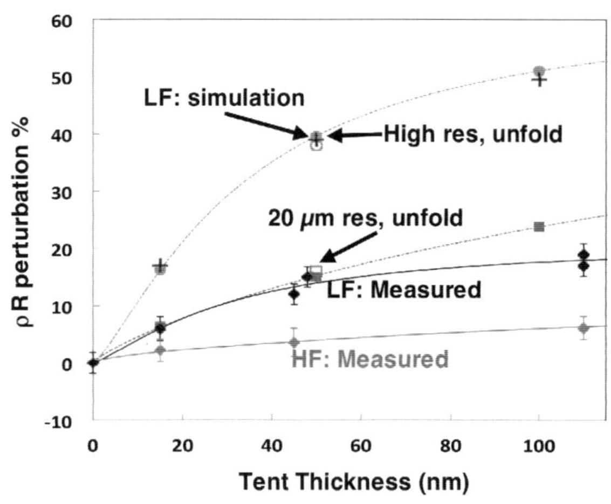

Figure 4. ρR perturbation (Peak-to-Average) *vs.* tent thicknesses. All with 14° OT angle. Black: LF drive, Red: HF drive. Closed circles and squares: DT implosion. Open circles and squares: Symcap implosion. Measurements: Symcap implosions.

As the capsules continue to compress, simulations indicate that the perturbation resulting from the tent has a significant impact on the hot-spot shape, and the resulting performance. Figure 5 shows density and temperature at peak compression for a DT target (N120321), driven with a LF pulse, with a 50-nm-thick tent at 14° OT angle. The region of low ρR results in escaping hot-spot mass, and the regions of high ρR result in a jet of cold DT (~1 μg) penetrating the hot-spot, causing local cooling. The resulting x-ray hot-spot shape has a strong (P_4/P_0 ~ 30%) distortion (Figure 6), in qualitative agreement with experiments.[9] For a 50-nm-thick tent, and no other sources of perturbation, the neutron yield is degraded to 18% of the 1D value. Although no CH enters the hot-spot in this case, simulations of N120321 that include drive asymmetry and a 100-nm-thick tent (n.b. tent on this shot was 110 nm) indicate that 1.5 μm of CH and 4 μg of cold DT enter the hot-spot.

We are exploring thinner tents and alternate mounting schemes. Simulations indicate that for a 15-nm-thick tent the ρR perturbation (Figure 4) and resulting jet are reduced. With only this perturbation, the neutron yield is degraded to 58% of the 1D value. Alternate mounting schemes under development include "polar-contact" tents that reduce the solid angle of the perturbation and the resulting mix.

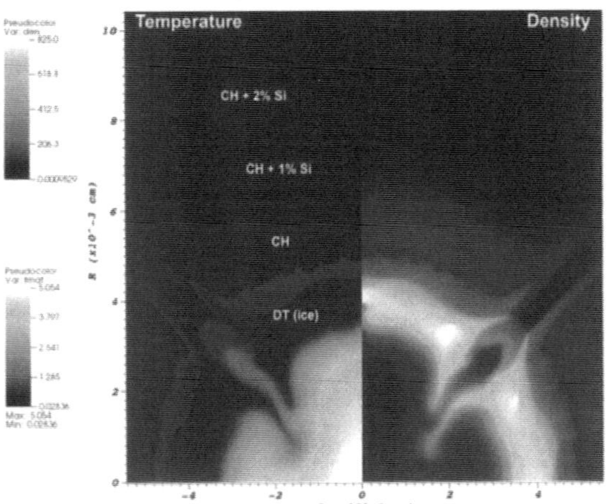

Figure 5. Simulation of DT target implosion near peak compression, driven by LF pulse, with 50-nm-thick tent at 14° OT angle. Right Frame: Density. Left Frame: Temperature. The tent perturbation results in a jet of cold DT that penetrates the hot-spot.

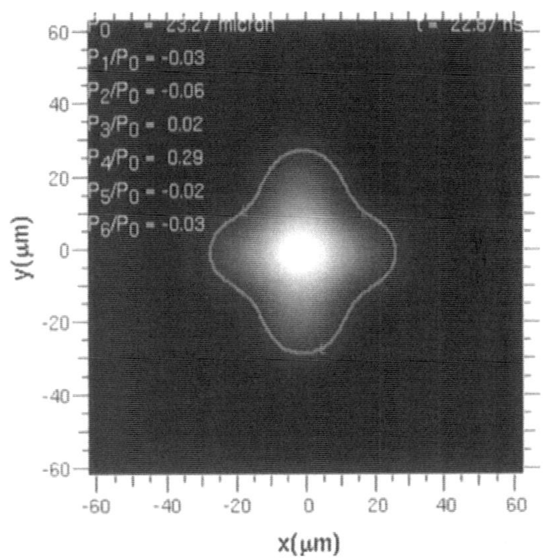

Figure 6. Simulated x-ray image. Legendre polynomial fit (15% contour) indicates a ~30% P_4/P_0 distortion.

References

[1] Baxamusa S H, et al., Langmuir **30**, 5126 (2014).
[2] Stadermann M, Letts S A, and Bhandarkar S, Fusion Sci. Technol. **59**, 58 (2011).
[3] Haan S W, Atherton J, Clark D S, and Hammel B A, Fusion Sci. Ignition **63**, 67–75 (2013).
[4] Clark D S et al., Physics of Plasmas, 2015. **22**(2).
[5] Lindl J, et al., Phys. Plasmas **21**, 129902 (2014)
[6] Capsule parameters and drive similar to shot N120321.
[7] Hurricane O A, et al., Nature **506,** 343–348 (2014).
[8] Tommasini R et al., Physics of Plasmas **22** (5), 2015.
[9] Nagel S R, et al., Phys. Plasmas **22**, 022704 (2015).

This work was performed under the auspices of the U.S. Department of Energy by Lawrence Livermore National Laboratory under Contract No. DE-AC52-07NA27344

Journal of Physics: Conference Series **717** (2016) 012022 doi:10.1088/1742-6596/717/1/012022

Reaction-in-Flight neutrons as a test of stopping power in degenerate plasmas

A C Hayes[1], C J Cerjan[2], G Jungman[1], M M Fowler[1], M E Gooden[1], G P Grim[2], E Henry[2], R S Rundberg[1], S M Sepke[2], D H G Schneider[2], R L Singleton[1], A P Tonchev[2], J B Wilhelmy[1] and C B Yeamans[2]

[1] Los Alamos National Laboratory, Los Alamos, NM, 87545 USA
[2] Lawrence Livermore National Laboratory, Livermore, CA, 94551, USA

Email: anna_hayes@lanl.gov

Abstract. Cryogenically cooled inertial confinement fusion capsule designs are suitable for studies of reaction-in-flight (RIF) neutrons. RIF neutrons occur when energetically up-scattered ions undergo DT reactions with a thermal ion in the plasma, producing neutrons in the energy range 9-30 MeV. The knock-on ions lose energy as they traverse the plasma, which directly affects the spectrum of the produced RIF neutrons. Here we present measurements from the National Ignition Facility (NIF) of RIF neutrons produced in cryogenic capsules, with energies above 15 MeV. We show that the measured RIFs probe stopping under previously unexplored degenerate plasma conditions and constrain stopping models in warm dense plasma conditions.

1. Introduction

RIF neutrons are produced in a DT plasma when a neutron or alpha particle born in the burn elastically scatters deuteron or triton ions up to MeV energies, and the energetic knock-on ion undergoes a DT reaction with a thermal ion in the plasma, producing RIF neutrons in the energy range 9.2-30 MeV. The knock-on ions lose energy through the plasma stopping power as they traverse the plasma. This energy loss directly affects the number and energy spectrum of the RIF neutrons. Thus, RIF measurements can be used to extract information on the plasma stopping. In the case of cryogenic NIF capsules, RIFs probe stopping under previously unexplored plasma conditions.

2. RIF production in NIF cryogenic capsules

The knock-on ion fluence responsible for RIF production is dominantly produced in the cold fuel surrounding the hotspot of a compressed cryogenic capsule. This is because the magnitude of the knock-on fluence (Q_0) is determined by the 14 MeV neutron fluence (Φ_{14}) and the number density of DT ions (n_{dt}) in the plasma, $Q_0=\Phi_{14}n_{dt}\,\sigma_{ko}$, where σ_{ko} is the neutron-ion knock-on cross section and DT density in the cold fuel is very high (10^{25} -10^{26} cm^{-3}). Transport of the knock-on ions through the plasma changes both the shape and magnitude of the knock-on spectrum, and hence also of the RIF spectrum. This spectrum change is determined by the stopping power, and a measurement of the RIF spectrum provides a constraint on stopping models used for the cold fuel. The detailed relationship between stopping power and the RIF spectrum is described in ref. [1].

At peak capsule compression, the outer dense DT fuel temperature is a fraction of a keV. When combined with its high density, this results in the cold fuel being degenerate, i.e, the temperature is less than the Fermi temperature, with a degeneracy parameter θ/θ_F~0.2-0.3, where θ (θ_F) is the electron (Fermi) temperature. The corresponding standard plasma coupling parameters $\Gamma=Ze^2/R_W\theta$ or $g= Ze^2/R_D\theta$, where R_W (R_D) is the Wigner (Debye) radius, are approaching or exceeding unity, Fig. 1. In the present case Z=1. Since the cold fuel is electron degenerate, the temperature determining the stopping power is an effective temperature, $\theta_{eff} = 3/5\ \theta_F F(\theta)$, where the function $F(\theta)$ is obtained by solving for the chemical potential of the system and has the property that in the zero temperature limit, $\theta_{eff} \rightarrow 3/5\theta_F$, while in the high temperature limit $\theta_{eff} \rightarrow \theta$. Thus, a better measure of the coupling in a degenerate system is the *ideality* $\xi= Ze^2/R_W\theta_{eff}$, and ξ=0.05-0.2 in cold fuel of cryogenic NIF capsules. Under these plasma conditions (g>1, ξ<1, and θ/θ_F<1), stopping powers have not been tested experimentally and existing models tend to disagree with one another. The predictions of the HYDRA code for degeneracy and the coupling parameters g and Γ as a function of the radius of the capsule are shown in the right panel of Fig.1.

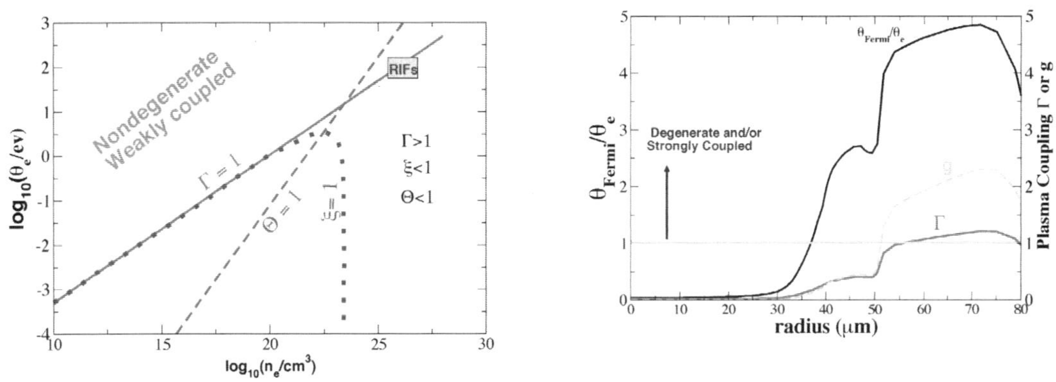

Fig. 1 The RIFs are dominantly produced in the cold fuel of a cryogenic NIF capsule, which is predicted by HYDRA calculations to be a degenerate plasma. The panel on the right shows the value of the degeneracy $\Theta=\theta_e/\theta_F$ and the coupling parameters g and Γ as a function of the capsule radius, as predicted by HYDRA. The relevant part of the (θ_e,n_e) phase diagram for the current experiments and the limits for strong coupling Γ=1, degeneracy Θ=1, and ideality ξ=1 are shown in the panel on the left. Stopping powers have not previously been tested under these plasma conditions (Γ>1, ξ<1, and θ/θ_F<1), where existing models tend to disagree with one another.

3. The RIF measurements

The RIF measurements are done by assaying neutrons with energies E_n>15 MeV using the activation of thulium foils. Thulium is chosen because the ^{169}Tm*(n,3n)* reaction has a threshold of 15 MeV, and produces ^{167}Tm with a half-life $t_{1/2}$= 9.25 days by electron capture to ^{167}Er, with the emission of a 207.8 keV gamma-ray 41% of the time. The Tm foil is part of a larger activation foil assembly that is mounted on a diagnostic instrument manipulator (DIM) at a nominal standoff of 50 cm from target chamber center, placed behind the solid debris collection diagnostic [2,3]. Activation of Zr, Al, and Au is used to characterize the neutron spectrum at and below the DT fusion primary neutron energy [4]. Activation assemblies are located both on the polar DIM, 14 degrees below the NIF north pole, and the equatorial DIM, 14 degrees below the NIF equator. Thulium samples were fielded on 14 shots with primary DT neutron yields of 1-8x10^{15} and down-scattered ratios [5] of 2.5-4.2%. The measurement of ^{168}Tm, produced in the ^{169}Tm*(n,2n)* reaction, provides a check of the primary 14 MeV neutron production, and is cross-compared with activation foils in the same holders.

The primary experimental challenge for the ^{167}Tm measurements is the huge background of γ-rays produced in the activation of the foils by the much higher fluence of primary 14 MeV neutrons, in particular those from ^{168}Tm. To suppress this background, the activated foils are shipped to Los

Alamos, where two clover detectors are deployed to assay the ^{168}Tm and ^{167}Tm activity in the foils. The clover system consists of two high-efficiency clover germanium detectors. Each clover consists of 4 HPGe crystals, surrounded by an active NaI Compton suppressor. The ^{167}Tm decays to a 208 keV isomeric level in ^{167}Er that has a half-life of 2.28 sec. This level decays 100% of the time to the ^{167}Er ground state. There is nothing in coincidence with the 208 keV γ-ray, in contrast to the complex decay scheme of ^{168}Tm. This allows us to use the clover array to identify ^{167}Tm above the background.

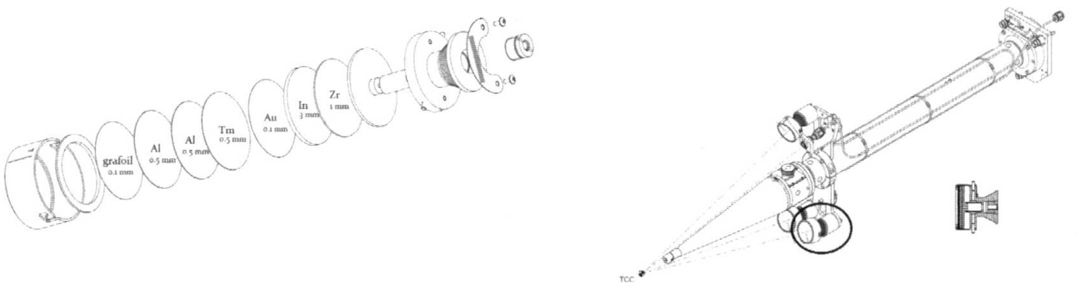

Figure 2 – (Left) Thulium activation assembly explosion drawing: The RIF neutron measurement thulium foil is held in an assembly containing materials for solid debris collection and measurement of the neutron spectrum. (Right) NIF diagnostic instrument manipulator (equatorial), showing snout-mounted Tm activation diagnostic.

4. Experimental Results and Analyses

The measured ratio of ^{167}Tm/^{168}Tm varies somewhat from shot to shot, but is typically measured to be on the 10^{-5} scale with 15% accuracy. As an example of how well the stopping power can be tested, we will restrict our discussion here to shot N140304 for which the measured ^{167}Tm/^{168}Tm ratio was 1.69+/- 0.24x 10^{-5}. The total RIF signal can be written as,

$$N_{RIF}^{total} = N_{14}\langle \rho R\rangle_{DT} \int \frac{d\bar{\psi}}{dE}\sigma_{DT}\, dE_{KO} \quad \textbf{(1)}$$

where $d\bar{\psi}/dE$ is the knock-on fluence per unit 14 MeV neutron. The knock-on fluence scales inversely with the stopping power, and our goal is to extract $d\bar{\psi}/dE$ from the RIF measurements using full three-dimensional (3-D) simulations of the capsules for N_{14} and $\langle\rho R\rangle$.

In our analyses of shot N140304 we used full 3-D simulations constrained [6] by a broad range of experimental data. The constraints involve a physically consistent description of the imploded capsule at stagnation. This method attempts to validate the model of the capsule through stringent comparisons between the radiation hydrodynamic code simulations and a suite of x-ray and neutron imaging data. We carried out a second independent analysis of the RIFs using the 1-D code CPT-Implode, which also focuses on describing the stagnation properties of the capsule correctly. In both cases, we varied the stopping models and examined the change in the shape of the predicted RIF signal. The results are summarized in Table 1.

Table 1	167Tm/168Tm
Experiment	1.69+/- 0.24x 10^{-5}
Maynard-Deutsch-Zimmerman [7]	1.59x10-5
Li-Petrasso [8]	1.16x10-5
Classical Model only (Grabowski) [9]	0.9x10^{-5}
BPS [10]	Model breaks down- g too large
BPS+ effective temperature	1.6x10^{-5}

5. Theoretical discussion

It has long been recognized [11] that quantum mechanical effects lower plasma stopping powers by about 40%, relative to classical model predictions. The situation is summarized in Fig. 3, where we compare the classical contribution to the stopping power, as predicted by Grabowski et al. [9] to the quantum mechanical predictions from the Maynard-Deutsch-Zimmerman model (MDZ) [7]. The Li-Petrasso stopping model [8] includes quantum effects and allows for degeneracy through the use of an effective temperature. But the latter model [8] leads to a stopping power that appears [12] to be too high. The BPS model [10] is only valid when g<1. However, with the use of an effective temperature derived from the chemical potential, $g \to g_{eff} < 1$, and the BPS results then agree with experiment. The leading dependence of the stopping power on the electron temperature and density scales as

$$\frac{dE}{dx} \propto \frac{n}{\theta^{3/2}} \to \frac{n}{\theta_{eff}^{3/2}}, \qquad (2)$$

where n is the electron density, and the degeneracy of the cold fuel results in a effective temperature (θ_{eff}~0.5 keV), which is be compared to the actual temperature (θ~0.2 keV). Thus, degeneracy further lowers the stopping power, as can be seen by comparing the red and dashed blue curve in left panel of Fig.3. The corresponding effect on the RIF spectrum is to increase the spectrum by about a factor of 2 at 15 MeV, but by less than a factor of 1.1 at 22 MeV, as seen in the right panel of Fig. 3.

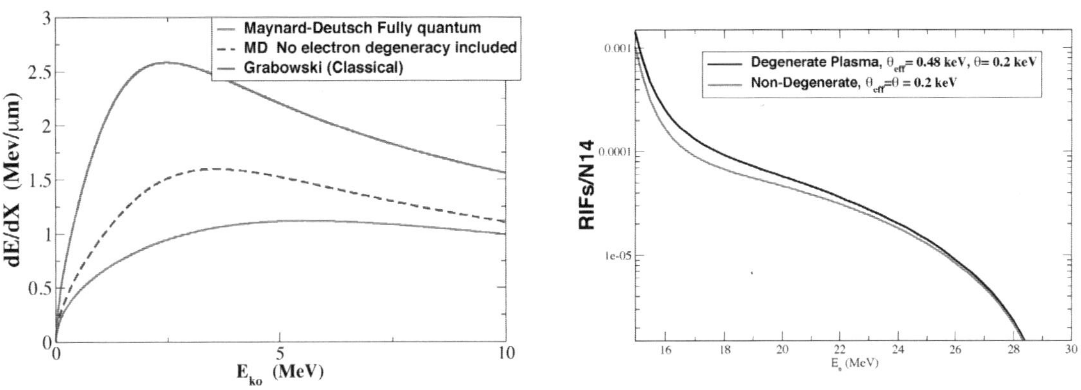

Fig. 3 The left panel shows the MDZ stopping model, with and without degeneracy, compared to the stopping power predicted for a purely classical model. The right panel shows the MDZ predicted RIF spectra, with and without degeneracy.

A more stringent test of stopping models in degenerate plasmas would be afforded by a measurement of the shape of the RIF spectrum. For this we are developing a second RIF diagnostic, using activation of bismuth foils. The ^{209}Bi*(n,4n)* reaction has an energy threshold of E_n=22.5 MeV and it anchors the RIF spectrum at high energies, where degeneracy has little effect. A comparison between the bismuth and thulium RIF diagnostics would place stringent constraints on the role of degeneracy on stopping.

References

[1] Hayes A.C. et al., , Physics of Plasmas **22 (8)** 082703 (2015)
[2] Gostic J. M. *et al.* 2012 *Rev. Sci. Instrum.* **83** 10D904.
[3] Shaughnessy D. A. *et al.* 2014 *Rev. Sci. Instrum.* **85** 063508.
[4] Bleuel D. L. *et al.* 2012 *Rev. Sci. Instrum.* **83** 10303.
[5] Wilson D C *et al.* 2002 *Nucl. Instrum Meth.* **488** S0168-9002(02)00474-6.
[6] Cerjan C., Springer P.T. , Sepke S.M. ,Phys. Plasmas 20, 056319 (2013).
[7] Zimmerman G.B., Recent Developments in Monte Carlo Techniques, Lawrence Livermore National Laboratory internal report, UCRL-JC-105616, (1990).
[8] Grabowski P.E., Surh M.P., Richards D.F. , Graziani F.R. , and Murillo M.S., Phys. Rev. Lett. **111**, 215002, (2013).
[9] Li C.K. and Petrasso R.D., Phys. Rev. Lett. **70**, 3059 (1993)
[10] Brown L.S., Preston D.L., and Singleton R.L., Jr.,Phys. Rep. **410**, 237 (2005).
[11] George E.P. and Hamada T., Phys. Letts. **67A**, 369 (1978).
[12] Ordonez C.A. and Molina M.I., Phys. Rev. Lett. **72**, 2407 (1994).

Implosion configurations for robust ignition using high-density carbon (diamond) ablator for indirect-drive ICF at the National Ignition Facility

D. D.-M. Ho, S. W. Haan, J. D. Salmonson, D. S. Clark, J. D. Lindl, J. L. Milovich, C. A. Thomas, L. F. Berzak Hopkins and N. B. Meezan

Lawrence Livermore National Laboratory, Livermore, CA U.S.A.

ho1@llnl.gov

Abstract. We present five ignition scale capsule designs using high-density carbon ablators with fuel adiabat (α) ranging from 1.5 to 4. All five have 1D yield > 1 MJ. The sensitivities of these capsules to surface roughness and P_2 radiation asymmetries were studied. The most robust configuration with respect to surface roughness depends on the amplitude of the surface spectrum. The most robust configuration with respect to P_2 asymmetry is the α = 1.5 configuration which has the highest 1D margin. We find that α = 2 and 2.5 configurations have the highest overall robustness. Further analysis is needed to study the effects of more complicated 3D behaviors.

1. Introduction
The two types of ablators used in most current indirect-drive implosion experiments at the National Ignition Facility (NIF) are plastic (CH) [1 - 4] and high-density carbon (HDC) or diamond [5 - 7]. The HDC ablators are made of micro-crystalline diamond with micron-scale crystal structure having grain size between one and a few microns [8].

The potential advantages of HDC ablator over other ablators, Be or CH, have been described previously [5]. The higher density and consequently thinner ablator allows better energy absorption for a given capsule outside diameter, a shorter pulse that provides hohlraum design flexibility, and less sensitivity to the perturbation seeded by the capsule support tent [Ref. 9].

The search for the most robust capsule configuration requires the tradeoff among key implosion parameters: fuel adiabat, implosion velocity, dopant level, hydrodynamic instability, and low-mode radiation asymmetries. Performing this tradeoff to obtain the most robust configuration, which is a fundamental exercise in ICF implosion design and physics, is the goal of this note. A detailed description of this tradeoff and HDC implosion physics will be given in a forthcoming paper [10].

In this note, we present five ignition capsule configurations and their performance with fuel adiabat α ranging from 1.5 to 4 in Sec. 2. The ignition cliff and robustness with respect to surface roughness are discussed in Sec. 3. The ignition cliff and robustness with respect to low-mode radiation flux asymmetries are discussed in Sec. 4. The selection of the overall most robust implosion configuration, according to the criteria assumed here, is presented in Sec. 5. Section 6 summaries this work.

2. Capsule configurations and performances
We present five capsule configurations with α ranging between 1.5 and 4. They have the same outer radius and ablator thickness (see Fig. 1). The fuel layer thickness is 55.6 μm for all except 46 μm for the α = 4.0 configuration. The radiation temperature Tr of the drive pulses are shown in Fig. 2. Note

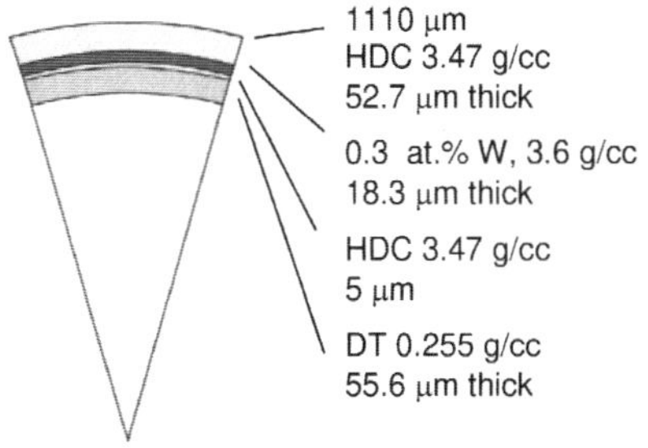

Figure 1. HDC capsule configuration with α = 2.05.

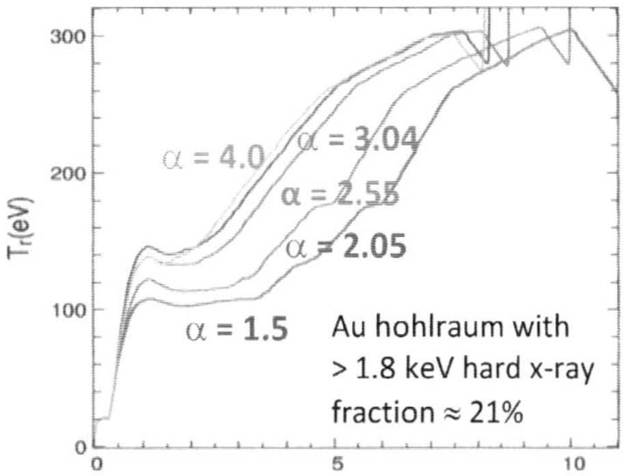

Figure 2. Channel Tr vs time for the α = 1.52, 2.05, 2.55, 3.04, and 4.0 configurations.

by the foot of the Tr profile of the α = 1.52 configuration, is close to the minimum required for HDC ablators. Experiments using the Omega laser showed that the velocity ripples of the first shock become unacceptably large if it is below 6 Mbar [11]. Shocks do not merge inside the fuel layer for all, except the α = 4 configuration in which the first two shocks merge inside the fuel layer.

The capsule performance parameters, computed using the radiation-hydrodynamics code HYDRA [12], are listed in Table 1. A doped layer with 0.3 at.% W, is used to control the Atwood number at the ablator-fuel interface. The amount of dopant is close to the highest level currently accessible to the fabrication technology, and appears to be suitable for these designs. The dopant layer thickness is determined by the requirement that the simulated clean fuel fraction (fraction of initial fuel mass contaminated by 5% or less by mass of ablator material, from 2D simulations) at peak velocity (vpeak) is between 80 and 85%. Ablation front Rayleigh-Taylor instability (RTI) growth decreases as α increases (as indicated by the single-mode growth factors shown in Fig. 3) and this reduces the feedthrough that enhances the mix at the ablator-fuel interface. Also, the in-flight fuel density is lower with higher α. The required dopant layer thickness therefore decreases with increasing α.

Table 1. 1D capsule performance

Fuel adiabat α	doped layer thickness (μm)	absorbed energy (kJ)	1D yield yield (MJ)	no-burn stagnation pressure	vpeak (km/s)	fuel ρr (g/cm^2)	Atwood number	percent ablator mass remaining	convergence ratio
1.52	25.5	209	20.5	515	382	1.41	0.074	7.7	34.6
2.05	18.3	215	17.8	441	400	1.29	0.087	5.5	32.7
2.55	15.1	216	13.6	376	417	1.14	0.11	4.3	31.1
3.04	14.0	216	10.4	341	424	1.08	0.134	3.6	29.4
4.0	15.5	214	1.7	259	429	0.8	0.06	4.2	26.0

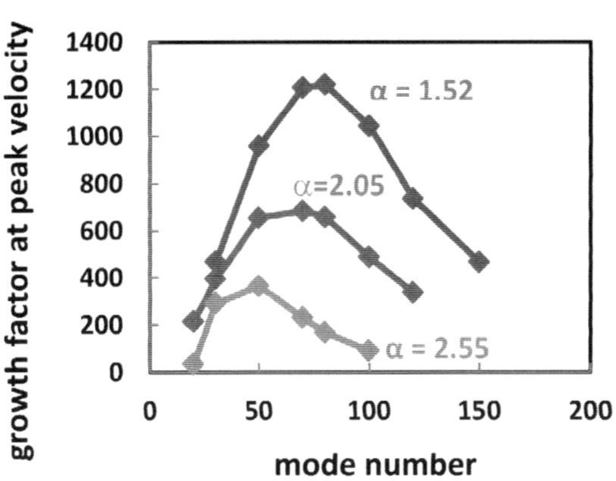

Figure 3. Ablation front growth factor at peak velocity vs mode number for α = 1.52, 2.05, 2.55 configurations.

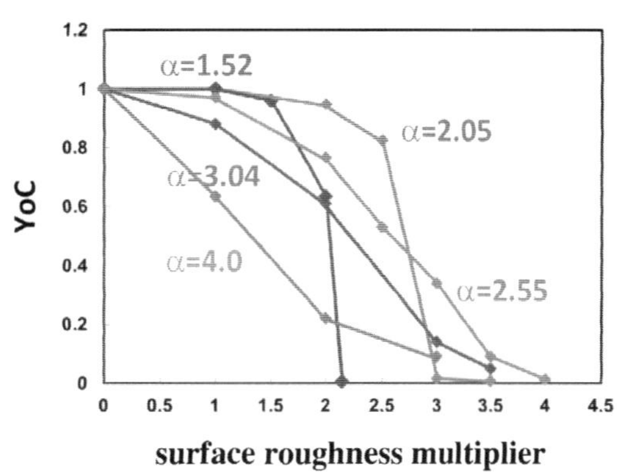

Figure 4. Yield cliff with respect to surface roughness.

Increasing α reduces the fuel compressibility which lowers the convergence ratio (initial outer radius to hotspot radius at ignition); consequently the no-burn stagnation pressure drops. This is the main disadvantage of high-α implosions. However, two factors partially offset this disadvantage: (i) thinner dopant layer improves ablation efficiency; (ii) lower RTI growth allows higher-α implosions to have less ablator remaining at while maintaining the same clean-fuel fraction. These two factors results in higher vpeak, as shown in Table 1. This is why 1D yield > 10 MJ is possible in these simulations for high α = 2.5 and 3 implosions. Doped layer thickness does not decrease further when α = 4 because of the thinner fuel layer. To maintain the same clean fuel fraction at vpeak for a thinner fuel layer, dopant layer thickness cannot be decreased further.

3. The ignition cliff and robustness with respect to surface roughness

The ignition cliffs for the five configurations with respect to surface roughness are shown in Fig. 4. The nominal ablator surface roughness used here is based on Atomic Force Microscope (AFM) spectra obtained prior to mid-2015. The indicated multiplier was put on the roughness of all surfaces.

The α = 2.5 design allows the largest surface roughness. At lower adiabat, the implosions fail at low surface roughness multiplier because larger ablation-front RTI. As α increase beyond 2.5, higher α reduces 1D margin and the ignition cliff becomes susceptible to perturbation growth at the hotspot boundary during deceleration phase RTI. Consequently, ignition becomes less robust for α > 2.5.

Substantial improvement on HDC surface roughness has been achieved for ablators fabricated after mid-2015. Simulations with smoother outer surface roughness show the ignition cliff of the α = 1.5configuration becomes more robust and approaches that of the α = 2. Therefore, improvement on the outer surface roughness improves the robustness more on low-α than on high-α implosions. The simulations shown in Fig. 4 do not explicitly include the effects of the capsule fill tube or support tent.

4. The ignition cliff and robustness with respect to P_2 radiation flux asvmmetrv

We found that with surface roughness, yield degradation is more susceptible to $+P_2$ than to $-P_2$ radiation flux asymmetry. To be conservative, $+P_2$ is therefore being used in our study.

The yield cliffs for P_2 asymmetry is shown in Fig. 5. Prior to the shocks merge near the inner fuel layer, P_2 = 0 and then in 0.5 ns, rises to a constant value thereafter. The performance becomes less robust as α increases. Even though the high-α implosions have lower convergence ratio, their low 1D margin makes them more susceptible to low-mode asymmetries. Therefore, this study leads us to the conclusion that the advantage of high 1D margin overrides the disadvantage of high convergence. The effect of the capsule support tents is not included in this analysis.

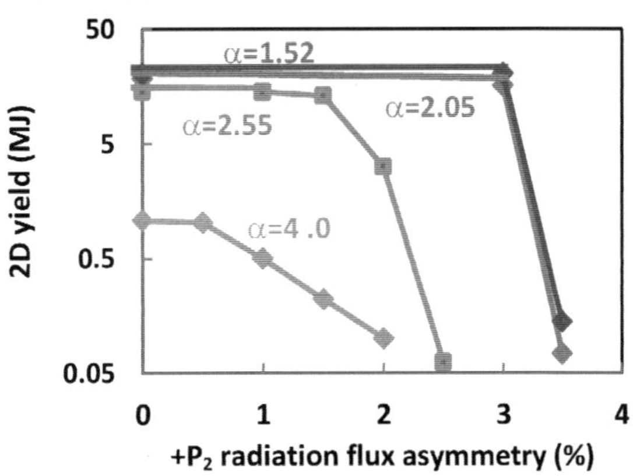

Figure 5. Yield cliffs for P_2 flux asymmetry for α = 1.52, 205, 2.55, and 4.0 configurations.

5. The most robust overall implosion configuration and relation to CH implosion experiments

Combining the results from these two sensitivity studies allows us to select the overall most robust implosion configuration for an ignition experiment. The logic for this selection is:

(a) For rougher outer surface spectrum

- Since the difference between the ignition cliff for $+P_2$ for α = 1.5 and 2.0 configuration is not large but α = 1.5 has the worst surface roughness sensitivity, is more susceptible to the tent, and is more difficult to obtain good radiation symmetry for its longer pulse, this eliminates α = 1.5 from the selection for the most overall robust configuration.
- The ignition cliff for $+P_2$ falls off rapidly for α > 2.5 and therefore configurations with for α > 2.5 can be eliminated from the selection.
- α = 2.5 is slightly more robust than α = 2.0 in terms of surface roughness but worse than α = 2.0 for

+P_2 robustness. Therefore, the choice between α = 2.0 and 2.5 depends on the following factors: the importance of tent perturbation and the quality of radiation symmetry provided by the hohlraum.

(b) For smoother outer surface spectrum

The robustness of the α = 1.5 becomes close to that of the α = 2.0 in terms of surface roughness if tent perturbation is not dominant (e.g., using alternate support strategy, or if the tent-seeded perturbation is acceptable for HDC capsules). If the hohlraum can accommodate the longer pulse length of the α = 1.5, then this design could be the most robust as it has the largest tolerance to P_2 asymmetries.

While the above description provides the first attempt to select the most overall robust configuration for ignition, some factors have not been included. For a detailed search of the most robust configuration, other factors to consider would include: P_4 radiation asymmetries, the effect of the tent, 3D phenomena, and the time-dependent radiation asymmetries for specific hohlraum designs.

The "high-foot" CH implosion campaign during the last two years shows higher α (~ 2.3), implosions perform better than low-α (~ 1.5) implosions [4] since higher α has RTI growth substantially lower than the low-foot. Because of this, high-foot is less sensitive to perturbations caused by the tent, which is the major factor in causing yield degradations in low-foot. The major sources of yield degradation for high-foot then becomes low-mode flux asymmetry as demonstrated by detailed 3D simulations by Clark [13]. This suggests the value of going to higher adiabat for stability. Furthermore, high-foot CH implosions typically have higher vpeak than low-foot, i.e., ~355 vs ~320 km/s and would make the high-foot even more robust than the low-foot implosions. In contrast, the ignition HDC capsules studied here have vpeak > 380 km/s and therefore the relative sensitivity to low modes are further reduced. Taking into account all these differences between CH experiments and our calculations, the CH experimental results are consistent with what we have presented here.

6. Conclusions

We have optimized the design of five DT-layered capsules with α between 1.5 and 4. By applying the appropriate amount of dopant so that the clean-fuel fraction at vpeak is between 80 and 85%, we are able to demonstrate that high-α, i.e., 2 to 3, implosions can achieve high yield. The most robust configuration with respect to surface roughness depends on the amplitude of the surface spectrum. For robustness with respect to P_2 flux asymmetry, the configuration with the lowest α is the most robust and the advantage of 1D margin overrides the advantage of low convergence. Combining the consideration of surface roughness and P_2 asymmetry, α between 2 and 2.5 appears to be the most robust overall configuration for ignition and high yield even taking into the consideration of the tent. Although the α = 4 configuration has low convergence ratio, it has less overall robustness, unless the corresponding hohlraum design can provide considerably better symmetry.

Acknowledgements

Prepared by LLNL under Contract No. DE-AC52-07NA27344.

Reference

[1] S. W. Haan, *et al.* 2011 Physics of Plasma **18**, 051001.
[2] M. J. Edwards, *et al.* 2013 Phys. Plasmas **20**, 070501.
[3] J. D. Lindl, *et al.* 2014 Phys. Plasmas **21**, 020501.
[4] O. A. Hurricane, *et al.* 2014 Nature **505**, 343.
[5] D. D. Ho, S. W. Haan, *et al.*, APS-DPP Philadelphia, Oct. 2006, Orlando Nov. 2007, Atlanta, Nov. 2009, Providence, Oct. 2012, Denver, Nov. 2013, New Orleans, Nov. 2014.
[6] J. S. Ross, *et al.* 2015 Phys. Rev. E **91**, 021101.
[7] L. F. Berzak Hopkins, *et al.* 2015 Phys. Plasmas **22**, 056318.
[8] J. Biener, *et al.* 2009 Nuclear fusion **49**, 112001; J. Biener, paper forthcoming.
[9] B. Hammel, *et al.*, this proceeding.
[10] D. D.-M. Ho, S. W. Haan, J. D. Salmonson, *et al.*, Phys. Plasmas, paper forthcoming.
[11] P. M. Celliers, *et al.* 2009 IFSA San Francisco.
[12] M. M. Marinak, *et al.* 2001 Phys. Plasmas **8**, 2275.
[13] D. S. Clark, *et al.*, this proceeding.

Journal of Physics: Conference Series **717** (2016) 012024 doi:10.1088/1742-6596/717/1/012024

Picosecond-petawatt laser-block ignition for avalanche fusion of boron by ultrahigh acceleration and ultrahigh magnetic fields

H Hora[*1], P Lalousis[2], L Giuffrida[3], D Margarone[3], G Korn[3], S Eliezer[4,5], G H Miley[6], S Moustaizis[7], G Mourou[8] and C P J Barty[9]

[1]Department of Theoretical Physics, University of New South Wales, Sydney 2052, Australia
[2]Institute of Electronic Structure and Lasers FORTH, Heraklion, Greece
[3]ELI-Beamline Project, Inst. Physics, ASCR, PALS Center, Prague, Czech Republic
[4]Institute of Nuclear Fusion, Polytechnic University of Madrid, Madrid, Spain
[5]Soreq Research Center, Yavne, Israel
[6]Deptartment of Nucluclear Plasma & Radiological Engineering University of Illinois, Urbana IL, USA,
[7] Technical University Crete, Laboratory of Matter Structure and Laser Physics, Chania, Greece
[8] DAER-IZEST, Ecole Polytechnique, Route de Saclay, 91128 Palaiseau Cedex, France
[9]Lawrence Livermore National Laboratory, Livermore, CA, USA
*h.hora@unsw.edu.au

Abstract. Fusion energy from reacting hydrogen (protons) with the boron isotope 11 (HB11) resulting in three stable helium nuclei, is without problem of nuclear radiation in contrast to DT fusion. But the HB11 reaction driven by nanosecond laser pulses with thermal compression and ignition by lasers is extremely difficult. This changed radically when irradiation with picosecond laser pulses produces a non-thermal plasma block ignition with ultrahigh acceleration. This uses the nonlinear (ponderomotive) force to surprizingly resulting in same thresholds as DT fusion even under pessimistic assumption of binary reactions. After evaluation of reactions trapped cylindrically by kilotesla magnetic fields and using the measured highly increased HB11 fusion gains for the proof of an avalanche of the three alphas in secondary reactions, possibilities for an absolutely clean energy source at comptitive costs were concluded.

1. Introduction

Energy generation from laser driven fusion of deuterium D with tritium T (DT) arrived at highest fusion gains at the NIF project [1][2] by using indirect drive using up to 2 MJ laser pulses of about nanosecond duration where the thermal processes with ablation-compression and spark ignition are involved. A similar option with direct drive and volume ignition is being studied [3] for thermal driven laser-fusion with nanosecond laser pulses.

This is in basic contrast with using picosecond (ps) laser pulses with needing energy fluxes E* of about 10^8J/cm^2 for igniting uncompressed solid density DT. Initiated by Chu [4], a non-thermal energy transfer from laser energy into plasma blocks is used to avoid complicate thermal determined mechanisms of the nanosecond interaction. This is possible now through the Chirped Pulse Amplification CPA [5] of laser pulses of ps and shorter duration with powers above petawatt (PW). The ultrahigh acceleration of the plasma blocks was theoretically-numerically predicted in 1978 and measured [6] in agreement with the theory [7].

The advantage of the ps-block-ignition arrived at the surprising result [8][9], that the environmentally clean fusion of protons with the boron isotope 11 (HB11) is possible at similar thresholds as DT, while thermal ignition was considered as impossible with needing compression of HB11 to more than 100000 times solid state density. Next steps of these developments focus on the combination of these results with cylindrical trapping of the fusion reaction with ultrahigh magnetic fields of about 10 kilotesla and by using the measured highly increased gains of HB11 fusion as an avalanche ignition process.

2. Block ignition by picosecond laser pulses

The drastic difference between the interactions of laser pulses of ps and ns duration is given by the force density **f** in the plasma being not only determined by the gas dynamic pressure p but also by the force $\mathbf{f}_{NL}$ due to electric **E** and magnetic **B** laser fields of frequency ω,

$$\mathbf{f} = -\nabla p + \mathbf{f}_{NL} \quad (1)$$

where the force $\mathbf{f}_{NL}$ is given by Maxwell's stress tensor as Lorentz and gauge invariant nonlinear force

$$\mathbf{f}_{NL} = \nabla\bullet[\mathbf{EE} + \mathbf{HH} - 0.5(\mathbf{E}^2 + \mathbf{H}^2)\mathbf{1} + (1+(\partial/\partial t)/\omega)(\mathbf{n}^2-1)\mathbf{EE}]/(4\pi) - (\partial/\partial t)\mathbf{E}\times\mathbf{H}/(4\pi c) \quad (2)$$

where $\mathbf{1}$ is the unity tensor and **n** is the complex optical constant of the plasma given by the plasma frequency ω_p. At plane laser wave interaction with a plane plasma front, the nonlinear force reduces to

$$\mathbf{f}_{NL} = -(\partial/\partial x)(\mathbf{E}^2+\mathbf{H}^2)/(8\pi) = -(\omega_p/\omega)^2(\partial/\partial x)(E_v^2/\mathbf{n})/(16\pi) \quad (3)$$

showing how the force density is given by the negative gradient of the electromagnetic laser-field energy-density including the magnetic laser field from Maxwell's equations. E_v is the amplitude of the electric laser field in vacuum after time averaging. The second expression in Eq. (3) is Kelvin's formulation of the ponderomotive force in electrostatics of 1845 [10].

The difference to laser interaction by ns thermal interaction against ps non-thermal nonlinear force driving is determined by $\mathbf{f}_{NL}$ interaction dominating in Eq. (1). For the ns interaction, the first term dominates at low laser intensities while with ps, the second term dominates in which case the quiver energy of the electrons of the laser field has to be higher than their thermal energy of motion. A numerical example about nonlinear force acceleration of a slab of deuterium plasma irradiated by a neodymium glass laser pulse is shown in Figure 1. During the 1.5 ps, the plasma reached velocities above 10^9cm/s by the ultrahigh acceleration above 10^{20}cm/s^2. The generation of the plasma blocks, one moving against the laser light and the other into the target is *a nonlinear force driven dielectric explosion as a non-thermal not collision-determined absorption* and should not be understood as radiation pressure acceleration.

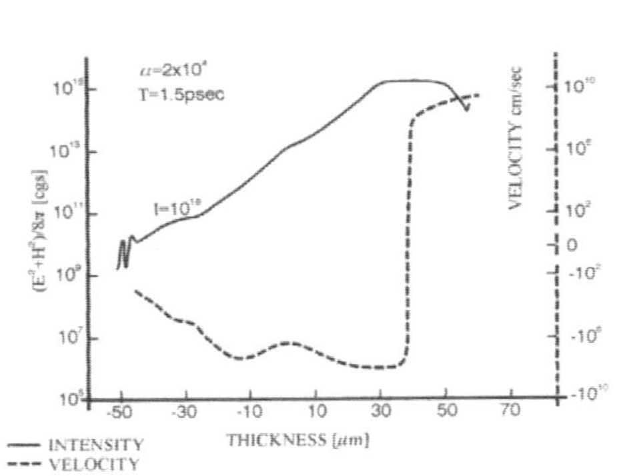

Fig. 1. 10^{18} W/cm^2 neodymium glass laser incident from the right hand side on an initially 100 eV hot deuterium plasma slab whose initial density has a very low reflecting bi-Rayleigh profile, resulting in a laser energy density and a velocity distribution from plasma hydrodynamic computations at time t=1.5 ps of interaction. The driving nonlinear force is the negative of the energy density gradient of the laser field ($\mathbf{E}^2$+$\mathbf{H}^2$)/8π. The dynamic development of temperature and density had accelerated the plasma block of about 15 vacuum wave length thickness of the dielectric enlarged skin layer moving against the laser (positive velocity) and another block into the plasma (negative velocity) showing ultrahigh >10^{20}cm/s^2 acceleration ([10]: figures 4.14 & 4.15) as computer result of 1978.

The experimental proof of the ultrahigh acceleration was possible [6] in full agreement with the results of computations in 1978, after laser pulses of higher than terawatt (TW) power and about ps duration were available only thanks to the Chirped Pulse Amplification CPA [5]. With these ps ultrahigh accelerations the plasma block ignition of solid density DT by the nonlinear force was possible and updated [8]. Computation of DT and HB11 fusion using the ps-block ignition [8] showed many details of the generated fusion flames with velocities of few 1000 km/s, the delayed generation of a Rankine-Hugoniot shock fronts, local distribution of reaction rates etc. [11], however, in one dimension plane wave computations. Using spherical laser irradiation needed laser powers in the exawatt range for gains of few hundred.

3. Secondary avalanche reactions

The results reported up to this stage were based on calculations of binary reactions as it is the case for DT. These binay reactions for protons with ^{11}B used in the computations, however, were a rather pessimistic assumption. The three 2.9 MeV alphas from an initial binary reaction can by elastic collisions transfer energy to boron nuclei or protons such that they secondarily produce each three further alphas with avalanching to a high energy gain. Computations were performed [13] confirming the secondary reactions.

Before considering the elastic collisions in view of stoping power and related processes it was fortunate that experiments became known [12] from which the avalanche process was evident [13]. Following the groundbreaking measurements of HB11 reactions of more than 10^6 under rather complicate experimental conditions [14], the transparent conditions of the measurement of 10^9 reactions per steradian [12] by the PALS iodine laser based on few hundred Joules laser pulses in the range of 100 ps duration were so strongly elevated even above comparable results with DT [13] that this could only be due to the secondary reactions with the avalanche.

The avalanche HB11 reaction can be explained in quantitative agreement with the experiment [12] numerically by evaluating an extremely non-equilibrium state of the plasma by elastic collisions of the alpha particles with the boron nuclei and protons where the broad energy range around 600keV fits with the most exceptionally increased HB11 fusion cross section [15].

Based on this clarification for the avalanche secondary processes, the following cylindrical trapping of the HB11 reaction with solid density fuel could be used [16] where the ultrahigh magnetic fields in the range of 10 kilotesla [17] were applied, Figure 2.

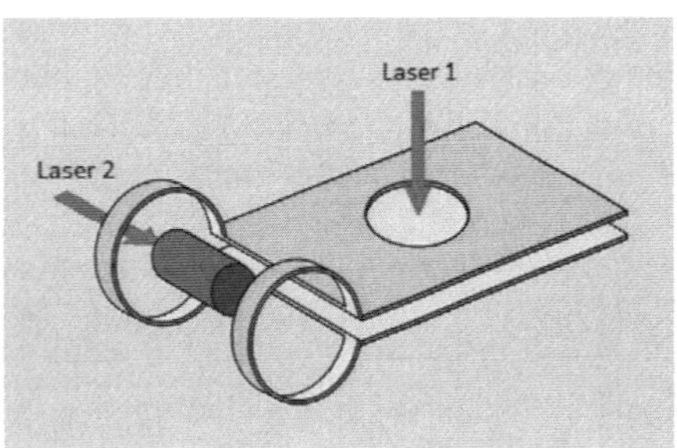

Figure 2. Generation of a 10 kiloTesla magnetic field of about 2ns duration in the coils by firing a >kilojoule nanosecond laser pulse 1 into the hole between the plates. The HB11 fusion fuel is a solid cylinder of 1cm length and 1 mm radius coaxially located in the coils and the block-ignition of the fusion flame is produced by a ps-30 PW laser pulse of 0.2 mm diameter for block ignition from laser 2.

Based on the computation [16] of block ignition by a ps laser pulse 2 in Figure 2 of 10^{20}W/cm^2 intensity on a solid HB11 cylinder of 0.2mm diameter within a 10 kilotesla magnetic field shows a slow radial expansion against the trapping and an axial propagation of the reaction with several thousands of km/s velocity during about a ns. With avalanche reaction, the cylinder of Fig. 2 is nearly completely reacting producing more than one gigajoule energy of alpha particles by the irradiated 30 kJ laser pulse 2.

4. Estimations about preliminary parameters for HB11 fusion generation.

Following the results [16] for a reaction scheme of Fig. 2, a ps laser pulse of 30 kilojoule energy (30 PW power) should produce more than a GJ energy in clean alpha particles. A technology path capable of producing 200-PW, 0.1-ps pulses based on NIF-like laser technology has recently been defined and could be extended to meet the needs of GJ alpha production [18].Figure 3 describes the scheme of a fusion reactor for converting the energy of the alpha particles into an electrostatic pulse in the range close to 1.4 megavolts using the technology of high voltage direct current HVDC power transmission electricity.

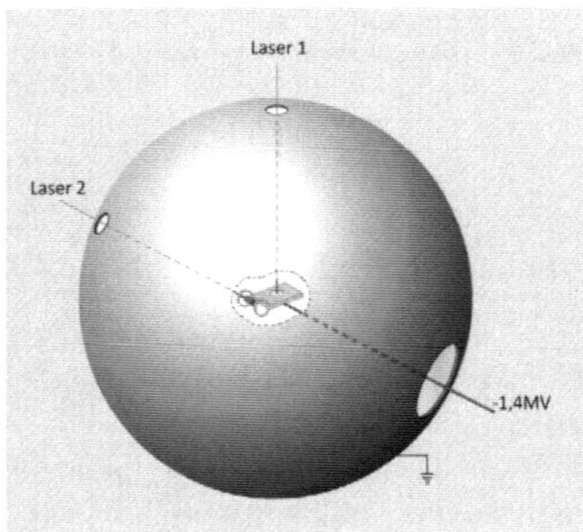

Figure 3. Scheme of a HB11 fusion reactor without any radioactive radiativon problems is based on non-thermal plasma block ignition by nonlinear forces using a 30kJ-picosecond laser pulse 2. The central reaction unit (Figure 2) located in the center of the reactor sphere is electric charged to the level of -1.4 million volts against the wall of the sphere such that the alpha particles (helium nuclei) produce more than a gigajoule energy, of which a small part is needed for the operation of the laser pulses. One part of the gained costs of electricity is needed for the reaction unit being destroyed and the HB11 fuel at each reaction [13][15].

It was estimated [13] for a power station working with one reaction per second, that after deducting the costs for investment and operation, the energy of the value of up to $300Million per year may be produced for the grid.

References:

[1]Hurricane, O.A, D. A. Callahan,' D. T. Casey, et al. 2014 Nature 506, 343
[2]Hinkel, D. E., Herzack-Hopkind, L.F. et al. IFSA Seattle 2015 Conf. Abstracts p. 163
[3]Hora. H. Extraodrinary jump of increasing laser fusion gains experienced at volume ignition for combination with NIF experiments. Laser and Particle Beams **31** 229-232 (2013)
[4]Chu, M.S. 1972 Thermonuclear reactions waves at high densities. Phys. Fluids 15, 412.
[5]Mourou, G., C.P.J. Barty & M.D. Perry. 1998 Ultrahigh-Intensity Lasers. Physics Today 51 No.1, 22.
[6]Sauerbrey, R. (1996) Acceleration of femtosecond laser produced plasmas. Physics of Plasmas 3, 4712-4716.
[7]Hora, H., J. Badziak, M.N. Read, Y.-T. Li, T.-J. Liang, H. Liu, Z.-M. Sheng, J. Zhang, F. Osman, G.H. Miley, W. Zhang, X. He, H. Peng, F. Osman, S. Glowacz, S. Jablonski, S. Wolowski, Z. Skladanowsli, K. Jungwirth, K. Rohlena & J. Ullschmied 2007 Phys. Plasmas 14, 072701
[8]Hora, H., Laser fusion with nonlinear force driven plasma blocks. 2009 *Laser and Particle Beams* **27**, 207-222
[9]Hora, H., G.H. Miley, M. Ghorannviss, H. Malekynia, N. Azizi & X-T. He 2010 Fusion energy without radioactivity: laser ignition of solid hydrogen-boron(11) fuel. Energy and Environment Science 3, 479-486.
[10]Hora, H., *Laser Plasma Physics. Forces and the Nonlinearity Princile* SPIE Book, Bellingham WA, 2000
[11]Lalousis, P., H. Hora, S. Eliezer, J.-M. Martinez-Val, Moustaizis, G.H. Miley & G. Mourou 2013 Phys. Letters A377, 885
[12]Picciotto, A., D. Margarone, A. Velyhan, P. Bellini, J. Krasa, A. Szydlowski, G. Bertuccio, Y. Shi, A. Margarone, J. Prokupek, A. Malinowska, E. Krouski, J. Ullschmied, L. Laska, M. Kucharik, and G. Korn. 2014 Boron-Proton Nuclear-Fusion Enhancement Induced in Boron-Doped Silicon Targets by Low-Contrast Pulsed Laser. Phys. Rev. X 4, 031030.
[13]Hora, H., G. Korn, L. Giuffrida, D. Margarone, A. Picciotto, J.Krasa, K. Jungwirth, J. Ullschmied, P. Lalousis, S. Eliezer, G.H. Miley, S. Moustaizis and G. Mourou, 2015 Fusion energy using avalanche increased boron reactions for block ignition by ultrahigh power picosecond laser pulses. Laser and particle Beams. 33, No. 4/2015, "First View" 14 July 2015. Cambridge University Press DOI: 10.1017/S026.03461.5000634
[14]Labaune, C., S. Deprierraux, S. Goyon, C. Loisel, G. Yahia & J. Rafelski. 2013 Fusion reactions initiated by laser accelerated particle beams in laser produced plasmas. Nature Communications 4, 2506
[15]Eliezer S., H. Hora, G. Korn, N. Nissim and J.M. Martinez-Val 2015 Nonequilibrium elastic nuclear collision model for avalanche generation at boron fusion, Phys. Lett. A submitted
[16]Lalousis, P., Hora and S. Moustaizis, Optimized boron fusion with magnetic trapping by laser driven shock initiation at nonlinear force driven ultrahigh acceleration. Laser and Particle Beams 32, 409-411 (2014).
[17]Fujioka S., Zhe Z., et al. (2013) Scientif. Rep. 3, 1170-1176.
[18]Barty, C.P.J., 2015, The Nexawatt: Strategies for Exawatt peak power laser based on NIF and NIF-like beam lines. IFSA Seattle Conf. 2015, Book of Abstracts, p. 459.

Journal of Physics: Conference Series **717** (2016) 012025 doi:10.1088/1742-6596/717/1/012025

Electron beam guiding by external magnetic fields in imploded fuel plasma

T Johzaki[1], Y Sentoku[2], H Nagatomo[3], A Sunahara[4], H Sakagami[5], S Fujioka[3], H Shiraga[3], T Endo[1] and FIREX project group

[1] Graduate School of Engineering, Hiroshima University, Higashi-Hiroshima, Hiroshima 739-8527, Japan
[2] Department of Physics, University of Nevada Reno, Reno, NV 89557, USA
[3] Institute of Laser Engineering, Osaka University, Suita, Osaka 565-0871, Japan
[4] Institute for Laser Technology, Suita, Osaka 565-0871, Japan
[5] National Institute for Fusion Science, Toki, Gifu 509-5292, Japan

E-mail: tjohzaki@hiroshima-u.ac.jp

Abstract. For enhancing the core heating efficiency in fast ignition laser fusion, we proposed the fast electron beam by externally-applied the kilo-tesla (kT) class longitudinal magnetic field. We evaluated the imploded core and the magnetic field profiles formed through the implosion dynamics by resistive MHD radiation hydro code. Using those profiles, the guiding effect was evaluated by fast electron transport simulations, which shows that in addition to the feasible field configuration (moderate mirror ratio), the kT-class magnetic field is required at the fast electron generation point. In this case, the significant enhancement in heating efficiency is expected.

1. Introduction

One of the most crucial issues of electron-driven fast ignition is efficient core heating by laser produced relativistic electron beam (REB). Two factors, which mainly degrade the efficient heating, are (1) too high electron energy, and (2) too large beam divergence. For improving the efficiency in FIREX project [1], we proposed the REB guiding by externally applied kilo-tesla (kT) -class longitudinal magnetic fields generated by a capacitor-coil target [2] and evaluated its effects by 2D PIC simulations[3,4]. It was demonstrated that the REB with a large divergence angle generated by a relativistic laser plasma interaction with a laser of intensity 3×10^{19} W/cm^2 and duration 1 ps can successfully focused by the moderately-converging fields (mirror ratio $R_M \leq 20$) [4]. In addition, formation of a magnetic-pipe like structure due to the resistive effects was shown, which indicates a possibility of the beam focusing even under the higher mirror fields that may be achieved in the ignition-scale or high-gain target implosion. However, in the implosion simulation [5] for a cone-attached CD shell target with externally-applied B-field by a one-turn coil, the ratio of B-field strength at the cone tip to that in the region between cone tip and dense core reaches $R_M > 100$ at the maximum compression. In such a case, the fast electrons could not pass the very high R_M region and could not reach the core, so that the core heating would be very inefficient. Hence it is critical to check the electron transport in more realistic configuration formed through the compression dynamics.

In the present paper, we discuss the core heating properties for the dense core with the converging magnetic field configuration formed through the implosion of cone-attached CD spherical solid target on the basis of integrated simulations [6, 7]. In the following, we present the integrated simulation results for core heating by REB with large beam divergence under the compressed core and magnetic fields.

2. Simulation model

To obtain the imploded core profile and magnetic field configuration, the implosion simulations were carried out for a Au cone attached CD spherical solid target using a 2D radiation-hydro code PINOCO [8], which is designed for single-fluid two-temperature hydrodynamics, thermal transport, radiation transport equations, equation of state and laser absorption, is extended to make a resistive magnetohydrodynamic code [5]. The implosion simulation conditions are based on the feasible conditions for the GEKKO-XII (GXII) laser system at ILE Osaka University. The total energy and the wavelength of the laser are 1.5 kJ on target and 0.53 μm, respectively. The pulse shape is Gaussian with 1.3 ns (FWHM) and the peak timing is 1.5 ns.

Using the obtained core and magnetic-field profiles, the core heating by the REB was simulated by FIBMET [9], where the transport of REB is solved by particle scheme instead of Fokker-Planck equation. In the present study, we did not consider the fast electron generation process. The REB was injected at the tip of cone. We used the experimentally observed energy spectrum [10], where $f(E) \propto 0.95\exp(-E/0.9[\text{MeV}]) + 0.05\exp(-E/5\ [\text{MeV}])$ (E is electron energy in MeV). The beam divergence angle of 45 degree half angle was assumed. The Gaussian profile with 1 ps duration (FWHM) and the Super Gaussian with 20 μm radius (HWHM) were assumed for temporal and spatial profiles. The peak REB intensity was assumed as $1.55\text{x}10^{19}\text{W/cm}^2$. The injected REB energy was 410 J.

3. Simulation results

3.1. *Implosion of cone-attached CD solid spherical target*

The implosion simulation was carried out for a cone-attached CD spherical solid target (Fig.1). In the simulation, the uniform magnetic field with the strength of 0.2kT was initially applied in z-direction. The obtained density and magnetic field profiles at the maximum compression are shown in Fig.2. In the conventional shell implosion simulation for GXII, the core density reached $\rho_{\text{core}} \sim 100$ g/cm^3 [5]. Contrary to this, in the present simulation for the solid target, the density is lower ($\rho_{\text{core}} \sim 30$ g/cm^3). The obtained fuel ρR at the maximum compression is 0.1 g/cm^2 in z-direction and 0.06 g/cm^2 in r-direction, which is comparable to that obtained in the shell implosion. The achievement of comparable ρR despite the lower density is due to the larger mass of the solid target than that for the shell target.

As for the magnetic field profile, the magnetic-field lines are smooth and the mirror ratio (the ratio of magnetic field strength at the cone tip to that in the compressed core) is $R_M \sim 4$. Such a moderate mirror configuration is feasible for REB guiding and focusing [4]. Such a moderate compression of magnetic field and mirror ratio is due to the lower compression of the fuel target and the high magnetic duffusivity in the cold CD solid region.

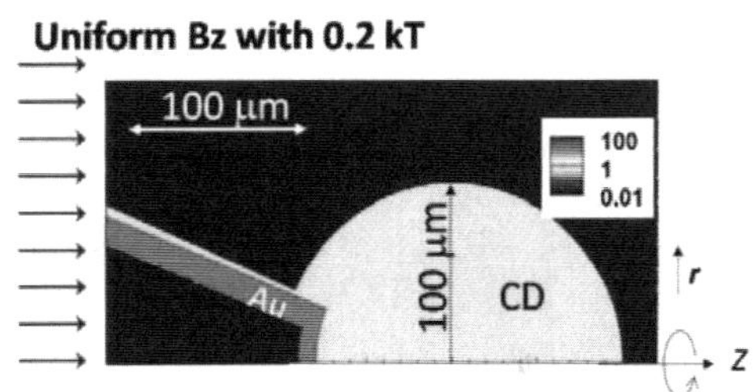

Figure 1. Initial target configuration for implosion simulation.

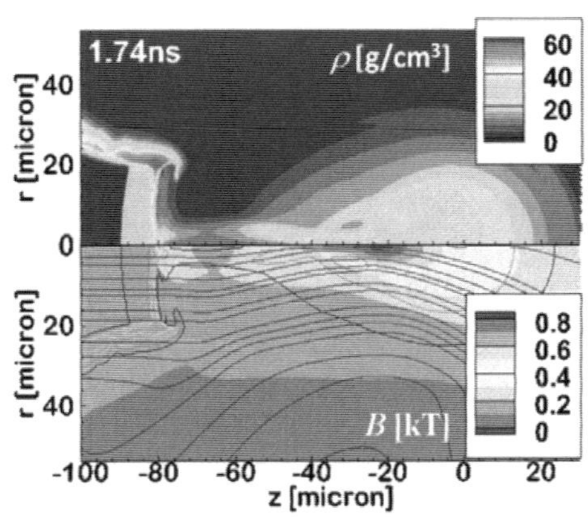

Figure 2. Spatial profiles of density (upper) and magnetic field strength (lower) at the maximum compression (t = 1.74 ns). The lines in the lower figure show the magnetic field lines.

3.2. Core heating of imploded core by fast electron beam

Using the fuel and magnetic field profiles at the maximum compression, we carried out the core heating simulations. The REB having above noted profiles is injected in the cone tip. To evaluate the dependence of magnetic field strength, the magnetic field strength shown in Fig.2 was multiplied by 0, 1, 2.5, 5, 10. Each multiplying factor corresponds to the magnetic field-strength at the injection point 0, 0.2, 0.5, 1.0, 2.0 kT, respectively.

Fig.3 shows the fast electron energy density profiles at the peak REB intensity. If the external field is neglected (upper figure), the REB is diverged due to the large beam divergence though the part of the fast electrons are focused at the central axis at z = 40 μm by the self-generated magnetic fields. Since the magnetic field strength of B_{z0} = 0.2 kT at the REB injection region is not strong enough to trap the MeV-class fast electrons, the REB is also diverged in this case (middle figure). When B_{z0} = 2 kT (lower figure), most of the fast electrons are trapped by the magnetic fields and moves along the magnetic field lines, then the REB is focused at the central axis at z = 60 μm. In Figure 4, the energy deposited by the fast electrons in the dense core region (ρ > 5 g/cm^3) is plotted as a function of B_{z0}. The energy coupling efficiency from the heating laser to the core is shown in the right axis, where the experimentally measured conversion efficiency from the laser energy to the REB energy, 31%, is used. The heating efficiency is monotonically increases in the sub-kT region, and the enhancement is saturated in the kT-class region. The heating efficiency for B_{z0} = 2 kT becomes about twice larger than that for the case without external field. It is found then that for effective guiding by external magnetic field, kT-class magnetic field is required at the injection point. This is because the MeV-class fast electrons should be confined within r = several ~ 10's μm. Otherwise, the fast electron escape from the strong magnetic field region (see Fig.2). This requirement is consist with the previous PIC simulations [3, 4].

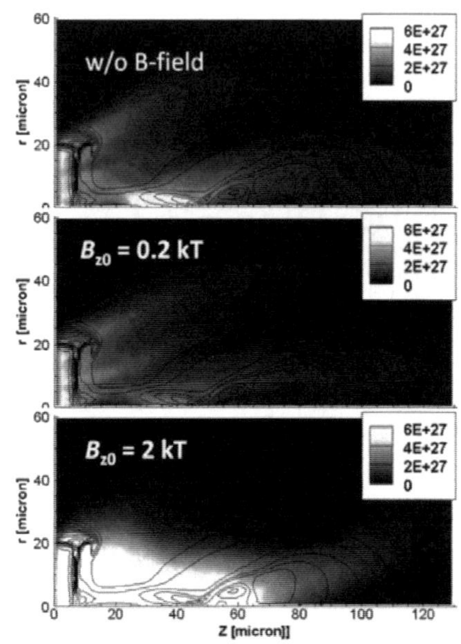

Figure 3. Spatial profiles of fast electron energy density [erg/cm^3] at the peak beam intensity for the cases 0 kT (upper), 0.2 kT (middle) and 2.0 kT (lower) at the injection point.

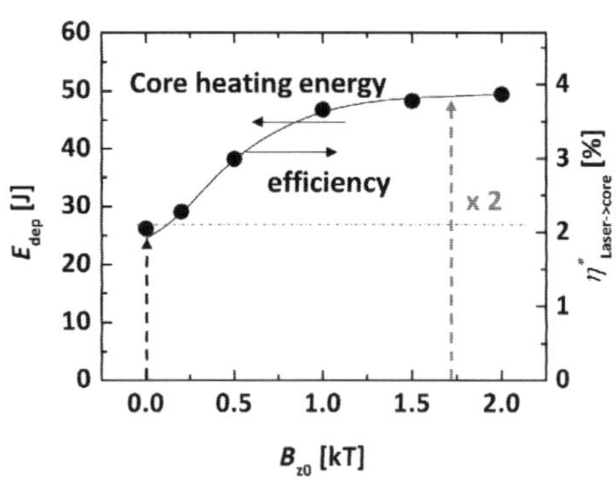

Figure 4. Core heating energy by REB as a function of magnetic field strength at the injection point B_{z0}. The right axis indicates the energy coupling efficiency from the heating laser to the fuel core, where the experimentally measured energy conversion ratio from the laser to REB, 31%, is assumed.

The direct measurement of energy coupling efficiency is difficult in the experiments. Instead, the neutron yield $Y_{n,dd}$ and neutron weighted ion temperature $\langle T_i \rangle_{n,dd}$ will be measured. So, the dependences of $Y_{n,dd}$ and $\langle T_i \rangle_{n,dd}$ on B_{z0} are evaluated and plotted in Fig.5. For the case without REB heating, $Y_{n,dd}$ is less than 10^5 and $\langle T_i \rangle_{n,dd}$ is about 0.5 keV. When the REB is injected without the external magnetic field or with sub-kT external field, $Y_{n,dd}$ and $\langle T_i \rangle_{n,dd}$ are increased slightly. However, such small enhancement cannot be observed experimentally due to the existence of large noise. If the kT-class field is applied, the heating effect becomes pronounced and then the clear enhancements in $Y_{n,dd}$ and $<T_i>_{,n,dd}$ can be observed. For B_{z0} = 2 kT, $Y_{n,dd}$ is 10^7, which is larger than that for the case without external field by two orders of magnitude, and $<T_i>_{n,dd}$ = 1.06 keV, which means 0.5 keV enhancement from the case without REB heating. These clear enhancements in $Y_{n,dd}$ and $<T_i>_{n,dd}$ could be measured in the experiments.

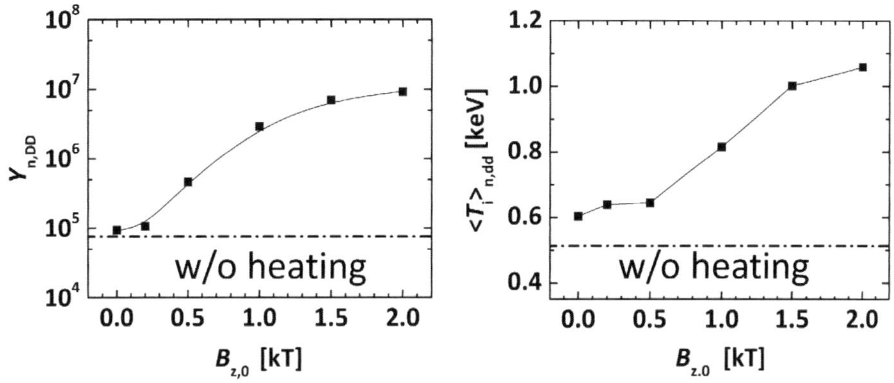

Figure 5. Neutron yield $Y_{n,DD}$ (left) and neutron-weighted ion temperature $<T_i>_{n,dd}$ as a f unction of magnetic field strength at the injection point B_{z0}

4. Conclusion

The integrated simulation for evaluation of fast electron guiding effect by external magnetic field showed that for the cone-attached CD solid spherical target case, the compression of magnetic field by implosion is not so high, then the feasible mirror ratio (R_M ~ 4) is obtained. From the transport simulations for REB using the imploded core and magnetic field profiles obtained from implosion simulation, it is found that not only the magnetic field configuration (mirror ratio), the strength of magnetic field at the fast electron generation region is important. The kT-class magnetic field is required around the fast electron generation point to confine them in the strong magnetic field region.

Acknowledgments

This work is partially conducted under the joint research projects of the ILE, Osaka University (FIREX-project), and with the supports of the NIFS Collaboration Research program (NIFS12KUGK057 and NIFS15KUGK093) and JSPS KAKENHI Grant Number 25400534, 25400539 and 26400532.

References

[1] Azechi H 2006 *Plasma Phys.Cont. Fusion* **48** B267-B275.
[2] Fujioka S, Zhang Z, Ishihara K, Shigemori K, Hironaka Y, Johzaki T, Sunahara A, Yamamoto N, Nakashima H, Watanabe H, Shiraga H, Nishimura H and Azechi H 2013 *Sci. Rep.* **3** 1170.
[3] Johzaki T, Mima K, Fujioka S, Sakagami H, Sunahara A, Nagatomo H and Shiraga H "Electron beam guiding by strong longitudinal magnetic fields" to be published in 2016 *J. Phys.: Conf. Seri.*
[4] Johzaki T, Taguchi T, Sentoku Y, Sunahara A, Nagatomo H, Sakagami H, Mima K, Fujioka S and Shiraga H 2005 *Nucl. Fusion* 2005 **55** 053022.
[5] Nagatomo H, Johzaki T, Asahina T, Sunahara A, Sano T, Sakagami H, Mima K, Fujioka S, Shiraga H and Azechi H 2015 *Nucl. Fusion*, 55 093028.
[6] Sakagami H, Johzaki T, Nagatomo H and Mima K 2006 *Laser Part. Beams* **24** 191-198.
[7] Johzaki T, Sakagami H, Nagatomo H and Mima K 2007 *Laser Part. Beams* **25** 621-629.
[8] Nagatomo H *et al* 2007 *Phys. Plasmas* **14** 056303.
[9] Johzaki T, Nakao Y and Mima K 2009 *Phys. Plasmas* **16** 062706.
[10] Arikawa Y et al., "Optimization of hit electron spectra by using plasma mirror for fast ignition" presented at IFSA2015, Seattle NE USA, Sept. 2015.

Towards a more universal understanding of radiation drive in gas-filled hohlraums

O S Jones[1], C A Thomas[1], P A Amendt[1], G N Hall[1], N Izumi[1], M A Barrios Garcia[1], L F Berzak Hopkins[1], H Chen[1], E L Dewald[1], D E Hinkel[1], A L Kritcher[1], M M Marinak[1], N B Meezan[1], J L Milovich[1], J D Moody[1], A S Moore[1], M V Patel[1], J E Ralph[1], S P Regan[2], M D Rosen[1], M B Schneider[1], S M Sepke[1], DJ Strozzi[1] and D P Turnbull[1]

[1]Lawrence Livermore National Laboratory, Livermore, CA, USA
[2]Laboratory for Laser Energetics, Rochester, NY, USA

oggie@llnl.gov

Abstract. We have found that radiation-hydrodynamic calculations that use the high flux model assumptions [1] can accurately predict the radiation drive produced by a laser-heated hohlraum under certain conditions, but can not predict drive over a broad range of parameters (pulse energy, hohlraum gas fill density, hohlraum case-to-capsule ratio). In particular, the model is accurate for ~7 ns long laser pulses used to implode capsules with high density carbon (HDC) ablators in hohlraums with helium fill gas densities of 0-0.6 mg/cc. By systematically varying the gas fill density from 0 to 1.6 mg/cc we found that the agreement with drive begins to diverge for fills > 0.85 mg/cc. This divergence from the model coincides with the onset of measureable SRS backscatter. In this same set of experiments the radiation drive symmetry inferred from the imploded shape of a gas-filled capsule is not predicted with this model. Finally, several possible fixes to the model to reduce the observed discrepancies are considered.

1. Introduction

To correctly model x-ray production and symmetry from a laser-heat hohlraum a number of physical processes must be included. These processes include laser absorption via inverse bremsstrahlung and heating of the plasma, thermal transport from the coronal plasma to the denser, cooler wall plasma, conversion of thermal energy to x-rays, expansion and dynamics of the ablated wall and ablator plasmas, the dynamics closure of the laser entrance hole, and laser plasma interactions, including backscatter and cross beam energy transfer (CBET).

We have developed a standard model in the radiation-hydrodynamics code Hydra [2] for calculating laser-driven hohlraum experiments. For an axisymmetric calculation we use about 34,000 zones, 85 photon energy groups, and 5 million IMC photonics particles. There is 2.5° polar zoning around the capsule, 60 radial zones in the DT (for ignition capsules), 220 radial zones in the ablator, and 70 radial zones in the hohlraum wall. In the wall zones STA tabular (LTE) wall opacities are used when the zone temperature is less than 300 eV. For wall zone temperatures greater than or equal to 300 eV the DCA model is used for NLTE opacity and equation of state. We use flux limited thermal transport with a limiter of 0.15.

When this model is applied to CH ablator implosions with ~15 ns ("high foot") or ~20 ns ("low foot") laser pulses and helium hohlraum fill densities ranging from 0.96 to 1.6 mg/cc it typically predicts about 30% too much radiation flux compared to the "measured" flux, which is usually inferred by matching the capsule bang time [3]. Experiments using open-ended "viewfactor" hohlraums confirmed that the discrepancy in bang time was primarily due to over-estimating the radiation flux compared to that measured through the open end [4].

2. Experimental results for 2-shock and 3-shock HDC ablator implosions

2.1. Results for hohlraum gas fill densities up to 0.6 mg/cc He

In contrast to the results for CH ablator implosions, when the model described above is applied to HDC ablator implosions with hohlraum gas fills up to 0.6 mg/cc it has been able to predict the total radiation flux measured by the Dante x-ray diode array to within the measurement errors, and has required about a 10% reduction in the absorbed peak power in calculations to match the observed bang times. The HDC radiation drives are shorter than those for CH capsules due to the higher density of the HDC. These experiments also have had low (< 3%) amounts of backscatter [5]. Figure 1 shows a comparison of the measured and calculated Dante flux out of the open end of a gold viewfactor vacuum (no gas fill) hohlraum (NIF experiment N150120-003-999). This experiment was done in a 5.75 mm diameter hohlraum, but with a 3-shock pulse shape that was scaled down by a factor of 0.8. That is, the scaled experiment is designed to produce the same radiation drive that the full-scale pulse would produce in a 6.72 mm diameter hohlraum. We see that the total calculated radiant intensity in GW/sr is within the experimental error bars, as is the gold "M-band" component, which is the radiant intensity for photon energies greater than 1.8 keV. Figure 2 shows similar agreement for a full-scale 6.72 hohlraum with a 0.6 mg/cc helium gas fill (NIF experiment N140701-002-999).

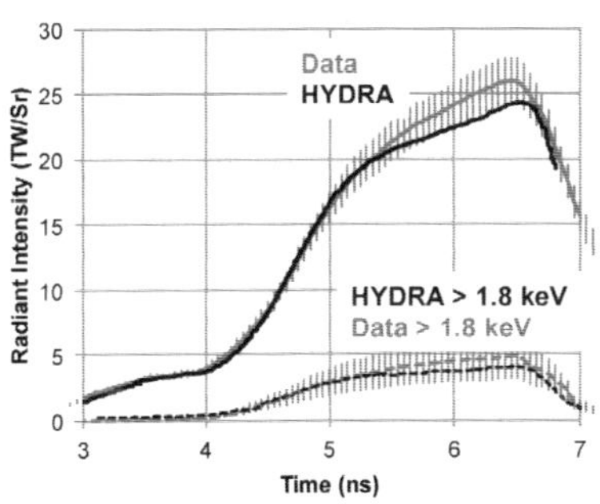

Figure 1. Measured and calculated Dante radiant intensity from the open end of a 5.75 mm diameter vacuum gold viewfactor hohlraum.

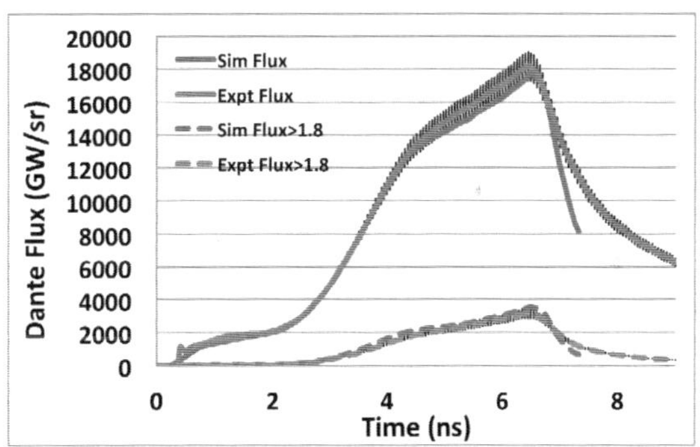

Figure 2. Measured and calculated Dante radiant intensity from the LEH of a 6.72 mm diameter gold hohlraum with 0.6 mg/cc He fill.

2.2. Hohlraum gas fill density scan experiments

While the agreement with our standard model is good for the two cases shown above, it is not a good drive predictor for all HDC ablator experiments. A number of HDC experiments were done with a 3-shock laser pulse, $\Delta\lambda$=0 A, and a hohlraum He fill density of 1.6 mg/cc (e.g. shot N140722-001-999). These were done in 5.75 mm diameter hohlraums with full-scale HDC capsules. This experiment series was more directly comparable to the CH ablator "high foot" series, which had the same hohlraum gas density (1.6 mg/cc). The relationship between measured and calculated drive was similar to what was observed for the CH ablator experiments. They had ~15% backscatter losses and required a further reduction of ~30% in absorbed power in the calculations to match bang time.

To investigate the drive discrepancy more systematically, we undertook a series of experiments in which we varied the hohlraum gas fill density only. These experiments were done in a 5.75 mm diameter hohlraum driven by a 2-shock laser pulse. The capsules were HDC shells filled with a mixture of 70% (atomic) D and 30% ^{3}He. The drive was measured using the Dante x-ray diode and also inferred from the bang time. Figure 3 is a plot of the multiplier applied to the peak of the laser pulse in the calculation in order to match the measured bang time. We see that for fill density less than or equal to 0.85 mg/cc, the required peak drive multiplier is > 0.9. For the two highest fill densities the required multiplier is significantly lower. Since the bang time is measured to about 50 ps accuracy, which corresponds to about 1% variation in the peak flux, the observed drive deviation from the model is well outside of the experimental error.

We also measured and calculated the shape of the imploded capsule at the time of peak x-ray brightness. The calculations have backscatter removed and include the calculated CBET without any ad hoc adjustments. The amount of calculated instantaneous power transfer is generally < 10%, and the total energy transfer is < 3%. Figure 4 shows that in the calculations the P2/P0 Legendre moment of the shape is expected to tend toward a more sausage shape as the gas fill is increased, whereas the data shows the opposite trend.

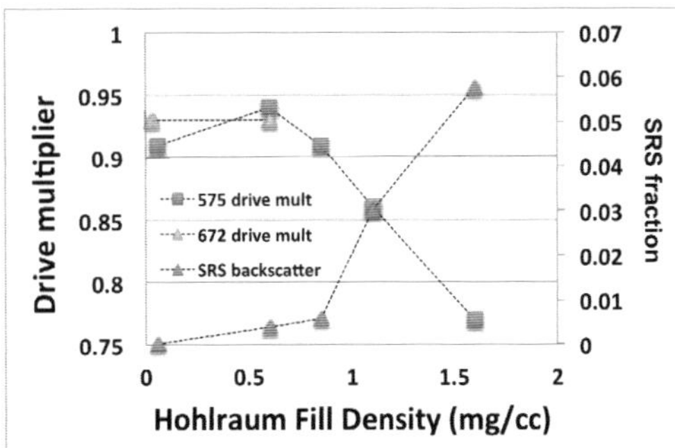

Figure 3. Peak drive multiplier required to make calculation match measured bang time (left) and SRS fraction (right) as a function of hohlraum fill density.

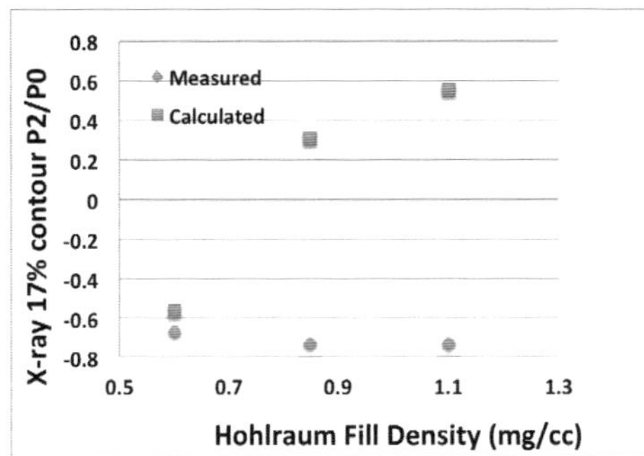

Figure 4. Measured and calculated x-ray P2/P0 as a function of hohlraum fill density.

3. Exploring potential avenues for improving hohlraum modelling capability

In this section we investigate some possible changes in the model that could lead to improved agreement at higher gas fill densities.

3.1. Sensitivity to DCA NLTE transition temperature

In the current model the wall zones transition at 300 eV from using tabular LTE opacities and tabular equations of state to using the DCA NLTE model for both. The reason for this choice is that at the peak of the ~300 eV radiation drive, the wall zones that are heated by x-rays should be in LTE and tabular LTE opacities should be more accurate than LTE opacities calculated using a simpler DCA having fewer transition levels. Wall zones that are above 300 eV are presumed to be heated directly by the laser and thus are probably in NLTE. This model has two problems. First, the transition from the tables to the DCA model is abrupt and leads to sudden jumps in opacity for small changes in temperature. Second, because it does not use a physical basis to decide if a zone is LTE or not, there can often be zones that are using LTE tables when in fact the conditions would drive it into NLTE, and vice versa. Figure 5 shows that the radiation temperature is sensitive to the choice of table transition temperature. For the 1D calculation of a near vacuum hohlraum that is shown, the radiation temperature is almost 10 eV cooler when we transition to DCA at 200 eV instead of 300 eV. This change is mostly due to fact that the DCA model LTE gold opacities are much lower than those from high-resolution tables. Thus, it would be beneficial to move towards a wall model that does not rely on an arbitrary temperature condition to decide when to transition between LTE and NLTE. And it would also help to have a more accurate DCA model that has opacities close to the tabular values when in LTE.

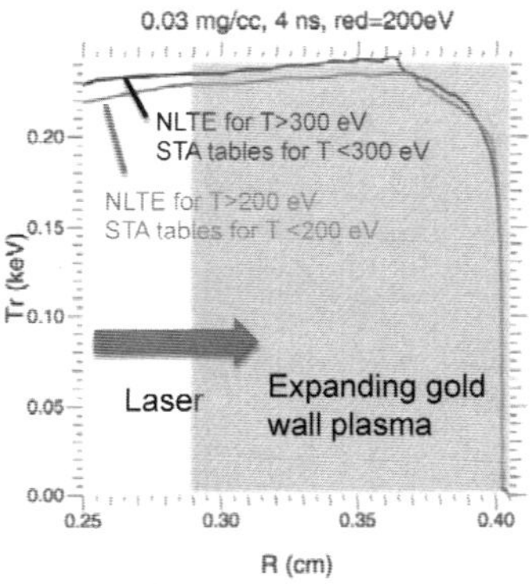

Figure 5. Radiation temperature spatial profile for two different LTE/NLTE transition temperatures.

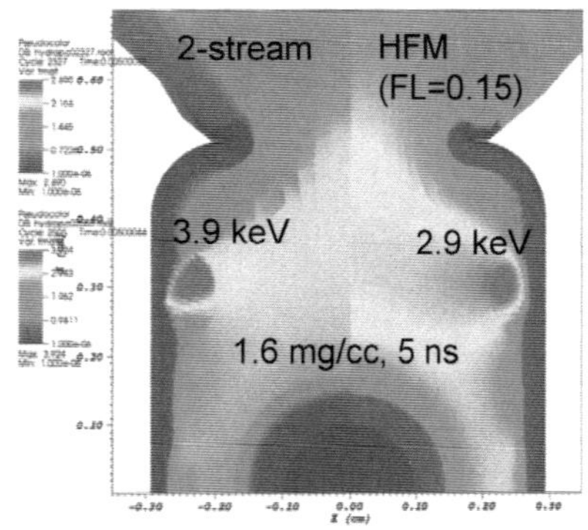

Figure 6. Plot of electron temperature for the 1.6 mg/cc fill shot with 2-stream limiter (left) and flux limiter = 0.15 (right).

3.2. Methods for inhibiting thermal transport to x-ray conversion layer

One hypothesis is that as the hohlraum gas fill is raised, other physics processes not currently included in the model are causing an inhibited heat transport from the coronal plasma to the wall, leading to reduced x-ray drive. One mechanism that has been proposed is the two-stream plasma instability [6]. If this instability is excited, it can quickly grow and saturate, leading to ion acoustic turbulence, which can lead to increased scattering and inhibited thermal transport [7]. In Hydra the two-stream instability is applied as a collision-dependent flux limiter, where the enhanced two-stream collisionality depends on the value of ZT_e/T_i in each zone. When this model is applied to the 1.6 mg/cc He fill case (see figure 3), figure 6 shows that the electron temperature in the coronal plasma heated by the outer beams is ~ 1 keV hotter than when the usual flux limiter of 0.15 is used. The higher temperature leads to a lower radiation drive.

4. Summary

We have described a Hydra model for laser-heated hohlraums that can predict the amount of radiation drive produced in hohlraums with gas fill densities between 0 and 0.6 mg/cc of helium for 2-shock and 3-shock pulses driving HDC ablator capsules. To match the measured bang time, 7-10% of the measured absorbed laser power must be removed from the peak of the pulse in the calculation. A series of 2-shock HDC implosions in which the hohlraum fill density was varied showed that the bang time discrepancy becomes larger for gas fill densities above 0.85 mg/cc. The discrepancy correlates with the amount of measured SRS backscatter. This model does not capture the variation in radiation drive symmetry as a function of the hohlraum fill density. We are currently investigating several different possible improvements to our model to improve its predictability over a broader range of conditions. The improvements include improved wall NLTE models and models that inhibit the thermal transport from the coronal plasma to the x-ray conversion layer. This work performed under the auspices of the U.S. Department of Energy by Lawrence Livermore National Laboratory under Contract DE-AC52-07NA27344.

References

[1] M. D. Rosen, H. A. Scott, D. E. Hinklel, et al., High Energy Density Physics **7**, 180 (2011)
[2] M. M. Marinak, G. D. Kerbel, N. A. Gentile, et al., Phys. Plasmas **8**, 2275 (2001)
[3] O. S. Jones, C. J. Cerjan, M. M. Marinak, et al., Phys. Plasmas **19**, 056315 (2012)
[4] S. A. MacLaren, M. B. Schneider, K. Widman, et al., PRL **112**, 105003 (2014)
[5] L. F. Berzak Hopkins, N. B. Meezan, S. LePape, et al., PRL **114**, 175001 (2015)
[6] B. D. Fried and R. W. Gould, Phys. of Fluids, **4**, 139 (1961)
[7] S. H. Glenzer, W. Rozmus, V. Yu. Bychenkov, et al., PRL **88**, 235002 (2002)

Kinetic studies of ICF implosions

Grigory Kagan[1], H. W. Herrmann[1], Y.-H. Kim[1], M. J. Schmitt[1], P. Hakel[1], S. C. Hsu[1], N. M. Hoffman[1], D. Svyatsky[1], S. D. Baalrud[2], J. O. Daligault[1], H. Sio[3] A. B. Zylstra[1], M. J. Rosenberg[4], H. G. Rinderknecht[5], M. Gatu Johnson[3], J. A. Frenje[3], F. H. Séguin[3], C. K. Li[3], R. D. Petrasso[3], B. J. Albright[1], W. Taitano[1], G.A. Kyrala[1], P. A. Bradley[1], C.-K. Huang[1], C. J. McDevitt[1], L. Chacon[1], B. Srinivasan[6], A. M. McEvoy[1], T. R. Joshi[1] and C. S. Adams[6]

[1]Los Alamos National Laboratory, Los Alamos, NM 87545, USA
[2]Department of Physics and Astronomy, University of Iowa, Iowa City, IA 52242, USA
[3]Massachusetts Institute of Technology, Cambridge, MA 02139, USA
[4]Laboratory for Laser Energetics, University of Rochester, Rochester, NY 14623, USA
[5]Lawrence Livermore National Laboratory, Livermore, California 94550, USA
[6]Department of Aerospace and Ocean Engineering, Virginia Tech, Blacksburg, VA 24061, USA

E-mail: kagan@lanl.gov

Abstract. Kinetic effects on inertial confinement fusion have been investigated. In particular, inter-ion-species diffusion and suprathermal ion distribution have been analyzed. The former drives separation of the fuel constituents in the hot reacting core and governs mix at the shell/fuel interface. The latter underlie measurements obtained with nuclear diagnostics, including the fusion yield and inferred ion burn temperatures. Basic mechanisms behind and practical consequences from these effects are discussed.

1. Introduction

In inertial confinement fusion (ICF) high temperature and density of the deuterium/tritium plasma are achieved through ablating the outer shell of the fuel capsule, which is then compressed by the rocket force. This process takes place over a short time scale, making large background gradients an inherent feature of ICF implosions. On the other hand, the major ICF modeling tool, rad-hydro codes, rely on the *single*-fluid framework. One should thus expect predictions of these codes to deviate from experimental observations with the plasma mean free path being comparable to the gradient length scale. Importance of the associated *kinetic effects* can be quantified with the Knudsen number $N_K = \lambda/L$, where λ and L are the characteristic mean free path and background scale, respectively.

Figure 1 shows N_K calculated by post-processing the temperature and density profiles from 1D HYDRA [1] simulations for two representative cases with L taken to be the hot-spot radius and λ the ion mean free path at the hot-spot center [2]. It can be observed that in the NIF case N_K does become of order unity during the course of implosion. Although it is predicted to drop down just before the burn, the initial conditions for this burn can thus be altered. Furthermore, in the thin shell Omega case N_K is predicted to be particularly large during the burn itself. Of course, one should not take these predictions at face value since the fluid simulations that give

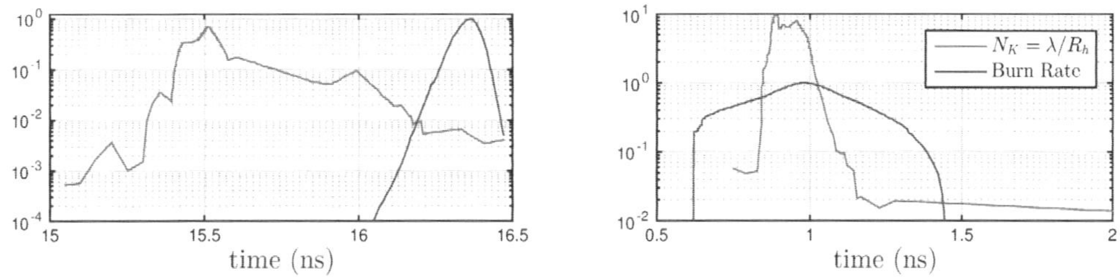

Figure 1. Knudsen number as a function of time, along with the burn rate normalized to its maximum value. Left: NIF high foot shot N130501, 1125 μm radius Si-doped CH capsule with 68.6 μm DT ice layer. Right: 865×7.4 μm CH capsule with 5 atm DT fill driven by a 27 kJ, 1 ns laser pulse.

them are no longer robust once N_K starts approaching even a fraction of unity. However, Fig. 1 does indicate that kinetic effects can play a role on ICF experiments.

In principle, such regimes can be investigated with fully kinetic simulations [3, 4, 5]. However, this path requires enormous computational resources that, in particular, makes it infeasible for modeling realistic 2D and 3D implosions. Neither is it applicable to dense plasmas with non-ideal equations of state and small Coulomb logarithms, as is the case at the fuel/shell interface. On the other hand, fully kinetic simulations can provide a benchmark for physics models, which can both deepen our understanding and allow describing practical implosions beyond the simplest 1D case.

A computationally inexpensive option is a mainline rad-hydro code with "kinetic patches"[6]. This framework involves free parameters, whose values are set through optimization process to fit experimental data, and so is somewhat phenomenological. In the present work we develop a first principles yet computationally efficient approach by noticing that a large class of kinetic effects can be encompassed through treating the thermal and suprathermal particles separately. For the former, the mean free path is smaller and distribution functions are often close to equilibrium, though the strong gradients can separate different ion species making the conventional *single-*fluid approach invalid. For the latter, the mean free path can be comparable to, or even larger than, L but these particles only constitute a small fraction by number and can be described by a reduced linear (as opposed to the full non-linear) kinetic equation. Sections to follow summarize our main theoretical results in these two research directions.

2. Inter-ion-species diffusion

The central hot plasma in a spherical implosion contains multiple ion species, which can include both low-Z and high-Z elements [7, 8, 9, 10, 11]. Their relative concentrations can evolve substantially over the course of implosion due to inter-ion-species diffusion [12, 13]. Observation of the resulting fuel stratification in experiments [14, 15, 16, 17] and fully kinetic simulations [3, 4] have recently been reported. The targets with high-Z dopants show particularly pronounced yield anomalies [8, 9, 18], suggesting that even stronger fuel stratification may take place. In addition, this is the same diffusion process that underlies the atomic mixing at the shell/fuel interface. This physics is not included in conventional rad-hydro codes, which therefore "force" the relative concentrations of the ion species to remain constant.

The concentration evolution of species α is governed by the so-called "diffusive flux" $\vec{i}_\alpha \equiv \rho_\alpha(\vec{u}_\alpha - \vec{u}_{\mathbf{c.m.}})$, where $\vec{u}_\alpha$ and $\vec{u}_{\mathbf{c.m.}}$ are fluid velocities of the species α and the overall plasma, respectively, and ρ_α is the mass density of species α. In the illustrative case of a plasma with two ion species the diffusive flux of the lighter species takes the form

$$\vec{i}_l = -\vec{i}_h = -\rho D\Big(\nabla c + k_p \nabla \log p_i + \frac{e k_E}{T_i}\nabla\Phi + k_T^{(i)}\nabla \log T_i + k_T^{(e)}\nabla \log T_e\Big),$$

where $p_i = p_l + p_h$ is the total ion pressure, Φ is the electrostatic potential, T_i and T_e are the ion and electron temperature, respectively, and subscripts "l" and "h" denote the light and heavy ion species, respectively. To get an idea of how fast the concentration changes one needs to know the classical diffusion coefficient D and the so-called "diffusion ratios" k_p, k_E, $k_T^{(i)}$ and $k_T^{(e)}$.

We have first conducted first principles transport calculation to obtain these coefficients in the conventional, or *weakly coupled*, plasmas [19, 20]. Unlike the well-known case of neutral gas mixtures, thermo-diffusion ratios $k_T^{(i)}$, $k_T^{(e)}$ were found to be comparable to, or even much larger than, the baro-diffusion ratio k_p. Equally important, in low-Z/high-Z mixtures, $k_T^{(i)}$ and $k_T^{(e)}$ were found to attain large absolute values and slopes at small concentrations of the high-Z component [20]. Plasma thermo-diffusion may thus be potentially responsible for the anomalous yield reduction in doped capsule implosions [8, 9] in addition to the laser absorption effects considered earlier [18].

This calculation relied on the binary collision model that breaks down for dense *coupled* plasmas, in which particle correlations are significant. Such plasmas naturally occur during various stages of the implosion and, in particular, in the mixing layer between the shell and the fuel. To gain an insight into diffusion issues in this regime we have utilized the recently proposed effective potential theory, whose predictions for transport in dense plasmas proved in remarkable agreement with molecular dynamic (MD) simulations [21, 22]. Taking this idea further we have evaluated the full set of transport coefficients in multi-component plasmas. Most interestingly, thermo-diffusion was found to vanish in substantially coupled plasmas, making it somewhat similar to thermo-diffusion in neutral gas mixtures [23].

3. Suprathermal ions

In fully ionized plasmas, the particle mean free path scales with the square of this particle's energy. Suprathermal ions can therefore be far from Maxwellian even if their thermal counterparts are nearly equilibrated. It is these suprathermal, or *tail*, ions that fuse in subignited implosions. Their distribution is thus the key to proper interpretation of nuclear diagnostics employed at ICF experiments in general and to correct fusion yield prediction in particular.

The possibility of the reactivity reduction due to ion tail depletion was originally pointed out by Henderson and Petschek [24, 25]. More recently Molvig et al. reexamined the problem in connection to experiments at Omega laser facility and suggested a simple upper bound model for the reactivity reduction [26]. This work was followed by more accurate calculations by Albright et al. [27], Schmit et al. [28], Tang et al. [29], Davidovits and Fisch [30] and Cohen et al. [31]. All these studies considered the perfectly planar or spherical models and relied either on direct numerical solution [28, 29, 30, 31] or phenomenological assumptions that affect the structure of the kinetic equation, thereby introducing uncontrollable modifications to the reactivity [26, 27].

Hydro-instabilities that commonly exist at the shell/fuel interface suggest that the 1D results need to be revisited for realistic implosions. Indeed, we have found that the reactivity reduction can be substantially aggravated by these instabilities [32]. This analysis utilized the self-similarity feature of the suprathermal ion distribution, which has been verified in the 1D planar and spherical cases with the newly developed code solving the reduced kinetic equation. On the practical side this feature effectively gives that, as far as the reactivity is concerned, the suprathermal ions "feel" the distance to the hot-spot boundary only, thereby providing a computationally expedient yet reasonably accurate tool for evaluating the tail depletion effects.

Finally, we have demonstrated that the inferred ion burn temperature, as evaluated from the standard Brysk prescription [33], is lower than the actual one due to the same ion tail depletion effect. This mechanism has a stronger impact on the DD reaction than on DT, thereby making the apparent DD temperature lower than DT [32]. Such a trend is often observed at NIF [34] and explanations available to date include the bulk fluid motion [35, 36], neutron scattering and burn weighting due to reactivity dependence on temperature [34]. None of these mechanisms alone seems capable of giving the observed 25% difference between the DD and DT temperatures, so the newly predicted effect may contribute.

This work is performed under the auspices of the U.S. Dept. of Energy by the Los Alamos National Security, LLC, Los Alamos National Laboratory under Contract No. DE-AC52-06NA25396.

References

[1] Marinak M, Kerbel G, Gentile N, Jones O, Munro D, Pollaine S, Dittrich T and Haan S 2001 *Physics of Plasmas* **8** 2275–2280
[2] Huba J D 2004 NRL: Plasma formulary Tech. rep. DTIC Document
[3] Larroche O 2012 *Physics of Plasmas* **19** 122706
[4] Bellei C *et al.* 2013 *Physics of Plasmas* **20** 012701
[5] Larroche O *et al.* 2016 *Physics of Plasmas* **23** 012701
[6] Hoffman N M *et al.* 2015 *Physics of Plasmas* **22** 052707
[7] Petrasso R *et al.* 2003 *Physical Review Letters* **90** 095002
[8] Wilson D *et al.* 2008 *Journal of Physics: Conference Series* **112** 022015
[9] Garbett W *et al.* 2008 *Journal of Physics: Conference Series* **112** 022016
[10] Yaakobi B, Epstein R, Hooper Jr C, Haynes Jr D and Su Q 1996 *Journal of X-ray science and technology* **6** 172–187
[11] Regan S *et al.* 2002 *Physics of Plasmas* **9** 1357–1365
[12] Amendt P, Landen O, Robey H, Li C and Petrasso R 2010 *Physical Review Letters* **105** 115005
[13] Amendt P, Wilks S, Bellei C, Li C and Petrasso R 2011 *Physics of Plasmas* **18** 056308
[14] Herrmann H, Langenbrunner J, Mack J, Cooley J, Wilson D, Evans S, Sedillo T, Kyrala G, Caldwell S, Young C *et al.* 2009 *Physics of Plasmas* **16** 056312
[15] Casey D *et al.* 2012 *Physical Review Letters* **108** 075002
[16] Rosenberg M *et al.* 2014 *Physical Review Letters* **112** 185001
[17] Rinderknecht H G, Rosenberg M, Li C, Hoffman N, Kagan G, Zylstra A, Sio H, Frenje J, Johnson M G, Séguin F *et al.* 2015 *Physical Review Letters* **114** 025001
[18] Dodd E *et al.* 2012 *Physics of Plasmas* **19** 042703
[19] Kagan G and Tang X Z 2012 *Physics of Plasmas* **19** 082709
[20] Kagan G and Tang X Z 2014 *Physics Letters A* **378** 1531–1535
[21] Baalrud S D and Daligault J 2013 *Physical Review Letters* **110** 235001
[22] Daligault J, Baalrud S D, Starrett C E, Saumon D and Sjostrom T 2016 *Physical Review Letters* **116** 075002
[23] Kagan G, Baalrud S D and Daligault J O 2016 *"Multi-component Plasma Transport Across Coupling Regimes" (to be submitted)*
[24] Henderson D B 1974 *Physical Review Letters* **33** 1142
[25] Petschek A and Henderson D 1979 *Nuclear Fusion* **19** 1678
[26] Molvig K, Hoffman N M, Albright B, Nelson E M and Webster R B 2012 *Physical Review Letters* **109** 095001
[27] Albright B, Molvig K, Huang C K, Simakov A, Dodd E, Hoffman N, Kagan G and Schmit P 2013 *Physics of Plasmas* **20** 122705
[28] Schmit P, Molvig K and Nakhleh C 2013 *Physics of Plasmas* **20** 112705
[29] Tang X Z, Berk H, Guo Z and McDevitt C 2014 *Physics of Plasmas* **21** 032707
[30] Davidovits S and Fisch N J 2014 *Physics of Plasmas* **21** 092114
[31] Cohen B I, Dimits A M, Zimmerman G B and Wilks S C 2014 *Physics of Plasmas* **21** 122701
[32] Kagan G, Svyatskiy D, Rinderknecht H, Rosenberg M, Zylstra A, Huang C K and McDevitt C 2015 *Physical Review Letters* **115** 105002
[33] Brysk H 1973 *Plasma Physics* **15** 611
[34] Gatu Johnson M *et al.* 2016 *"Indications of flow near maximum compression in layered DT implosions at the National Ignition Facility", Physical Review Letters (in review)*
[35] Appelbe B and Chittenden J 2011 *Plasma Physics and Controlled Fusion* **53** 045002
[36] Murphy T 2014 *Physics of Plasmas* **21** 072701

Journal of Physics: Conference Series **717** (2016) 012028 doi:10.1088/1742-6596/717/1/012028

Implosion uniformity improvement of fuel target in heavy ion fusion

T Karino[1], S Kawata[1], T Suzuki[1], S Kondo[1], T Iinuma[1], D Barada[1] and A I Ogoyski[2]

[1] *Utsunomiya University, Utsunomiya 321-8585, Japan*

[2] *Varna Technical University, Varna 9010, Bulgaria*

E-mail: mt146615@cc.utsunomiya-u.ac.jp, kwt@cc.utsunomiya-u.ac.jp

Abstract. A dynamic mitigation mechanism for instability growth was proposed and discussed in the paper [Phys. Plasmas 19, 024503 (2012)]. In the present paper the robustness of the dynamic instability mitigation mechanism is discussed further. The results presented here show
that the mechanism of the dynamic instability mitigation is rather robust against changes in the phase, the amplitude and the wavelength of the wobbling perturbation applied.

1. Introduction

The dynamic stabilization for the Rayleigh-Taylor instability (RTI) has been studied for a uniform compression of a fusion fuel target in inertial confinement fusion [1-7]. The RTI dynamic stabilization was proposed many years ago in inertial fusion; the oscillation amplitude of the driving acceleration should be sufficiently large to stabilize RTI. In inertial fusion, the fusion fuel compression is essentially important to reduce an input driver energy, and the implosion uniformity is one of critical issues to release the fusion energy stably [8-9].

The heavy ion accelerator could provide a beam axis wobbling with a high frequency. The wobbling heavy ion beams also define the perturbation phase. This means that the perturbation phase is known, and so the perturbations successively imposed are superposed in the plasma. The heavy ion beams accelerate the fusion target fuel with a large acceleration in inertial fusion. The wobbling heavy ion beams would provide a small oscillating acceleration perturbation in an inertial fusion fuel target during the target implosion. So the RTI growth would be reduced by the phase-controlled superposition of perturbations in heavy ion inertial fusion.

If there are other sources of perturbations in the physical system except the driven-beam illumination non-uniformity, the dynamic mitigation mechanism proposed here does not work. For example, when the shell thickness of an inertial fusion fuel target is not uniform at the target fabrication process, the dynamic mitigation mechanism does not work. In this sense, the dynamic mitigation mechanism is not almighty but works well to mitigate the RTI growth originated by the driver beam non-uniformity itself.

In this paper we discuss the robustness of the dynamic mitigation mechanism for instabilities presented in Refs. [10-12]. The results presented here show that the mechanism of the dynamic instability mitigation is rather robust against changes in the phase, amplitude and wavelength of the wobbling perturbation applied. The promising results presented in this paper ensure the viability of the mechanism of the dynamic instability mitigation.

2. Dynamic instability mitigation

Let us consider an unstable system, which has one mode of $a = a_0 e^{ikx+\gamma t}$. Here a_0 is the amplitude, $k = 2\pi/\lambda$ is the wave number, λ the wavelength and γ the growth rate of the instability. An example initial perturbation is shown in Fig. 1(a). At t=0 the perturbation is imposed. The initial perturbation grows with γ. After Δt, if another perturbation, which has an inverse phase, is actively imposed (see Fig. 1(b)), the overall amplitude is the superposition of all the perturbations, and so the actual perturbation amplitude is well mitigated as shown in Fig. 1(c). This is an ideal example for the dynamic instability mitigation [10-12].

In plasmas the perturbation phase and amplitude cannot be measured. So the perfect feedback control cannot be realized in plasmas and fluids. However, a heavy ion beam accelerator can also provide a controlled wobbling or oscillating beam with a high frequency. It would provide the defined phase and amplitude of perturbations [13, 14].

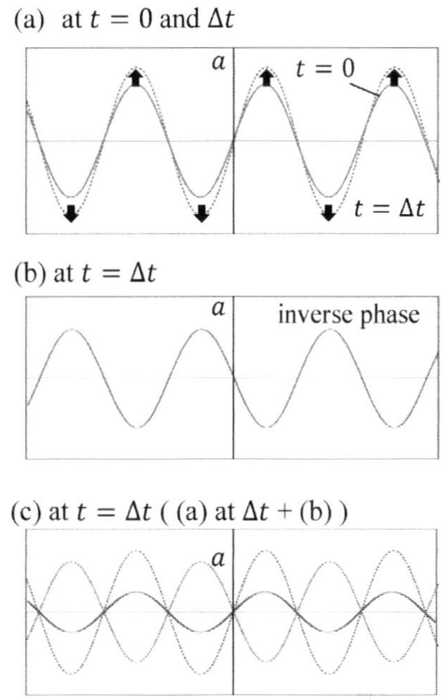

Figure 1. An ideal example concept of the dynamic mitigation.

When the instability driver wobbles uniformly in time, the imposed perturbation for a physical quantity of F at $t = \tau$ may be written as

$$F = \delta F e^{i\Omega\tau} e^{\gamma(t-\tau)+i\vec{k}\cdot\vec{x}} \quad (1)$$

Here δF is the amplitude, Ω the wobbling or oscillation frequency defined actively by the driving wobbler, and $\Omega\tau$ the phase shift of superimposed perturbations. At each time t, the wobbler or the modulated driver provides a new perturbation with the phase and the amplitude actively defined by the driving wobbler itself. The superposition of the perturbations provides the actual perturbation at t as follows:

$$\int_0^t d\tau \ \delta F e^{i\Omega\tau} e^{\gamma(t-\tau)+i\vec{k}\cdot\vec{x}} \propto \frac{\gamma+i\Omega}{\gamma^2+\Omega^2} \delta F e^{\gamma t} e^{i\vec{k}\cdot\vec{x}} \quad (2)$$

When $\Omega \geq \gamma$, the perturbation amplitude is reduced by the factor of γ/Ω, compared with the pure instability growth ($\Omega = 0$) based on the energy deposition non-uniformity[10,11]. The result in Eq. (2) presents that the perturbation phase should oscillate with $\Omega \geq \gamma$ for the effective amplitude reduction.

Figure 2 shows an example simulation for RTI, which has a single mode. In this example, two stratified fluids are superimposed under an acceleration of $g = g_0 + \delta g$. The density jump ratio between the two fluids is 10/3. In this specific case the wobbling frequency Ω is γ, the amplitude of δg is $0.1g_0$, and the results shown in Figs. 2 are those at $t = 5/\gamma$. In Fig. 2(a) δg is constant and drives the RTI as usual, and in Fig. 2(b) the phase of δg is shifted or oscillates with the frequency of Ω as stated above for the dynamic instability mitigation. The RTI growth mitigation ratio is 72.9% in Fig. 2. The growth mitigation ratio is defined by $(H_0 - H_{mitigate})/H_0 \times 100\%$. Here H is defined as shown in Fig. 2(a), H_0 shows the deviation amplitude of the two-fluid interface in the case in Fig. 2(a) without the oscillation ($\Omega = 0$), and $H_{mitigate}$ presents the deviation for the other cases with the oscillation ($\Omega \neq 0$). The example simulation results support well the effect of the dynamic mitigation mechanism.

3. Dynamic instability mitigation

In order to check the robustness of the dynamic instability mitigation mechanism, here we study the effects of the change in the phase, the amplitude and the wavelength of the wobbling perturbation δF, that is, δg in Fig. 2 on the dynamic instability mitigation.

When the perturbation amplitude $\delta F = \delta F(t)$ depends on time or oscillates slightly in time, the dynamic mitigation mechanism is examined first. We consider $\delta F(t) = \delta F_0\left(1 + \Delta e^{i\Omega' t}\right)$ in Eq. (1).

Here $\Delta \ll 1$. In this case, Eq. (2) is modified as follows:

$$\int_0^t d\tau\ \delta F e^{i\Omega\tau} e^{\gamma(t-\tau)+i\vec{k}\cdot\vec{x}} \propto \left\{\frac{\gamma+i\Omega}{\gamma^2+\Omega^2} + \Delta\frac{\gamma+i(\Omega+\Omega')}{\gamma^2+(\Omega+\Omega')^2}\right\}\delta F_0 e^{\gamma t} e^{i\vec{k}\cdot\vec{x}} \quad (3)$$

When $\Delta \ll 1$ in Eq. (3), just a minor effect appears on the dynamic mitigation of the instability.

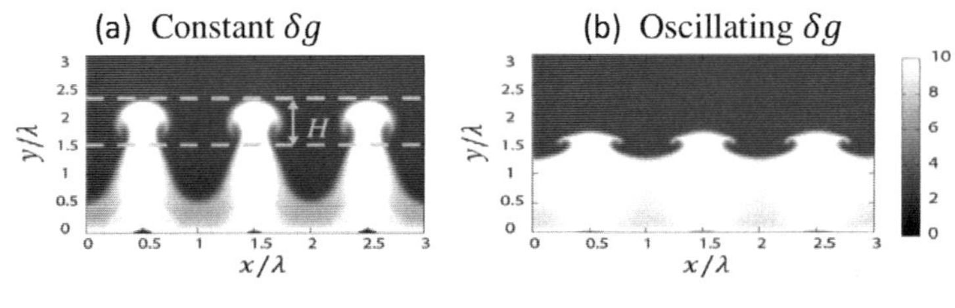

Figure 2. Example simulation results for the Rayleigh-Taylor instability (RTI) mitigation.

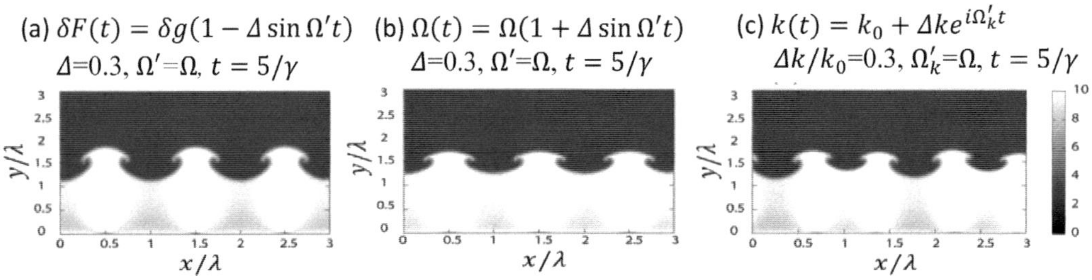

Figure 3. Fluid simulation results for the RTI mitigation for the time-dependent amplitude, oscillation frequency and wave number.

We also performed the fluid simulations. In the simulations $\delta F(t) = \delta g(1 - \Delta \sin \Omega' t)$. The RTI is simulated again based on the same parameter values shown in Fig. 2 except the perturbation amplitude oscillation $\delta F(t)$. In the simulations we employ Ω'=3Ω, Ω and Ω/3 in Eq. (3). For Δ=0.1 and 0.3, and for Ω'=3Ω, Ω and Ω/3, the RTI growth reduction ratio is 54.9~72.9% at $t = 5/\gamma$. The results by the fluid simulations and Eq. (3) demonstrate that the perturbation amplitude oscillation $\delta F = \delta F(t)$ is uninfluential as long as $\Delta \ll 1$. Figure 3(a) shows the RTI growth reduction ratio of 61.3% for Ω'=Ω.

When the oscillation frequency Ω of the perturbation δF depends on time ($\Omega = \Omega(t)$), the time-dependent frequency means that $\Omega(t)$ would consist of multiple frequencies: $e^{i\Omega t} = \sum_i \Delta_i e^{i\Omega_i t}$. In this case Eq. (3) becomes

$$\int_0^t d\tau\ \delta F e^{i\Omega\tau} e^{\gamma(t-\tau)+i\vec{k}\cdot\vec{x}} \propto \sum_i \Delta_i \frac{\gamma+i\Omega_i}{\gamma^2+\Omega_i^2}\delta F e^{\gamma t} e^{i\vec{k}\cdot\vec{x}} \quad (4)$$

The result in Eq. (4) shows that the highest frequency of Ω_i contributes to the instability mitigation. In a real system the highest frequency would be the original wobbling frequency Ω or so, and the largest amplitude of Δ_i is also that for the original wobbling mode. So when the frequency change is slow, the original wobbler frequency of Ω contributes to the mitigation.

The fluid simulations are also done for the RTI with $\Omega(t) = \Omega(1 + \Delta \sin \Omega' t)$ together with the same parameter values employed in Fig. 2. In this case Δ=0.1 and 0.3, and Ω'=3Ω, Ω and Ω/3. The growth reduction ratio was 66.9~74.0% at $t = 5/\gamma$. The little oscillation of the imposed perturbation oscillation frequency $\Omega(t)$ has a minor effect on the dynamic instability mitigation. Figure 3(b) shows the RTI growth reduction ratio of 68.8% for Ω'=Ω.

When the wobbling wavelength $\lambda = 2\pi/k$ depends on time, one can expect as follows in a real system: $k(t) = k_0 + \Delta k e^{i\Omega'_k t}$ and $k_0 \gg \Delta k$. In this case the wobbling wavelength changes slightly in time, and Eq. (3) becomes as follows:

$$\int_0^t d\tau \delta F e^{i\Omega\tau} e^{\gamma(t-\tau)+ik\cdot x} \propto \delta F e^{\gamma t + ik_0\cdot x} \int_0^t d\tau\, e^{(i\Omega-\gamma)\tau} \sum_{m=-\infty}^{\infty} i^m J_m(\Delta k\cdot x) e^{im\Omega'_k \tau}$$

$$\propto \sum_{m=-\infty}^{\infty} i^m J_m(\Delta k \cdot x) \int_0^t d\tau\, e^{i(\Omega+m\Omega_k')\tau-\gamma\tau} \propto \sum_{m=-\infty}^{\infty} i^m J_m(\Delta k \cdot x) \frac{\gamma+i(\Omega+m\Omega_k')}{\gamma^2+(\Omega+m\Omega_k')^2} \quad (5)$$

Here J_m is the Bessel function of the first kind. The result in Eq. (5) demonstrates that the instability growth reduction effect is not degraded by the small change in the wobbling wavelength. In actual situations the mode $m = 0$ contributes mostly to the instability mitigation, and in this case the original reduction effect shown in Eq. (2) is recovered.

The fluid simulations are also performed for this case $k(t) = k_0 + \Delta k e^{i\Omega_k' t}$ together with the same parameter values employed in Fig. 2. In this case Δ=0.1 and 0.3, and Ω'=3Ω, Ω and Ω/3. The growth reduction ratio was 68.8~93.3% at $t = 5/\gamma$. The little oscillation of the imposed perturbation oscillation frequency $\Omega(t)$ has a minor effect on the dynamic instability mitigation. For a realistic situation Ω_k'~ Ω, where Ω is the wobbling or modulation frequency. Figure 3(c) shows the RTI growth reduction ratio of 93.3% for Ω_k'=Ω.

All the results shown above demonstrate that the dynamic instability mitigation mechanism proposed is rather robust against the changes in the amplitude, the phase and the wavelength of the wobbling or modulating perturbation of δF in general or δg in RTI.

4. Conclusions

We have discussed the dynamic mitigation method, in which the perturbation growth is controlled and mitigated by the wobbler or the driver modulation. In this paper we focus on the robustness of the dynamic mitigation mechanism of the instability. The changes in the perturbation frequency, amplitude and wavelength were examined. The theoretical and simulation results demonstrate that the dynamic instability mitigation mechanism is rather robust against the changes in the perturbation frequency, the amplitude and the wavelength of the driver wobbling motion or the driver modulation. The results in this paper show the viability of the mechanism of the dynamic instability mitigation.

The wobbling or the modulation defines the imposed perturbation phase and amplitude at each time. Consequently the overall superposition of all the perturbations mitigates its growth through the control of the wobbling or modulating motion. The dynamic mitigation would work for the mitigation of instabilities in which the imposed perturbation phase is controlled actively.

Acknowledgements

This work was partly supported by MEXT, JSPS, ASHULA, ILE/Osaka Univ., CORE/Utsunomiya Univ., CDI/Utsunomiya Univ., and Japan/U.S. Fusion Research Collaboration Program, Japan.

References

[1] Wolf G H 1970 *Phys. Rev. Lett.* **24** 444
[2] Troyon F and Gruber R 1971 *Phys. Fluids* **14** 2069
[3] Boris J P 1977 *Comments Plasma Phys. Cont. Fusion* **3** 1
[4] Betti R, McCrory R L and Verdon C P 1993 *Phys. Rev. Lett.* **71** 3131
[5] Piriz A R, Prieto G R, Diaz I M and Cela J J L 2010 *Phys. Rev.* E **82** 026317
[6] Piriz A R, Piriz S A and Tahir N A 2011 *Phys. Plasmas* **18** 092705
[7] Nuckolls J, Wood L, Thiessen A and Zimmmerman G 1972 *Nature* **239** 139
[8] Emery M H, Orens J H, Gardner J H and Boris J P 1982 *Phys. Rev. Lett.* **48** 253
[9] Kawata S and Niu K 1984 *J. Phys.* Soc. Jpn. **53** 3416
[10] Kawata S, Sato T, Teramoto T, Bandoh E, Masubichi Y, Watanabe H and Takahashi I 1993 *Laser Part. Beams* **11** 757
[11] Kawata S, Iizuka Y, Kodera Y, Ogoyski A I and Kikuchi T 2009 *Nucl. Inst. Meth. Phys. Res.* A **606** 152
[12] Kawata S 2012 *Phys. Plasmas* **19** 024503
[13] Qin H, Davidson R C and Logan B G 2010 *Phys. Rev. Lett.* **104** 254801
[14] Arnold R C, Colton E, Fenster S, Foss M, Magelssen G and Moretti A 1982 *Nucl. Inst. Meth.* **199**, 557

Uniform fuel target implosion in heavy ion inertial fusion

S Kawata[1], T Karino[1], S Kondo[1], T Iinuma[1], D Barada[1], Y Y Ma[1,2] and A I Ogoyski[3]

[1]Utsunomiya University, Utsunomiya 321-8585, Japan
[2]National University of Defence Technology, Changsha, China
[3]Techinical University of Varna 9010, Varna, Bulgaria

E-mail: kwt@cc.utsunomiya-u.ac.jp

Abstract. For a steady operation of a fusion power plant the target implosion should be robust against the implosion non-uniformities. In this paper the non-uniformity mitigation mechanisms in the heavy ion beam (HIB) illumination are discussed in heavy ion inertial fusion (HIF). A density valley appears in the energy absorber, and the large-scale density valley also works as a radiation energy confinement layer, which contributes to the radiation energy smoothing for the HIB illumination non-uniformity. The large density-gradient scale, which is typically ~500μm in HIF targets, also contributes to a reduction of the Rayleigh-Taylor instability growth rate. In HIF a wobbling HIBs illumination would also reduce the Rayleigh-Taylor instability growth and to realize a uniform implosion.

1. Introduction

In inertial confinement fusion the target implosion non-uniformity is introduced by a driver beams' illumination non-uniformity, a fuel target alignment error in a fusion reactor, the target fabrication defect, et al. For a steady operation of a fusion power plant the target implosion should be robust against the implosion non-uniformities. The requirement for the implosion uniformity is stringent, and the hotspot dynamics is also essentially important. It was confirmed that the implosion uniformity should be less than a few % [1]. In this paper the non-uniformity mitigation mechanisms in the heavy ion beam (HIB) illumination are discussed in heavy ion inertial fusion (HIF) [2-5]. In HIF targets a density valley appears in the energy absorber, and the large-scale density valley also works as a radiation energy confinement layer, which contributes to a radiation energy smoothing for the HIBs illumination non-uniformity [5]. In HIF fuel targets the density gradient scale length becomes large (typically ~500μm or so), and it also contributes to a reduction of the Rayleigh-Taylor instability (RTI) growth rate. In HIF the wobbling HIBs illumination also contributes to reduce the RTI growth [2-4]. In the paper compressible-fluid computer simulations are performed to investigate the mitigation mechanism of the HIB illumination non-uniformity. In the fluid code the three-temperature model is employed to simulate the HIF target implosion and the RTI growth with the CIP (Cubic Interpolated Profile Scheme)[6]-based ALE (Arbitrary Lagrangian-Eulerian) method [7].

Figure 1 shows a typical structure for a HIF direct drive target. The HIB ions deposit their energy inside the material mainly in the Al layer in Fig.1. The HIBs deposition stopping length is long (typically ~1mm or longer) depending on a HIB particle energy and the target material. The density gradient scale length L in a HIF target is the order of ~500 micron m or so. The large scale L of the density gradient comes from the long scale length of the HIBs energy deposition in a material.

By the expression of the RTI growth rate of $\gamma \approx \sqrt{gk/(1+kL)}$ [8, 9], the RTI modes with the long wavelength $2\pi/k$ become dominant in the HIF targets. On the other hand, inside the density valley the radiation energy would be confined, and the radiation energy is transported in the lateral direction along with the density valley to smooth the HIBs energy deposition non-uniformity.

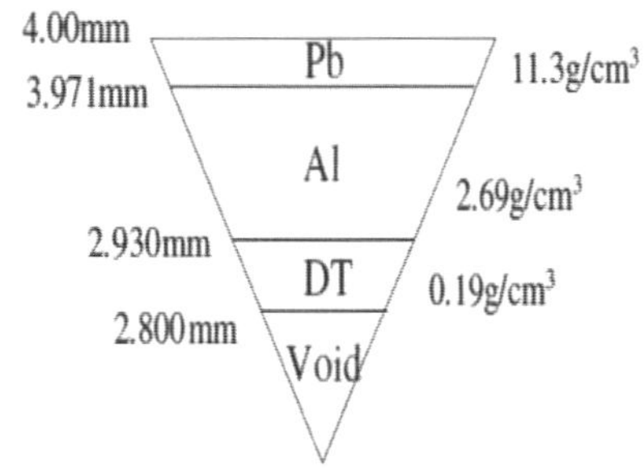

Figure 1. An example fuel target structure in heavy ion inertial fusion.

The HIB accelerator has a capability to control and rotate the HIB's axis precisely with a high frequency [10]. The wobbling HIBs would induce the dynamic mitigation of the RTI in HIF. Normally the perturbation phase is unknown so that the instability growth rate is discussed. However, if the perturbation phase is known, the instability growth can be controlled by a superposition of perturbations imposed actively: if the perturbation is induced by, for example, a driving beam axis oscillation or wobbling, the perturbation phase could be controlled and the instability growth is mitigated by the superposition of the growing perturbations [2-5]. So far the dynamic stabilization mechanism has been discussed in order to obtain a uniform compression of a fusion fuel in inertial confinement fusion. The RTI dynamic stabilization was found many years ago [9, 10] and is important in inertial fusion. It was found that the oscillation amplitude of the driving acceleration should be sufficiently large to stabilize the R-T instability [11-16].

2. Smoothing mechanisms of HIBs illumination non-uniformity

In inertial confinement fusion, the driver beam illumination non-uniformity leads to a degradation of fusion energy output. Therefore, it is important to reduce the illumination non-uniformity. In this section two smoothing mechanisms are presented for the beam illumination non-uniformity in HIF.

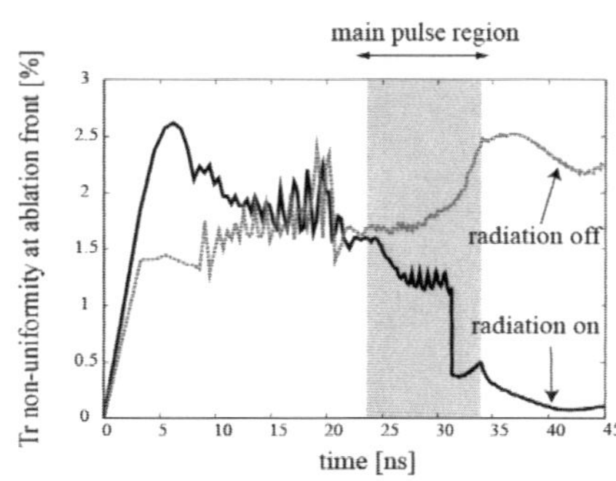

Figure 2. Histories of the non-uniformity of the radiation temperature at the ablation front in the cases of the radiation transport ON and OFF in the direct drive HIF target.

2.1. Radiation Smoothing in a HIF target implosion

The HIB ions deposit their energy in a deep layer of the energy absorber, and so a density valley appears inside the energy absorber. Even in a direct driven fuel target shown in Fig. 1, a part of the HIB energy is converted to a radiation energy confined in the density valley. The density valley plays a role to confine the radiation energy and to smooth the HIB deposition energy partly. The total HIB energy is 4.0MJ. We employ a 32-HIBs illumination system [18].

Figure 2 presents the time histories of the RMS non-uniformity of the radiation temperature at the ablation front in the cases of the radiation transport ON and OFF. In Fig. 2 we see that during the main pulse, the implosion non-uniformity is well smoothed by the radiation transport effect.

In the direct drive HIF target implosion the peak conversion efficiencies of the HIB total energy to the radiation energy are ~ 4.5%. The result means that a few hundred kJ of the radiation energy is confined in the density valley and contributes to the non-uniformity mitigation. The smoothing mechanism demonstrates that the direct drive HIF target naturally serves a direct-indirect hybrid mode of the implosion through the large density valley, which appears in the thick HIBs energy deposition layer.

2.2. *Wobbling HIBs smoothing effect of target implosion non-uniformity*

We introduce another smoothing mechanism [2-5] by the wobbling HIBs. In general a perturbation of physical quantity would be an origin of instability. Normally the perturbation phase is unknown so that the instability growth is discussed with the growth rate. However, if the perturbation phase is known, the instability growth can be controlled by a superposition of perturbations (see Fig. 3). If the perturbation is induced by, for example, a particle beam axis oscillation or wobbling [8], the perturbation phase could be controlled and the instability growth is mitigated by the superposition of the growing perturbations.

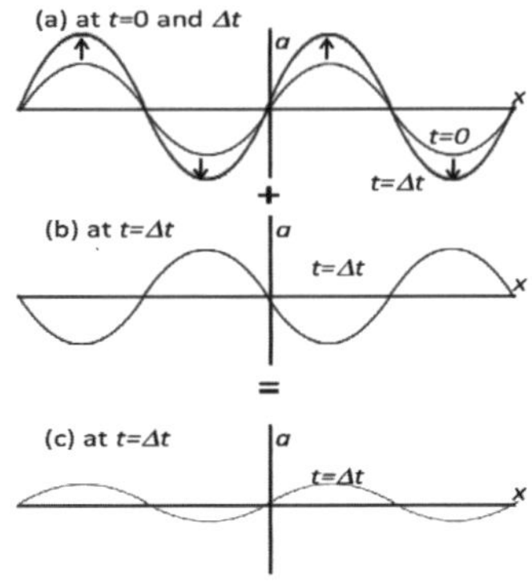

Figure 3. An ideal example concept of the dynamic mitigation. (a)At t=0 a perturbation is imposed. The initial perturbation grows with γ. (b) After Δt another perturbation, which has an inverse phase, is actively imposed, so that (c) the actual perturbation amplitude is mitigated very well after the superposition of the initial and additional perturbations.

When the instability driver wobbles uniformly in time, the imposed perturbation for a physical quantity of F at $t=\tau$ may be written as $F=\delta F e^{i\Omega\tau}e^{\gamma(t-\tau)+i\vec{k}\cdot\vec{x}}$. Here δF is the amplitude, Ω the wobbling or oscillation frequency defined actively by the driving wobbler, and $\Omega\tau$ the phase shift of superimposed perturbations. At each time t, the wobbler or the modulated driver provides a new perturbation with the phase and the amplitude actively defined by the driving wobbler itself. The superposition of the perturbations provides the actual perturbation at t as follows: $\int_0^t d\tau\ \delta F e^{i\Omega\tau}e^{\gamma(t-\tau)+i\vec{k}\cdot\vec{x}} \propto \frac{\gamma+i\Omega}{\gamma^2+\Omega^2}\delta F e^{\gamma t}e^{i\vec{k}\cdot\vec{x}}$. When $\Omega \gg \gamma$, the perturbation amplitude is reduced by the factor of γ/Ω, compared with the pure instability growth ($\Omega=0$) based on the energy deposition nonuniformity[11, 12]. When $\Omega \cong \gamma$, the amplitude mitigation factor is still about 50%. The result presents that the perturbation phase should oscillate with $\Omega \gtrsim \gamma$ for the effective amplitude reduction. Figure 4 shows an example simulation for RTI, which has two modes. In this example, two stratified fluids are superimposed under an acceleration of $g=g_0+\delta g$. The density jump ratio between the two fluids is 10/3, which corresponds to the Atwood number of 0.539. In this specific case the wobbling frequency Ω is γ, the amplitude of δg is $0.1g_0$, and the results shown in Figs. 4 are those at $t=5/\gamma$. In Fig. 4(a) δg is constant and drives the RTI as usual, and in Fig. 4(b) the phase of δg is shifted or oscillates with the frequency of Ω as stated above for the dynamic instability mitigation. The RTI growth mitigation ratio is 76.0% in Fig. 4. The growth

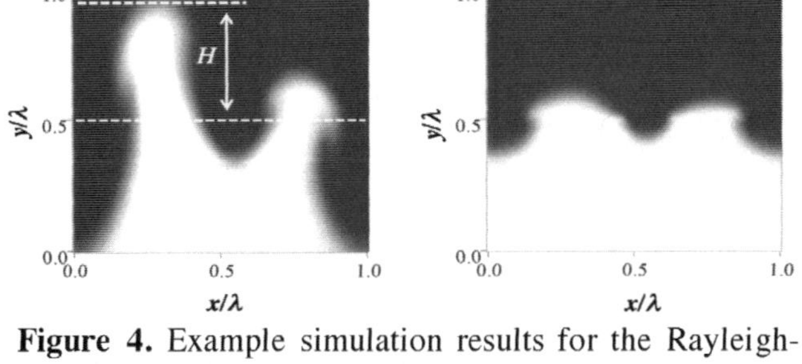

Figure 4. Example simulation results for the Rayleigh-Taylor instability (RTI) mitigation. δg is 10% of the acceleration g_0 and oscillates with the frequency of Ω=γ. The dynamic instability mitigation mechanism works well to mitigate the instability growth.

mitigation ratio is defined by $(H_0 - H_{mitigate})/H_0 \times 100\%$. Here H is defined as shown in Fig. 4(a), H_0 shows the deviation amplitude of the two-fluid interface in the case in Fig. 4(a) without the oscillation ($\Omega = 0$), and $H_{mitigate}$ presents the deviation for the other cases with the oscillation ($\Omega \neq 0$). The example simulation results support well the effect of the dynamic mitigation mechanism.

In addition, the results presented in Ref. [18] show that the mechanism of the dynamic instability mitigation is rather robust against changes in the phase, the amplitude and the wavelength of the wobbling perturbation applied.

3. Summary

The target fuel implosion dynamics in HIF demonstrated that the density gradient scale length is rather large, about several hundred μm. In order to reduce the HIB illumination non-uniformity, two non-uniformity mitigation mechanisms were presented: the first one is the direct-indirect hybrid target, in which a part of the input HIBs energy is converted to the radiation energy in the density valley inside the target and the radiation energy contributes to the non-uniformity smoothing in HIF. The dynamic mitigation mechanism of instabilities is also introduced, and would contribute to smooth the beam driver non-uniformity. The HIB accelerator has a capability to control and rotate the HIB's axis precisely with a high frequency. The wobbling HIBs would induce the dynamic mitigation of the R-T instability in HIF. In our previous paper [5], we also found that the frequency spectrum of the HIBs illumination non-uniformity is synchronized with the rotation frequency of the wobbling beams. This result would work to reduce the growth of the R-T instability originated from the HIBs illumination non-uniformity.

Acknowledgements

This work is partly supported by MEXT, JSPS, JSPS KAKENHI Grant Number 15K05359, ASHULA project, ILE/Osaka University, CORE/Utsunomiya University, Collaboration Center for Research and Development of Utsunomiya University, and Japan/U.S. Fusion Research Collaboration Program conducted by MEXT, Japan. The authors would like to express their appreciations to our collaborative friends, Friends in U.S. HIF-VNL for their fruitful discussions on this work.

References

[1] Kawata S and Niu K 1984 J. Phys. Soc. Jpn. **53** 3416.
[2] Kawata S, Sato T, et al. 1993 Laser Part. Beams **11** 757.
[3] Kawata S, Iizuka Y, et al. 2009 Nucl. Inst. Meth. Phys. Res. A **606** 152.
[4] Kawata S 2012 Phys. Plasmas **19** 024503.
[5] Kawata S, Noguchi K, et al. 2014 Physica Scripta **89** 088001.
[6] Yabe T, Ishikawa T, et al. 1991 Comput. Phys. Comm. **66** 233.
[7] Hirt C W, Amsden A A, and Cook J L 1997 J. Comput. Phys. **135** 203.
[8] Bodner S E 1974 Phys. Rev. Lett. **33** 761.
[9] Takabe H, Mima K, et al. 1985 Phys. Fluids **28**, 3676.
[10] Qin H, Davidson R C, et al. 2010 Phys. Rev. Lett. **104** 254801.
[11] Wolf G H 1970 Phys. Rev. Lett. **24** 444.
[12] Troyon F, and Gruber R 1971 Phys. Fluids **14** 2069.
[13] Boris J P 1977 Comments Plasma Phys. Cont. Fusion **3** 1.
[14] Betti R, McCrory R L and Verdon C P 1993 Phys. Rev. Lett. **71** 3131.
[15] Piriz A R, Prieto G R, Diaz I M and Cela J J L 2010 Phys. Rev. E **82** 026317.
[16] Piriz A R, Piriz S A and Tahir N A 2011 Phys. Plasmas **18** 092705.
[17] Skupsky S and Lee K 1983 J. Appl. Phys. **54** 3662.
[18] Kawata S and Kario T 2015 Phys. Plasmas, **22** 042106.

Direct-drive DT implosions with Knudsen number variations

Y Kim[1], H W Herrmann[1], N M Hoffman[1], M J Schmitt[1], P A Bradley[1], S Gales[2], C J Horsfield[2], M Rubery[2], A Leatherland[2], M Gatu Johnson[3], J A Frenje[3] and V Yu Glebov[4]

[1]Los Alamos National Laboratory, Los Alamos, New Mexico 87545, USA
[2]Atomic Weapons Establishment, Aldermaston, Reading, Berkshire RG7 4PR, UK
[3]Massachusetts Institute of Technology, Cambridge, Massachusetts 02139, USA
[4]Laboratory for Laser Energetics, Rochester, New York 14623, USA

Email: yhkim@lanl.gov

Abstract. Direct-drive implosions of DT-filled plastic-shells have been conducted at the Omega laser facility, measuring nuclear yields while varying Knudsen numbers (i.e., the ratio of mean free path of fusing ions to the length of fuel region) by adjusting both shell thickness (e.g., 7.5, 15, 20, 30 μm) and fill pressure (e.g., 2, 5, 15 atm). The fusion reactivity reduction model showed a stronger effect on yield as the Knudsen number increases (or the shell thickness decreases). The Reduced-Ion-Kinetic (RIK) simulation which includes both fusion reactivity reduction and mix model was necessary to provide a better match between the observed neutron yields and those simulated.

1. Motivation

Recent work by Molvig *et al.* [1] examined how fusion reactivity may be reduced by the loss of fast ions to the imploding capsule-wall. Their theory of fusion reactivity reduction was formulated as a function of Knudsen number (i.e., the ratio of mean free path of ions to the length of fuel region) and implemented in a 1-D radiation-hydrodynamic code. In addition, their fusion reactivity reduction + mix model was benchmarked against existing direct-drive deuterium-tritium (DT) implosion data collected at the Omega laser facility between 2005 and 2011. Although Molvig *et al.* demonstrated that the fusion reactivity reduction + mix model provided a better match between the observed neutron yields and those simulated, the Omega experimental data had either large uncertainty in the initial target conditions (e.g., D/T mixing ratio and total DT gas pressure at shot time) or incomplete diagnostic information (e.g., no Knudsen number and shell area-density have been measured) to confirm that Knudsen effects are the cause of the discrepancies.

A platform for the study of Knudsen effects has been developed with the goal of measuring burn-averaged Knudsen numbers and yields as a function of the Knudsen number controlled via shell thickness and fill pressure variations.

2. Experimental setup

Nuclear and X-ray diagnostics were used to measure the performance of implosions and to infer the Knudsen number during implosions. The Knudsen number (N_K) in DT plasma is given by [1]

$$N_K = \frac{\lambda_i}{L} \approx 3.86 \times 10^{-2} \frac{1}{\ln\Lambda} \frac{T_i\,[keV]^2}{\rho L\,[mg/cm^2]},$$

where, λ_i is the mean free path of fusing ions, L is the system length (i.e., radius of fuel hot spot), and $\ln\Lambda = 3 + \ln\left(\frac{T_i\,[keV]^{\frac{3}{2}}}{\rho\,[g/cm^3]^{\frac{1}{2}}}\right)$ is the Coulomb logarithm [2]. In these expressions, T_i is the burn-averaged ion temperature in units of keV as measured by neutron time-of-flight (nTOF) diagnostics, and ρL is the fuel area-density in units of mg/cm^2 as measured by the knock-on deuteron spectrum from a charged particle spectrometer (CPS). Fuel density ρ is obtained by the measured ρL divided by the hot spot radius L captured using an X-ray framing camera.

Two different methods were used to vary the Knudsen number between shots. Since $N_K \sim T_i^2$, and T_i ~ implosion velocity ~ inverse of shell mass [3], we used four different thicknesses of plastic shells (7.5, 15, 20, 30 μm) at fixed inner diameter to vary the Knudsen number. The second way used to vary N_K was to have three different fill pressures (2, 5, 15 atm). Direct-drive implosions of these targets were conducted at OMEGA using 60 beams of frequency-tripled (351 nm) UV light in a 1.0 ns square pulse and a total energy of 27 kJ with smoothing by spectral dispersion and with SG4 pulse shape.

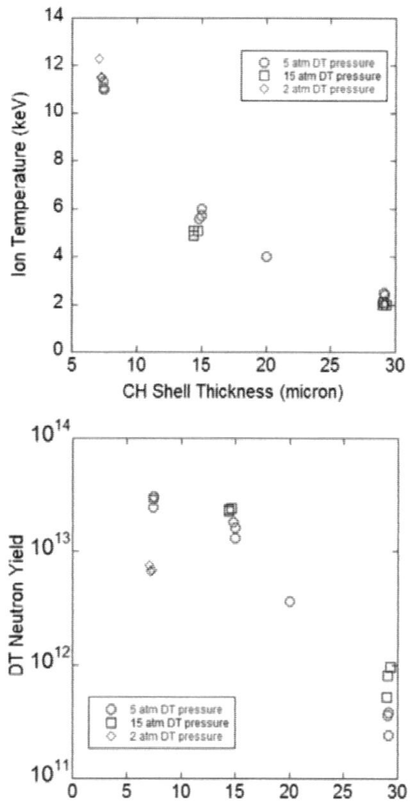

Figure 1: (a) Burn-averaged ion temperature and (b) DT neutron yields as a function of plastic shell thickness from 7.5 to 30 μm.

3. Experimental results

Time-integrated, burn-averaged ion temperature measurements using nTOF are shown in figure 1(a). As the wall thickness of plastic-shell increases from 7.5 to 30 μm, the burn-averaged ion temperature decreases significantly from ~ 12 to ~ 2 keV. There does not appear to be a strong temperature dependence with the DT fill pressure, although a slight decrease in ion temperature was observed going from 2 to 5 atm or from 5 to 15 atm, respectively. Figure 1(b) shows the DT neutron yield as a function of shell-thickness and DT fill pressure. Maximum neutron yield of (2-3)×10^{13} was achieved at 7.5 μm shell-thickness and 5 atm fill conditions. As the plastic-shell thickness increases from 7.5 to 30 μm, yield decreases nearly two order of magnitude. At 7.5 μm and 2 atm fill case, significant yield drop is observed compared to the 5 atm case. However, fill pressure variations at other shell thicknesses (e.g., 15 – 30 μm) do not affect yield as strongly.

Table I summarizes a subset of the data (i.e., 5 atm fill pressure and 7.5, 15, 20 μm thickness only) with shot number, capsule type, T_i, fuel ρL, hot spot radius, fuel mass density, $ln\Lambda$, N_K, fusion reactivity at $N_K = 0$ (i.e., $\langle\sigma v\rangle_0$), fusion reactivity at non-zero N_K (i.e., $\langle\sigma v\rangle_{N_K}$), and the resultant reduction factor for fusion reactivity expressed as $\langle\sigma v\rangle_{N_K}/\langle\sigma v\rangle_0$. No fuel ρL was recorded at 30 μm thickness capsule due to insufficient yield. As the shell thickness decreased from 20 to 7.5 μm, (1) the fuel ρL decreased from ~ 6 to ~ 2 mg/cm^2, (2) the hot spot radius increased from ~ 42 to ~ 60 μm, and (3) the Coulomb logarithm $ln\Lambda$ increased from ~ 5 to ~ 7. As a result, we achieved a minimum empirical Knudsen number N_K of ~ 0.025 at 20 μm shell thickness and a maximum N_K of ~ 0.36 at 7.5 μm shell thickness, which is nearly a factor of 14 variation in N_K in this data subset.

Table I: As-shot conditions for a subset of the data (i.e., 5 atm fill pressure and 7.5, 15, 20 μm thickness only) are shown with shot number, capsule type, T_i, fuel ρL, hot spot radius, fuel density, $ln\Lambda$, N_K, fusion reactivity at $N_K = 0$ (i.e., $\langle\sigma v\rangle_0$), fusion reactivity at non-zero N_K (i.e., $\langle\sigma v\rangle_{N_K}$), and resultant reduction factor for fusion reactivity expressed as $\langle\sigma v\rangle_{N_K}/\langle\sigma v\rangle_0$

Shot#	DT(atm)CH[μm]OD[μm]	T_i (keV)	Fuel ρL (mg/cm^2)	Hot spot radius (μm)	Fuel mass density (g/cc)	$ln\Lambda$	N_K	$\langle\sigma v\rangle_0$	$\langle\sigma v\rangle_{NK}$	$\langle\sigma v\rangle_{NK}$ / $\langle\sigma v\rangle_0$
70858	DT(5.1)CH[20]883	4.0	4.9	42	1.17	5.0	0.025	6.00E-18	5.70E-18	0.95
70861	DT(5.4)CH[14.8]872	5.6	6.2	42	1.48	5.39	0.036	2.03E-17	1.90E-17	0.94
70860	DT(5.4)CH[15]868	5.7	5.3	42	1.26	5.49	0.043	2.15E-17	1.98E-17	0.92
70849	DT(5)CH[15.0]873	6.0	4.5	42	1.07	5.65	0.055	2.55E-17	2.28E-17	0.90
70882	DT(5)CH[7.5]865	11.0	2.5	60	0.42	7.03	0.266	1.44E-16	9.02E-17	0.63
70868	DT(5)CH[7.4]860	11.1	2.3	60	0.38	7.09	0.292	1.48E-16	8.85E-17	0.60
70862	DT(5.2)CH[7.4]866	11.3	2.0	60	0.33	7.19	0.343	1.54E-16	8.56E-17	0.56
70852	DT(5)CH[7.4]870	11.0	1.8	60	0.30	7.20	0.360	1.44E-16	7.70E-17	0.53

The $\langle\sigma v\rangle$ as a function of T_i and N_K provided by A. Simakov [4] are plotted in figure 2. The $\langle\sigma v\rangle$ increases as T_i increases and more interestingly, it decreases as N_K increases. The last three columns in Table I (i.e., $\langle\sigma v\rangle_0$, $\langle\sigma v\rangle_{N_K}$, $\langle\sigma v\rangle_{N_K}/\langle\sigma v\rangle_0$) were obtained from figure 2 using the measured T_i and N_K. The $\langle\sigma v\rangle_{N_K}/\langle\sigma v\rangle_0$ indicates how much the fusion reactivity decreases at given N_K condition and can be used to estimate the yield reduction. In this work, a nearly ~ 0.5 reduction factor was obtained at 7.5 μm shell thickness, whereas a 15-20 μm shell thickness provided only 0.90 – 0.95 reduction factor.

4. Discussion

1-D radiation-hydrodynamic simulations incorporated "reduced" (i.e., simplified or approximate) ion-kinetic (RIK) models [5] were performed to assess the relative importance of (1) mix and (2) fusion reactivity reduction. In this paper, the electron thermal flux limiter was equal to 0.06. To account for laser refraction past the capsule, the incident laser energy was reduced by a factor of 0.66. Two additional parameters affecting ion mass transport (f_{idiff} used in [5]) and ion thermal conduction (f_{icnd} used in [5]) are fixed to f_{idiff} = 1, f_{icnd} = 4, respectively. Turbulent mixing was accounted for using the buoyancy-drag model of Dimonte with a fixed drag coefficient of 2.5 and an adjustable initial scale length l [5]. An improved fusion-reactivity reduction model [4] was implemented in the simulation code, where the coefficient f_{KNU} can be varied.

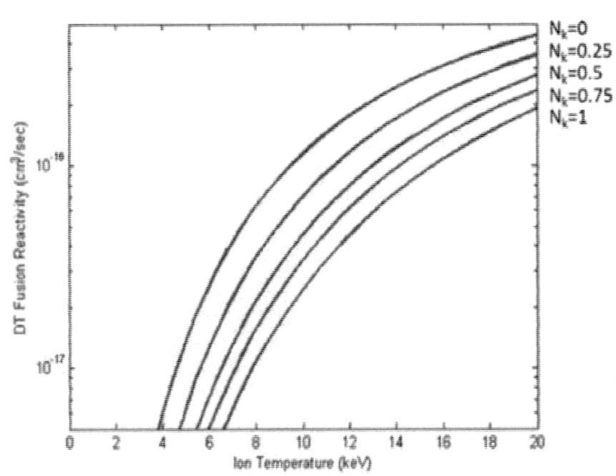

Figure 2: DT fusion reactivity as a function of ion temperature and Knudsen number [4]

In figure 3, DT neutron yields from 8 shots listed in Table I and additional 3 shots performed by 30 μm shell thickness and 5 atm fill pressure conditions are re-plotted as a function of T_i (red x mark). Three simulations were performed for each shot: first of which (the "nominal" simulation) did not invoke either the mix model (l = 0) nor fusion reactivity reduction (f_{KNU} = 0),

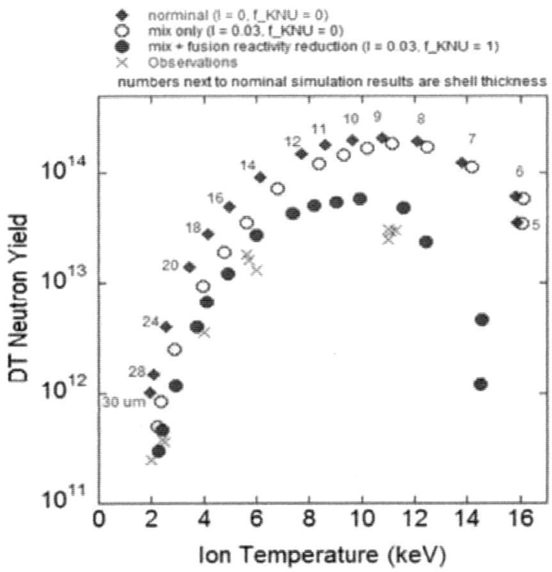

Figure 3: DT neutron yield as function of ion temperature for a norminal simulation (l = 0, f_{KNU} = 0), a mix only (l = 0.03, f_{KNU} = 0), a mix + fusion reactivity reduction (l = 0.03, f_{KNU} = 1), and experimental data

second of which (the mix only) invoked the mix model only (l = 0.03, f_{KNU} = 0), and third of which (mix + fusion reactivity reduction) invoked both (l = 0.03, f_{KNU} = 1). Turbulent mix shows a stronger effect on yield as shell thickness increases (or T_i and N_K decreases). However, the fusion reactivity reduction is more important than turbulent mix as T_i and N_K increases (or shell thickness decreases). However, the fusion reactivity reduction is more important than turbulent mix as T_i and N_K increases (or shell thickness decreases). For example, for an implosion with ~ 11 keV T_i, the observed yield over simulation (YOS) was ~ 0.19 by the norminal simulation, ~ 0.21 by the mix alone, and ~ 0.83 by the mix + fusion reactivity reduction. The Reduced-Ion-Kinetic (RIK) simulation which includes both fusion reactivity reduction and mix model was necessary to provide a better match between the observed neutron yields and those simulated.

5. Conclusion

Unlike the previous Omega DT experiments, empirical Knudsen number (N_K) was inferred by using nuclear diagnostics (i.e., nToF, CPS, X-ray framing camera). Systematic variation of N_K by a factor of 14 (0.025 - 0.36) was achieved by varying shell thickness (ion temperature) and fill pressure. As the N_K increases, the fusion reactivity reduction model was more important than turbulent mix to explain the discrepancies between simulated yield and measured yield. Turbulent mix showed a stronger effect on yield as shell thickness increases (or T_i and N_K decreases). The Reduced-Ion-Kinetic (RIK) simulation which includes both fusion reactivity reduction and mix model was necessary to provide a better match between the observed neutron yields and those simulated. Two future experiments may provide additional information to qualify fusion reactivity reduction [6]. First, a N_K higher than 0.36 may provide further qualification of the fusion reactivity reduction. Second, DT mixing ratio variation would separate fusion reactivity reduction physics and possible multi-species diffusion physics [7].

References

[1] Molvig K, *et al.* Phys. Rev. Lett. **109**, 095001 (2012).
[2] Huba J D, NRL Plasma Formulary (2009).
[3] Rosen M D, Phys. Plasmas **6**, 1690 (1999).
[4] Albright B J, *et al.* Phys. Plasmas **20**, 122705 (2013).
[5] Hoffman N M, *et al.* Phys. Plasmas **22**, 052707 (2015).
[6] Kagan G, *et al.* Phys. Rev. Lett. **115**, 105002 (2015).
[7] Kagan G and Tang X, Physics Letters A **378**, 1531 (2014).

Relativistic soliton formation in laser magnetized plasma interactions

W Feng, J Q Li[1] and Y Kishimoto

Graduate School of Energy Science, Kyoto University, Gokasho, Uji, Kyoto 611-0011, Japan

[1]Emai address: lijq@energy.kyoto-u.ac.jp

Abstract. The laser plasma interactions in the presence of strong magnetic field are studied by employing particle-in-cell simulations. Simulations show that the energy absorption of strong laser pulse is mainly characterized by the electron cyclotron resonance heating (ECRH) when the magnetic field is large enough. However, it is found that for a weaker magnetic field, a standing or moving soliton can be generated in some moderate laser intensity regions, greatly enhancing the laser absorption. The laser intensity for the soliton heating decreases as the magnetic field increases. Furthermore, the soliton position moves towards the front boundary when the laser intensity or magnetic field strength increases.

1. Introduction

Recently, the study of laser plasma interactions in the presence of strong magnetic field has become a hot research topic in fast ignition. Kilo-tesla magnetic field has been proposed to guide the relativistic electron beams (REBs) and enhance the coupling efficiency of the REBs to the fuel core. However, due to the conversional laser pulse contrast, the under-dense pre-plasma formed by the pre-pulse is inevitable. Early particle-in-cell (PIC) simulations and experiments showed that a bright soliton can be generated during the interactions of a strong laser pulse with an under-dense plasma and up to 40% of the laser energy can be converted into the soliton structures, which makes the soliton formation an important energy conversion process in the laser plasma interactions [1, 2]. Hence, the formation and properties of the solitons should be reconsidered in the presence of strong magnetic field.

The dynamics of soliton formation in unmagnetized plasmas has been studied intensively. Kozlov et al first studied the relativistic solitons in a cold plasma by an envelope approximation [3]. Their work were extended by several authors to include the ion's motion and warm temperature effects [4-12]. Different types of solitons under different soliton velocities have been reviewed by D. Farina [12]. Compared with the systematic and abundant research results on the unmagnetized plasmas, the study of solitons in magnetized plasma is much less and insufficient [13,14]. In this paper, we present the numerical simulations of the effects of magnetic field on the soliton formation and properties. The analytical analysis of the soliton dynamics is left for another publication.

2. Simulation Results

One dimensional PIC simulations are carried out by employing the extended particle based integrated code (EPIC3D) [15,16]. The simulation is set up as follows. The system size is $15.36\mu m$, which corresponds to $L_y = 3072$ in normalized unit of $0.005\mu m$. A Gaussian shaped laser pulse is irradiated from the left boundary with the wavelength $0.82\mu m$ and pulse duration 40fs. The peak intensity of the laser varies from $2.1 \times 10^{16} W/cm^2$ to $8.2 \times 10^{18} W/cm^2$ or $a_0 = 0.1$ to $a_0 = 2.0$ in normalized values. The laser is linearly polarized in the x direction. A fully ionized carbon plasma with electron and ion temperature $0.5kev$ is initially set between $y = 500$ to $y = 2500$ and a pre-plasma of density layer $n_e(y) = 0.63 n_c \exp[(y - 1500)/164]$ is assumed from $y = 500$ to $y = 1500$. Here $n_c = 1.66 \times 10^{21} cm^{-3}$ is the cut off density for the laser wave. The uniform magnetic field is also applied in the y direction. In the simulations, periodic boundary conditions are employed for the particles while the laser waves are outgoing.

The laser absorption rate is used to evaluate the laser heating efficiency during the propagation of the laser pulse in different magnetized plasmas, as summarized in figure 1 [17]. The absorption rate is calculated by the ratio of energy increment of the electrons and ions to the laser energy. It shows that the dependence of heating efficiency on laser intensities are quite different for different magnetic field strength. Generally, larger amplitude laser can effectively heat the plasma, however, for not so strong magnetic field strength, there exists a moderate laser intensity region where enhanced heating occurs. The enhanced heating region moves to the lower laser intensity area as the magnetic field increases. When the magnetic field approaches the strength where electron cyclotron resonance happens, the enhanced heating disappears and the heating efficiency is almost independent of laser intensity, as shown by the line with upper triangles ($B_0 = 15.0kT$) in figure 1. Hence, the absorption is mainly characterized by the electron cyclotron resonance heating (ECRH) in the high magnetic field regime. The peak of the enhanced heating is about 37%, a little smaller than the ECRH heating efficiency.

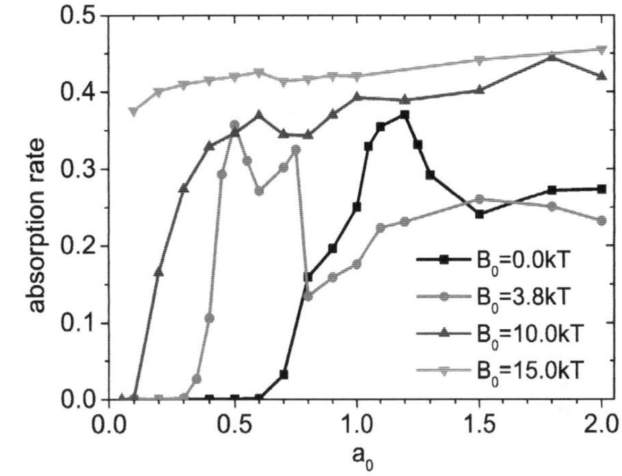

Figure 1. Energy absorption rate with laser intensity a_0 for different magnetic field strength. The magnetic field strength for ECRH is about 13kT.

The details on the laser pulse propagation in the enhanced heating region are depicted in figure 2, where a localized mode, namely, a bright soliton is observed, which co-exists with a plasma cavity. The bright soliton is the wave which the wave intensity is maximum at the center and vanishes at the infinity. The soliton formed in the magnetized plasma has the same properties as in the unmagnetized plasma [1], for example, long-lived and short width of about one laser wavelength. The mechanism of the soliton generation is as follows [12]. The density perturbation is generated after the laser pulse due to the ponderomotive force and the laser starts to lose energy. Meanwhile, the laser frequency decreases because this energy transfer is an adiabatic process and the ratio between the wave energy density and the frequency is conserved. Actually, if one checks the frequency spectra of the laser field at $y = 2000$, where the soliton is generated, a lower frequency spectrum appears besides the incident laser frequency. More importantly, it has a higher intensity than the laser frequency one. Once the laser frequency falls below the plasma frequency, the laser wave becomes trapped in the density cavity, forming a solitary wave. Since the group velocity of the laser pulse decays with the frequency in the weak magnetic field regime, the soliton wave propagates with almost zero velocity. As a result, the laser can be absorbed efficiently in the standing density well. In the case of weakly relativistic limit, the frequency downshift is approximately proportional to the density perturbation and the square of laser intensity. Meanwhile it has a positive correlation with the magnetic field strength [18]. Hence, as the magnetic field strength increases, the laser intensity region needed for the soliton formation reduces. If the density perturbation is not large enough to trap the laser wave, the soliton will collapse soon after its generation and moves away. Note that there is strong electrostatic field inside the soliton in the initial phase of the soliton formation. However, afterwards, the ponderomotive force starts to dig a hole in the ion density and the quasi-neutrality state can be achieved.

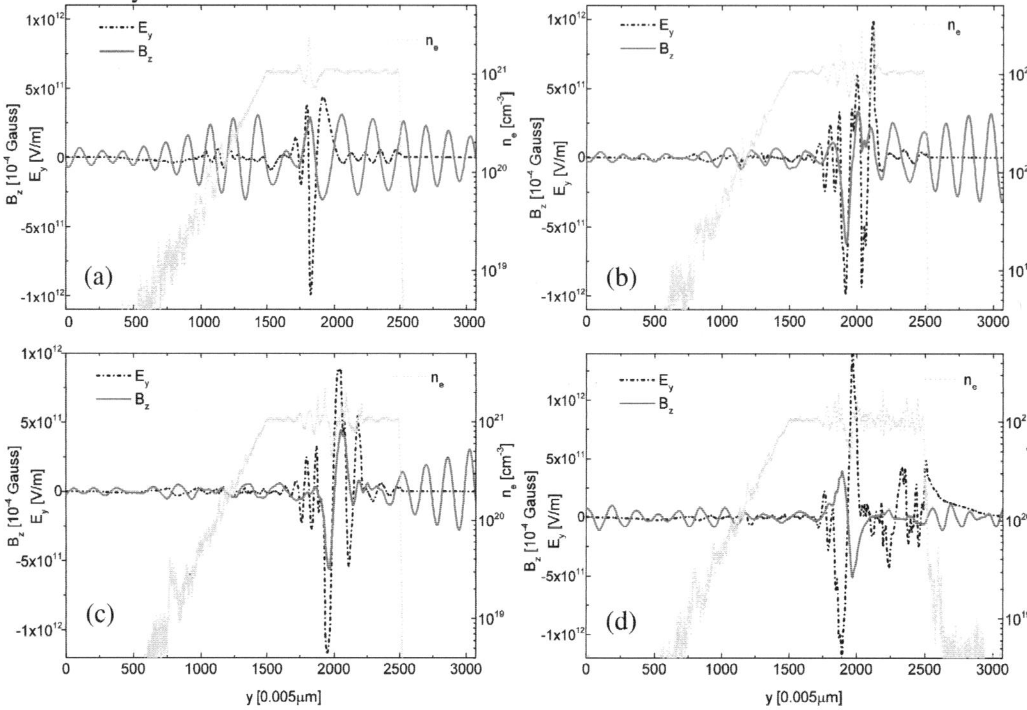

Figure 2. The time evolution of a standing soliton at $t = 119.6fs$ (a: upper-left), $t = 135.6fs$ (b: upper-right), $t = 139.1fs$ (c: bottom-left) and $t = 161.1fs$ (d: bottom-right) with $a_0 = 0.5$ and $B_0 = 3.8kT$.

The soliton position in the uniform density area is an interesting and meaningful issue, since we may potentially use the soliton generation to control the energy deposition position in the plasma. Some factors relevant to the soliton position are studied as shown in figures 3 and 4. Fig. 3 shows that, the soliton position shifts from the rear boundary to the front one when one increases the laser intensity or magnetic field strength. This may result from the dependence of the frequency downshift on the magnetic field strength and laser intensity, which may be characterized by a threshold for the soliton formation: one of the two factors increases, the other needed for soliton generation decreases. On the other hand, we have extended the system size to $L_y = 4096$ and changed the characteristic length of the pre-plasma, as shown in Figure 4. It is observed that the soliton position is independent of the system size, but shifts to the front boundary as the characteristic length of the pre-plasma increases. The detailed mechanism of the soliton positon on the density profile still remains unknown and needs to be resolved in the future.

3. Summary

We have used the EPIC3D code to study the soliton formation and their properties in the strongly magnetized plasmas. It is found that, the absorption is dominated by the ECRH when the magnetic field strength is strong enough, e.g. comparable to the strength where ECRH occurs. In the case of weak magnetic field, standing or moving solitons can be generated in some moderate laser intensity regions, which can greatly enhance the absorption rate. 2D simulations also confirmed this enhancing effect [17]. The laser intensity for the soliton generation decreases as the magnetic field increases. The soliton is of long-living time with a very short width and a lower frequency compared to the incident laser wave. The relevant factors to soliton position are also studied. It is found that the soliton position depends on the characteristic length of the pre-plasma rather than the uniform plasma length. Furthermore, the soliton position moves from the rear boundary to the front one as the laser intensity or magnetic field increases. Hence, considering the high efficiency of energy transfer by the solitons, the soliton generation in strongly magnetized plasmas may play an important role in fast ignition, laser-plasma based particle acceleration as well as other potential applications.

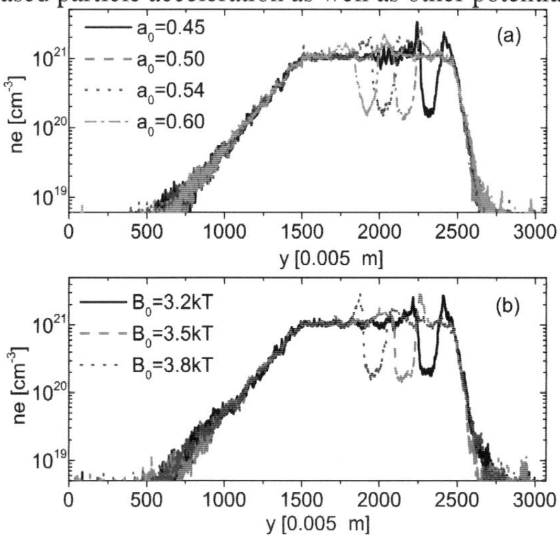

Figure 3. The soliton positon dependence on laser intensities (a) and magnetic field strength (b). In (a) $B_0 = 3.5kT$, and in (b) $a_0 = 0.5$.

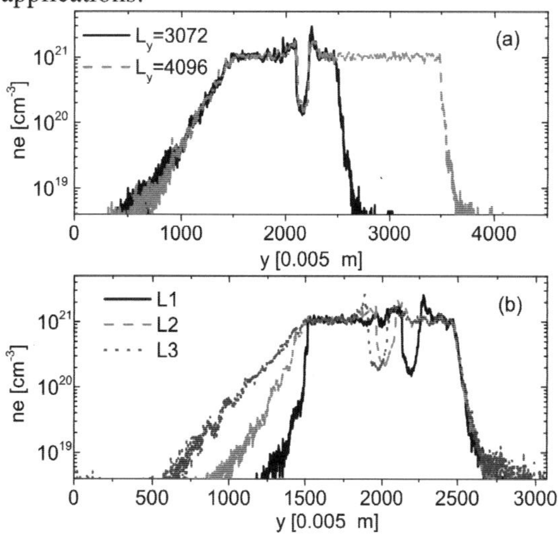

Figure 4. The soliton positon dependence on system size (a) and characteristic length of the pre-plasma (b). The parameters are $a_0 = 0.5$, $B_0 = 3.5kT$ for (a) and $B_0 = 3.8kT$ for (b). “L1”, “L2” and “L3” in (b) represent the case without pre-plasma, with pre-plasma of characteristic length 82 and 164, respectively.

References

[1] Bulanov S V, Inovenkov I N, Kirsanov V I, Naumova N M and Sakharov A 1992 *Phys. Fluids B* **4** 1935

[2] Borghesi M, Bulanov S, Campbell D H, Clarke R J, Esirkepov T Zh, Galimberti M, Gizzi L A, MacKinnon A J, Naumova N M, Pegoraro F, Ruhl H, Schiavi A and Willi O 2002 *Phys. Rev. Lett.* **88** 135002

[3] Kozlov V A, Litvak A G and Suvorov E V 1979 *JETP* **49** 75

[4] Kaw P K, Sen A and Katsouleas T 1992 *Phys. Rev. Lett.* **68** 3172

[5] Esirkepov T Zh, Kamenets F F, Bulanov S V and Naumova N M 1998 *JETP Lett.* **68** 36

[6] Poornakala S, Das A Sen A and Kaw P K 2002 *Phys. Plasmas.* **9** 1820

[7] Farina D and Bulanov S V 2001 *Phys. Rev. Lett.* **86** 5289

[8] Lontano M, Bulanov S and Koga J 2001 *Phys. Plasmas.* **8** 5113

[9] Poornakala S, Das A, Kaw P K, Sen A, Sheng Z M, Sentoku Y, Mima K and Nishikawa K 2002 *Phys. Plasmas.* **9** 3802

[10] Sanchez-Arriaga G, Siminos E and Lefebvre E 2011 *Plasma Physics and Controlled Fusion* 53 045011

[11] Sanchez-Arriaga G, Siminos E and Lefebvre E 2011 *Phys. Plasmas.* **18** 082304

[12] Farina D and Bulanov S V 2005 *Plasma Physics and Controlled Fusion* **47** A73

[13] Farina D, Lontano M and Bulanov S 2002 *Phys. Rev. E* **62** 4146

[14] Borhanian J, Kourakis I and Sobhanian S 2009 *Phys. Lett. A* **373** 2667

[15] Masaki T and Kishimoto Y 2005 *Plasma Fusion Res.* **81** 843

[16] Kishimoto Y and Masaki T 2006 *J. Plasmas Phys.* **72** 971

[17] Feng W, Li J Q and Kishimoto Y 2016 *Phys. Plasmas.* **23** 032102

[18] Shukla P K and Stenflo L 1984 *Rhys. Rev. A* **30** 2110

Analysis of the symmetry of the direct-drive target implosion under laser pulse of a megajoule scale

N N Demchenko[1], S Yu Gus'kov[1,3], P A Kuchugov[1,2], V B Rozanov[1], R V Stepanov[1], G A Vergunova[1], R A Yakhin[1] and N V Zmitrenko[2]

[1] P.N. Lebedev Physical Institute of RAS, 53 Leninskiy Ave., 119991, Moscow, Russia
[2] Keldysh Institute of Applied Mathematics of RAS, 4 Miusskaya Sq., 125047, Moscow, Russia
[3] National Research Nuclear University MEPhI, 31 Kashirskoe Hwy., 115409, Moscow, Russia

E-mail: pkuchugov@gmail.com

Abstract. The class of direct-drive targets for the laser pulse of a megajoule scale is considered in the work. A distinctive feature of the design of these targets is a relatively low aspect ratio (outer DT-shell radius/shell thickness, $A \sim 10$) to provide greater compression stability. The irradiation of the target surface is carried out by the shaped laser pulse on 2ω of Nd-laser. The influence of laser energy absorption inhomogeneity on the parameters of compression and burning of thermonuclear fuel on the final stage of implosion is studied in the work based on 1D and 2D numerical calculations. The results of 2D modeling show that in the case of target's offset from the point of beams crossing significantly greater reduction in the neutron yield is observed than in the case when only irregularities caused by the geometry of irradiation by the finite number of beams are taken into account.

1. Introduction

It is considered that the main obstacle to the achievement of high gain thermonuclear burning is a possible growth of various types of perturbations [1]. This is due to the fact that the precise compliance with the conditions for the irradiation symmetry, as well as capsule manufacturing without deviations from the nominal values of the parameters is difficult. In these conditions the possible approach is to look for such targets designs, which would provide maximum compression stability, accompanied by efficient DT-fuel burning. The class of such low aspect targets was proposed in [2–4]. In this work the efficiency of these targets for the different variants of inhomogeneity of the absorption of incident laser flux is evaluated. The cases which correspond to target irradiation by specific multi-beam laser system and target offset from chamber center will be considered. It should be noted that problem of target positioning i.e. placing it at the desired point is especially relevant for the prospects of the future fusion reactors where the capsules should be put in the core with a frequency of 1-10 Hz [5].

2. Target design and scheme of irradiation

The considered target is a system of nested shells and comprises an inner cavity with a radius of 1414 μm filled by DT-gas of density 10^{-3} g/cc (the pressure is 0.2–0.4 atm at cryogenic temperatures). Then, the target consists of a layer of frozen DT-ice of 149 μm thick and a density of 0.253 g/cc. Finally, the target outside the DT-shell is surrounded by a polystyrene

$(C_8H_8)_n$ of 34 μm thick and a density of 1.05 g/cc. Schematic representation of the target is shown on Fig. 1a. Irradiation of the target is carried out by 192 beams, united into clusters by 4 beams, which are located on the sides of the cube, as shown on Fig. 1b, the optical axes intersect at the center of the target. The total energy of the laser pulse is 2 MJ, a shape of the pulse is shown on Fig. 1c, the duration is 10 ns, the working wavelength is the second harmonic of the Nd-laser ($\lambda_L = 0.53$ μm). Under the conditions outlined above, total absorbed laser energy

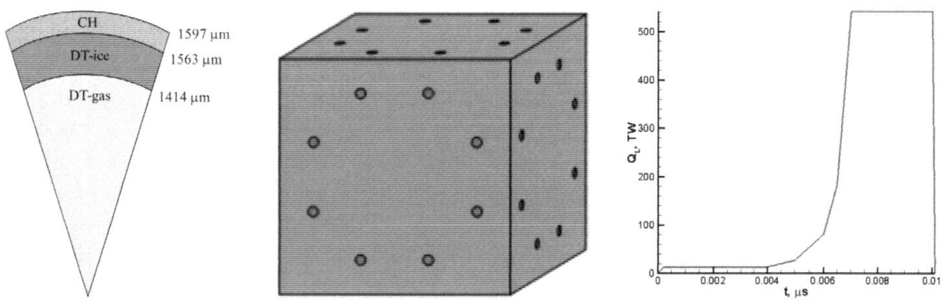

(a) (b) (c)

Figure 1: (a) Target dimensions, (b) scheme of target irradiation by multi-beam laser system, (c) shaped incident laser pulse.

about 1.5 MJ and without taking into account angular dependency of the absorbed laser flux thermonuclear gain $G = 20.8$ and maximal areal density in DT-fuel $(\rho R)_{\max} = 0.92$ g/cm^2 [3] was obtained in 1D calculation.

3. Modeling of the inhomogeneity of laser energy absorption in the target

An assessment of inhomogeneity of laser energy absorption was done by using RAPID [6] and SEND [3, 7] numerical codes. The first one solves equations of one-dimensional two-temperature hydrodynamics together with Maxwell's equations to determine a part of incident laser energy absorbed in the target, the second one is responsible for the calculation of target heating uniformity by multi-beam laser system. On Fig. 2 the angular distributions of the total absorbed laser energy for the two cases described in the introduction are shown. It is clear that they differ

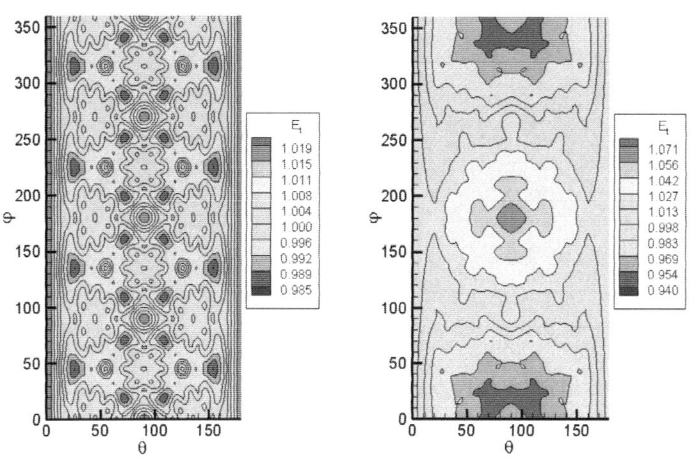

(a) (b)

Figure 2: The angular distributions of the total absorbed laser energy: (a) inhomogeneity due to target irradiation geometry, (b) inhomogeneity due to target offset along X-direction on 80 μm.

significantly. Firstly, the uniformity of the absorption of laser energy $\eta_E = E_t^{\min}/E_t^{\max}$ takes value 0.9663 in the first case and 0.8714 in the second one. Secondly, the dominant modes l_d of induced perturbations are $8 \div 10$ and 1, respectively. It should be noted that when the target is displaced from the chamber center on 80 μm the efficiency of laser energy absorption slightly decreases in comparison with the standard conditions of target irradiation while the rate of uniformity drops significantly. This is due to the fact that on the opposite sides of the target located on the axis of offset, the values of absorbed energy change in opposite directions, creating a large difference between the target heating.

4. Target implosion efficiency

Investigations of target implosion efficiency were carried out using numerical programs DIANA [8] and NUTCY [9, 10] in the 1D and 2D geometry, respectively. One-dimensional calculations can be represented as points on the distributions shown on Fig. 2, two-dimensional – in the form of straight line segments. Modeling of the compression and burning of thermonuclear fuel with the help of DIANA program, i.e. spherically symmetric formulation, for different pairs of angles (θ, φ) led to the ranges of variation of the thermonuclear gain G from 8 to 14 for standard irradiation conditions and from 4 to 16 in the case of target offset by 80 μm along one of the axes if the center of coordinate system is placed in the center of the chamber. Omitting details, we only note that in the case of the shifted target variation of obtained values is significantly higher than without considering such an offset.

The next step of the study of the target performance is investigation of the efficiency of target compression and burning in the 2D geometry. The initial state for 2D calculations is formed from a set of one-dimensional profiles, taken at time $t = 10$ ns, which corresponds to the end of the laser pulse. Modeling for different segments on the maps of absorbed energy leads to the following ranges of variations of G: from 2.82 to 5.85 under standard target irradiation conditions and from 0.07 to 5.4 in the case of target offset on 80 μm along the X-direction. Thus, in the first case, a two-fold reduction of thermonuclear gain is observed, and in the second case worsening of the burning takes place until its termination. On Fig. 3 the distributions of density and temperature at the time of peak compression for standard irradiation conditions and for the case of shifted target with different resulting values of the gain which correspond to different line segments on Fig. 2 are shown.

Analysis of 2D distribution shows that in the case of target offset from the point of beam crossing the temperature of central area is lower than at standard irradiation conditions due to the fact that because of presence of significant low mode perturbations the transform of kinetic energy of accelerated DT-shell into internal energy doesn't occur effectively. This leads to a reduction in the rate of main fusion reaction. This circumstance, together with a reduction in the maximal value of areal density in the DT-fuel is the reason for decreasing DT burnout which directly affects the value of thermonuclear gain degrading it.

5. Conclusion

The influence of inhomogeneities of laser radiation absorption associated with both the geometry of the irradiation by the laser system and the offset of the target from the point of beams focusing on the implosion efficiency was investigated. It is shown that the geometry of the target irradiation has no significant effect on the values of the thermonuclear gain compared with a spherically symmetric case. Target offset from the chamber center on the contrary can lead to situations with not burning DT-fuel even when the shift amount is not more than 5% of the initial radius of the target. Thus low mode perturbations have much greater impact on the thermonuclear yield than ones associated with the geometry of the irradiation by the laser system.

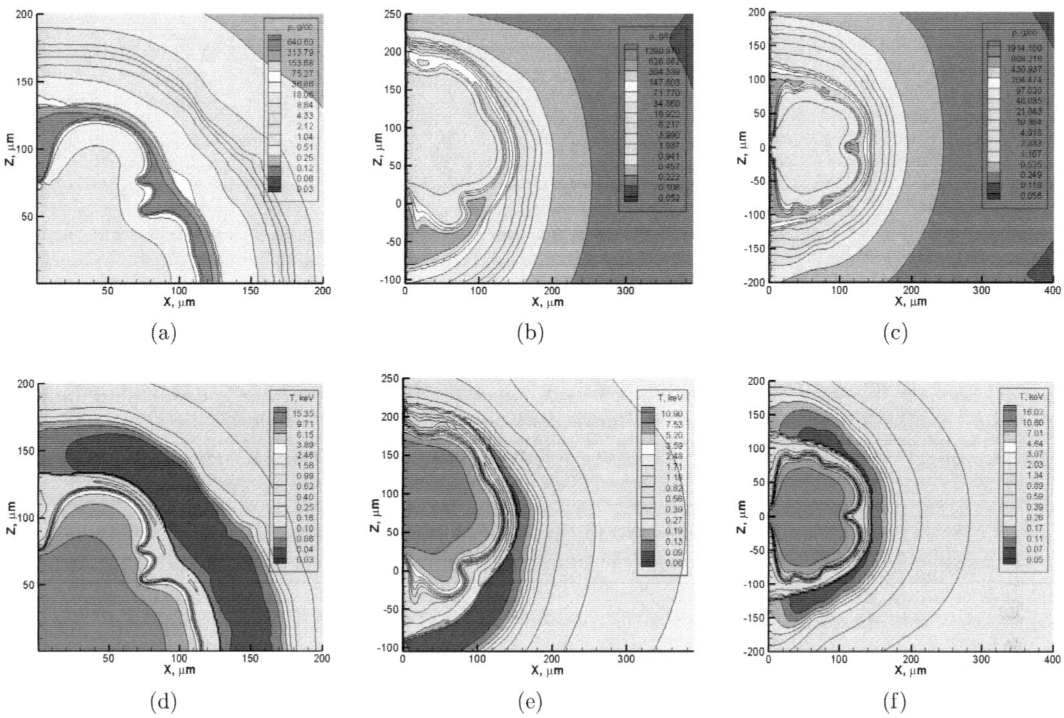

Figure 3: Distributions of density (a), (b), (c) and temperature (d), (e), (f). The first column refers to standard irradiation conditions at time moment $t = 11.1$ ns with resulting gain 2.82 ($\theta = 0° - 90°$, $\varphi = 46°$), the second and third columns – to the case of shifted target at time moment $t = 11.14$ ns with resulting gain 0.07 ($\theta = 90°$, $\varphi = 0° - 180°$), and at time $t = 11.24$ ns with resulting gain 5.4 ($\theta = 0° - 180°$, $\varphi = 0°$) respectively.

Acknowledgments

The work was partially supported by Russian Foundation for Basic Research, projects No. 14-01-00828-a, 14-02-00270-a and 16-31-60101-mol_a_dk.

References

[1] Edwards M, Patel P, Lindl J, Atherton L, Glenzer S, Haan S, Kilkenny J, Landen O, Moses E and Nikroo A et al 2013 *Phys. Plasmas* **20** 070501

[2] Rozanov V, Gus'kov S, Vergunova G, Demchenko N, Stepanov R, Doskoch I, Yakhin R, Bel'kov S, Bondarenko S and Zmitrenko N 2013 *Book of Abstracts of the* 8th *International Conference on Inertial Fusion Sciences and Applications, Nara, Japan, September 8–13, 2013* p 236

[3] Bel'kov S, Bondarenko S, Vergunova G, Garanin S, Gus'kov S, Demchenko N, Doskoch I, Kuchugov P, Zmitrenko N and Rozanov V et al 2015 *JETP* **121** 686–698

[4] Demchenko N, Doskoch I, Gus'kov S, Kuchugov P, Rozanov V, Stepanov R, Vergunova G, Yakhin R and Zmitrenko N 2015 *Laser and Particle Beams* **33** 655–666

[5] Hogan W, Coutant J, Nakai S, Rozanov V and Velarde G (eds) 1995 *Energy from inertial fusion* (Vienna: IAEA)

[6] Afanas'ev Y, Gamalii E, Demchenko N, Krokhin O and Rozanov V 1980 *Sov. Phys. JETP* **52** 425–431

[7] Rozanov V and Demchenko N 1985 *Sov. J. of Quantum Electronics* **15** 1251

[8] Zmitrenko N, Karpov V, Fadeev A, Shelaputin I and Shpatakovskaya G 1983 *VANT. Series: Methods and Software for Numerical Solution of Problems of Mathematical Physics* **2** 34–37

[9] Tishkin V, Nikishin V, Popov I and Favorskii A 1995 *Matem. Mod.* **7** 15–25

[10] Lebo I and Tishkin V 2006 *The Study of Hydrodynamic Instability in Problems of Laser Fusion by Methods of Mathematical Modeling* (Moscow: FIZMATLIT)

Control of Be capsule low mode implosions symmetry at the National Ignition Facility

G A Kyrala[1], J L Kline[1], S Yi[1], A N Simakov[1], R E Olson[1], D C Wilson[1], S Batha[1], E L Dewald[2], R Tommasini[2], J E Ralph[2], A G MacPhee[2], D A Callahan[2], O A Hurricane[2], D E Hinkel[2], S F Khan[2], T Ma[2], N Izumi[2], S Nagel[2] and J R Rygg[2]

[1]Los Alamos National Laboratory, Los Alamos, NM, USA
[2]Lawrence Livermore National Laboratory, Livermore, CA, USA

Kyrala@lanl.gov

Abstract. We present results of the beryllium experimental campaign on the implosion symmetry properties of beryllium capsules at the National Ignition Facility (NIF) [1]. These indirect drive experiments measure both the inflight and core self-emission implosion symmetry. The inflight symmetry of the ablator before stagnation is measured using a backlight imaging technique. A copper backlighter was used to measure the transmissions (or backlit absorption) of the copper doped beryllium shells. Images of the x-ray emission from the core around bang time provide a measure of the symmetry near peak compression. Both pieces of information about the 2D symmetry are used to infer the drive and velocity uniformity enabling us to predictably adjust the properties of the incident laser, mainly the time dependent ratio of the inner beam cone power to the outer laser beam powers, to achieve proper symmetry of the implosion. Results from these experiments show inner beam propagation is not degraded compared to similar implosions with CH ablators. Variations in the shape compared with implosions using CH ablators also provides information about the cross beam energy transfer used to adjust the equatorial shape and thus infer information about the differences in plasma conditions near the laser entrance holes. Experimental results of the implosion shape for beryllium capsules will be presented along with comparisons relative to CH ablators.

1. Introduction

Symmetry of implosion systems has been currently identified as one of the main culprits in not achieving ignition at the National Ignition Facility (NIF) using indirectly driven cryogenically layered DT ignition fuels. [2 ,3 ,4] Although low order spherical symmetry has been achieved with symcaps, where the layered DT fuel is replaced with an equivalent mass of the shell material [5], the same conditions did not achieve the same goals necessary for ignition with cryogenically layered targets. Differences due to low order symmetry have been explored, but the incapability of accurately calculating the symmetry of the hohraum radiation limits conclusions about the effect. Higher order symmetries, due to instability growth from either objects that carry the capsule, or from the perturbations on the various surfaces in the capsule, offers a slightly better explanation. All the

Journal of Physics: Conference Series **717** (2016) 012033 doi:10.1088/1742-6596/717/1/012033

previous studies have used various CH or Diamond capsules with different levels of doping. Beryllium (Be) ablators have been recognized for quite a while as more efficient ablators, with different instability growth characteristics, requiring different drive characteristics from CH capsules [6,7]. The Be low opacity leads to higher ablation rates, [8] pressure, and velocity.[9] Thus a Be campaign was started to examine the use of Be and find what similar or different issues from those of CH or Diamond occur. Initial studies, in this first Be campaign, have been made with hohlraum conditions as close to those from the high foot campaign [10] to quickly establish differences and discover any issues. In this study, part of the first Be capsule campaign, we concentrate on the study of symmetry of both the imploding non-layered Be capsule shell, as well as the symmetry of the x-ray emission from the imploding fuel. In this paper we concentrate on the results and techniques as used on a 2d-ConA shot techniques [11, 12] that are similar to those used on other shots, and we summarize the results of the symmetry tuning.

2. Experimental Conditions

For this study we used a copper doped Be capsule (Cu:Be) in a 575 Au hohlraum using a 2D-ConA x-ray backlit configuration [11] to measure the symmetry of the in-flight shape of the capsule. The capsule dimensions and doping are shown in Figure 1, while the hohraum arrangement, target mounting, and backlighter configuration are shown in Figure 2. Basically the capsule is a symcap of the layered capsule, where the DT fuel layer is replaced by an equivalent mass of Be to preserve the hydrodynamic, rocket like, motion of the ablator layer. The capsule also has three copper doped layers, that protects the fuel from the >1.8 keV preheat from the hohlraum and helps controls the ablation front position in the shell. The doped layer, having a different density from the rest of the Beryllium in the capsule, is graded in three densities to reduce the spatial gradient of the density jumps necessary for the copper doping. The surfaces of the Be capsule are polished to NIF specs [13] to reduce the seeding of instabilities. Inherent spatial structures within the Be remain to be controlled in the future. The capsule is held in the center of the hohraum to an accuracy of 30 μm using a tent, Figure 2, with a thickness of 45 nm enclosing the capsule. Two square windows are placed along the axial line of sight to allow the backlighter radiation, and the x-ray emission form the implosion, to reach a gated x-ray detector [10]. These windows do perturb the capsule implosion, producing an M_2 mode as seen from the polar direction, but are not expected to affect the Legendre mode decomposition in the axial line of sight used in this study. Filtered pinhole x-ray images were acquired using time integrated image plates as well as a gated x-ray camera with a gating time of 80 ps per image [14]. A spatial resolution of 11 microns was achieved using a nominal 10 μm diameter pinholes, and a magnification 6X.

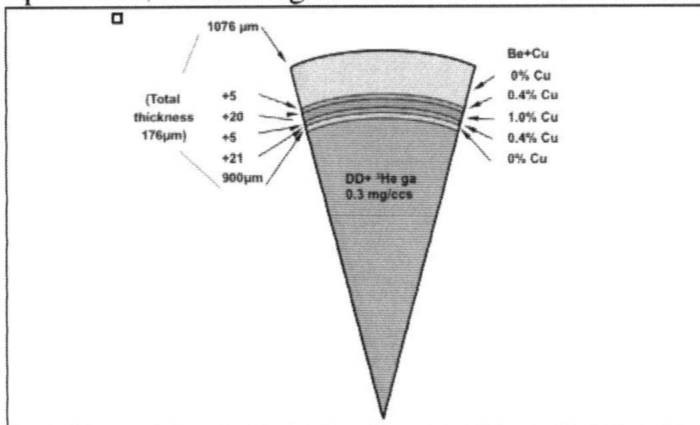

Figure 1, Diagram of Capsule.

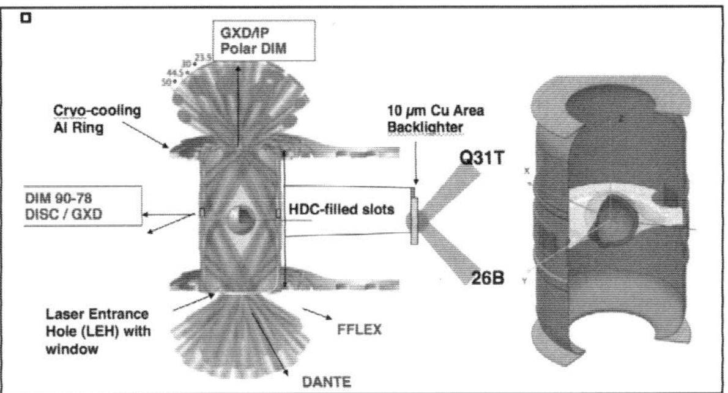

Figure2 , Hohlraum, Lasers and mounting

1.44 MJ of laser energy was converted to an x-ray drive in a hohraum to implode the Cu:Be capsule. Eight laser beams of the NIF facility were pointed at a copper backlighter target placed outside the hohlraum, while the rest of the 184 beams were directed to the same positions as in a cryo hohlraum. The laser had different wavelengths for each of the cones incident to the hohlraum axis: inner 23.5°

10530.70 Å, 30° 10530.0 Å, outer cones at 45° and 50° 10524.3 Å. The choice of wavelengths was made to allow for cross beam energy transfer [15] in the hohlraum gas (1.6 mg/cc of ^{4}He at 24 K). The laser pulse shape is shown in figure 3, where the inner and outer beams had slightly different pulse shapes and energy per beam to allow for better control of the symmetry and drive in the hohlraum. The laser had 1.32353 Ghz SS modulation. The backlighter beams (BL) turned on from 15.5 to 18 ns to capture the backlit transmission images of the imploding shell before stagnation. Turning off BL before stagnation allowed us to record the X-ray emission around bang time as well.

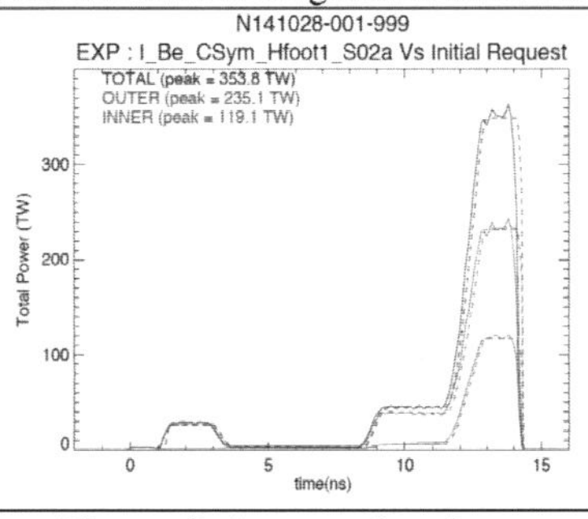

Figure 3. Laser pulse shape

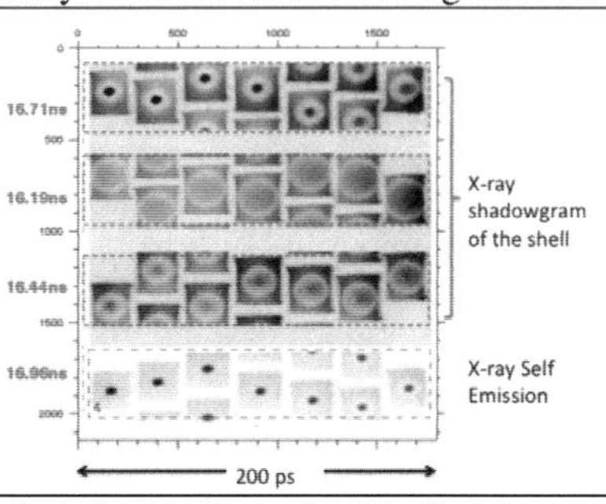

Figure 4. A frame from the 2D-ConA experiment

3. Data

A typical record of the data is shown in figure 4, where the delays at the center of each strip are shown. The data is analysed using two different techniques for different features. At early times, we use the transmission images to extract the position of the Limb (the ring-like light region in the images) as a function of angle and time (Ref 10), and the position of the maximum slope on the outside of the capsule. The angular data is used to extract a Legendre decomposition of the position of the limb as a function of time giving an indication of the symmetry of the imploding shell. At Bang time, we use the x-ray self emission images, shown in the lowest strip in Figure 4, and to extract symmetry properties, [4] also characterized by Legendre Polynomial amplitudes.

A summary of the data for shot N150420-003 is shown in figure 5. Two graphs are shown. The left graph shows the measure size of different parts of the capsule implosion at different times as well as the emitted x-ray flux from the core emission. The radius for the limbs from the shell decreases in time, similar to that from the emission for the edges (17%) of the x-ray emission. The figure also shows the stagnation of the core just before maximum x-ray emission. The graph to the right shows the history of the Legendre P_2 amplitudes from the Limb minimum and from the x-ray emission. The P_2 shows a slight swing as a function of time, but also shows that the absolute P_2 amplitude has the same sign, and is within the statistics, of the Limb amplitudes. Currently, calculations are not able to reproduce these shapes, but are able to reproduce (figure 6) the temporal variation of the Limb radii, provided a multiplier of 0.86X on the laser power was used after subtracting backscatter.

4. Summary

When we compare the results of this measurement to other measurements, figure 7, we note that the equatorial symmetry of the implosion varies monotonically, from prolate to oblate, with the wavelength difference between the inner and outer beams. The figure shows the progression of the tuning of the P_2 amplitude for different shots. The points at the right side of the figure are from baseline High-Foot CH shots used to start tuning the Hohraum. The three points at the left are for Be shots. They show the symmetry data from the 2DConA at 6 angstroms, and 1D cona at 5 Angstroms, and the final results of tuning a cryo capsule at 5.5 Angstroms. While the laser pulse shape changed little in between these shots, progress on tuning the symmetry with wavelength difference change is evident. We note that a similar behavior for the Polar M modes was harder to achieve since there are necessary differences in hohlraum symmetry between the 1dConA, 2dconA shots and the cryo implosions due to the use of some laser beams for backlighting.

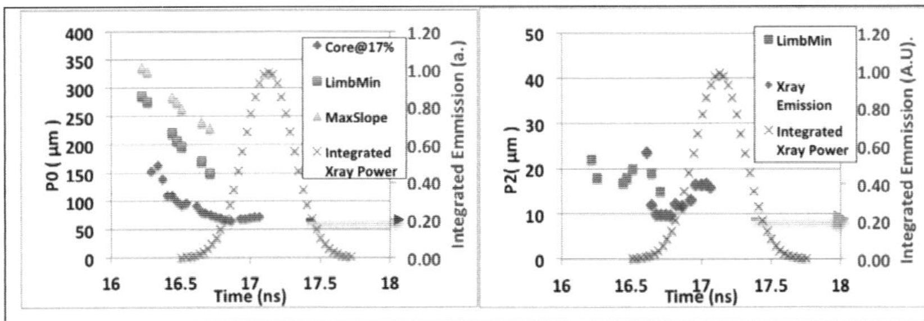

Figure 5. Summary of the 2DConA. The Limbmin, and Maxslope are the locations of the minimum and max radial slope of the transmission.

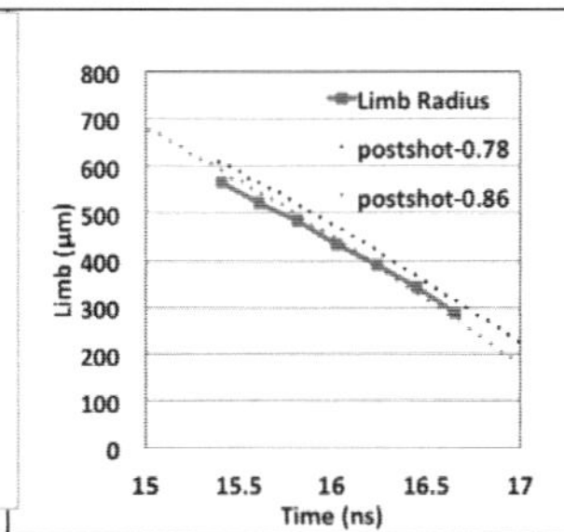

Figure 6. Predictions of the shell implosion

In a few shots, we were able to demonstrate that we can tune the equatorial symmetry of the Be implosions, that the shell and the core have roughly the same P_2 contribution, and that the effects of the tent and fill tube in the implosion images were not observed. We also show that different Δλs are needed to get to the same P_2/P_0 for different materials in the same hohlraum and laser drives.

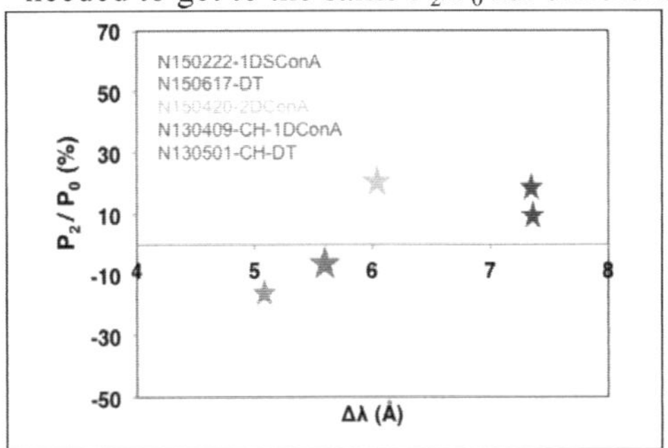

Figure 7. Summary of the tuning effort

5. Acknowledgements

We wish to thank the NIF operations team and General Atomics for which these experiments would not have been possible. This work performed under the auspices of the U.S. Department of Energy by LANL under contract DE-AC52-06NA25396 and by LLNL under Contract DE-AC5 2-07NA27344.

References

[1] E. I. Moses et al., *Phys. Plasmas* 16, 041006 (2009).
[2] D. S. Clark, et al., *Phys. Plasmas* 22 (2015).
[3] C. R. Weber, et al., *Phys. Plasmas* 22 (2015).
[4] A. L. Kritcher and e. al., *Phys. Plasmas* (submitted (2012).
[5] G. A. Kyrala, *et al*, *Phys. Plasmas* Volume: 18 , 056307 (2011).
[6] T. R. Dittrich, *et al*, *Fusion Technology* **31**, 402 (1997).
[7] D. C. Wilson, et al., *Phys. Plasmas* **5**, 1953 (1998).
[8] R. E. Olson, G. A. Rochau, O. L. Landen, and R. J. Leeper, *Phys. Plasmas* **18** (2011).
[9] A. N. Simakov, *et al.*, *Phys. Plasmas* **21** (2014).
[10] O. A. Hurricane, *et al.*, *Nature* 506, 343 (2014).
[11] J. R. Rygg, *et al.*, *Physical Review Letters*, 112, 195001 (2014).
[12] J. E. Field, *et al.*, *Review of scientific Instruments* 85, 11E503 (2014).
[13] S. W. Haan, *et al.* , *Phys. Plasmas* 18, 051001 (2011)
[14] G. A. Kyrala, *et al.* , *Rev. Sci. Instrum.* 81, 10E316 (2010).
[15] P. Michel, et al., *Phys. Plasmas* 17, 056305 (2010).

Journal of Physics: Conference Series 717 (2016) 012034 doi:10.1088/1742-6596/717/1/012034

Indirect-drive ablative Richtmyer Meshkov node scaling

O L Landen[1], K L Baker[1], D S Clark[1], V N Goncharov[2], B A Hammel[1], D D Ho[1], O A Hurricane[1], J D Lindl[1], E N Loomis[3], L Masse[1], C Mauche[1], J L Milovich[1], J L Peterson[1], V A Smalyuk[1], S A Yi[3], A L Velikovich[4], and C. Weber[1]

[1]Lawrence Livermore National Laboratory, Livermore, CA, USA
[2]Laboratory for Laser Energetics, Rochester, NY, USA
[3]Los Alamos National Laboratory, Los Alamos, NM, USA
[4]Naval Research Laboratory, Washington, DC, USA

landen1@llnl.gov

Abstract. The ablation front Rayleigh Taylor hydroinstability growth dispersion curve for indirect-drive implosions has been shown to be dependent on the Richtmyer Meshkov growth during the first shock transit phase. In this paper, a simplified treatment of the first shock ablative Richtmyer-Meshkov (ARM) growth dispersion curve is used to extract differences in ablation front perturbation growth behavior as function of foot pulse shape and ablator material for comparing the merits of various ICF design option.

1. Simplified Indirect-Drive Ablative Richtmyer Meshkov Model

Ignition requires a pulse shape with a low power foot designed (see figure 1) to send a carefully timed series of shocks through the DT shell such that they overtake each other soon after they travel into the enclosed DT gas [1]. This minimizes the in-flight adiabat of the fuel and hence increases its compressibility and the final fuel areal density that can be achieved.

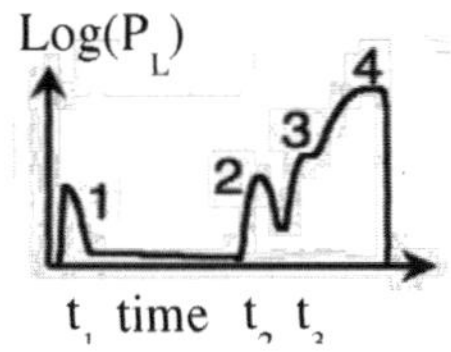

Figure 1. Schematic of laser pulse profile for the example of a 4-shock drive with launch times of shocks identified. The low power section between the first shock picket and 2nd shock launch is denoted the trough.

However, the amplitude of any perturbations of transverse wavenumber k at the capsule surface will begin evolving as a function of k after first shock launch. Extensively evaluated both theoretically and computationally for direct-drive [2-4] where labeled an ablative Richtmyer-Meshkov (RM) instability, the shorter wavelengths even have enough time to oscillate in sign before second shock launch after time $t = t_2 - t_1$. Specifically, the ablation front perturbation growth after shock transit but before acceleration [5] that is seeded over a depth $\approx 1/k$ will initially grow, then decay and eventually reverse phase [6] under sustained drive and ablation at a rate V_a. The drive during the picket and trough, by virtue of its long duration compared to the successive shock phases, dominates this ablative RM growth phase. We define the longest transverse wavenumber (= shortest wavelength) that does not undergo phase reversal as k_0, which also makes it by definition the first node on a plot of growth factor vs wavenumber. We seek to understand the dependence of the perturbation node k_0 reached at time t_2 on foot pulse shape and ablator material since that largely sets the later acceleration-driven Rayleigh-Taylor (RT) dispersion curve for the case of indirect-drive [7]. In particular, it is advantageous if the node k_0 can be set at ≈ 1/in-flight shell thickness ΔR, the most potentially damaging ablation front mode based on the combination of RT growth and feedthrough [8].

Noting that for all the indirect-drive ICF foot drive cases of interest shown in Table 1, k_0V_at is < 1, so we can approximate the dominant terms in [4,7] as follows:

$$\eta \approx e^{-2kV_at}\frac{c_s}{\sqrt{V_aV_{bl}}}f(C)sin\left(k\sqrt{V_aV_{bl}}t+0.3\right)-\frac{c_s}{V_{bl}}g(C)sin\left(1.1\sqrt{C/3}kV_at+1\right) \quad (1)$$

where η is the linear amplitude growth factor, c_s is the postshock sound speed, V_{bl} is the k dependent blow-off velocity ≈ $V_a/(2.4kL/\nu)^{1/\nu}$, L is the ablation front width, ν is the thermal conductivity exponent > 1, and C is the first shock compression jump ρ/ρ_0. Conveniently, the arguments ($\sqrt{(C/3)}k_0V_at + 1$) and ($k_0\sqrt{(V_aV_{bl})}t + 0.3$) cluster around π/2, so setting the sine terms = 1 and η = 0 to find the mode number node $\ell_0 = Rk_0$, where R is the average capsule outer radius up to time t_2 (> 95% of R_0), yields:

$$\ell_0 \approx \frac{R\,ln\left[\left(2.1/(C-1)^{0.9}\right)\left(R\nu/2.4\ell_0L\right)^{1/2\nu}\right]}{2V_at} \quad (2)$$

This is reasonably consistent with the more exact equation (2) in [7] where for the specific case C = 3, $\ell_0 \sim V_a^{-1.5}t^{-0.94}c_s^{0.2}\nu^{-0.1}L^{-.05}$. Ignoring for the moment the weaker log term dependence, equation (2) shows that the node mode number will not depend on capsule scale if the shock merge depth x ~ t is scaled with R. The additional scaling here of interest is with first shock compression, accurate to 5% over relevant C = 2 – 4. We note that as one approaches the incompressible limit (C = 1) corresponding to near zero particle speed behind the shock, there will be little transverse motion and RM growth consistent with simulations [9] and hence no time for phase reversal at any k consistent with equation (2).

2. Model Results

Table 1 lists 8 prior and current implosion designs using either partially Si-doped CH polymer capsules (CH) [10,11], undoped High Density Carbon (HDC) [12] or partially Cu-doped beryllium (Be) [13]. The designs with lower trough T_r have lower fuel adiabat. The drives are designated by the number of shocks, by SS if sub-scale capsule, and by adiabat-shaped (AS) [10,11] in the CH cases where picket and/or trough T_r are modified relative to the regular NIC 4-shock and High-Foot 3-shock designs to create a more strongly decaying first shock [14,15]. Because V_a is typically only 10% of the first shock speed u_s, we can assume the ablation front never reaches the inner mass elements that are either doped or sensing the decaying portion of the first shock. The first shock compression ratios C vs T_r can then be extracted from undoped material shock Hugoniot experiments and simulations of shocked ρ vs pressure P for CH [16], HDC [17] and Be [18,19] and simulations relating P to picket T_r. The conductivity exponent ν as inferred from code-calculated ablation front density profiles [7] is 1.3 and approximated as the same for all designs since a 30% uncertainty in ν translates to only a 10% (3% per [7]) change in ℓ_0. We note that this value of ν = 1.3 is well below the radiative diffusion value of 3 because low Z x-ray ablation is only a weakly diffusive process even for NIF timescales. It is also below the electron thermal conduction exponent of ≈ 1.5-2 predicted [20] for dense partially ionized low Z plasmas, suggesting volumetric x-ray bleaching is probably dominant.

The key parameter, the ablation rate V_a = dm/dt/ρ, is scaled either from simulations for CH [7] or higher T_r planar HDC and Be data [21]. Defining an exhaust velocity v_{ex} and ablator albedo α, we can write the generalized form of the mass ablation rate assuming ablated $T_e \sim T_r$ as:

$$\frac{dm}{dt} \sim \frac{T_r^4(1-\alpha)}{v_{ex}^2} \sim \frac{T_r^4(1-\alpha)}{(Z+1)T_e/A} \sim \frac{T_r^3(1-\alpha)}{(Z+1)/A} \quad (3)$$

C is not fully ionized at these foot T_r drives, leading to $(1 - \alpha_C)$ scaling as $T_r^{-0.3}$ [22]. By contrast, we assume fully ionized Be for a relevant ablated $T_e \approx T_r > 80$ eV and hence a T_r independent lower average albedo ($\approx$15 vs 30%). The specific heat (Z+1)/A of Be is taken to be 4/9, while for CH it is assumed to be between 8/13 and 8.5/13 and for the higher T_r HDC designs, 6.5/12. Combining this information while ensuring consistency with existing data and simulations, V_a becomes $4.4T_r^{2.7}/\rho$, $5.3T_r^{2.7}/\rho$ and $8.0T_r^{3.0}/\rho$ μm/ns with T_r in heV and ρ in g/cc for CH, HDC and Be, respectively. The ordering in mass ablation rate follows from the lower (Z+1)/A of HDC and Be vs CH and the lower albedo of Be [23]. We also assume isentropic decompression at the ablation front $\rho_{trough}/\rho_{shock} = (P_{trough}/P_{picket})^{3/5}$ when the trough T_r < picket T_r in calculating an average V_a. Since $V_a \sim 1/C$, current 4% uncertainties in first shock compression C lead to only a 1-1.5% uncertainty in ℓ_0 through largely cancelling C dependent terms in equation (2). The ablation front widths L are based on CH implosion simulations [7], scaling as $T_r^{1.9}/\rho$. We approximate the multiplicative constant on L as the same between C and Be since the Planck mean free path at relevant $50 < T_r < 100$ eV is an average below and above the more absorptive C K edge while it predominantly senses above the less absorptive Be K edge [24]. Moreover, a 30% uncertainty in L only translates to a 5% uncertainty in ℓ_0.

Table 1. Capsule and foot drive design values, calculated first shock RM relevant parameters and calculated first RM node by 2nd shock launch and first RT node mode numbers at peak acceleration.

Ablator	Design	R (μm)	Picket T_r (eV)	Trough T_r (eV)	C	<L> (μm)	<V_a> (μm/ns)	t (ns)	l_0 RM	l_0 RT
CH	NIC	1126	63	63	2.4x	0.7	0.5	11.5	116	160
CH	HF	1132	93	85	3.1x	1.2	1.1	7	65	90
HDC	4-Sh	1110	105	105	1.8x	0.7	1	2.5	298	260±15
HDC	3-Sh	1110	123	115	2x	0.9	1.3	2	242	200±15
HDC	3-Sh SS	910	123	115	2x	0.9	1.3	1.5	242	235
Be	4-Sh AS	1051	100	80	2.3x	1	1.6	7.7	62	73
Be	3-Sh AS	1050	120	103	2.7x	1.3	2.6	5.5	43	28
CH	4-Sh AS	1126	85	63	3x	1	0.8	8	85	75
CH	3-Sh AS	1132	93	70	3.1x	1.1	1	9	63	75

A comparison of the last 2 columns on Table 1 shows that the RM node l_0 analytic scaling from equation (2) quantitatively tracks the calculated RT growth node [10-12,25], confirming the importance of the first shock in setting the hydroinstability dispersion curve. The RT node locations for the four CH designs have already been confirmed by RT experiments [15,25] and all but the Be 4-shock design have been shot as DT implosions. The capsule hydrodynamics simulations used the 2D version of the radiation-hydrodynamics code Hydra [26] with the hohlraum drive inferred from measured Dante spectra, shock strengths, timings and backlit shell trajectories as inputs [27]. We caution that we only expect rough one-to-one correspondence between RM and RT modes since simulations and a more accurate analytic model [7] show that the CH node shifts 20-50% to higher mode number between shock breakout and peak acceleration. Nevertheless, we can understand the trends in l_0 relative to the CH 4-shock NIC l_0 as follows: HDC l_0 is substantially greater as the product of $V_a \sim T_r^{2.7}/C\rho_0$ and $t \approx \Delta R/u_s \sim 1/\rho_0 u_s \sim 1/T_r^{1.4}\sqrt{\rho_0}$ yields $V_a t \sim T_r^{1.3}/C\rho_0^{1.5}$ which is dominated by the 3.5x larger ρ_0 of HDC. The Be l_0 is less as has $\approx$ 2x higher mass ablation rate at a given T_r. The CH 3-shock High-foot and AS l_0 with higher picket and trough T_r are less as $V_a \sim T_r^{2.1}$ (substituting for the

CH Hugoniot scaling [16] $C\rho_0 \sim T_r^{0.6}$) increases faster than $t \sim 1/u_s \sim 1/T_r^{1.4}$ decreases. This also explains why all 3-shock designs have lower predicted l_0 than the companion lower foot T_r 4-shock designs. The 4-shock AS l_0 is less as the larger picket V_a, L and C overcomes a shorter t.

3. Conclusions

We have derived a simple, approximate formula for the first shock driven ablation front dispersion node l_0 of indirect-drive implosions for arbitrary foot pulse shape and ablator material. The scaling explains why the 3-shock and 4-shock AS CH and all Be ignition designs can provide lower l_0. Such designs will be crucial in mitigating the RT mode of greatest concern for feedthrough $l \approx R/\Delta R \approx 40$ which is also the dominant mode seeded by the contact discontinuity of the thin plastic membranes [28,29] currently used for capsule support. The scaling also explains why all current and conceivable HDC-only implosion designs (adiabat-shaped or not) are susceptible to higher mode 100-150 growth, those seeded by small isolated surface features, for example, the 10 μm diameter filltube [30]. From the standpoint of minimizing l_0 while keeping the fuel on a low adiabat, the most favorable design is the 4-shock AS Be.

Acknowledgments

This work performed under the auspices of the U. S. Department of Energy by Lawrence Livermore National Laboratory under Contract No. DE-AC52-07NA27344.

References

[1] Munro D et al 2001 *Phys. Plasmas* **8** 2245
[2] Goncharov V 1999 *Phys. Rev. Lett.* **82** 2091
[3] Velikovich A et al 2000 *Phys. Plasmas* **7** 1662
[4] Goncharov V et al 2006 *Phys. Plasmas* **13** 012702
[5] Loomis E, Braun D, Batha S, Sorce C and Landen O 2011 *Phys. Plasmas* **18** 092702
[6] Aglitskiy Y et al 2002 *Phys. Plasmas* **9** 2264
[7] Peterson J, Clark D, Masse L and Suter L 2014 J *Phys. Plasmas* **21** 092710
[8] Lindl J et al 2004 *Phys. Plasmas* **11** 339
[9] Hammel B et al 2010 *High Energy Density Physics* **6** 171
[10] Clark D et al 2014 *Phys. Plasmas* **21** 112705
[11] Milovich J et al 2015 *Phys. Plasmas* **21** 112702
[12] Mackinnon A et al 2014 *Phys. Plasmas* **21** 056318
[13] Yi S et al 2014 *Phys. Plasmas* **21** 092701
[14] Baker K et al 2015 *Phys. Plasmas* **22** 052702
[15] Macphee A et al 2015 *Phys. Plasmas* **22** 080702
[16] Barrios M et al 2012 *J. Appl. Phys.* **111** 093515
[17] Hicks D et al 2008 *Phys. Rev.* B **78** 174102
[18] Cauble R et al 1998 *Phys. Rev. Lett.* **80** 1248
[19] Wilson B, Sonnad V, Sterne P and Isaacs W 2006 *J. Quant. Spectrosc. Radiat. Transfer* **99** 658
[20] Lee Y and More R 1984 *Phys. Fluids* **27** 173
[21] Olson R, Rochau G, Landen O and Leeper R 2011 *Phys. Plasmas* **18** 032706
[22] Eidmann K et al 1995 *Phys. Rev.* E **52** 6703
[23] Wilson D et al 1998 *Phys. Plasmas* **5** 1953
[24] Iglesias C and Rogers F 1993 *Astrophys. J* **412** 752
[25] Casey D et al 2014 *Phys. Rev.* E **90** 011102
[26] Marinak M et al 2001 *Phys. Plasmas* **8** 2275
[27] Clark D et al 2013 *Phys. Plasmas* **20** 056318
[28] Haan S et al 2013 *Fus. Sci. Tech.* **63** 67
[29] Nagel S et al 2015 *Phys. Plasmas* **22** 022704
[30] Edwards J et al 2005 *Phys. Plasmas* **12** 056318

Journal of Physics: Conference Series **717** (2016) 012035 doi:10.1088/1742-6596/717/1/012035

NIF Rugby High Foot Campaign from the design side

J-P Leidinger[1], D A Callahan[2], L F Berzak-Hopkins[2], J E Ralph[2], P Amendt[2], D E Hinkel[2], P Michel[2], J D Moody[2], J S Ross[2], J R Rygg[2], P Celliers[2], J-F Clouët[1], E L Dewald[2], P Kaiser[1], S Khan[2], A L Kritcher[2], S Liberatore[1], D Marion[1], P-E Masson-Laborde[1], J L Milovich[2], O Morice[1], A E Pak[2], O Poujade[1], D Strozzi[2] and O A Hurricane[2]

[1] CEA-DAM, DIF, F-91297 Arpajon, France
[2] Lawrence Livermore National Laboratory, Livermore, CA-94551, USA

E-mail: jean-pierre.leidinger@cea.fr

Abstract. The NIF Rugby High Foot campaign results, with 8 shots to date, are compared with the 2D FCI2 design simulations. A special emphasis is placed on the predictive features and on those areas where some work is still required to achieve the best possible modelling of these MJ-class experiments.

1. Introduction

Part of the NIF 2014-2015 High Foot Campaign was carried out with the same "Rugby 700" hohlraum (figure 1, waist diameter 7.0 mm) used in 2013 for 3 Low Foot shots [1]. Due to the high neutron yield obtained with the high foot pulse in a cylindrical hohlraum [2], it was of interest to test this laser pulse in a parabolic hohlraum geometry such as the rugby one.

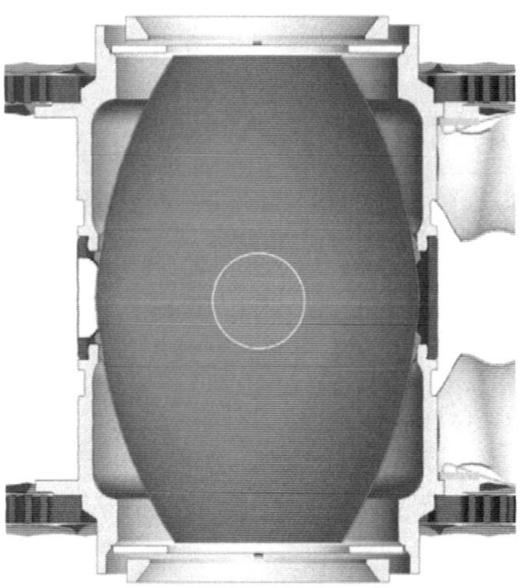

Figure 1. "Rugby 700" hohlraum with higher case-to-capsule ratio than "cylinder 575" but 10% larger surface.

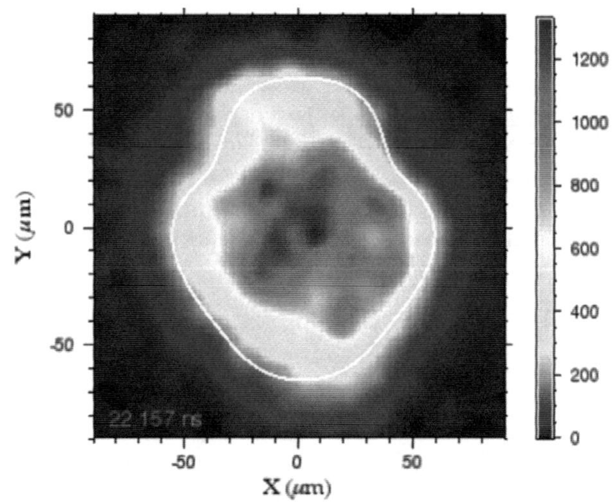

Figure 2. N131011 Rugby Low Foot shot (SymCap) GXD hot spot. The Legendre mode 2 (P2/P0) is 6% and P0 = 55 μm, achieved by repointing outer cones.

Moreover the last low foot rugby shot, N131011, which was the first shot in the CEA/LLNL rugby collaboration, showed that repointing the outer cones inwards by 500 µm could effectively achieve a round hot spot (figure 2) in a 1.2 mg/cc He filled rugby hohlraum. The same pointing was then selected for the high foot campaign, with a higher fill density of 1.6 mg/cc to counteract the faster outer gold bubble expansion.

The campaign started with 2 Keyholes shots followed with a Reemit and 3 2dConAs. Finally 2 additional Keyholes were shot in August 15. We detail the whole campaign hereafter.

2. First Keyholes N140822 (0.75 MJ) and N141102 (1.1 MJ)

The NIF Keyhole platform's main goal is to allow precise shock-timing using a Visar diagnostic and to tune the laser pulse for an optimal DT adiabat. The high foot pulse is characterised by 3 shocks. Figure 3 shows the shocks inside the ablator (top) and the surrogate liquid deuterium (center) in a Lagrangian plot.

The pulse was calculated with the CEA FCI2 rad-hydro code in order to allow the first and second shocks to coalesce at a depth from the ablator/deuterium interface of about 80 µm. The capsule is the regular LLNL "T-1" with 175 µm ablator thickness. Some power multipliers [3] are taken into account to simulate laser losses. For instance, the first shock is computed with a multiplier of 0.9 (same value used in the LLNL cylinder simulations) applied to the laser foot in order to fit the data.

The N140822 Visar shock speeds are within 1% of the FCI2 prediction, but the 2nd shock arrived 700 ps earlier than predicted (figure 4).

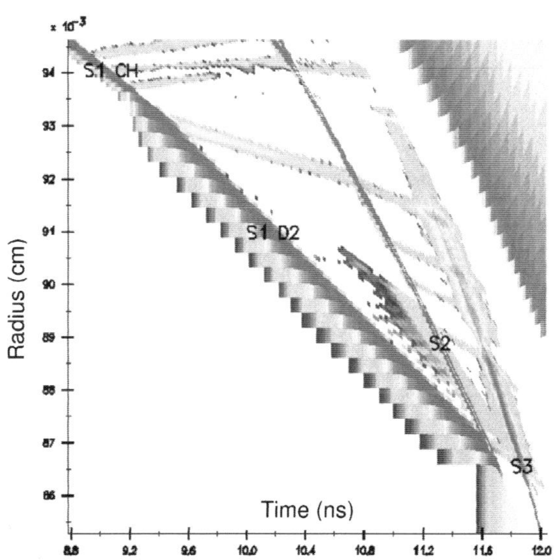

Figure 3. High foot 3 shocks Lagrangian visualisation in ablator (CH) and deuterium (D2) between 8.8 and 12 ns.

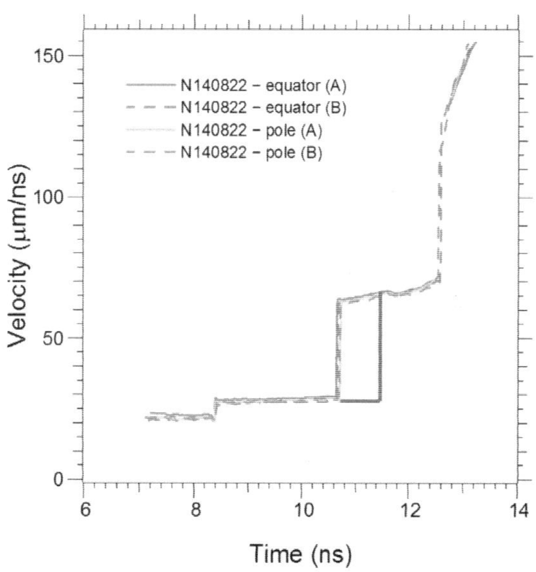

Figure 4. N140822 measured velocities. The plain red line is the predicted 2nd shock delayed by 0.7 ns.

The likely explanation for this delay was inferred from the post-shot calculations using a new NLTE model for the wall and especially the ablator. With the CEA standard ablator LTE model, the 2nd shock is created by the coalescence of two separate shocks. The extra shock forms around 100 µm in front of the main ablation front and is due to K-shell C opacity at 300 eV. Launching the shock 100 µm outside the main ablation front causes a 700 ps delay in the 2nd shock arrival time [4].

Modeling the ablator in NLTE using Nohel [5], the fully ionized ablator no longer absorbs the 300 eV photons, allowing them to reach the main ablation front and launching a single shock that arrives 700 ps earlier than for the LTE calculation, thus matching the data.

N141102 then used a delayed 2nd picket and validated the full high foot shock-timing.

Journal of Physics: Conference Series **717** (2016) 012035 doi:10.1088/1742-6596/717/1/012035

3. Reemit N150104

The aim of the Reemit platform is to measure the early-time flux symmetry around the capsule. The design picket cone fraction at 0 Å of 0.5 is very different from the NIF high foot cylinder shots, which rely on cross-beam energy transfer (CBET) to achieve hot spot symmetry.

This shot measured a pole-hot flux (mode 2 +5.6%) inside the 7.5% specification. A slight discrepancy was found with the simulations predicting rather a waist-hot flux at early time. One post-shot way to close the gap calls for extra outwards refraction in the hohlraum : with a 400 μm refraction for all cones, the calculated mode 2 evolution becomes similar to the measured one but the physical cause behind this extra-refraction, if true, is still unknown.

4. 2dConAs N150127, N150329 and N150504

The 3 ConAs shots, all done at 1.3 MJ, differed from each other by the peak cone fraction (0.33 vs 0.4 resp. for N150127 and N150329) and the cone colors. N150504 was a "3-color" shot that used a 2 Å separation for the 23.5° cone and a 5 Å separation for the 30° cone relative to the outer cones.

The first shot N150127 hot spot came out 34% pancake, despite the prediction stating a close-to-round hot spot (P2/P0 -5%). The initial strategy to mitigate any out-of-round shape was to vary the peak cone fraction, see figure 5. The 2nd 2dConA therefore used a cone fraction of 0.4 instead of 0.33 previously (inner energy +20%, outer energy -10%). The hot spot shape result was unfortunately unchanged, both SXIs showing the same relative outer/inner intensities (figure 6). The most likely explanation for this surprising outcome is that a higher inner cone SRS and a lower outer cone SBS compensated for the cone fraction change, the useful laser energy deposited in the wall being the same.

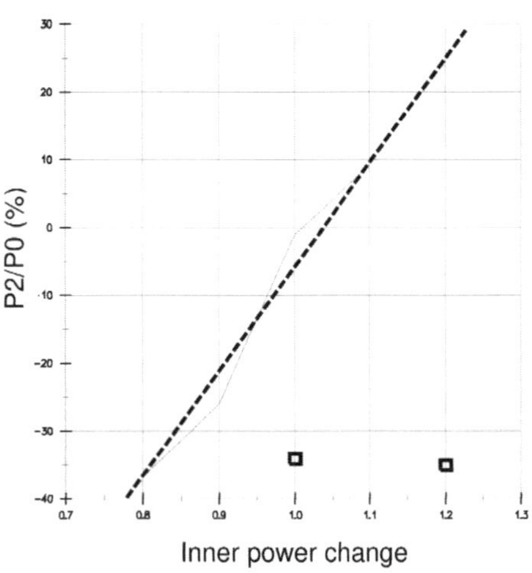

Figure 5. 2dConA mode 2 FCI2 prediction as a function of the inner cone power change. Black squares are N150127 (center) and N150329 (right) results.

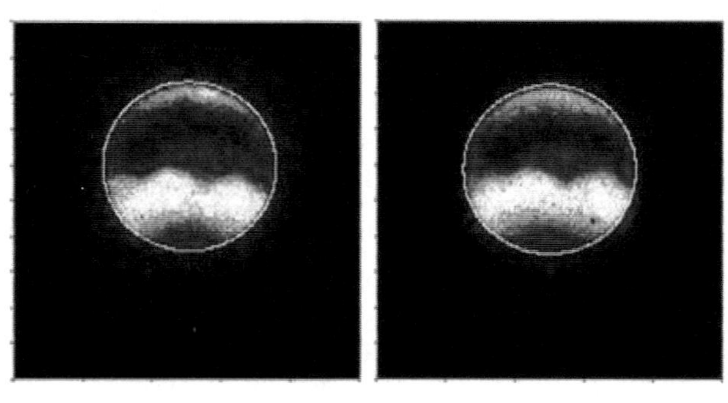

Figure 6. Comparison of N150127 (right) and N150329 (left) SXI images. Inner cones are on top and outer cones are the main bright coronal areas.

The third 2dConA N150504 used the 3-color approach to see a drastic hot spot change, maximising the laser energy absorbed at the hohlraum waist with a designed high CBET towards the 30° cone, the closest one to the equator. The transfer prediction, using either P. Michel's script [6] or the FCI2 CBET model (figure 7 (a)), was +80% laser energy applied to the 30° cone from the 44.5° one mainly and no change on the other 23.5° inner cone.

The shot result was more satisfactory than the previous one, achieving a "sausage" hot spot shape (mode 2 +30%) and demonstrating that round symmetry can be obtained with a high foot pulse in this rugby hohlraum. However we obtained a waist-high flux because the CBET was likely higher than calculated, possibly doubled according to the in-flight radiography analysis (figure 7 (b), (c)).

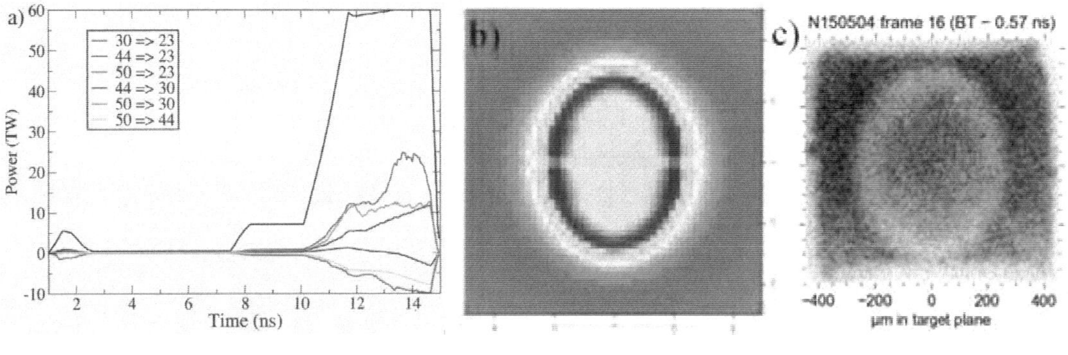

Figure 7. (a) Cone to cone CBET FCI2 prediction with the laser pulse (black). Negative values (red curve for instance) mean the transfer goes from 23 to 30°. (b) FCI2 synthetic radio with +160% energy on 30° cone. (c) N150504 radio at the same time showing the same P2 as the FCI2 picture.

5. Keyholes N150810 (1.1 MJ) and N150823 (0.8 MJ)

The campaign ended with 2 new Keyholes due to platform availability, the first one still using a 3-color scheme with half-N150504 wavelengths (1 Å and 2.5 Å) and the last one testing a quad unsplitting + repointing proposal to draw closer to the 2D modelling for the 44° cone.

The best overall coupling was achieved with the 2nd shot, whose shock-timing waist/pole symmetry analysis indicates an effective 400 μm inwards repointing is optimal for hot spot symmetry. N150823 SXI also showed for the first time a bright 44° spot (figure 8). A future 2dConA shot should allow us to check for a symmetric hot spot shape with this new 44° cone configuration.

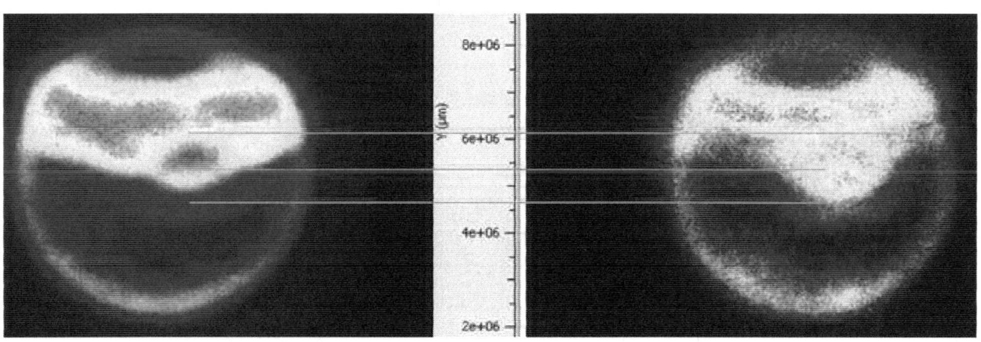

Figure 8. SXI pictures from N140822 (left) and N150823 (right) showing new 44° cone.

Acknowledgments

We thank the whole NIF staff and operations people who made this LLNL/CEA rugby campaign possible.

[1] Amendt P *et al* 2014 *Phys. Plasmas* **21** 112703
[2] Hurricane O A *et al* 2014 *Phys. Plasmas* **21** 056314
[3] Jones O S *et al* 2013 *EPJ Conf.* **59** 02009
[4] Poujade O, Bonnefille M and Vandenboomgaerde M 2015 *Phys. Rev. E* **92** 053105
[5] Decoster A, *in* 2003 *JQSRT* **81** 71-84
[6] Michel P *et al* 2009 *Phys. Rev. Lett.* **102** 025004

High-power laser-plasma interaction in nanosecond regimes 'at a glance' using proton deflectometry

P Loiseau[1], A Castan[1,2], J-R Marquès[2], L Lancia[3], T Gangolf[2], J Fuchs[2], P-E Masson-Laborde[1], D Teychenné[1], A Debayle[1], M-C Monteil[1], M Casanova[1], C Rousseaux[1], S Lemaire[1] and D Riz[1]

[1]CEA, DAM, DIF, F-91297 Arpajon, France
[2]LULI - CNRS, École Polytechnique, CEA: Université Paris-Saclay; UPMC Université Paris 6: Sorbonne Universités - F-91128 Palaiseau Cedex, France
[3]Dipartimento SBAI, Univ. Roma La Sapienza, Via Antonio. Scarpa 14, 00161 Rome, Italy

E-mail: pascal.loiseau@cea.fr

Abstract. Recent experiments indicate that controlling the propagation of high-power laser beams through millimeter long and low-density plasmas still remains challenging. In such plasma conditions, it is equally important to consider the impact of the plasma on laser propagation and laser properties, and the impact of the laser on plasma conditions. These complex phenomena are still difficult to implement in fluid models owing to the highly non-linear physics at play. Yet, electromagnetic fields prove to be good signatures of most of these low frequency phenomena. In particular, local pressure gradients and electron transport can be inferred from the electric fields. Such in-depth plasma characterization can be achieved through proton deflectometry. For that purpose, we have developed a three-dimensional simulation capability in order to compute protons' trajectories modified by the local electric fields.

1. Introduction

Detailed knowledge and understanding of the laser-plasma interaction is of fundamental interest in many high energy density (HED) configurations, and specifically in inertial confinement fusion (ICF). This is the case both in the indirect drive approach of ICF, where high-power laser beams have to propagate through several millimeters of plasma before reaching the hohlraum walls, and in the direct drive approach, where the reduction of the imprint on the fuel capsule could be achieved through plasma induced smoothing [1,2]. Recent experiments on the LIL [3,4], NIF (NEL) [5], LULI [6] and OMEGA EP [7] facilities brought to light the difficulty in modeling laser plasma propagation. Setting aside ionization that involves atomic physics, laser propagation in plasmas depends mainly on the local electron temperatures and densities that can trigger instabilities such as self-focusing and stimulated scattering off plasma waves. An interesting tool to investigate this interplay is to probe the electric fields driven by electron transport and ponderomotive effects inside the plasma through proton deflectometry [6]. In addition to the usual optical measurements, this diagnostic provides unique and novel characterization of the laser-plasma interaction by allowing to measure the electric fields within the plasma. In a joint effort, we explore this new tool both experimentally at the LULI 2000 facility and numerically, by using the transport code Diane [8] in order to compute the proton deflectometry due to the local electric fields calculated by our 3D paraxial code Hera [9]. In addition, we use the radiative-hydrodynamics code FCI2 to compute initial values for Hera.

2. Proton deflectometry in underdense plasmas

Multi-MeV proton beams generated by high-power, typically greater than 100 TW, laser beams are commonly used on numerous facilities for radiography purposes, see reference [10] for a review. For underdense plasmas, that is plasmas whose electron densities are lower than the critical density $n_c \sim 10^{21}/\lambda_0^2$ cm^{-3}, with λ_0 the laser wavelength in μm, protons undergo mainly electromagnetic deflections. Proton deflectometry consists in recording proton's deflections after their propagation through a probed object, onto a pack of radiochromic films (RCF), as illustrated Fig. 1 for a typical image. The resulting patterns give an indication about the strength of the electromagnetic fields encountered, and indirectly about the physics under consideration. Due to the protons' high velocities, proton deflectometry probes the plasma on a picosecond time-scale, and temporally resolved, depending on the proton energy spectrum, the more energetic protons arriving first. The proton beam traverses a mm^3 volume, and impinges onto a cm-size RCF with a few μm spatial resolution. Then, dose modulations recorded on each RCF allow to estimate an amplitude of the electric and magnetic fields generated during the laser-plasma interaction. In hot underdense plasmas, magnetic field amplitudes are too low for deflecting protons [6] and the electrostatic fields E mostly stem from the electron pressure gradient, first term in the right hand side (RHS) of the following equation, and the ponderomotive force, second term in the RHS,

$$eE \sim -\frac{1}{n_e}\nabla(n_e T_e) - \frac{1}{2cn_c}\nabla I\,, \tag{1}$$

where I is the laser intensity, n_e the electron density, T_e the electron temperature and e the electron charge. Plasma modifications due to those two forces lead to electron density modifications with a characteristic electrostatic signature. Thus, proton deflectometry is a unique tool for probing in depth ICF plasmas, giving quantitatively access to electrostatic field amplitude arising from the electron thermal and ponderomotive effects. In principle, such a diagnostic could discriminate between local or non local electron transport regimes when compared to hydrodynamic simulations [6].

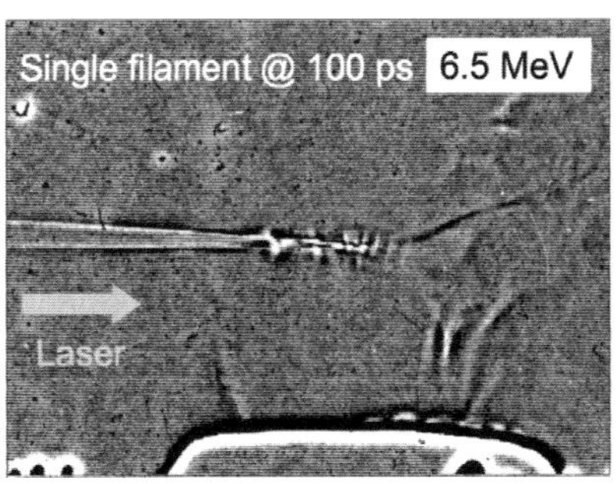

Figure 1. RCF pattern produced by 6.5 MeV protons from a recent LULI experiment. The laser beam propagates from the left to the right in a helium gas jet (we see clearly the 1 mm diameter nozzle at the bottom of the image), whose initial density is 1% of the critical density for a 1 μm laser wavelength. The protons arrive 100 ps after the beginning of the pulse.

Although many experiments have demonstrated the interest of proton deflectometry, exhibiting specific patterns attributed to the existence of specific fields, it is still difficult to analyse quantitatively RCF without knowing *a priori* the field structures. This aspect arises from the two-dimensional (2D) representation of a three-dimensional (3D) interaction. The inversion problem points to the need of a simulation tool that calculates the 3D electric field structures resulting from the two forces, and then the proton's deviations into those structures. For this, we use our 3D massively parallel laser-plasma interaction code Hera to compute the electric fields, coupled to the 3D Monte-Carlo proton transport code Diane to simulate the dose received by the RCF. Hera includes a paraxial laser propagation solver, coupled to the mass and momentum conservation equations and taking into account Brillouin backscattering. Diane solves the multigroup Boltzmann-Fokker-Planck equation, including the Lorentz force in stationary regime. Our numerical tool was successfully compared to the analytical expressions detailed in Ref. [10].

3. Application to beam self-focusing

We use the Hera and Diane codes in two configurations. First, we consider a gaussian laser beam ($\lambda_0 = 1\,\mu m$, $f_{\#} = 10$ and $I = 10^{14}\,W/cm^2$) propagating in a 1-mm long, 160-μm wide and 3.2%n_c He plasma. The electron temperature is around 180 eV. The laser beam has a constant power and propagates in the x direction and the proton beam in the y direction, see insert e) Fig. 3 for the geometry. We present, see Fig. 2, longitudinal maps of a) the laser intensity, b) electron density, c) electric field z-component and d) proton dose 140 ps and 250 ps after the beginning of the pulse.

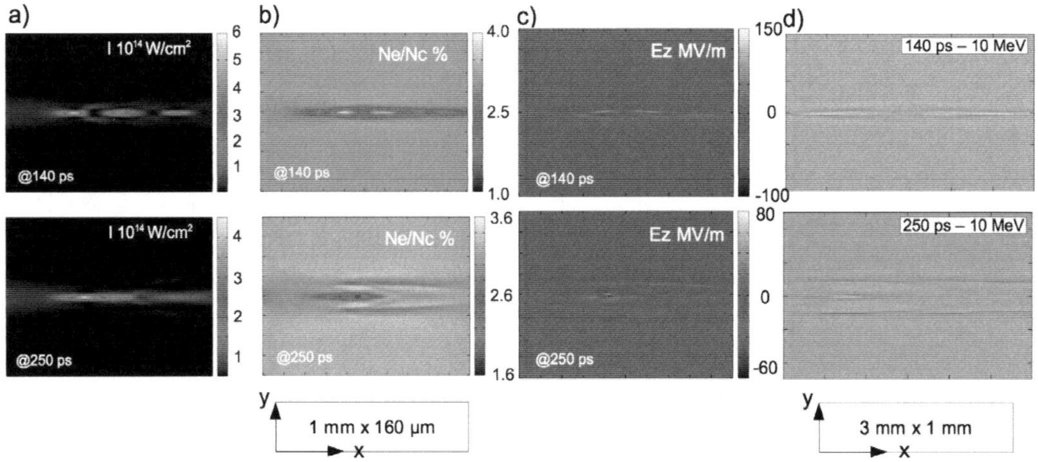

Figure 2. Propagation of a single filament in a 3.2%n_c He plasma: a) laser intensity, b) electron density, c) electric field z-component and d) simulated proton dose. Hera simulation box is 1 mm long and 160 μm wide. The magnification factor is 15 on Diane's simulated detector.

At 140 ps, the laser exhibits a strong self-focusing. The secular effect of the ponderomotive pressure digs plasma channels and results in high amplitude electric fields. Later, at 250 ps, the laser is less perturbed. The electron density map keeps the hydrodynamical history of the laser plasma interaction: transversal density bumps come from the successive self-focusing. The simulated proton dose matches this time history.

It is quite straightforward that proton deflectometry is useful for a single beam configuration. It is challenging to consider a more complex situation where the lasers are smoothed by a random phase plate (RPP), producing a speckle pattern in the focal plane. The second configuration corresponds to $1\,\mu m$ wavelength, 1.5 ns long and 20 J RPP beam propagating in a 3.2%n_c He plasma and focused over a 200 μm FWHM focal spot (see Fig. 3 a). The plasma is 2-mm long and 500-μm wide. We needed up to $\sim 5.4 \times 10^8$ cells (512 x 1024 x 1024) for the total domain discretization on 1024 cores. The focal spot in vacuum is shown top of Fig. 3 a), it can be compared to the focal spot in the transverse median (yz) plane after 500 ps of propagation Fig. 3 b). In this simulation, the RPP beam undergoes a smoothing induced by the plasma response and becomes more and more spatially incoherent [1,2]. This is illustrated at the bottom of Fig. 3 c) and d) where, due to the ponderomotive pressure, strong density fluctuations appear, generating high amplitude electric fields, around 100 MV/m. We use the Diane code in order to simulate proton deviations through the probe volume. For this simulation, the proton beam is modeled by a 5 degrees half angle cone, positioned 7 mm off the plasma and aligned to the center of the simulation box. The detector is positioned 10 cm after the plasma, giving a x15 magnification. Due to the cone aperture, we select a $1 \times 0.5 \times 0.5$ mm^3 volume centered to the middle of the simulation box. For this simulation, the Diane code generated 2.5×10^8 protons in 5h, using 32 parallel runs, each with 32 cores. The simulated proton dose on the detector is

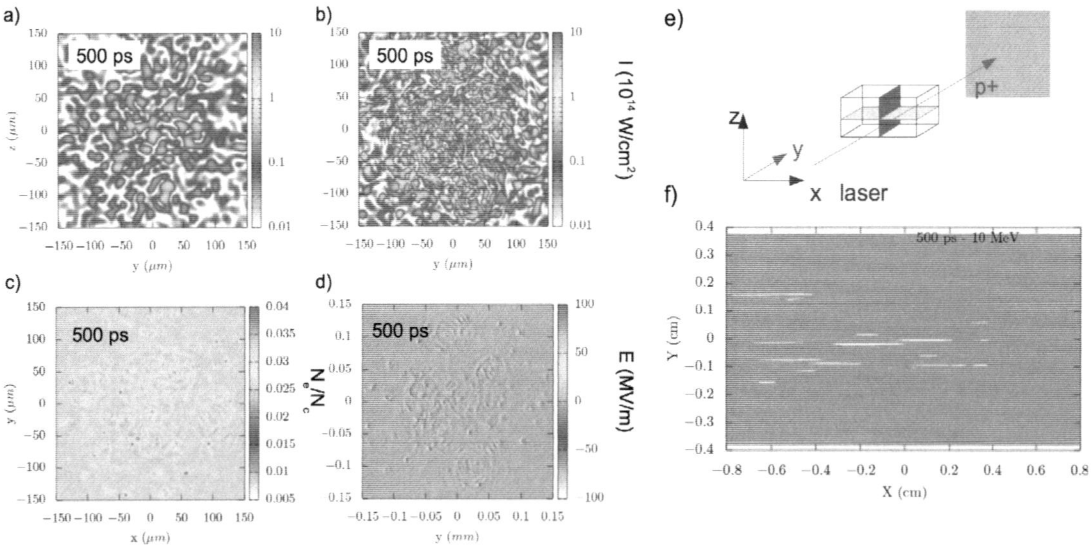

Figure 3. Propagation of a laser beam with a speckle pattern in a $3.2\%n_c$ He plasma. Maps a) of the laser intensity in vacuum, b) the laser intensity in the transverse median plane, c) electron density perturbations and d) the corresponding E_z. The geometry is shown in e) where the transverse median plane is colored in dark blue and the RCF in grey, and f) the simulated RCF for 10 MeV protons.

shown Fig. 3 f) for a 10 MeV proton beam, exhibiting in white strong density depletions. Such structures are linked to the speckles induced plasma modifications and their individual and dynamical self-focusing. It demonstrates that proton deflectometry applied to realistic laser-plasma interaction leads to discernable proton dose modulations, but a statistical approach (for example: measuring the characteristic filament's length) is necessary for a quantative comparison to experiments. Probing plasmas by proton deflectometry provides a way to characterize the laser beam's properties during its propagation.

4. Conclusion

Proton deflectometry is a promising tool for studying laser-plasma interaction in underdense plasmas. Complementary to the usual optical diagnostics, proton deflectometry, with multi-MeV beams, allows in situ characterization on mm-scale with very high temporal and spatial resolution from a single laser shot. Although evaluating an electric field amplitude is conceivable given an analytic form of the electric potential, this is not the case in general. For this reason, we have developed simulation capabilities using a laser-plasma interaction code coupled to a charged-particle transport code. Because intense speckles may lead to stronger local magnetization than anticipated, magnetic fields will be soon included in Hera simulations. Experiments on HED facilities such as LMJ with PETAL are envisioned.

[1] Fuchs J, Labaune C, Depierreux S, Baldis H A, Michard A and James G 2001 *Phys. Rev. Lett.* **86** 432
[2] Depierreux S *et al.* 2009 *Phys. Rev. Lett.* **102** 195005
[3] Loiseau P *et al.* Accepted in *IOP Conf. Series*
[4] Rousseaux C, Huser G, Loiseau P, Casanova M, Alozy E, Villette B, Wrobel R, Henry 0 and Raffestin D 2015 *Phys. Plasmas* **22** 022706
[5] Glenzer S *et al.* 2007 *Nat. Phys.* **3** 716
[6] Lancia L *et al.* 2011 *Phys. Plasmas* **18** 030705
[7] Harvey-Thompson A J *et al. Proc. of IFSA 2015* 149
[8] Riz D and Chiche M 2003 *Int. Conf. on Supercomputing in Nuclear Applications (Paris) Am. Nucl. Soc.*
[9] Ph. Ballereau *et al.* 2007 *J. Scient. Comput.* **33** 1
[10] Kugland N L, Ryutov D D, Plechaty C, Ross J S and Park H S 2012 *Rev. Sci. Instrum.* **83** 101301

Journal of Physics: Conference Series **717** (2016) 012037 doi:10.1088/1742-6596/717/1/012037

Study of fast electron generation using multi beam of LFEX-class laser

M Hata[1], H Sakagami[2], T Johzaki[3], Y Sentoku[4] and H Nagatomo[1]

[1]Institute of Laser Engineering, Osaka University, Suita, Osaka 565-0871, Japan
[2]Fundamental Physics Simulation Research Division, National Institute for Fusion Science, Toki, Gifu 509-5292, Japan
[3]Graduate School of Engineering, Hiroshima University, Higashihiroshima, Hiroshima 739-8527, Japan
[4]Department of Physics, University of Nevada, Reno, Nevada 89557, USA

E-mail: masayasuhata@gmail.com

Abstract. Fast Ignition Realization Experiment project phase-I (FIREX-I) is being performed at Institute of Laser Engineering, Osaka University. In this project, the four-beam bundled high-energy Petawatt laser (LFEX) is being operated. LFEX laser provides great multi-beam irradiation flexibility, with the possibility of arrange the pulses in temporal sequence, spatially separate them in distinct spots of focus them in a single spot. In this paper, we study the two-beam interference effects on high-intensity picosecond laser-plasma interaction (LPI) by two-dimensional relativistic Particle-In-Cell simulations. The interference causes surface perturbation, which enhances laser absorption and underdense plasma generation, increasing the accelerated electron number and their slope temperature. The laser-to-electron energy conversion efficiency for two-beam interference case is suitable for Fast Ignition (FI) compared to the single beam case, but the increment of fast electron divergence leads to lower energy coupling. To optimize the target design for FI, these interference effects should be taken into consideration.

1. Introduction

Fast Ignition Realization Experiment project phase-I (FIREX-I) is being performed at Institute of Laser Engineering [1,2], Osaka University. In this project, the four-beam bundled high-energy petawatt laser (LFEX) is being operated [3]. LFEX laser provides great multi-beam irradiation flexibility, with the possibility of arrange the pulses in temporal sequence, spatially separate them in distinct spots of focus them in a single spot. In case of no beam overlapping, the maximum beam energy, and consequently intensity, is limited by the maximum allowed energy for a single beam. In contrast, in case of beam overlapping the laser intensity on target is increased, but the beams undergo interference. According to previous computational research [4], the spatially overlapped irradiation causes the generation of structures on the target plasma surface according to the interference pattern and therefore affecting the laser-plasma interaction. However, scale-down parameters are used in this study, e.g. the pulse duration is less than 700 femtoseconds and the spot diameter (FWHM) is set to 10 ∝m. Features of the LFEX, which are high power, long pulse, and spot diameter much larger than the diffraction limit, are considered to provide interesting results that differ from many usual research works, where the laser is ultrahigh intensity, but much shorter pulses and close to diffraction limit spot sizes. Kemp et al. [5] reported that a high-power long-pulse laser such as LFEX creates a large underdense plasma during irradiation which in turn affects the LPI resulting in a increment of the hot electron temperature even with no initial pre-plasma. Therefore, large-scale simulations using realistic laser and target parameters are strongly desired. In this paper, we study the two-beam interference effect on high-intensity LPI by two-dimensional relativistic Particle-In-Cell simulations using PICLS2D code [6].

2. Simulation of multi beam irradiation to planar target

Two-dimensional Particle-In-Cell simulations of Petawatt-class laser irradiation of a planar target have been carried out. The target is simulated as a simple 20 μm Au plasma with uniform electron density of $40n_{cr}$ and an exponentially decreasing, 1μm scale-length pre-plasma extending from 0.1 to $40n_{cr}$, where n_{cr} is the critical electron density.

The choice of such short pre-plasma scale-length is justified because we are assuming a high contrast laser pulse, following the recent upgrades of LFEX laser. Fixing the total laser energy, two laser profile conditions are chosen: one-beam irradiation and two-beam irradiation which intensity per beam is half the intensity of the single beam case. The temporal profile of each beam at the focus is expressed by $I(r,t) = I_0 \exp\left(-\frac{r^2}{r_0^2}\right) \exp\left[-\frac{(t-3t_0)^2}{2t_0^2}\right]$, where $r_0 = 25\lambda_L$ and $t_0 = 191(2\pi/\omega_L)$, corresponding to 41.6 μm FWHM spot size and 1500 fs FWHM pulse duration for 1 μm wavelength. The peak intensities of the laser beam are set to be 1.36×10^{19} $\mathrm{Wcm^{-2}}$ for one-beam irradiation and 6.8×10^{18} $\mathrm{Wcm^{-2}}$ for two-beam irradiation, keeping the same total laser energy. The laser is focused at the center of the target surface, namely (x,y) = (105 μm,72 μm) in fig 1(b). Laser incidence angle is normal for one-beam irradiation and tilted by ± 2.86 degrees for two-beam irradiation.

2.1. *Effects of interference*

Figure 1 shows two-dimensional spatial profile of (a) square of the electric field and (b) electron density for (1) one- and (2) two-beam irradiation at 2228 fs , close to the time of peak laser irradiation on the plasma surface. In fig. 1(a-2), apparent interference pattern can be observed with interval of about 10 μm. Considering interference of two plane waves, the interference pattern interval is expressed by $l_{\text{interference}} = \frac{1}{2}\frac{\lambda_L}{\sin\theta}$, where λ_L is the wavelength and θ is the half of angle between two waves, and it is calculated to be 10.0 μm using simulated parameters of λ_L = 1 μm and θ = 2.86 degrees, in agreement with the observed interval. Laser interference generates a periodic intensity pattern at specific points with hole boring occurring at high-intensity irradiation areas, correspondingly generating a periodic surface structure as shown in fig. 1(b-2).

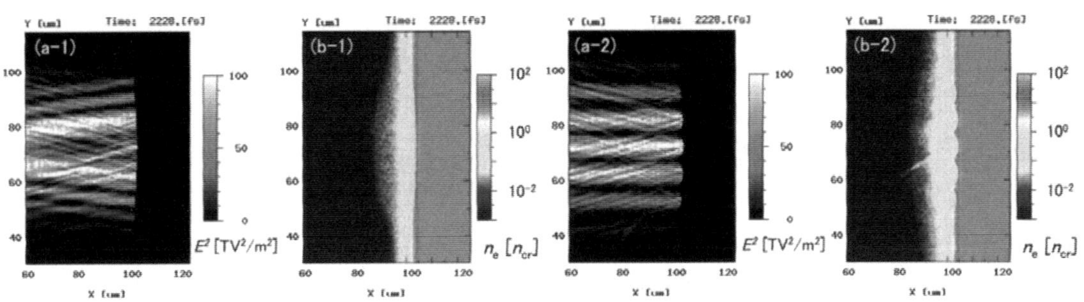

Figure 1. 2-D spatial profile of (a) square of electric field and (b) electron density in the cases of (1) one-beam and (2) two-beam irradiation at 2228 fs, namely around the time of laser peak intensity.

Figure 2 shows cross sections of the electric field (E_y) profile along x-axis at 2228 fs, where red and green lines represent the cross sections in case of one and two-beam irradiation and solid and dashed lines indicate those of Y = 72 and 67 ∝m, respectively. For two-beam irradiation, maximum electric field is around 20 TV/m, twice the maximum electric field for one-beam irradiation. Figure 3 shows cross sections of the electron density profile along the x-axis, where black solid line indicates initial density, red and green solid lines represent the profiles at 2228 fs respectively in case of one- and two-beam irradiation along the focal spot centerline (Y = 72∝m), while dashed red and green lines indicate the profiles 5 ∝m below centerline (Y = 67 ∝m) respectively. For one-beam irradiation, electron density profiles are similar both at Y = 72 and 67 ∝m. On the other hand, a large difference is observed for electron density profiles at Y = 72 and 67 ∝m in two-beam irradiation. For the latter case in fact the laser irradiated surface is prominently compressed inwards at Y = 72 ∝m compared to the one-beam case. In contrast, at Y = 67 ∝m a larger underdense plasma blown-off is observed for the two beam case compared to the one beam case, although the electron profile above critical density does not significantly differ from initial density profile. These features may affect the parameters of the generated fast electron distribution.

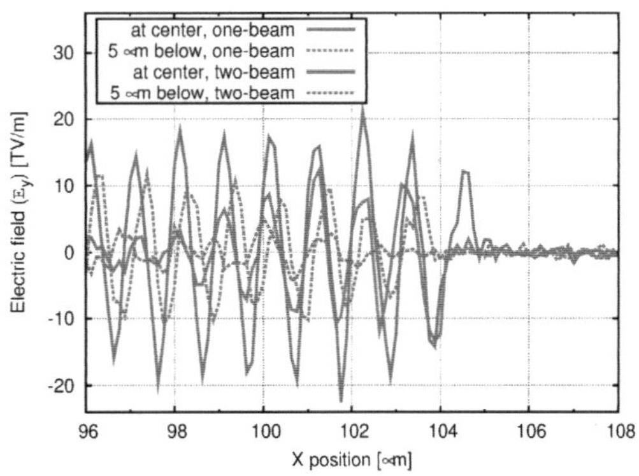

Figure 2. Cross sections of electric field (E_y) profile along x axis at 2228 fs, where red and green lines represent respectively the cross sections for one- and two-beam irradiation case, solid and dashed lines indicate respectively profiles taken at Y = 72 and 67 ∝m.

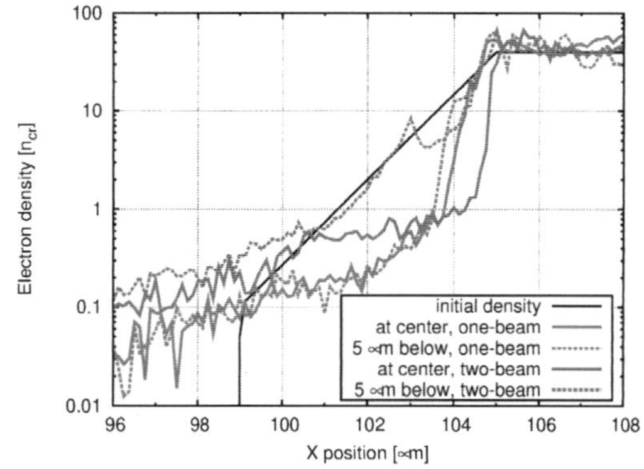

Figure 3. Cross sections of electron density profile along x axis, where black solid line indicates initial density, red and green lines represent profiles at 2228 fs in case of one- and two-beam irradiation, solid and dashed lines indicate profiles taken at Y = 72 and 67 ∝m, respectively.

2.2. Fast electron characteristics

Fast electron parameters are measured at the X=115 ∝m plane. Figure 4 shows the time-integrated fast electron energy spectrum. A larger number of fast electrons is generated in the two-beam irradiation compared to the one-beam irradiation case, with slope temperature estimated to be 2.4 and 2.8 MeV for one- and two-beam irradiation respectively. This result can be explained by the generation of large pre-plasma and surface structure in the two-beam irradiation case. The surface modulation at relativistic critical density increases the area where the laser incidence is oblique. Reflected laser light is measured on the plane at X = 0 ∝m and the calculated absorption rates are about 47% and 80 % respectively for the one and two beam irradiation case at the end of simulation. Consequently enhanced fast electron generation for two beam interference case is observed, as shown in Figure 4.

The half-angle fast electron angular divergence, (HWHM of the fast electron angular distribution) as function of the electron energy is shown in figure 5. The fast electron divergence for two-beam irradiation is larger than for one-beam case over whole energy range. Because of surface modulation at relativistic critical density, the number of obliquely propagating fast electrons increase. In addition, the hgh intensity fields modulated over the interference pattern locally generate high currents and large magnetic fields. These magnetic fields enhance the divergence of relatively low-energy electrons.

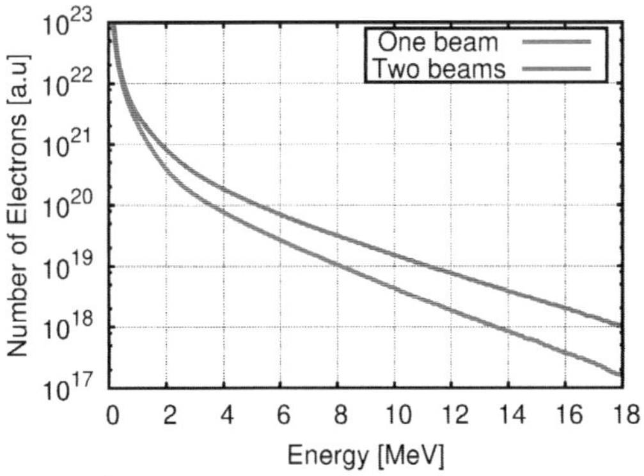

Figure 4. Time-integrated energy spectrum of fast electrons passing through X = 115 ∝m.

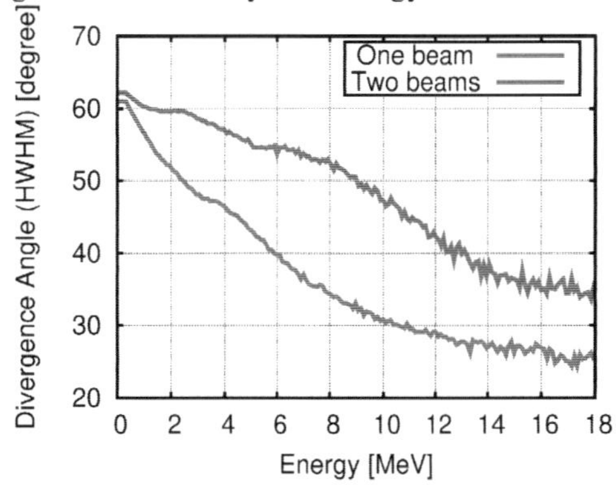

Figure 5. Divergence angle (HWHM) of observed fast electrons in momentum space.

3. Summary

Two-dimensional PIC simulation using realistic parameters of LFEX laser were performed and effects of beam interference were investigated. The interference induces surface perturbations, which enhance the laser absorption and the creation of underdense plasma, increasing the generated fast electron number and the relative slope temperature. On the other hand, the surface perturbation also cause wider fast electrons divergence. From a fast ignition point of view, large fast electron divergence leads to lower coupling efficiency to the fuel core, but at the same time high absorption leads to higher coupling efficiency. Simply, a large divergence angle does not significantly affect the coupling efficiency if the core is located near the fast electrons generation point, but it strongly affects it if the core is far from the fast electron generation point. Fast electron guiding via external, laser-generated magnetic fields may solve the disadvantage of interference effects [7]. To optimize the target design for fast ignition, these conflicting effects, namely effects of higher absorption and wider fast electron divergence due to interference, should be clarified via full-scale PIC-Hybrid-Hydro simulation under the realistic conditions, including implosion, fast electron generation, and core heating processes.

Acknowledgments

This work was supported by JSPS Grant-in-Aid (No. 15K17798) and NIFS Collaboration Research program (NIFS12KUGK057 and NIFS14KNSS054).

References

[1] H. Azechi and the FIREX Project, Plasma Phys. Controlled Fusion **48**, B267 (2006).
[2] H. Azechi, et al., Nucl. Fusion **53**, 104021 (2013).
[3] N. Miyanaga *et al., J. Physique IV France* **133**, 81 (2006)
[4] H. Sakagami, et al., Proc. 39th EPS Conf. on Plasma Physics P2.122 (2012)
[5] A. J. Kemp and L. Divol, Phys. Rev. Lett. **109**, 195005 (2012).
[6] Y. Sentoku and A. J. Kemp, J. Comput. Phys. **227**, 6846 (2008).
[7] T. Johzaki, et al., Nucl. Fusion **55**, 053022 (2015).

Off-Hugoniot characterization of alternative inertial confinement fusion ablator materials.

Alastair S Moore[1], **Shon Prisbrey**[1], **Kevin L. Baker**[1], **Peter M. Celliers**[1], **Jonathan Fry**[1], **Thomas R. Dittrich**[1], **Kuang-Jen J. Wu**[1], **Margaret L. Kervin**[1], **Michael E. Schoff**[2], **Mike Farrell**[2], **Abbas Nikroo**[2], **Omar A. Hurricane**[1]

[1] Lawrence Livermore National Laboratory, P.O. Box 808, Livermore, CA 94551-0808 USA
[2] General Atomics, San Diego CA 92121 USA

E-mail: alastair.moore@physics.org

Abstract. The ablation material used during the National Ignition Campaign, a glow-discharge polymer (GDP), does not couple as efficiently as simulations indicated to the multiple-shock inducing radiation drive environment created by laser power profile [1]. We investigate the performance of two other ablators, boron carbide (B_4C) and high-density carbon (HDC) and compare with GDP under the same hohlraum conditions. Ablation performance is determined through measurement of the shock speed produced in planar samples of the ablator subjected to the identical multiple-shock inducing radiation drive environments that are similar to a generic three-shock ignition drive. Simulations are in better agreement with the off-Hugoniot performance of B_4C than either HDC or GDP.

1. Introduction

The ablation of the outer capsule surface by x-rays created in a hohlraum cavity is a key element to the indirect-drive scheme being pursued on the NIF. X-rays generated by the interaction of a purposefully shaped laser pulse incident inside a high-Z, Au or depleted uranium hohlraum result in the ablation of material from the capsule. The rocket-like blow-off of material generates multiple shocks that accelerate the capsule to velocities of > 350 km/s and implode it from an initial diameter of ≈ 1.0 mm to < 35 μm. The efficiency with which this rocket effect occurs is a result of the x-ray ablation properties of the capsule material, and so is a key-parameter in the quest for ignition. An approximate analytic expression for the ablation pressure can be found by considering the conservation of momentum and energy, and constant specific heat [2]:

$$P_{abl} = (1-\alpha)(\gamma-1)\sigma T^4 \sqrt{\frac{Am_p}{(Z+1)k_B T}} \quad (1)$$

where σ is the Stefan-Boltzmann constant, A is the atomic number and m_p is the proton mass. Assuming an albedo (α) of approx 0.3 since the ablator thickness is typically tuned to optimize for the material opacity and $\gamma = 5/3$, we calculate the average ionization (Z) using FlyCHK for a temperature of 110 and 170eV corresponding to typical 1st and 2nd shock conditions [4]. Table 1 shows typical ablation pressures for the different materials, motivating the interest in HDC, B_4C and Be, where 15-20% enhancement in P_{abl} might be expected over GDP.

Table 1. Estimated ablation pressures for GDP, B_4C, HDC and Be

	ρ (g/cm^3)	A	T=110eV		T=170eV	
			Z	P_{abl} (Mbar)	Z	P_{abl} (Mbar)
GDP	1.1	13	7	8.2	8	35.8
HDC	3.5	12	5	9.7	6	41.1
B_4C	2.4	55	21	10.0	26	41.9
Be	1.9	9	4	9.2	4	42.2

To-date experiments have concentrated on GDP which has the motivating advantage of a long history of development and production of high-quality capsules, together with the ease with which a high-z dopant can be added to reduce fuel pre-heat. Using Be, B_4C as an ablator actually offers greater margin to reach ignition conditions since they all absorb more energy than CH, having a lower albedo, and therefore offer a higher ablation rate for a given temperature. In addition, since B_4C and HDC are higher density, they enable a thinner ablator shell to be used while maintaining the same shell mass. This has a two-fold advantage: (1) shorter laser pulses can be used reducing filling of the hohlraum and allowing the use of near-vacuum hohlraums that suffer less backscatter, and (2) for an equivalent outer diameter, the thinner shell can contain more fuel. In addition, higher mass ablation rates provide better ablative stabilization of mix seeded by the outer surface roughness.

The ablation rates of GDP under single-shock ignition relevant conditions have previously been compared to a range of materials including Be and diamond-like HDC at temperatures between 160-260 eV and indicate that better ablation performance can be acheived with Be and HDC to that of GDP [3]. However the standard ignition laser pulse design uses three or four shocks to accelerate the capsule, so by the time that the final shock is launched to set the final implosion velocity (v_{imp}), the ablator has experienced two or three previous shocks, so it is the ablation performance under these off-Hugoniot conditions that is of real importance in determining v_{imp}. To investigate this we compare the relative performance of B_4C and HDC to GDP under multiple-shock, ignition-relevant conditions and for simplicity of manufacture we make measurements in a planar geometry.

2. Methods

We measure the absolute shock velocity obtained for a given ablator material using the NIF VISAR diagnostic [6] and characterize the x-ray energy incident on the ablator by measuring the hohlraum temperature through the laser entrance hole (LEH). The x-ray emission is measured by the NIF Dante diagnostic [7]. The same 5.75 mm diameter by 9.43 mm long hohraum and laser-beam pointing as in other fusion experiments [5] is used, Figure 1, but we replace the fusion capsule with a 2.0 mm planar disc of the ablator material. The hohlraum is filled to 400 torr with He_4 gas and cooled to 21 K to create a gas-fill density of 1.45 mg/cc. The ablator disc is situated 0.4 mm from the hohlraum axis to creaste a similar viewfactor of the gold hohlraum wall as a normal ignition capsule. The laser pulse shape is fixed regardless of ablator material to, as-much-as-possible, create a 'universal' x-ray drive. The ablator densities were measured by simple weight and volume characterization to within $\pm$0.01 g/cc; the B_4C was 2.51 g/cc, HDC was 3.50 g/cc, and GDP was 1.08 g/cc. The Rq surface roughness was typically 500-800 nm rms for the B_4C samples 40-60 nm rms for HDC and 140-180 nm for the GDP, all of which are below the minimum cell-size used in calculations - typically <0.5 μm.

The shock pressure history of the ablator is measured by adding either a 10 μm gold pre-heat shield and 400 μm of α-quartz followed by liquid deuterium (D_2), or only liquid D_2. The α-quartz acts as a reference material so that the shock velocity can be used to infer the ablation

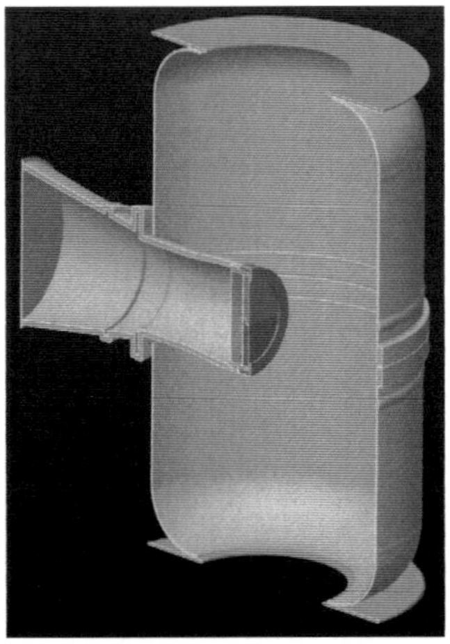

Figure 1. A planar ablator package replaces the capsule inside an ignition scale hohlraum.

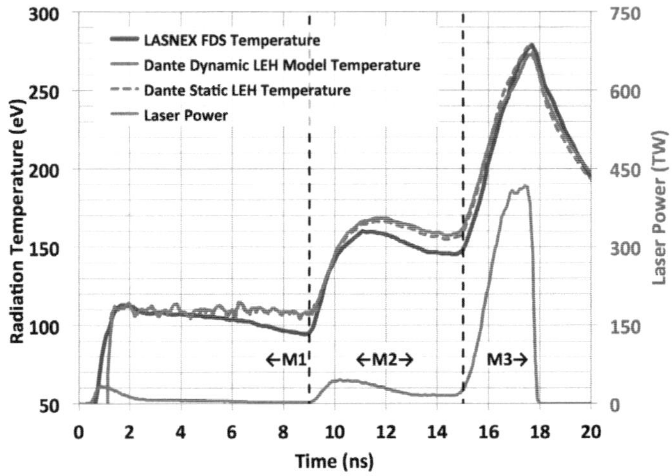

Figure 2. Hohlraum radiation temperature measured by Dante: (red - - - -) assumes no LEH motion , (red ——) dynamic LEH model, (blue) temperature source used in LASNEX simulations, and (green) laser pulse.

pressure from the well-understood equation of state of α-quartz [8]. Unfortunately experiments using the α-quartz also necessitate the Au preheat shield. This introduces a shock-impedance mis-match and reflected shock that disrupts the ablation front meaning that the results are less relevant to a fusion capsule. To mitigate this we also perform experiments in which the ablator is backed by liquid deuterium, which is a good surrogate for a DT ice-layer. The equation of state of deuterium under these conditions is less well understood, and also more easily ionized by hot-electrons or few-keV photons generated in the hohlraum, but the shock and release behavior is directly applicable to a fusion capsule performance.

3. Results and Discussion

Soft x-ray images of the hohlraum laser-entrance hole (LEH) were used to correct the x-ray flux measured by the Dante instrument in Figure 2 and calculate the radiation temperature for comparison with simulation. Simulations were performed using the 'High-flux' model in the radiation hydrodynamics code LASNEX [9]. Once post-processed to simulate the Dante measurement these match the measured temperature very well and within the 5% x-ray flux uncertainty of Dante at the peak. The simulations are tuned to match the shock velocity and overtake times measured by the VISAR diagnostic by applying multipliers (M1, M2 and M3) to the laser power in the three regions shown: $t < 9$ ns, $9 \text{ ns} < t < 15$ ns, and $t > 15$ ns.

The VISAR data in figure 3 shows nuanced differences in the shock and ablator material reflectivity for the different ablators. In addition to the shock signatures, in figure 3(a) and (c) when the peak laser power is reached at 17.5 ns, the data disappears. This is also coincident with the peak in the hot electron signal inferred from hard x-rays measured by FFLEX and also the greatest production of Au M-band radiation inside the hohlraum. It is postulated that these factors cause the either the D_2 and/or quartz window at the rear of the VISAR cone to be weakly ionized and non transmissive to the VISAR laser. In figure 3(b),(d) and (e) the behavior is different. A reduction in fringe visibility is observed at the same time as in (a) and (c) which is again hypothesized to be the result of hot electrons or hard x-rays from the hohlraum causing the quartz to 'blank'. However, in these experiments the Au preheat shield limits the duration

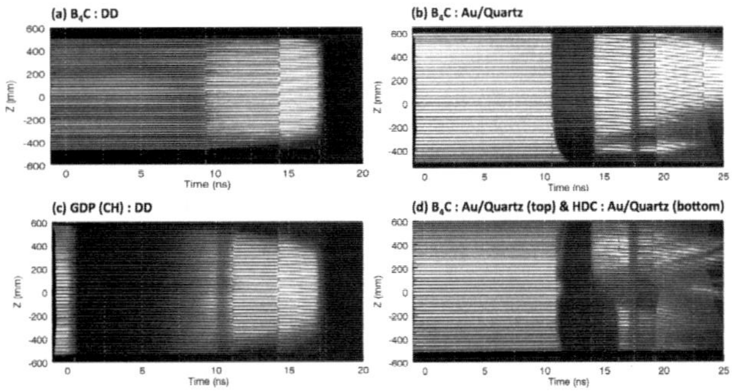

Figure 3. VISAR data for four experiments (a) N130218: 152.9 μm B_4C backed by liquid D_2, (b) N130114: 149.4 μm B_4C, 9 μm Au and 402 μm α-quartz, (c) N140719: 250.0 μm GDP backed by liquid D_2, and (d) N130524: top: 148.3 μm B_4C and bottom: 148.6 μm HDC, 11 μm Au and 402 μm α-quartz.

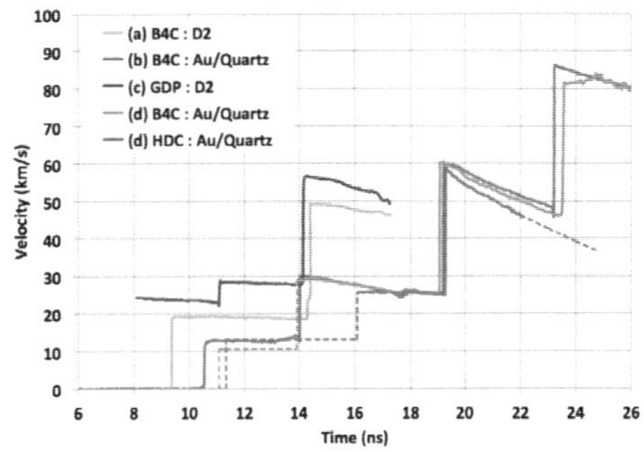

Figure 4. Shock velocities: Dotted and dashed lines are calculated based on the time the shock broke out of the quartz and assume that the shock has a constant velocity during the periods missing data. The velocity uncertainty is 0.8 μm/ns

of this effect to about 1 ns, after which the quartz recovers. It is also of note that the data in (b) and the top of (d) are essentially repeated, and so the 0.4 ns later arrival of S1 in (d) is a result of the 2% lower laser energy prior to 9 ns and the 2 μm thicker Au layer.

Obviously M3 cannot be accurately infer from the data in figure 3(a) and (c) due lack of data after the laser power peaks, however, M1 and M2 in figure 3(a) and (c) are directly relatable to an ignition capsule and indicate better agreement between the simulated and measured ablation performance of B_4C than that of GDP. For all experiments the value of M1 inferred from the tuned simulations was about 1.00, however an M2 multiplier of 1.06, 1.00 and 0.95 was required to match the B_4C experiments (a), (b) and (d), while M2=0.85 was required to match the HDC measurement (e) and and even greater reduction M2=0.60 for GDP (c).

4. Conclusions

To conclude, we have investigated the off-Hugoniot ablation properties of three candidate ablator materials for ICF capsules. By studying the ablation properties away from the principle Hugoniot we rate how well the current equation of state models for the different materials reproduce the measured performance. Results indicate that the measured off-Hugoniot shock velocities of GDP and HDC fall below that predicted by simulations, while the equation of state of B_4C better predicts the shock pressure after multiple shocks. We are grateful to the dedication of the NIF operation team. This work was supported by UK Ministry of Defence and performed under the auspices of Lawrence Livermore National Laboratory under contract DE-AC52-07NA273444.

References

[1] Robey H F *et al.* 2012 *Phys. Rev. lett.* **108** 215004
[2] Wilson D C *et al.* 1998 *Phys. Plasmas* **5** 1953
[3] Olsen R E *et al.* 2011 *Phys. Plasmas* **18** 032706
[4] Chung H-K, Chen M H, Morgan W L, Ralchenko Yu and Lee R W 2005 *High Energy Density Physics* **1** 3
[5] Hurricane O A *et al.* 2014 *Nature* **506** 343
[6] Celliers PM *et al.* 2004 *Rev. Sci. Instrum.* **75** 4916
[7] Kline J L *et al.* 2010 *Rev. Sci. Instrum.* **81** 10E321
[8] Knudsen M D and Desjarlais M P 2009 *Phys Rev Lett* **103** 225501
[9] Rosen M D *et al.* 2011 *High Energy Density Physics* **7** 180

Analysis of plasma channels in mm-scale plasmas formed by high intensity laser beams

R Murakami[1], H Habara[1], S Ivancic[2,3], K Anderson[2], D Haberberger[2], T Iwawaki[1], H Sakagami[4], C Stoeckl[2], W Theobald[2], Y Uematsu[1] and K A Tanaka[1]

[1]Osaka University, Suita, Osaka, Japan, [2]LLE, U. Rochester, Rochester, NY, USA, [3]Dep. Mech. Eng., U. Rochester, Rochester, NY, USA, [4]NIFS, Toki, Gifu, Japan

rmurakami@eie.eng.osaka-u.ac.jp

Abstract. A plasma channel created by a high intensity infrared laser beam was observed in a long scale-length plasma (L ~ 240 μm) with the angular filter refractometry technique, which indicated a stable channel formation up to the critical density. We analyzed the observed plasma channel using a rigorous ray-tracing technique, which provides a deep understanding of the evolution of the channel formation.

1. Introduction

For fast ignition of inertial confinement fusion, the effective conversion of Petawatt laser power into ~MeV electrons is required to heat a high-density fuel core to ignition temperatures. We use a scheme making use of relativistic laser self-focusing to penetrate through the surrounding plasma corona to the core. However, during the laser propagation in the underdense plasma, the beam loses energy through several laser-plasma instabilities. Therefore, a pre-formed low-density plasma channel may be a good choice for efficient energy transport to the core. In previous experiments [1] with the Omega EP laser at the Laboratory for Laser Energetics at the University of Rochester, we observed the plasma channel formation using the angular filter refractometry (AFR) technique [2]. In this paper, we provide a quantitative characterization of the plasma channel created inside near-critical-density (~10^{21}/cm^3), mm-size plasmas by ray-trace calculation of the short-pulse, ultraviolet probe beam.

2. Experiment

Figure 1 shows the experimental setup. High-density plasma plume is created from a CH flat target (3 mm × 3 mm × 0.125 mm) irradiated by two ultraviolet (UV) beams (λ_{UV} = 0.351 μm) with a total energy of 1 kJ. After the UV beam irradiation, an infrared (IR) channeling laser (λ = 1.054 μm, pulse length = 10 ps or 100 ps, pulse energy = 1 kJ or 2.6 kJ, peak intensity = 4×10^{19} W/cm^2 or 1×10^{19} W/cm^2, respectively) was injected into the plasma and focused to $n_c/4$. The AFR diagnostic utilized a fourth-harmonic probe pulse (λ = 0.263 μm, 10 ps, 10 mJ), which observes the plasma density profile by measuring the refraction angle of the probe rays. The refracted probe beam was collected through a lens. On the way to the image plane, an angular filter is inserted in order to measure the refraction angles. At the image plane, contours of constant refraction angle can be observed. The angular filter consists of 2-mm-wide opaque rings blocking the light and 2-mm-wide transparent rings and a 500-

μm-diameter central opaque dot. The spatial resolution of the diagnostic is ~4 μm, and the field of view extends over 4 mm.

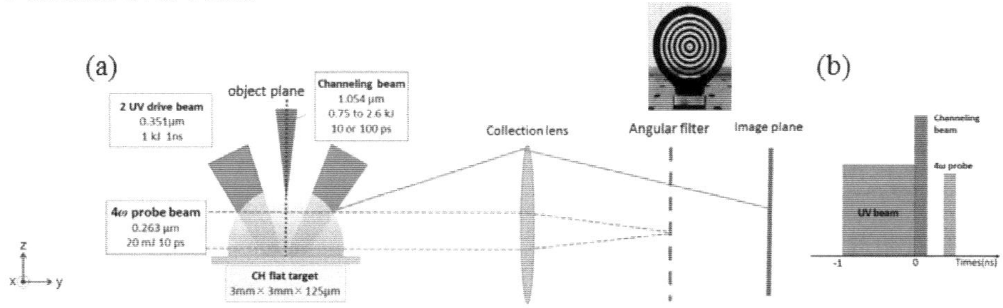

FIG. 1(a) The experimental setup and the model of the AFR diagnostic. The dotted lines represent the unrefracted probe beam and the solid line represents the refracted probe beam. (b) Shot timing diagram.

The delay time of 0 ps of the probe beam is defined as the channelling laser injection timing. The ring collapse at the center of the image (Figs. 2(a) and 3(a)) shows the IR laser propagation and the resultant plasma channel formation. The 100 ps pulse laser propagated deeper into the plasma to higher density region than the 10 ps pulse laser. In the case of the 100 ps pulse, a strong 4ω self-emission generated at the critical density was observed at the second ring from the bottom (Fig. 3(a)). For the 10 ps laser, a filament structure was observed at early probe times that extended from the tip of the plasma channel into the dense plasma. The width of the filament structure was ~200 - 300 μm. Details of the experimental results are presented in Ref. 1.

3. Analysis

3.1 Ray tracing calculation

The plasma density profiles are analysed by ray trace calculations for various plasma parameters such as plasma scale length or channel diameter. In the calculations, the plasma density profile is given as

$$\boldsymbol{n_e = n_0 \exp\left[-\frac{x^2+y^2}{L_g{}^2}\right] \exp\left[-\frac{z}{L_n}\right]} \tag{1}$$

where L_n is the longitudinal plasma scale length along the propagation direction of the IR beam and L_g is a transversal plasma scale length. The low-density plasma channel and the high-density channel wall are approximated by a quadratic function and are then additionally applied inside the profile. The refraction angle of probe rays is related to the spatial gradient of phase shifts.

$$\delta\theta_\alpha = \frac{\lambda_p}{2\pi}\frac{\partial\phi}{\partial\alpha} \tag{2}$$

where λ_p is the wave length of the probe ray, ϕ is the phase shift of the probe ray passing through the plasma, y is the propagation direction of the probe, and α denotes x or z. The summation of the phase shift is calculated along the optical path for spatially divided rays in the calculation.

$$\delta\phi(x,z) = \frac{2\pi}{\lambda_p}\int_{-L/2}^{L/2}\left[1-(1-\frac{n_e(x,z)}{2n_c})\right]dy \tag{3}$$

In the calculation, each ray propagates in one step for a distance that is equal to λ_p and the refraction angles are updated by Eq. (3). The density gradient in Eq. (3) is calculated by central difference method ($\Delta\alpha=\lambda$).

Journal of Physics: Conference Series **717** (2016) 012039 doi:10.1088/1742-6596/717/1/012039

3.2 Calculation results

Figures 2 and 3 show the experimental and calculation results for the 10 ps and 100 ps pulses. The calculated images agree very well with the experimental AFR images. Figs. 2(c) and 3(c) render the inferred plasma profiles producing the calculated AFR images in Figs. 2(b) and 3(b). The scale length of background plasma can be estimated by the width and separation of the rings, and was $L_n = 240$ μm in this experiment. The 10 ps and 100 ps pulses reach an electron density of 0.8 n_c and >1.5 n_c, respectively. The critical density corresponds to the second band from the bottom of the experimental or calculated AFR images, where the strong self-emission signal is observed for 100 ps pulse shot (Fig. 3(a)). Under current plasma conditions, the AFR images exhibit the near actual size of plasma channel at TCC. However, if the channel diameter becomes large (300 μm~), the size may have an error.

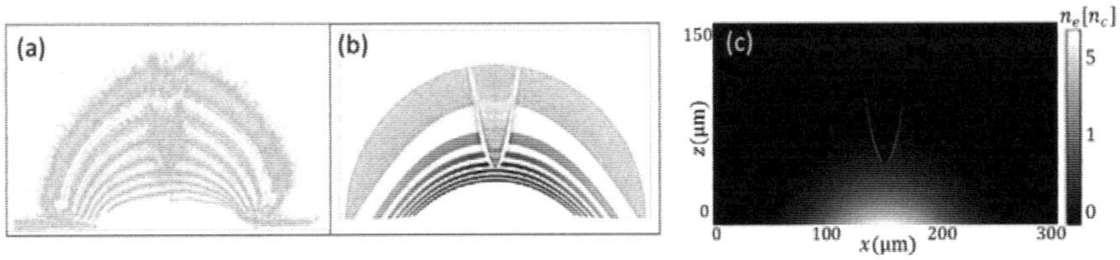

FIG. 2 (a) AFR image of 10 ps (1 kJ) channeling pulse. (t = 15 ps) (b) Well-fitted calculated image. (c) A cross section of the plasma density profile used in the calculation.

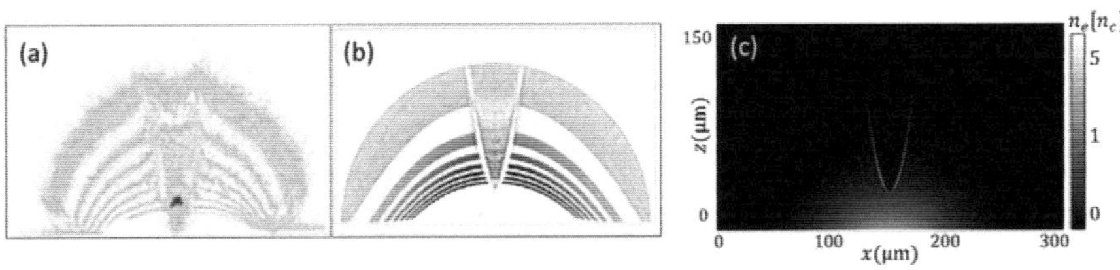

FIG. 3 (a) AFR image of 100 ps (2.6 kJ) channeling pulse. (t = 65 ps) (b) Well-fitted calculated image. (c) A cross section of the plasma density profile used in the calclation.

3.3 The effect of the channel profile on the AFR image

Here, we conducted more detailed analysis of the channel by changing the channel parameters. We found (a) that the density inside the plasma channel affects the ring curvature, (b) the high density plasma wall collapses the ring pattern inside the channel, and (c) filamented patterns are reproduced by very small density gap.

Figures 4 (a) and (b) show the magnified experimental and calculated images for the 10 ps pulse. In the calculation, the density inside the plasma channel significantly changes the ring curvature inside the channel. From the comparison of both images, the channel density is estimated to be 0.2 of its background density. In the case of the 100 ps pulse (Fig. 4(c,d)) the channel density is estimated to be 0.5 of its background density. This result indicates that the higher intensity laser pulse forms a lower density plasma channel as predicted by PIC simulations.

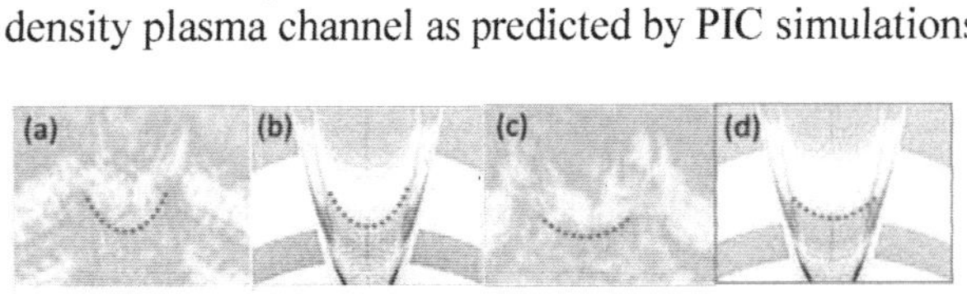

FIG. 4 Experimental images for (a) 10 ps pulse shot (t = 15 ps) and (c) 100 ps pulse shot (t = 65 ps). (b), (d) Calculation results. Channel density is (b) 0.2 times back ground and (d) 0.5 times back ground.

The effect of the plasma channel wall density on the AFR image was also studied. Figures 5(a) and 5(b,c) show experimental and calculated images for wall densities of 10 times and 2 times of the background density, respectively. The ring pattern inside the channel disappears when the wall density becomes higher, which is independent of the inner density of the plasma channel. Therefore, a loss of the ring pattern does not necessarily mean the collapse of the plasma channel structure.

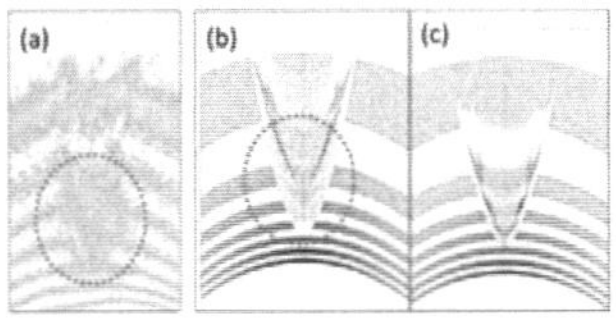

FIG. 5 (a) Experimental image for the 10 ps pulse shot (t = 15 ps). Calculation results of (b) high-density (10×) wall and (c) low-density (2×) wall.

Figure 6 indicates (a) filamented structures in the experimental image, and the ray trace images for two different cases of filament structure, (b) micro plasma channel (0.1 times background density) and (c) fast electron filaments (10 times background density). We fixed the diameters of filaments to 5 μm, which is about the spatial resolution of the diagnostic. In both cases the filament structure is clearly observed and there is no difference between both. We will conduct a further analysis using a large-scale PIC code to identify the origin of the filamented structure.

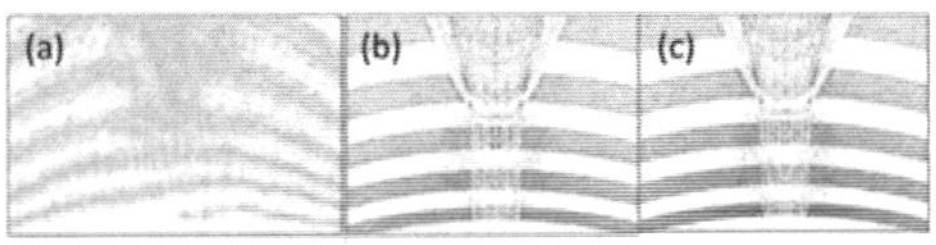

FIG. 6 (a) Filament structure in the experimental image and calculated images of (b) micro plasma channel and (c) electron filaments.

3.4 Error evaluating

The plasma density corresponding to each ring can be obtained by Abel inversion [3]

$$n_e\left(R = \sqrt{x^2 + y^2}\right) = -\frac{\lambda_p n_c}{\pi^2} \int_{-\infty}^{\infty} \frac{\partial \phi}{\partial x} \frac{ds}{\sqrt{s^2 + R^2}} \tag{4}$$

where $s = \sqrt{x^2 - R^2}$ is substituted to eliminate the singularity at $x = R$. However in Eq. (4), the ray refractions are completely ignored. Therefore, in the case of large-scale plasma, Abel inversion might give large errors because the refracted ray passes through a lower density region than along its original direction. Figure 8 indicates the plasma density corresponding to each of the rings obtained by Abel inversion (green triangles) and ray tracing (blue dots). From the comparison, the maximum error of the plasma density between Abel inversion and ray trace calculation is less than $0.2n_c$.

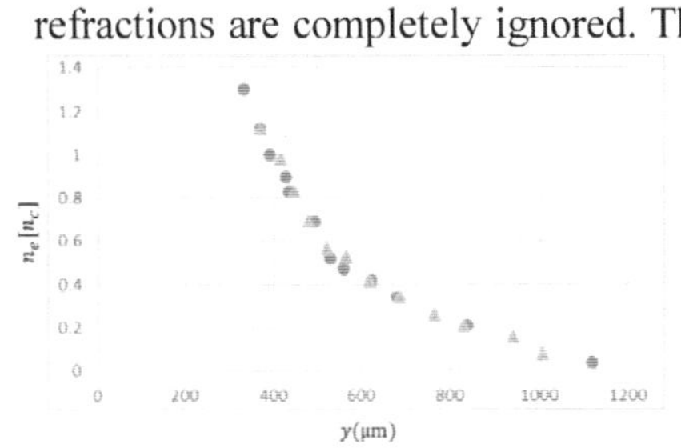

FIG. 7 Plasma densities calculated by the ring positions. The blue dots and green triangles are from the ray trace analysis and the Abel inversion, respectively.

4. Summary

The effect of the plasma channel profile on AFR images was studied with detailed ray trace calculations. The analysis indicates that the plasma channel front propagated up to a background electron density of 0.8 n_c for the 10 ps, 125 TW laser pulse and >1.5 n_c for the 100 ps, 20 TW laser pulse, respectively. The residual density inside the channel was found to be ~20% and ~50% of the background density for the 10 ps and 100 ps pulses, respectively.

References

[1] S. Ivancic et al., Phys. Rev. E 91, O51101 (2015).
[2] D. Haberberger et al., Phys. Plasmas 21, 056304 (2014).
[3] M. Seki and S. Himeno, Technical Digest of Hokkaido University, 2(2), 1-9 (1991).

Acknowledgments

A part of this research has been supported by ASPS S-class(15H05751) Grants - in - aid for Scientific Research and JSPS Core-to-Core Program on International Alliance for Material Science in Extreme States with High Power Laser and XFEL.

Laser–plasma interaction in direct-drive inertial confinement fusion

J F Myatt,[1,2] J Shaw,[1] V N Goncharov,[1,2] J Zhang,[1,2] A V Maximov,[1,2] R W Short,[1] R K Follett,[3] W Seka,[1] D H Edgell,[1] and D H Froula[1,3]

[1] Laboratory for Laser Energetics, University of Rochester, 250 East River Road, Rochester NY 14623-1299, United States
[2] Department of Mechanical Engineering, University of Rochester
[3] Department of Physics, University of Rochester

E-mail: jmya@lle.rochester.edu

Abstract. Hot electrons generated by the two-plasmon–decay instability in direct-drive targets are a preheat concern. A mitigation strategy that employs a layered ablator [V. N. Goncharov *et al.*, Phys. Plasmas **21**, 056315 (2014)] has been investigated both numerically and experimentally. The numerical simulations described here predict reduced hot-electron production compared with similar targets using either a solid CH or Be ablator. These findings are shown to be consistent with experimental observations

1. Introduction

Of the many challenges facing laser-driven inertial confinement fusion (ICF) [1, 2], controlling the impact of laser–plasma interactions [3] has shown itself to be one of the most difficult and uncertain. In the direct-drive approach [4, 5] the collective interaction of multiple beams [6] leads to two important instabilities: multiple-beam two-plasmon decay (TPD) [7, 8] and cross-beam energy transfer (CBET) [9, 10, 11]. Two-plasmon decay has the potential to create enough hot electrons to preheat the cold fuel, while CBET reduces the efficiency of laser coupling, i.e., an $\approx 30\%$ loss of drive energy on OMEGA.

While TPD is known to be active in spherical implosions on OMEGA [12, 13], the preheat associated with the instability is not currently thought to be sufficient to impair target performance [14]. It has the potential to become so when CBET is mitigated since the mitigation of CBET will increase the laser intensity at the quarter-critical density (the active region for TPD) by as much as a factor of two. Beam zooming is planned as a mitigation strategy for CBET [15, 16]. Two-plasmon–decay preheat may be detrimental at the longer plasma scale lengths obtainable in NIF-scale implosions [17] that are required for ignition.

For these reasons it is important to understand the effectiveness of mitigation strategies for TPD. The strategy described here employs a mid-Z (silicon) layer buried within the ablator. The effectiveness of the mid-Z layer can arise because of hydrodynamic effects (higher coronal temperature and shorter density scale length) and/or collisional effects that modify the plasmon physics [18, 19]

2. Simulating TPD with LPSE

Each of three targets was simulated with the 1-D hydrodynamics code LILAC [20] to obtain hydrodynamic variables as a function of time. These simulations incorporated the effects of CBET using a ray-based in-line model [10]. The targets corresponded to those shot in a recent OMEGA experimental campaign (shots 77388, 77391, and 77392). A schematic of a multilayer target is shown in Fig. 1. In the other two targets, the ablator was either solid CH (shot 77388) or solid beryllium (shot 77391).

The predicted coronal temperature, at a radius corresponding to the quarter-critical density, is shown for the three target designs as a function of time in Fig. 2. Notice that the coronal temperature in the multiple-ablator design is higher, particularly when the Si layer is present at the quarter-critical surface (shaded region in the figure). The increased temperature combined with a shorter density scale length (not shown) in these targets corresponds to a reduced TPD threshold factor [13] (shown in green). This suggests that TPD should be below threshold for the multiple-ablator experiment.

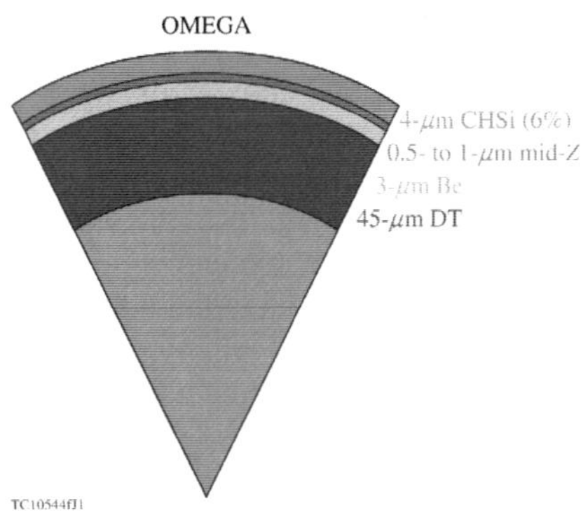

Figure 1. Schematic for a multi-ablator target design. The purpose of the thin Si layer is to mitigate the two-plasmon–decay (TPD) instability.

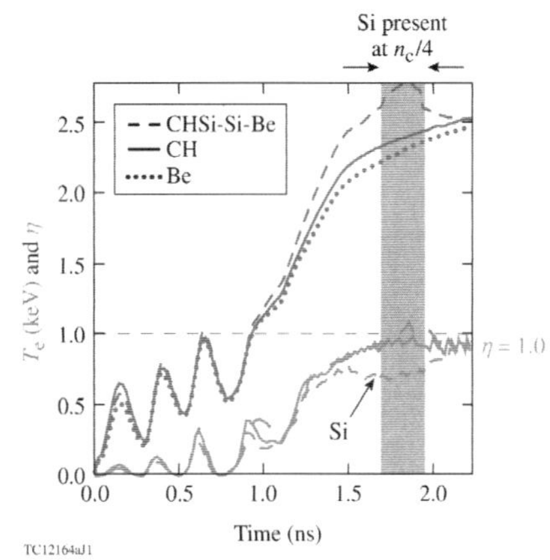

Figure 2. Quarter-critical temperature and empirical TPD threshold factor for three ablator designs, as simulated by LILAC.

While the TPD threshold factor $\eta = I_{14} L_{\mu m}/(230\ T_{e,keV})$ is based on the instability threshold for a single plane wave [21, 22], its application to spherical implosion experiments (substituting total overlapped intensity) is empirical. It is therefore of interest to compute the effects on the modified target hydrodynamics on TPD hot-electron production using a physical model that takes into account the details of the multiple-beam irradiation on the threshold and nonlinear evolution, including hot-electron production.

The LPSE code (laser plasma simulation environment) does this by combining a physical model of TPD with an established model of plasma-wave turbulence [23, 24] in three spatial dimensions. The incident laser light in the simulations reflects the actual illumination — i.e., it takes into account the angles of the overlapping beams and beam smoothing. Kinetic effects and hot-electron production are incorporated in a way that has similarities with the quasi-linear model described in Ref. [18]. Instead of solving a quasi-linear diffusion equation to evolve the spatially averaged distribution function, however, electron particle trajectories are numerically integrated in the electrostatic field using a novel algorithm that exploits hardware (GPU) acceleration.

The LPSE model takes advantage of the separation between the hydrodynamic and plasma-wave time scales. The duration of the ∼2-ns implosion is sampled at several points in time and the hydrodynamics are imported into LPSE. These variables are then "frozen" over the duration of the LPSE simulation (with the exception of density profile modification), which is performed over a time sufficient to obtain a statistical steady state for the hot-electron production (typically $\sim$ 20 ps). The volume of the calculation does not include the whole $n_c/4$ surface but only a small ($30 \times 30 \times 70$-μm^3) fraction of it. This is therefore a local analysis in the neighborhood straddling a point (r, θ, ϕ) on the surface. Several angular locations are simulated, confirming that electrons are preferentially generated at hex centers [25].

Figure 3 shows the simulated hot-electron production as a function of time (each marker corresponds to a sample time) for three simulated implosions.

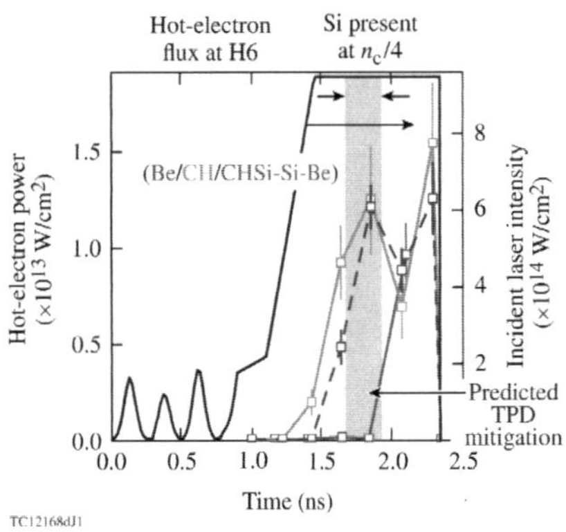

Figure 3. Hot-electron production as simulated by LPSE for three targets with different ablator types.

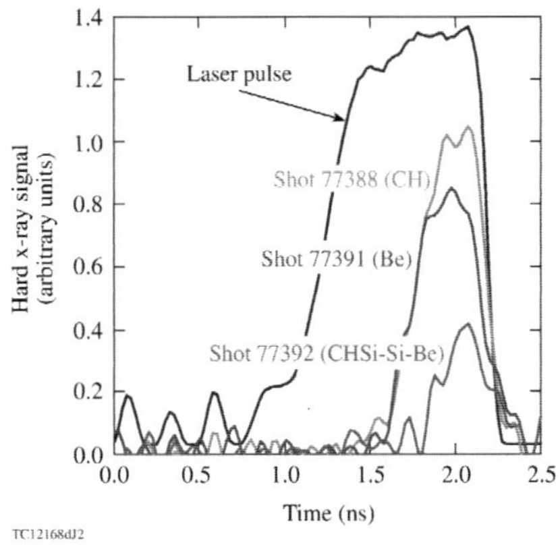

Figure 4. Measured hard x-ray yield from the three corresponding implosion experiments.

The simulation results for the CH ablator, shown in green in Fig. 3, compare well with experimental results (Fig. 4). The error bars in hot electron fraction reflect fluctuations about the statistical steady state seen in the simulations. Uncertainties in the hydroynamic conditions are not accounted for. While the comparison is complicated because the experiment measured hard x rays (in the approximate range of 40-80 keV) and not hot-electron power, a good agreement is observed between the time of onset. This suggests that the LPSE model makes a good prediction for the threshold. The Be simulation is predicted to be very similar to CH (blue markers). Again, the experimental x-ray signatures for Be and CH are very close. For all cases, the simulated hot electron temperatures are close to those observed experimentally (30-40 keV).

Importantly, the simulations confirm the mitigating effect of the Si layer. Very few hot electrons are predicted to occur while the Si is present at the $n_c/4$ surface (magenta points). Hot electrons are produced at later times and at a level that is similar to the other ablators. Experimentally, the late onset is similar to the predictions, but the mitigating effect appears to be even stronger than predicted by LPSE. This may be due to mixing of the Si layer.

3. Summary

Simulations of TPD mitigation experiments using the LPI code LPSE confirm the effectiveness of buried mid-Z layers. Furthermore, several aspects of these simulations are consistent with experimental observations. It remains to be seen how effective this strategy will be at the ignition scale when the instability is anticipated to be more strongly driven.

Acknowledgments

The authors are grateful for the contributions from H. X. Vu (unaffiliated), D. A. Russell, and D. F. DuBois (Lodestar Research Corporation).

This material is based upon work supported by the Department of Energy National Nuclear Security Administration under Award Number DE-NA0001944, the University of Rochester, and the New York State Energy Research and Development Authority. The support of DOE does not constitute an endorsement by DOE of the views expressed in this article.

References

[1] Nuckolls J, Wood L, Thiessen A and Zimmerman G 1972 *Nature* **329** 139
[2] Lindl J D 1998 *Inertial Confinement Fusion: The Quest for Ignition and Energy Gain Using Indirect Drive* (New York: Springer-Verlag)
[3] Kruer W L 1988 *The Physics of Laser Plasma Interactions* Frontiers in Physics, Vol. 73 Pines D ed (Redwood City, CA: Addison-Wesley)
[4] Cratxton R S, Anderson K S, Boehly T R, Goncharov V N, Harding D R, Knauer J P, McCrory R L, McKenty P W, Meyerhofer D D, Myatt J F *et al.* *Direct-Drive Inertial Confinement Fusion: A Review* to be published in Physics of Plasmas
[5] McCrory R L, Betti R, Boehly T R, Casey D T, Collins T J B, Craxton R S, Delettrez J A, Edgell D H, Epstein R, Frenje J A *et al.* 2013 *Nucl. Fusion* **53** 113021
[6] Myatt J F, Zhang J, Short R W, Maximov A V, Seka W, Froula D H, Edgell D H, Michel D T, Igumenshchev I V, Hinkel D E and Moody J D 2014 *Phys. Plasmas* **21** 055501
[7] Michel D T, Maximov A V, Short R W, Hu S X, Myatt J F, Seka W, Solodov A A, Yaakobi B and Froula D H 2012 *Phys. Rev. Lett.* **109** 155007
[8] Zhang J, Myatt J F, Short R W, Maximov A V, Vu H X, DuBois D F and Russell D A 2014 *Phys. Rev. Lett.* **113** 105001
[9] Igumenshchev I V, Edgell D H, Goncharov V N, Delttrez J A, Maximov A V, Myatt J F, Seka W, Shvydky A, Skupsky S and Stoeckl C 2010 *Phys. Plasmas* **17** 122708
[10] Igumenshchev I V, Seka W, Edgell D H, Michel D T, Froula D H, Goncharov V N, Craxton R S, Divol L, Epstein R, Follet R *et al.* 2012 *Phys. Plasmas* **19** 056314
[11] Froula D H, Igumenshchev I V, Michel D T, Edgell D H, Follet R, Glebov V Yu, Goncharov V N, Kwiatkowsi J, Marshall F J, Radha P B *et al.* 2012 *Phys. Rev. Lett.* **108** 125003
[12] Stoeckl C, Bahr R E, Yaakobi B, Seka W, Regan S P, Craxton R S, Delettrez J A, Short R W, Myatt J, Maximov A V and Baldis H 2003 *Phys. Rev. Lett.* **90** 235002
[13] Seka W, Edgell D H, Myatt J F, Maximov A V, Short R W, Goncharov V N and Baldis H A 2009 *Phys. Plasmas* **16** 052701
[14] Goncharov V N, Sangster T C, Betti R, Boehly T R, Bonino M J, Collins T J B, Craxton R S, Delettrez J A, Edgell D H, Epstein R *et al.* 2014 *Physics of Plasmas* **21** 056315
[15] Igumenshchev I V, Froula D H, Edgell D H, Goncharov V N, Kessler T J, Marshall F J, McCrory R L, McKenty P W, Meyerhofer D D, Michel D T *et al.* 2013 *Phys. Rev. Lett.* **110** 145001
[16] Froula D H, Kessler T J, Igumenshchev I V, Betti R, Goncharov V N, Huang H, Hu S X, Hill E, Kelly J H, Meyerhofer D D *et al.* 2013 *Phys. Plasmas* **20** 082704
[17] Hohenberger M, Radha P B, Myatt J F, LePape S, Marozas J A, Marshall F J, Michel D T, Regan S P, Seka W, Shvydky A *et al.* 2015 *Phys. Plasmas* **22** 056308
[18] Myatt J F, Vu H X, DuBois D F, Russell D A, Zhang J, Short R W and Maximov A V 2013 *Phys. Plasmas* **20** 052705
[19] Lafon M, Betti R, Anderson K S, Collins T J B, Epstein R, McKenty P W, Myatt J F, Shvydky A and Skupsky S 2015 *Phys. Plasmas* **22** 032703
[20] Delettrez J A, Epstein R, Richardson M C, Jaanimagi P A and Henke B L 1987 *Phys. Rev. A* **36** 3926
[21] Simon A, Short R W, Williams E A and Dewandre T 1983 *Phys. Fluids* **26** 3107
[22] Seka W, Edgell D H, Knauer J P, Myatt J F, Maximov A V, Short R W, Sangster T C, Stoeckl C, Bahr R E, Craxton R S *et al.* 2008 *Phys. Plasmas* **15** 056312
[23] DuBois D F, Bezzerides B and Rose H A 1992 *Phys. Fluids B* **4** 241
[24] Russell D A, DuBois D F and Vu H X presented at the Workshop on the Nonlinear Saturation of Stimulated Raman and Brillouin Instabilities for Plasma and Laser Parameters of Interest to Inertial Confinement and High Energy Density Science, Livermore, CA, 3-5 April 2002
[25] Seka W, Myatt J F, Short R W, Froula D H, Katz J, Goncharov V N and Igumenshchev I V 2014 *Phys. Rev. Lett.* **112** 145001

An optimum design of implosion with external magnetic field for electron beam guiding in fast ignition

H Nagatomo[1], T Johzaki[2], T Asahina[1], K Matsuo[1], A Sunahara[3], H Sakagami[4], T Sano[1], K Mima[5], A Morace[1], Z Zhang[6], S Fujioka[1], K Shigemori[1], H Shiraga[1] and FIREX project group

[1]Institute of Laser Engineering, Osaka University, Suita, Osaka, Japan
[2]Graduate School of Engineering, Hiroshima University, Higashi-Hiroshima, Japan
[3]Institute for Laser Technology, Suita, Osaka, Japan
[4]National Institute for Fusion Science, Toki, Gifu, Japan
[5]Graduate School for the Creation of New Photonics Industries, Hamamatsu, Japan
[6]Institute of Physics, Chinese Academy of Science, Beijing, P. R. China

naga@ile.osaka-u.ac.jp

Abstract. Compression of solid spherical target under the strong external magnetic field is studied using two dimensional radiative magneto-hydrodynamic (MHD) simulation code for fast ignition. The simulation results show that a compression of a solid sphere target is stable, and it is possible to achieve a high areal density core plasma. Assuming the GXII scale laser, it will be ρR=60-80mg/cm^2. Due to the magnetic diffusion in the solid target, the magnetic mirror ratio is less than 4, which does not reflect most of the hot electrons for heating core. These properties are preferable for fast ignition scheme.

1. Introduction

In the first ignition scheme, formation of a high dense core plasma is required in implosion process, as well as the high heating efficiency in heating the core process. Recent studies show that heating efficiency from laser to core is very low due to the large divergence angle of the hot electron beam generated by an ultra-intense laser. A reliable beam convergence method is necessary to improve it. External magnetic field is a possible candidate for the purpose. Recently, the beam divergence method is confirmed computationally and experimentally [1, 2].

However, the strong magnetic field may affect the implosion dynamics where electron heat conduction is anisotropic depending on the angle between the ablation surface direction and the magnetic field line [3]. In our previous work, we showed that imploding CD shell was broken during the acceleration phase in an implosion [4]. Controlling the implosion dynamics in order to avoid or delay the fatal break is extremely difficult by changing the target thickness or laser irradiation laser patterns. Furthermore, at the maximum compression time, the compressed magnetic field formed a mirror structure between the tip of the cone and dense core. Its mirror ratio was high enough to reflect the hot electrons. Therefore, we should introduce an alternative scheme to achieve high dense core plasma which has preferable properties for fast ignition under these conditions.

Here, we investigate a possibility of the solid target for fast ignition. The compression of solid target is hydrodynamic instability free and the life time of the core is longer than shell implosion. Therefore, it is assumed to be more reliable target to achieve high dense core under such extreme conditions.

However, maximum density is not high enough as ideal conventional shell implosion, and intensity of magnetic field may be reduced by the diffusion in the solid target. 2-D radiative MHD simulation is conducted to confirm the possibility of the target in this study.

2. Computational simulations

2.1. Simulation method

We have conducted computational simulations using 2-D radiative magneto-hydrodynamics simulation code (PINOCO-MHD) [4] for this study. PINOCO-MHD is designed for single-fluid two-temperature hydrodynamics with thermal transport, radiation transport equations, equation of state (QEOS), laser ray-trace and absorption, resistive MHD models and so on. The diameter of the deuterated polystyrene (CD) solid target is 200 μm (Fig.1). The solid target is inserted by a gold conical target of which open angle, and thickness are 45 degree and 15 μm respectively. The distance between tip of the cone and center of the solid target is 60 μm. The laser energy, wavelength, and pulse shape are 2.0 kJ with 9 beams of GXII laser, 0.53 μm, and Gaussian (pulse width: 1.3 ns) respectively. Several simulations are conducted for different initial magnetic field strength, 0, 0.2, 0.5 1.0, 1.5, and 2.0 kilo-Telsa (kT), of which directions are parallel to the symmetric axis. Braginskii's electron heat conduction coefficient [3] is applied in the MHD simulation. Magnitude of magnetic pressure is estimated. In the result, it is very small at whole region during the compression. The ratio of the pressure plasma to the magnetic pressure β is β>100. Therefore magnetic pressure is not taken account into this study.

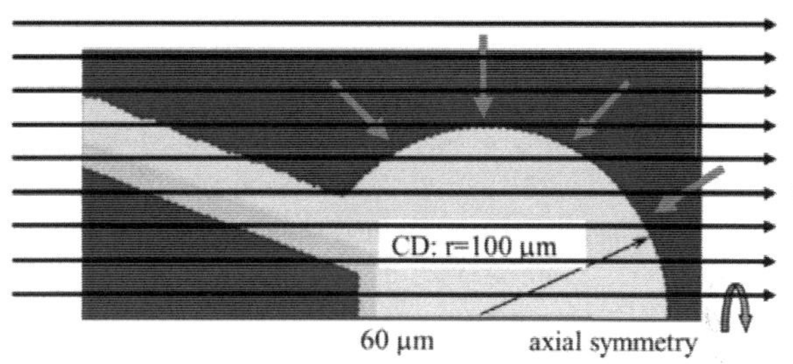

Fig. 1. Target configuration.

2.2. Simulation results

Figures 2 show mass density contours at the maximum compression time for B_0=0, 0.5 and 2.0 kT. The targets are fairly compressed in spite of the strong magnetic field and the existence of the gold cone. The maximum core densities are reached ρ_{core} are 60-80 g/cm^3 for all cases. Due to the suppression of the electron heat conduction which cross the magnetic field lines, core is compressed asymmetrically, and the core becomes cigarette shape, of which radius is about 20 μm, whereas it is about 30 μm in without magnetic field case. This narrow cross section leads the low heating efficiency. However it will be improved if elliptic solid target and/or sophisticatedly controlled non-uniform laser irradiation is applied.

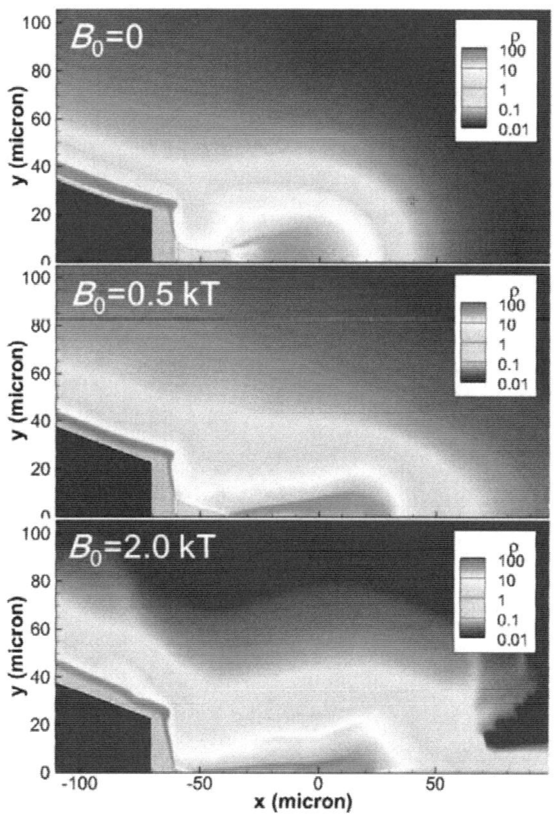

Fig. 2. Mass density contours at the maximum compression time for B_0=0, 0.5, 2.0 kT

Figures 3 show the magnetic field lines at

the maximum compression time for B_0=0.5 and 2.0 kT cases. The line colours indicate the magnetic field strength. For ideal MHD assumption, magnetic field at the compressed core will be B_{core} =3-4 and 13-17 kT for B_0=0.5 and 2.0 kT cases respectively. However, it is only a quarter to the ideal cases, due to the magnetic diffusion transport at solid target area. Because the solid target is kept low temperature and solid density until the shockwave pass through, the magnetic diffusion is dominant during the compression process. This is preferable fact for heating process, and finally the magnetic mirror ratio is limited in $R_M < 4$ for all the cases. In these conditions, most of the electron beam for heating can be transported into the compressed core.

Temporal averaged areal densities for B_0=0, 0.5, 1.0 and 2.0 kT are plotted in Fig. 4. The maximum areal densities increase as a function of the B_0, and they are higher than the case without magnetic field, because the suppression of the thermal conduction enhances the compression toward the radial direction. The maximum times of the case with magnetic field are 170-190 picoseconds earlier than the case without magnetic field also. This difference should be paid attention in experiment. The life time of the high dense core are about 100 picosecond, which is longer than shell implosion case which is a few tens of picosecond. This means that heating triming is less sensitive for solid target case. However, the tip of the cone is exposed to the high pressure for longer period. Optimization of the heating timing is still significant issue.

In the result we conclude that a solid target is probable target for electron beam heating fast ignition. This is preliminary simulation with Gaussian shaped pulse. The solid target has high potential to obtain higher compression if we can use sophisticated tailor laser pulse [5].

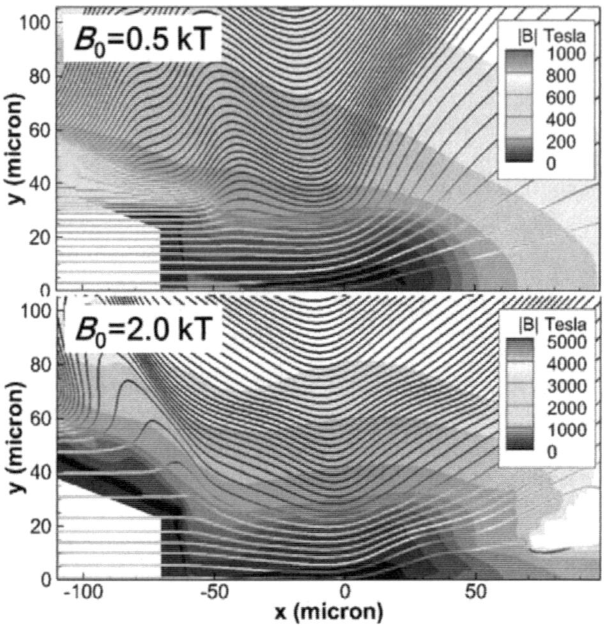

Fig. 3. Magnetic field lines at the maximum compression time for B_0=0.5 and 2.0 kT. The line colours indicate the magnetic field strength.

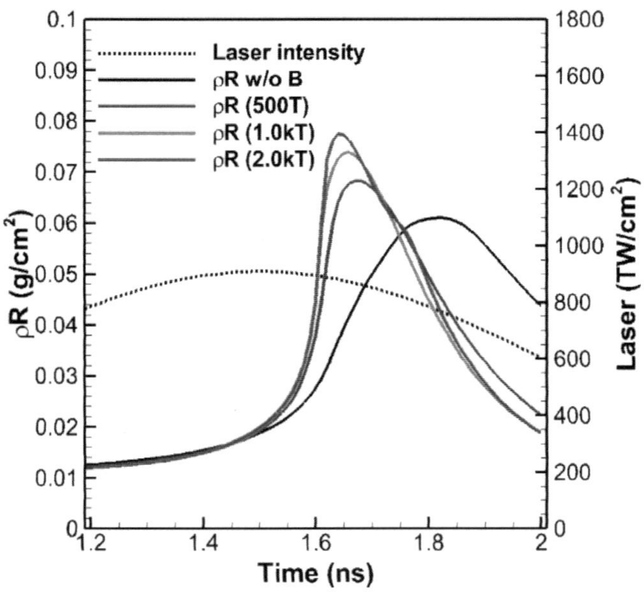

Fig. 4. Temporal averaged areal density (left axis; g/cm^2) and laser intensity (right axis; TW/cm^2).

In order to investigate the detail performance of the solid target, integrated simulations of whole process of fast ignition [6] is necessary. Some preliminary simulations show that 2 kT of the magnetic field at the injection point of the hot electron to guide them toward the high dense core [7].

3. Summary

We have studied the compression of solid target with gold conical target under the strong external magnetic field using two dimensional radiative MHD simulation code. The simulation results show that a compression of a solid sphere target is stable, and it is possible to achieve a high areal density core plasma. Assuming the GXII scale laser, it will be ρR=60-80 mg/cm^2. In the solid target magnetic diffusion is dominant and it is diminished. The magnetic mirror ratio is less than 4 which does not reflect most of the hot electrons for heating core at the maximum compression time.

In this study laser pulse shape is Gaussian shape only. With a sophisticated tailored pulse and target we can achieve higher compression with the solid target.

Acknowledgment

This work is partially conducted under the joibnt research projects of the Institute of Laser Engineering, Osaka University (FIREX-project), and partially supported by JSPS KAKENHI Grant Number 26400532. These simulations were conducted at the Cyber Media Center (CMC), Osaka University, and the ILE, Osaka University.

References

[1] S. Fujioka *et al.*, *Plasma Phys. Control. Fusion* **54** 124042, 2012.
[2] T. Johzaki *et al.*, *Nuclear Fusion* **55**, 053022, 2015.
[3] S. I. Braginskii, *Reviews of Plasma Physics* 1, 1965.
[4] H. Nagatomo *et al.*, *Nuclear Fusion* **55**, 093028, 2015.
[5] R.E. Kidder, *Nuclear Fusion* **14**, pp.53-60, 1974.
[6] H. Sakagami *et al.*, *Laser and Particle Beams* **24** 191-198, 2006.
[7] T. Johzaki *et al.*, "Electron beam guiding by external magnetic fields in imploded fuel plasma" presented at IFSA2015, Seattle NE USA, Sept. 2015.

Wetted foam liquid fuel ICF target experiments

R E Olson[1], R J Leeper[1], S A Yi[1], J L Kline[1], A B Zylstra[1], R R Peterson[1], R Shah[1], T Braun[2], J Biener[2], B J Kozioziemski[2], J D Sater[2], M M Biener[2], A V Hamza[2], A Nikroo[2], L Berzak Hopkins[2], D Ho[2], S LePape[2] and N B Meezan[2]

[1]Los Alamos National Laboratory, Los Alamos, New Mexico, USA
[2]Lawrence Livermore National Laboratory, Livermore, California, USA

E-mail: reolson@lanl.gov

Abstract. We are developing a new NIF experimental platform that employs wetted foam liquid fuel layer ICF capsules. We will use the liquid fuel layer capsules in a NIF sub-scale experimental campaign to explore the relationship between hot spot convergence ratio (CR) and the predictability of hot spot formation. DT liquid layer ICF capsules allow for flexibility in hot spot CR via the adjustment of the initial cryogenic capsule temperature and, hence, DT vapor density. Our hypothesis is that the predictive capability of hot spot formation is robust and 1D-like for a relatively low CR hot spot (CR~15), but will become less reliable as hot spot CR is increased to CR>20. Simulations indicate that backing off on hot spot CR is an excellent way to reduce capsule instability growth and to improve robustness to low-mode x-ray flux asymmetries. In the initial experiments, we will test our hypothesis by measuring hot spot size, neutron yield, ion temperature, and burn width to infer hot spot pressure and compare to predictions for implosions with hot spot CR's in the range of 12 to 25. Larger scale experiments are also being designed, and we will advance from sub-scale to full-scale NIF experiments to determine if 1D-like behavior at low CR is retained as the scale-size is increased. The long-term objective is to develop a liquid fuel layer ICF capsule platform with robust thermonuclear burn, modest CR, and significant α-heating with burn propagation.

1. Background and motivation

The hot spot formation processes in DT ice layer and DT liquid layer ICF capsules are quite different. The National Ignition Campaign (NIC) baseline DT ice layer ICF ignition capsule design [1] requires a hot spot convergence ratio (CR) of 35 with a hot spot that is dynamically formed from DT mass originally residing in a very thin layer (a few microns) at the inner DT ice surface. In contrast, DT liquid layer capsules can have the hot spot formed from mass originating within a spherical volume of DT vapor. Liquid DT (or D2) layer capsule designs allow for flexibility in hot spot CR through the adjustment of the initial cryogenic temperature with significant hot spot energy gain occurring at relatively large radius. Additional benefits of the liquid layer approach include potentially improved robustness to low-mode x-ray flux asymmetry and ablator surface roughness. A detailed comparison of the performance of DT liquid and DT ice layer capsules can be found in Reference 2.

We are developing a new NIF experimental platform that employs wetted foam liquid fuel layer ICF capsules. We plan to use the liquid fuel layer capsules in a NIF experimental campaign to explore the relationship between hot spot CR and the predictability and robustness of hot spot formation. Our hypothesis is that hot spot formation will be robust and 1D-like for a relatively low convergence ratio hot spot (CR~15) in which the hot spot is formed largely from the vapor, but will deviate from 1D-like behavior as vapor pressure is reduced and hot spot CR is increased to CR>20. We will begin with NIF sub-scale experiments to determine a CR limit at which burn truncation begins to occur due to 3D effects. When we discover a CR threshold for 1D-like behavior in the NIF sub-scale experiments, we will advance from sub-scale to full-scale NIF experiments to determine if the good predictive capability is retained as the scale-size is increased. The wetted foam liquid layer platform would then be used for hot spot diagnostic applications (eg., perhaps via dopants in the foam) to understand and repair departures from 1D-like behavior as CR is gradually increased.

2. Liquid fuel layer NIF sub-scale experimental platform

The NIF sub-scale liquid layer platform builds upon some recent innovations – 1) a new approach to lining the interior of HDC capsules with an ultra-low density CH foam that can survive wetting with liquid hydrogen [3,4]; and 2) NIF near-vacuum hohlraums (NVH) for driving implosions with HDC ablators [5]. Our initial NIF experiments will use sub-scale HDC capsules with liquid DT (or D2) layers fielded in a scale 575 NVH (5.75 mm diameter, 10.13 mm length, 3.375 mm diameter LEH's, and 0.032 mg/cm^3 He gas fill) [5,6]. The detailed sub-scale HDC wetted foam capsule dimensions and densities are shown in Fig. 1.

In the initial DT liquid layer sub-scale experimental series, we will use a 900 kJ NIF laser pulse of ~ 7 ns duration with peak power of 300 TW to produce a capsule drive with peak Tr ~ 285 eV. The laser pulse shape is a slight variation on a 3-shock pulse shape that has previously been used to achieve a symmetric implosion of a HDC sub-scale "symcap" capsule in a 575 NVH [5,6]. As shown in the plot in Fig. 2, DT liquid layer capsules can provide flexibility in hot spot CR via the adjustment of the initial vapor density. The initial goal of the experiments will be to measure hot spot size, neutron yield, ion temperature, and burn width to infer hot spot pressure [7] for liquid layer implosions. We will begin with a CR ~ 15 (initial vapor density of 3 mg/cm^3) experiment, and then gradually increase the hot spot CR (by decreasing the initial vapor density) in succeeding experiments. Our hypothesis is that the hot spot pressure will be predictable and 1D-like for a relatively low CR hot spot in which the hot spot is created largely from the vapor, but will deviate from 1D-like behavior as vapor pressure is reduced and hot spot CR is increased. Simulations indicate that sensitivity to capsule instability growth and low-mode x-ray flux asymmetries will increase as CR is increased. An example of this effect is shown in Fig. 3, in which 2D simulations of liquid layer capsules with initial vapor density of 3.0 mg/cm^3 are compared to ice layer capsules with initial vapor density of 0.3 mg/cm^3. One caveat here is that the presence of the foam matrix in the fuel layer might introduce perturbations that are not currently included in the simulations.

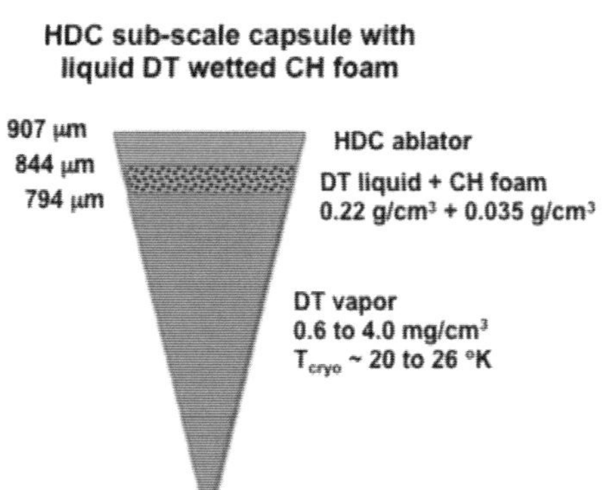

Figure 1. HDC sub-scale capsule with a liquid DT wetted CH foam layer.

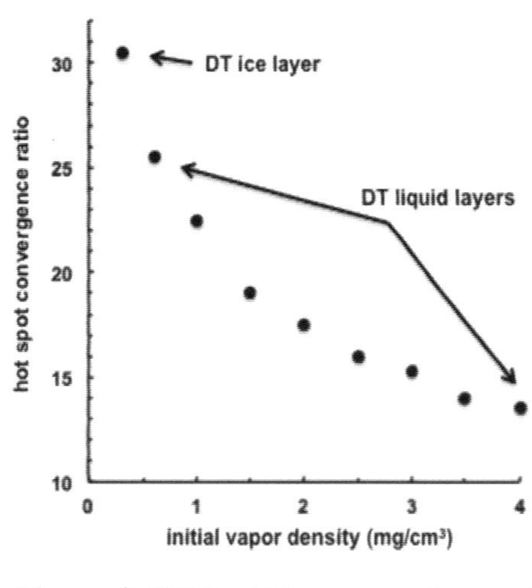

Figure 2. DT liquid layer capsules can be used to access hot spot CR's in the range of 12 to 25.

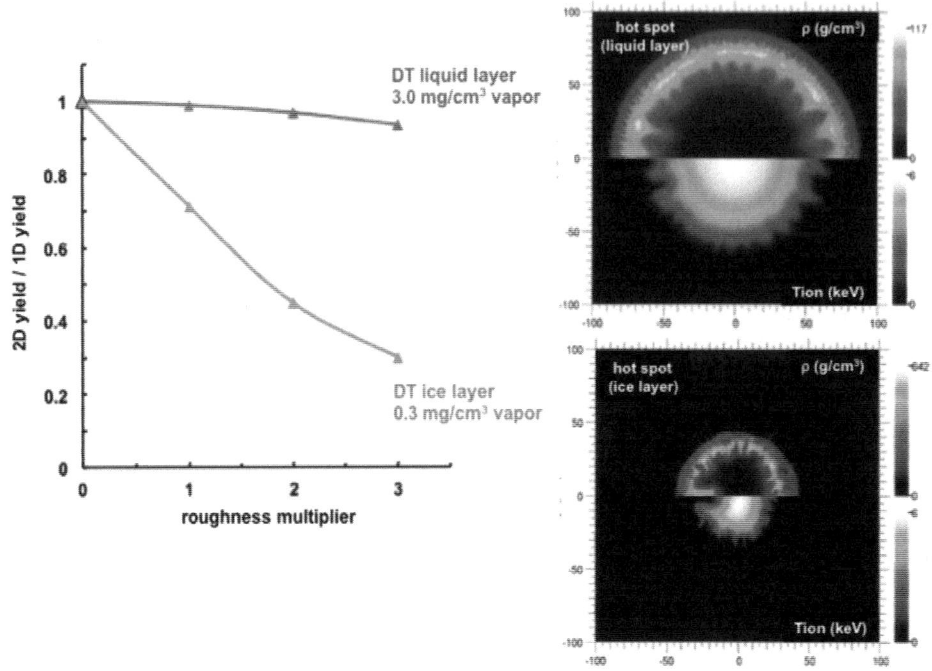

Figure 3. Variations on 2D simulations of NIF sub-scale HDC wetted foam capsules. Roughness has been applied to inner and outer ablator surfaces, as well as to the DT vapor / liquid (or ice) surface.

3. Liquid fuel layer NIF full-scale experimental platform

When we discover a CR threshold with 1D-like behavior in the NIF sub-scale experiments, we will advance from sub-scale to full-scale NIF experiments to determine if the 1D-like predictive capability is retained as the scale-size is increased. An example full-scale HDC capsule with liquid DT wetted CH foam is shown in Fig. 4. We envision that a full-scale wetted foam capsule would be fielded in a NIF scale 672 NVH [6] and driven with a 1.7 MJ laser pulse of ~9 ns duration and 420 TW peak power. The resulting peak radiation drive is predicted to be ~300 eV. Of course, the 1D capsule performance will depend upon the initial vapor density and resulting hot spot CR as determined by the "threshold CR" for 1D-like behavior from the sub-scale campaign.

The red, green, and blue plots in Fig. 5 show the cumulative neutron yield as time progresses in a series of 1D simulations of the full-scale HDC wetted foam capsule design at different initial vapor densities of 1.0, 2.0, and 3.0 mg/cm^3. The result for a comparable DT ice layer simulation with vapor density of 0.3 mg/cm^3 is plotted in black. In all cases, as the capsule implodes a hot spot radius is defined, beginning at shock flash, as the radius of the zone at which TN burn is 10% of the peak value. Then, as the implosion progresses, the maximum hot spot CR increases to a value that is relatively low ($15 < CR < 20$) for the liquid layer simulations compared to the simulation employing an ice layer. In all cases, after achieving peak CR, the cumulative neutron yield continues to increase as the hot spot disassembles. Looking at times prior to peak CR, it can be seen that, for a given CR, the liquid layer implosions have achieved a larger 1D neutron yield than the ice layer implosion. This is important, since, for high CR implosions, we expect experimental measurements to show that the 1D-like burn will be truncated, and that the achievable yield will be set by a limiting CR caused by 3D effects. If, for example, 1D-like behavior is limited to CR's of < 20, the black plot (ice layer) will have achieved a yield of ~ 10^{14}-10^{15} prior to the onset of burn truncation, while the liquid layer capsules will be able to achieve full 1D yields of up to $3x10^{17}$. If our hypothesis is correct, we would expect that liquid layer capsules will have higher achievable yields at lower limiting CR's.

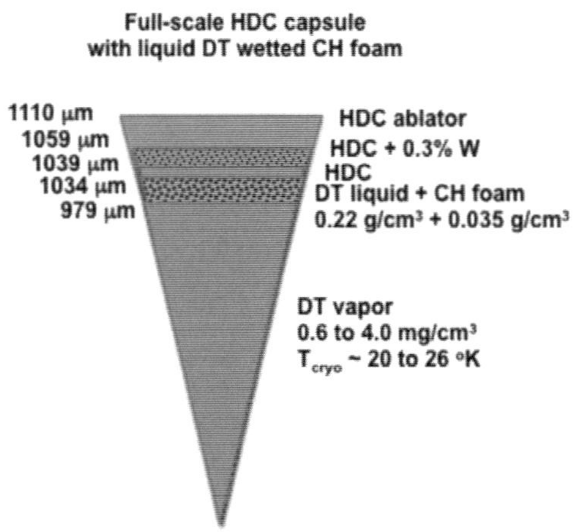

Figure 4. HDC full-scale capsule with a liquid DT wetted CH foam layer.

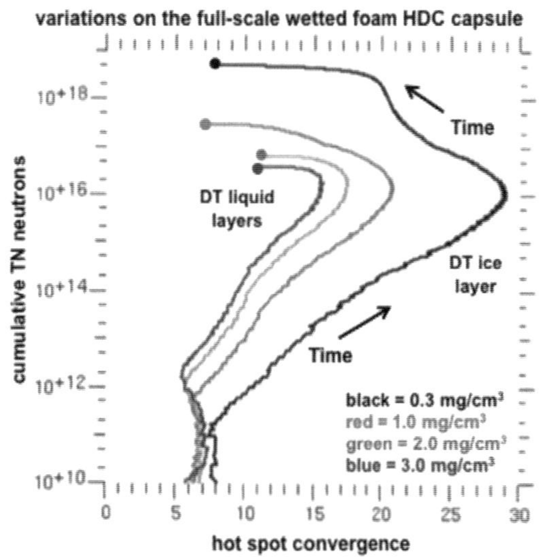

Figure 5. Cumulative neutron yield as time progresses in 1D wetted foam simulations.

4. Summary

We will use DT liquid layer capsule implosions to explore the relationship between CR and robustness of hot spot formation. We will begin by using sub-scale HDC capsules with wetted CH foam layers to implode capsules with vapor densities in the range of 0.6 to 4.0 mg/cm^3 and predicted hot spot CR's in the range of 12 to 25. The HDC sub-scale wetted foam capsules will be fielded in 575 NVH's. Our hypothesis is that hot spot formation will be robust and 1D-like for a relatively low convergence ratio hot spot in which the hot spot is formed largely from the vapor, but will deviate from 1D-like behavior as vapor pressure is reduced and hot spot CR is increased. A goal of the sub-scale experiments is to measure hot spot size, neutron yield, ion temperature, and thermonuclear burn width to infer hot spot pressure for liquid layer implosions and determine a CR limit at which burn truncation occurs due to 3D effects. When we discover a CR threshold with 1D-like behavior in the NIF sub-scale experiments, we will advance from sub-scale to full-scale NIF experiments to determine if the good predictive capability is retained as the scale-size is increased. The wetted foam liquid layer platform would then be used for hot spot diagnostic applications (eg., perhaps via dopants in the foam) to understand and repair departures from 1D-like behavior as CR is gradually increased. The long term objective is to develop a liquid fuel layer ICF capsule platform with robust thermonuclear burn, modest CR, and significant α-heating with burn propagation.

References

[1] S. W. Haan *et al.* 2011 *Phys. Plasmas* **18** 051001.
[2] R. E. Olson and R. J. Leeper 2013 *Phys. Plasmas* **20** 092705.
[3] J. Biener *et al.* 2012 *Nucl. Fusion* **52** 062001.
[4] T. Braun *et al.* 2016 *ACS Applied Materials and Interfaces* **8** 2600.
[5] L. Berzak Hopkins *et al.* 2015 *Phys. Rev. Lett.* **114** 175001.
[6] S. LePape *et al.* 2016 *Phys. Plasmas*, to be published.
[7] C. Cerjan, P. T. Springer, S. M. Sepke 2013 *Phys. Plasmas* **20** 056319.
[8] L. Berzak Hopkins *et al.* 2015 *Phys. Plasmas* **22** 056318.

Electron transport estimated from electron spectra using electron spectrometer in LFEX laser target experiments

T Ozaki[1], M Hata[2], K Matsuo[2], S Kojima[2], Y Arikawa[2], S Fujioka[2], H Sakagami[1], A Sunahara[2], H Nagatomo[2], T Johzaki[3], A Yogo[2], A Morace[2], Z Zhang[2,4], H Shiraga[2], S Sakata[2], T Nagai[2], Y Abe[2], S Lee[2], M Nakai[2], H Nishimura[2], H Azechi[2], FIREX group and GXII-LFEX group

[1]National Institute for Fusion Science, 322-6 Oroshi Toki, Gifu 509-5292, Japan
[2]Institute of Laser Engineering, Osaka Univ., 2-6 Yamada-Oka, Suita, Japan
[3]Hiroshima Univ., 1-4-1, Kagamiyama, Higashi-Hiroshima, 739-8527, Japan
[4]Institute of Physics, Chinese Academy of Sciences, Beijing 100080, P.R. China

E-mail: ozaki@nifs.ac.jp

Abstract. Hot electrons which are generated from targets irradiated by a high-intense laser are measured by two electron spectrometers (ESMs). However, total electron energy observed by the ESM is only less than 1 %. Hot electrons are confined by self-fields due to the huge current. When an external magnetic field of several hundred Tesla is applied during the laser irradiation on targets, the ESM signals always increase. In the simulation, the same result can be obtained. The reason is that the Alfvén limit can be mitigated due to the external longitudinal magnetic field.

1. Introduction

In the fast ignition (FI) [1,2], it is a key issue whether enough energy to ignite can be deposited to the imploded core. The most traditional scheme of FI is to utilize hot electrons, which are generated by the interaction between the guiding cone and the ultra-high intense laser. We have measured the hot electrons, which are emitted to the vacuum region by using the electron spectrometer (ESM) [3] in various fundamental and integrated target experiments. However, the observed total energy of the hot electrons is less than 1 % of the input laser energy. The reason is that the electron current up to the Alfvén limit can be only emitted to the vacuum because the electron current is huge and the self-magnetic field is too large. Additionally, the sheath potential [4] created by escaped electrons from targets pulls the electrons to the targets.

The electron behavior has been investigated by the PIC simulation [5]. The injected laser is partially reflected on the target surface. The remaining laser accelerates electrons in pre-formed plasma, created by the pre-pulse of the heating laser. However, the electrons are trapped by the self-magnetic field B since the return current is not enough on the front surface although the electron current is huge. Electrons run on the surface by $E \times B$ drift (similar to the skin effect), where E is the sheath electric field. Residual electrons enter to the target if those energies are high enough. In the target, the electrons go almost straight because sufficient return currents are supplied. However, when the electrons reach the rare surface, the electrons are trapped by the self-fields again. The electrons can be observed by the ESM when they overtake those fields. Therefore, when two ESMs are located in different ports, the electrons are observed even at the front surface side. Those tendencies are in agreement with the simulation result.

Here we discuss the electron behavior from the laser reflection, the result of the ESM, and the X-ray measurement. To investigate the behavior precisely, we explain the ESM results (increasing electron/ion fluxes) when the external magnetic field is applied. Finally, we discuss about the proper target in which the effective electron deposition can be expected.

2. Laser Reflection

Target experiments have been performed in the target chamber I of the Gekko XII (GXII) facility, where GXII imploding laser and LFEX heating laser are settled. The detail of the experimental setup is shown elsewhere [6]. The part of the laser is reflected on the front surface according to the simulation. We measured the LFEX laser reflection at various targets. In the experiments, two beams in LFEX (four beams) were used. A reflection plate was located at the lack of the beam inlet. The reflection light was observed by a CCD camera with a filter. The results are shown in Fig. 1. There is the target material dependence in the laser reflection. The reflection in a diamond-like-carbon (DLC) is higher than that in Au. The reflection decreases at the higher laser energy. According to the simulation by Cai *et al.* [7], the reflection decreases at the longer scale length of the pre-formed plasma. The pre-pulse, which creates the pre-formed plasma, is a positive correlation to the laser energy if the beam contrast is maintained. The scale length in the Au is smaller than that in DLC experimentally. The experimental result is agreed with the simulation result. If we remove the pre-pulse and obtain the lower effective hot electron temperature (T_{eff}) in order to obtain the high target coupling efficiency, the reflection may increase. Therefore, the beam contrast may be the trade-off between the laser absorption and the hot electron deposition.

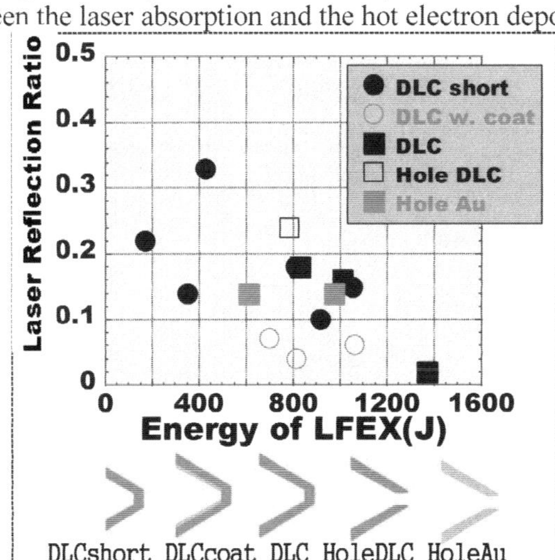

Figure 1. Laser reflection on targets.

Reflection in Au is smaller than that in DLC(5mm length), DLC short (2mm) because the scale length in Au is shorter than that in DLC. DLC w. coat(Au coated DLC); Hole DLC(Au):(DLC(Au) with a hole instead of cone tip).

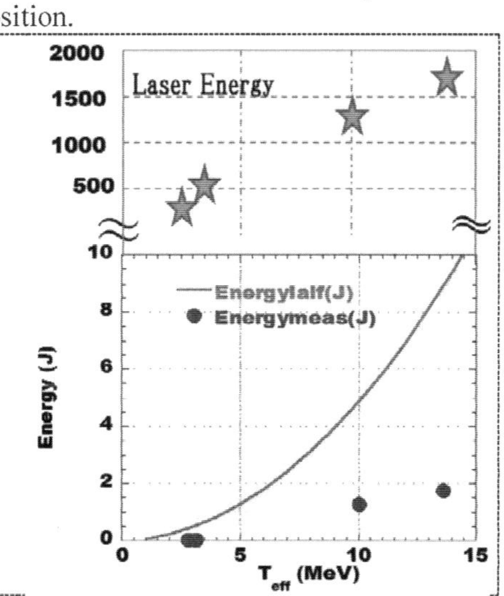

Figure 2. Electron energy measured by ESM (Energymeas).

Energylalf is the energy assuming Alfvén limit. Upper is the incident laser energy.

3. Electron behavior

Part of the electrons, which are accelerated in the pre-formed plasma, run on the surface by $E \times B$ drift. Here the self-B appears due to insufficient cancellation of the return current on the surface. Residual electrons enter to the target if those energies are high enough. In the target, the electrons move straight because there are sufficient return currents. According to the electron coupling efficiency measurement using Ta Kα by Zhang *et al.* [8], about 40 % of the electrons can enter the thick Ta block, which has a solid angle of 2π. However, many electrons may move around the row dense ablation plasma region same as was observed in the simulation, but not in the core because the realistic core is much smaller than the thick Ta block. When the electrons reach the end of the target, the electrons are trapped by the self-fields again on the rare surface. Therefore, less than 1 % of the hot electrons can emit to the vacuum due to the Alfvén limit and the sheath electric field. Figure 2 shows the comparison between the electron energy observed by the ESM and the Alfvén limit. The discrepancy between the Alfvén limit and the observed electron current may mean the effect of the sheath potential.

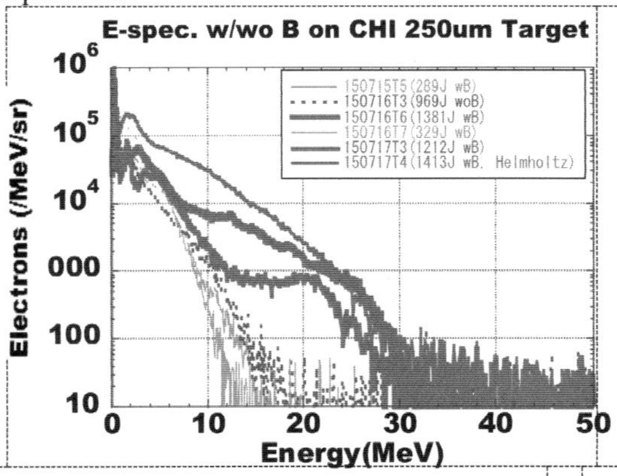

Figure 3. Electron flux with/without the B_{ext} in plain targets. (lines: with B_{ext}; dot: without B_{ext}.)

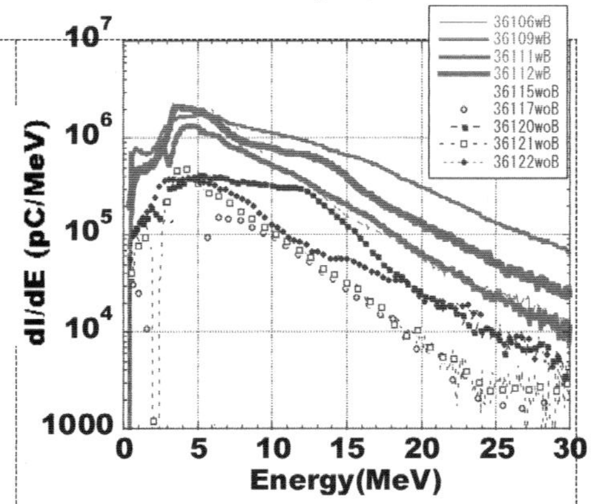

Figure 4. Electron flux with/without the B_{ext} in integrated targets. (lines: with B_{ext}; dots: without B_{ext}.)

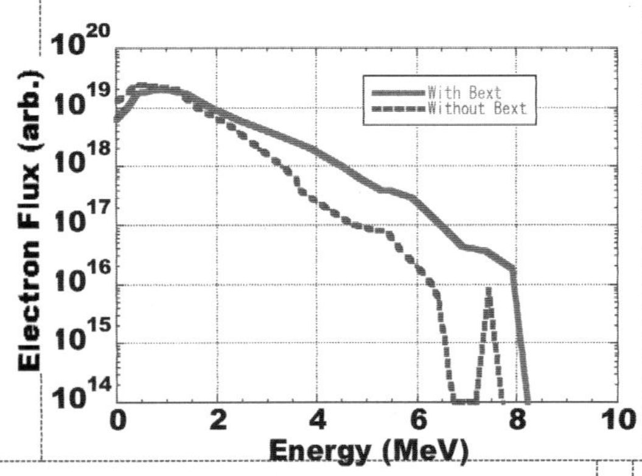

Figure 5. Simulation results with/without the B_{ext}.

Electron flux increases by the B_{ext}.

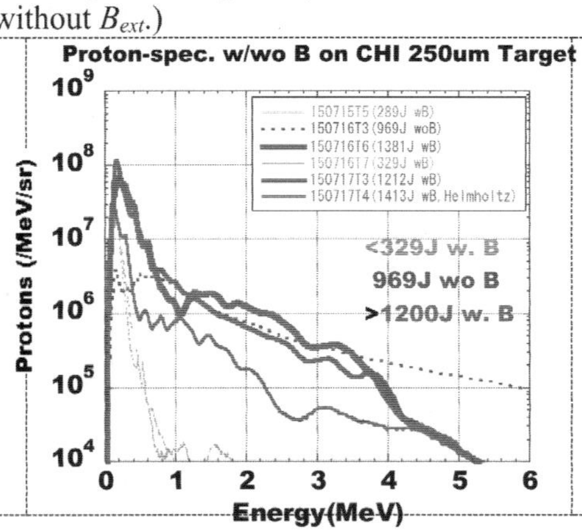

Figure 6. Ion spectra with/without the B_{ext}.

Low energy ion flux increases by the B_{ext}.

4. External Magnetic Field

The increase of the electron flux measured by the ESM has been studied when the external longitudinal magnetic field B_{ext} is applied along the LFEX laser direction. The B_{ext} of several hundreds Tesla can be generated by the one-turn coil connected with two condenser plates. Three beams of GXII (not used for implosion) irradiates those plates and induces the huge current. The detailed setup is mentioned elsewhere [9]. The B_{ext} was introduced originally to converge the hot electron and to improve the electron-target coupling. In this concept, the B_{ext} is created by laser induced current on the capacitor plates before the target implosion and the LFEX injection. The magnetic field, which stays in the shell target, is converged with the imploding shell during the implosion phase. The hot electron are trapped and focused by the converged magnetic field line of the B_{ext} and heat the core effectively. Even if there is a large laser spot size (in order to prevent energetic tail of hot electrons which reduces the core deposition), the high coupling efficiency can be expected.

The electron flux just behind the target should increase. However, the electron flux measured by the ESM should not be changed because the ESM is located far from the target (90 cm). Figures 3 and 4 show the electron flux with/without the B_{ext} on the plain and on the integrated targets, respectively. When the B_{ext} is applied, the electron fluxes increase both on plain and integrated targets. In PIC simulation at the same condition, the flux clearly increases when the B_{ext} is applied, as shown in Fig. 5. This means that the vertical magnetic field line against the target surface, which can easily extract the trapped electrons to the vacuum, can be created. Therefore, the electron flux increases because the Alfvén limit is mitigated by the vertical magnetic field. The ESM has the ability to measure the ion spectrum (mainly proton) at the same time. As shown in Fig. 6, low energy ions remarkably increase. The sheath potential created by the escaped electron also increases because the electron flux increased when the B_{ext} is applied. Low energy ion is extracted by the sheath potential.

From these results, the hot electrons stay around the target by the self-magnetic field addition to the sheath potential if the B_{ext} is not applied. Unfortunately, those electrons are not effectively contributed to heat the core from the results of integration experiments. Those electron energies are lost due to the radiation induced by collision between the electrons and the low density plasma surrounding the core.

5. Summary

Most of the hot electrons are trapped by the self-magnetic field and sheath potential around the target. They lose their energy by the radiation, without heating the core. Residual electrons enter to the target and heat it only once. In the FI, the high coupling efficiency can be expected if the hot electrons effectively irradiate the core geometrically and if there is enough areal density (ρR) in the targets. However, much effort to converge the hot electrons within the core diameter is necessary. The large ρR requires much heating laser energy in order to heat the targets because the fuel amount is also large. Therefore, the laser intensity on targets becomes higher, and higher T_{eff} reduces the target coupling efficiency as a result. One of the procedures to achieve FI is by using wide laser spot so as not to produce the energetic electron tail. The high coupling efficiency can be obtained by converging the hot electrons by the B_{ext}. However, this process does not promise the success of the FI at the moment. Based on the above experimental results, we propose another candidate of electron-based FI, where the hot electrons are used to heat the target many times. If the heating laser irradiates the inside of the hole-shell just before the implosion, the hot electrons may be able to be used effectively [10].

6. Acknowledgements

This work was supported by the Japanese Ministry of Education, Science, Sports and Culture by a Grant-in-Aid (No.24561028), Collaboration Research Program by NIFS (NIFS14KUGK081 and NIFS-URSX302).

7. References

[1] Tabak M, *et al.*, *Phys. Plasmas* **1** (1994) 1626.
[2] Kodama R, *et al.*, *Nature* **418** (2002) 933.
[3] Ozaki T, *et al.*, *Rev. of Sci. Instrum.* **83** (2012) 10D920-1.
[4] Gurevich, A.V., et al., Sov. Phys. JETP 22, 449 (1966).
[5] Sakagami H, *et al.*, *Proc. 39th EPS conf. on Plasma Phys.* **p2.**122 (2012).
[6] Shiraga H, *et al.*, *Plasma Phys. Cont. Fus.* **53** (2011) 124029
[7] Cai HB, *et al.*, *Phys. of Plasma* **13** (2006) 094504.
[8] Zhang Z, *et al.*, *High Density Ene. Phys.* **9** (2013) 435-438.
[9] Fujioka S, *et al.*, *Sci. Rep.* **3** (2013) 1170.
[10] Ozaki T, *et al.*, *Physica Scripta* **2014** T161 (2014) 014025.

Measurement of inflight shell areal density near peak velocity using a self backlighting technique

L A Pickworth[1], B A Hammel[1], V A Smalyuk[1], A G MacPhee[1], H A Scott[1], H F Robey[1], O L Landen[1], M A Barrios[1], S P Regan[2], M B Schneider[1], M Hoppe Jr[3], T Kohut[1], D Holunga[1], C Walters[1], B Haid[1] and M Dayton[1]
[1]Lawrence Livermore National Laboratory, 7000 East Ave, Livermore, CA, USA
[2]University of Rochester, Laboratory for Laser Energetics, Rochester, NY, USA
[3]General Atomics, San Diego, CA, USA

pickworth1@llnl.gov

Abstract. The growth of perturbations in inertial confinement fusion (ICF) capsules can lead to significant variation of inflight shell areal density (ϱR), ultimately resulting in poor compression and ablator material mixing into the hotspot. As the capsule is accelerated inward, the perturbation growth results from the initial shock-transit through the shell and then amplification by Rayleigh-Taylor as the shell accelerates inwards. Measurements of ϱR perturbations near peak implosion velocity (PV) are essential to our understanding of ICF implosions because they reflect the integrity of the capsule, after the inward acceleration growth is complete, of the actual shell perturbations including native capsule surface roughness and "isolated defects". Quantitative measurements of shell-ϱR perturbations in capsules near PV are challenging, requiring a new method with which to radiograph the shell. An innovative method, utilized in this paper, is to use the self-emission from the hotspot to "self- backlight" the shell inflight. However, with nominal capsule fills there is insufficient self-emission for this method until the capsule nears peak compression (PC). We produce a sufficiently bright continuum self-emission backlighter through the addition of a high-Z gas (~ 1% Ar) to the capsule fill. This provides a significant (~10x) increase in emission at hυ~8 keV over nominal fills. "Self backlit" radiographs are obtained for times when the shock is rebounding from the capsule center, expanding out to meet the incoming shell, providing a means to sample the capsule optical density though only one side, as it converges through PV.

Recent experiments at the National Ignition Facility (NIF)[1] have demonstrated sensitivity to instabilities seeded through surface density perturbations on the capsule ablator to the laser pulse shape. The growth of perturbations in ICF[2] capsules can lead to significant variation of inflight shell areal density (ϱR) and ultimately result in poor compression and ablator material mixing into the hotspot degrading the neutron yield[3], [4].

This paper presents the initial measurements of growth from the first self backlighter experiments on the NIF measuring growth from a pre-imposed perturbation, mode 40 and amplitude 100 nm, at radii <200 μm driven by a low-adiabatic laser pulse shape[5].

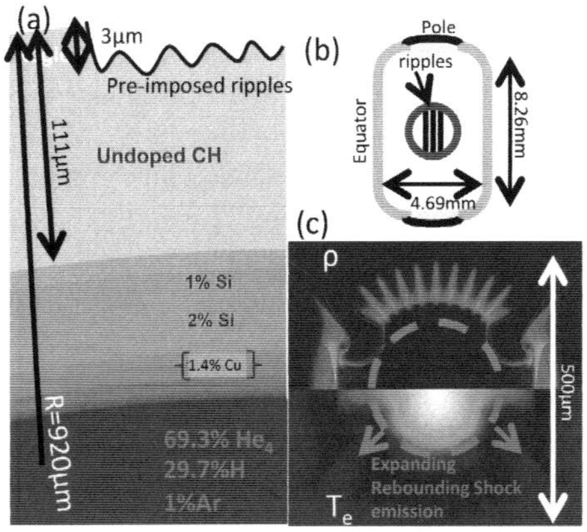

Figure 1: (a) A plastic (CH) capsule is doped with Si and 1.4% Cu. A mode 40 ripple is machined in to the ablator with a 220nm peak-to-valley. The initial radius of the capsule is 920μm. (b) Cartoon showing the orientation of the ripples in the hohlraum, visible on both the polar and the equatorial lines of sight. (c) When the shock rebounds through the capsule gas fill the presence of ~1% Ar enhances the self-emission[5], shown here by a HYDRA[6] simulation indicating the enhancement of T_e at R~ 150 μm, near to PV.

The primary aim of the experiment is to observe a snapshot of the total growth arising from the native perturbations at the surface of the capsule. We wanted to qualify the technique with a known perturbation that would be visible at the time of the peak emission from the rebounding shock. A mode 40 sinusoid ripple was machined into the ablator surface as considerable growth was expected when the capsule had reduced to radii less than 300 μm or convergences greater than 3x.

To boost the self-emission at PV a high Z gas is added into the fill of the capsule. This provides an additional experimental challenge as higher Z gases condense at a higher temperature than a cryogenic target is normally fielded at (~24 K). Ar was chosen at the 1% level as this allowed the target to be fielded at a cryogenic temperature, 75 K, with minimal risk of Ar condensation. The enhancement of emission from the Ar begins ~400 ps prior to PC providing a temporal window in which a series of self-radiographs can be taken.

Figure 1(a) shows the dopant quantities in the CH capsule and the location of the 1.4% Cu doped layer. From simulations we calculate that during the peak Ar emission (~PC -350 ps) the ablator will have predominantly burnt off, yet enough will remain so that the perturbation has not penetrated into the Cu doped layer. Figure 1(c) shows a prediction of the perturbation in the shell when it is at a radius of ~150 μm, from a capsule only 2D simulation using HYDRA[6]. The ripples are clearly visible in the density map, grown to a peak-to-valley of ~55 μm, and the enhanced T_e from the added Ar provides sufficient self-emission to radiograph them.

The added Cu dopant in the shell absorbs the enhanced continuum emission from the rebounding shock at the K-absorption edge, 8.9keV. At this energy the emission from the rebounding shock is a continuum with a distribution determined by the electron temperature (T_e) of the capsule gas. A spectroscopic measurement can be made of the relative emission above and below the absorption edge to find the Cu dopant ρR as a function of time.

The self-radiographs are taken in a 0.8 scale ignition design with a modified low adiabatic laser pulse shape. The smaller scale of the capsule and hohlraum with the addition of the Ar-gas does not allow direct comparison to an ICF experiment designed to achieve ignition, however they show the validity of the technique to observe growth arising from the initial conditions of the capsule surface that may not be measurable with current target fabrication techniques. The target has been used to measure the largest areal density growth factors arising from acceleration Rayligh-Taylor of ~7000x[5].

The ripples on the capsule were orientated such that they were visible from the polar and equatorial lines of sight as shown in figure 1(b). A pinhole imaging system coupled to an Micro-channel plate (MCP) gated x ray detector (GXD)[7] captured the self emission of the implosion, a sample of the polar images are shown in Figure 2. The GXD provided a large number of images spanning the period between PC -400 ps and -100 ps that corresponds to the time when the shell reaches peak velocity. The images are observed though a 10 μm Cu filter which provided a pseudo narrow energy band response allowing the modulation in optical depth to be found. The orientation of the ripples and identical setup of the two detectors allows a simultaneous measurement of growth from the mode 40 perturbations to be made on both the pole and equator. The growth observed from these measurements is discussed in detail in L.A. Pickworth, PRL 2016. Figure 2 shows the (vertical) ripples very clearly, and in addition some three-dimensional features are visible, clear as two regions of brighter emission on the central two ripples. These two spots can be followed frame-to-frame coming closer together as the capsule converges.

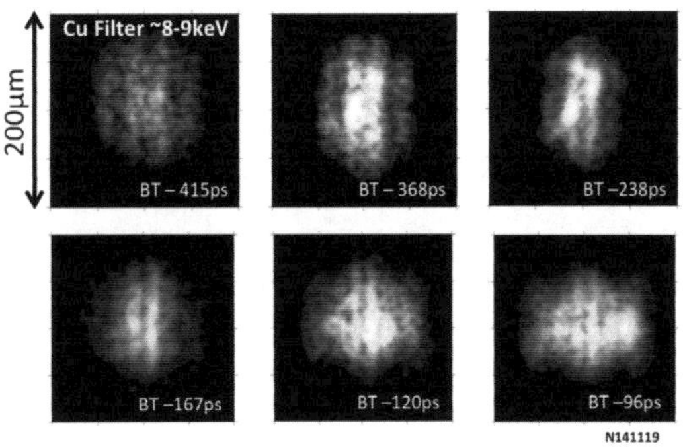

Figure 2: A series of self-radiographs taken with a 12 x magnification, 12 μm pinhole imaging system onto a gated x-ray detector with ~100 ps time resolution, from the polar line of sight.

Figure 3 shows the self-emission from the capsule on a 'streaked' spectrometer[8]. The record shown in figure 3(a) covers 1 ns around PC, viewing the equatorial region of the shell unperturbed by the ripples. Two bands of self-emission are clearly visible, and highlighted in the emission time history shown in figure 3(b), the enhanced emission from the rebounding shock near to PV and the self-emission from the 'hot spot' at PC. Figure 3(b) compares the experimental time history to two simulations on a relative intensity scale, one with additional Ar and one without. The order of magnitude enhancement in self-emission near PV is clearly visible. It can also be seen that the timing of the enhanced self-emission in the simulation matches well to the experiment. The exact intensity of the simulation and experiment is not matched, this is interesting and further work is ongoing to improve the diagnostic analysis and the predicative capability that will allow us to compare absolute emissivity during the shock rebound phase in future.

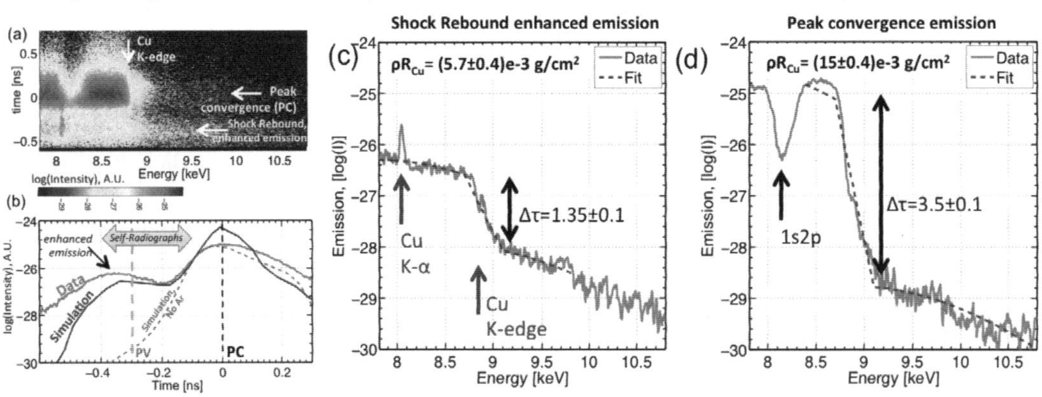

Figure 3. (a) Streaked spectrum, viewed equatorially through a region of the capsule unperturbed by the pre-imposed ripples. (b) Time history of the self-emission below the Cu edge compared to simulation, with and without Ar, on a relative intensity scale. (c) Spectrum averaged over 100 ps at PV (d) Spectrum averaged over 100 ps at PC from the spectrum shown in (a).

Two spectra are shown in Figure 3 averaging over 100 ps at the rebounding shock and 100 ps at PC. The ϱR of the Cu dopant can be inferred by measurement of the change in self-emission at the Cu dopant K-edge. For the time of the rebounding shock self emission we estimate the density of the Cu to be (5.7±0.4)E-3 g/cm^3. To arrive at this conclusion we used the cold opacity of Cu [9]. This corresponds to a capsule convergence of 5.1±0.2x, consistent with the expected capsule radius near PV.

There are other notable features visible in the spectrum that change over the course of the implosion. At the time of PV there is visible emission from the Cu K-alpha line emission. At the time of PC this is no longer clearly visible in the spectrum, however the Cu 1s-2p absorption feature is highly visible. The presence of the 1s-2p feature indicates the Cu is both ionized and dense in a similar manner as previously investigated in a direct drive configuration using Ti dopant[10-12] . We intend to utilize this feature in future analysis to better infer the shell T_e and ϱR. The whole spectrum can be carefully compared to a model for the T_e and average shell opacity allowing the T_e to be more accurately inferred. Initial application of this simple method (dashed blue lines in Figure 3(c, d)) highlights the need of a more detailed comparison to simulation that we are currently perusing.

Areal density of the shell can be inferred from time resolved images (Figure 2) through Ross pair filter imaging[13] that was fielded during these experiments; these will be discussed in a separate publication[5]. In future we intend to apply this technique to observe the 'native' shell surface features that have 'grown up' at the time of PV. In order to make these measurements as accurately as possible an imaging system with higher spatial resolution, improved signal to the detector and spectral selectivity is desired. NIF has recently commissioned a Kirkpatrick-Baez Microscope[14], [15] which has <8 μm spatial resolution and a narrow band spectral response that we hope to apply to this experimental platform in the near future, replacing the use of Ross-Pair filtering.

The authors thank the engineering, target fabrication, and operations teams at the National Ignition Facility who made these experiments possible. This work was performed under the auspices of the U.S. Department of Energy by Lawrence Livermore National Laboratory under Contract No. DE-AC52-07NA27344.

References

[1] G. H. Miller et. al. "The national ignition facility", *Optical Engineering*, 43, no. 12, p. . 2841-2853, 2004.
[2] J. D. Lindl, *Inertial confinement fusion : the quest for ignition and energy gain using indirect drive*]: AIP Press ; New York, NY, USA : Springer, 1998.
[3] J. L. Milovich et. al. *Physics of Plasmas (1994-present)*, 22, no. 12, p. . 122702, 2015.
[4] S. P. Regan et. al. *Phys Rev Lett*, 111, no. 4, p. . 045001, 2013.
[5] L. A. Pickworth et. al *Physical Review letters*, in review 2016.
[6] M. M. Marinak et. al. *Physics of Plasmas*, 8, no. 5, p. . 2275, 2001.
[7] J. Park et. al. , *Review of Scientific Instruments*, 81, no. 10, p. . 10E319, 2010.
[8] F. Pérez et. al., *Review of Scientific Instruments*, 85, no. 11, p. . 11D613, 2014.
[9] B. L. Henke et. al. , *Atomic Data and Nuclear Data Tables*, 54, no. 2, p. . 181-342, 1993.
[10] B. Yaakobi et. al. , *Optics Communications*, 34, no. 2, p. . 213-217, 1980.
[11] V. A. Smalyuk, et. al. , *Physics of Plasmas (1994-present)*, 10, no. 3, p. . 830-834, 2003.
[12] H. M. Johns et. al. , *Physics of Plasmas (1994-present)*, 21, no. 8, p. . 082711, 2014.
[13] N. Izumi et. al. , *Review of Scientific Instruments*, 83, no. 10, p. . 10E121, 2012.
[14] N. F. Brejnholt et. al. , *in SPIE Optical Engineering+ Applications*, 2015, no., p. . 95910J-95910J
[15] L. A. Pickworth et. al., *Review of Scientific Instruments*, 85, no. 11, p. . 11D611, 2014.

Performance of indirectly driven capsule implosions on NIF using adiabat-shaping

H F Robey[1], V A Smalyuk[1], J L Milovich[1], T Döppner[1], D T Casey[1], K L Baker[1], J L Peterson[1], B Bachmann[1], L F Berzak Hopkins[1], E Bond[1], J A Caggiano[1], D A Callahan[1], P M Celliers[1], C Cerjan[1], D S Clark[1], S N Dixit[1], M J Edwards[1], N Gharibyan[1], S W Haan[1], B A Hammel[1], A V Hamza[1], R Hatarik[1], O A Hurricane[1], K S Jancaitis[1], O S Jones[1], G D Kerbel[1], J J Kroll[1], K N Lafortune[1], O L Landen[1], T Ma[1], M M Marinak[1], B J MacGowan[1], A G MacPhee[1], A Pak[1], M Patel[1], P K Patel[1], L J Perkins[1], D B Sayre[1], S M Sepke[1], B K Spears[1], R Tommasini[1], C R Weber[1], C C Widmayer[1], C Yeamans[1], E Giraldez[2], D Hoover[2], A Nikroo[1], M Hohenberger[3] and M Gatu Johnson[4]

1) Lawrence Livermore National Laboratory, Livermore, CA USA 94551
2) General Atomics, San Diego, CA USA 92186
3) Laboratory for Laser Energetics, Rochester, NY USA 14623
4) *Plasma Science and Fusion Center, Massachusetts Institute of Technology*, Cambridge, MA *USA* 02139

robey1@llnl.gov

Abstract. A series of indirectly driven capsule implosions has been performed on the National Ignition Facility to assess the relative contributions of ablation-front instability growth vs. fuel compression on implosion performance. Laser pulse shapes for both low and high-foot pulses were modified to vary ablation-front growth & fuel adiabat, separately and controllably. Two principal conclusions are drawn from this study: 1) It is shown that an increase in laser picket energy reduces ablation-front instability growth in low-foot implosions resulting in a substantial (3-10X) increase in neutron yield with no loss of fuel compression. 2.) It is shown that a decrease in laser trough power reduces the fuel adiabat in high-foot implosions results in a significant (36%) increase in fuel compression together with no reduction in neutron yield. These results taken collectively bridge the space between the higher compression low-foot results and the higher yield high-foot results.

1. Introduction

Inertial confinement fusion (ICF) implosions using indirect-drive are being conducted on the National Ignition Facility (NIF). In the indirect-drive technique, a precisely tailored sequence of laser pulses is used to compress a cryogenic layer of deuterium-tritium (DT) fuel to the high temperature plasma conditions required to initiate DT fusion reactions in the central hot spot core. Details of the driving laser pulse shapes are important for maintaining a low fuel adiabat [1] and also have been demonstrated to strongly influence the stability of perturbations growing on the ablation front [2-6]. Initial experiments conducted during the National Ignition Campaign [7] used a relatively low power and energy (15TW/15kJ) initial picket pulse (0-2 ns) followed by a low power (1 TW) trough (2-8 ns). These "low-foot" (LF) implosions were successful in achieving a high total (including ablator and fuel) areal density (ρR_{max} = 1.3 g/cm^2), but the measured neutron yield was significantly lower than expectations. This reduced yield was attributed to poor ablation-front stability. Subsequent "high-foot" (HF) implosions [8] using an increased picket (38TW/39kJ) and increased trough power (4TW) resulted in a 10X increase in neutron yield, though at the expense of reduced fuel compression. The adiabat-shaping campaign was conducted to test the performance of pulse shapes designed to achieved the best of the LF and HF implosions, i.e. both good fuel compression and good ablation-front stability.

2. Quantifying the fuel adiabat

As was shown in Clark [9], modifications to the "foot" of the laser pulse can have a significant impact. An increase in the picket power was found to significantly improve ablation-front stability, while a decrease in the trough power was shown to decrease the fuel adiabat. Figure 1(a) shows the specific changes that were made to the foot of both the LF and HF laser pulses in this study. The standard low-foot (black) and high-foot (red) pulses are shown with dashed lines. The adiabat-shaped (AS) versions of the low-foot (blue) and high-foot (green, orange) are shown with solid lines. The two AS HF pulses (AS HF #1 green, AS HF #2 orange) are nominally identical with the exception that the 2nd and 3rd pulses (not shown here) for AS HF #2 were advanced in time by 700 ps. This was done to achieve proper shock merger timing for the 20 μm thinner ablator used in AS HF #2.

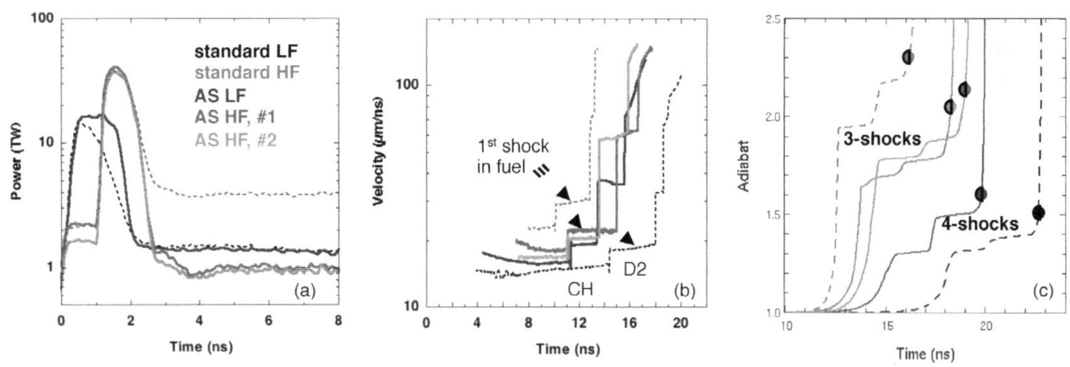

Figure 1. (a) Foot laser power histories for the five shots considered in this study. The standard LF (black) and HF (red) pulses are shown with dashed lines. The adiabat-shaped versions of the LF (blue) and HF (green, orange) are shown with solid lines. (b) Measured VISAR shock velocity histories for all five pulses with the same colors as the laser pulses of (a). (c) Plot of simulated fuel adiabat time history for all five pulses. The symbols indicate the adiabat at the time of peak fuel velocity.

The picket energy of the AS LF pulse was increased from 15 to 23 kJ by extending the duration of the picket by ~400 ps. In contrast, no change was made to the picket of the AS HF pulses. The trough power of the AS HF pulses was decreased from 4.0 to 1.0 TW. The timing of subsequent higher-power portions of all AS pulses (not shown here) was advanced to achieve proper shock merger timing. This results in all three AS pulses being of similar duration (~16-17 ns), a value that is intermediate between the standard LF and HF pulses. Note that the total changes in pulse energy that have been made are very small, with a total energy difference less than 0.5% from the standard LF and HF companion pulses.

The corresponding shock velocity histories measured by VISAR [10] for each of these pulses are shown in Fig. 1(b). The entire shock velocity history for all AS pulses falls between those of the standard HF and LF. The velocity of the 1st shock is important, as it adds most of the entropy to the fuel. The first shock velocity for the AS LF (blue, 19.0 μm/ns) is very comparable to that of the standard LF (black, 18.5 μm/ns) as intended, since no change was made to the trough of the laser pulse, which controls the 1st shock velocity [1]. The average first shock velocity for the two AS HF pulses are a little higher than the two LF pulses at 22.2 (green) and 20.45 μm/ns (orange) for AS HF #1 and #2, respectively. Both are substantially lower than the standard HF pulse, though, which has a first shock velocity of 29.7 μm/ns (red).

One-dimensional simulations using the radiation-hydrodynamics code HYDRA [11], tuned to precisely match the VISAR-measured velocity histories of Figure 1(b), can be used to extract the mass averaged entropy of the fuel layer, from which the adiabat can be obtained. The temporal history of the simulated adiabat is plotted in Figure 1(c). The various steps seen in these adiabat time histories correspond to the entropy added during the traversal of the sequential shocks. The circle symbols indicate the value of the fuel adiabat at the time of peak fuel velocity. The adiabat values are 2.3 (HF, red), 2.1 (AS HF #1, green), 2.0 (AS HF #2, orange), 1.6 (AS LF), and 1.5 (LF, black). Note that a significant difference remains in the adiabats of the LF and HF cases. This is simply due to the difference in the number of driving pulses (3 for all HF-based pulses and 4 for all LF-based pulses), and therefore the number of shocks and corresponding pressure jumps seen in the fuel.

3. Quantifying the ablation-front stability

Having discussed the effect of these small changes to the foot of the laser pulse on the fuel adiabat, we now turn our attention to the corresponding effect of these same modifications on ablation-front stability. Figure 2(a) shows a schematic of the experimental geometry that is used on NIF in the Hydro-Growth Radiography (HGR) target platform. This geometry and the initial experiments are described in detail in Refs. [2,3]. A standard NIF hohlraum and capsule are used. A Au diagnostic cone, similar to that used in keyhole experiments [12], is used to provide a diagnostic line-of-sight for the 4.3 keV backlighter x rays, which are produced by interaction of two quads incident on a Vanadium foil located outside the hohlraum. The capsule is precisely machined with sinusoidal perturbations of very small initial amplitude on the outer surface within the field-of-view of the diagnostic x-ray framing camera. A comparison of the resulting optical depth modulations in the x-ray images at a capsule radius of ~ 650 μm (convergence of ~2X) is shown in Figure 2(b) for all four pulses. The agreement between simulated and measured growth is very good. Growth in the HF implosions is measured to be ~4X less than that of the LF, and both AS pulses show similar growth to that of the HF pulse. Further comparison of data with simulations is given in [13].

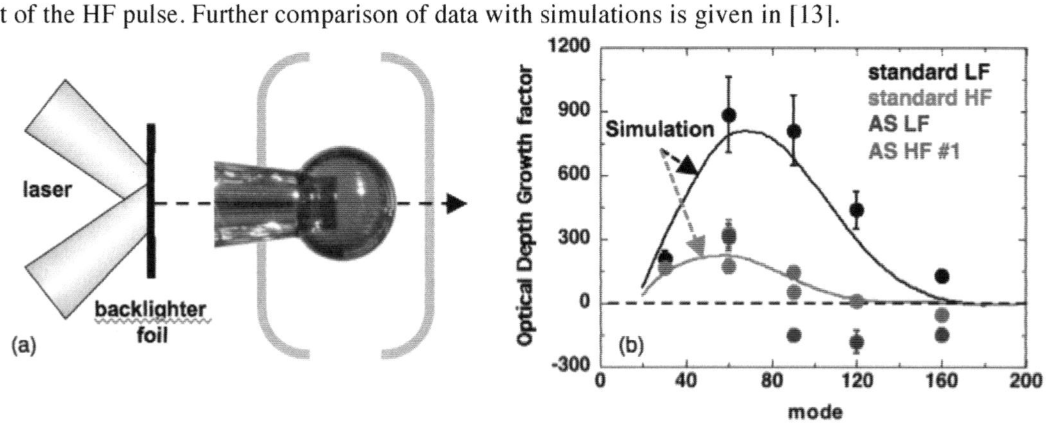

Figure 2. (a) Schematic of the Hydro-Growth Radiography (HGR) experimental platform. (b) Measured optical depth growth factor dispersion curves for all four pulses.

4. Effect of adiabat-shaping on implosion performance

In this Section, the performance of the three adiabat-shaped implosions of Figure 1 is discussed. We begin with the AS LF pulse (blue) of Figure 1(a). To summarize earlier observations, this pulse showed a very similar adiabat in Figure 1(c) but significantly improved stability in comparison to the standard LF pulse in Figure 2(b). Figure 3 shows the parameter space of neutron yield vs. DSR for all NIF shots. Shot N141123 (magenta) was an AS LF shot with a total laser energy of 1.6 MJ and a peak laser power of 336 TW. It is compared in Figure 3 with 5 companion shots from the NIC database that had very similar laser power and energy. Complete experimental details of this shot are given in Casey [14]. This shot exhibited a 3-10X increase in neutron yield with no loss of compression (blue vertical arrow) as compared to its companion shots. These results strongly indicate that ablation-front instability growth was indeed a major factor in degrading the yield of NIC LF implosions.

We now turn to the performance of the two AS HF pulses of Fig. 1(a). To review the previous observations, these pulses showed a moderately reduced adiabat in Figure 1(c) and similar stability to the standard HF in Figure 2(b). Results for the lower power shot N150115 (pulse AS HF #1) showed a significant increase in DSR (36%) and a modest (10%) increase in yield over similar power high-foot companion shot N150610. This is a clear consequence of the decreased adiabat of Figure 1(c) allowing increased fuel compression.

Increasing laser power in the AS HF pulse, however, resulted in a more modest increase in DSR. In shot N150416, a 20 μm thinner capsule was used, peak laser power was increased from 328 to 388 TW, and total laser energy was increased from 1.58 to 1.74 MJ. The foot of the pulse remained the same, however, and the adiabat was expected from Figure 1(c) to decrease slightly from the lower power AS HF shot N150115. This should correspond to increased fuel ρR and a higher measured DSR. As Figure 3 shows, however, the DSR increased over companion high-foot shots N140520, N150121, and N150409 by only 14%, as opposed to the 36% increase seen at lower power. The DSR on shot N150416 (4.65%) actually decreased relative to that of shot N15015 (5.04%), contrary to expectations from the measured adiabats of Figure 1(c). Total neutron yield (including the down-scattered component) was 8.41e15, very comparable to similar high-foot shots N140520 (8.98e15), N150121 (7.33e15), and N150409 (8.07e15). This decrease in fuel compression with increasing laser power strongly suggests that additional physics is playing a role in offsetting the increased fuel compressibility. The possible role of increased electron preheat is discussed in Ref. [13].

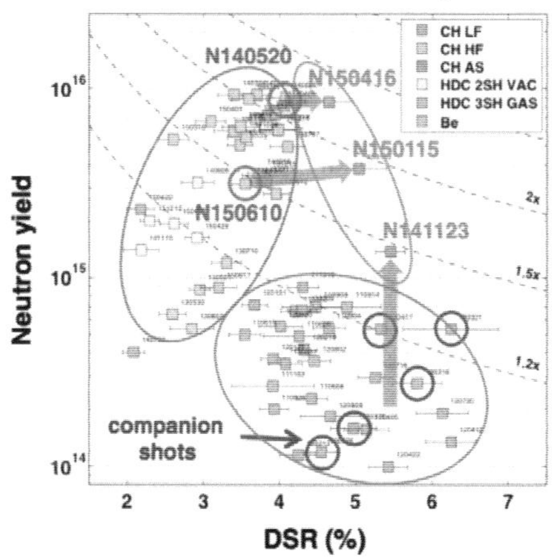

Figure 3. Plot of measured neutron yield vs. DSR. Low-foot shots are shown in blue, high-foot shots in green, and adiabat-shaped shots in magenta. Similar power and energy companion shots are circled.

References

[1] Robey H F, *et al.*, 2012 *Phys. Rev. Lett.* **108** 215004
[2] Raman K S, *et al.*, 2014 *Phys. Plasmas* **21**, 072710
[3] Smalyuk V A, *et al.*, 2014 *PRL* **112**, 185003
[4] Casey D T, *et al.*, 2014 *Phys. Rev. E* **90**, 011102
[5] Peterson J L, *et al.*, 2015 *Phys. Plasmas* **22**, 056309
[6] MacPhee A G, *et al.*, 2015 *Phys. Plasmas* **22**, 080702
[7] Edwards J M, *et al.*, 2013 *Phys. Plasmas* **20**, 070501
[8] Hurricane O A, *et al.*, 2014 *Nature* **506**, 343
[9] Clark D S, *et al.*, 2014 *Phys. Plasmas* **21**, 112705
[10] Barker L M and Hollenbach R E, 1972 *J. Appl. Phys.* **43**, 4669
[11] Marinak M M, *et al.*, 2001 *Phys. Plasmas* **8**, 2275
[12] Robey H F, *et al.*, 2012 *Phys. Plasmas* **19**, 042706
[13] Robey H F, *et al.*, 2016 *Phys. Plasmas* **23**, 056303
[14] Casey D T, *et al.*, 2015 *PRL* **115**, 105001

Simulation analysis for ion assisted fast ignition using structured targets

H Sakagami [1], T Johzaki [2], A Sunahara [3] and H Nagatomo [4]

[1] Fundamental Physics Simulation Research Division, National Institute for Fusion Science, 322-6 Oroshi-cho, Toki 509-5292, Japan
[2] Graduate school of Engineering, Hiroshima University, 1-3-2 Kagamiyama, Higashi-Hiroshima 739-8511, Japan
[3] Institute for Laser Technology, 2-6 Yamada-oka, Suita 565-0871, Japan
[4] Institute of Laser Engineering, Osaka University, 2-6 Yamada-oka, Suita 565-0871, Japan

sakagami.hitoshi@nifs.ac.jp

Abstract. As the heating efficiency by fast electrons in the fast ignition scheme is estimated to be very low due to their large divergence angle and high energy. To mitigate this problem, low-density plastic foam, which can generate not only proton (H^+) but also carbon (C^{6+}) beams, can be introduced to currently used cone-guided targets and additional core heating by ions is expected. According to 2D PIC simulations, it is found that the ion beams also diverge by the static electric field and concave surface deformation. Thus structured targets are suggested to optimize ion beam characteristics, and their improvement and core heating enhancement by ion beams are confirmed.

1. Introduction

In the fast ignition scheme, first a fuel target is imploded by long-pulse implosion lasers, and then its compressed core is heated by a short-pulse ultrahigh-intense laser. The FIREX project has started at ILE, Osaka University to demonstrate core heating up to 5 keV with the fast ignition scheme using Au cone-guided targets. Results of both fundamental experiment campaigns and 2D PIC simulations indicate that the coupling efficiency from fast electrons to the core is quite low because the divergence angle of fast electrons is too large to efficiently hit the core and their slope temperature is too high to deposit the energy into the core even the energy conversion efficiency from the heating laser to fast electrons is generally high [1]. Thus a relatively small enhancement in neutron yield was achieved in integrated experiment campaigns.

To mitigate this critical issue, the ion assisted fast ignition scheme is suggested, where a plastic (CH) ion beam generator is introduced into the cone-guided target and additional core heating is expected by proton (H^+) and carbon (C^{6+}) beams [2,3]. To adapt the Ponderomotive force ion acceleration [4], the low-density CH foam is installed in front of the cone tip surface as the ion beam generator in order to shorten the distance between the ion generation point and the compressed core. This simple design can reduce core-arrival time lags due to different ion energies or between electrons and ions, and not only fast electrons but also energetic ions with wide energy range can be used to heat the core. In addition, this design makes it easy to introduce the ion beam generator into FIREX experiments combining with currently used Au cone-guided targets. Ion beam characteristics are investigated by 2D PIC simulations, and found that the divergence angle of energetic ions is smaller than that of fast electrons but it is still too large to efficiently heat the core using the simple unstructured targets. Thus structured targets for the ion generator are needed to optimize the ion beam properties, and improvements for the divergence angle of energetic ions are confirmed.

2. Unstructured targets

Initial electron density profile of the unstructured target is shown in figure 1 (a). The Au cone plasma (Z=30, A=197, 20n_{cr}, 60 degree open angle, 10 μm tip width, yellow in figure) is introduced and the low-density CH foam (Z=6, A=12, 7.37n_{cr} and Z=1, A=1, 1.23n_{cr}, 8 μm thickness, light blue in figure) is placed in front of the cone tip surface. The total electron density of the foam, 8.60n_{cr}, is selected to be the relativistic critical density to maximize the ion acceleration. The heating laser is set to λ_L=1.06 μm, I_L=10^{20} W/cm^2, τ_{rise}=τ_{fall}=50 fs, τ_{flat}=400 fs and ϕ_{FWHM}=10 μm. Super-Gaussian with α=5 laser is injected to flatly push plasmas than the Gaussian laser beam. Energetic ions are observed at 1 μm behind of the cone tip surface with 30 (±15) μm width. To ignore a circulation of fast electrons, we introduce an artificial cooling region (1 μm width), in which fast electrons are gradually cooled down to the initial temperature, behind the observation region, top and bottom regions of the Au plasma.

Electron density profile at t=500 fs is shown in figure 2 (a). It is clearly seen that the low-density CH foam is pushed by the Ponderomotive force of the heating laser and ions in the center of the foam (±5 μm) are accelerated forward. On the other hand, electron density along the cone wall increases, but ion density at the same place is relatively preserved. Thus charge separation occurs and this induces static electric field in y-direction that can diverge ion beams. Time evolution of normalized angular distributions of (top) C^{6+} and (bottom) H^+ are shown in figure 3 (a). Angular distributions are normalized by the instantaneous maximum value at each time. At an early stage, both ions are accelerated along the laser axis with a small divergence angle. On the other hand, the surface of the low-density CH foam is concavely deformed at a later time. As ions are accelerated to the normal direction of the surface, both ion beams begin to split into two beams with 60 degree oblique angle after 800 fs for C^{6+} or 650 fs for H^+ due to the concave surface, and efficient core heating cannot be expected.

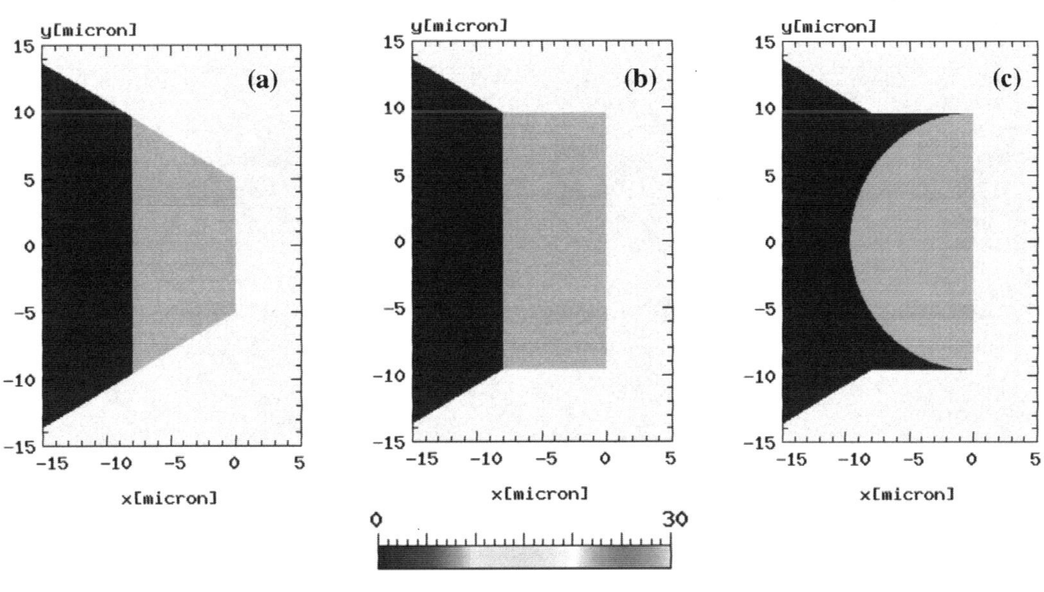

Figure 1. Initial electron density profiles of (a) unstructured, (b) non-tapered and (c) convex foam targets. Density scale is normalized by n_{cr}.

Journal of Physics: Conference Series **717** (2016) 012046 doi:10.1088/1742-6596/717/1/012046

3. Structured targets

To prevent from forming the static electric field, non-tapered cone tip is introduced. Electron density profiles at t=0 fs and 500 fs are shown in figure 1 (b) and figure 2 (b), respectively. The non-tapered target can successfully suppress electron density increment along the cone wall. Time evolution of normalized angular distributions of (top) C^{6+} and (bottom) H^{+} are shown in figure 3 (b). Although beam splitting for both ions are also suppressed before 1100 fs for C^{6+} or 1000 fs for H^{+}, the angular distributions are somewhat flat and ion beams have the large divergence angle because the foam surface is still concave shape. To compensate the concave surface deformation, convex foam target is introduced as the structured target. Electron density profiles at t=0 fs and 500 fs are shown in figure 1 (c) and figure 2 (c), respectively, it is found that the convex foam target can preserve surface flatness

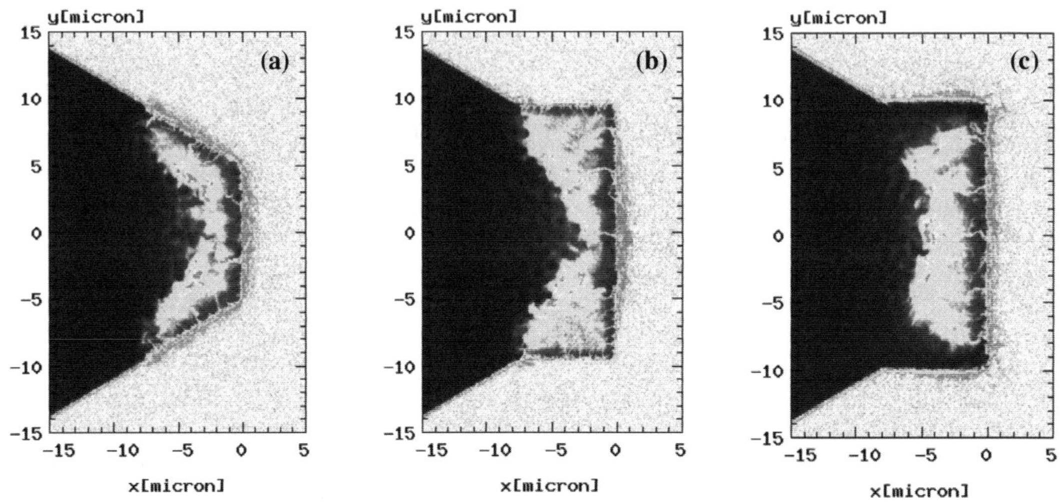

Figure 2. Electron density profiles at t=500 fs of (a) unstructured, (b) non-tapered and (c) convex foam targets. Density scale is as same as in figure 1.

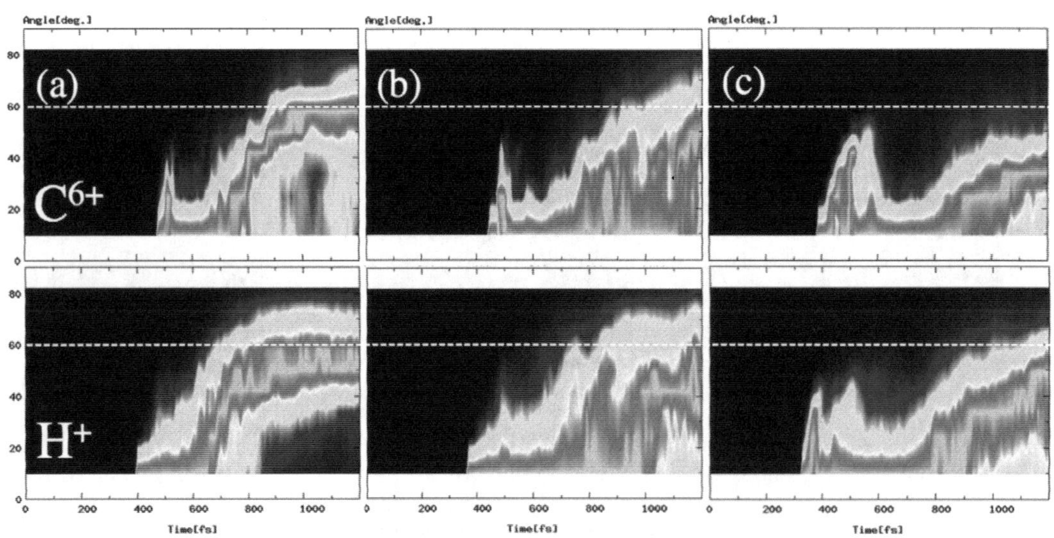

Figure 3. Time evolution of normalized angular distributions of (top) C^{6+} and (bottom) H^{+} for (a) unstructured, (b) non-tapered and (c) convex foam targets. Angular distributions are normalized by the instantaneous maximum value at each time.

for a long time. Time evolution of normalized angular distributions of (top) C^{6+} and (bottom) H^{+} are shown in figure 3 (c), and improvement for the divergence angle of ion beams is clearly seen.

4. Integrated simulations

Transferring 2D PIC data of fast electrons and energetic ions, which are observed in $0 < y < 3$ μm into 1D RFP-Hydro code, we carried out integrated simulations to evaluate the core heating properties [5], including 7 μm transport in the Au cone tip. Time evolutions of averaged core electron temperatures are shown in figure 4 for all 3 cases. As electrons finish to heat the core but ions do not reach the core yet due to slow speed, electrons play a dominant role in core heating before 2 ps. After that, ions finally reach and dominantly heat the core instead of electrons. In the case of the non-tapered target, enhancement to temperature increment by ions is larger than that of the unstructured target, but electron heating is somewhat reduced. On the other hand, ion heating is significantly enhanced with the convex foam target, but electron heating is seriously reduced. Fast electron generation can be enhanced by using a relativistically underdense multi-layer foam in front of the solid target [6]. So target structures should be well optimized for fast electron generation and it's left for future work. Additionally, 2D core heating simulations are needed to accurately estimate the heating efficiency in the ion assisted fast ignition as current heating simulations are performed by 1D RFP-Hydro code and divergence effects are not properly calculated.

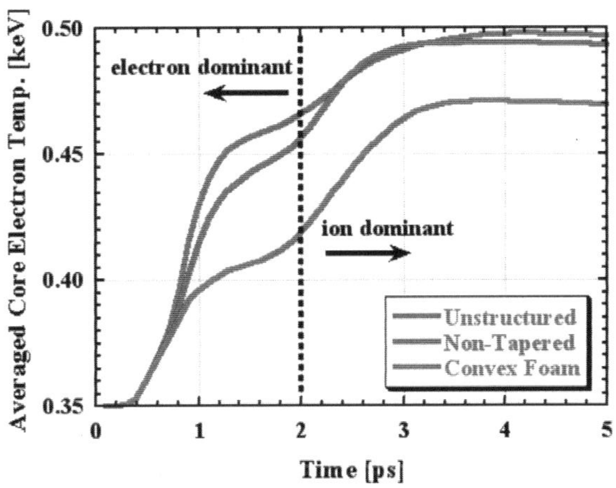

Figure 4. Time evolutions of averaged core electron temperatures for all 3 cases.

Acknowledgements

This work is partially supported by JSPS KAKENHI Grant-in-Aid for Scientific Research (C)(25400539) and (S)(15H05751), and is performed with the support and under the auspices of the NIFS Collaboration Research program (NIFS12KUGK057 and NIFS15KUGK093).

References

[1] Johzaki T, Nagatomo H, Sunahara A, Cai H B, Sakagami H, Nakao Y and Mima K 2011 *Nucl. Fusion* **51** 073022

[2] Sakagami H, Johzaki T, Sunahara A and Nagatomo H 2013 *Proc. 8th Int. Conf. on Inertial Fusion Sciences and Applications* (Nara, Japan, September 8-13) P.Tu_14

[3] Sakagami H, Johzaki T, Sunahara A and Nagatomo H 2014 *Proc. 41st EPS Conf. on Plasma Physics* (Berlin, Germany, June 23-27) P1.102

[4] Wilks S C, Kruer W L, Tabak M and Langdon A B 1992 *Phys. Rev. Lett.* **69** 1383

[5] Johzaki T, Sakagami H, Nagatomo H and Mima K 2007 *Laser Part. Beams* **25** 621

[6] Yazdani E, Sadighi-Bonabi R, Afarideh H, Yazdanpanah J and Hora H 2014 *Laser Part. Beams* **32** 509

Journal of Physics: Conference Series **717** (2016) 012047 doi:10.1088/1742-6596/717/1/012047

Hot spot formation and stagnation properties in simulations of direct-drive NIF implosions

Andrew J. Schmitt and Stephen P. Obenschain

Plasma Physics Division, Naval Research Laboratory, Washington DC 20375, USA

E-mail: andrew.schmitt@nrl.navy.mil

Abstract. We investigate different proposed methods of increasing the hot spot energy and radius in inertial confinement fusion implosions. In particular, shock mistiming (preferentially heating the inner edge of the target's fuel) and increasing the initial vapor gas density are investigated as possible control mechanisms. We find that only the latter is effective in substantially increasing the hot spot energy and dimensions while achieving ignition. In all cases an increase in the hot spot energy is accompanied by a decrease in the hot spot energy density (pressure) and both the yield and the gain of the target drop substantially. 2D simulations of increased vapor density targets predict an increase in the robustness of the target with respect to surface perturbations but are accompanied by significant yield degradation.

1. Introduction

In ICF, spherical implosions concentrate the kinetic energy of the imploding fuel shell and convert it to internal energy of a hot spot and compressed fuel layer when the shell stagnates at its center. This process is inefficient: only $\sim 2\%$ [1] of the absorbed energy ends up in the stagnated hot spot. As the hot spot is solely responsible for igniting the rest of the fuel, its energy content is crucial in determining the success and gain of the target. In conventional central-ignition icf, the hot spot is heated primarily by pdV work of the imploding fuel shell. The gas vapor initially inside the target makes up only a small fraction of the hot spot ($\sim 5\%$), the rest being contributed by the inner edge of the initial fuel shell.

Models describing the scaling and evolution of a pre-established hot spot have been offered [2, 3], but do not address its origin. For instance, they are silent about which parts of the inner fuel edge participate in the hot spot as opposed to comprising the compressing fuel. These models of the hot spot energy predict a scaling like $E \sim \alpha^a p^{-b} V^{-c}$ where α is the fuel shell adiabat, p is the drive pressure, V is the implosion velocity, and $a, b,$ and c are model-dependent (positive) constants. We will assume that the parameters α, p and V are fixed here, as well as the mass and geometry of the target and the laser energy. We instead are interested in the parameters that form the "constant of proportionality" in that scaling; we'd like to know what other factors can control the hot spot in a stagnated target. We will investigate two approaches: changing the timing of the drive laser pulse to preferentially shock heat the inner edge of the fuel layer, and changing the initial gas density inside the target.

Altering the pulse timing is a common procedure for most laser facilities and is limited only by the accuracy of the timing system. The idea behind this is to cause shock collisions in the compressed fuel or shocks in the rear rarefaction, and induce more heating of just the innermost fuel shell which forms the bulk of the hot spot. The hope is that a significant change in the hot spot assembly might be induced, while changing the rest of the implosion in only a minor way. In contrast, adjusting the vapor density is usually done by fielding the target at a different initial (cryogenic) temperature. Such an approach presents implementation challenges, since the highest gas density that can be reached while still using a solid-DT fuel layer (at the $19.79°K$ triple point) is about $6.2\,mg/cm^3$ [4] a higher fielding temperature, and the DT-ice layer melts. A low-density foam matrix can be used to support such a liquid layer in place [5]. The decrease in the DT fuel density in melting is countered by the additional density contributed by the foam matrix and the resulting fuel-foam matrix has about the same density as the solid DT ice. A previous similar study investigating indirectly-driven targets concluded that this substitution is not a major design consideration [6]. Here we ignore this effect and consider the ice layer to remain the same (in composition and density) through the range of vapor densities considered. As a result the yields reported in this analysis may be viewed as an upper-limit.

We will address two questions: the first is whether we can increase the radius, energy and/or pressure of the hot spot while still achieving ignition. Using the FASTRAD3D radiation hydrocode [7], we investigate this with a series of 1D studies varying the laser pulse characteristics or the vapor density. Secondly, we'd like to know if the changes induced in the first step have a postive effects on the robustness, stability and/or performance of the target. 2D simulations using FASTRAD3D will provide this answer.

It is worth noting that the results can depend on how the hot spot is defined. Because we are primarily interested in finding a robust ignition, we define the hot spot as the central region where the ion temperature exceeds $4\,keV$ when it either exceeds $0.3\,gm/cm^2$ or reaches a maximum areal mass less than 0.3.

2. Hot spot formation: 1D results

We start with a tuned design that is consistent with the NIF laser parameters. A relatively low initial aspect ratio target previously studied for shock-ignition [9] was converted to conventional central ignition by removing the final shock spike pulse and raising the compression power. Fig. 1 shows the target with the $1\,MJ$, $\lambda = 0.351\,\mu m$ drive laser pulse ($\sim 740\,kJ$ with $0.248\,\mu m$ light) and in-flight adiabat ~ 2 which produces about 73 MJ in yield; the pulse timings shown were found to maximize the target yield.

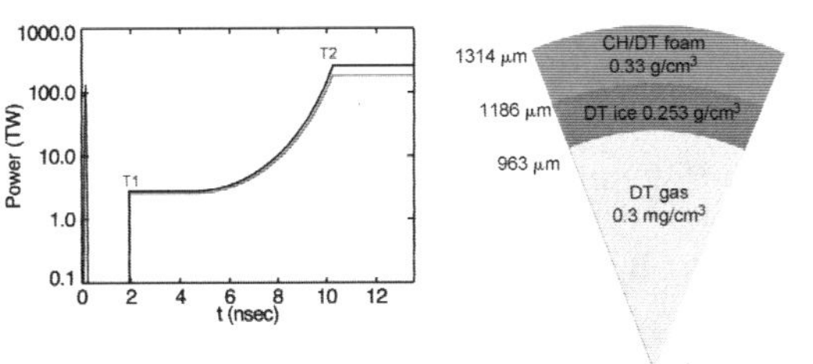

Figure 1. The baseline target design for this study. The pulse shown is for laser wavelengths $0.351\,\mu m$ (black) and $0.248\,\mu m$ (red). Times T1 ($1.94ns$) and T2 ($10.25ns$) mark the onset of the pulse foot and the main power, respectively. A single zoom in the laser spot diameter to the concurrent critical surface radius occurs at T2.

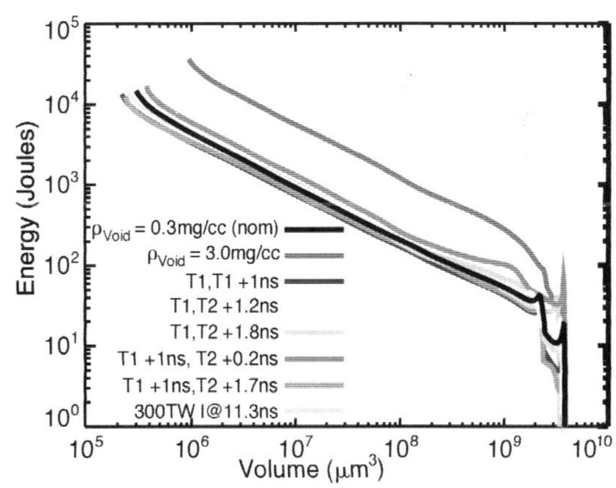

Figure 2. The Energy of the hot spot as it shrinks in volume. The black line is the baseline design.

First we consider the consequence of increasing the initial fill gas density by over an order of magnitude. The additional gas results in a substantial increase in the radius and energy contained in the hot spot (Fig. 3). Conversely, the hot spot pressure ($\sim E_{hs}R_{hs}^{-3}$) and the target yield decrease monotonically. A major portion of the yield decrease is due to the smaller assembled ρR caused by the larger hot spot. However there is still ignition and appreciable yield when the gas density is an order of magnitude larger than nominal, and the hot spot radius has increased about 50%. Here we might expect a favorable tradeoff between yield and target robustness.

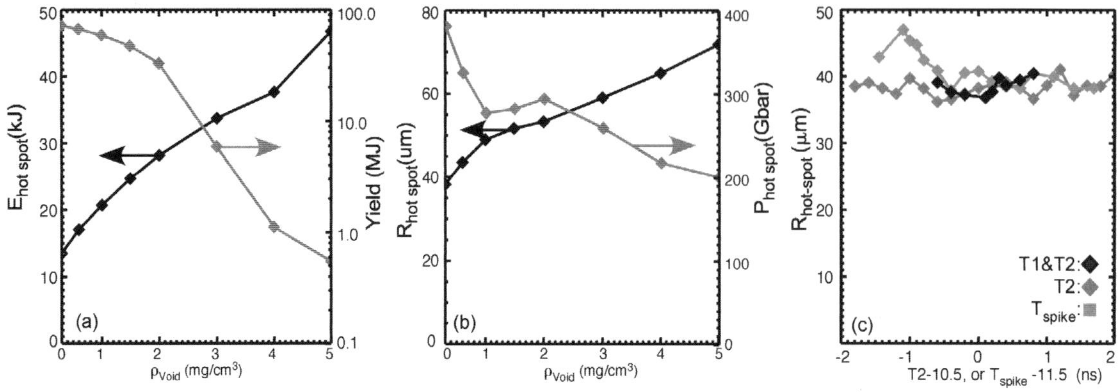

Figure 3. The response of the hot spot parameters to the increased vapor density in the initial target interior: (a) and (b). (c) shows the resulting hot spot parameters when pulse timing is changed or a spike pulse is added. In (c), data is shown at all points where the yield is at least 10% of the baseline.

Next we investigate the consequences of deliberately mistiming and creating shocks. Three variations were tried: (1) the foot (at time T1) and the main drive (onset at T2) were together delayed or advanced relative to the initial spike shock; (2) the main drive (T2) was delayed / advanced relative to the spike and foot onset; and (3) an additional picket pulse (250 psec FWHM, 300 TW) was added in addition to the baseline pulse at times T_{spike} between 8.8 and 13.2 nsec. In all cases, the simulations predict the resulting hot spot radius to vary less than $\sim$ 10% (Fig. 3c) when significant yield occurs. Other hot spot parameters vary in a similar minor way. Although the fuel from the affected inner edge forms the bulk ($>$ 95%) of the subsequent hot spot, little effect is seen from the attempt to preferentially heat it just before the implosion progresses. This is not just an artifact of ignition truncating a growing hot spot: similar minor effects are seen in an unignited hot spot when TN burn is artificially turned off.

The reason for this is shown in Fig. 2, which plots the total energy of the mass that forms the hot spot as a function of its volume. This energy is due to the PdV work done on the hot spot by the boundary pressure. In all the cases of mistiming the pulse, there is an initial jump to a higher pressure at the hot spot boundary as the material is shock heated. However, the material at the boundary then expands due to the subsequent rarefaction wave, and the pressure soon drops to about the same level as it would be without the shock mistiming. The net effect over most of the implosion is little difference in this energy. In contrast, raising the initial vapor density by a factor of ten (the upper red line in the figure) allows the pressure to maintain a higher level throughout the implosion, (the rarefaction is more contained by the higher density), and the energy put into the hot spot can be significantly increased.

3. Multidimensional robustness

In the cases where we were successful in increasing the hot spot dimensions (by increasing the initial vapor density), we then checked the robustness of the larger hot spot radius / lower convergence ratio by using high-resolution 2D simulations. In all cases, a 1 μm RMS roughness was put on the inner DT ice layer, while we varied the outer surface roughness from 0.1 μm to 0.4 μm. The spectral distribution of these perturbations is the same as in Ref. [1] and is meant to emulate the "NIF spec". The laser drive is uniform. Sample images of the density and temperature at stagnation are shown in Figs. 4. For this perturbation level (and others simulated here) the larger hot spot does not correlate with a more symmetric stagnation. Over the full set of simulations, we find that increasing the target vapor density always results in a lower yield (c.f. Fig. 5a). As the perturbation level increases, both the yield and its perturbation variance (slope) are significantly smaller. However as a ratio of the 1D yield (Fig. 5b), the decrease in both is milder at higher vapor densities.

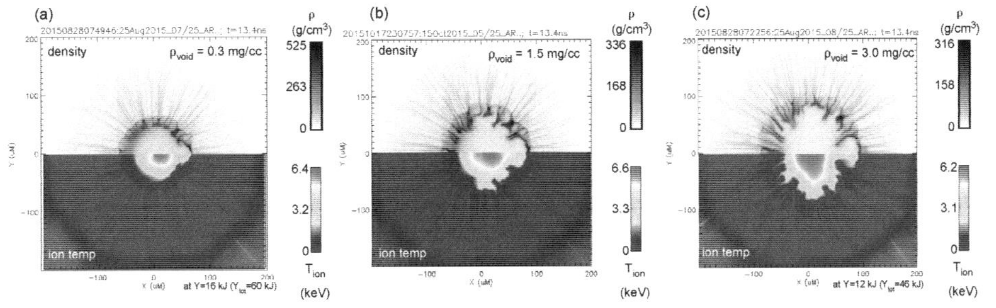

Figure 4. The mass density (top, grayscale) and ion temperature (bottom, color) images from the 2D simulations at the approximate time of hot-spot ignition for increasing vapor density: (a) $0.3\,mg/cm^2$; (b) $1.5\,mg/cm^2$; and (c) $3\,mg/cm^2$. All simulations started with $1\,\mu m$ inner ice surface roughness and $0.4\,\mu m$ outer surface perturbations.

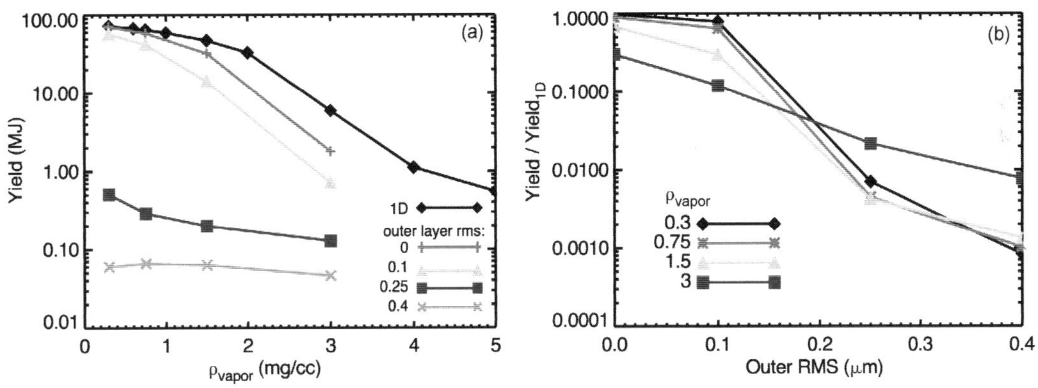

Figure 5. (a) The 2D yield degradation as a function of vapor density for $1\,\mu m$ RMS inner surface ice roughness and different outer surface perturbations; (b) the yield/1D-yield as a function of the outer surface perturbation level.

Acknowledgments
This work supported by the U.S. D.O.E / NNSA. The authors acknowledge helpful discussions with Drs. R. Leeper and R. Olson as well as helpful comments from the anonymous reviewer.

References

[1] Schmitt A J, Bates J W, Obenschain S P, Zalesak S T and Fyfe D E 2010 *Phys. Plasmas* **17** 042701
[2] Betti R, Anderson K, Goncharov V N, *et al.* 2002 *Phys. Plasmas* **5** 2277.
[3] Kemp A, Meyer-ter-Vehn J and Atzeni S 2001 *Phys. Rev. Lett.* **15** 3336
[4] Souers P C 1986 *"Hydrogen Properties for Fusion Energy"* (Berkeley: University of California Press)
[5] Biener J *et al.* 2012 *Nucl. Fusion* **52** 062001
[6] Olson R E and Leeper R 2012 *Phys. Plasmas* **20** 092705
[7] Gardner J H, Schmitt A J, Dahlburg, J P, *et al.* 1998 *Phys. Plasmas* **5** 1935; in addition to the models described therein, FASTRAD3D also has a thermonuclear burn package with multigroup diffusive alpha particle diffusion, as in Corman E G, Loewe W E, Cooper G E, and Winsolw A M 1975 *Nucl. Fusion* **15** 377.
[8] Schmitt A J, Colombant D G, Velikovich A L, *et al.* 2004 *Phys. Plasmas* **11** 2716.
[9] Schmitt A J, Obenschain S P, Zalesak S T and Fyfe D E 2011 *Bull. Am. Phys. Soc.* **56** 221

Journal of Physics: Conference Series **717** (2016) 012048 doi:10.1088/1742-6596/717/1/012048

Using HT and DT gamma rays to diagnose mix in Omega capsule implosions

M J Schmitt[1], H W Herrmann[1], Y H Kim[1], A M McEvoy[1], A Zylstra[1], B A Hammel[2], S M Sepke[2], A Leatherland[3] and S Gales[3]

[1]Los Alamos National Laboratory, MS F699, Los Alamos NM 87545
[2]Lawrence Livermore National Laboratory, 7000 East Ave, Livermore, CA 94550
[3]Atomic Weapons Establishment, Aldermaston, Reading, Berkshire RG7 4PR, UK

Email: mjs@lanl.gov

Abstract. Experimental evidence [1] indicates that shell material can be driven into the core of Omega capsule implosions on the same time scale as the initial convergent shock. It has been hypothesized that shock-generated temperatures at the fuel/shell interface in thin exploding pusher capsules diffusively drives shell material into the gas core between the time of shock passage and bang time. We propose a method to temporally resolve and observe the evolution of shell material into the capsule core as a function of fuel/shell interface temperature (which can be varied by varying the capsule shell thickness). Our proposed method uses a CD plastic capsule filled with 50/50 HT gas and diagnosed using gas Cherenkov detection (GCD) to temporally resolve both the HT "clean" and DT "mix" gamma ray burn histories. Simulations using Hydra [2] for an Omega CD-lined capsule with a sub-micron layer of the inside surface of the shell pre-mixed into a fraction of the gas region produce gamma reaction history profiles that are sensitive to the depth to which this material is mixed. An experiment to observe these differences as a function of capsule shell thickness is proposed to determine if interface mixing is consistent with thermal diffusion $\lambda_{ii}\sim T^2/Z^2\rho$ at the gas/shell interface. Since hydrodynamic mixing from shell perturbations, such as the mounting stalk and glue, could complicate these types of capsule-averaged temporal measurements, simulations including their effects also have been performed showing minimal perturbation of the hot spot geometry.

1. Introduction and background

Suggestive evidence for the detection of 19.8MeV gamma ray emission from the fusion of hydrogen (H) and tritium (T) was recently obtained from experiments on Omega [3] containing a 44.7/5.3/50 percentage mixture of H/D/T. Although the cross section for HT fusion is small, it can still be the dominant (>10 MeV) gamma reaction for capsules filled primarily with HT. Owing to the small branching ratio for DT fusion gamma rays (4×10^{-5}), calculations indicate that equal numbers of HT and DT gamma rays will be obtained in Omega implosions when approximately 1% of the HT fill gas is replaced by deuterium (D). This percentage of D is an order of magnitude greater than the D contamination level (0.1%) present in current 50/50 HT gas supplies for ICF experiments. Consequently, a 49.95/0.1/49.95 capsule gas fill should produce gamma reaction history (GRH) signals where HT-γs outnumber DT-γs by an order-of-magnitude or more. Further energy discrimination by gas Cherenkov detection (between the 19.8MeV HT-γs and the ≤16.75MeV DT-γs) can increase the relative measurement signal levels to 30 or more.

Journal of Physics: Conference Series **717** (2016) 012048 doi:10.1088/1742-6596/717/1/012048

The ability to measure both HT and DT gamma rays in ICF implosions provides an opportunity to observe the effects of varying distributions and amounts of shell mix on fusion burn in separated reactant experiments, where a CD-lined capsule is filled with HT gas (containing 0.1% contamination D). Here HT gamma rays will be produced by fusion in the clean gas, while DT gamma rays will primarily be produced only through the mixing of the CD shell with the HT gas. This concept has been assessed via simulation of the gamma ray time histories of such implosions using the radiation hydrodynamics code Hydra [2] for an Omega CD capsule where a sub-micron layer of the inside surface of the shell is pre-mixed into the outer half of the gas region. An example is shown in Figure 1 where one can see that the HT gamma signal rises first owing to clean HT fusion burn when the shock reaches the axis, followed by DT burn of the out-going shock in the in-flowing CD/HT mix, and finally by both HT and DT fusion at full hot spot compression.

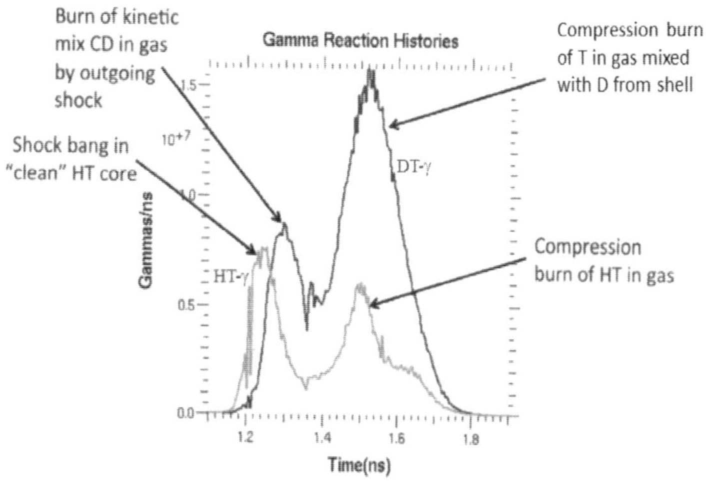

Figure 1 Gamma reaction histories for both HT (red) and DT (blue) gamma rays from a separated reactants capsule with 0.5 um of CD pre-mixed into the outer half of the gas region.

2. Experiment definition

To probe how shell interface temperature (and its commensurate diffusion) affects early implosion mix, one must construct a capsule platform where the interface temperature can be varied. This can be done by varying the shell thickness. Figure 2 shows the time histories of the interface temperatures for both 9 μm and 15 μm CH capsules (driven by the Omega laser) simulated using Hydra. Here the interface temperature differs by a factor of two between the two capsule thicknesses. Using the standard scaling [4] for the ion mean free path, λ_{ii}, given by

$$\lambda_{ii} \propto T_i^2/Z_i^2\rho$$

where T_i is the ion temperature, Z_i is the ion ionization state and ρ is the ion density, we should expect a factor of 4 difference in the diffusion of shell material into the gas.

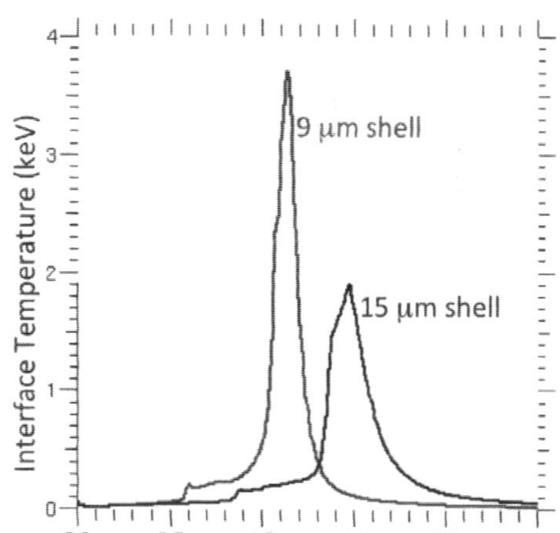

Figure 2 Termperature at the inside gas/shell interface for 1ns laser pulse Omega CH capsules.

3. Simulation design

Optimization of an experiment to use gamma reaction histories to characterize mix requires simulation of signal time histories for different levels of mix. Here we use Hydra for both 1D simulations (to assess gamma signal levels) and 2D simulations (to look at perturbations from mounting the capsules). Simulations assume an 870 μm x 15μm plastic CD-lined CH capsule (preheated on inner half to 70eV to reduce compression) filled with 10 atmospheres of HT gas (including 0.1% D contamination) and driven by a 27kJ, 1 ns laser pulse (scaled by 0.6 to emulate the experimentally absorbed laser fraction).

Journal of Physics: Conference Series **717** (2016) 012048 doi:10.1088/1742-6596/717/1/012048

3.1 1-D Hydra Particle Monte Carlo burn simulations with pre-mixed gas layers

Simulations were performed to assess the relative magnitudes of the HT and DT gamma ray signals for various amounts of shell pre-mix in the gas. For these calculations a 0.5 mm thick CD layer of the inner shell surface was assumed to atomically mix into the HT gas fill. The results of three simulations, each with different depths of pre-mix, are shown in Figure 3.

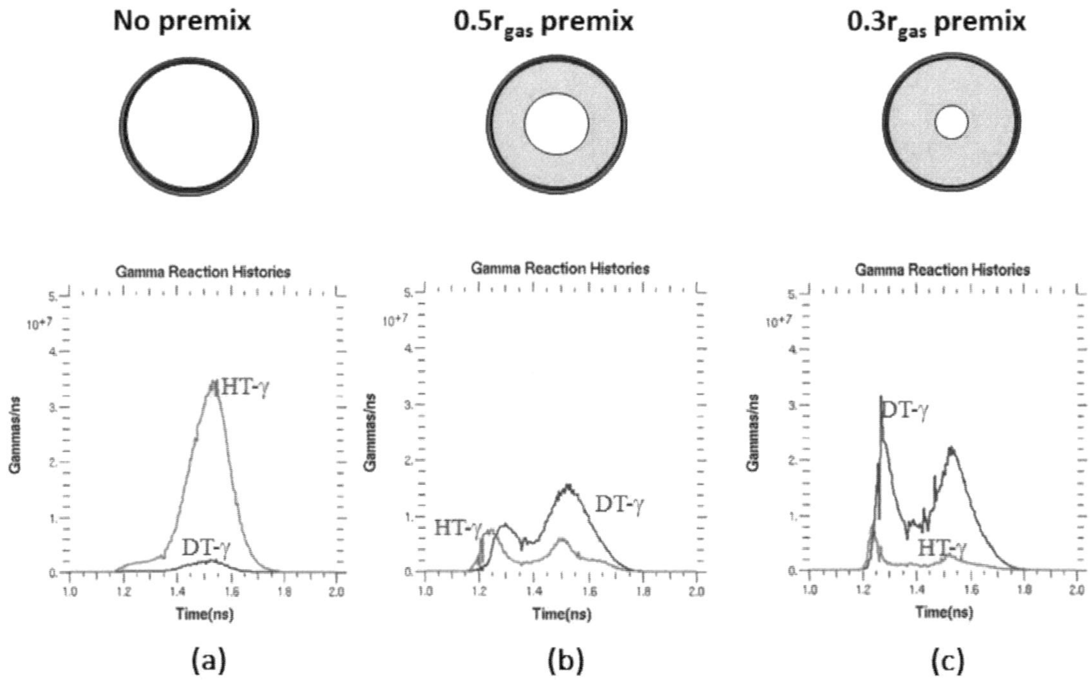

Figure 3 Gamma reaction history profiles for both HT (red) and DT (blue) gamma rays shown for a HT-filled CD capsule with (a) no CD pre-mix case, (b) CD shell pre-mixed half way to the core, and (c) CD pre-mixed 70% into the core.

Note that there are characteristic changes in the shape of the reaction time histories dependent on the variations of the pre-mixed material depth, making this a sensitive diagnostic for the mix conditions in the capsule during burn. A plot of the total HT-γ and DT-γ flux as a function of pre-mix depth is shown in Figure 4. Interestingly, both signal levels change inversely by an order-of-magnitude over the mix range. The DT-γ signal at 0% pre-mix is caused by the 0.1% D contamination in the HT gas fill. Reduction of the DT-γ flux increase with pre-mix depth could be achieved by either reducing the thickness of the CD layer on the inside of capsule, or by reducing the isotopic percentage of D in the CD layer, for example by having a layer that is 90% CH and 10 % CD. This would aid in constraining the dynamic range of the combined GRH signals.

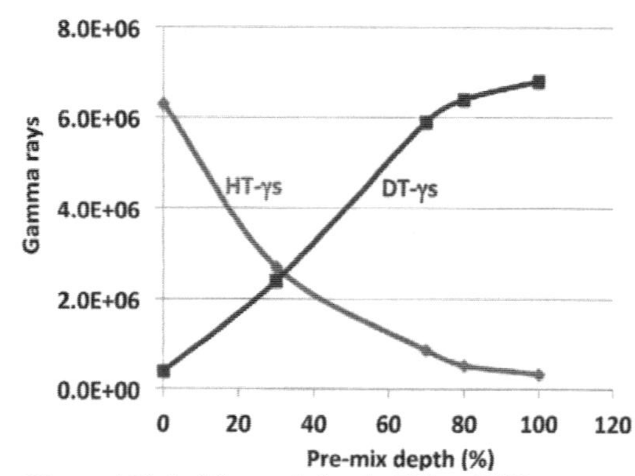

Figure 4 Plot of the variation in HT and DT gamma fluxes as a function of the pre-mix deth percentage.

3.2 Hydra 2-D simulations including stalk and glue perturbations

A potential issue affecting separated reactant capsule experiments is the effect of the mounting stalk and its glue on the implosion dynamics. Hydra 2D simulations of a plastic capsule (full 180°) were performed using 0.2° angular resolution (chosen to obtain several angular zones in the stalk). The geometry is depicted in Figure 5(a) for the 15 μm thick CH capsule driven by a 1ns, scaled 27kJ laser pulse. Simulations were performed for 4 cases: (1) no stalk nor glue, (2) stalk only, (3) stalk plus nominal 80x100μm glue fillet, and (4) stalk plus a double wide, 160x100μm, glue fillet. Elongation of the hot spot in the direction of the stalk was observed at bang time, with more elongation seen as more glue was added. This can be observed in Figure 5(b) where the density of capsule at bang time (1.5ns) is shown for the nominal glue fillet case (where inner shell preheat of 70eV was used). Note that there is a P2 low-mode perturbation of the hot spot, but no significant jetting of material into the core is present. When the same simulation was run without preheat, Rayleigh-Taylor spike growth at the gas-shell interface and the beginnings of a stalk jet are seen at bang time. Since Hydra does not have a predictive mix model, only the variation of the DD burn in the shell was analysed which showed only a small, 20% change in yield over all stalk/glue cases run.

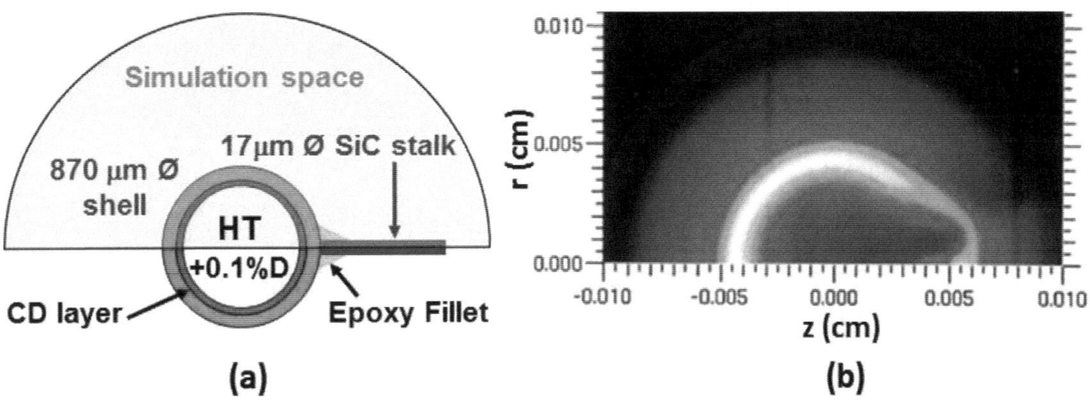

Figure 5(a) Geometry for the 2D capsule simulations with a SiC stalk and epoxy glue fillet, and (b) capsule density at bang time (1.5ns) for the stalk with nominal glue fillet case.

4. Summary

Variations in the yields and temporal histories of HT and DT fusion γ-rays in separated reactant Omega capsule implosions have been characterized via Hydra "pre-mix" simulations indicating that the amplitudes of the HT-γ and DT-γ burn histories are of similar magnitude and have temporal variations consistent with measurements using existing GRH diagnostics. Simulated yields show order-of-magnitude variations versus pre-mix depth that can provide evidence of the burn/mix morphology during Omega capsule implosions. Experimental results will be important for validating current theories on mix evolution described in the literature. Hydra 2D simulations indicate only small changes in yield due to mounting stalk & glue perturbations for separated reactant capsules using 1 ns laser drive (with a pre-heated inner capsule shell to limit convergence). CD mix layer experiments on Omega are planned for early 2016.

Acknowledgments

This work was performed under the auspices of the U.S. DOE by LANL under contract DE-AC52-06NA25396.

References

[1] H.G. Rinderknecht, et al., Phys Rev Lett **112** 135001 (2014).
[2] M.M. Marinak, et al., Phys Plasmas **3** 2070 (1996).
[3] Y.H. Kim, et al., *Kinetic plasma and interspecies-ion diffusion studies using DT, DT/3He, DT/H implosions*, IFSA 2015 these proceedings.
[4] N.M. Hoffman, et al, Phys Plasmas **22** 052707 (2015).

Images of the gold bubble feature in NIF Gas-Filled Ignition Hohlraums

M B Schneider[1], S A MacLaren[1], K Widmann[1], N B Meezan[1], J H Hammer[1], B E Yoxall[1], P M Bell[1], D K Bradley[1], D A Callahan[1], M J Edwards[1], T M Guymer[2], D E Hinkel[1], W W Hsing[1], M L Kervin[1], O L Landen[1], J D Moody[1], A S Moore[1], N E Palmer[1] and A T Teruya[1]

[1]Lawrence Livermore National Laboratory, P.O. Box 808, Livermore, CA 94550, USA
[2]Atomic Weapons Establishment, Aldermaston, Reading RG7 4PR, UK

E-mail: `schneider5@llnl.gov`

Abstract.
The ViewFactor experiments at the National Ignition Facility use a truncated hohlraum to allow excellent diagnostic views of the interior of the hohlraum. Time-integrated, hard x-ray (3 - 5 keV) images show the region where the laser deposits its energy at peak power. These images show a three-dimensional structure in the region where the outer beams deposit their energy (the "gold bubble") which varies with pulse shape and cross beam energy transfer. The images from two-dimensional simulations have similar trends but show some discrepancies.

1. Introduction

The National Ignition Facility (NIF)[1] at Lawrence Livermore National Laboratory is currently engaged in the pursuit of fusion ignition using indirect drive inertial confinement fusion (ICF)[2]. A shaped pulse of up to 1.8 MJ and 500 TW of laser energy and power from the NIF, a 192 beam, 351 nm Nd glass laser, is converted to an x-ray drive in a cryogenic gas-filled gold hohlraum. The x-ray drive implodes the capsule which has a doped plastic (CH) ablator surrounding the deuterium-tritium fuel.[3] Measured ablator implosion velocity is observed to be slow relative to simulations[4],[5] but these measurements could not determine if this was due to an x-ray drive deficit or capsule physics.

The ViewFactor experiment was designed[6] to determine if the slower implosion velocity was due to an x-ray drive deficit. These experiments use an ignition hohlraum with one end truncated (see Figure 1a) to allow views of the hohlraum interior from the capsule point-of-view.[7, 8] The drive irradiance measured through the open end, which needs no source size correction was 70-85% of that predicted by the model[7] and thus explained nearly all of the velocity deficit.

The open end allow diagnostics a complete view of the hohlraum interior, and we focus this paper on the time-integrated images of the hohlraum interior at 18° to the hohlraum axis, (see Figure 1b) in an x-ray energy band of 3-5 keV. These are images of the regions in which the laser deposits its energy at peak power.

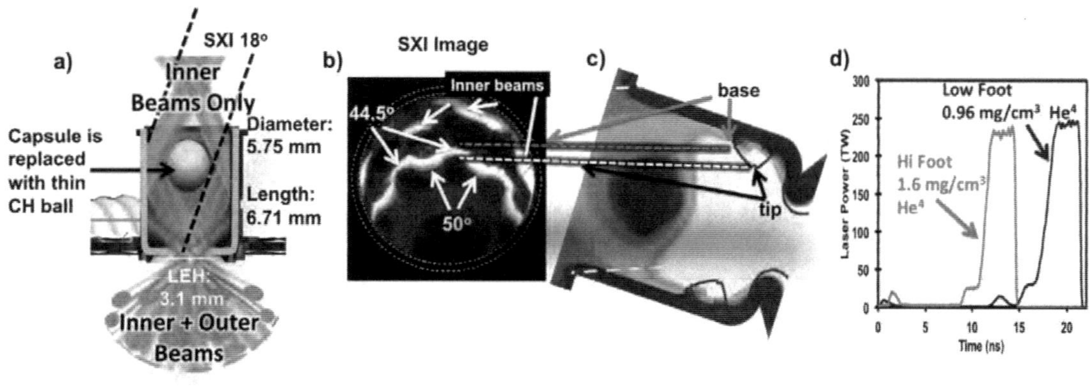

Figure 1. The experimental configuration. (a) The ViewFactor target showing the truncated "open" end facing up and the LEH end facing down. (b) Typcial time-integrated SXI image taken in an x-ray energy band of 3-5 keV and viewing the hohlraum at 18° to its axis. (c) Map of electron temperature at time of maximum power from a HYDRA[9] simulation. The map is tilted 18° to identify features in the SXI image. Horizontal lines show the gold bubble features. The tip and base of the bubble are defined. (d) Plot of laser power vs. time.

2. Experimental Description

The nominal ignition hohlraum is 5.75 mm in diameter, 9.425 mm long, with 3.1 mm diameter laser entrance holes (LEHs). The ViewFactor target has one half identical to the ignition hohlraum but the other half extends only 2 mm beyond the hohlraum center, with an opening equal to the full diameter of the hohlraum as shown in Figure 1a. The ViewFactor target is shot at cryogenic temperature and is filled with He gas at the same mass density as in the corresponding ignition design. The fuel capsule is replaced by a thin plastic spherical shell (3 mm diameter with 20 or 30 μm shell thickness). There are several holes in the CH shell so that its interior has same mass density as the hohlraum. The thin shell is almost completely ablated by the peak of the drive. ViewFactor experiments were done in pairs, open-end down and open-end up, so all diagnostics obtained data looking into both the open end and LEH end.

The hohlraum is driven (at wavelength 351 nm) with one of two standard ignition pulse shapes, a low foot pulse or a high foot pulse, which are shown in Figure 1d. The full complement of 96 beams (64 outer beams at 44.5^o and 50° to the hohlraum axis) and 32 inner beams (23.5^o and 30° to the hohlraum axis) is used at the LEH side but only the 32 inner beams are used at the open end. The relative power between the inner and outer beams at the LEH end can be adjusted by by transferring energy from the outer to the inner beams by tuning the laser wavelength difference between them.[10] This is called cross beam energy transfer (CBET).

Four experiments are discussed here. The first experiment used a 20 μm thick CH shell, a low-foot laser pulse, and a wavelength separation of 7.9A (9.1A) between the 30° (23.5^o) and outer beams. The second experiment was identical to this except the CH shell was 30 μm thick. The third experiment was identical to the second experiment except the wavelength separations were reduced to 1.5 A to turn off CBET. The fourth experiment used a hi-foot laser pulse with 7.3A (8.5A) wavelength separation. (The wavelength separations are for the fundamental wavelength, 1053 nm, before frequency tripling to 351 nm.)

This paper focuses on the images obtained with the the Static X-ray Imager (SXI), a pinhole camera with an x-ray CCD detector.[11] It views the hohlraum at 18° to the its axis from the upper hemisphere. It has a hard channel sensitive to 3 to 5 keV x-rays defined by a titanium

Journal of Physics: Conference Series **717** (2016) 012049 doi:10.1088/1742-6596/717/1/012049

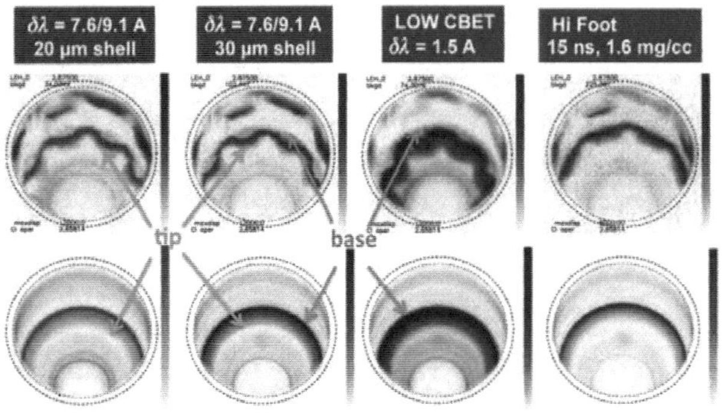

Figure 2. SXI images of data (top row) and model (bottom row). The tip and base of bubble are indicated.

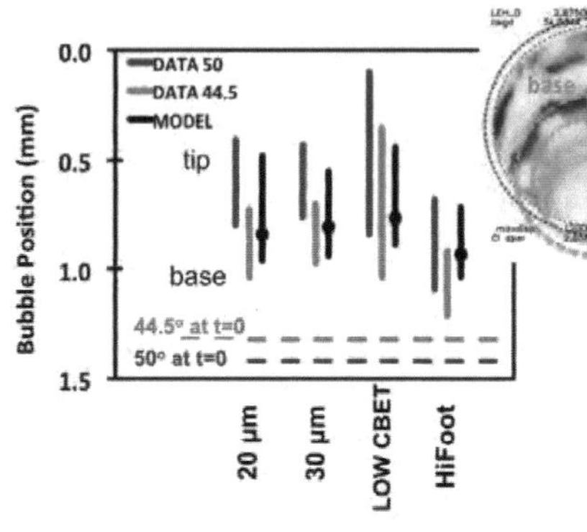

Figure 3. Plot of location of bubble (base to tip) in SXI image. The brightest part of bubble in model is marked.

filter. The filters in SXI are not calibrated, so the images are normalized[8] to the signal levels in the 4 keV x-ray channel of the Dante diagnostic[12], a time-resolved, low resolution x-ray spectrometer which views the hohlraum from the lower hemisphere at 37° to its axis. Figure 1b and c show a typical SXI image and a simulated map of electron temperature. The map is used to identify the gold bubble feature formed by the outer beams. Figure 2, top row, shows the normalized images from the four experiments.

3. Modelling

The ViewFactor experiments were modeled[7] with the LASNEX code[13] using the High Flux Model[14]. This model uses a flux limiter based on the Spitzer formula for electron conduction of $f = 0.15$ and it uses the detailed configuration accounting (DCA) non-local thermodynamic equilibrium (NLTE) atomic physics model[15] to generate the opacities responsible for emission. The simulation is first run using the measured laser energy to establish plasma conditions. The CBET is calculated for these plasma conditions, the laser pulses are modified using this CBET and the measured backscatter, and the model is run again[16].

To compare to the data, a post shot simulation was run using the measured laser drive minus the backscatter. These simulations have been post-processed to produce DANTE voltages and SXI images. The normalized simulated SXI images are shown in Figure 2, bottom row.

4. Results

The data in Figure 2 top row shows that the gold bubble pattern is three-dimensional, with the bubble from the 50° beams growing more than that from the 44.5^o beams. This cannot be captured in the images from the two-dimensional models, shown in the bottom row. However, the modeled images do look similar to the data. For instance, comparing the second and third experiments whose only difference is CBET vs LOW CBET, respectively, the bubble regions are wider in LOW CBET images because energy is not lost from the outer beams.

More insight can be gained by measuring the position of the bubble tips and bases, as defined in Figure 1c. The position is taken at the point of half the maximum intensity. Each series of tips or bases in the SXI image is fit to a circle foreshortened by 18°, as shown in the insert to Figure 3 for the 50° bubble by the cyan (base) and orange (tip) circles. The position of the tip (or base) is defined as the vertical distance from the top of the circle to the middle of the SXI image, as shown by the vertical lines. The positions of the bubbles (base to tip) are plotted in

Figure 3 for all the images shown in Figure 2. The model predicts the position of the bubble correctly, especially for the first two experiments (low foot pulse with CBET). However, the model predicts the LOW CBET bubble width to be only 65% of the measured width.

The inner beams in the data images in Figure 2 look brighter than those in the model, except for the LOW CBET shot. To quantify this, the total counts in the inner and outer beam regions are measured. Figure 4 shows the definition of the region of inner beams (outlined in red) and outer beams (outlined in blue). The ratio of the total counts in the inner beam region to that in the outer beam region is calculated and plotted. Both data and model see a decrease in inner to outer beam ratio in going to the thicker capsule. This, plus the increase in laser backscatter[7], is an indication of the influence of capsule blow-off on plasma conditions. Figure 4 shows that the model underestimates the ratio for the two low foot experiments with CBET and the HiFoot experiment but agrees with the data for the LOW CBET case. This may indicate an issue in the way the model implements CBET.

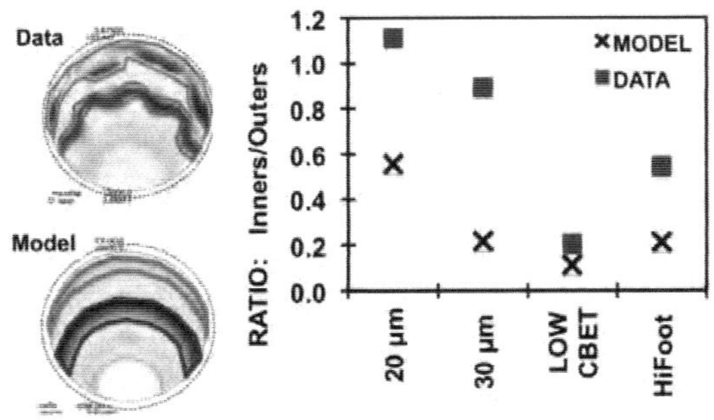

Figure 4. Ratio inner/outer beam brightness.

5. Conclusion

The ViewFactor experiment allows excellent diagnostic views of the interior of an ignition hohlraum. Time-integrated images in the 3-5 keV x-ray range at 18° to the hohlraum axis show that the region in which the outer beams deposit their energy at peak power, the gold "bubble" is three-dimensional. The two-dimensional models produce images that are similar to the data but underestimate the size of the bubble when there is LOW CBET and they underestimate the brightness of the inner beams when there is CBET. Three-dimensional models are in progress[17] as are enhancements to the current models, such as an in-line CBET package[18].

6. Acknowledgments

The authors gratefully acknowledge the efforts of the NIF Operations, Laser Performance, Target Diagnostics, and Target Fabrication Teams. This work was performed under the auspices of the Lawrence Livermore National Security, LLC, (LLNS) under Contract No. DE-AC52-07NA27344.

[1] Moses E, Boyd R N, Remington B A *et al.* 2009 *Phys. Plasmas* **16** 041006
[2] Lindl J, Amendt P, Berger R L *et al.* 2004 *Phys. Plasmas* **11** 339
[3] Haan S W, Lindl J D, Callahan D A *et al.* 2011 *Phys Plasmas* **18** 051001
[4] Hicks D G, Meezan N B, Dewald E L *et al.* 2012 *Phys Plasmas* **19** 122702
[5] Meezan N, MacKinnon A J, Hicks D G *et al.* 2013 *Phys.Plasmas* **20** 056311
[6] Hammer J H 2013 *LLNL-PROC-646357*
[7] Maclaren S A, Schneider M B, Widmann K *et al.* 2014 *Phys. Rev. Lett.* **112** 105003
[8] Schneider M B, MacLaren S A, Widmann K *et al.* 2015 *Phys Plasmas* **22** 122705
[9] Marinak M M, Kerbel G D, Gentile N A *et al.* 2001 *Phys Plasmas* **8** 2275
[10] Michel P, Divol L, Williams E A *et al.* 2009 *Phys. Plasmas* **16** 042702
[11] Schneider M B, Jones O S, Meezan N B *et al.* 2010 *Rev. Sci. Instrum* **81** 10E358
[12] Kline J, Widmann K, Warrick A *et al.* 2010 *Rev. Sci. Instrum.* **81** 10E538
[13] Zimmerman G B and Kruer W L 1975 *Comments Plasma Phys. Controlled Fusion* **2** 51
[14] Rosen M D, Scott H A, HInkel D E *et al.* 2011 *High Energy Density Physics* **7** 180
[15] Scott H and Hansen S 2010 *High Energy Density Phys.* **6** 39
[16] Town R P J, Rosen M D, Michel P A *et al.* 2011 *Phys. Plasmas* **18** 056302
[17] Meezan N 2013 personal communication
[18] Strozzi D J, Sepke S M, Kerbel G D *et al.* 2014 *LLNL-CONF-662797*

Journal of Physics: Conference Series **717** (2016) 012050 doi:10.1088/1742-6596/717/1/012050

Converging shock generation with cone target filled with low density foam

K Shigemori[1], T Yamamoto[1], Y Hironaka[1], T Kawashima[1], S Hattori[1], H Nagatomo[1], H Kato[1], N Sato[2], T Watari[2] and M Takagi[2]

[1]Institute of Laser Engineering, Osaka University, Suita, Osaka, Japan
[2]Hamamatsu Photonics KK, Hamamatsu, Shizuoka, Japan

shige@ile.osaka-u.ac.jp

Abstract. We have been developing an alternative scheme of fast ignition of inertial confinement targets with converging shock. Experiments were carried out on GEKKO-XII laser facility at ILE, Osaka University. We employed Au cone targets filled with low-density RF foam (2 mg/cm^3). The foam-filled cone targets were irradiated by three beams of the GEKKO-XII, with pulse duration of 1.3 ns, intensity of ~ 10^{14} W/cm^2 in 2 :λ(ω0.527μ m). Self-emission at the tip of cone was observed by one-dimensional streaked optical pyrometer (SOP) and two-dimensional images.

1. Introduction

Fast ignition scheme is an attractive approach for fusion ignition with lower laser energy because hot spark and high-density fuel should not be generated simultaneously with one laser condition. Many experimental and theoretical works have been done in order to find optimum condition for ignition by means of hot electrons [1,2], ions [3], strong shock wave by "spike pulse" [4], and high velocity impactor [5]. Here we propose the hot spark generation by a different scheme with fast converging shock wave. The high-temperature spark generation with converging shock wave is very similar to high-velocity implosion [6-8], which generates high temperature and low density plasmas. Schematic picture of ignition by the converging shock wave is shown in Fig. 1. The target configuration is combination of slow and fast implosions. The slow implosion of the main hollow shell is to produce high-density and low-temperature fuel with isochoric compression whereas the fast implosion of inside the cone target filled with low density foam is to produce hot spark with shock multiplexing. By using cone-in-shell target, these two implosions can be separated spatially and temporally. In our preliminary target design, the fast implosion is guided to inside of the cone. The strong shock wave converges towards the tip of the cone in which low-density foam is filled in the cone. The advantage of the use of foam is for higher initial density than gas medium. In the

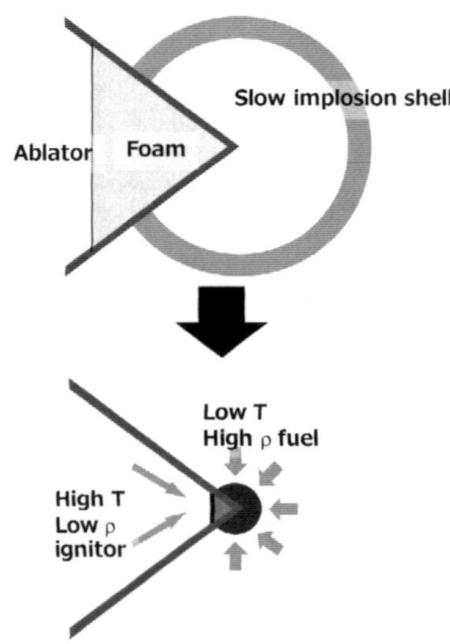

Figure 1 Schematics of ignition by converging shock wave

previous high-velocity implosion experiments, the gas pressure was around 10 to 50 Torr in deuterium or tritium.

In this paper, we report on preliminary experiments on shock propagation for converging geometry using cone targets filled with low-density foam. Temporal evolution of self emission at the top of cone was observed by optical diagnostics in order to verify the uniformity of the converging shock.

2. Experiments

Experiments were carried out on GEKKO-XII laser facility at ILE, Osaka University. Figure 2 shows schematic view of the target irradiation configuration. We employed Au cone targets filled with low-density RF foam. The density of the RF foam (CHO) was 2 mg/cc. The thickness and opening angle of the cone were ~7 μm and 90 degrees, respectively. A polystyrene foil (5-μm thickness) was attached on the entrance of the cone, and an Al foil (5-μm thickness) was attached on the tip of the cone. The foam-filled cone targets were irradiated by three beams of the GEKKO-XII, with pulse duration of 1.3 ns (Gaussian), energy of ~ 1kJ in 2ω (λ: 0.527 μm). The focal spot diameter of each beam on the cone entrance was around 400 μm. For the reference data in planar geometry characteristics, we also irradiated planar low-density foam of 300-μm thickness.

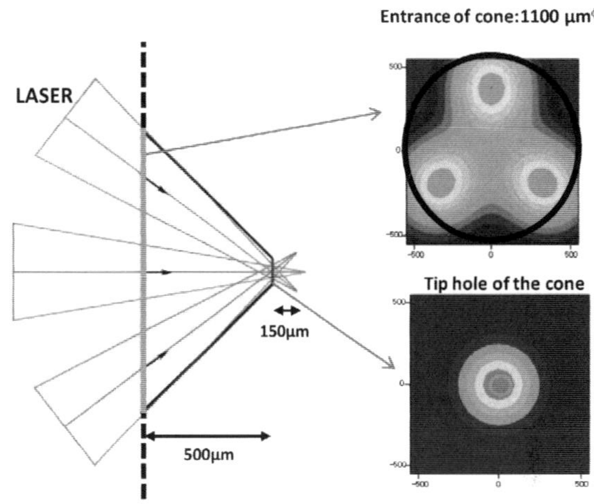

Figure 2 Irradiation configuration of cone target filled with low-density foam.

On the optical measurements, we employed streaked optical pyrometers (SOPs) in order to estimate the temperature of converging shock front at the tip of the cone. One of the SOPs was conventional one-dimensional SOP. We also employed HISAC (high-speed sampling camera) [9] in order to obtain time-resolved two-dimensional self-emission images at the top of the cone. For the planar geometry experiment, x-ray shadowgraph measurements with an x-ray streak camera were also done simultaneously with the optical diagnostics.

3. Results and Discussions

The basic parameters on shock propagation in the low-density foam were taken with planar low-density foam target. Figure 3 (a) shows a raw streaked image by the SOP. From the raw streak image, temporal profile of the self emission at the centre of the foam was plotted as in Fig. 3 (b). Also overlayed

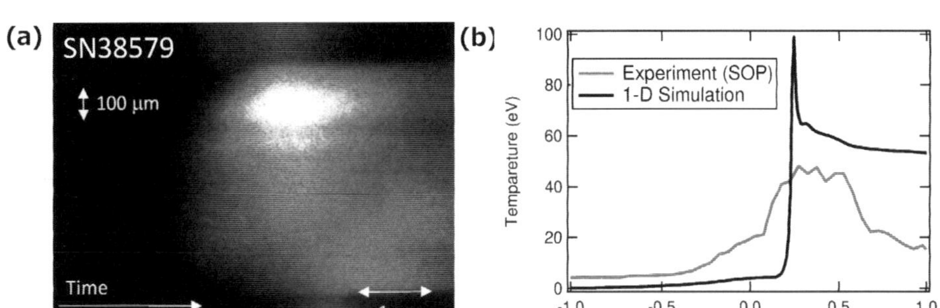

Figure 3 (a) Raw streaked image with SOP for planar low-density target. (b) Temperature from SOP data and from ILESTA-1D calculation.

in Fig. 3 (b) is the temperature at the rear surface (Al) calculated by one-dimensional hydrodynamics code ILESTA-1D [10]. The experimental shock breakout timing shows good agreements with the simulation results. The experimental data and the simulation calculation also indicates a small preheating prior to the shock breakout. The temperature in the foam at uncompressed region from the simulation calculation shows more than 5 eV (not shown here), suggesting that the foam is not no longer foam material but partly ionized gas which is compressed by the shock wave.

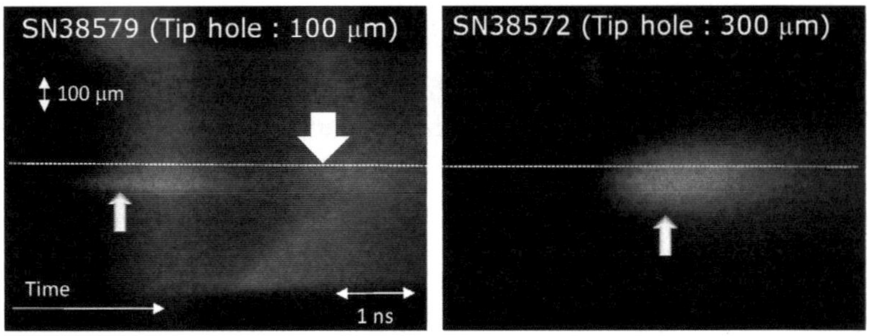

Figure 4 One-dimensional SOP images for the tip size of 100- and 300-μm. Dotted lines show the position of the TCC (target chamber centre). Yellow arrows are timing of laser peak timing for each shot.

In order to observe the behavior of the shock converging, we measured the self-emission at the tip of the cone by changing the hole size of the tip. Since the shape of the cone was same, the shock propagation distance was slightly different by changing the hole size. Figure 4 shows raw streaked images of cone targets taken by the one-dimensional SOP. These images are for the cone tip diameters of 100 and 300 μm. The raw data clearly indicates collimated emissions in accordance with the hole size of the tip. The shock breakout timing of each shot shows in good agreements with the ILESTA-1D simulation. On the other hand, the position of the self-emission is not at exactly center of the target chamber center (TCC). That might be due to the accuracy of the target alignment.

In order to evaluate the two-dimensional self-emission image, we employed the HISAC with a bundled fiber system. The bundle fiber has 225 ch (15x15), with the field of view of ~500 μm. Figure 5 shows the re-constructed two-dimensional self-emission for the tip size for 100 μm. The HISAC data suggests that the peak emission was at around 220-ps after the laser peak timing, where the position of the peak emission was slightly sifted ~ 60 μm from the TCC. The shift of the self-emission position is in good agreements with the one-dimensional SOP data.

Both from the one-dimensional SOP and the HISAC, the self emission from the tip hole starts much earlier than calculations with ILESTA 1-D calculations for all target conditions. The self-emission starts at more than 500-ps earlier than the laser peak emission timing which indicates that strong preheating is a probable source of the self emission. The preheating would be due to hot electrons which penetrates the low density foam region. Also x-ray radiation is source of the preheating because a part of the laser irradiated inner portion of the cone as shown in Fig. 2. The measured temperature of the preheating was

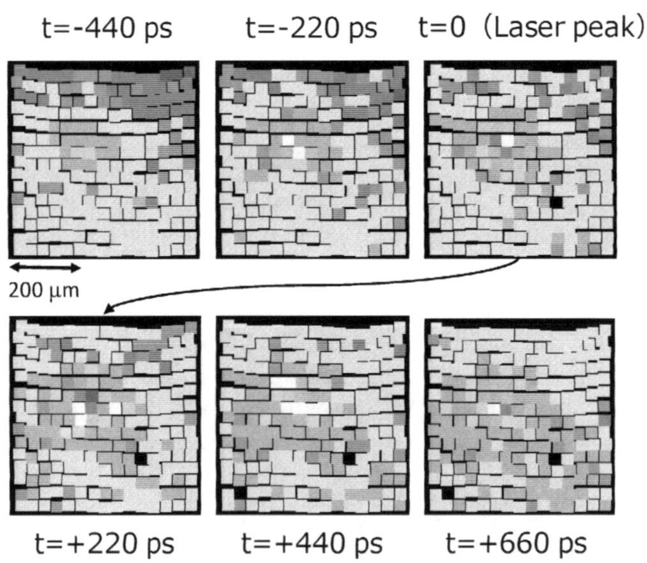

Figure 5 Two-dimensional self-emission image with the HISAC for the 100-μm cone tip target. The TCC position is the centre of each image.

around 20 eV, which is from calibrated value of the SOP. Please note that there is a second flash of the self emission (white arrow in Fig. 4) at later timing. The timing of the second flash is around ~500 ps later from the calculated timing with ILESTA 1-D simulation. We suspect that the second flash is from shock breakout at the tip of the cone after the expansion by the preheating, resulting later timing and weak emission. The irradiation configuration and the target design should be revised and improved for observation of the converging shock front in future experiments.

4. Summary

We have performed experiments on converging shock wave with cone targets filled with low-density form. Optical measurements were carried out in order to verify the shock convergence at the top of the cone. The experimental observations with optical measurements suggest clear shock conversion inside of the cone targets. We observed the self emission due to the shock breakout from the tip of the cone. However, preheating prior to the shock breakout is significant, which should be improved with modifying target design and drive configurations.

Acknowledgments

The authors would like to acknowledge the dedicated technical support by the staff at the GEKKO-XII facility for the laser operation, target fabrication, and plasma diagnostics. This work was performed under the joint research project of the Institute of Laser Engineering, Osaka University. his work was supported by JSPS KAKENHI Grant Number 26630472.

References

[1] Kodama R 2002 Fast heating scalable to laser fusion ignition. *Nature* **418** 933–4

[2] Tabak M, Hammer J, Glinsky M E, Kruer W L, Wilks S C, Woodworth J, Campbell E M, Perry M D and Mason R J 1994 Ignition and high gain with ultrapowerful lasers *Phys. Plasmas* **1** 1626

[3] Roth M, Cowan T E, Key M H, Hatchett S P, Brown C, Fountain W, Johnson J, Pennington D M, Snavely R A, Wilks S C, Yasuike K, Ruhl H, Pegoraro F, Bulanov S V., Campbell E M, Perry M D and Powell H 2001 Fast Ignition by Intense Laser-Accelerated Proton Beams *Phys. Rev. Lett.* **86** 436–9

[4] Betti R, Zhou C D, Anderson K S, Perkins L J, Theobald W and Solodov a. a. 2007 Shock ignition of thermonuclear fuel with high areal density *Phys. Rev. Lett.* **98** 155001

[5] Murakami M, Nagatomo H, Azechi H, Ogando F, Perlado M and Eliezer S 2006 Innovative ignition scheme for ICF—impact fast ignition *Nucl. Fusion* **46** 99–103

[6] Yamanaka C, Nakai S, Yabe T, Nishimura H, Uchida S, Izawa Y, Norimatsu T, Miyanaga N, Azechi H, Nakai M, Takabe H, Jitsuno J, Mima K, Nakatsuka M, Sasaki T, Yamanaka M, Kato Y, Mochizuki T, Kitagawa Y, Yamanaka T and Yoshida K 1986 Laser implosion of high-aspect-ratio targets produces thermonuclear neutron yields exceeding 10^{12} by use of shock multiplexing *Phys. Rev. Lett.* **56** 1575–8

[7] Richardson M, McKenty P, Keck R, Marshall F, Roback D, Verdon C, McCrory R, Soures J and Lane S 1986 High-aspect-ratio laser-fusion targets driven by 24-beam uv laser radiation. *Phys. Rev. Lett.* **56** 2048–51

[8] Piriz a R and Wouchuk G 2000 Laser driven implosion of high aspect ratio spherical shell target *Plasma Phys. Control. Fusion* **32** 469–74

[9] Kodama R, Okada K and Kato Y 1999 Development of a two-dimensional space-resolved high speed sampling camera *Rev. Sci. Instrum.* **70** 625

[10] Takabe H, Yamanaka M, Mima K, Yamanaka C, Azechi H, Miyanaga N, Nakatsuka M, Jitsuno T, Norimatsu T, Takagi M, Nishimura H, Nakai M, Yabe T, Sasaki T, Yoshida K, Nishihara K, Kato Y, Izawa Y, Yamanaka T and Nakai S 1988 Scalings of implosion experiments for high neutron yield *Phys. Fluids* **31** 2884

High-density implosion via suppression of Rayleigh–Taylor instability

Takashi Shiroto, Naofumi Ohnishi, Atsushi Sunahara[1], Shinsuke Fujioka[2], and Akira Sasaki[3]

Tohoku University, 6-6-01 Aramaki-Aza-Aoba, Aoba-ku, Sendai, Miyagi, 980-8579, Japan
[1]Institute for Laser Technology, 2-6 Yamada-oka, Suita, Osaka, 565-0871, Japan
[2]Osaka University, 2-6 Yamada-oka, Suita, Osaka, 565-0871, Japan
[3]Japan Atomic Energy Agency, 8-1-7 Umemidai, Kizugawa, Kyoto, 619-0215, Japan

E-mail: tshiroto@rhd.mech.tohoku.ac.jp

Abstract. Radiation hydrodynamic simulations of ICF capsules assuming a kJ-class laser facility were conducted to evaluate the hydrodynamic stability of a brominated plastic shell. An opacity table based on a detailed atomic model was employed so that more quantitative forecast of the implosion dynamics could be performed. A lightly doped shell could form a high-density core at the maximum compression by suppressing the hydrodynamic instability.

1. Introduction

Until today, inertial confinement fusion (ICF) has not generated the extra amount of energy because implosion asymmetry induced by Rayleigh–Taylor instability (RTI) fails to ignite a thermonuclear fuel. The RTI prevents from obtaining the ICF energy mainly in two ways. First one is contamination of the hotspot by higher-Z elements [1]. Perturbation is seeded on the gas/solid interface of a thermonuclear fuel at the acceleration phase; then it is amplified by the RTI on the deceleration phase. Thus, the hotspot is contaminated by higher-Z elements included in the ablator, resulting in a reduction of hotspot temperature due to radiation cooling. Second is reduction of density–radius product (ρR) which regulates the thermonuclear chain reaction [2].

A number of investigations on the ablative RTI have been conducted in theoretical [3, 4, 5] and experimental [6, 7] manners. One of the most famous outcomes of linear stability analysis is so-called modified Takabe formula [3, 4];

$$\gamma(k) = \alpha\sqrt{\frac{kg}{1+kL}} - \beta k v_{\mathrm{a}}, \qquad (1)$$

where γ is the linear growth rate for the wavenumber of k, g the acceleration, L the density scale length of an ablation front, v_{a} the ablation velocity, α the square root of Atwood number, and β a fitting parameter for the ablative stabilization. A high-Z doping scheme [7] was proposed to suppress the ablative RTI. Emission from the high-Z ions may enhance v_a in the case of the direct-drive, so it is expected that the RTI is suppressed according to Eq. (1).

In this study, we investigated whether a brominated plastic shell can suppress the RTI and can achieve greater ρR than the non-doped one. Implosion dynamics with a large non-uniformity

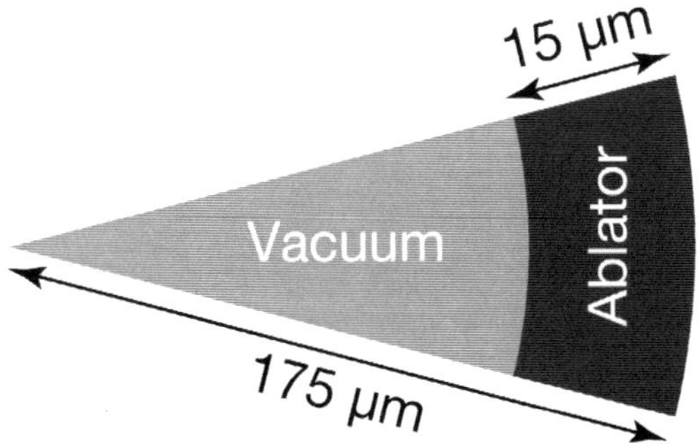

Figure 1. Schematic of the shell target used in this study.

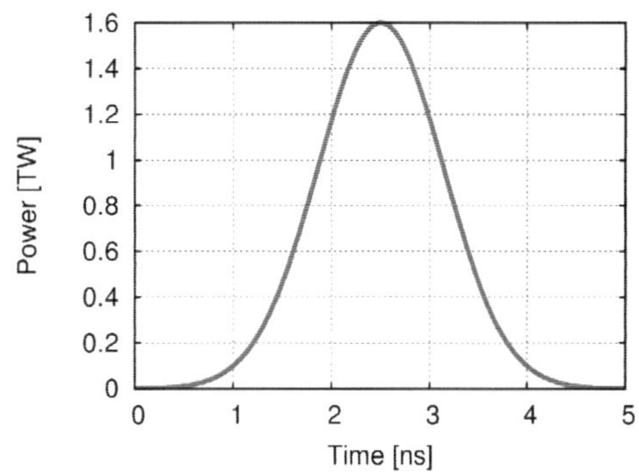

Figure 2. Pulse profile to simulate the 12-beam illumination with GEKKO-XII.

of the incident lasers was studied to evaluate the effect of suppression of the RTI. The temporal profile of ρR was then analyzed to discuss the mechanism that the high-Z doping can enhance ρR at the maximum compression. Moreover, robustness of the high-Z doped target for the hydrodynamic instability was investigated by 2D parametric study.

2. Numerical methods and conditions

Numerical simulations in this study were conducted with a radiation hydrodynamic code, RAICHO [8]. Governing equations are discretized in (r, θ) coordinates whose pole is an axis of rotation. Implosion dynamics was described by Euler equation with energy sources and energy equations for ions and electrons. Laser energy was deposited on the electrons by one-dimensional ray-tracing in the radial direction with free–free absorption coefficient. Thermal conduction was approximated by the flux limited diffusion (FLD) manner with Spitzer–Härm conductivity [9]. The flux limiter of FLD was set to 0.1. Radiative transfer equation was reduced to the multi-group diffusion form with variable Eddington factor of Minerbo [10]. Atomic process of bromine ions was calculated by HULLAC code [11] which is based on the detailed configuration accounting (DCA) model, and opacity table was composed by JATOM code [12] with collisional radiative (CR) model. Although the DCA can generate the better opacity table than the average-ion (AI) model, the HULLAC code cannot handle low-Z plasmas. Thus, the opacity was calculated by the AI and CR models for hydrogen and carbon ions.

Schematic of the shell target is shown in Fig. 1. The ablator was composed of polystyrene (CH) or 0.05%-doped brominated one (CHBr). In order to simulate the laser incidence of GEKKO-XII, the pulse shape was set like Fig. 2. The pulse energy was 2.5 kJ, the pulse width was 1.3 ns in full width half maximum, and the wavelength was 527 nm. The shell surface was genuinely spherical, and perturbation was added in the angular distribution of the incident laser as following;

$$\delta a/a = (\delta a/a)_{\max} \cos(l\theta), \tag{2}$$

where $\delta a/a$ is the ratio of the perturbation to the uniform component. The wave mode l was set to 32 in this paper. Note that this is not a fundamental wave in the axial symmetric (r, θ) coordinates, i.e., the Legendre mode perturbation. Computational domain D was set as the below expression;

$$D = \{(r, \theta) \mid 0 \le r\ [\mu\text{m}] \le 300,\ 0 \le \theta \le \pi/2\}. \tag{3}$$

The domain D was partitioned into computational cells which have the equal interval in the (r, θ) space. The numbers of grids were 513 in the radial direction and 257 in the polar one. Boundary condition was the outflow condition at $r = 300$ μm while the others were the mirror boundaries.

3. Results

Figure 3 shows the implosion dynamics of the CH and CHBr shells. At the beginning of acceleration (Figs. 3(a) and 3(b)), density of the CHBr is less than the CH since bromine ions emit high-energy photons and preheat inner side of the shell. Additionally, amplitude of the perturbation is not different so much between them, so the capability to suppress the RTI can be estimated by the perturbation growth after this time. Figures 3(c) and 3(d) suggest that the CHBr target shows shorter length of the spike than the CH one, and the inner wall of the CH is completely broken. As a result, only the CHBr target could result in high-density core at the maximum compression (Figs. 3(e) and 3(f)).

Figure 4 depicts the temporal profile of ρR with the amplitude of $\delta a/a = 10\%$. The CHBr target achieved nearly twice greater ρR than the CH one. Note that the two peaks of ρR was found for the CH target at $t = 2.5$ and 2.65 ns. This phenomenon corresponds to the Bubble–spike (BS) structure generated by the RTI. When the BS structure is developed like Fig. 3, a reflected shock wave is generated at the inner wall of the target. Thus, the first peak corresponds to the shock compression. The CH target has longer spikes than the CHBr one, so the shock compression arrives at the earlier time. On the other hand, the second peak corresponds to the stagnation of the massive region. The CHBr target has a stationary point around $t = 2.65$ ns, while the CH one has the second peak at the same time. This is because the acceleration of both targets seems to be the same for such a lightly doped case. Hence, it is important to shorten the time interval between the shock compression and the stagnation in order to generate a high-density core.

Moreover, we conducted a parametric study to obtain the dependency of ρR on the amplitude of laser non-uniformity. Figure 5 shows the maximum ρR obtained by the 2D simulations except the 1D simulations for the uniform case. At the cases of nearly uniform, the CH target has higher ρR than the 0.05% doped one, and the 0.3% one has lower ρR than the CH target at the most perturbed case. This is because strong emission from the bromine ions preheats the shell and compression at the stagnation is weaken due to the higher adiabat. Therefore, brominated targets should not be used if the hydrodynamic instability is not the most crucial cause to spoil the maximum ρR. Laser non-uniformity of GEKKO-XII seems to be several tens of percent, so the high-Z doped target might support the core generation.

4. Conclusion

RHD simulations of ICF implosions were conducted to analyze the hydrodynamics of a brominated shell. In our previous work, opacity tables were generated with the AI model which is one of the most primitive models for atomic process. The simulations with the DCA opacity revealed that the lightly doped target can suppress the RTI resulting in the core generation at the maximum compression. In the case of large perturbation, the CHBr shell achieved greater ρR because the time interval between the shock compression and stagnation was shortened. The brominated target has more robust characteristics for disturbance while the 1D performance is inferior due to the higher adiabat. However, intensity of the lasers on this study is sub-ignition class which is much less than the mega-joule class reactor. There are some arguments that the RTI can not be suppressed with mid-Z ablators in the ignition-relevant implosion [13]. It is our future work to demonstrate that the high-Z doping scheme is available in such a scale.

Journal of Physics: Conference Series **717** (2016) 012051 doi:10.1088/1742-6596/717/1/012051

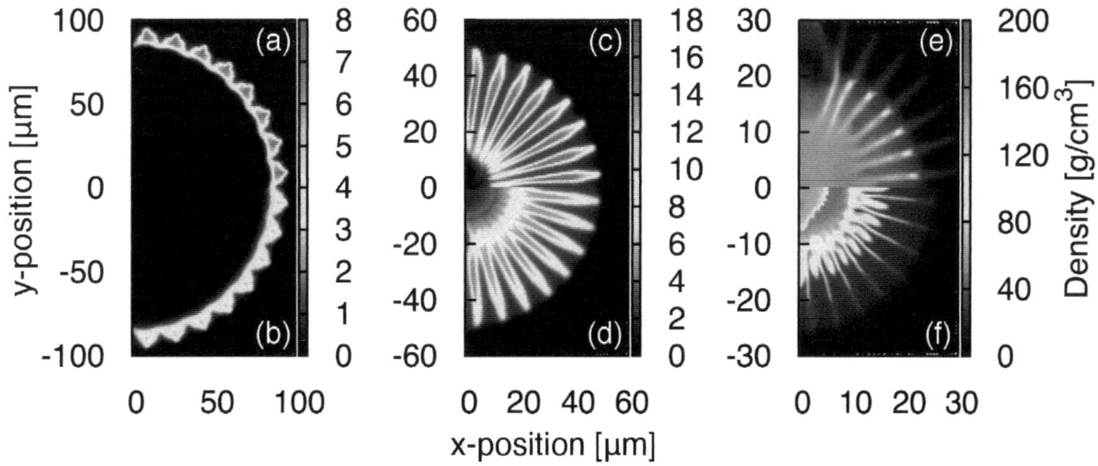

Figure 3. Density contours of simulations with perturbation amplitude of $\delta a/a = 10\%$. Panels consist of (a) CH and (b) CHBr at the beginning of acceleration ($t \simeq 2.1$ ns), (c) CH and (d) CHBr at the end of acceleration ($t \simeq 2.5$ ns), and (e) CH and (f) CHBr at the stagnation phase ($t \simeq 2.6$ ns).

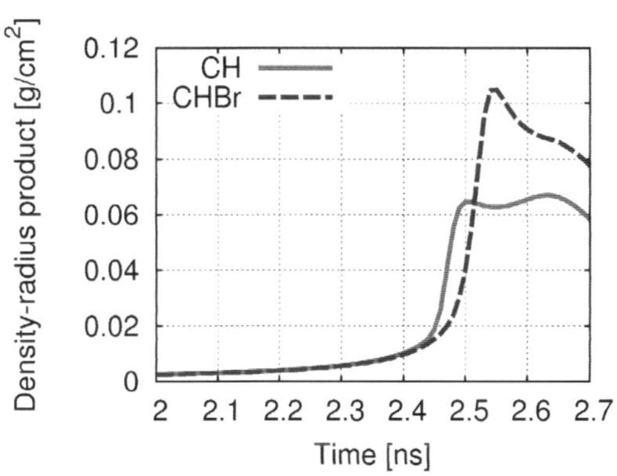

Figure 4. Temporal profile of ρR for $\delta a/a = 10\%$. ρR was averaged over the polar direction.

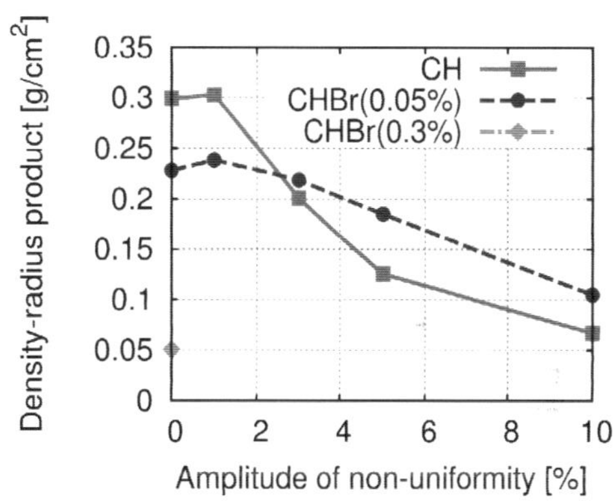

Figure 5. Dependency of ρR at the maximum compression on the laser non-uniformity.

References

[1] Ma T et al. 2013 *Phys. Rev. Lett.* **111** 085004
[2] Haan S W et al. 2005 *Phys. Plasmas* **12** 056316
[3] Takabe H et al. 1985 *Phys. Fluids* **28** 3676
[4] Budil K S et al. 1996 *Phys. Rev. Lett.* **76** 4536
[5] Betti R et al. 1998 *Phys. Plasmas* **5** 1446
[6] Glendinning S G et al. 1992 *Phys. Rev. Lett.* **69** 1201
[7] Fujioka S et al. 2004 *Phys. Rev. Lett.* **92** 195001
[8] Ohnishi N 2012 *High Energy Density Phys.* **8** 341
[9] Spitzer L and Härm R 1953 *Phys. Rev.* **89** 977
[10] Minerbo G N 1978 *J. Quant. Spectrosc. Radiat. Transfer* **20** 541
[11] Bar-Shalom A et al. 2001 *J. Quant. Spectrosc. Radiat. Transfer* **71** 169
[12] Sasaki A and Murakami I 2013 *Plasma Fusion Res.* **8** 2401021
[13] Lafon M et al. 2015 *Phys. Plasmas* **22** 032703

Journal of Physics: Conference Series **717** (2016) 012052 doi:10.1088/1742-6596/717/1/012052

Hydrodynamic growth experiments with the 3-D, "native-roughness" modulations on NIF

V A Smalyuk, S V Weber, D Casey, D S Clark, F Coppari, J E Field, S W Haan, B A Hammel, A Hamza, W Hsing, O Landen, A Nikroo, H F Robey, and C R Weber

Lawrence Livermore National Laboratory, Livermore, CA 94550, USA

Abstract. Hydrodynamic instability growth experiments with three-dimensional (3-D) surface-roughness modulations were performed on plastic (CH) shell spherical implosions at the National Ignition Facility (NIF). The initial capsule outer-surface roughness was similar to the standard specifications ("native roughness") used in a majority of implosions on NIF. At a convergence ratio of ~3, the measured tent modulations were close to those predicted by 3-D simulations (within ~15-20%), while measured 3-D, broadband modulations were ~3-4 times larger than those simulated based on the growth of the known imposed initial surface modulations. One of the hypotheses to explain the results is based on the increased instability amplitudes due to modulations of the oxygen content in the bulk of the capsule. These new experiments results have prompted looking for ways to reduce UV light exposure during target fabrication.

1. Introduction

Hydrodynamic instabilities, including Rayleigh-Taylor, Richtmyer-Meshkov, and Kelvin-Helmholtz instabilities, and mix play a central role in the implosion performance degradation. Any drive asymmetries and surface imperfections are amplified by the hydrodynamic instabilities during implosion resulting in a distorted shell with reduced hot-spot temperature, volume, and pressure [1]. In addition, small-scale turbulence seeded by these instabilities can mix ablator material into the DT hot spot [1].

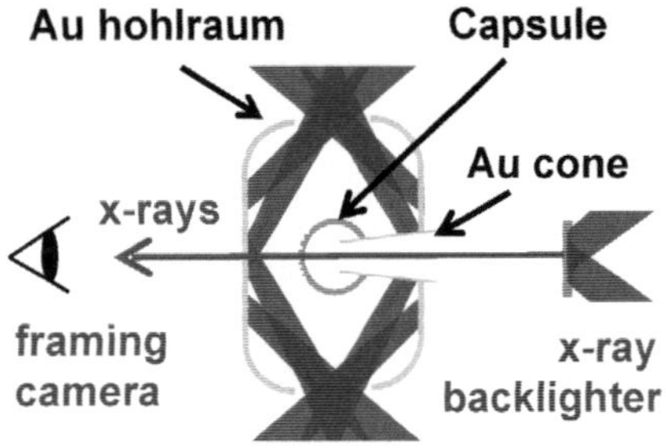

Fig. 1. Experimental configuration Au hohlraum, plastic (CH) capsule, Au cone, and an iron backlighter.

The presence of mixed ablator material was correlated with reduced experimental yields and temperatures in high-compression layered DT implosions [2] in experiments during National Ignition Campaign (NIC). Several experimental platforms have been developed to measure and understand various aspects of the instability growth on NIF. Recent experiments with the Hydrodynamic Growth Radiography (HGR) platform were used to study instability growth using large-amplitude, two-dimensional (2-D) pre-imposed modulations [3].

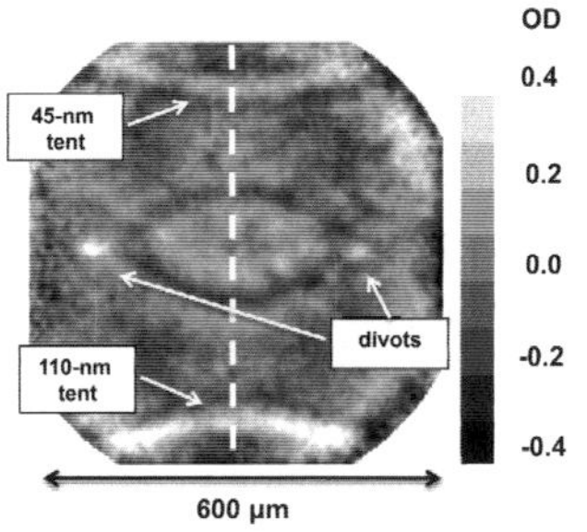

Fig. 2. Measured image of optical-depth (OD) modulations shows locations of the two divots, as well as 45-nm and 110-nm thick tents. The dark color represents areas of stronger backlighter x-ray absorption (larger column density in the shell), while the light color represents areas of weaker absorption (thin regions). The dashed vertical curve shows the location of the vertical lineout used in Fig. 3.

These experiments showed that the measured instability growth of sinusoidal outer surface ripples were modeled well with 2-D simulations [3]. This article describes experiments with the 3-D, "native-roughness" modulations. They were measured at ignition relevant conditions on NIF and compared with 3-D simulations. In addition to the 3-D surface-roughness growth, these measurements included imaging of the modulations caused by a support membrane (the "tent") used to hold capsules in a hohlraum, and pre-imposed divots. The experimental configuration is described in Sec. 2; while the modulation growth data are discussed in Sec. 3. The results are summarised in Sec. 4.

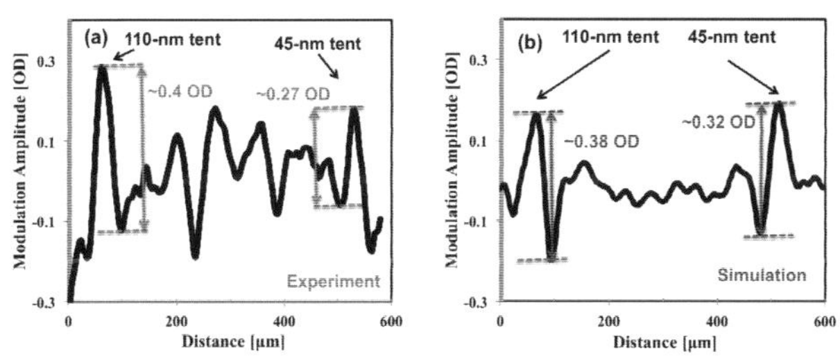

Fig. 3. Vertical lineouts show a comparison of (a) experimental and (b) simulated optical-depth modulations

2. Experimental configuration

The experimental configuration used in the experiments is shown in Fig. 1. It is similar to the HGR platform used in earlier experiments [3] to measure instability growth with 2-D modulations during acceleration phase. The configuration includes a Au hohlraum, a plastic (CH) capsule, a Au cone, and an iron backlighter. The experiments were conducted with laser drives and conditions similar to those used in high-compression layered DT implosions [4]. They were designed to test the acceleration phase hydrodynamic growth predictions used to model these DT layered implosions that achieved fuel areal densities of ~1.2 g/cm^2, peak fuel velocities of ~320-330 km/s, and peak radiation temperatures of ~300 eV [4].

The nominal 209-μm thick plastic capsule with nominal 1120-μm outer radius had the same Si-doped layers as used in the previous DT layered implosions [4]. These capsules used an extra 20-μm thick CH layer to replace the 69-μm thick DT ice layer, thereby maintaining approximately the same shell mass as in the layered DT implosions. The capsule had two divots that were imprinted on the surface by laser ablation, to be spatial fiducials for the convergence measurements. The divots were both ~40-μm wide, with peak-to-valley amplitudes of ~260 nm and ~280 nm.

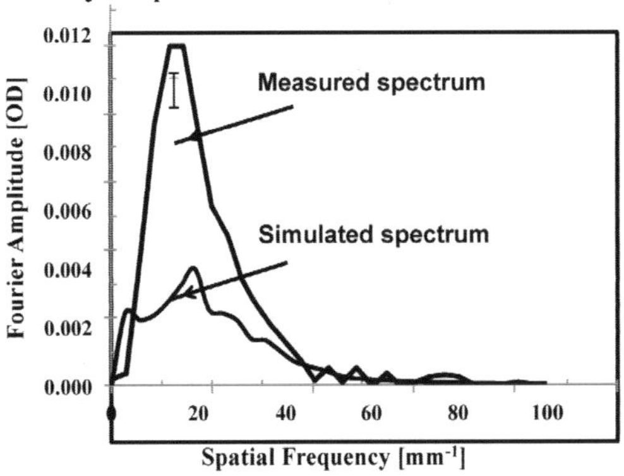

Fig. 4. Measured Fourier amplitudes of optical-depth modulations in the central, 360-μm square region (thick solid curve), compared to the simulated optical-depth modulations (thin solid curve).

The target also used two different support membranes ("tents"), with 110-nm and 45-nm thickness, located at the lower and upper parts of the target, respectively. The modulation growth was measured with through-foil x-ray radiography [5] using ~6.7-keV x-rays generated by the iron backlighter, which was located 12 mm from the target center [5]. X-ray images of the growing capsule modulations were formed using 20-μm diameter pinholes located 80 mm from the capsule. The images were captured using a framing camera [5] with a magnification of ~8 for the imaging system.

3. Ablation-front Rayleigh-Taylor instability experiments

Figure 2 shows measured capsule optical-depth (OD) x-ray image on a framing camera. It represents an average of 30 independent intensity images captured during a ~240-ps temporal range. The dark color in Fig. 2 represents areas of stronger backlighter x-ray absorption (or spikes), while the light color represents areas of weaker backlighter x-ray absorption (or bubbles). One can see the locations of both 45-nm and 110-nm tents, two divots, and non-systematic 3-D modulations in this image. In addition, the "eye" feature is also prominent in the central part of this OD image. Such a feature was not measured in outer-surface capsule characterization, indicating it could have been present in the bulk of the shell, or created during target assembly subsequent to characterization, as will be discussed later. Figure 3 shows a comparison of the measured tent data with the simulations. In the experiment, the peak-to-valley of the 110-nm tent modulation of ~0.4 OD was larger than that of the 45-nm tent by a factor of 1.5. Such large tent modulations would be expected to significantly degrade the performance of the layered DT implosions. In the simulation, the predicted modulations were in fair agreement with the measured ones, as shown in Fig. 3(b). The widths of the modulations were similar in experiment and simulations, however the shapes of the modulations were slightly different. The spikes on sides towards capsule center are more evident in simulation than data, indicating the difference in the modulation shapes. On average, the simulated tent amplitudes were ~20% lower than the measured amplitudes [5].

The modulation growth of the randomly seeded 3-D modulations was quantified by analyzing the central 360-μm square parts of the measured radiographs, away from the tent and divot modulations. This is quantitatively demonstrated in Fig. 4 by comparing the optical-depth Fourier spectra of the measured and predicted modulations. The measured modulations do not resemble the simulated modulations, indicating that they do not originate from the outer-surface modulations, but come from some other seeds, possibly from modulations in the bulk of the capsule [5]. These results are consistent with the hypothesis that spatially non-uniform oxygen contamination in the bulk of the capsule could seed significant modulation growth [6]. Oxygen absorption is increased by exposure to UV or to visible light [6]. The targets used for layered DT implosions routinely experience UV and visible light during handling and assembly procedures, and modulations in exposure are very likely. Coherent features like the "eye" in Fig. 2 could very plausibly arise from differential exposure to light or UV. Experiments with reduced UV exposure were recently performed and the measured image is shown in Fig. 5. The lack of "eye" feature in Fig. 5 suggests that that reduced UV exposure can eliminate these oxygen-induced modulations.

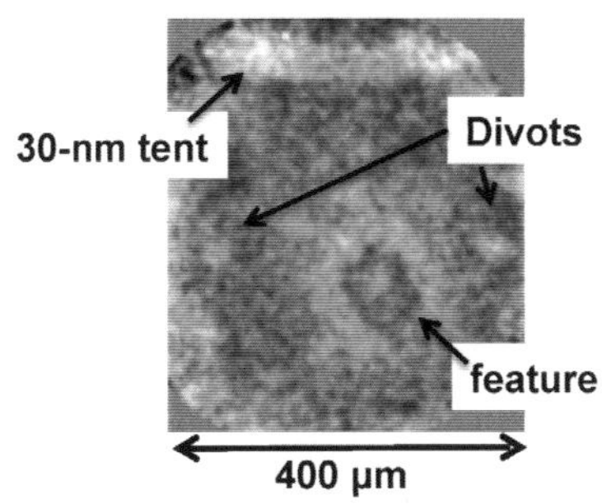

Fig. 5. Optical-depth image of the modulations in a shot with reduced UV exposure during target assembly. The round feature in the middle of the image was expected to grow from the feature measured on the capsule before the experiment. The "eye" feature was not present in this shot, suggesting the new assembly procedure was able to eliminate it.

4. Conclusions

Hydrodynamic instability growth of three-dimensional (3-D) surface-roughness modulations was studied in spherical implosions at the National Ignition Facility. The measured 3-D broadband modulations were ~3-4 times larger than those simulated based on the growth of the known imposed initial surface modulations. These modulations included a striking systematic feature, presumably originating from some part of the target fabrication process. The hypothesis to explain the results is the seeding of instability growth by modulations of the oxygen contamination in the bulk of the capsule.

This work was performed under the auspices of the U.S. Department of Energy by Lawrence Livermore National Laboratory under Contract DE-AC52-07NA27344.

References

[1] S. W. Haan et al., Phys. Plasmas **16**, 051001 (2011)
[2] T. Ma et al., Phys. Rev. Lett. **111**, 085004 (2013)
[3] K. Raman et al., Phys. Plasmas **21**, 072710 (2014)
[4] V. A. Smalyuk et al., Phys. Rev. Lett. **111**, 125001 (2013)
[5] V. A. Smalyuk et al., Phys. Plasmas **21**, 056301 (2014)
[6] S. W. Haan et al., Phys. Plasmas **22**, 032708 (2015)

Hydrodynamic simulations of long-scale-length plasmas for two-plasmon–decay planar-target experiments on the NIF

A A Solodov[1,4], M J Rosenberg[1], J F Myatt[1], R Epstein[1], S P Regan[1], W Seka[1], J Shaw[1], M Hohenberger[1], J W Bates[2], J D Moody[3], J E Ralph[3], D P Turnbull[3] and M A Barrios[3]

[1]Laboratory for Laser Energetics, University of Rochester, Rochester
250 East River Road, Rochester, NY 14623-1299, USA
[2]Naval Research Laboratory, Washington, DC, USA
[3]Lawrence Livermore National Laboratory, Livermore, CA, USA

asol@lle.rochester.edu

Abstract. The two-plasmon–decay (TPD) instability can be detrimental for direct-drive inertial confinement fusion because it generates high-energy electrons that can preheat the target, thereby reducing target performance. Hydrodynamic simulations to design a new experimental platform to investigate TPD and other laser–plasma instabilities relevant to direct-drive–ignition implosions at the National Ignition Facility are presented. The proposed experiments utilize planar plastic targets with an embedded Mo layer to characterize generation of hot electrons through Mo K_α fluorescence and hard x-ray emission. Different laser-irradiation geometries approximate conditions near both the equator and the pole of a polar-direct-drive implosion.

1. Introduction

Coronal plasmas of direct-drive–ignition designs with a baseline plastic ablator are characterized by long-density scale lengths $L_n \sim$ (500 to 600) μm. Understanding and controlling the impact of laser–plasma interaction (LPI) instabilities in such plasmas are key requirements of inertial confinement fusion (ICF) research. One of the instabilities driven by multiple laser beams that can exceed the instability threshold is two-plasmon decay (TPD) [1–4]. In TPD, the overlapping intense laser beams excite large-amplitude electron plasma waves in the region near the quarter-critical density (n_{qc}) surface, leading to extra laser absorption and hot-electron production. The extra absorption at n_{qc} may reduce, however, the laser intensity reaching critical density, thereby decreasing the hydroefficiency. The hot-electron generation from TPD may negatively affect target compression because of the possible preheat of the imploding shell, which must remain on a low adiabat for efficient compression. Other LPI instabilities, such as multibeam stimulated Raman scattering (SRS), can also lead to anomalous laser-energy dissipation before the n_{qc} surface and/or hot-electron generation.

To support direct-drive ICF experiments at the National Ignition Facility (NIF) in its indirect-drive beam configuration, the polar-direct-drive (PDD) concept was proposed [5]. The impact of laser parametric instabilities on the PDD implosions has been recently tested in experiments on the NIF [6]. To investigate the scaling of TPD-induced hot electrons to the laser intensity and plasma conditions, a series of planar experiments have also been conducted at the Omega Laser Facility [7]. The plasma parameters at the n_{qc} surface achieved in those experiments (as predicted by the simulations using the

[4] Author to whom any correspondence should be addressed.

code *DRACO* [8]) are summarized in Table 1. The coronal plasma parameters in the NIF PDD–ignition design developed at the Laboratory for Laser Energetics [9] are also shown. Table 1 shows that all the parameters in the previous experiments [the overlapped laser intensity (I), density scale length (L_n), and electron temperature (T_e) at the n_{qc} surface] are still ~1.5 to 2× below that in the ignition design. The empirical TPD threshold parameter $\eta = I_{14}L_{n,\mu m}/(230\ T_{e,keV})$, often used to evaluate the effect of TPD [2,10], is ~2× less than in the ignition design. Importantly, cross-beam energy transfer (CBET) reduces the laser beam energy, reaching the n_{qc} surface in current NIF implosion experiments, so that current implosion experiments do not achieve ignition-relevant coronal plasma conditions.

Table 1. Plasma parameters at the n_{qc} surface and TPD threshold in OMEGA and NIF experiments, ignition NIF PDD design, and planar targets in this paper, as predicted by *DRACO* simulations.

Parameters at n_{qc} surface	OMEGA	Current NIF PDD	Ignition NIF PDD	Planar NIF
I_L (W/cm^2)	<4 × 10^{14}	<4.5 × 10^{14}	8 to 10 × 10^{14}	6 to 8 × 10^{14}
L_n (μm)	<350 μm	<350 μm	600 μm	550 to 600 μm
T_e (keV)	<2.5 keV	<3.5 keV	5 keV	3.2 keV
η	<2.4	<2	4.7	4.5 to 5.5

In this paper, hydrodynamic simulations using the code *DRACO* are presented to show that coronal plasma conditions in the ignition PDD design can be approached in planar-target experiments on the NIF (Table 1). Since planar targets exhibit a very high absorption efficiency, CBET seeded by backscattered light represents a negligible source of losses in laser energy. It is speculated [6] that the TPD instability will be able to share decay waves most effectively along the polar axis and around the equatorial region of a PDD implosion. We present two planar-target simulations that differ by the NIF beam irradiation geometry: (1) irradiation by the NIF inner-cone beams from the south pole (23.5° and 30° incidence angle with respect to the target normal) and (2) irradiation by the outer-cone beams (44.5° and 50°). The higher-angle cones approximate irradiation conditions near the equator of a PDD implosion (where the beams come at similar higher angles), while the lower-angle cones correspond to those near the poles.

2. Proposed experimental configuration

The planar-target design is shown schematically in figure 1. The target is an ~5-mm-diam, ~500-μm-thick plastic CH disk. A 30-μm-thick, ~4-mm-diam Mo disk is buried 40 μm behind the target's front surface. The target is oriented in the equatorial plane of the NIF chamber and irradiated by a subset of NIF beams from the south pole. Hot electrons generated by the LPI instability in the coronal plasma flow into the target. Time-resolved K_α line emission and the hard x-ray bremsstrahlung from Mo are used as the main hot-electron diagnostics. The Mo thickness is equal to the range of electrons of a typical energy ~120 keV. Hot electrons that are not stopped in Mo are slowed down in the back CH, so that electron recirculation is minimal in this experiment. The front CH layer is chosen to be thick enough to avoid a burnthrough to the Mo layer, while sufficiently thin to reduce collisional energy losses of hot electrons on the way toward the Mo. For the proposed target thickness, the laser-induced shock does not reach the back of the target and the target is not accelerated when the laser pulse is on. The simulations use laser pulses with a 2-ns linear power rise from zero to the maximum value and flattop after that, with a total duration of 5.5 to 7.5 ns.

The measurements can be performed using the NIF x-ray spectrometer [11] to measure the time-resolved Mo K_α emission and the filter-fluorescer x-ray diagnostic [12] to measure the time-resolved hard x-ray emission. NIF optical spectrometers can measure the half-frequency ($\omega_L/2$) harmonic of the incident light, which is a characteristic signature of TPD, and SRS spectra.

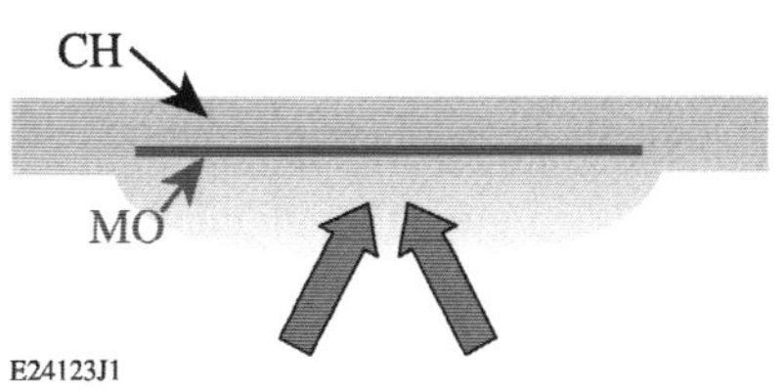

Figure 1. The proposed target design.

3. *DRACO* simulations

The simulations were performed using the Eulerian version of *DRACO* [8] in cylindrically symmetric geometry. *DRACO* includes a full 3-D laser ray trace, a flux-limited heat-conduction model (with a flux limiter $f = 0.1$), multigroup diffusive radiation transport, and *SESAME* equation of state. For the low-*Z* plastic CH ablator, the Astrophysics Opacity Table was applied; the average-ion model, which is a collisional-radiative-equilibrium model, was used for the high-*Z* Mo.

The simulations used the actual measured focal spot shapes of the indirect-drive NIF beams. The beams are used at best focus and pointed at the averaged-over-time longitudinal position of the n_{qc} surface: 320 μm in front of the target surface. The simulations are designed to have similar plasma parameters at the n_{qc} surface, with the flattop total powers of 17 TW and 15 TW, durations of 5.5 ns and 7.5 ns in the inner- (32 beams at 23.5° and 30°) and outer-cone beam (64 beams at 44.5° and 50°) simulations, respectively. The duration of the flattop used in the inner-cone beam simulation is chosen to be the longest allowable while still avoiding laser damage on the NIF.

Figure 2 presents the electron density and electron temperature in the coronal plasma at $t = 4.5$ ns in the inner-cone beam simulation. The outer-cone beam simulation predicts similar results (with a slightly higher peak temperature of 3.15 keV) and is not shown.

Figure 3 shows the time evolution of the plasma parameters at the n_{qc} surface and TPD threshold parameter at $r = 0$ in both simulations. The density scale length and electron temperature are almost stationary at $t > 2.5$ ns with $L_n = 500$ to 600 μm and $T_e \sim (3$ to $3.3)$ keV. Laser intensity slowly decreases with time, with $I = (5$ to $6.5) \times 10^{14}$ W/cm^2 in the inner-cone beam and $I = (6$ to $8) \times 10^{14}$ W/cm^2 in the outer-cone beam simulations. Notably, the empirical TPD threshold is greatly exceeded in these simulations: $\eta \sim 4$ to 5.

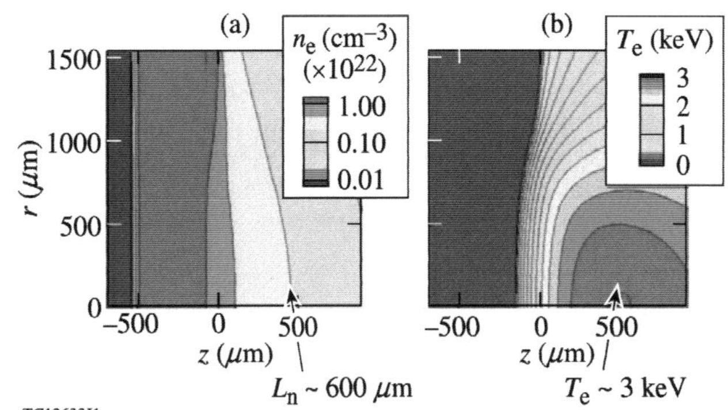

Figure 2. The electron density and electron temperature in the coronal plasma at $t = 4.5$ ns in the inner-cone beam simulation.

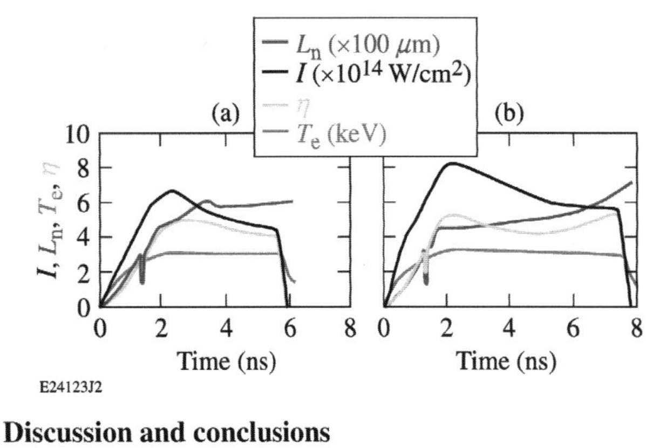

Figure 3. Time evolution of the plasma parameters at the n_{qc} surface and TPD threshold parameter at $r = 0$ in the (a) inner- and (b) outer-cone beam simulations.

4. Discussion and conclusions

Overall, a similar evolution of the plasma parameters at the n_{qc} surface is predicted by *DRACO* simulations for the inner- and outer-cone beams. Planar-target experiments on the NIF, therefore, can study the effect of a beam's incidence angle on TPD instability and hot-electron generation. Simulations of TPD using the three-dimensional laser–plasma interaction code *LPSE* [13] have been performed using the NIF irradiation geometry and plasma parameters at the n_{qc} surface predicted by *DRACO*. *LPSE* models the TPD instability in a small volume of plasma (200 μm × 30 μm × 30 μm) close to the n_{qc} surface. *LPSE* simulations confirm the onset of TPD instability when the TPD threshold (η) exceeds unity in *DRACO* simulations, for both irradiation geometries. The mechanisms of saturation of the TPD instability (such as pump depletion) are currently under implementation in *LPSE*. *LPSE* will be used to study hot-electron production and laser absorption at the nonlinear stage of TPD.

Table 1 shows that the plasma parameters at the n_{qc} surface in the present simulations are closer to the PDD-ignition design than in the OMEGA and current NIF PDD implosion experiments, with the exception of the electron temperature. Particularly, the plasma density scale length is as long as in the ignition design. A relatively low temperature is explained by higher transversal thermoconduction losses in planar experiments compared to the spherical implosions. The laser power can be further increased, provided the optics damage threshold is not exceeded. This can allow one to study TPD at higher overlapped laser intensity (equal or exceeding that in the ignition design) and electron temperature at the n_{qc} surface. The power can be increased by up to a factor of 2 in the outer-beam configuration. The power can also be increased at the expense of decreased pulse duration in the inner-beam configuration.

In conclusion, hydrodynamic simulations suggest that planar-target experiments on the NIF can be a powerful tool in the study of TPD and other LPI processes in the plasma conditions relevant to the ignition direct-drive designs. While current NIF PDD experiments suffer from CBET, which reduces the laser absorption, planar NIF experiments can provide a first look at the effect of TPD in NIF PDD implosions when CBET has been mitigated. Subsequently, the NIF planar platform can be used to study TPD mitigation strategies by using different ablator materials.

Acknowledgment
This material is based upon work supported by the Department of Energy National Nuclear Security Administration under Award Number DE-NA0001944, the University of Rochester, and the New York State Energy Research and Development Authority. The support of DOE does not constitute an endorsement by DOE of the views expressed in this article.

References

[1] Rosenbluth M N 1972 *Phys. Rev. Lett.* **29** 565
[2] Simon A, Short R W, Williams E A and Dewandre T 1983 *Phys. Fluids* **26** 3107
[3] Michel D T, Maximov A V, Short R W, Hu S X, Myatt J F, Seka W, Solodov A A, Yaakobi B and Froula D H 2012 *Phys. Rev. Lett.* **109** 155007
[4] Zhang J, Myatt J F, Short R W, Maximov A V, Vu H X, DuBois D F and Russell D A 2014 *Phys. Rev. Lett.* **113** 105001
[5] Skupsky S, Marozas J A, Craxton R S, Betti R, Collins T J B, Delettrez J A, Goncharov V N, McKenty P W, Radha P B, Boehly T R *et al.* 2004 *Phys. Plasmas* **11** 2763
[6] Hohenberger M, Radha P B, Myatt J F, LePape S, Marozas J A, Marshall F J, Michel D T, Regan S P, Seka W, Shvydky A *et al.* 2015 *Phys. Plasmas* **22** 056308
[7] Yaakobi B, Chang P-Y, Solodov A A, Stoeckl C, Edgell D H, Craxton R S, Hu S X, Myatt J F, Marshall F J, Seka W *et al.* 2012 *Phys. Plasmas* **19** 012704
[8] Radha P B, Collins T J B, Delettrez J A, Elbaz Y, Epstein R, Glebov V Y, Goncharov V N, Keck R L, Knauer J P, Marozas J A *et al.* 2005 *Phys. Plasmas* **12** 056307
[9] Collins T J B, Marozas J A, Anderson K S, Betti R, Craxton R S, Delettrez J A, Goncharov V N, Harding D R, Marshall F J, McCrory R L *et al.* 2012 *Phys. Plasmas* **19** 056308
[10] Seka W, Edgell D H, Myatt J F, Maximov A V, Short R W, Goncharov V N and Baldis H A 2009 *Phys. Plasmas* **16** 052701
[11] Regan S P, Fournier K B, Bedzyk M J, Agliata A, Ayers S L, Barrios M A, Bell P M, Bradley D K, Chen H, Emig J A *et al.* 17–21 August 2014 presented at Optics and Photonics, San Diego, CA abstract 9211-15
[12] Hohenberger M, Albert F, Palmer N E, Lee J J, Döppner T, Divol L, Dewald E L, Bachmann B, MacPhee A G, LaCaille G *et al.* 2014 *Rev. Sci. Instrum.* **85** 11D501
[13] Myatt J F, Solodov A A, Shaw J G, Maximov A V, Short R W, Zhang J, Hohenberger M, Regan S P, Rosenberg M J and Seka W 20–25 September 2015 presented at the Ninth International Conference on Inertial Fusion Sciences and Applications IFSA 2015 Seattle WA paper Fr.O.5.3

Journal of Physics: Conference Series **717** (2016) 012054 doi:10.1088/1742-6596/717/1/012054

Multi-fluid studies of plasma shocks relevant to inertial confinement fusion

B Srinivasan[1,2], **G Kagan**[2] **and C S Adams**[1]
[1]Department of Aerospace & Ocean Engineering, Virginia Tech, Blacksburg, VA 24061
[2]Theoretical Division, Los Alamos National Laboratory, Los Alamos, NM 87545

E-mail: srinbhu@vt.edu

Abstract. Results from inertial confinement fusion (ICF) experiments performed at the Omega laser facility suggest the potential role of kinetic effects in plasmas during implosion. Recent theoretical and numerical work has indicated the importance of diffusion effects in the presence of multiple ion species as well as the importance of ion viscosity. This provides the motivation to adequately develop multi-fluid plasma models capable of capturing kinetic physics including concentration diffusion and ion species separation driven by the ion concentration gradient, the ion pressure gradient, the electron and ion temperature gradients, and the electric field. Benchmarks between the newly developed code and analytical results are presented for multi-fluid plasma shocks.

1. Introduction

Recent results from inertial confinement fusion (ICF) experiments performed at the Omega laser facility suggest the potential role of kinetic effects in plasmas during implosion [1, 2, 3]. In particular, inter-ion-species diffusion can degrade yields in an ICF implosion via fuel stratification in the implosion core. This has been supported by theoretical and numerical results which have highlighted the importance of inter-species transport diffusion effects as well as viscosity in the presence of multiple ion species [4, 5, 6, 7, 8, 9, 10, 11, 12]. This motivates the development of multi-fluid plasma modeling capabilities that capture adequate kinetic physics unlike the majority of radiation-hydrodynamic codes that are traditionally used for ICF simulations. Concentration diffusion and ion species separation are driven by the ion concentration gradient, the ion pressure gradient, the electron and ion temperature gradients, and the electric field. Multi-fluid codes are a valuable tool to understand subtle physics phenomena such as fuel stratification that can significantly impact ICF implosion performance. The purpose of this paper is to benchmark the newly-developed code against a known physics-motivated solution. Diffusion is likely to affect many ICF-relevant phenomena such as mix at the interface. The benchmark demonstrated in this paper is motivated by the separation of the fuel constituents in the hot-spot by a shock wave hence, the simulations presented here focus on the structure of a plasma shock wave.

The WARPX (Washington Approximate Riemann Plasma) code [13, 14] is used for the results presented here. WARPX is a two-fluid plasma code, the two fluids being ions and electrons. A second order MUSCL [15] scheme is used for the results presented here with a third order accurate Runge-Kutta time-stepping scheme. In order to study multi-ion diffusion effects relevant to ICF, the WARPX code is updated to include multiple ion species and an electron species. The work presented in this paper focuses on a single ion and single electron species multi-fluid implementation with relevant kinetic terms included [16].

Multi-component plasmas can be modeled in two basic ways. The first is using center-of-mass dynamics along with an equation accounting for concentration evolution. The second approach involves evolving individual momentum equations for each species. Except for the viscous effect on the diffusive flux, the two approaches are equivalent [5]. Transport properties of multi-component plasmas ranging from weakly-coupled to strongly-coupled have been decribed in Refs. [8, 17, 12] respectively. The electrons are assumed to be massless. The collisional, electrostatic, two-fluid plasma equations with relevant kinetic effects are

$$\frac{\partial \rho_i}{\partial t} + \nabla \cdot (\rho_i \mathbf{u}_i) = 0 \tag{1}$$

$$\frac{\partial \rho_i \mathbf{u}_i}{\partial t} + \nabla \cdot (\rho_i \mathbf{u}_i \mathbf{u}_i + p_i \mathbf{I}) = n_i Z_i e \mathbf{E} - \nabla \cdot \mathbf{\Pi}_i - \mathbf{R} \tag{2}$$

$$\frac{\partial \epsilon_i}{\partial t} + \nabla \cdot ((\epsilon_i + p_i)\mathbf{u}_i) = \mathbf{J}_i \cdot \mathbf{E} + \nabla \cdot \mathbf{\Pi}_i \cdot \mathbf{u}_i + Q_i \tag{3}$$

$$\frac{\partial \epsilon_e}{\partial t} + \nabla \cdot ((\epsilon_e + p_e)\mathbf{u}_e) = \mathbf{J}_e \cdot \mathbf{E} - \nabla \cdot q_e + Q_e \tag{4}$$

$$n_e e \mathbf{E} = \frac{\rho_i}{m_i} e \mathbf{E} = -\nabla p_e + \mathbf{R} \tag{5}$$

with mass density ρ_i for ion species i, ion/electron number density $n_{i,e}$ such that $n_e = \sum_N Z_i n_i$ with Z_i being the charge state and N representing all ion species present. Other variables include velocity of ion species i given by $\mathbf{u}_i$, ion/electron current density $\mathbf{J}_{i,e}$, ion pressure p_i, electron pressure p_e, ion viscous stress tensor $\mathbf{\Pi}_i$, electron thermal conduction q_e, electric field $\mathbf{E}$, ion-electron thermal equilibration $Q_{i,e}$, and ion-electron friction and thermal forces $\mathbf{R}$. The ion viscosity, electron thermal conductivity, friction and thermal forces, and the thermal equilibration terms are described in Ref. [16]. The total ion energy ϵ_i includes the ion kinetic and thermal energies whereas the electron energy ϵ_e is solely the thermal energy due to the assumption of massless electrons. The electric field is determined from Ohm's law using the electron pressure gradient and the ion-electron friction and thermal forces. These equations enforce quasineutrality and ambipolar flow. A separate set of continuity, momentum, and energy equations is solved for each ion species and $\mathbf{R}$ is modified to include the summation of friction and thermal forces over all ion species present [5].

Simulations of stationary, two-fluid, collisional shocks are presented here and are compared to the semi-analytical model presented by Jaffrin and Probstein [18]. The purpose of the results presented here is two-fold. Firstly, the multi-fluid plasma model is benchmarked to known analytical results which is necessary prior to applying it to dynamic shock problems. Secondly, the two-fluid plasma shocks presented here go beyond the analytical work of Jaffrin and Probstein by including the ion-electron thermal forces that were absent in the analytical work. Any qualitative and quantitative differences are highlighted. In the presence of multiple ion species, the thermal force term has a significant impact on shock structure and species separation hence, it cannot be neglected for ICF applications.

2. Two-fluid shock results

Simulations are initialized with Hugoniot conditions for a two-fluid plasma shock at a plasma Mach number of 3. This corresponds to an ion Mach number of 4.2. The analytical results from the Jaffrin and Probstein model are reproduced here for arbitrary Mach numbers and are compared to the numerical results. For the electron collision frequency, an artificial ion-to-electron mass ratio of 100 is used to accelerate the computational time. The initial conditions

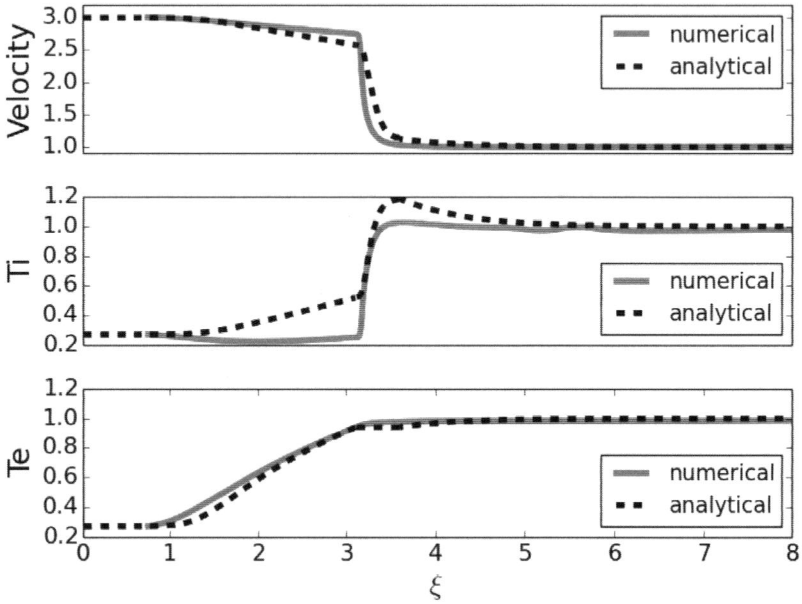

Figure 1. Comparing numerical and analytical [18] solutions of two-fluid electrostatic stationary shocks at a plasma Mach number of 3. ξ refers to the spatial scale normalized to the mean-free-path downstream of the shock. The velocities and temperatures are normalized to their respective downstream values.

are then allowed to relax to the stationary shock solutions via collisional effects. The boundary conditions for density, velocity, and temperature are maintained throughout the simulation based on the prescribed Hugoniot conditions. The simulation domain is chosen so as to provide sufficient spatial extent so that the ion and electron temperatures equilibrate at both domain boundaries. The simulations are run to approximately 100 electron collision times to allow for adequate relaxation of the shock.

Figure 1 presents results of ion velocity, ion temperature, and electron temperature along with corresponding analytical solutions for a plasma Mach 3 shock. ξ is the spatial scale normalized to the mean-free-path (λ_{mfp}) downstream of the shock. Consistent with theory, note the formation of three distinct regions in the shock: the pre-heat region, the ion viscous jump, and the post-shock region. Due to the electron thermal conductivity being much larger than the ion thermal conductivity, the electrons are heated in the pre-heat layer resulting in a shallow gradient of the electron temperature across the upstream region of the shock. The second region is the viscous jump layer that corresponds to the heating and compression of the ions which contributes to a sharp gradient of the ion temperature across a few mean-free-paths. The analytical results assume a flat electron temperature profile across this layer since there is insufficient time for ions and electrons to exchange energy in this region, but the numerical results contain a small gradient in the electron temperature across this region. In the third layer, the post-shocked region, the ion and electron temperatures equilibrate after approximately $\sqrt{m_i/m_e}\lambda_{mfp}$ due to the electron heat conduction and energy exchange between ions and electrons.

Slight quantitative differences are observed in the ion temperature profile. In the pre-heat region, the analytical solution gives rise to a shallow gradient in the ion temperature whereas the numerical solution has a fairly flat ion temperature profile across this region. In the viscous jump, the analytical solution reaches a higher ion temperature compared to the numerical results. This is due to having a shallow temperature gradient in the pre-heat region that produces a higher ion temperature just upstream of the viscous jump for the analytical solution. Corresponding differences are also seen in the ion velocity profile in the viscous jump layer. The quantitative and qualitative differences between the numerical and analytical results may be attributed to the inclusion of the ion-electron thermal forces in the numerical solutions which were not present in the analytical solutions of Jaffrin and Probstein. Furthermore, the analytical work assumes a constant Coulomb logarithm since it is a very slow function of temperature and density. In the numerical simulations, the Coulomb logarithm is computed based on the density and temperature at the given spatial location so it is allowed to vary through the shock, however, the effect of this is expected to be relatively minor.

3. Summary

Results are presented for collisional two-fluid plasma shocks and comparisons are made to theoretical results. The simulation framework has been extended to multiple ion species and simulations involving two or more ions will be presented in future publications. Due to the need to resolve the mean-free-path within the shock and the need to resolve the electron collision frequency, explicit time-stepping is used for the results presented here. The choice of plasma Mach number of 3 used here is primarily due to the large computational effort involved when using higher Mach numbers. Higher Mach numbers correspond to larger mean-free-paths and longer collision times hence, solutions requiring higher Mach numbers increase the size of the simulation domain and the run-time significantly. Furthermore, at higher Mach numbers, the fluid description of plasma shock waves breaks down [19]. Accelerated time-stepping schemes and the use of emerging computing architectures will potentially allow access to higher Mach number regimes for multi-dimensional problems which will be explored in future work.

References

[1] Casey D T, Frenje J A, Gatu Johnson M, Manuel M J E, Rinderknecht H G, Sinenian N, Séguin F H, Li C K, Petrasso R D, Radha P B, Delettrez J A, Glebov V Y, Meyerhofer D D, Sangster T C, McNabb D P, Amendt P A, Boyd R N, Rygg J R, Herrmann H W, Kim Y H and Bacher A D 2012 *Phys. Rev. Lett.* **108**(7) 075002
[2] Rinderknecht H G, Rosenberg M J, Li C K, Hoffman N M, Kagan G, Zylstra A B, Sio H, Frenje J A, Gatu Johnson M, Séguin F H, Petrasso R D, Amendt P, Bellei C, Wilks S, Delettrez J, Glebov V Y, Stoeckl C, Sangster T C, Meyerhofer D D and Nikroo A 2015 *Phys. Rev. Lett.* **114**(2) 025001
[3] Rosenberg M, Rinderknecht H, Hoffman N, Amendt P, Atzeni S, Zylstra A, Li C, Séguin F, Sio H, Johnson M G *et al.* 2014 *Physical review letters* **112** 185001
[4] Amendt P, Landen O L, Robey H F, Li C K and Petrasso R D 2010 *Phys. Rev. Lett.* **105**(11) 115005
[5] Kagan G and Tang X Z 2012 *Physics of Plasmas* **19**
[6] Srinivasan B and Tang X Z 2014 *Europhysics Letters* **107**
[7] Haines B M, Vold E L, Molvig K, Aldrich C and Rauenzahn R 2014 *Physics of Plasmas* **21** 092306
[8] Kagan G and Tang X Z 2014 *Physics Letters A* **378** 1531 – 1535 ISSN 0375-9601
[9] Kagan G and Tang X Z 2014 *Physics of Plasmas* **21** 022708
[10] Bellei C, Amendt P A, Wilks S C, Haines M G, Casey D T, Li C K, Petrasso R and Welch D R 2013 *Physics of Plasmas* **20** 012701
[11] Bellei C and Amendt P A 2014 *Phys. Rev. E* **90**(1) 013101
[12] Kagan G, Baalrud S D and Daligault J O 2016 *"Fluid description and transport properties of multi-component plasmas across coupling regimes"*
[13] Shumlak U, Lilly R, Reddell N, Sousa E and Srinivasan B 2011 *Computer Physics Communications* **182** 1767–1770
[14] Srinivasan B, Hakim A and Shumlak U 2011 *Communications in Computational Physics* **10** 183–215
[15] Colella P 1985 *SIAM Journal on Scientific and Statistical Computing* **6** 104–117
[16] Braginskii S I 1965 *Reviews of Plasma Physics* **1** 205
[17] Zhdanov V M 2002 *Plasma Physics and Controlled Fusion* **44** 2283
[18] Jaffrin M Y and Probstein R F 1964 *Physics of Fluids* **7** 1658–1674
[19] Casanova M, Larroche O and Matte J P 1991 *Physical review letters* **67** 2143

Direct heating of compressed core by ultra-intense laser

A Sunahara[1], T Johzaki[2], H Sakagami[3], H Nagatomo[4], K Mima[5], Y Abe[4], Y Arikawa[4], S Fujioka[4], H Shiraga[4], H Azechi[4], Y Mori[5], Y Sentoku[6] and Y Kitagawa[5]

[1] Institute for Laser Technology, 2-6 Yamadaoka Suita Osaka 565-0871, Japan
[2] Graduate school of Engineering, Hiroshima University,
1-4-1 Kagamiyama, Higashi-Hiroshima, 739-8527, Japan
[3] National Institute for Fusion Science, 322-6 Oroshi-cho, Toki 509-5292, Japan
[4] Institute of Laser Engineering, Osaka University,
2-6 Yamadaoka Suita, Osaka 565-0871, Japan
[5] The Graduate School for the Creation of New Photonics Industries,
Kurematsuchou, 1955-1 Nishi-ku, Hamamatsu 431-1202, Japan
[6] Department of Physics, University of Nevada, Reno
1664 N. Virginia St. Reno, NV 89557, USA

E-mail: suna@ilt.or.jp,

Abstract. We propose a new scheme for heating an imploded core in the fast-ignition scheme. In this method, a heating laser irradiates an imploded core plasma directly. The accelerated fast-ions as well as fast-electrons heat the core. Two-dimensional particle in cell (PIC) simulation confirmed that carbon C^{6+} and deuteron D^{+} ions were accelerated as well as fast electrons when ultra-intense laser irradiates the CD plasma. In order to estimate the temperature scaling of the heated core in this scheme, we conducted transport simulations in the one-dimensional conical geometry. Our results show that 5 keV of ignition temperature can be achieved at the intensity of 10^{21} W/cm^2, and 1.5 ps pulse for the compressed CD plasma with 10g/cm^3 density.

1. Introduction

In the fast ignition scheme of inertial confinement fusion (ICF), the conventional method uses laser-accelerated fast electrons only to heat the imploded core plasma. The plasma core is heated by the energy deposition of fast electrons up to the temperature required for initiating the thermonuclear ignition [1]. The final temperature of the compressed core plasma should reach at least 5 keV in order to ignite the thermonuclear burn. At the institute of Laser Engineering at Osaka University, we have been conducting the Fast Ignition Realization Experiment (FIREX) [2], demonstrating an increase of the neutron yield with increasing heating laser energy [3]. In order to ignite the thermonuclear burn with minimal heating energy, the energy coupling efficiency from heating laser to the core heating should be large. Therefore, we proposed an alternative method that included a fast ion based ignition scheme [4, 5, 6, 7, 8]. In this presentation, we propose a new approach, in which the imploded core is directly heated by an ultra-intense laser to increase the energy coupling as shown in Fig. 1. As in the conventional fast ignition, we use a cone attached shell target to prevent the interaction of heating laser with the long scale coronal plasma generated in the implosion. When the high pressure of imploded core plasma breaks up the tip of the cone, the heating laser can directly irradiate the imploded plasma expanding into the cone. As the result, the heating laser directly irradiates the imploded core. This is a straightforward way to improve the energy coupling from the laser to the core heating. Although this scheme requires the exact simultaneous occurrence of the break up of the tip and irradiation of the heating laser, it allows generation of fast-ions, as well as the fast electrons, and both fast electrons and fast-ions also contribute to heating of the core. In order to estimate the temperature scaling of the core in this scheme, we simulated the temperature of the core plasma heated by the fast-electron and ions. In Sec. II, we estimate the generation of fast ions and electrons. The heating model is described in Sec. III. In Sec. IV, we estimate the core temperature. Our summary and conclusion are in Sec. V.

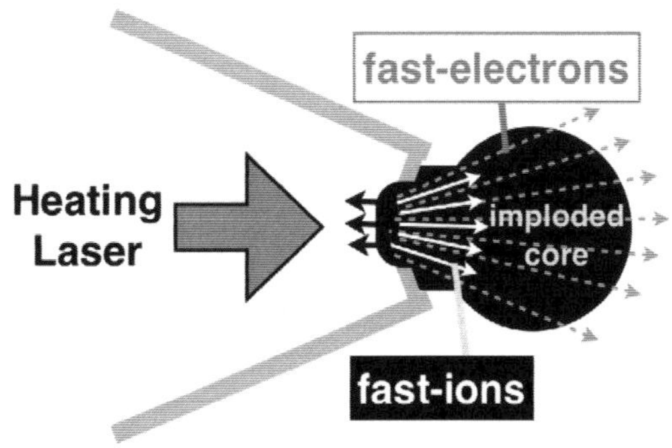

Fig.1 Schematic of our heating concept. Heating laser irradiate the imploded CD core directly through the hole of cone target. White solid lines show accelerated fast-ions. Dotted lines show fast electrons.

2. Generation of fast ions and electrons

When an ultra-intense laser irradiates an imploded plasma core, the laser interacts with the critical density plasma that is a relatively lower density part of the plasma surrounding the high-density core. In the laser-irradiated region, the fast electrons are accelerated by the ponderomotive force of the ultra-intense laser, while the fast ions are accelerated by the electric field caused by charge separation [9]. In order to estimate generation of fast-ions and electrons, we have conducted 2D Particle in Cell (PIC) simulations. In the simulation, a 10 μm thick and 10 μm wide fully ionized CD target with pre-formed plasma having the scale length of 5 μm, and the electron density n_e of $0.1n_c$ to $10n_c$ is irradiated by $\lambda_L = 1\,\mu m$ wavelength laser. Here n_c is the critical density. The laser temporal profile is a 1 ps flat pulse. The grid size is set to $\lambda_L/64$. The maximum electron density is 100 n_c. The estimated conversion efficiency from the ultra-intense laser to the fast ions and electrons at varying laser intensities are shown in Table. I. The conversion efficiency η_{C6+} of fast-carbon is limited to be of the order of 1-3% of incident laser energy, and η_{e-} of fast electrons is over 70%. However, even a small number of the fast ions can contribute the core heating due to its shorter stopping range compared to that of fast electrons.

Table. I
PIC calculated conversion efficiency of fast- electron, carbon ion, and deuteron, in function of laser intensity, respectively.

I_L (W/cm²)	η_{e^-} (%)	$\eta_{C^{6+}}$ (%)	η_{D^+} (%)
1e19	71.4	1.2	0.2
3e19	69.4	2.2	0.6
1e20	84.8	3.0	0.7

3. Core heating model

In order to estimate the temperature of the core plasma heated by both fast-ions and fast-electrons, we constructed 1D integrated heating model, based on the 1D fluid equation coupled with the transport of ions, electrons, and the radiation. For the transport of the fast electrons, we use the ray-tracing method. This treatment is valid in the condition where the change of the electrons kinetic energy ΔE during transport is negligible compared to the kinetic energy E. We assume the energy spectrum of fast electrons at the injection point as

$$f_e(E) = n_{e0}\exp(-E/T_{slope}). \quad (1)$$

Here, E and T_{slope} is the kinetic energy and the slope temperature of fast electrons, respectively. n_{eo} is the electron number density at velocity = 0. We divided energy spectrum in 100 bins of rays in the energy range of 0 to 50 MeV. For fast electrons, we calculated the drag-heating and ohmic-heating with Spitzer resistivity at each spatial mesh. For the ion heating, we inject the ion particles having the monotonic energy estimated from the momentum balance between laser and accelerated ions [9]. The ion-stopping range is estimated by Mehlhorn [10]. The ionization degree and heat capacity of CD plasma is calculated using Thomas-Fermi model. We also include the electron-ion temperature relaxation and the radiation energy loss due to the brems-starahlung from CD plasma.

4. Simulation

In order to carry out simulation in the realistic condition by our 1D model, we assume the conical plasma region in the calculation shown in Fig.2. Here, the diameter D of fast electrons and ions at the injection point are assumed to be 50μm of the current LFEX laser spot diameter [2]. Here, we use the constant plasma density with immobile bulk CD plasmas. In order to set different divergence angle of fast electron and ions, we set the angles of the conical regions to be θ_e = 35 deg (half angle) for the fast electrons and θ_i = 20 deg for the fast ions, respectively. We set the plasma size, ℓ is set to 200μm, and the initial temperature of the core plasma to $T_i(x) = T_e(x)$ = 500eV. The plasma density is set to 100g/cm^2. The wavelength of the heating laser is 1.05μm, and the laser pulse duration is 1.5 ps. The slope temperature of fast electrons T_{slope} is given by the scaling law of Wilks [9] and Haines [11] as a function of the laser intensity. The conversion efficiency from the laser to the forward-moving fast electron is set to 0.4, estimated by PIC simulation. T_{slope} may be higher than that expected by Wilks and Haines scaling, due to the existence of coronal plasma with the long density scale length. A way of avoiding this is to use a membrane.

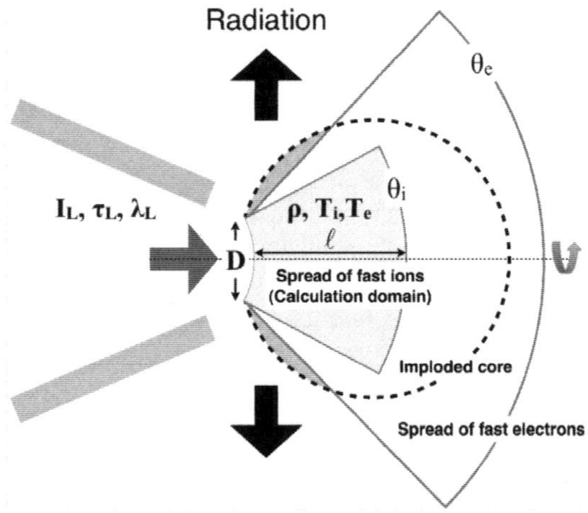

Fig. 2. Simulation condition

In Figs. 3(a) and (b), we show the temperature profiles of the heated core of 100g/cm^3, using Wilks model for the electron slope temperature shown in Fig. 3(a), and those using Haines model in Fig. 3(b), for the heating laser intensity of 10^{19}W/cm^2, 10^{20}W/cm^2, and 10^{21}W/cm^2, respectively. These profiles correspond to the timing when the ion temperature is maximum just after the relaxation time between electrons and ions of several ps. Here the energy of the heating laser at the intensity of 10^{21}W/cm^2 corresponds to 30 kJ. In Figs. 3(a) and 3(b), we see the temperature rise, which is composed of three different components: for long scale heating component is due to fast electrons. middle-range component due to D^+ ion heating, and the Short-range temperature rise due to C^{6+}, respectively. Figure 3(a) shows that the maximum temperature is 1 keV at 10^{19}W/cm^2 corresponding to the laser intensity of the current FIREX experiment [2]. Also, as the laser intensity increases, contribution from the ion heating increases. In comparison of the temperature with or without the ion heating, the condition of fast electron heating only gives the maximum ion temperature of 2 keV at the intensity of 10^{21}W/cm^2. However, the case including both fast electrons and ions at the same intensity yields the maximum ion temperature of 6 keV, showing that the contribution of the fast ions is significant. In Fig. 3(b), we observe that the contribution of fast electrons to the core heating is larger than that shown in Fig. 3(a). This is because of the difference of the slope temperature of fast electrons

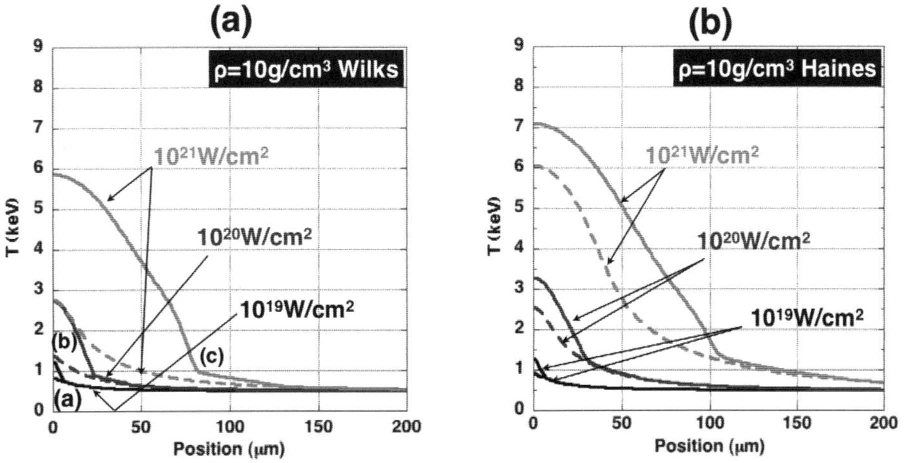

Figs. 3(a) and (b) Temperature profile of heated core with density of 10g/cm^3. (a) Wilks model is used for the electron slope temperature, (b) Haines slope temperature is used. In each figure, solid lines include both of electron and ion heating. Broken lines show electron heating (ohmic + drag) only. The heating laser intensity (a) 10^{19}W/cm^2, (b) 10^{20}W/cm^2, and (c) 10^{21}W/cm^2, respectively.

given by Wilkes and Haines. Haines scaling law shown in Fig. 3(b) gives lower slope temperatures of fast electrons. Therefore, Fig 3(b) using Haines slope temperature shows a more efficient heating, compared to that of Fig. 3(a) for the same initial condition.

5. Summary

We propose a new scheme for heating the plasma in the fast-ignition scheme. In this method, a heating laser *directly* irradiates the imploded core through the hole of a cone attached to the shell. Consequently, we can use both the fast ion and the fast electrons for heating of the core. We estimated the temperature scaling of the CD core, using our conical-1D integrated model. The contribution of fast ion is not negligible in the range of the laser intensity from 10^{19} to 10^{21} W/cm^2. Our results show that 5 keV of ignition temperature at the intensity of 10^{21} W/cm^2 and 1.5 ps pulse in CD plasma is achievable.

Acknowledgements

This work is partially conducted under the joint research of the ILE, Osaka University (FIREX-project), and JSPS KAKENHI Grant Number 24540545 and 15K05368), and Japan-US Joint Institute for Fusion Theory (JIFT) 2014, 2015.

References

[1] Tabak, M. et al., *Phys. Plasmas* **1** (1994) 1626-1634.
[2] Azechi H. et al., *Plasma Phys. Control. Fusion* **48** (2006) B267.
[3] Kodama R. et al., *Nature* **418**, (2002) 933-934.
[4] Roth M. et al., *Phys. Rev. Lett.* **86** (2001) 436.
[5] Temporal M. et al., *Phys. Plasmas.* **9** (2001) 3098.
[6] Honrubia J. J. et al., *Phys. Plasmas.* **16** (2009) 102701.
[7] Naumova N. et al., *Phys. Rev. Lett.* **102** (2009) 025002.
[8] Fernandez J. C. et al., *Nuclear Fusion* **54** (2014) 054006.
[9] Wilks S. C. et al., *Phys. Rev. Lett.* **69** (1992) 1383.
[10] Mehlhorn T. A., *J. Appl. Phys.* **52** (1981) 6522.
[11] Haines M. G. et al., *Phys. Rev. Lett.* **102** (2009) 045008.

Journal of Physics: Conference Series **717** (2016) 012056 doi:10.1088/1742-6596/717/1/012056

Plasma viscosity in spherical ICF implosion simulations

E Vold[1], A Joglekar[2], M Ortega[3], R Moll[4], D Fenn[5] and K Molvig[1]

[1]Los Alamos National Laboratory, Los Alamos, NM, USA
[2]University of Michigan, Ann Arbor, MI, USA
[3]University of New Mexico, Albuquerque, NM, USA
[4]University of California, Santa Cruz, CA, USA
[5]Florida State University, Tallahassee, FL, USA

E-mail: elv@lanl.gov

Abstract. Inertial confinement fusion (ICF) hydrodynamic codes often ignore the effects of viscosity though recent research indicates plasma viscosity and mixing by classical transport processes may have a substantial impact on implosion dynamics. A Lagrangian hydrodynamic code in one-dimensional spherical geometry with plasma viscosity and mass transport, and including a three temperature model for ions, electrons, and radiation treated in a gray radiation diffusion approximation, is used to study differences between ICF implosions with and without plasma viscosity and to examine the role of artificial viscosity in a Lagrangian implosion simulation. It was found that plasma viscosity has substantial impacts on ICF shock dynamics characterized by shock burn timing, maximum burn temperatures, fuel compression, and time history of neutron production rates. Plasma viscosity reduces the need for artificial viscosity to maintain numerical stability in the Lagrangian formulation and this study suggests that artificial viscosity may provide an unphysical stability in implosion simulations.

1. Introduction

Direct-drive inertial confinement fusion (ICF) refers to laser heating of a spherical shell that contains fusion fuel, resulting in compression and fusion [1, 2]. Species plasma viscosity is usually assumed to be negligible [1] and may be overwhelmed by artificial viscosity used in many Lagrangian-based hydrodynamic codes for numerical stability and to capture the shock discontinuities. Recently, the role of plasma viscosity in smoothing small scale fluctuations in 2D ICF simulations has been reported [3]. Plasma viscosity and diffusion have been shown in 2D simulations to play a significant role in attenuating RT and KH interfacial instabilities for ICF relevant conditions [4]. These studies suggest that plasma dissipative processes are important at micron scale lengths for ICF relevant conditions. In a study examining plasma viscosity [5], it was found that the plasma viscosity in conditions relevant to ICF is likely to be important in degrading implosion performance. Computational studies examining plasma viscosity in ICF were summarized in [6].

2. Computations

We explore the dynamic effects of viscosity on ICF implosions through the use of a one dimensional, three temperature, Lagrangian hydrodynamics model that includes plasma viscosity and a self-consistent plasma mass transport [7] for treating fuel-plastic mass mixing. The present study focuses on 'Omega-facility' ICF experiments with simplified capsules of a CH shell and a deuterium fuel region. Lagrangian hydrodynamic equations model the mixture averaged quantities, and distinguish T_i, T_e and a grey-body approximation to the radiation temperature, T_r. A radiative diffusion coefficient uses a simple fit to approximate tabular data for Rossland averaged opacity.

The viscous stress tensor in spherical coordinates simplifies in 1D with the implied symmetries, $u_\theta = u_\phi = 0$ and $\partial/\partial\theta = \partial/\partial\phi = 0$. Evaluating the divergence, $\nabla \cdot u = (1/r^2)\partial(r^2 u)/\partial r = \partial u/\partial r + 2u/r$, and for $u \equiv u_r$, the radial component of the viscous stress is

$$(\nabla : \tau)_r[1D] = \frac{1}{r^2}\frac{\partial}{\partial r}\left(r^2 \frac{4}{3}\eta_o \left(\frac{\partial u}{\partial r} - \frac{u}{r}\right)\right) + \frac{4\eta_o}{3r}\left(\frac{\partial u}{\partial r} - \frac{u}{r}\right) \tag{1}$$

These four terms are interpreted as a diffusion of velocity, a compressible correction to the diffusion, a spherical metrics correction and the compressible contribution to the spherical coordinate correction. The form for viscous dissipation in the energy equation with the assumed radial symmetry, $[u_\theta = u_\phi = \partial/\partial\theta = \partial/\partial\phi = 0]$, becomes

$$\Phi_{visc} = \eta_o \left(2\left(\left(\frac{\partial u}{\partial r}\right)^2 + 2\left(\frac{u}{r}\right)^2\right) - \frac{2}{3}\left(\frac{1}{r^2}\frac{\partial}{\partial r}(r^2 u)\right)^2\right) \tag{2}$$

The numerical scheme is based on a standard Lagrange methodology with velocity evaluated at mesh cell faces or nodes, located between thermodynamic quantity cell centers. The three temperatures are updated semi-implicitly through an iterative procedure until converged at each time step. The 3T model was first tested in a 0-D model, the hydrodynamic and plasma physics underwent extensive testing as a one temperature code and then in the 3T version. Details of the model, numerical methods, and testing are provided in [6] and references therein.

3. Results

An ICF implosion of a plastic, CH, capsule of 25 micron thickness with inner radius of 400 microns imploding on a DD fuel was selected to validate the 1D code model and algorithms. The direct-drive laser power is 22 TW with an absorption fraction (default set to 60%) delivered to the electrons in the outer CH zones over a 1 ns square pulse assumed to be representative of an Omega-scale ICF implosion. Setup details were previously described [6] and ICF test results from the code were found to be in reasonable agreement with the HELIOS ICF code [8].

A radius-time plot of the Lagrangian interface between the fuel and plastic capsule is shown in Fig. 1(a) for the inviscid and for the viscous solutions and for a third case, where the plasma viscosity is on, and the artificial viscosity is turned off (labelled VISCwQavOFF_150607) immediately after the main shock converges on center ($t \approx 1.4ns$). Figure 1 also shows surfaces on a log scale for plasma density for these three cases: (b) inviscid (artificial viscosity is ON), (c) viscous (plasma viscosity and artificial viscosity are ON), and (d) plasma viscosity is ON with artificial viscosity turned OFF after first shock convergence. In the density plots, the maximum values follow a 'ridge' showing the location of the inner-most layer of the plastic capsule as it undergoes compression in each case. While artificial viscosity, Q_{av}, is essential for Lagrangian numerical stability at early times, we find that once the entire domain becomes an ionized plasma (after the shock convergence at center) then the simulations can proceed with the plasma viscosity as the only dissipative mechanism and artificial viscosity is not needed for numerical stability. The capsule convergence for the viscous cases are less than the inviscid case (about 6% in radius or 20% in volume).

The incoming shock front is sharp in the inviscid case and remains so after convergence on center and during subsequent reflections in the fuel. The viscous shock has a less distinct density front location as it converges on center and after reflection. The time of shock arriving at center is approximately 0.1 ns later for the viscous case than the inviscid. With artificial viscosity turned off after shock convergence, there is considerably more wave structure or fluctuations in the 1D density solutions during compression. Zeroing out the artificial viscosity does not appear to greatly modify the 1D compression as seen in Fig.1(a), however, it is hypothesized that the increased fluctuations, apparent in the absence of artificial viscosity, lead to increases in entropy which will reduce implosion efficiency in the full 3D geometry.

The plots of burn weighted temperatures and the neutron production rates in Fig. 2 illustrate the key differences in shock timings between the inviscid and viscous cases. The calculation of burn weighted ion temperatures in Fig. 2 (a) shows for each case that the burn weighted ion temperatures exhibit two local maximum values, at the times of shock heating and compression burn. The maximum burn weighted ion temperature in the inviscid case is approximately 7 keV, while the viscous case maximum burn weighted ion temperature is about 3 keV, at first shock or at maximum compression. The difference between viscid and inviscid cases is large only during the first shock convergence and reflection from the center. The time averaged burn weighted temperature in either case was similar, about 2.6 keV, and comparable to experimental results and other 1D ICF simulations [8, 9]. Neutron production rates verses time in

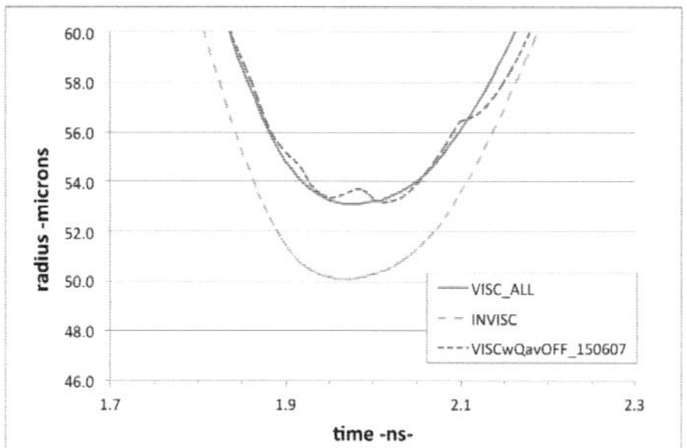

(a) fuel-capsule radius(t) - 3 cases

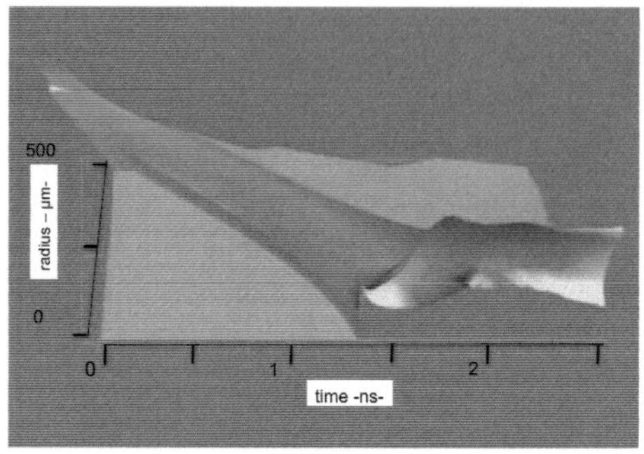

(b) density - inviscid case

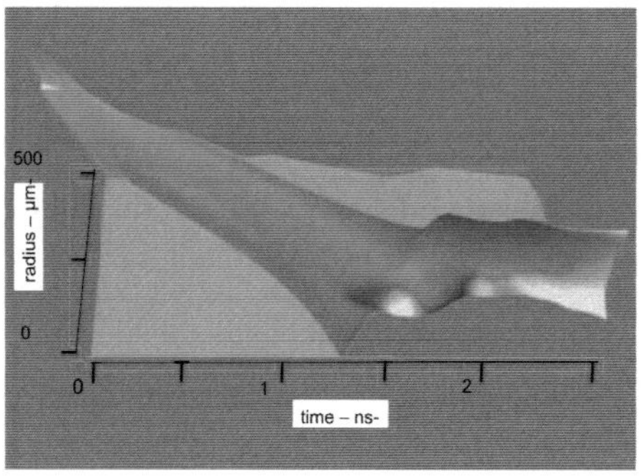

(c) density - viscous plasma case

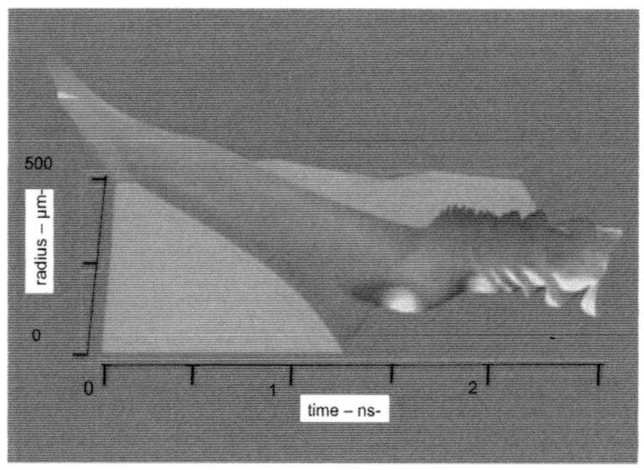

(d) density - viscosity ON - Q_{av} OFF

Figure 1: Radius of the fuel-capsule interfaces near time of maximum compression for three cases (a). Plasma density surfaces in radius-time for Lagrangian artificial viscosity only (b), with plasma viscosity added (c), and with plasma viscosity ON and artificial viscosity OFF after shock convergence (d).

Fig. 2 (b) exhibit a local maximum value at the time of compression burn. During shock heating, neutron production is about 10% of the maximum for the inviscid case and significantly less for the viscous case, attributed to the low density at shock arrival time.

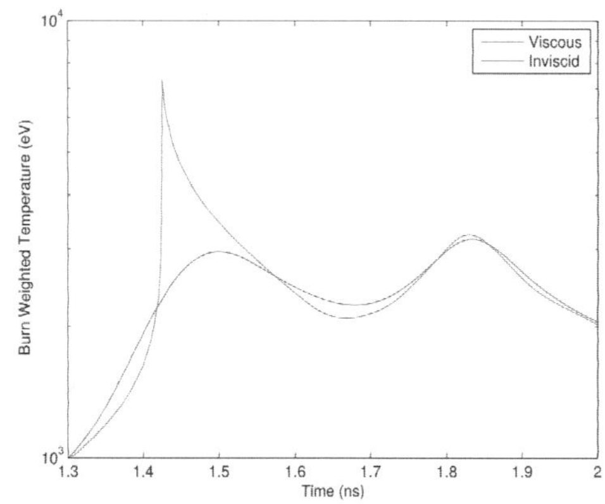

(a)

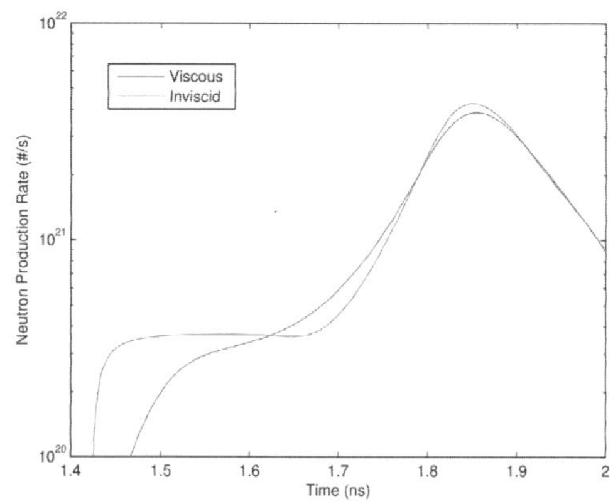

(b)

Figure 2: Burn weighted ion temperature (a) and neutron production rates (b) over time for both cases, viscid and inviscid.

4. Summary

A one dimensional staggered grid Lagrangian hydrodynamic code with a three temperature model for ions, electrons, and radiation was written to include plasma transport with viscous diffusion of momentum, viscous heating of ions, and species mass transport, in addition to thermal conductivities and temperature coupling through collisions. The code was used to examine ICF implosions typical of 'Omega-scale' direct-drive DD-CH fuel-capsules and it recovers reasonable results compared to experiments and to other codes for this class of ICF implosions. Overall, our examination of viscosity in ICF implosions, shows the plasma viscosity and viscous heating of the ions resulted in differences in maximum fuel compression, neutron production rate, and the time dependent burn weighted ion temperatures, with additional detail in [6].

We examined differences in results for simulations with or without plasma viscosity, and also in the viscous case with artificial viscosity turned off after shock convergence. The timing of the first shock on target center agrees with previous work [2, 8] and shows a timing lag ($\approx 0.1ns$) in the viscous case. The neutron production rate and the maximum and burn weighted ion temperatures are consistent with expected results [9, 10]. The ion temperatures are consistently higher for the inviscid case at first shock convergence by a factor of approximately two and it follows that the neutron production rate is augmented during shock convergence. The time averaged burn weighted ion temperatures agree within a few percent, and the neutron yields are comparable during compression between inviscid and viscous cases.

The minimum fuel radius during compression for the viscous case is greater by 6%, compared to the inviscid case. This corresponds to a 20% increase in fuel volume for the viscous case, and combined with a slightly wider mix width for the viscous case there is a 40% increase in the mix volume for the viscous case [6]. It is evident from the results that viscous effects cause a decrease in the compression.

After the first shock converges, the entire region is ionized and thus plasma viscosity is significant. After that time, the plasma viscosity and viscous dissipation eliminate the need for artificial viscosity to stabilize the numerics, and the artificial viscosity can be turned off. Results show that the subsequent solutions introduce greater shock-wave fluctuations in density (and in kinetic energy, temperatures and pressure) within the fuel and especially in the capsule material compared to cases when artificial viscosity and its corresponding energy dissipation are included. Eulerian methods can also introduce numerical dissipation and may present similar limitations when trying to accurately model physical viscosity in a shocked plasma.

Viscosity influences shock timings, fuel compression, neutron production rates, and heat fluxes especially at the time of first shock convergence. Viscous effects during the later compression phase appear to be small in this 1D example, but are likely to play a more important role in 2D and 3D as suggested in previous studies [3, 4]. It is expected that viscosity and plasma species mass transport will have a significant influence in other ICF implosions.

Acknowledgments

Los Alamos National Laboratory is operated by Los Alamos National Security, LLC for the U.S. Department of Energy NNSA under Contract No. DE-AC52-06NA25396.

References

[1] S. Atzeni, and J. Meyer-ter-Vahn. The Physics of Inertial Fusion: Beam Plasma Interaction, Hydrodynamics, Hot Dense Matter. *Clarendon Press*, Gloucestershire, UK (2004).
[2] J.M. Soures, R.L. McCrory, C.P. Verdon, A. Babushkin, R.E. Bahr, T.R. Boehly, et.al., *Phys.Plasmas* **21(2)** 2108 (1996).
[3] C. R. Weber, D. S. Clark, A. W. Cook, L. E. Busby, and H. F. Robey. *Phys. Rev. E*, **89**:053106, (2014).
[4] B.H. Haines, E.L. Vold, K. Molvig, R. Rauenzahn, and C. Aldrich. *Phys. Plasmas*, **21(9)**, 092306, DOI: 10.1063/1.4895502 (2014).
[5] R.J. Mason, R.C. Kirkpatrick, and R.J. Faehl, *Phys. Plasmas* **21** 022705 (2014), and *Phys. Plasmas* **21** 039902 (2014).
[6] E. L. Vold, A. S. Joglekar, M. I. Ortega, R. Moll, D. Fenn, K. Molvig, *Phys. Plasmas* **22** 112708; doi.org/10.1063/1.4935906 (2015).
[7] K. Molvig, A. N. Simakov, and E. L. Vold, *Phys. Plasmas* **21** 092709 (2014).
[8] I.E. Golovkin, J.J. MacFarlane, P.R. Woodruff, L.A. Welser, D.L. McCrorey, R.C. Mancini, and J.A. Koch, *Inertial Fusion Science and Applications 2003* 166-169 (2004).
[9] D. C. Wilson, C. W. Cranfill, C. Christensen, R. A. Forster, et.al., G. J. Schmid, N. Izumi, and P. Amendt, *Phys. Plasmas* **11(5)** 2723 (2004).
[10] P. Bradley, *Phys. Plasmas* **21** 062703; doi: 10.1063/1.4882247 (2014).

Journal of Physics: Conference Series **717** (2016) 012057 doi:10.1088/1742-6596/717/1/012057

Hydrodynamic instability measurements in DT-layered ICF capsules using the layered-HGR platform

C Weber[1], T Döppner[1], D Casey[1], T Bunn[1], L Carlson[2], R Dylla-Spears[1], B Kozioziemski[1], A G MacPhee[1], J Sater[1], A Nikroo[1,2], H Robey[1] and V Smalyuk[1]

[1]Lawrence Livermore National Laboratory, Livermore, California, USA
[2]General Atomics, San Diego, California, USA

E-mail: `weber30@llnl.gov`

Abstract. The first measurements of hydrodynamic instability growth at the fuel-ablator interface in an ICF implosion are reported. Previous instability measurements on the National Ignition Facility have used plastic capsules to measure ablation front Rayleigh-Taylor growth with the Hydro.-Growth Radiography (HGR) platform. These capsules substituted an additional thickness of plastic ablator material in place of the cryogenic layer of Deuterium-Tritium (DT) fuel. The present experiments are the first to include a DT ice layer, which enables measurements of the instability growth occurring at the fuel-ablator interface. Instability growth at the fuel-ablator interface is seeded differently in two independent NIF experiments. In the first case, a perturbation on the outside of the capsule feeds through and grows on the interface. Comparisons to an implosion without a fuel layer produce a measure of the fuel's modulation. In the second case, a modulation was directly machined on the inner ablator before the fuel layer was added. The measurement of growth in these two scenarios are compared to 2D rad-hydro modeling.

1. Introduction

Understanding and mitigating the hydrodynamic instability growth that occurs during the implosion of an inertial confinement fusion (ICF) capsule is important for improving performance in experiments on the National Ignition Facility (NIF). This has been the focus of an experimental campaign using face-on radiography to measure the in-flight growth of pre-imposed perturbations[1, 2]. Thus far, good agreement between simulations and the observed growth has been found at several mode numbers and using several pulse shapes. These hydro-growth radiography (HGR) measurements were made on gas-filled plastic capsules (called symmetry capsules or "symcaps"), where the thickness of the ablator is increased to compensate for the missing DT fuel mass. It is expected that ablation front growth is similar between a symcap and a capsule with a DT ice layer, but this has not been experimentally verified. Additionally, the fuel-ablator interface can be unstable and mix ablator material into the fuel. Recently the first experiments directly measuring instability growth in the presence of a DT ice layer have been conducted[3] and are reported here.

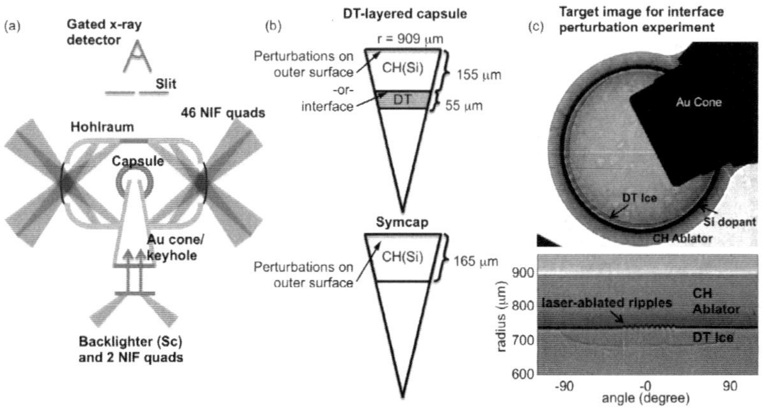

Figure 1. (a) Experimental setup of the hydro.-growth radiography platform. (b) Layered capsules are used with perturbations either on the outer surface or on the fuel-ablator interface. A symcap was also used with identical perturbations on the outer surface. (c) Target image for the experiment with an interface perturbation viewed through the laser entrance hole during cryogenic layering. The capsule, the cone, and the ice layer can be seen. Below shows the same image unrolled into radius vs. angle.

2. Experimental Setup

These experiments have perturbations on two separate surfaces. The first set of experiments uses perturbations on the outer surface of the ablator. This perturbation feeds through to the fuel-ablator interface and grows through the Rayleigh-Taylor and Richtmyer-Meshkov instabilities. The effect on the interface is identified by comparison to a separate experiment using a symcap with identical outer-surface perturbations. In the second experiment, a perturbation is machined directly on the inner ablator surface, between the DT ice and the CH plastic.

The setup for these experiments is shown in Fig. 1(a). The capsule is placed on an Au cone, centering it within the Au hohlraum. Of NIF's 192 laser beams, 184 are directed into the hohlraum, with the remaining 8 laser beams pointed towards a scandium backlighter, creating 4.5 keV x-rays. The backlighter x-rays pass down the axis of the cone, through half the shell of the capsule as it is imploding, out a high-density carbon (HDC) window in the wall of the hohlraum, through a 12 μm wide slit, and finally are recorded by the gated x-ray detector (GXD). The GXD is timed to record images at four times as the capsule is imploding.

These targets are similar to the ignition-scale design[4] used during the National Ignition Campaign (NIC), but scaled down by 0.8× in order to be hydrodynamically similar yet operate at reduced laser energy (and reduced laser damage). The laser pulse has 0.9 MJ of energy and a peak power of 230 TW. The capsule for the DT layered experiment is 909 μm in radius with a 153 μm thick CH ablator and a DT ice layer that is 55 μm thick in the field of view. The ablator contains graded silicon dopant of up to 2.5% to block high-energy x-ray preheat. The symcap contains 10 μm of additional CH to account for the absent DT mass.

The target for the inner-surface perturbation experiment is shown in the Fig. 1(c). The ice layering process was precisely controlled so that the ice would be a uniform 55 μm thickness over ±90° and have minimal grooves within the field of view. The perturbation at the interface can be seen in the radiograph. This perturbation was imposed through a new technique using laser ablation[5]. After the hole for the Au cone was cut into the capsule, a UV laser was used to remove individual spots to create a sinusoidal mode 60 pattern. This technique also left higher-mode features the size of the laser spot (mode ~500). Because of these higher mode

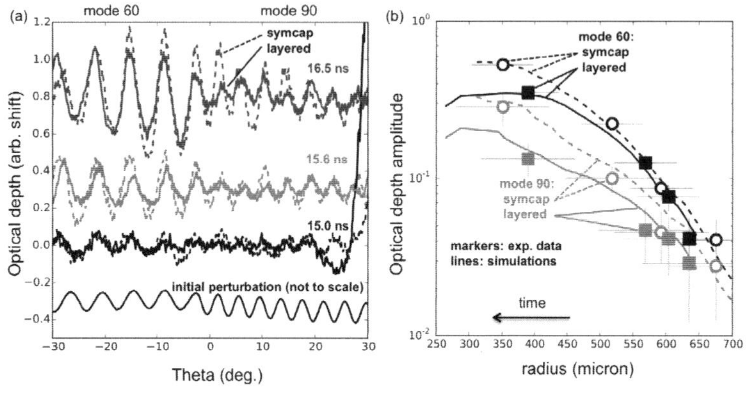

Figure 2. Outer-surface perturbation HGR experiments, comparing a layered capsule (solid) to a symcap (dashed). (a) Radiograph results converted to optical depth. The initial condition amplitude is shown at the bottom. (b) Single mode amplitudes vs. radius. Markers represent experimental data and lines represent simulations.

features and the small amplitude growth expected at the interface, only a single wavelength was included on this experiment to simplify the analysis. While high modes (~1000) are of the most interest at this interface because they reflect the in-flight Atwood number, those small scales are challenging to measure. Mode 60 was chosen because it is expected to grow to a measurable size and can be compared to the outer-surface experiment.

The outer-surface perturbations for the comparison between the symcap and the layered capsule were made using a lathe technique, similar to past HGR experiments. Side-by-side perturbations of modes 60 and 90 and amplitude of 0.7 μm are used, chosen because modes near 60 are expected to grow the fastest. These perturbations are shown at the bottom of Fig. 2(a). The perturbation amplitudes for all experiments are chosen so that the radiograph has sufficient contrast (optical depth modulation of ~0.1 during the measurement), requiring perturbations that are much larger than those that natively appear on NIF capsules.

3. Results

The results from outer-surface perturbation experiments are shown in Fig. 2(a). The images from the GXD were processed to show optical depth and the spatial dimension was converted to angle around the capsule by aligning individual perturbations. A similar amplitude is observed between the symcap (dashed) and layered capsule (solid) at the earliest time, but later in time the symcap appears to grow larger. This effect is reproduced in post-shot modeling and occurs when the ablation front growth feeds through to the interface. In the simulations, both capsules have the same ablation front perturbation amplitude and the same ρR ($\int \rho\, dr$) modulation, but the low opacity of the DT fuel to the x-ray backlighter compared to the CH results in less observed optical-depth ($\int \kappa\rho\, dr$) modulation.

The single mode amplitudes from these experiments are shown in Fig. 2(b) and compared with post-shot simulations using HYDRA[6]. Here the transfer function of the imaging system has been corrected for. Good agreement is found between the simulations and the data, with the simulation falling within the error bars of most data points. Both the experiments and the simulations show the optical depth modulation amplitude growing larger in the symcap than in the layered capsule. As mentioned previously, this is consistent with both the ablation front amplitude and the ρR amplitude being similar between the two capsules, but a lower optical

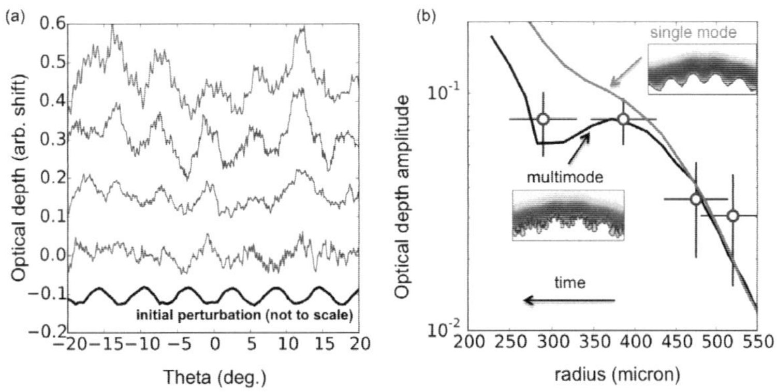

Figure 3. Inner-surface perturbation HGR experiment. (a) The initial condition is shown in black (not to scale) and the optical depth at four times is shown in blue. (b) Single mode amplitudes vs. radius. Simulations with and without the high-mode roughness are shown and images of the opacity ($\kappa\rho$) are inset.

depth modulation due to feed-through of growth to the interface and the low DT opacity.

Results from the experiment with a perturbation at the interface are shown in Fig. 3(a). The 2.2 µm amplitude mode 60 perturbation is shown at the bottom of this figure. The phase of the perturbation was expected to invert from the initial condition, but this cannot be constrained given the uncertainty in angle between the initial condition and the in-flight radiographs. In the data, moderate amplitude growth can be observed between the earliest and latest time. There are also variations in the modulations further from the center of the target (the spike at θ=10-15° appears larger than in the center). This is an effect also observed in simulations and appears to be due to the curvature of the capsule - rays going at an angle (rather than radially, as with θ=0°) have a longer path length (and optical depth).

The sinusoidal amplitude from the center three waves are shown in Fig. 3(b). Two post-shot calculations are also shown in this figure. When only mode 60 is included in the calculation and higher modes are filtered out, the simulation shows good agreement with the earlier time data but over-predicts the late-time growth. The high-mode structure, resulting from the laser-ablation process used to create the perturbation, is included in the "multimode" calculation. The growth in this case saturates later in time, with high modes coupling to lower modes. This case shows good agreement with all of the experimental data. This suggests that high-mode features may be influencing current measurements. Future experiments will seek to directly measure the growth of these small wavelengths.

Acknowledgements
This work was performed under the auspices of the U.S. Department of Energy by Lawrence Livermore National Laboratory under Contract No. DE-AC52-07NA27344.

References
[1] Raman K S *et al.* 2014 *Phys. Plasmas* **21** 072710
[2] Smalyuk V A *et al.* 2014 *Phys. Rev. Lett.* **112** 185003
[3] Weber C R *et al.* 2016 *Phys. Rev. Lett.* in review
[4] Haan S W *et al.* 2011 *Phys. Plasmas* **18** 051001
[5] Carlson L *et al.* 2015 *Fusion Sci. Tech.* **67**
[6] Marinak M M *et al.* 2001 *Phys. Plasmas* **8** 2275

9th International Conference on Inertial Fusion Sciences and Applications (IFSA 2015) IOP Publishing
Journal of Physics: Conference Series **717** (2016) 012058 doi:10.1088/1742-6596/717/1/012058

X-ray drive of beryllium capsule implosions at the National Ignition Facility

D C Wilson[1], S A Yi[1], A N Simakov[1], J L Kline[1], G A Kyrala[1], E L Dewald[2], R Tommasini[2], J E Ralph[2], R E Olson[1], D J Strozzi[2], P M Celliers[2], M B Schneider[2], A G MacPhee[2], A B Zylstra[1], D A Callahan[2], O A Hurricane[2], J L Milovich[2], D E Hinkel[2], J R Rygg[2], H G Rinderknecht[2], H Sio[3], T S Perry[1] and S Batha[1]

[1]Los Alamos National Laboratory, Los Alamos, NM, USA
[2]Lawrence Livermore National Laboratory, Livermore, CA, USA
[3]Massachusetts Institute of Technology, Cambridge, MA, USA
dcw@lanl.gov

Abstract. National Ignition Facility experiments with beryllium capsules have followed a path begun with "high-foot" plastic capsule implosions. Three shock timing keyhole targets, one symmetry capsule, a streaked backlit capsule, and a 2D backlit capsule were fielded before the DT layered shot. After backscatter subtraction, laser drive degradation is needed to match observed X-ray drives. VISAR measurements determined drive degradation for the picket, trough, and second pulse. Time dependence of the total Dante flux reflects degradation of the of the third laser pulse. The same drive degradation that matches Dante data for three beryllium shots matches Dante and bangtimes for plastic shots N130501 and N130812. In the picket of both Be and CH hohlraums, calculations over-estimate the x-ray flux > 1.8 keV by ~100X, while calculating the total flux correctly. In beryllium calculations these X-rays cause an early expansion of the beryllium/fuel interface at ~3 km/s. VISAR measurements gave only ~0.3 km/s. The X-ray drive on the Be DT capsule was further degraded by an unplanned decrease of 9% in the total picket flux. This small change caused the fuel adiabat to rise from 1.8 to 2.3. The first NIF beryllium DT implosion achieved 29% of calculated yield, compared to CH capsules with 68% and 21%.

1. Introduction

Beryllium ablators [1] were chosen for the first ignition target design for the National Ignition Facility [2], but the challenges of handling beryllium contamination in the target chamber and diagnostics led to selecting glow discharge polymer (GDP). The first implosion of a DT filled beryllium capsule at NIF required a sequence of shots to understand the interaction of the beryllium capsule with the hohlraum and to experimentally tune the timing and velocity of shocks within the capsule. The first beryllium DT capsule experiment was designed to be similar to the first high-foot CH DT capsule (N130501). Both were fielded in a now standard, 5750 μm diameter Au hohlraum, supported by a 45 nm thick tent, and filled through a similar glass tube glued to a 10 μm dia. hole in the capsule. This similarity in laser drive, hohlraum dimensions, and capsule parameters allows us to compare these two experiments, and the performance of beryllium vs. plastic capsules in general.

Journal of Physics: Conference Series **717** (2016) 012058 doi:10.1088/1742-6596/717/1/012058

2. Drive reduction for keyhole, 1D and 2D convergent ablator targets

Our HYDRA [3] calculations began with subtracting the backscattered energy, assuming it never reached the hohlraum. The plastic capsules have ~ 2% less backscatter than beryllium: less SRS in the 30° beams and no SBS in the 50° beams. Keyhole targets have a gold cone penetrating the hohlraum wall and into a capsule filled with liquid D_2. A VISAR (Velocity Interferometer System for Any Reflector) reflected off the moving shock front to measure its velocity versus time [5] at both the equator and pole. Ad hoc reduction to laser power and energy transfer between the laser cones is needed to bring the calculations into agreement with experiment. Implosions are backlighted with copper x-rays to give streaked radius vs. time (1DconA) [6] or time-gated images of transmission (2DconA)[7]. Measurements of the time of peak X-ray emission at the south pole [8] (bangtime) further constrain power reduction. Jones *et al.* [4] introduced time dependent multipliers on the laser power incident into the hohlraum as a calculational tool to match better the measured shock speeds, shock merger times, peak implosion velocity, capsule bangtime, and other experimental data. Experiments have shown that the radiation flux within the hohlraum is actually depleted [9]. The hohlraum physics in the calculations [10] is thought to be in error. Figure 1 shows the drive degradation derived from keyhole, 1D and 2DconA targets and by matching Dante flux measurements. The 0.3 value at 12ns is required to explain the low VISAR 3rd shock velocity and the late bangtime with 0.9 MJ.

N150222 was a 1DconA target, using ~8.3 keV emission from a copper backlighter foil to measure in transmission the radius vs. time of the imploding Be/Cu shell. From this the imploding velocity of the unablated mass is derived [6]. The observed velocities those calculated for the un-ablated mass agree within experimental error bars. Table I summarizes the results.

Table I- Post Shot Simulations of N150222 vs. Observations

	Bangtime (ns)	Yield (e+11)	X-ray fwhm (ps)	DD Tion (keV)	ConA peak V (km/s)	P0 (um)	P2/P0 (%)	Mass Remaining (%)
Calculated	**17.437**	**6.28**	**300**	**2.17**	**285**	**67.9**	**-10.3**	**11**
Data	**17.44±0.02**	**2.31±0.16**	**377±20**	**2.11±0.2**	**282±9**	**~60**	**-16**	**10±1.5**

N150420 was a 2DconA target, again using a copper backlighter to image the implosion within about 1ns of peak X-ray emission (bangtime) [7]. From these time-dependent images the radii of minimum transmission, and of the points of maximum slope in the transmission are derived. The former corresponds roughly to the radius of the inner copper doped beryllium layer, and the latter to its outer radius. The calculated and measured points differ by an ~30ps shift of the measured points to earlier times, within the 80ps uncertainty. The velocity derived from the measured minimum radii is 285 ± 20 um/ns compared with 270 um/ns calculated 0.62ns before bangtime. As with the 1DconA, N150222, the observed implosion is slightly earlier than calculated. A calculated velocity for the maximum slope point is 245 km/s compared to the measured 223 km/s.

2.1. Dante Flux modelling in the second and third laser pulses

Dante [11], an 18 channel soft x-ray spectrometer, uses filters, mirrors, and X-ray diodes to measure the absolute flux emitted out the hohlraum laser entrance hole between 50 eV and 20 keV. The power and spectrum are sensitive to laser drive degradation. Figure 2 compares calculated and observed Dante spectrally integrated flux, and in figure 3 the flux >1.8 keV for the 1DConA. The 2DConA and the DT capsule also agree with calculations.

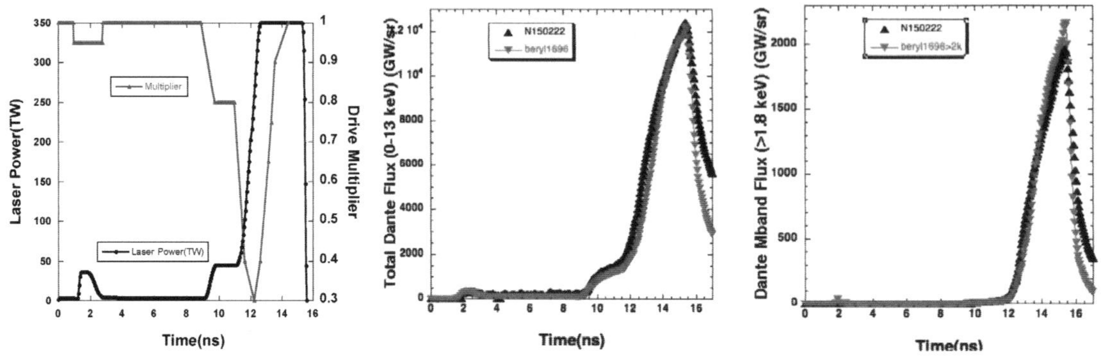

Figures 1. Drive degradation and laser pulse; 2. Dante total flux; and 3. Dante flux >1.8 keV

Since calculated drive degradation is thought to be a consequence of inadequate hohlraum, not capsule, modeling, the same degradation could apply to hohlraums with plastic (CH) as well as with beryllium capsules. Applying the degradation in figure 1 in the third pulse to CH capsules also brings both total and m-band fluxes into agreement with the observations.

The Dante instrument is not spatially resolved, but integrates all the X-ray emission within its field of view. Typically this emission comes from within the LEH, but the effective aperture of the LEH to X-rays changes with time. Images of the LEH with the Static X-ray Imager (SXI) [12] can measure the amount of this closure. The calculated image has a slightly smaller diameter (2750 vs. 2950μm) than the experiment. Thus the hohlraum brightness (flux/unit area) is about 13% less than calculated from the spatially integrated flux, or, in a time integrated sense, the drive degradation is ~13% too high.

2.2. *Laser picket modelling*

One of the deficiencies in HYDRA hohlraum modeling is the over-prediction of M-band flux in the picket of the laser pulse. A drive multiplier in the picket is thought to represent errors in the equation of state. It is set to obtain the 1st shock velocity observed by the VISAR: 0.93 for beryllium capsules and 0.85 for CH. The M-band is then over-predicted by about a factor of 100 compared to Dante data for both CH and Be capsules, suggesting an error in hohlraum modeling. HYDRA calculates that a thin hohlraum wall layer is being heated to high temperature at low density. The data suggest that softer X-rays are being emitted at lower temperatures. The total Dante flux is correctly calculated. We don't know how to correct this error. For beryllium capsules this over-prediction has a measureable consequence in VISAR measurements. Unlike CH, which transmits VISAR light, the metallic beryllium/DD surface reflects the VISAR. M-band X-rays can penetrate through the ablator, preferentially heating the copper doped layers that expand. HYDRA calculations predict an interface velocity of 2-4 km/s, but VISAR measurements show only 0.2-0.3 km/s, reflecting about 1/100 the kinetic energy.

3. Sensitivity to the laser picket

With the drive degradation determined from VISAR, 1DconA and 2DconA experiments, the hohlraum radiation drive could be calculated for the DT layered beryllium shot N150617. However a small but important difference occurred between the intended laser picket and that delivered by the NIF laser. The delivered energy in the picket was 9% low (within NIF specifications), but the second and third pulses were as requested. This caused a slower first shock in the beryllium. The second and third shocks then over took the first shock 20 μm inside the DT ice/gas interface, instead of just inside the gas. This increased the entropy of some of the fuel, changing the total fuel adiabat from an intended 1.9 to 2.3 times the Fermi degenerate. The final ρR was decreased. The expected down-scattered ratio

(DSR) dropped from 5% to 4%, and the calculated yield was reduced a factor of ~2. This unfortunate picket deficit also affected the first CH DT layered capsule as well, N130501, allowing us to still compare the relative performance of beryllium to plastic.

4. Conclusions

Using the discussed drive degradation and saturated cross beam energy transfer [13], we calculated the performance of both beryllium and plastic DT layered implosions. In the absence of heating from alpha deposition asymmetric radiation drive alone accounts approximately a factor of two degradation. This leads to less alpha particle deposition and less heating, which accounts for N130812 not igniting. Driven by 1.37 MJ (vs. 1.27 and 1.69 for CH), the performance of the beryllium ablator DT layered capsule gave a 3.8% measured down-scattered ratio (vs. 3.1 and 4.1%), a lower implosion velocity 285km/s (vs. 310 and 350), a lower inferred un-ablated mass, 200 μg (vs. 280 and 420), and an intermediate 29% ratio of observed yield/ calculated 2D asymmetric (vs. 68 and 21%). Within the uncertain degradations from drive and fabrication asymmetry, such as M1 asymmetry in the DT ice, the beryllium and plastic implosions are similar. The laser-hohlraum interaction is only slightly different with Be than with CH capsules. Drive degradation for the 3rd pulse is similar, perhaps identical for Be and CH. Similar saturation of crossbeam energy transfer is needed for both Be and CH capsule hohlraums to replicate observed x-ray and neutron images. In summary when driven similarly the Be capsules performed much like a CH capsules. Performance degradation for both seems dominated by drive and target fabrication asymmetries. The advantages of beryllium are its low opacity, higher mass ablation rate, and greater stability to high frequency surface perturbations. These advantages were not emphasized in the first experiments comparing beryllium and CH.

Acknowledgements

Execution of NIF experiments requires the participation and collaboration of large teams of individuals. We thank all of those who make the NIF laser a success: the experimentalists who created and fielded the diagnostics, and the target fabrication specialists who performed the myriad tasks involved in designing, building, and fielding the complex holraums and beryllium capsules in our experiments. This work was performed at LANL and LLNL, funded by the US Department of Energy.

References

[1] Krauser W J Hoffman N M Wilson D C Wilde B H Varnum W S Harris D B Swenson F J Bradley P A Haan S W, Pollaine S M *et al.* 1996 *Phys. Plasmas* **3** 2084
[2] Moses E I, Boyd R N, Remington B A Keane C J and Al-Ayat R 2009 *Phys. Plasmas* **16** 041006
[3] Marinak M Kerbel G D Gentile N A Gentile N A Jones O Munro D Pollaine S Dittrich T R and Haan S W 2001 *Phys. Plasmas* **8**(5) 2275-2280
[4] Jones O S Cerjan C J Marinak M M *et al.*, 2012 *Phys. Plasmas* **19** 056315
[5] Robey H F Boehly T R Olson R E *et al.* 2010 *Phys. Plasmas* **17** 012703
[6] Hicks D G Spears B K Braun D G *et al.* 2010 *Rev. Sci. Instrum.* **81** 10E304
[7] Rygg J R Jones O S Field J E *et al.* 2014 *Phys. Rev. Lett.* 2014 **112** 195001
[8] Edgell D H Bradley D K Bond E J *et al.* 2012 *Rev. Sci Instrum.* **83** 10E119
[9] MacLaren S A Schneider M B Widmann K *et al.* 2014 *Phys. Rev. Lett.* **112** 105003
[10] Rosen M D Scott H A Hinkel D E *et al.* 2011 *High Energy Density Physics* **7** (3) 180-190
[11] Kornblum H N Kauffman R L and Smith J A 1986 *Rev. Sci. Instrum.* **57** 2179 also Dewald E L Campbell K M Turner R E *et al.* 2004 *Rev. Sci. Instrum.* **75** 3759 and Kline J L Widmann K Warrick A *et al.* 2010 *Rev. Sci. Instrum.* **81**, 10E321
[12] Schneider M B Meezan N B Alvarez S S *et al.* 2012 *Rev. Sci. Instrum.* **83**(10) 10E525
[13] Michel P Divol L Williams E A *et al.* 2009 Phys. Rev. Lett. **102** 025004

Increasing shot and data collection rates of the Shock/Shear experiment at the National Ignition Facility

F. W. Doss, K. A. Flippo, D. Capelli, T. Cardenas, B. DeVolder, J. Kline, L. Kot, S. Kurien, E. Loomis, E. C. Merritt, T. Perry, D. Schmidt, and C. Di Stefano

Los Alamos National Laboratory, Los Alamos, NM 87545

E-mail: `fdoss@lanl.gov`

Abstract. Updates to the Los Alamos laser-driven high-energy-density Shock/Shear mixing-layer experiment are reported, which have collectively increased the platform's shot and data acquisition rates. The strategies employed have included a move from two-strip to four-strip imagers (allowing four times to be recorded per shot instead of two), the implementation of physics-informed rules of engagements allowing for the maximum flexibility in a shot's total energy and symmetry performance, and splitting the laser's main drive pulse from a monolithic single pulse equal to all beams into a triply-segmented pulse which minimizes optics damage.

Figure 1. A LANL Shock/Shear target for NIF, with two indirect-drive halfraums on top and bottom and a BABL to the left. The visible portion of the shock tube is 5 mm tall.

1. Introduction

The Los Alamos National Laboratory (LANL) high-energy-density Shock/Shear experiment has used the National Ignition Facility (NIF) [1] at Lawrence Livermore National Laboratory (LLNL) to obtain shear-driven mixing data at high Atwood and Reynolds numbers. The Shock/Shear platform (Figure 1) is a shock tube filled by foam and separated by a dense material (typically aluminum) driven at each end by a halfraum, which launches shocks and drives hydrodynamic flow into the shock tube. When the shocks cross in the center of the tube, extremely strong shear is created, and the development of a temporally growing mixing layer is subsequently observed.

Journal of Physics: Conference Series **717** (2016) 012059 doi:10.1088/1742-6596/717/1/012059

The Big Area BackLighter (BABL) radiography source [2], a foil mounted from the halfraums covering one side of the experiment, is then driven to provide x-ray illumination and diagnose the experiment. The platform development campaign covering the first five shots, along with many other details of the design and analysis methods discussed below, is fully documented in Doss et al [3].

Four major innovations have contributed to an increased rate of data acquisition:

- Moving from a two-strip radiography scheme to a four-strip, doubling the number of mixing-layer images which can be taken in a single shot.
- The implementation of physics-based constraints on drive energy and symmetry.
- Breaking up the main drive pulse into a three-part segmented drive to reduce the impact on laser optics of each shot.

Together with the facility's efforts to increase the overall NIF shot rate [4] by a factor of 1.5 from 2014 to 2015, the rate of usable data acquisition of the Shock/Shear experiment was increased by these improvements by a factor of ~6. Below we will summarize the methods and impact of increasing the platform data output with these methods. In addition to increasing the quantity of radiography, the quality of data taken was also improved by a change from plastic to beryllium tubes for superior x-ray transmissibility. Improvements to target fabrication which allowed the move to beryllium tubes have been documented in Capelli et al. [5]. Statistics on the shots taken so far using the platform will also be given.

2. Four-strip radiography

During its initial platform development campaign, the Shock/Shear experiment could obtain two frames of data per experiment. The hGXD, a class of Gated X-ray Detector [6] which images onto film [7], can be outfitted with gated micro-channel plate stages which split the image into either two or four images taken at separate times.

For the platform development, shock timing and drive characterization required a full view of the target geometry. In addition to grids which can be used to determine magnification and, with preshot target metrology, absolute positioning of features with respect to the target ends, location-orienting notch fiducials were placed on the outer edge of the target volume and the shock's interaction with the wall was used to locate the shock. The radiography produced by the BABL used in this experiment was very spatially flat in intensity, which allowed this method to accurately locate the shocks.

The tradeoff in moving to the four-strip camera is that it reduces the imageable area to only that around the mixing layer. The drive conditions therefore had to be accurately characterized before this switch. Figure 2 shows an example of the reduced lateral span of the four-strip imaging scheme compared to the 2-strip shot N140923-2. The shock tube walls, and correspondingly the opportunity to register the shock location, are cut out. However, as can be seen in a subsequent image from shot N150415, the grid is positioned such that it is now imaged in an adjacent strip and can still be used in the four-strip configuration.

To fully take advantage of the additional data which can be taken in the four-strip configuration, the radiography scheme also needed to be extended to create a longer usable pulse of x-rays. This work is described in detail in Kirk Flippo's paper (this volume) [8].

3. Physics-based limits

As shots continued, it became necessary to define rules of engagement for each shot which allowed the maximum permissible variation in drive conditions while still maintaining physics fidelity. Compared to many other experiments on the NIF, the drive requirements in total energy are relatively loose for these shots, and have been quantified. Design calculations performed using the LANL hydrocode RAGE [9] and subsequently with the LLNL code KULL [10] identified a

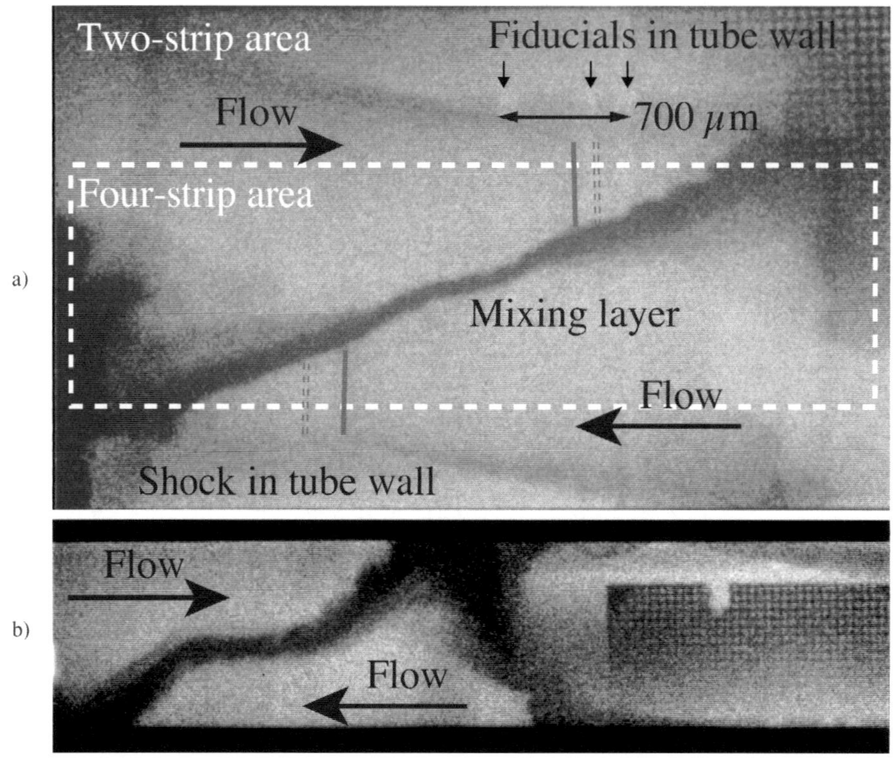

Figure 2. (a) Data from a single strip of Shot N140923-2 at 23 ns. Analysis of the radiography uses the machined fiducials and refracted components of shocks in the tube walls to determine the inferred shock locations (shown in red). The white rectangle shows the equivalent usable imaging area allowed by a single strip of the four-strip imager instead of the two-strip, discussed in Section 2. The dashed red lines show the shock locations measured in the cognate shot N141017 which fired with 611 kJ compared to this shot's 569 kJ (see Section 3 for discussion). Underneath, (b) shows a single strip from a four-strip shot (N150415 at 29.5 ns).

10% end-to-end drive asymmetry as sufficient to substantially modify the mixing dynamics at the center of the tube. This level of 10% is taken as the limiting factor past which the data has been corrupted and is likely unusable for precision mix measurements.

Figure 3 shows a collection of data from shots taken through January 2016. With the exception of several shots in which entire beams, quads (clusters of four beams), or bundles (pairs of quads linked between the top and bottom of the chamber) were lost, the data cluster near the design energy of 602 kJ (607 ± 7 kJ). Even in the cases of the experiments in which significant beam energy was lost (including the six named shots to the left of the main drive cluster), drive symmetry can be preserved. It is typically known leading up to the experiment if entire quads will be unable to fire; our rules of engagement, informed by the simulated limits at which the experiment can function, allow us to then drop a corresponding amount of energy from the opposite side. This has allowed us to keep drive symmetry between the top and the bottom halfraums to within 5% over all shots (providing a factor-of-two safety margin to the 10% limit at which it affects the mixing layer) at the cost of additional energy loss (but also keeping all shots within, to date, 8% of the nominal drive energy). The efficient use of these limits has allowed us to field experiments rapidly.

The second-to-left-most shot in Figure 3, shot N140923-2, was the last of the named "low-

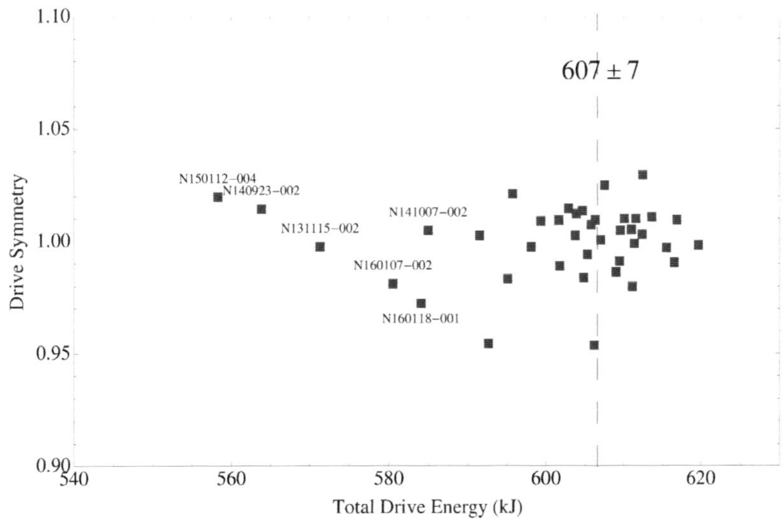

Figure 3. Drive energies in kilojoules for all fully-driven Shock/Shear NIF experiments through January 2016. The vertical axis shows drive symmetry between the top and bottom halfraums. The main cluster of shots is centered around 607 kJ with a spread of ± 7 kJ. Six shots in which the drive varied strongly due to entire dropped beams have been named explicitly to the left.

drive" shots to be imaged with the two-strip instead of the four-strip diagnostic discussed in Section 2 and the first shot designed to field the triply-segmented pulse described in Section 4. It is therefore an interesting shot for comparing the impact both of total energy and of the other changes discussed here. A full-power cognate shot, in which only the drive was varied, is N141017, which fired with 611 kJ. The shocks in N141017 were located (measured from the bottom surface of the shock tube) at 3073±13 μm and 1860±60 for the top shock (which travels from 0 to 5 mm) and bottom shock (which travels in the opposite direction), respectively. In N140923-2, driven with 569 kJ, for data taken at the same time, we instead find positions of 2952±44 μm and 2067±21 μm, respectively. This indicates a delay in shock position of 100-200 μm, or 2-4% of the length of the tube. This shock displacement is illustrated in Figure 2. For drive variations at this level, the final data analysis to be compared with simulation output should be expected to require comparable scaling of the shock and flow velocities. However, this was an extreme case, with a 7% discrepancy in delivered power; the vast majority of shots, which fell within 2% of the design energy, should exhibit no noticeable effects from their small changes in laser energy.

4. Pulse splitting

A major component in the scheduling and fielding of a NIF campaign is the burden of the shots to the NIF optics recycling program [11, 12]. The accumulation of optics damage is sensitive to the structure of the laser pulse [13], and the laser-optics modeling code LPOM [14] predicted that we could lower the costs of the laser drive with no impact (ideally) to its performance by using the modified drive pulse shown in Figure 4. This allowed the shot rate of the Shock/Shear experiment to nearly triple, by fielding more shots per given allocation of optics usage.

In Figure 4, the original single-shaped laser pulse which used 60 beams to drive each halfraum (available in [3]) has been split into three shorter component pulses, using 32, 16, and then 12 beams which sum to recreate the original platform drive pulse (shown in black). In order to maintain the same total power entering the halfraum, the power of each component beam was

Journal of Physics: Conference Series **717** (2016) 012059 doi:10.1088/1742-6596/717/1/012059

increased, but this is more than compensated by the shortening of pulse duration during which the damage accumulates.

The tradeoff here is that optics damage has been exchanged for robustness against drive fluctuations. Previously, in the 60-beam pulse in which each beam participated equally, the loss of a four-beam quad would lead to an even 6% loss of power on that side over the entire pulse. With the split pulse, the loss of the same four beams can be as much as 30% of the power of the drive group. Worse, if it the beams are lost from the second or third group, what was originally a flat or slowly decreasing beam profile becomes a modulated one possibly containing local increases, which can affect the mixing layer data if the imaging time of the experiment is sufficiently late. Without modifying the rules of engagement, based on the energy delivery statistics reported in Figure 3 we could expect loss of data due to loss of drive energy of 5-10%. Consequently, the triply-segmented pulse, while allowing for an increased shot rate, requires additional specification to the total energy and symmetry rules above; typically a quad loss is only permitted on the first, earliest group. For images sufficiently early in time (before 30 ns) it was determined by simulation that a single quad may be lost from the last pulse and the loss of energy will not propagate to the layer before the experiment ends.

The as-shot laser profiles for two shots are shown in Figure 4, averaging the top and bottom hohlraums together. The lower of the two curves is the record of shot N140923-2, the drive characterization shot for the triply-segmented pulse, which we have seen in Figure 2 and studied above as the lowest total energy shot (569 kJ) which was fired with the two-strip GXD. As can be seen, a significant drop occurred in the first pulse on each side, leading to a 7% overall loss to the total integrated energy, reflected in the delayed shock locations discussed above. We found however that the drive was consistently described just by the effects of this lowered energy, with no indication of transients or anomalous hydrodynamics caused by handing off one group of beams to the next.

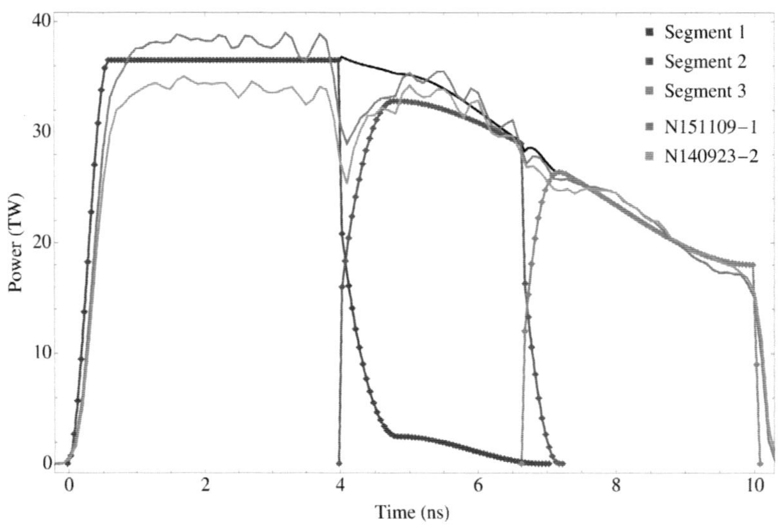

Figure 4. The three-stepped pulse used by the Shock/Shear experiments in 2015. The blue, purple, and tan dots show the drive pulse segments with 32, 16, and 12 beams respectively contributing to each segment on each side of the target. The black line shows the sum of the segments. The red curve shows the as-shot laser drive for a shot which performed well (N151109-1) and the orange curve shows a shot which underperformed (N140923-2). The total integrated design power of this pulse is 301 kJ from each side.

5. Summary

Following the platform development and scaling study reported in Doss et al [3], modifications to the Shock/Shear platform were made to increase the campaign's shot and data throughput. These included moving from two-strip to four-strip radiography (at the penalty of losing additional drive calibration information with each shot), implementing specific pyhsics-informed rules of engagement (obtaining data which may require simulation-informed scaling in total energy to be properly folded into the complete shot sequence), and fielding a triply-segmented drive pulse to mitigate the campaign's optics damage budget (at the cost of increased risk to beam losses from the later segments). Through these improvements, the campaign has increased its data-taking rate by a factor of ~6 (three times as many shots for a given assignment of laser time, acquiring twice as much data per shot, reduced slightly by the need for tighter rules of engagement), which will be important to complete the high-Atwood and roughness scaling studies in progress in a timely fashion.

Acknowledgments

This work was supported by the U.S. Department of Energy and executed by Los Alamos National Laboratory under Contract DE-AC52-06NA25396. Experiments performed on the National Ignition Facility additionally reflect facility development and operations performed under the U.S. Department of Energy by Lawrence Livermore National Laboratory under Contract DE-AC52-07NA27344. This manuscript is numbered LAUR-16-20676.

References

[1] Moses E I and Wuest C R 2003 *Fusion Science and Technology* **43** 420–427
[2] Flippo K A, Kline J L, Doss F W, Loomis E N, Emerich M, Devolder B, Murphy T J, Fournier K B, Kalantar D H, Merritt E, Perry T S, Tregillis I, Welser-Sherrill L and Fincke J R 2014 *Review of Scientific Instruments* **85** 093501
[3] Doss F W, Kline J L, Flippo K A, Perry T S, DeVolder B G, Tregillis I, Loomis E N, Merritt E C, Murphy T J, Welser-Sherrill L and Fincke J R 2015 *Physics of Plasmas* **22** 056303
[4] Brunton G, Bowers G, Conder A, Nicola J D, Nicola P D, Fedorov M, Fishler B, Fleming R, Lau G, Kalantar D, Mathisen D, Miller-Kamm V, Pacheu V, Paul M, Reed R, Rouse J, Sanchez R, Shaw M, Stout E, Tang Y, Weaver S and Wilson R 2016 *Proceedings of ICALEPCS 2015, Melbourne, Australia* (Joint Accelerator Conferences Website) p MOD3O03
[5] Capelli D, Schmidt D W, Cardenas T, Rivera G, Randolph R B, Fierro F, Merritt E C, Flippo K A, Doss F W and Kline J L 2016 *Fusion Sciences and Technology* **in press**
[6] Oertel J A, Aragonez R, Archuleta T, Barnes C, Casper L, Fatherley V, Heinrichs T, King R, Landers D, Lopez F, Sanchez P, Sandoval G, Schrank L, Walsh P, Bell P, Brown M, Costa R, Holder J, Montelongo S and Pederson N 2006 *Review of Scientific Instruments* **77** 10E308
[7] Glenn S, Bell P M, Benedetti L R, Bradley D K, Celeste J, Heeter R, Hagmann C, Holder J, Izumi N, Kilkenny J D, Kimbrough J, Kyrala G A, Simanovskaia N and Tommasini R 2011 *Penetrating Radiation Systems and Applications* (*Proceedings of SPIE* vol 8144) ed Grim G P and Schirato R C p 814409
[8] Flippo K, DeVolder B, Doss F, Kline J, Merritt E, Loomis E, Capelli D, Schmidt D and Schmitt M J 2014 *J. Phys.: Conf. Ser.* **IFSA 2015** to be published
[9] Gittings M, Weaver R, Clover M, Betlach T, Byrne N, Coker R, Dendy E, Hueckstaedt R, New K, Oakes W R, Ranta D and Stefan R 2008 *Computational Science and Discovery* **1** 015005
[10] Rathkopf J A, Miller D S, Owen J M, Stuart L M, Zika M R, Eltgroth P G, Madsen N K, McCandless K P, Nowak P F, Nemanic M K, Gentile N A, and Keen N D 2000 KULL: LLNL's ASCI inertial confinement fusion simulation code Tech. Rep. UCRL-JL-137053 Lawrence Livermore National Laboratory
[11] NIF and Photon Science 2016 National Ignition Facility User Guide Tech. Rep. LLNL-TM-681123-P2103437-WO15466-NUG Lawrence Livermore National Laboratory
[12] Spaeth M L, Wegner P J, Suratwala T I, Nostrand M C, Bude J D, Conder A D, Folta J A, Heebner J E, Kegelmeyer L M, MacGowan B J, Mason D C, Matthews M J and Whitman P K 2016 *Fusion Science and Technology* **29** FST15–119
[13] Carr C W, Trenholme J B and Spaeth M L 2007 *Applied Physics Letters* **90** 041110
[14] Shaw M, House R, Williams W, Haynam C, White R, Orth C and Sacks R 2008 *Journal of Physics: Conference Series* **112** 032022

Journal of Physics: Conference Series **717** (2016) 012060 doi:10.1088/1742-6596/717/1/012060

Suprathermal electrons in kJ-laser produced plasmas: M-shell resolved high-resolution x-ray spectroscopic study of transient matter evolution

F P Condamine[1,2], M Šmíd[3], O Renner[3], M Dozières[4], F Thais[4], P Angelo[1,2], J-L Bobin[1,2] and F B Rosmej[1,2]

[1] Sorbonne Universités, Pierre et Marie Curie UPMC, UMR7605, LULI, case 128, 4 Place Jussieu, F-75252 Paris Cedex 05, France

[2] Ecole Polytechnique, LULI, CEA-CNRS-UPMC-UPS, Palaiseau, France

[3] Institute of Physics of the ASCR, ELI Beamlines project, Prague, Czech Republic

[4] CEA/DSM/IRAMIS/LIDyL, Saclay, Gif-Sur-Yvette, France

Email: florian.condamine@etu.upmc.fr, frank.rosmej@upmc.fr

Abstract. Hot electrons are of key importance to understand many physical processes in plasma physics. They impact strongly on atomic physics as almost all radiative properties are seriously modified. X-ray spectroscopy is of particular interest due to reduced photoabsorption in dense matter. We report on a study of the copper Kα X-ray emission conducted at the ns, kJ laser facility PALS, Prague, Czech Republic. Thin copper foils have been irradiated with 1ω pulses. Two spherically bent quartz Bragg crystal spectrometers with high spectral and spatial resolution have been set up simultaneously to achieve a high level of confidence in the spectral distribution. In particular, an emission on the red wing of the $K\alpha_2$ transition ($\lambda = 1.5444$ Å) could be identified with complex atomic structure calculations. We discuss possible implications for the analysis of non-equilibrium phenomena and present first atomic physics simulations.

1. Introduction

The study of suprathermal electrons in hot and dense matter is of fundamental interest for the understanding of many physical processes, namely in fusion sciences. Simulations are challenging due to the non-equilibrium physics involved.

Hot electrons have also large impact in atomic physics as almost all radiative properties of the matter are seriously modified. This in turn provides the possibility for their detailed characterization. Hot electrons ionize atomic inner-shells as described in equation (1.1). The use of a high-resolution spectrometer allows hot electrons tracing *via* the induced Kα X-ray emission (see equation (1.2)). Time evolution is related to the different ionization stages (different values of X and Y). Thus, the line intensity of the Kα transitions provide potentially information about the transient hot electron fraction (f_{hot}).

$$K^2L^XM^Y + e^-_{hot} \rightarrow K^1L^XM^Y + 2e^- \qquad (1.1)$$

$$K^1L^XM^Y \rightarrow K^2L^{X-1}M^Y + h\nu_{K\alpha} \qquad (1.2)$$

2. Experimental Setup

The PALS is an iodine gas laser capable to deliver about 0.7kJ of energy on target at 1ω. Its pulse duration is approximately equal to 350ps with a 80μm focal spot diameter. In the present experiment the laser has been used in 1ω mode at λ=1315nm. With intensity of about 10^{16} W/cm^2 on the target, the laser beam can create hot plasmas and K-shell X-ray emission observable in a single shot.

2.1 Target

Copper (thickness: 1 to 6μm) has been chosen as the target for this experiment. Mid- Z-elements, like copper (Z=29), offer the possibility to observe almost all ionization states in X-ray spectra during the transient plasma evolution. Copper ionization energies are located in a range which allows the bulk plasma electrons with temperature (0.1 < E (keV) < 10) to successively ionize outer-shell electrons. Furthermore, the hot electron temperature ($T_{hot} \approx 85$ keV) is largely sufficient to ionize the K-shell. Kα emission is then created.

2.2 Geometry

In this experiment, the spherical crystal has been chosen to increase the signal level on a detector and in the same time to provide spatial resolution without slit. Two quartz Bragg crystals $22\bar{4}3$ (2d=2.024Å) with a curvature radius R=150mm have been employed as dispersive elements. Photographic films (Kodak Industrex AA-400) were used as detectors. The spectral resolution is $\Delta\lambda/\lambda \approx 5000$ and the spatial resolution is about 15μm.

Two X-ray spectrometers (called FSSR) are implemented at two different observation angles as shown in Figure 1. One spectrometer is placed almost parallel to the target surface (± 0.5°) in order to provide spatial resolution along z-axis. The second one is placed at an angle of 57° to simplify its implementation inside the chamber. Having two high-resolution spectrometers with two detectors inside the chamber gives a very good confidence level about the obtained results.

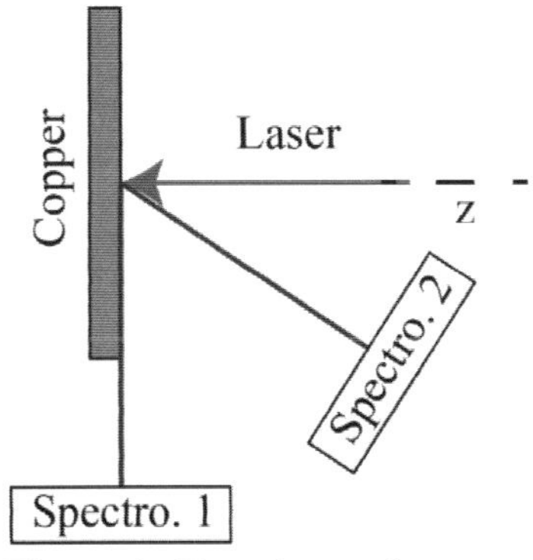

Figure 1. Experimental setup

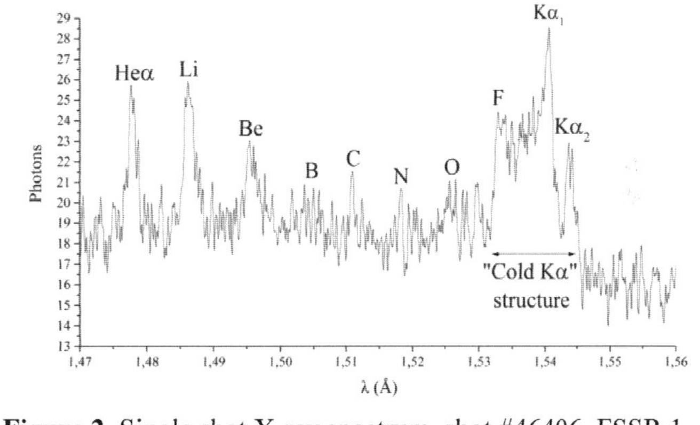

Figure 2. Single shot X-ray spectrum, shot #46406, FSSR 1

Figure 2 shows a X-ray spectrum taken in a single shot (#46406). As can be seen, a strong overlap of many transitions emitted by M-shell ionization states between 1.545 > λ(Å) > 1.53 creates the "cold Kα structure". In addition to this structure, all lines of L-shell ionization states of copper can be identified (indicated by the letters above the different visible peaks, eg., C for C-like etc.).

3. Data analysis

A LSJ-split Hartree-Fock calculation (including intermediate coupling and configuration interaction) allows the correlation of the experimental data. This code based on the work of Cowan [1] associated with a spectrum synthesizer module of the MARIA-code [2] has been used. Here, a gaussian profile has been employed with a line width that mates the experimental observation. Populations have been described by a Boltzmann distribution law. Line identification can be realized via equation (3.1).

$$I_{\omega}^{a} = \sum_k f_k \sum_{i,j} g_j^{(k)} A_{ji}^{(k)} e^{-E_j^{(k)}/kT_e} \phi_{ji}^{(k)}(\omega, \omega_{ji}^{k}) \quad (3.1)$$

The index (k) indicates different ionization states (k=0,1...,29) and f_k is a relative population factor for each ionization state "k" which is associated with the population of levels $n_k = const.\cdot f_k$. The index j corresponds to the upper-level and i to the lower-level, g_j is the statistical weight, E_j the energy, A_{ji} the spontaneous transition probability and ϕ_{ji} the line profile.

3.1 High-charge states

Figure 3 compares the theoretical calculation for Kα L-shell (and He-like) copper ionization states (and satellites) emission with the experiment. Correlation of this data is generally good and the presence of the lines corresponding to all ionization states of copper and their Kα emission in the spectrum indicates that hot electrons are present during the whole lifetime of the plasma. Then, line intensity ratios can give detailed information on the evolution of the hot electron fraction inside the plasma.

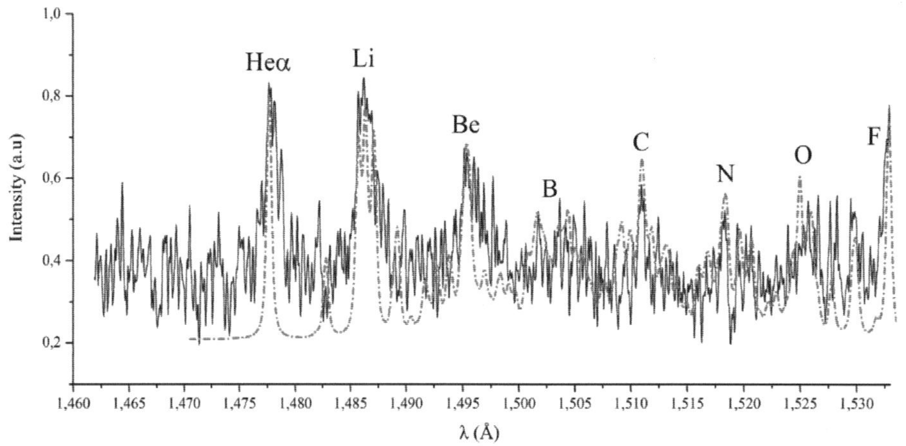

Figure 3. 1ω experimental (solid) and theoretical (dash) spectra (shot #46406, FSSR 1)

3.2 "Cold" structure

The Kα structure corresponds to lines generated by low ionization states and is designated as "cold structure" as these states originate from low temperature plasmas. However, the study of these transitions is challenging due to the complexity of configurations involving M-shell electrons. Figure 4a shows the comparison between the positions of $K\alpha_1$ and $K\alpha_2$ lines for all M-shell ionization states. Two groups can be separated. Ionization states involving 3d-shell electrons have almost all the same $K\alpha_1$ wavelengths (group D). On the contrary, 3p/3s-shell lines are much more separated (group P/S).

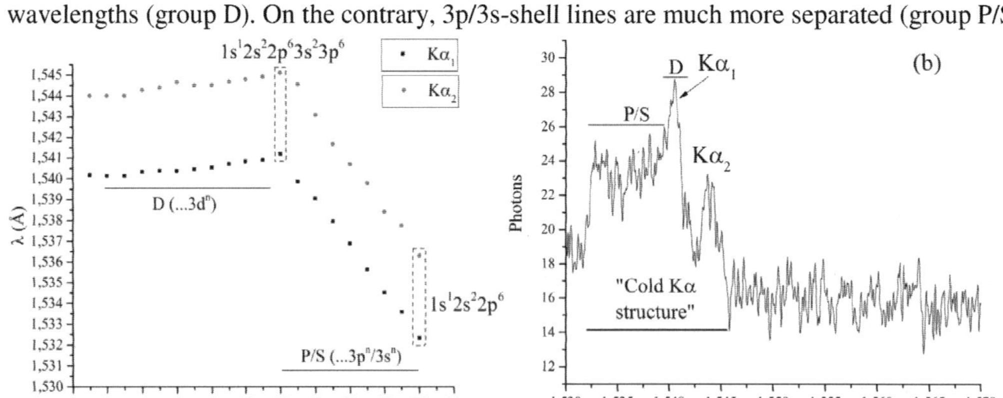

Figure 4. (a) Position of $K\alpha_1$ and $K\alpha_2$ lines for M-shell ionization states, (b) "cold structure" on experimental spectrum (#46406, FSSR 1)

Figure 4b reflects this identification in the experimental spectrum (shot #46406). The strong and narrow peak inside the structure corresponds to a very strong overlap of all lines from 3d-shell (group D) whereas the more extended structure inside the "cold structure" can be associated to the 3p/3s-shell emission (group P/S).

Figure 5a shows the simulation of emission from the states with 3d-shell configurations. States with an open shell generate lines on the red wing of the $K\alpha_2$ line. This "red wing" can be associated to the emission originating from configuration interaction and intermediary coupling. Figure 5b shows the correlation between an experimental spectrum and the simulation (each shell emission is separated for more clarity). The "red wing" seems to be present on experimental data via the enhanced slope (see arrow in figure 5) in the $K\alpha_2$ emission line.

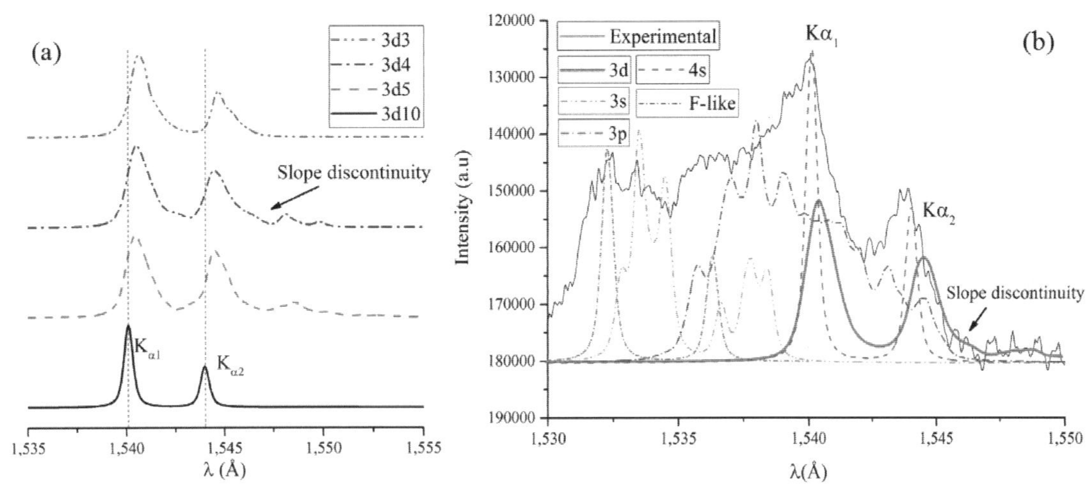

Figure 5. (a) simulations of states possessing a $K_{\alpha 2}$ red wing, (b) Sum of 5 experimental spectra (FSSR 1) and their comparison with simulations for the emission from different ionization stages

4. Conclusion

$K\alpha$ lines of different ionization states (emitted at different times) can be used to visualize hot electrons during the plasma lifetime. However, this study is challenging for low ionization states due to a strong overlap between their emission lines. High-resolution spectroscopy allows to differentiate some groups of lines and their contribution inside the $K\alpha$ "cold structure". Furthermore, the red wing identified in the spectra provides the possibility to resolve the contribution of some states inside the M-shell that could not be resolved otherwise.

Further analysis and experiments are envisaged to study the $K\alpha$ "cold structure". Finally, these next studies should allow a verification of atomic codes describing the time evolution of hot electrons inside the plasma.

Acknowlegments

We thank all the PALS technical team for their support.

This research was performed within the LASERLAB project PALS001990, and supported by MSMT ELI Project No. CZ.1.05/1.1.00/02.0061 and CZ COST Action MP1208 Developing the Physics and the Scientific Community for Inertial Fusion.

References

[1] Cowan R D 1981 *The Theory of Atomic Structure and Spectra* (Berkeley, Univ. California Press)

[2] Rosmej F 1997 *Journal. of Physics B: Atomic Molecular and Optical Physics* **30**(22) L819-L828

Journal of Physics: Conference Series **717** (2016) 012061 doi:10.1088/1742-6596/717/1/012061

Filamentation due to the Weibel Instability in two counter-streaming laser ablated plasmas

Quan-Li Dong[1,*], Dawei Yuan[2,3], Lan Gao[4], Xun Liu[2], Yangao Chen[4], Qing Jia[4], Neng Hua[5], Zhanfeng Qiao[5], Ming Chen[5], Baoqiang Zhu[5], Jianqiang Zhu[5], Gang Zhao[3], Hantao Ji[4], Zheng-Ming Sheng[6], Jie Zhang[6]

1 School of physics and optoengineerings, Ludong University, Yantai 264025, China;
2 Institute of Physics, CAS, Beijing 100190, China;
3 National Astronomical Observatories, CAS, Beijing 100012, China;
4 PPPL, Princeton University, Princeton, New Jersey 08543, USA;
5 National Laboratory on High Power Lasers and Physics, Shanghai, 201800, China;
6 Key Laboratory for Laser Plasmas (MoE) and Department of Physics, Shanghai Jiao Tong University, Shanghai 200240, China

* qldong@aphy.iphy.ac.cn

Abstract. Weibel-type filamentation instability was observed in the interaction of two counter streaming laser ablated plasma flows, which were supersonic, collisionless, and closely relevant to astrophysical conditions. The plasma flows were created by irradiating a pair of oppositely standing plastic (CH) foils with 1ns-pulsed laser beams of total energy of 1.7 kJ in two laser spots. With characteristics diagnosed in experiments, the calculated features of Weibel-type filaments are in good agreement with measurements.

1. Introduction

The physical mechanisms of magnetic field generation, amplification and annihilation play important roles in many astrophysical phenomena. Take the SN1006 as example. One explanation for surrounding shock structures is that the magnetic field are generated through Weibel Instability (WI) during interaction between outflow plasmas, reorganized and amplified, forming transverse structure (perpendicular to plasma flow direction) at large scales and capturing/stagnating incident charged particles to form shocks within which they accelerate and irradiate the synchrotron emission, manifesting magnetic field configuration they are gyrating around [1,2]. Following the above hyperthesis for shocks and ultrahigh energy cosmic rays from supernova, laboratory laser-plasma experiments designed for astrophysics study have to demonstrate, step by step, the magnetic field generation and amplification, charged particle acceleration and gyration to radiate synchrotron emission [3-6].

This paper reports density observations of Weibel-type filamentation of counter streaming laser plasmas, which act as the supplement for the magnetic filaments observed in [7]. However the filamentation instability is still far from nonlinear regime, indicating large laser facility is needed for WI mediated collisionless shock study.

2. Experiment setup

Experiments were conducted on the SG-II laser facility platform in September, 2011. The setup is shown in Figure 1. Two face-to-face standing CH foil targets of 5×5 mm^2 were used to generate a pair of counterstreaming plasma flows. To produce asymmetrical plasma flows, seven 1ns-pulsed laser beams were launched and divided into two groups: the 4 southern laser beams of 1 kJ were focused on the northern target, and the 3 northern of 0.7 kJ on the southern target. The southern bunch of laser beams was delayed by 1ns making sure that two plasma flows have different properties when they encounter between two CH foils separated by 4.5 mm. The laser spots on CH foils were 150 um in diameter, generating an average laser intensity of 10^{14}–10^{15} W/cm^2. The 9th laser beam of SG-II platform was used as a probe to measure the plasma density profile through Normaski Interferometry (NI) and shadowgraphy techniques. With the probe of 2ω (527 nm in wavelength) and 30 ps duration was applied, the upper measurement limit of the electron density is around 5-8×$10^{19}$$cm^{-3}$ in our experiments. At different delay times with respect to the southern bunch, the recorded interferometry can provide the temporal evolution of the electron density. One interferometry image and one shadowgraphy image are given as examples, which are also the main objectives for our analysis. For all shots, x-ray pinhole cameras of 10 time magnification were also used to measure the uniformity of the laser spots.

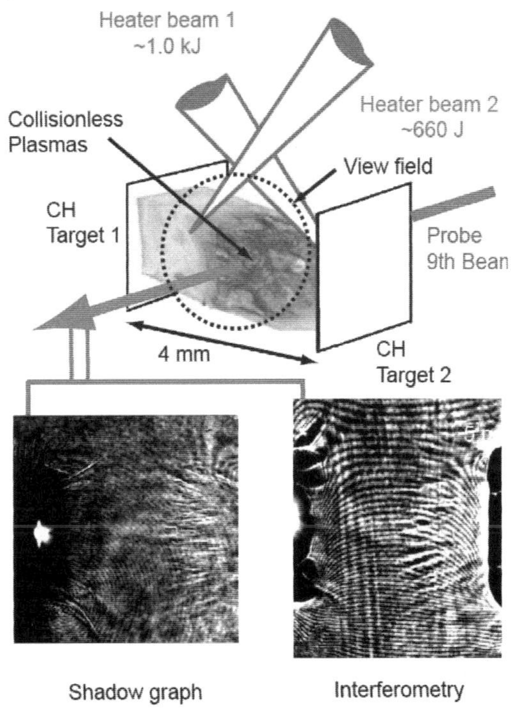

Figure 1. Experimental setup. Two oppositely standing CH foil targets were irradiated by 1ns-pulsed laser bunches with energy 1.0 kJ and 0.7 kJ, respectively. The 9th beam of SG II was used to probe the colliding plasmas, producing interferometry and shadowgraphy images of which two examples were given for 4.5 ns delay measurements.

3. Experimental result analysis

Figure 2 shows the recorded shadowgraphy at time delay of 4.5 ns. The main feature is the filaments that follow the expansion direction of the right plasma ball. The filament location near the side of the right (northern) laser spot rather than the center is attributed to 1-ns-delay of launch time of the Southern bunch lasers. The shot time arrangement of as well as the different energy contained in two laser spots produced a condition with tenuous background plasma for the hotter plasma from the right target, while avoiding the Southern laser heating of the preformed plasma. The indicated over dense plasma region shows the edge of plasma with density higher than 5×$10^{19}$$cm^{-3}$. The filaments can be divided into two groups according to their lengths. The longer filaments are supposed to be located in or parallel to face-on planes containing two laser spots, while shorter filaments are supposed to be the projections of those filaments with larger angles to face-on planes. The dashed line gives the position where the filament distribution pattern is analyzed in Figure 3.

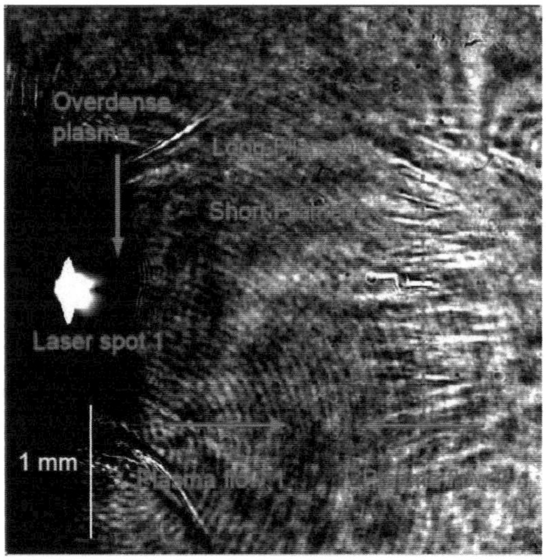

Figure 2. Shadowgraphy taken at 4.5 ns delay with respect to the tail of the laser pulse. The width distribution of the spaces between filaments along the dashed line will be analyzed in Figure 3.

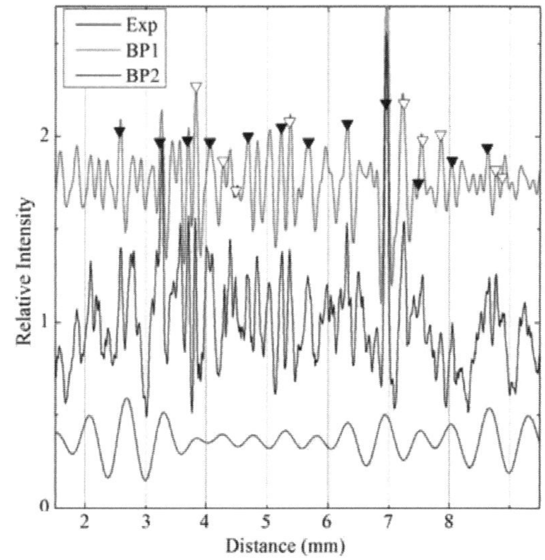

Figure 3. Profile of the filament distribution along the dashed line in Figure 1. The black line is the experimental result. With different band FFT (Fast Fourier Transformation) filters, the red line presents all (short and long) filaments spaced by 90-250 um, while the blue line presents only long filaments spaced by 400-600 um as indicated by the solid triangles.

The profile of the filament distribution is analyzed by using different band FFT (Fast Fourier Transformation) filters. The red line presents the filtered profile involving all (short and long) filaments spaced by 90-250 um, while the blue line presents only long filaments spaced by 400-600 um as indicated by the solid triangles. Both FFT-filtered distributions reproduce the corresponding peaks /filaments in the experimental profile. It can be expected that the spaces between real filaments in 3D is 400-600 um, and the smaller value of 90-250 um is due to the "forest-effect" the projection process usually has.

To explore the mechanisms underlying the filamentation, growth rates were calculated for various instabilities with the diagnosed plasma parameters. It is found that for the filamentation parallel to the flow velocity, Weibel instability is the most likely mechanism. The dispersion relation of the beam-Weibel instability is given by:

$$\omega^2 = (kc)^2 - \sum_s \omega_{ps}^2 \left[\frac{\upsilon_{sth}^2 + 2(V_s \sin(\theta))^2)}{\upsilon_{sth}}\left(1 + \zeta_s Z(\zeta_s)\right) - 1\right]$$

where $Z(\zeta) \equiv \pi^{-1/2} \int_{-\infty}^{\infty} \frac{e^{-z^2}}{z-\zeta} dz$ is the dispersion function, $\zeta_s = \zeta_s(\omega, k, \theta) \equiv (\omega/k - V_s \cos(\theta))/\upsilon_{sth}$, with V_s the bulk velocity, and $\upsilon_{sth} = \sqrt{2T/m}$ is the average thermal velocity at temperature T. The calculation result is shown in Figure 4, indicating the Weibel instability grew at $\gamma \sim 4 \times 10^{-5} \omega_{pe}$ and $kc/\omega_{pe} \sim 0.17$, which means, for plasmas with electron density of $5\times10^{18}cm^{-3}$ that can be measured by the interferometry technique, the space between filaments is around 480 um, agreeing well with the measurements. With the calculated growth rate, the amplitude of filamentations will grow by hundreds of times in several nanoseconds, making it observable but still in weak linear regime as indicated by the ratios (>> 1) of the filament length (or the parallel wavelength) and the space (or the perpendicular wavelength) to the diameter of filaments.

Contributions to growth rates from different plasma properties were also studied. As CH plasma contains various carbon ions, to simplify the calculation $C^{4+}$and H^{+} were adopted. Growth rates of following cases were calculated: plasma-plasma interactions, plasma-ion-beam interaction, and plasma-electron-beam interaction. The growth rates changed little for three different cases with the same bulk velocity. But for higher bulk velocity, the growth rate increased significantly.

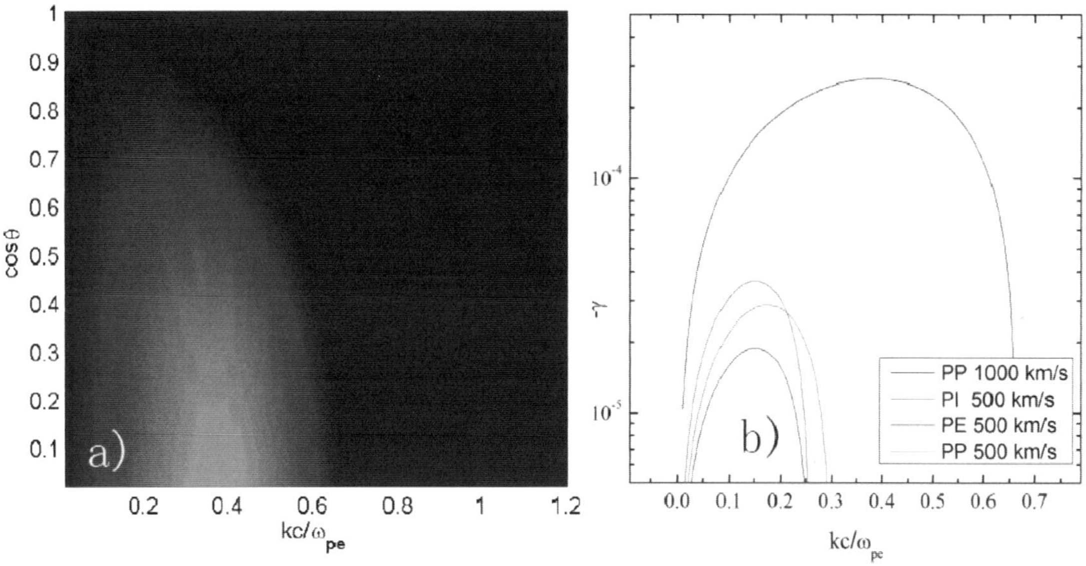

Figure 4. a) Growth rate of Weibel-type filamentation instability in the pair of counterstreaming laser plasmas with experimental conditions. **b)** Growth rate γ of filamentations parallel to plasma bulk velocity υ_b. Effects of different components to γ were compared. PP 1000km/s represents for plasma-plasma interaction with $\upsilon_b = 1000\text{km/s}$; PI for plasma-ion-beam interaction, and PE for plasma-electron-beam interaction.

4. Summary

A pair of counterstreaming laser plasmas were found to have filamentation behavior which were attributed to Weibel instability. However, the instability is still in linear regime, and the magnetic field is weak, far from the strength needed for stagnating charged particles to form collisionless shocks as observed in SN remnants. Larger plasma systems allowing for longer interaction time might help Weibel instability evolve into nonlinear regime necessary for laboratory study of collisionless shock related ultra high energy cosmic rays.

This work is jointly supported by the National Natural Science Foundation of China (Grants No. 11274152 and No. 11220101002).

References

[1] G. Cassam-Chenai et al., Astrophys. J. **680**, 1180 (2008);
[2] Francisco Suzuki-Vidal et al., Nature Physics **11,** 98 (2015);
[3] Hideaki Takabe et al., Plasma Phys. Control. Fusion **50**, 124057 (2008);
[4] R. P. Drake and G. Gregori, Astrophys. J. **749**, 171 (2012);
[5] Will Fox et al., Phys. Rev. Lett. **111**, 225002 (2013);
[6] K. Quinn et al., Phys. Rev. Lett. **108**, 135001 (2012);
[7] C. M. Huntington et al., Nature Physics **11**, 173 (2015;

The Laser-Driven X-ray Big Area Backlighter (BABL): Design, Optimization, and Evolution

Kirk Flippo, Barbara DeVolder, Forrest Doss, John Kline, Elizabeth Merritt, Eric Loomis, Deanna Capelli, Derek Schmidt, and Mark J. Schmitt
Los Alamos National Laboratory, PO Box 1663, Los Alamos, NM, USA

E-mail: kflippo@lanl.gov

Abstract. The Big Area BackLigher (BABL) has been developed for large area laser-driven x-ray backlighting on the National Ignition Facility (NIF), which can be used for general High Energy Density (HED) experiments. The BABL has been optimized via hydrodynamic simulations to produce laser-to-x-ray conversion efficiencies of up to nearly 5%. Four BABL foil materials, Zn, Fe, V, and Cu, have been used for He-α x ray production.

1. Introduction

The Big Area Backlighter (BABL)[1] is a large area (2.5 x 3 mm^2) x-ray backlighter designed for radiographing large volume experiments on the National Ignition Facility (NIF) [2]. Many considerations went into the development of the BABL, some unique to the facility, and others for optimal performance [3]. Essentially the facility configuration dictated the position and dimensions of the BABL [1] foil, and target material limitations (CH tubes) dictated the energy had to be above 6 keV for good contrast. With those criteria and with NIF allowing for an arbitrary single valued pulse shape in a 35 ns window (extendable with fiber optics), a pulse shape was developed such that only the maximum energy and fluence constrained the pulse power. The original BABL used a 2 ns flat-top pulse with a sharp ~4 ns picket in front of the flat-top as a way of generating a small amount of preplasma to increase the laser absorption and hence the x-ray conversion efficiency, which achieved a conversion of ~3.5% of the 132kJ incident laser energy into x rays between 6.4 at 7 keV, the Fe He-α and associated lines (a total of 4.6 kJ)[1]. However, due to optics damage considerations this

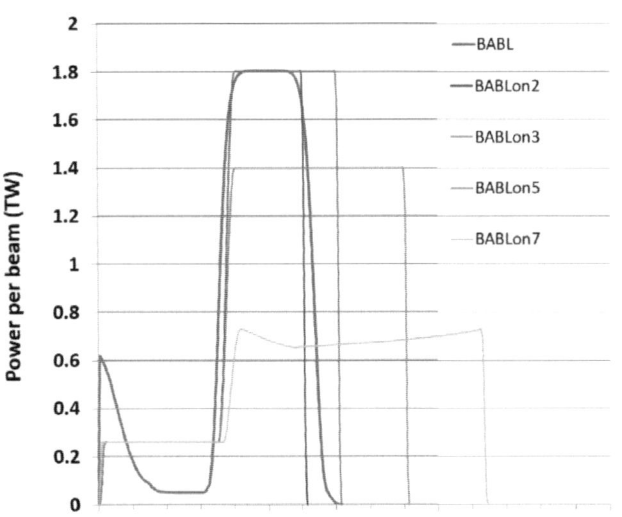

Figure 1. BABL pulses shapes, original BABL (blue) had a sharp picket, newer versions (purple, red, green yellow) have a plateau of nearly equivalent energy, with sharper edges on the flat-top and trailing edge.

pulse needed to be modified. Instead of the fast rising sharp picket, a new plateau was added that stayed below the laser damage growth power threshold of 0.26 TW/beam. The plateau was extended to have nearly the equivalent integrated energy as the original picket to keep pre-plasma generation at around the same level. Also, using different target materials (a Be tube) the energy criteria were relaxed due to greater transmission, and to date we have been able to use the BABL to produce V, Fe, Cu, and Zn He-α x rays. Figure 1 shows the original BABL pulse with the sharper picket in blue and the subsequent variations intended to save optics damage, including the quick all-off at the end of the pulse, such that it can pushed up against the edge of the 35 ns laser timing window. The plateau has increased the conversion efficiency to about 4.7% using the BABLon3 pulse shape.

2. Optimization Simulations

To optimize the flux of x rays reaching the x-ray camera the target generally needs to be thin enough to be transparent to its own He-α line and also be thick enough to withstand the laser ablation pressure throughout the pulse duration. A Lagrangian hydrodynamics code was used to model the BABL's performance with the various laser pulses to optimize the BABL foil thickness. The highest power pulse are the BABLon2 and BABLon3 pulses, since the BABLon3 is on for a total of 7.15 ns, this is the highest energy pulse as well. The longest pulse is the BABLon7 which lasts for 11.15 ns, and has a power of ~0.75 TW for 7 ns, but less total energy. Thus, these two pulses were used as the extremes in power and duration to test the burn-through of the foils from 50 μm down to 5 μm.

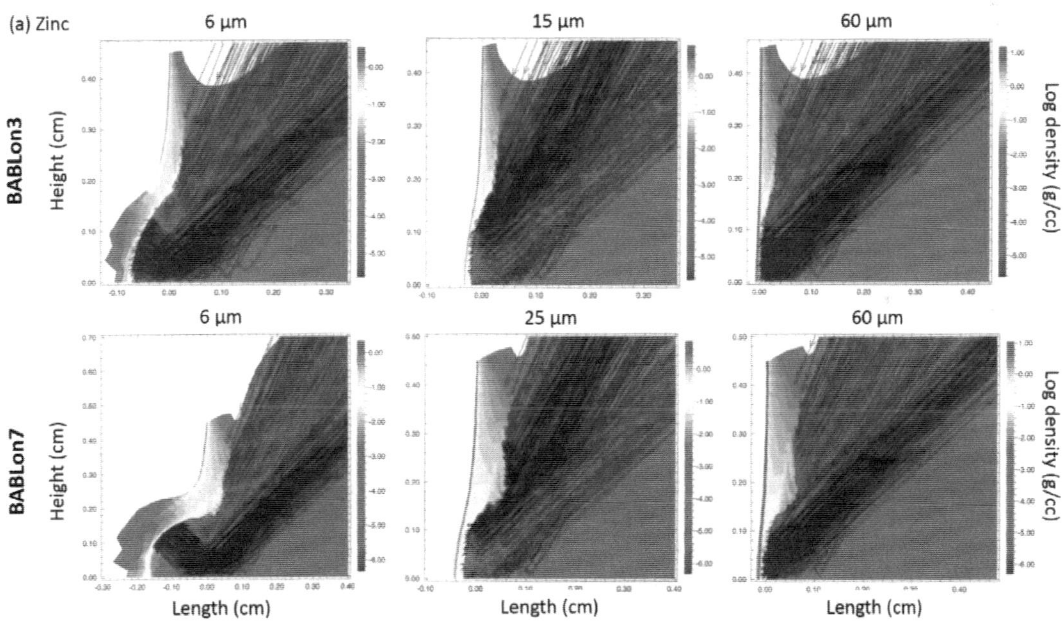

Figure 2. Burn-through hydro simulations for Zn BABL foils using the BABLon3 (top) and BABLon7 (bottom) pulse shapes with various thicknesses of BABL foils. Colorscale is log density in g/cc and the red, pink and dark-blue rays are the laser ray traces for differing angles.

Figure 2 shows the results of 6 simulations run for the duration of two pulse shapes: either the BABLon3 (7.15 ns) or the BABLon7 (11.15 ns) using 24 NIF beams in 6 Quads from 3 angles (red, pink and dark-blue tray traces). The simulations are symmetric around the length (x) axis, which is a satisfactory assumption, as half the beams come from the bottom and half come from the top in the NIF chamber onto the BABL foil. The simulations show that a 6 μm Zn BABL foil is sufficient to keep the laser from burning through the foil over the pulse duration. Though, by the end of the pulse the bulk of the foil for the thinnest (6 μm) case is deformed by nearly 100 μm in the BABLon3 case and almost 200 µm in the BABLon7 case. In the actually experiment this is likely less as the axial symmetry of the reflecting boundary condition puts more energy on half of the target than would actually be present in the experiment; the energy would be more evenly spread out between the top and bottom halves. Thus, this should represent the worst case.

Figure 3 shows a similar plot for the V BABL for both pulse shapes and several thicknesses. Here too, the 5 µm V foil is sufficient to stop the light from burning through the foil, though the foil is now displaced by more than 100 µm for the BABLon3 case and 300 µm for the BABLon7 case.

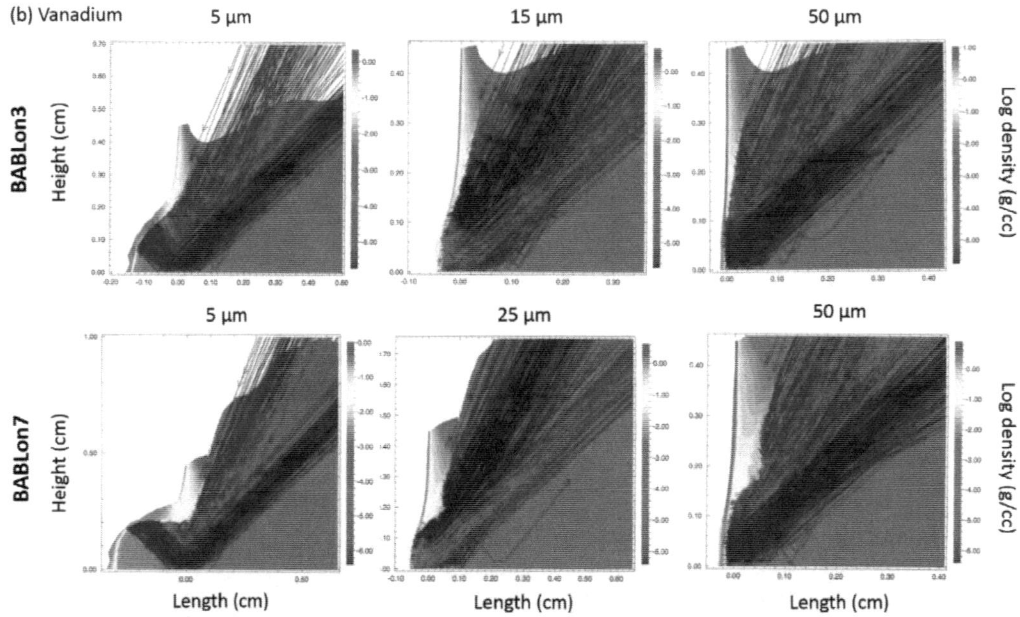

Figure 3. Simulations for V BABL foils using the BABLon3 (top) and BABLon7 (bottom) pulse shapes with various thicknesses of BABL foils. Colorscale is log density in g/cc and the red, pink and dark-blue rays are the laser ray traces for differing angles.

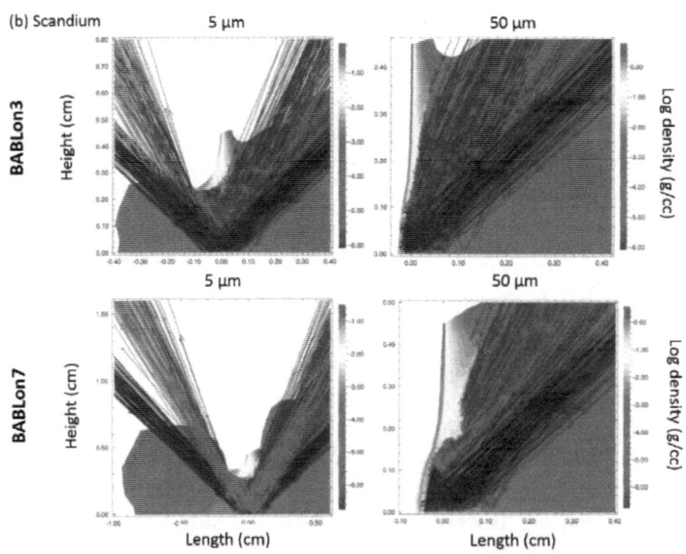

Figure 4. Simulations for Sc BABL foils using the BABLon3 (top) and BABLon7 (bottom) pulse shapes with various thicknesses of BABL foils. Colorscale is log density in g/cc and the red, pink and dark-blue rays are the laser ray traces for differing angles.

Figure 4 presents the case for a Sc BABL for both pulse shapes. It is immediately apparent that for Sc the thin foils present a very possible burn-through scenario. In this scenario the Sc foils will need to be thicker than 10 microns to be sure to prevent laser burn-through and degradation of the experiment.

3. Summary

The BABL has been optimized using hydrodynamic simulations of thin foils (5-6 μm) to increase the x-ray through-put for radiography. To date, four materials (V, Fe, Cu, and Zn) have been used successfully at the NIF to produce He-α x rays. The BABL yields a conversion efficiency for the thinner foils of nearly 5% into He-α x rays, and is driven using between 120 to 200 kJ of laser energy depending on pulse duration and number of beams, so nearly 10 kJ of x rays are produced per shot. These x rays range from 4.95 keV for V and 8.6 keV for Zn. There is no reason to believe that this range cannot be extended down to Sc (4 keV) or up to Ge (9.88 keV) in the near future with similar conversion efficiencies.

References

[1] Flippo, K.A., et al., *Development of a Big Area BackLighter for high energy density experiments.* Review of Scientific Instruments, 2014. **85**(9).

[2] Moses, E.I. and W.R. Meier, *Preparing for ignition experiments on the National Ignition Facility.* Fusion Engineering and Design, 2008. **83**(7-9): p. 997-1000.

[3] Flippo, K.A., et al., *Investigating Turbulent Mix in HEDLP Experiments.* J. of Phys.: Conf. Series, 2016 (in press). **8th International Fusion Science and Applications 2013**.

Journal of Physics: Conference Series **717** (2016) 012063 doi:10.1088/1742-6596/717/1/012063

Input energy measurement toward warm dense matter generation using intense pulsed power generator

R Hayashi[1], T Ito[1], T Ishitani[1], F Tamura[1], T Kudo[1], N Takakura[1], K Kashine[2], K Takahashi[1], T Sasaki[1], T Kikuchi[1], Nob. Harada[1], W Jiang[1] and A Tokuchi[1,3]

[1]Nagaoka University of Technology, 1603-1, Kamitomioka, Nagaoka, 940-2188, Japan
[2]National Institute of Technology, Kagoshima College, 1460-1, Shinkou, Hayato, Kirishima, 899-5193, Japan
[3]Pulsed Power Japan Laboratory Ltd., 2-3-16, Oji, Kusatsu, Shiga, 525-0058, Japan

E-mail: r_hayashi@stn.nagaokaut.ac.jp

Abstract. In order to investigate properties of warm dense matter (WDM) in inertial confinement fusion (ICF), evaluation method for the WDM with isochoric heating on the implosion time-scale using an intense pulsed power generator ETIGO-II (~1 TW, ~50 ns) has been considered. In this study, the history of input energy into the sample is measured from the voltage and the current waveforms. To achieve isochoric heating, a foamed aluminum with pore sizes 600 μm and with 90% porosity was packed into a hollow glass capillary (ø 5 mm × 10 mm). The temperature of the sample is calculated from the numerical calculation using the measured input power. According to the above measurements, the input energy into a sample and the achievable temperature are estimated to be 300 J and 6000 K. It indicates that the WDM state is generated using the proposed method with ICF implosion time-scale.

1. Introduction

In inertial confinement fusion (ICF), understanding implosion dynamics of fuel pellet is a key issue to obtain effective nuclear fusion reactions. To suppress the implosion non-uniformity, a foamed metal such as a pusher and a radiator is considered as structural materials in the fuel pellet [1]. Implosion of the fuel shell is driven by beam energy deposiotion with an energy driver, and the target material becomes dense plasma through warm dense matter (WDM) region with the implosion time-scale (several-10 ns). The WDM region pertains to densities from 10^{21} to 10^{24} cm^{-3} and for temperature varying from 10^3 to 10^5 K. However, the WDM is complex regime, because of unclear theoretical model and lacked experimental evaluations.

To understand the implosion dynamics, numerical simulations are crucial approach. However, the implosion dynamics from the numerical simulation result depends on EOS model in WDM [2]. There is no accurate and useable EOS model for the numerical simulation. The achievable parameters such as the confinement parameter and the moment of stagnation are affected by the properties of WDM in early stage of implosion. Therefore, to predict the implosion dynamics, the properties in WDM such as equation of state should be cleared.

Journal of Physics: Conference Series **717** (2016) 012063 doi:10.1088/1742-6596/717/1/012063

In previous studies, a short pulse laser and a pulsed power discharge were used for the measurement for properties of WDM [3, 4]. In the case of that foamed metal with a cellular structure is heated with several-10 ns, which is similar to the time-scale of hydrodynamic behavior of material, the relationship between heating time-scale and properties of matter is not well known. In order to estimate the more accurate temperature of the fuel pellet during implosion, the specific heat evaluation method for the WDM with isochoric heating [5] on the implosion time-scale using an intense pulsed power generator ETIGO-II ($\sim$1 TW, $\sim$50 ns) [6] has been considered [7]. To measure the specific heat, the evaluation method for history of the internal energy and the temperature of the sample should be established. The internal energy need to be measured with the input energy and the sample mass. The temperature need to be measured with an optical system using spectroscopic measurement. In this study, the input energy into a sample is measured from the voltage and current waveforms. As a result of establishing the measurement method of input energy, the internal energy for the specific heat is obtained.

2. Measurement for input energy into sample

To investigate the properties of WDM as a function of temperature, the current control into the sample with an electron beam diode as an impedance controller has been proposed [8]. In this study, the input energy measurement is performed with short circuit condition without the electron beam diode.

Figure 1 shows the experimental setup for history of input energy into the sample. To achieve isochoric heating, the sample, which is a foamed aluminum with pore sizes $\sim$600 μm and with $\sim$90% porosity, was packed into a hollow glass capillary (ϕ 5 mm $\times$ 10 mm). The average mass and the average density of the sample in the capillary are $\sim$60 mg and $\sim 0.1\rho_s$ (ρ_s is solid density of aluminum: 2699 kg/m^3), respectively. history of the current $I(t)$ needs to be measured with a Rogowski coil. The voltages need to be measured with using resistive dividers before and behind the sample. The resistive divider consists of two-stage non-inductive resistors of $R_1 = 1.5$ kΩ and $R_2 = 5$ Ω. When a transmission line is terminated in its characteristic impedance $Z_0 = 50$ Ω, the resistive voltage divider's ratio was designed 615 times. history of the voltage $V(t)$ is measured to be $V(t) = V_1(t) - V_2(t) - LdI(t)/dt$, where $V_1(t)$ is the voltage at the ground, $V_2(t)$ is the voltage between the sample and ground, the stray inductance is estimated as L = 98 nH. The input power $P(t)$ and the input energy $E(t)$ into the sample is measured to be $P(t) = V(t)I(t)$, $E(t) = \int P(t)dt$.

Figure 2 shows the current and voltage waveforms and input power and energy into the sample with a short circuit condition. Figure 2, to compare with the heating time-scale and the implosion time-scale, t_o and t_f are defined as the rise time of the input power and the time 50 ns from t_o. The pulse width was 100 ns (FWHM), with peak voltage and current values of 80

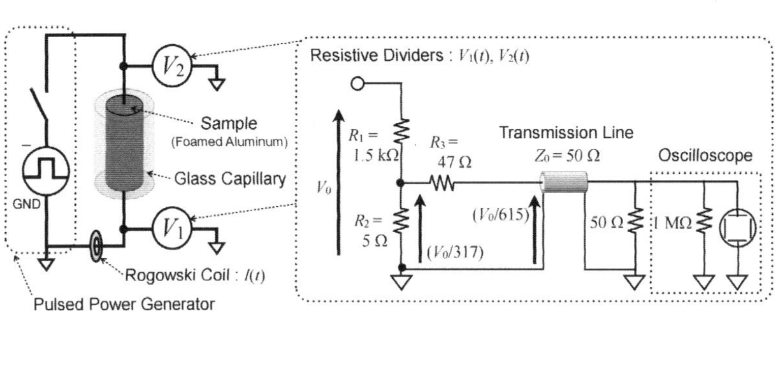

Figure 1. Experimental setup for history of input energy into sample.

kV and 140 kA, respectively. Peak input power and the input energy into the sample achieved 10 GW and 400 J at t_f.

3. Numerical calculation for temperature of sample

To estimate the achievable temperature, the history of the temperature of sample $T(t)$ is estimated from numerical calculation using measured input power $P(t)$.

The history of the temperature of sample $T(t)$ is estimated to be

$$T(t) = \int \frac{P(t)dt}{r^2\pi \cdot l \cdot \rho_{\text{foam}} \cdot c(\rho, T)}, \tag{1}$$

where the input power $P(t)$ is given by the corresponding experimental data as shown in Fig. 2, l is the length of the sample with 10 mm, r is the radius of the sample having 2.5 mm, the density of foamed aluminum sample is $\rho_{\text{foam}} = 0.1\ \rho_s$ and $c(\rho, T)$ is the specific heat that the data of aluminum in solid, liquid, and gas phases are given by Refs. [9, 10]. The initial temperature is set at 300 K.

Figure 3 shows the calculation result of the temperature as a function of the time. The temperature of the sample achieved 7000 K at t_f. Figure 4 shows the peak current as a function of gap distance of electron beam diode. In short circuit condition (gap distance: 0 mm), the peak current increased that is compared with the case of adjustment of gap distance in electron beam diode [8]. Figure 5 shows the temperature as a function of input energy at the t_f. In short circuit condition, the averaged input energy and temperature were 300 J and 6000 K, respectively.

4. Conclusion

In order to investigate the properties of WDM in ICF, input energy measurement toward WDM generation using intense pulsed power generator was performed.

The input power and the input energy were estimated with experimentally obtained current and voltage waveforms. Peak input power and the input energy into the sample achieved 10 GW and 400 J at the time t_f. The achievable temperature of the sample was estimated from calculation using the input power. The temperature for the sample of foamed aluminum ($0.1\rho_s$)

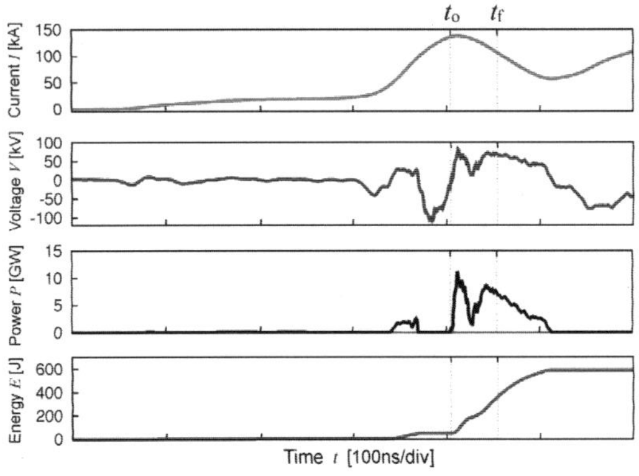

Figure 2. Current and voltage waveforms and input power and energy into the sample with short circuit condition.

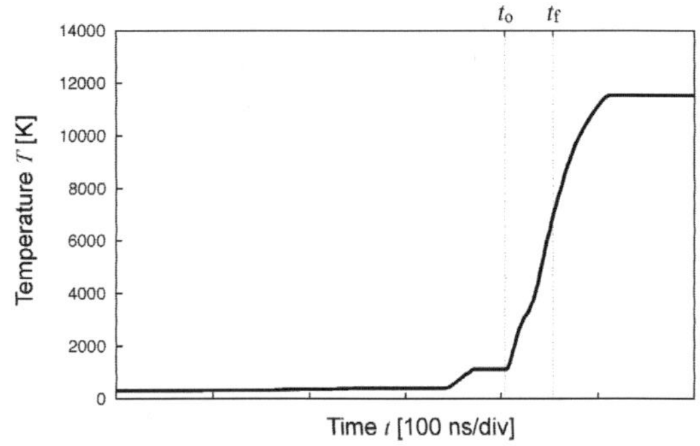

Figure 3. history of sample temperature. (t_o and t_f are defined as the rise time of the input power and the time 50 ns from t_o.)

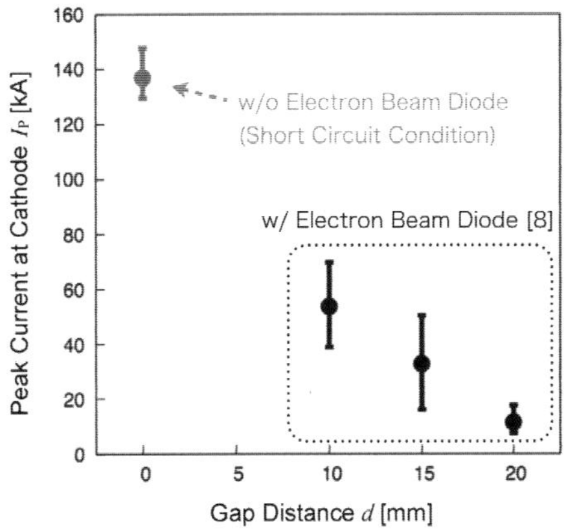

Figure 4. Peak current as a function of gap distance of electron beam diode. The error bars indicate maximum and minimum values.

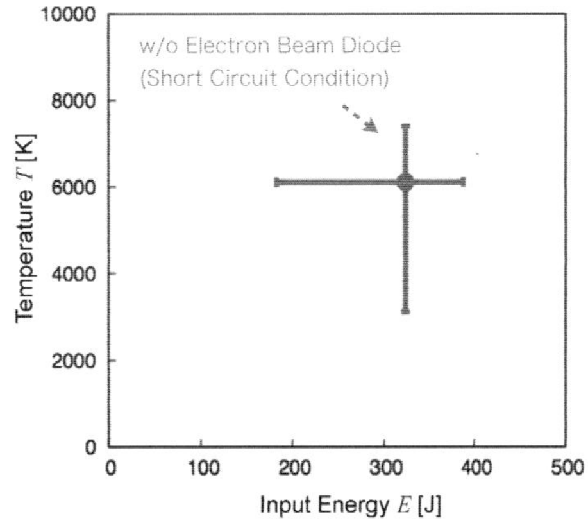

Figure 5. Temperature at t_f as a function of input energy. The error bars indicate maximum and minimum values.

was several-1000 K at the time t_f. In short circuit condition, the peak current increased that is compared with the case of adjustment of gap distance in electron beam diode.

According to above results, the measurement method of input energy with the current and voltage waveforms for the estimasion of internal energy was established.

Acknowledgments

This work was supported from JSPS Grant-in-Aid for Challenging Exploratory Research No.25630418.

References

[1] Atzeni S and Meyer-ter-vehn J 2004 *The Physics of Inertial Fusion: Beam Plasma Interaction, Hydrodynamics, Hot Dense Matter* (New York: Oxford University Press) Chapter 9
[2] Komatsu Y, Sasaki T, Kikuchi T, Harada N and Nagatomo H 2013 *EPJ Web of Conferences* **59** 04010
[3] Glenzer S H, *et al* 2007 *Phys. Rev. Lett.* **98** 065002
[4] Clérouin J, Noiret P, Blottiau P, Recoules V, Siberchicot B, Renaudin P, Blancard C, Faussurier G, Holst B and Starrett C E 2012 *Phys. Plasmas* **19** 082702
[5] Amano Y, Miki Y, Takahashi T, Sasaki T, Kikuchi T and Harada N 2012 *Rev. Sci. Instrum.* **83** 085107
[6] Jiang W, Sakagami T, Masugata K and Yatsui K 1993 *Jpn. J. Appl. Phys* **32** pp L752–L754
[7] Hayashi R, *et al Proc.* IFSA2013 to be published in *J. Phys.: Conf. Series*
[8] Hayashi R, *et al* 2015 *Laser Part. Beams* **33** pp 163–167
[9] NIST Standard Reference Database Number **69**
[10] Chase M W 1998 *J. Phys. Chem. Ref. Data* NIST-JANAF Thermochemical Tables Fourth Edition Monograph 9

9th International Conference on Inertial Fusion Sciences and Applications (IFSA 2015) IOP Publishing
Journal of Physics: Conference Series **717** (2016) 012064 doi:10.1088/1742-6596/717/1/012064

First-principles studies on the equation of state, thermal conductivity, and opacity of deuterium–tritium (DT) and polystyrene (CH) for inertial confinement fusion applications

S X Hu[1,3], L A Collins[2], V N Goncharov[1], J D Kress[2], T R Boehly[1], R Epstein[1], R L McCrory[1] and S Skupsky[1]

[1]Laboratory for Laser Energetics, University of Rochester,
250 East River Road, Rochester, NY 14623-1299, USA

[2]Theoretical Division, Los Alamos National Laboratory,
Los Alamos, NM 87545, USA

shu@lle.rochester.edu

Abstract. Using first-principles (FP) methods, we have performed *ab initio* compute for the equation of state (EOS), thermal conductivity, and opacity of deuterium–tritium (DT) in a wide range of densities and temperatures for inertial confinement fusion (ICF) applications. These systematic investigations have recently been expanded to accurately compute the plasma properties of CH ablators under extreme conditions. In particular, the first-principles EOS and thermal-conductivity tables of CH are self-consistently built from such FP calculations, which are benchmarked by experimental measurements. When compared with the traditional models used for these plasma properties in hydrocodes, significant differences have been identified in the warm dense plasma regime. When these FP-calculated properties of DT and CH were used in our hydrodynamic simulations of ICF implosions, we found that the target performance in terms of neutron yield and energy gain can vary by a factor of 2 to 3, relative to traditional model simulations.

1. Introduction

Accurate plasma properties of deuterium–tritium (DT) fuel and ablators under extreme conditions are crucial for simulating and understanding inertial confinement fusion (ICF) implosions. Knowledge of these physical properties, such as equation of state (EOS), thermal conductivities, viscosity, and opacity, is required beforehand to simulate ICF. Historically, these quantities have been estimated with approximated physics models such as the *SESAME* EOS model [1] the Lee–More [2] or Purgatorio [3] models for thermal conductivity κ, and the Astrophysics Opacity Table (AOT) [4] for radiation transport. Since the ignition designs generally have small margins, it is legitimate to ask whether or not the high-energy-density plasma properties given by these physics models are accurate enough for reliable ICF target designs. To that end, we have employed first-principles (FP) methods to systematically investigate the EOS, κ, and opacity of ICF-relevant materials (DT and CH ablator) in a wide range of plasma conditions through state-of-the-art *ab initio* calculations. Through these

[3]Author to whom any correspondence should be addressed

systematic studies, we have established the first-principles equation-of-state (FPEOS) tables [5–8] and thermal conductivities [9,10] (κ_{QMD}) for both DT and CH, as well as the first-principles opacity table (FPOT) [11] for DT. Implementing these FP-based plasma properties [FPEOS, κ_{QMD}, and FPOT] into our hydrocodes, we have examined their effects on ICF simulations. It was found that neutron yield and target gains from hydrodynamic simulations using these FP-based plasma properties can differ by a factor of 2 to 3 with respect to the traditional model predictions [12].

2. The first-principles methods

These FP methods include the quantum-molecular dynamics (QMD) and the path-integral Monte Carlo (PIMC) simulations. Both methods treat the electrons quantum-mechanically, while the ions are described by classical Newtonian mechanics (in QMD). In the PIMC method [13], the density matrix of a many-body system at high temperatures is built from the plane waves that describe the free particles. The density matrix at low temperatures can be obtained by performing the path-integral along the temperature path, using the known high-T density matrix. This method can handle plasmas with temperatures down to ~0.1 T_F (T_F is the Fermi temperature); however, for lower-T plasmas, the Fermi-sign problem renders the PIMC method impractical.

For complex plasmas (beyond DT) and transport properties of DT, we switched to the QMD method, which is based on the density functional theory (DFT) [14]. The QMD method combines the Kohn–Sham orbital-based molecular dynamics (KSMD) and the orbital-free molecular dynamics (OFMD) [15]. Both methods apply the DFT to find the thermodynamic-equilibrium electron density for each evolving ion configuration in the plasma. Subsequently, the motion of ions is then driven by the electronic force (derived from the obtained electron density) and the ion–ion interaction force, through the classical Newton equation. Once the plasma evolution is simulated, the various plasma properties can be extracted self-consistently. Combining both the KSMD and the OFMD methods, we can sample a wide range of temperature and density conditions relevant to both shell and coronal plasmas in ICF implosions.

3. FP-based plasma properties of DT and CH

In this section, we compare the plasma properties extracted from our first-principles calculations with both experiments and traditional model predictions. To be specific, we have conducted both PIMC and QMD simulations of D_2 plasmas of ρ = 0.001 to 1596 g/cm^3 and T = 1000 K to 64,000,000 K, and CH plasmas of ρ = 0.1 to 100 g/cm^3 and T = 1000 K to 4,000,000 K with QMD calculations. These ρ,T conditions fully cover the ICF plasma conditions. By doing mass/density scaling to the D_2/CH results, we can obtain the plasma properties of DT and deuterated plastic (CD). It is noted that the same QMD method can also be directly used for DT/CD property calculations, which should be equivalent to what obtained by mass/density scaling from the D_2/CH results.

Next, in figure 1, we compare the FP-predicted Hugoniots with experimental measurements [16–21] and the *SESAME* and Kerley03 models [1,22], respectively, for (a) D_2 and (b) CH.

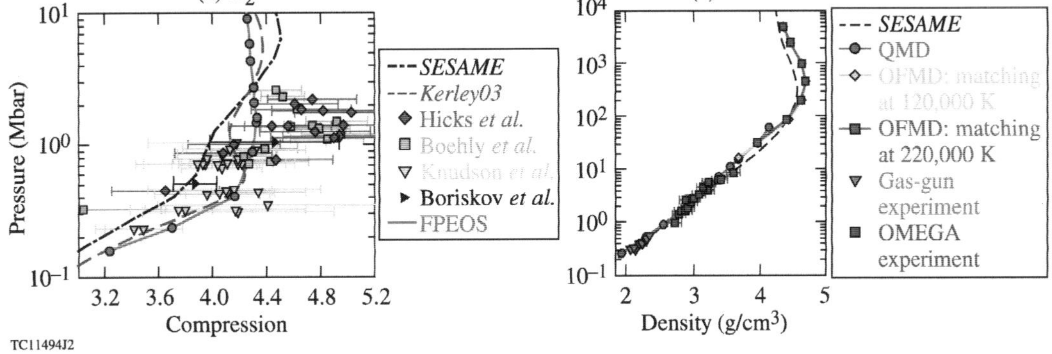

Figure 1. Comparison of FPEOS-predicted Hugoniot with both experiments and models.

Figure 1 shows that the FP calculations of shock Hugoniot of both D2 and CH agree well with experiments, while the traditional EOS models deviate from the FP predictions in certain pressure ranges. The off-Hugoniot EOS has been compared with the EOS models in previous publications [5–8], which indicated significant differences in warm dense plasma conditions where strong-coupling and electron degeneracy effects are important.

Figure 2 compares the resulting (a) opacity and (b) κ_{QMD} of D_2 and (c) the κ_{QMD} of CH with traditional models used in our hydrocodes. We see from figure 2 that the thermal conductivities of warm dense D_2 and CH plasmas are higher than the Lee–More predictions by a factor of 2 to 10; also, the total Rosseland opacities of D_2 are higher than the cold-opacity patched AOT by a factor of 10 to 100 at plasma temperatures below 20eV.

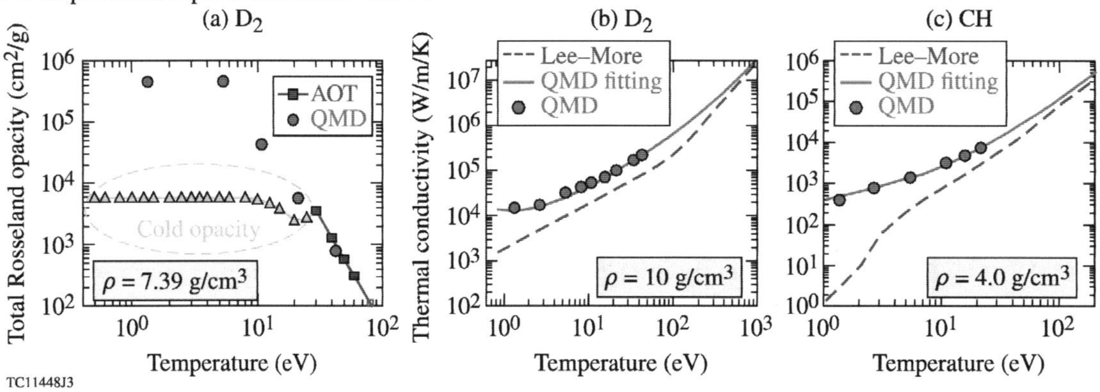

Figure 2. Comparisons of FP-predicted κ and opacity with traditional model predictions.

4. Effects of FP-based plasma properties on ICF simulations

Incorporating these FP-based properties [FPEOS, κ_{QMD}, and FPOT] of D_2 and CH into our hydrocodes, we can examine their effects on direct-drive ICF designs. One example is shown in figure 3, where a direct-drive ICF target design (a) is simulated with our 1-D hydrocode *LILAC*. The simulation results are compared figure 3(b), in which the dashed line represents the traditional simulations in contrast to the solid lines using the FP-based plasma properties of DT, while the high-density carbon ablator is still simulated with traditional models. The final target gain varies from G = 28 in the traditional models to the more-reliable value of G = 11 using the accurate FP-based plasma properties of DT. The effects of CH FPEOS and κ_{QMD} of CH further decrease the target performance [8,10] in CH-ablator based ICF implosions because of the reduction in the implosion velocity. To recover high gains, the FP-based plasma properties should be used to retune the laser pulse shape [12].

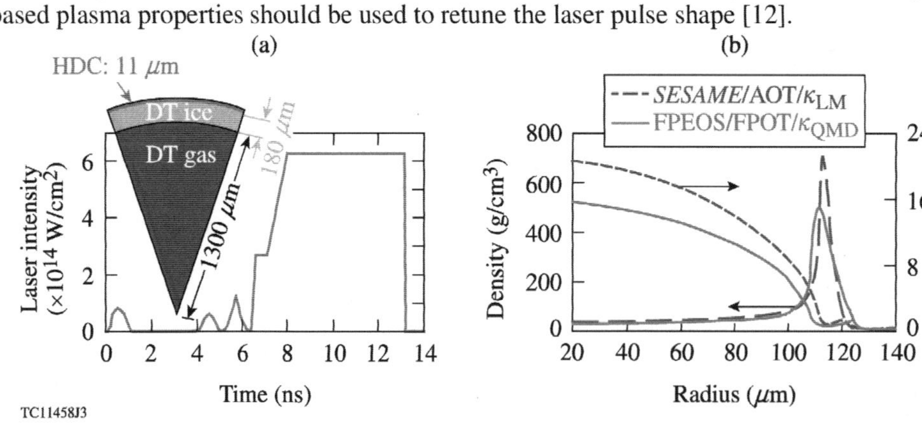

Figure 3. Comparisons of hydro simulations of a direct-drive ICF target design, using the FP-predicted properties of DT versus traditional models.

5. Conclusion

Using first-principles methods, we have systematically calculated plasma properties for a wide range of densities and temperatures of DT and CH for ICF applications. The calculated properties have shown significant differences when compared to traditional model predictions. Hydro simulations of ICF implosions with these FP-based plasma properties have shown a gain that is a factor of 2× to 3× smaller with respect to traditional model simulations. These studies indicated that FP-based plasma properties of DT and ablators should be used for reliable ICF target designs.

Acknowledgments

This material is based upon work supported by the Department of Energy National Nuclear Security Administration under Award No. DE-NA0001944, the University of Rochester, and the New York State Energy Research and Development Authority. The support of DOE does not constitute an endorsement by DOE of the views expressed in this article. This work was also supported by Scientific Campaign 10 at the Los Alamos National Laboratory, operated by Los Alamos National Security, LLC for the National Nuclear Security Administration of the U.S. Department of Energy under Contract No. DE AC52-06NA25396.

References

[1] Bennett B I, Johnson J D, Kerley G I and Rood G T 1978 Los Alamos, NM: Los Alamos National Laboratory Report LA-7130
[2] Lee Y T and More R M 1984 *Phys. Fluids* **27** 1273
[3] Wilson B, Sonnad V, Sterne P and Isaacs W 2006 *J. Quant. Spectrosc. Radiat. Transf.* **99** 658
[4] Huebner W F, Merts A L, Magee N H, Jr. and Argo M F 1977 Los Alamos, NM: Los Alamos National Laboratory Report LA-6760-M
[5] Hu S X, Militzer B, Goncharov V N and Skupsky S. 2010 *Phys. Rev. Lett.* **104** 235003
[6] Hu S X, Militzer B, Goncharov V N and Skupsky S 2011 *Phys. Rev. B* **84** 224109
[7] Hu S X, Boehly T R and Collins L A 2014 *Phys. Rev. E* **89** 063104
[8] Hu S X, Collins L A, Goncharov V N *et al.* 2015 *Phys. Rev. E* **92** 043104
[9] Hu S X, Collins L A, Boehly T R *et al.* 2014 *Phys. Rev. E* **89** 043105
[10] Hu S X, Collins L A, Goncharov V N *et al.* "First-principles investigations on ionization and thermal conductivity of polystyrene (CH) for inertial confinement fusion applications" to be published in *Phys. Plasmas*
[11] Hu S X, Collins L A, Goncharov V N *et al.* 2014 *Phys. Rev. E* **90** 033111
[12] Hu S X, Goncharov V N, Boehly T R *et al.* 2015 *Phys. Plasmas* **22** 056304
[13] Militzer B and Ceperley D M 2000 *Phys. Rev. Lett.* **85** 1890
[14] Kohn W and Sham L J 1965 *Phys. Rev.* **140** A1133
[15] Lambert F, Clérouin J and Zérah G 2006 *Phys. Rev. E* **73** 016403
[16] Hicks D G, Boehly T R, Celliers P M 2009 *Phys. Rev. B* **79** 014112
[17] Boehly T R, Hicks D G, Celliers P M, *et al.* 2004 *Phys. Plasmas* **11** L49
[18] Boriskov G V, Bykov A I, Il'kaev R I *et al.* 2003 *Dokl. Phys.* **48** 553
[19] Knudson M D and Desjarlais M P 2009 *Phys. Rev. Lett.* **103** 225501
[20] Marsh S P ed 1980 *LASL Shock Hugoniot Data* Los Alamos Series on Dynamic Material Properties (Berkeley, CA: University of California Press)
[21] Barrios M A, Hicks D G, Boehly T R *et al.* 2010 *Phys. Plasmas* **17** 056307
[22] Kerley G I 2003 Albuquerque, NM: Sandia National Laboratory Report SAND2003-3613

Controllable laser ion beam generation

D Kamiyama[1], M Takano[1], T Nagashima[1], D Barada[1], Y J Gu[2], X F Li[2], Q Kong[2], P X Wang[2] and S Kawata[1]

[1] Department of Advanced Interdisciplinary Sciences, Utsunomiya University, Utsunomiya, Japan
[2] Institute of Modern Physics, Fudan University, Shanghai, China

E-mail: mt146614@cc.utsunomiya-u.ac.jp, kwt@cc.utsunomiya-u.ac.jp

Abstract. In intense-laser plasma interaction, several issues still remain to be solved for a future laser particle acceleration. In this paper we focus on a bunching of ion beam, which is pre-accelerated by a strong electric field generated in a laser plasma interaction. In this study, a near-critical-density plasma target is illuminated by an intense short laser pulse. A moving strong inductive electric field is generated inside of the target. We have successfully obtained a bunched ion beam in our particle-in-cell simulations in this paper.

1. Introduction

Intense short-pulse lasers are now available. Based on the new laser technology, new mechanisms for particle acceleration have been proposed by the laser pulse, as an alternative of the conventional accelerators [1-3]. However, in the laser particle acceleration method there are issues, which include the controllability for the particle beam quality [3, 4] and for the particle energy [5-12]. Figure 1 shows a concept of an example future laser ion accelerator, which consists of post-accelerators, collimators and bunchers to improve the laser-produced particle beam quality [3]. In this study we focus on a bunching of the laser-produced proton beam. The buncher is a device to prevent the ion beam elongation during its propagation by reducing a velocity difference in the ion beam.

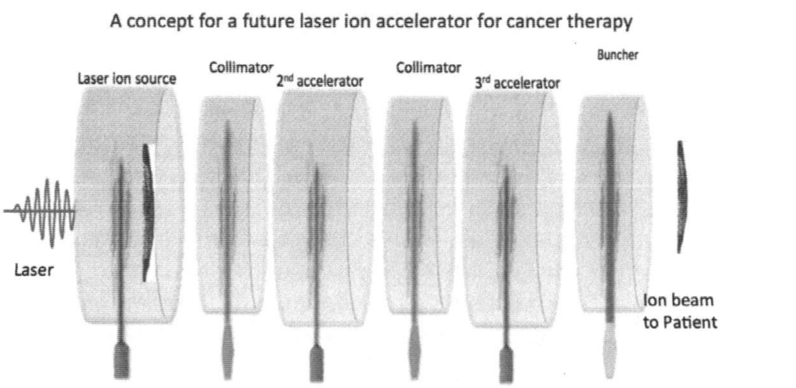

Figure 1. A concept of an example future laser ion accelerator.

Ion beam has been used in cancer treatment, ion beam inertial fusion, material sciences and other purposes, and has a unique preferable feature to deposit its main energy inside of a material [2-3]. In this paper a bunching of a laser produced ion beam is discussed for a future compact intense-laser ion accelerator [3].

In the laser ion acceleration, first electrons are accelerated in the near-critical-density plasma, and generate a current flow, which creates a strong magnetic field. The strong magnetic field moves together with the electron movement inside of the plasma. The inductive electric field is generated at the head of the electron current. The ion beam is accelerated by the inductive field. The inductive field would be also used for the ion post-acceleration and for the ion beam bunching. Figures 2 show the mechanism of the inductive field generation.

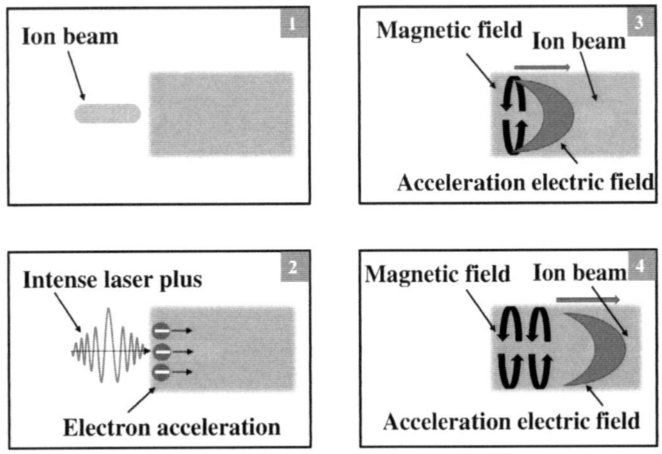

Figure 2. Mechanism of the moving inductive electric field generation in a near-critical density plasma.

In this study, we perform 2.5-dimensional PIC (Particle-In-Cell) [13] simulations to study the ion beam bunching.

2. Ion beam bunching

In order to reduce the proton energy divergence and to reduce the ion beam longitudinal length, a bunching device would be required. When a near-critical-density plasma target is illuminated by an intense laser, the strong inductive electric field is generated inside of the plasma target. The inductive electric field contributes to the ion beam tail acceleration to reduce the ion beam energy divergence. The moving speed of the inductive electric field becomes higher as the plasma density decreases.

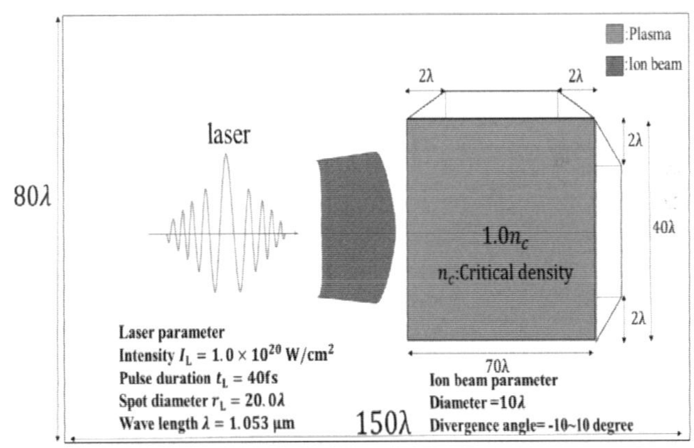

Figure 3. Simulation model for the ion beam bunching.

Figure 3 presents our particle simulation model for the bunching. The laser intensity is 1×10^{20} W/cm^2, the laser spot size 20λ, and the laser pulse length 40fs. The laser transverse profile is in the Gaussian distribution, and the laser temporal profile is also Gaussian. The laser wavelength is λ=1.053μm. The simulation box is 150λ in the longitudinal direction and 80λ in the transverse direction.

Figures 4 show the simulation results for the inductive electric field inside of the plasma target. The inductive field for the acceleration moves in the *x* direction. The electric field accelerates the tail of the ion beam. Figures 5 show the spatial distributions of the proton beams. When the buncher is employed, the longitudinal length of the proton beam becomes relatively shorter as shown in Fig. 5(b). If the buncher is not used, the length of the proton beam (Δ*x*) is 17.8λ. However, the buncher is employed, the proton beam spatial distribution is kept short as shown in Fig. 5(b), and the beam length is 14.8λ.

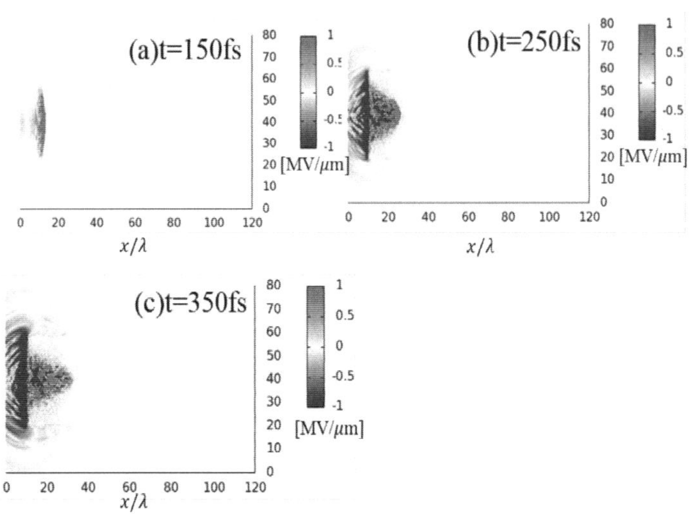

Figure 4. Inductive electric field inside of the plasma target at t=150fs, 250fs and 350fs. The inductive electric field accelerates the tail of the ion beam.

Figures 6 show the results for the energy distributions of the proton beams. When the buncher is not used, the proton beam is elongated as shown in Fig. 6(a), and the longitudinal velocity tilt ΔV_x is 0.121*c*. However, when the buncher is employed, the proton beam velocity divergence is reduced successful (ΔV_x=0.071*c*) and the proton beam length is kept short as shown in Fig. 5(b).

Based on the results presented here, we have succeeded to reduce the longitudinal spread of the entire proton beam by the bunching device.

3. Conclusions

In this study, we focused on the ion beam bunching using the near-critical-density plasma target. The inductive field acceleration also contributed to reduce the ion energy spread in this study inside of the target. The results presented in this paper contribute to the high-quality proton beam generation in laser plasma interaction. We have succeeded to reduce the energy spread and so to shorten the ion beam longitudinal length through the bunching device. The laser-illuminated near-critical-density plasma target also provides the new bunching device for the laser ion acceleration system.

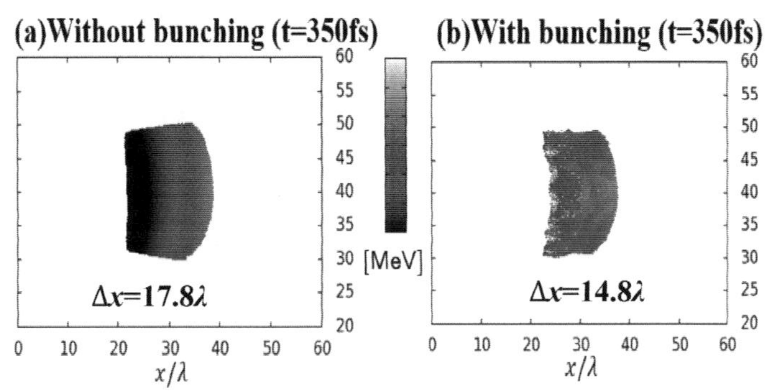

Figure 5. Spatial distributions of the proton beams. If the buncher is not used, the length of the proton beam (Δ*x*) is 17.8λ in Fig. 5(a). However, when the buncher is employed, the proton beam spatial distribution is kept short as shown in Fig. 5(b), and the beam length is 14.8λ.

The remaining issues in the laser ion accelerator include the energy efficiency from the laser to the ions, the ion beam spread in the *y* direction, the ion energy spectrum control, the ion particle energy control, the repetitive operation of the laser particle acceleration system, the laser generation efficiency, the laser target alignment, etc., as well as, the ion beam spread in the x direction. These issues should be studied further toward a real laser particle accelerator.

Acknowledgements
This work was partly supported by MEXT, JSPS, JSPS KAKENHI Grant Number 15K05359, ASHULA, ILE/Osaka University, CORE/Utsunomiya University, CDI/Utsunomiya University, Collaboration Center for Research and Development of Utsunomiya University, and Japan/U.S. Fusion Research Collaboration Program conducted by MEXT, Japan.

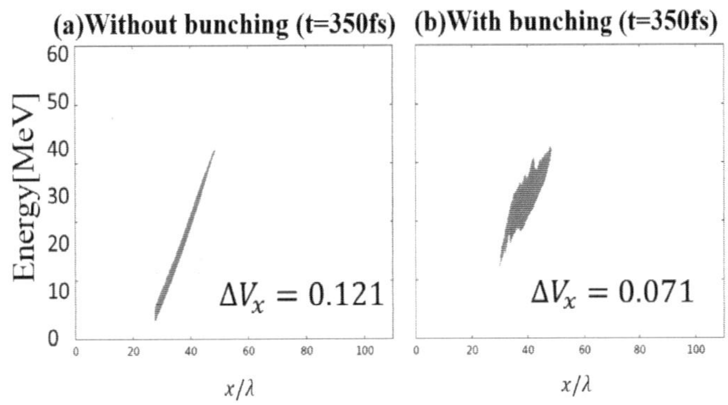

Figure 6. Ion beam energy spatial distributions. When the buncher is not used, the proton beam is elongated as shown in Fig. 6(a) and the longitudinal velocity is 0.121c. However, when the buncher is employed, the proton beam velocity divergence is reduced successfully (ΔVx=0.071c) as shown in Fig. 6(b).

References

[1] Tajima T and Dawson J M 1979 *Phys. Rev. Lett.* **43** 267
[2] Bulanov S V and Khoroshkov V S 2002 *Plasma Phys. Rep.* **28** 453
[3] Kawata S, Izumiyama T, Nagashima T, Takanao M, Barada D, Kong Q, Gu Y J, Wang P X, Ma Y Y and Wang W M 2013 *Laser Therapy* **22** 103
[4] Nakamura M, Kawata S, Sonobe R, Kong Q, Miyazaki S, Onuma N and Kikuchi T 2007 *J. Applied Phys.* **101** 113305
[5] Wang W M, Sheng Z M, Li Y T, Chen L M, Kawata S and Zhang J 2010 *Phys. Rev. STAB.* **13** 071301
[6] Nodera Y, Kawata S, Onuma N, Limpouch J, Klimo O and Kikuchi T 2008 *Phys. Rev.* E **78** 046401
[7] Hatchett S P, Brown C G, Cowan T E, Henry E A, Johnson J S Key M H, Koch J A, Langdon A B, Lasinski B F, Lee R W, Mackinnon A J, Pennington D M, Perry M D, Phillips T W, Roth M, Sangster T C, Singh M S, Snavely R A, Stoyer M A, Wilks S C and Yasuike K 2000 *Phys. Plasmas* **7** 2076
[8] Kaluza M, Schreiber J, Santala M I, Tsakiris G D, Eidmann K, Meyer T V and Witte K J 2004 *Phys. Rev. Lett.* **93** 045003
[9] Borghesi M, Mackinnon A J Campbell D H, Hicks D G, Kar S, Patel P K, Price D, Romagnani L, Schiavi A and Willi O 2004 *Phys. Rev. Lett.* **92** 055003
[10] Robinson A P L, Bell A R and Kingham R J 2006 *Phys. Rev. Lett.* **96** 035005
[11] Miyazaki S, Kawata S, Sonobe R and Kikuchi T 2005 *Phys. Rev.* E **71** 056403
[12] Takahashi K, Kawata S, Satoh D, Ma Y Y, Barada D, Kong Q and Wang P X 2010 *Phys. Plasmas* **17** 093102
[13] Lasinski B F, Langdon A B, Hatchett S P, Key M H and Tabak M 1999 *Phys. Plasmas*, **6** 2041

Ionization and acceleration of heavy ions in high-Z solid target irradiated by high intensity laser

D Kawahito and Y Kishimoto

Graduate School of Energy Science, Kyoto University, Uji, Kyoto, 611-0011

Abstract. In the interaction between high intensity laser and solid film, an ionization dynamics inside the solid is dominated by fast time scale convective propagation of the internal sheath field and the slow one by impact ionization due to heated high energy electrons coupled with nonlocal heat transport. Furthermore, ionization and acceleration due to the localized external sheath field which co- propagates with Al ions constituting the high energy front in the vacuum region. Through this process, the maximum charge state and then q/A increase in the rear side, so that ions near the front are further accelerated to high energy.

1. Introduction

The development of short pulse high power lasers which intensity reaches to 10^{21}W/cm^2 have opened up new schemes for ion acceleration and associated applications such as hadron radiotherapy, fusion neutron generation, ion driven fast ignition, etc., by choosing the material state properly [1,2]. The relatively lighter elements such as hydrogen, carbon, etc., which easily become fully ionized state with the largest charge-to-mass ratio q/A, and then are accelerated efficiently, have been widely used for such applications. On the basis of these experiences, it is worthwhile to extend the schemes to high energy acceleration of heavy elements, which explores a wider class of application especially related to nuclear physics, e.g., highly charged heavy ion beam, nucleus-nucleus collision and nuclear transmutation, extraction and acceleration of short lived exotic nuclei [3]. In contrast to such lighter elements, heavier ones irradiated by high power laser exhibit complex structure and dynamics dominated by plural physical processes such as ionization and recombination, collision and relaxation, radiation and the transport, etc. These processes have to be taken into account in the analysis simultaneously and self-consistently.

2. Simulation model

We carry out the 2-dimensional simulation for evaluating the ionization and ion acceleration by using EPIC3D (Extended Particle based Integrated Code [4]) which includes atomic processes such as a field ionization and an impact ionization. The field ionization is calculated as tunnel a field ionization process by calculating the electric field per particle based on ADK model [5], and the impact ionization is calculated by the cross-section through a binary collisional process between ion and electron based on BEB model [6]. The field energy loss can be estimated from the energy conservation with the ionization energy, so that we introduce the ionization current calculated from the energy as Joule heating. The simulation capability is set as L_y=14.4 μm (18λ) in the laser propagation direction (-y) and L_x=28.8 μm (36λ) in the transverse direction (-x), where λ (=0.8μm) is a wave length of laser field. In the simulation, the laser peak intensity is assumed to be I=1.0×10^{21} W/cm^2 (a_0=21). The laser field is emitted with P-polarization from the left hand boundary of the simulation domain during the gaussian pulse duration τ=40 fsec and focused into 3 micron diameter to the solid surface. The solid film is based on an Al (Z=13) atom which is initially located in 3.2 μm < y < 4.0 μm with a thickness l_{Al}=0.8 μm (=λ) and distributed uniformly on the transverse direction. The mass density of the film is 0.675 g/cm^3 which value is 1/4 for the real mass density of Al material, i.e., 2.7 g/cm^3. The initial charge state is set as Al^{1+} assumed the free electrons in the film, so that the initial electron density is n_e/n_{cr}=8.6, where n_{cr} is the cut-off density, i.e., 1.74×10^{21} cm^{-3}. Therefore, the initial penetration length is $\delta_{q=1}=4.35\times10^{-2}$ μm, and that at the fully ionized plasma reaches to $\delta_{q=13}=1.21\times10^{-2}$ μm. In the simulation model, we investigate the ionization dynamics and ion acceleration of Al.

3. Simulation result

3.1 Time history of the ionization process

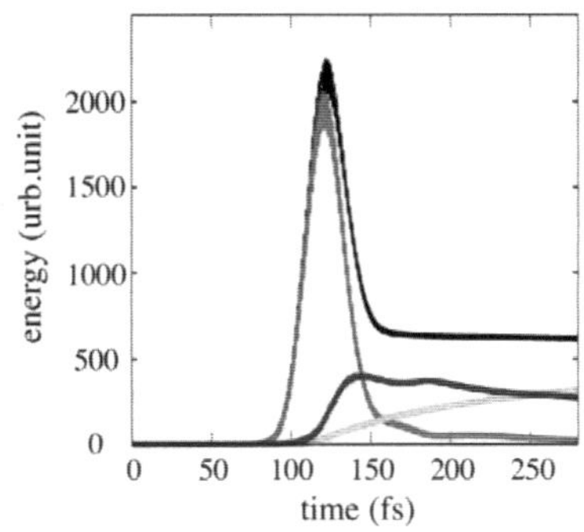

Figure 1. Time history of field (red), electron (blue), ion (gren), and total energy (black).

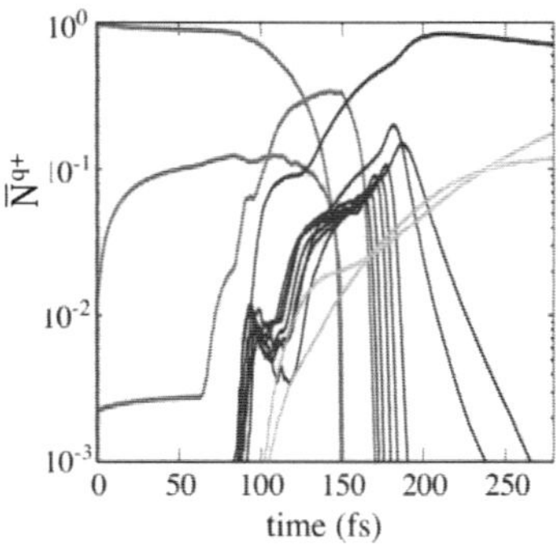

Figure 2. Time history of ion abundance of each charge state, where each lines are M-shell(red), L-shell(blue), and K-shell(green).

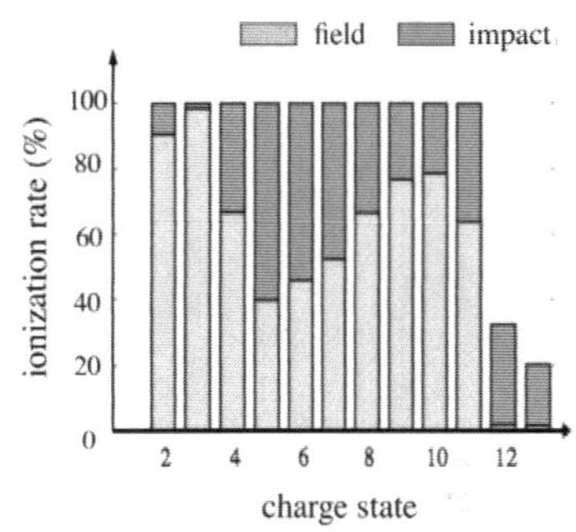

Figure 3. Ionization rate of the the field(blue) and impact(red) for each charge state.

Figure 1 represents the time history of field energy E_f, electron kinetic energy E_e, ion kinetic energy E_i summed up each charge state q (1-13), and total energy E_t (=E_f+E_e+E_i) in the simulation domain, respectively. The laser field is irradiated from the antenna, and then it applies the energy into the simulation domain and the plasma. From t=100 fsec, the both energies of electron and ion start to increase due to the absorption from the laser field with approaching the time of the peak laser intensity (t=120 fsec). After the laser ceases, the electron energy saturates at t=135 fsec and then decrease gradually. A part of field energy remains in the simulation domain as the sheath field which generated by the charge separation between the high energy electrons and ion surface. Then, the ion energy keeps increasing due to the energy transfer from that of the electrons and the field to that of the ions, i.e., sheath ion acceleration.

In this process, we found the two kind of rapid ionization dynamics as seen in Figure 2. Fig.2 shows the ion abundance $\bar{N}^{q+}$ for each charge state of Al in logarithmic scale. Here $\bar{N}^{q+}$ is determined as N^{q+}/N_{ion}, where N^{q+} is ion number for each charge state, and N_{ion} is total ion number. After the laser field hits the surface of the film, the ionization in M-shell electrons proceeds gradually from the initial charge state Al^{1+} until t=80 fsec. Then, the first rapid ionization starts to proceed for L-shell electrons, and the abundance of Al^{11+} which is the final state of the L-shell reaches to $\bar{N}^{11+}$=0.08. The increase of the abundance with the L-shell electrons saturates from t=100 fsec, and the inner ionization to Al^{12+} and Al^{13+} starts to increase gradually. At t=115 fsec, the second rapid ionization takes place from the lower charge state, and the population of ions with charge state in M- and L-shell mostly transits to Al^{11+}, i.e., the ionization state is held in the final charge state of L-shell due to the ionization potential gap between L- and K- shell. The abundance of Al^{11+} reaches $\bar{N}^{11+}$=0.9 at t=200 fsec and then decrease gradually while that of inner charge state: Al^{12+} and Al^{13+} keeps increasing with slower time scale than the other ionization process.

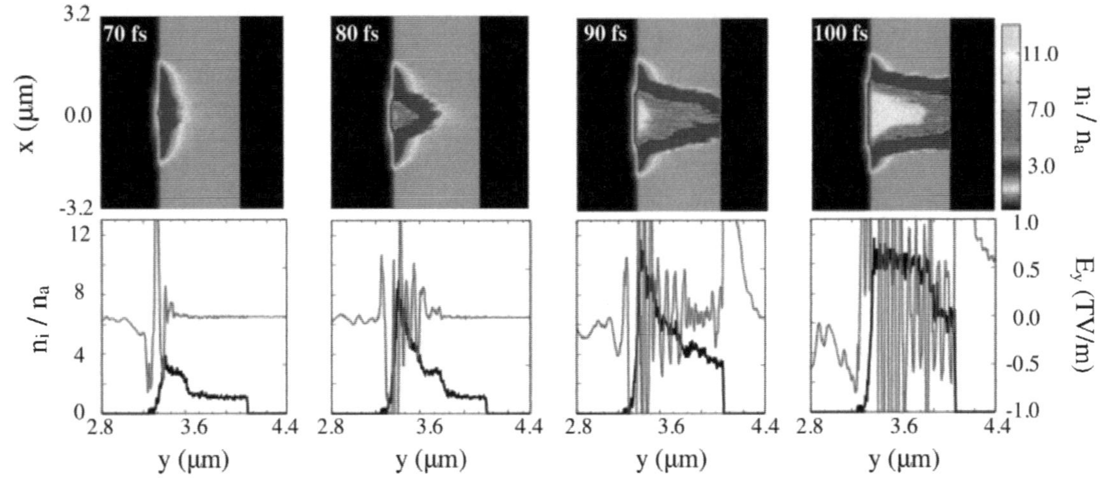

Figure 4. (a) Normalized ion density n_i/n_ain 2-dimentional domain at representative times, where n_i is the ion charge density, and n_a is the initial atom density, and (b) the profile of n_i/n_a and the electric field Ey on x=0 at same times to (a)

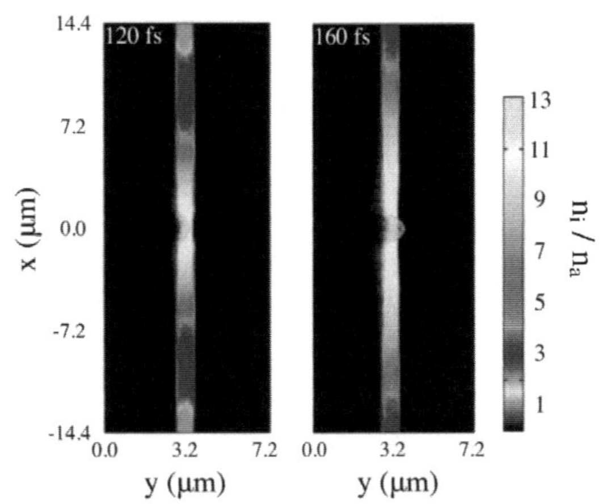

Figure 5. Normalized ion density n_i/n_a in 2-dimentional domain at t=120 fs and t=160 fs.

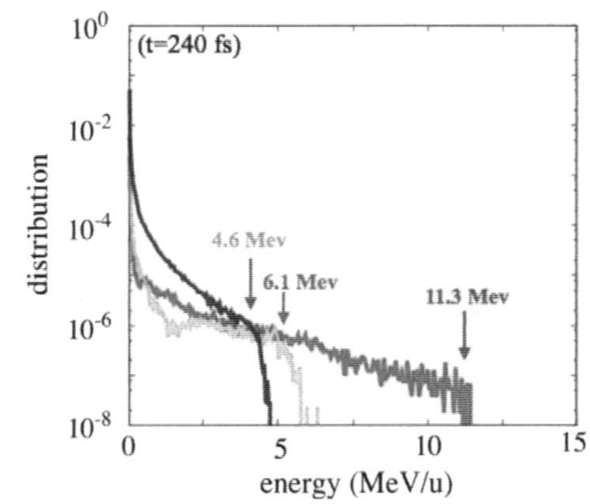

Figure 6. Ion energy distributions for the charge state q=11(blue). q=12(green), and q=13(red)

We also evaluate the difference between the field and the impact as seen in figure 3. Fig.3 shows the ionization rate of the field and the impact component for each charge state until t=250 fsec when the laser has already ceases from the simulation domain. In outer shell, i.e., M-shell, the field ionization takes place dominantly due to the laser field, and the rate reaches to 94.6 % which value is averaged from q=2 to q=3. The impact ionization proceeds only 5.6 %, i.e., the additional rate is 100. In L-shell, the field ionization rate decreases to averagely 40.4 %. Instead of that, the impact ionization rate increases due to the increase of the collisional frequency depended on the ionization degree, and the rate reaches to 59.6 % with slower time scale than that of the field. Finally, in inner-shell, i.e., K-shell, the impact components become dominant rate while that of the field is less than the 1%, however the total ionization proceed to only 25 % of the film. In this case, the ionization loss is estimated from the ionization degree, and the rate is only 0.1% compared with the laser energy. Thus, the ionization loss is negligible for the investigating the dynamics.

3.2 Spatial ionization structure and ion acceleration

Figure 4 represents (a) the normalized ion density n_i/n_a in the region of -3.2 μm < x < 3.2 μm and 2.8 μm < y < 4.4 at representative times, where n_i is the ion charge density, and n_a is the initial atom density, i.e., that shows the charge state as far as ions remain, and (b) the profile of n_i/n_a and the electric field E_y on x=0 at same times to (a). The ionization proceeds only around the front surface due to the cut-off of the laser until t=70 fsec. Then, the ionization front starts to propagate to the inside of the film with the electric field E_y as seen in Fig.4 (a) and (b). This transition is driven by the propagation of the fast electron bunches generated by the J×B heating [7]. The electron bunch is accelerated to the relativistic region at the front surface and injected to the inside of the film every twice of the laser frequency. The return current is also excited for the current neutralization with the bunch, and the solitary electric field at the front of bunch and the plasma wave behind that are generated for propagation direction by the charge separation inside the film [8]. Therefore, the field ionization to Al^{11+} takes places with the propagation of the electric fields E_y with the bunch as the first ionization dynamics. The ionization velocity is v_f=0.8c which is almost same as the averagely bunch velocity. After the electrons reach to the rear surface, the sheath field is generated by the charge separation between the ion surface and the electron bunches. The electrons behind of that are reflected by the sheath field to the inside of the film and recirculate between the front and the rear surface with the initial scattering angle. So that, the electrons spread to the transverse direction over the laser spot area with the ionization to Al^{11+} as seen in Figure 5, which is the same as Fig. 4 (a) in the whole region. Around the laser spot area, the field ionization proceeds due to the electron bunches with the electric field as discussed in the above. However, the amplitude of the electric field gradual decreases due to the electron diffusion with the spread to the transverse direction, and the impact ionization rate changes larger than that of the field as second rapid ionization dynamics. Therefore, the velocity of the ionization to Al^{11+} for transverse direction is decelerated to averagely v_f=0.6c compared with the first one. Additional to these rapid ionization processes, the ionization to Al^{12+} and Al^{13+} takes place due to the sheath field on the each surface and the collision around the laser spot area. So that, the charge states of accelerated ion are decided more than Al^{11+} through these ionization processes and extracted from the film.

The multi charged ions are accelerated by the sheath field generated on the both surfaces. The peak amplitude of the sheath field reaches to 35 TV/m which is same order to the laser peak amplitude, so that the ionizations to Al^{12+} and Al^{13+} proceed due to the field ionization from the charge state of the bulk Al^{11+}. Then, the ions are accelerated and extracted to the vacuum region with the dependence of its charge to mass ratio as seen in Figure 6 which represents the ion energy distribution for each charge state. The ions of fully ionized state (Al^{13+}) with the largest charge to mass ratio is accelerated most efficiently in the Al, and then the maximum ion energy reaches to 11.3 MeV/u. On the other hand, the ion energy of Al^{11+} and Al^{12+} reaches to 4.6 MeV/u and 6.1 MeV/u which is less than the dependence of the charge to mass ratio compared with that of Al^{13+}. Therefore, the sheath amplitude for a lower charge state becomes smaller than that for Al^{13+} in the outside than the laser spot area.

In the interaction between such high intensity laser and the solid film of Al, the ionization dynamics with the field and the impact is mainly driven by the electron transport and decide the charge state of accelerated ions. Therefore, we will perform the simulation for higher-Z elements including the other atomic process such as recombination, radiation, etc.

Referencce

[1] H. Daido, M. Nishiuchi, AS.Pirozhkov, Rep.Proq.Phys., **75**, 056401(2012)
[2] Y. Kishimoto, T. Masaki, *et al.*, Phys.Plasmas. **9**, 589(2002)
[3] M. Nishiuchi, H. Sakaki, *et al.*, Phys. Plasmas. **22**, 033107(2015)
[4] T. Masaki, Y. Kishimoto, J. Plasma Fusion Res., **81**, 643(2005)
[5] M. V. Ammosov, et al, Sov. Phys. JETP **64**, 1191 (1986)
[6] Y. K. Kim and M. E. Rudd, Phys. Rev. A **50**, 3954 (1994)
[7] W. L. Kruer and Kent Estabrook., Phys. Fluid **28**,430(1985)
[8] V. T. Tickhonchuk, Phys. plasmas **9**, 4 (2002)

Platform for spectrally resolved x-ray scattering from imploding capsules at the National Ignition Facility

D Kraus[1], T Döppner[2], A L Kritcher[2], A Yi[3], K Boehm[2], B Bachmann[2], L Divol[2], L B Fletcher[4], S H Glenzer[4], O L Landen[2], N Masters[2], A M Saunders[1], C Weber[2], R W Falcone[1] and P Neumayer[5]

[1]Department of Physics, University of California, Berkeley, CA 94720, USA
[2]Lawrence Livermore National Laboratory, Livermore, CA 94550, USA
[3]Los Alamos National Laboratory, Los Alamos, NM 87545, USA
[4]SLAC National Accelerator Laboratory, Menlo Park, CA 94309, USA
[5]GSI Helmholtzzentrum für Schwerionenforschung, 64291 Darmstadt, Germany

E-mail: dominik.kraus@berkeley.edu

Abstract. We present a new experimental platform to perform spectrally resolved x-ray scattering measurements of ionization, density and temperature in imploding CH or beryllium capsules at the National Ignition Facility. Scattered x-rays at 9 keV from a zinc He-alpha plasma source at a scattering angle of 120 degrees are highly sensitive to K-shell ionization, while at the same time constraining density and temperature. This platform will allow for x-ray scattering studies of dense plasmas with free electron densities up to 10^{25} cm^{-3} giving the possibility to investigate effects of pressure ionization and Pauli blocking on the ablator ionization state right before or shortly after stagnation of the implosion.

1. Introduction

A precise knowledge of ionization at given temperature and density is crucial in order to properly model compressibility and heat capacity of inertial confinement fusion (ICF) ablator materials for efficient implosions producing energy gain [1]. Spectrally resolved x-ray scattering [2] has been proven as outstanding tool to characterize such dense plasma parameters and experiments at the OMEGA laser facility of University of Rochester on directly driven beryllium and plastic (CH) capsules could investigate electron densities up to $10^{24}\,\mathrm{cm}^{-3}$ [3, 4]. With the availability of the National Ignition Facility (NIF), these studies can be extended to much higher densities: The experimental platform presented here aims for densities approaching and eventually exceeding $10^{25}\,\mathrm{cm}^{-3}$. These states of matter are strongly affected by pressure ionization and, besides ICF, highly relevant for modeling astrophysical objects like brown dwarfs and stars [5]. These conditions can currently only be accessed experimentally at NIF [6, 7].

2. Experiment

Our approach utilizes all 192 beams of the NIF laser: 184 beams are focused into a standard ICF hohlraum to implode and compress a CH or beryllium capsule and 8 beams are directed to a zinc foil in order to create intense Zn He-alpha line emission at 9 keV as x-ray source for

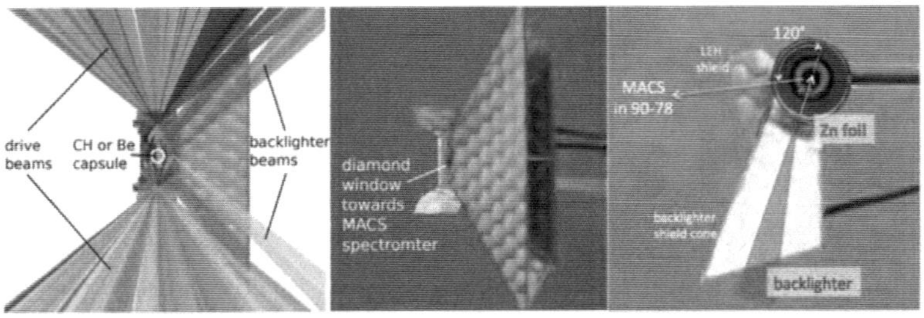

Figure 1. Target design of the new experimental platform to measure x-ray Thomson scattering from imploding capsules on the National Ignition Facility. X-rays created by irradiation of a Zn foil attached to the hohlraum wall are guided by a gold cone towards the imploded sample. The highly efficient MACS spectrometer records the scattered photons in the equatorial plane. Left: schematic showing the laser pathways onto the target. Middle: side-view of finished target. Right: top-view of finished target.

the scattering diagnostics (see Fig. 1). The x-rays are guided onto the imploding capsule by a gold cone which sticks into the hohlraum and initially holds the capsule [8]. Scattered x-rays can escape from the hohlraum through a diamond window of 80 μm thickness at a scattering angle of 120 degrees in the equatorial plane. The areal size of the window can be adjusted to the expected size of the imploding capsule at the time of probing. An additional larger diamond window on the opposite side ensures that the scattered x-rays are not overlaid by continuum or line emission from the gold hohlraum wall. The scattered x-rays are collected by the highly efficient Mono Angle Crystal Spectrometer (MACS) which can apply different HOPG crystals as dispersive element towards a 4-strip gated x-ray detector (GXD) [9]. A cylindrically curved crystal focusing the signal on a single strip allows for maximum collection efficiency, but is limited to a single probe time in the experiment. A flat crystal illuminates all four strips and thus can probe four different times in the implosion while providing 10x less scattered intensity on a single strip compared to the cylindrically curved crystal. An intermediate solution can be obtained by a two-segment cylindrically curved crystal focusing the collected x-rays on two strips of the GXD and thus giving the possibility to probe the implosion at two different times while delivering reasonable signal levels. The spectrometer is protected from a direct line-of-sight to the Zn source plasma by a gold cone that is mounted onto the hohlraum wall. It extends 12.5 mm from target chamber center in order to prevent the plasma plume from being imaged onto relevant areas of the MACS detector.

For the analysis of the x-ray scattering signal, it needs to be considered that the whole plasma volume that is reached by the source radiation and is visible to the spectrometer contributes to the measured spectrum. Thus, a three-dimensional analysis scheme has been developed which accounts for spatial gradients by volumetric and density weight as well as absorption losses inside the probe volume [10]. In previous work, we found that density-weighted averages of density, temperature and ionization as single parameters can describe the scattering spectrum of such an inhomogeneous target reasonably well as long as opacity losses in the imploded capsule are not significant [11].

The left panel of Fig. 2 shows conditions which can be achieved in a standard low-adiabat ICF implosion at the maximum density in comparison to previous experiments at the OMEGA laser facility. Highest densities approaching $10^{25}\,\mathrm{cm}^{-3}$ are present in the final convergence phase very close to stagnation. The right panel of Fig. 2 shows radial profiles of density and temperature close to stagnation of the implosion. In order to limit the sensitivity of the x-ray scattering diagnostics to the most dense regions, scattering from the low density coronal plasma needs to

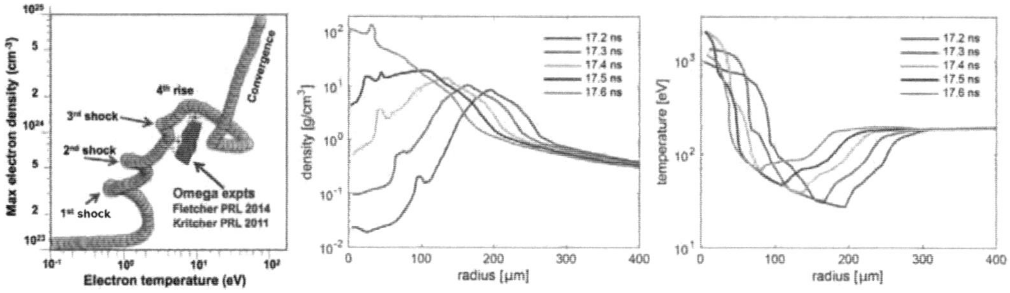

Figure 2. Left: conditions at maximum electron density in an exemplary CH four-shock low adiabat ICF implosion. Middle: radial density lineouts approaching stagnation. Right: radial temperature lineouts approaching stagnation.

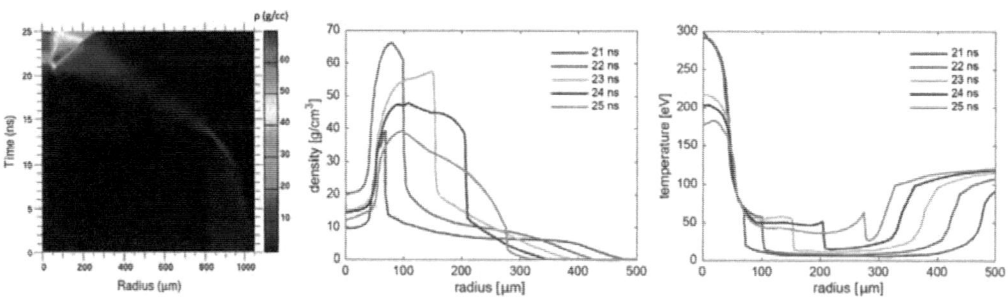

Figure 3. Left: one-dimensional simulation results of density at given radius and time of a beryllium capsule driven by using a truncated ICF pulse shape. Middle: radial lineouts of electron temperature at given times around stagnation. Right: radial lineouts of density at given times around stagnation.

be shielded. This is achieved by a corresponding opening of the cone which holds the capsule as well as the diamond window dimensions in the hohlraum wall. These dimensions can be adjusted depending on the exact timing during the implosion that will be probed in the experiment.

Figure 3 shows simulations for beryllium capsules using a drive which is optimized for achieving relatively homogeneous conditions at high densities and moderate temperatures. In order to realize this, the driving pulse is strongly truncated resulting in a reduced implosion velocity. The inward moving material is then accumulating around the central hot spot which does not exceed temperatures of 300 eV due to the slow implosion. The outgoing shock produces relatively homogeneous conditions in density and temperature.

Beryllium is a useful material to study pressure ionization of K-shell electrons: Since inner-shell electrons are more weakly bound than in carbon, ionization potential depression will have a stronger effect on K-shell ionization for given density and temperature compared to carbon. Moreover, as a single component system, modeling of scattering spectra is generally simpler than for CH. Figure 4 shows the effect of ionization potential depression with increasing density at zero temperature using the interpolation formula by Stewart and Pyatt (ΔI_{SP}). At electron densities around 10^{25} cm^{-3}, ΔI_{SP} exceeds the single atom ionization potential for Be^{3+}, meaning that beryllium would be fully ionized at this density. At the same time, the Fermi energy of the free electrons reaches comparable values. This starts to suppress ionization since the state, where an additional free electron would end up after an ionization process, might already be occupied

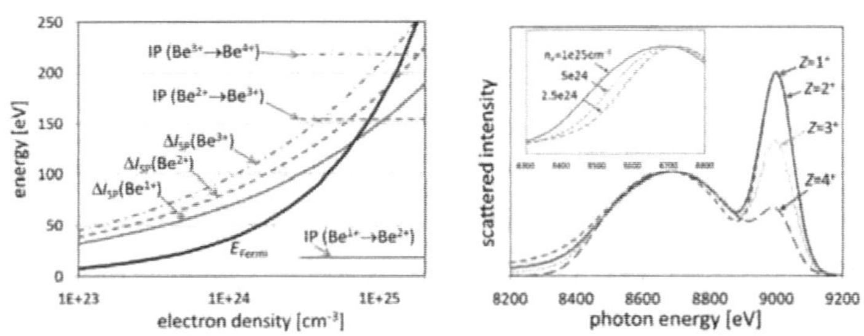

Figure 4. Left: comparison of ionization potentials, ionization potential depression given by the Stewart-Pyatt formula, and Fermi energy for beryllium. Right: Simulated scattering spectra for various charge states and for different electron densities (inset).

(Pauli-blocking). The right panel of Fig. 4 illustrates the sensitivity of spectrally resolved x-ray scattering to ionization and density of dense degenerate beryllium plasma. The elastic scattering is dominated by scattering from K-shell electrons and thus highly sensitive to K-shell ionization. The inelastic scattering from the electrons is shifted by the Compton energy and broadened by the energy distribution of the electrons. For a dense degenerate plasma, this is given by a Fermi distribution and hence, the width of the inelastic feature is strongly sensitive to density. For higher temperatures, the distribution steadily transforms into a Boltzmann distribution where the width is set by temperature. Thus, in the intermediate regime of dense degenerate plasmas with finite temperature, sensitivity to both density and temperature can be achieved by spectrally resolved x-ray scattering.

3. Conclusions

The presented platform allows for unique studies of pressure ionization effects in ICF ablator materials at extreme densities. The applied scattering diagnostics is strongly sensitive to K-shell ionization while at the same time constraining density and pressure. First experiments will be performed using well-tested standard ICF drives for CH capsules to demonstrate the functionality of the platform. Further experiments will then use an optimized drive for beryllium capsules in order to achieve relatively homogeneous conditions at extreme densities and moderate temperatures.

Acknowledgments

This work was performed with the assistance of Lawrence Livermore National Laboratory (LLNL) under Contract DE-AC52-07NA27344 and supported by Laboratory Directed Research and Development (LDRD) grants 11-ER-050 and 13-ERD-073. RWF and DK acknowledge support from SSAA program Contract No. DE-FG52-06NA26212.

References

[1] O. A. Hurricane et al., Nature **506**, 343 (2014).
[2] S.H. Glenzer and R. Redmer, Rev. Mod. Phys. **81**, 1625 (2009).
[3] A. L. Kritcher et al., Phys. Rev. Lett. **107**, 015002 (2011).
[4] L. B. Fletcher et al., Phys. Rev. Lett. **112**, 145004 (2014).
[5] W. B. Hubbard et al., Phys. Plasmas **4**, 2011 (1997).
[6] A. L. Kritcher et al., High Energy Density Phys. **10**, 27 (2014).
[7] T. Döppner et al., J. Phys.: Conf. Ser. **500**, 192019 (2014).
[8] K. Boehm et al., *Design and engineering of a target for x-ray Thomson scattering measurements on matter at extreme densities*, submitted (2015).
[9] T. Döppner et al., Rev. Sci. Instrum. **85**, 11D617 (2014).
[10] D. A. Chapman et al., Phys. Plasmas **21**, 082709 (2014).
[11] D. Kraus et al., *Increased K-shell ionization of carbon in imploding CH spheres at the National Ignition Facility*, submitted (2015).

Anomalous inverse bremsstrahlung heating of laser-driven plasmas

Mrityunjay Kundu
Institute for Plasma Research, Bhat, Gandhinagar - 382 428, Gujarat, India
E-mail: mkundu@ipr.res.in

Abstract. Absorption of laser light in plasma via electron-ion collision (inverse bremsstrahlung) is known to decrease with the laser intensity as $I_0^{-3/2}$ or with the electron temperature as $T_e^{-3/2}$ where Coulomb logarithm $\ln\Lambda = 0.5\ln(1+k_{\rm min}^2/k_{\rm max}^2)$ in the expression of electron-ion collision frequency $\nu_{\rm ei}$ is assumed to be independent of ponderomotive velocity $v_0 = E_0/\omega$ which is unjustified. Here $k_{\rm min}^{-1} = v_{\rm th}/\max(\omega, \omega_{\rm p})$, and $k_{\rm max}^{-1} = Z/v_{\rm th}^2$ are maximum and minimum cut-off distances of the colliding electron from the ion, $v_{\rm th} = \sqrt{T_e}$ is its thermal velocity, $\omega, \omega_{\rm p}$ are laser and plasma frequency. Earlier with a total velocity $v = (v_0^2 + v_{\rm th}^2)^{1/2}$ dependent $\ln\Lambda(v)$ it was reported that $\nu_{\rm ei}$ and corresponding fractional laser absorption (α) initially increases with increasing intensity, reaches a maximum value, and then fall according to the conventional $I_0^{-3/2}$ scaling. This anomalous increase in $\nu_{\rm ei}$ and α may be objected due to an artifact introduced in $\ln\Lambda(v)$ through $k_{\rm min}^{-1} \propto v$. Here we show similar anomalous increase of $\nu_{\rm ei}$ and α versus I_0 (in the low temperature and under-dense density regime) with quantum and classical kinetic models of $\nu_{\rm ei}$ without using $\ln\Lambda$, but a proper choice of the total velocity dependent inverse cut-off length $k_{\rm max} \propto v^2$ (in classical case) or $k_{\rm max} \propto v$ (in quantum case). For a given $I_0 < 5\times10^{14}\,{\rm Wcm}^{-2}$, $\nu_{\rm ei}$ versus T_e also exhibits so far unnoticed identical anomalous increase as $\nu_{\rm ei}$ versus I_0, *even* if the conventional $k_{\rm max} \propto v_{\rm th}^2$, or $k_{\rm max} \propto v_{\rm th}$ is chosen. However, for higher $T_e > 15$ eV, anomalous growth of $\nu_{\rm ei}$ and α disappear. The total velocity dependent $k_{\rm max}$ in kinetic models, as proposed here, may explain anomalous increase of α with I_0 measured in some earlier laser-plasma experiments. This work may be important to understand collisional absorption in the under-dense pre-plasma region due to low intensity pre-pulses and amplified spontaneous emission (ASE) pedestal in the context of laser induced inertial confinement fusion.

1. Introduction

Collisional absorption [1, 2, 3, 4] through electron-ion collision (inverse bremsstrahlung) occurs almost all the time in the sub-relativistic laser field. However, there is no universal agreement for the electron-ion collision frequency ($\nu_{\rm ei}$) in the laser field [1, 5, 6, 7]. In the conventional scenario, Coulomb logarithm $\ln\Lambda = 0.5\ln(1+k_{\rm min}^2/k_{\rm max}^2)$ in the expression of $\nu_{\rm ei}$ is assumed to be independent of the ponderomotive velocity $v_0 = E_0/\omega$ and the laser field strength E_0 which is clearly unjustified. Here $k_{\rm min}^{-1} = v_{\rm th}/\max(\omega, \omega_{\rm p})$, and $k_{\rm max}^{-1} = Z/v_{\rm th}^2$ are maximum and minimum cut-off distances of the colliding electron from the ion; $\omega, \omega_{\rm p}$ are laser and plasma frequency, and $v_{\rm th} = \sqrt{T_e}$ is the thermal velocity of an electron. Unless explicitly noted, we use atomic units (a.u.) i.e., $|e| = m_e = \hbar = 4\pi\varepsilon_0 = 1$, where e and m_e are charge and mass of an electron, $\hbar$ is the reduced Planck constant, and ε_0 is the permittivity of the free space. As a consequence, for a given $\omega_{\rm p}$, T_e, and ω, one finds that $\nu_{\rm ei}$ initially remains almost constant [6, 8, 9] with increasing $v_0/v_{\rm th}$ up to some value of $v_0/v_{\rm th}$, then decreases as v_0^{-3} when $v_0/v_{\rm th} \gg 1$ (see Fig.1 in Ref.[6]). In terms of the laser intensity $I_0 \propto v_0^2$, $\nu_{\rm ei}$ remains almost constant up to an intensity I_c, then it decreases as $I_0^{-3/2}$ when $I_0 \gg I_c$. The corresponding fractional absorption α (defined as the ratio of the absorbed laser energy to the incident laser energy) of the laser pulse vary with I_0 in the same way as ν_{ei} versus I_0 since $\alpha \propto \nu_{ei}$. In some experiments [1, 10], however, with normally incident s-polarized laser light, α was found to increase with I_0 up to a maximum value corresponding to an intensity I_c, and a decrease when $I_0 > I_c$. This anomalous increase of α versus I_0 was explained [11] with a classical scattering model (CSM) [7] using total velocity $v = (v_0^2 + v_{th}^2)^{1/2}$ dependent $\ln\Lambda(v)$. But, CSMs which use $\ln\Lambda$ have deficiencies. They do not match with more accurate kinetic models [6, 8, 9, 12, 13]) of ν_{ei} which has no $\ln\Lambda$ dependence. Kinetic models use only one cut-off k_{max}. Therefore, the artifact in $\ln\Lambda$ through k_{min} is questionable for CSMs. A large unphysical value of ν_{ei} may result with CSM compared to the kinetic models in the low temperature regime of our interest. To overcome those limitations of CSM, we use kinetic models and show that anomalous variation of ν_{ei} and α with I_0 are possible to explain in the regime of low temperature (< 15 eV) and under-dense densities for a given laser wavelength, provided k_{max} is chosen in some appropriate form. By choosing the conventional classical cut-off $k_{max} = v_{th}^2/Z$ (see Refs.[6, 8]) or the quantum cut-off $k_{max} = 2v_{th}$, we find anomalous increase of ν_{ei} versus T_e for a given I_0. Then we propose a modified $k_{max} = v^2/Z$ (for the classical case) or $k_{max} = 2v$ (for the quantum case) which depends on E_0 and ω, and show anomalous increase of ν_{ei} vs I_0 for a given $T_e < 15$ eV as in our earlier work [11] using CSM. The proposed k_{max} is useful to explain the low intensity growing part of the experimental absorption curve in Ref.[10]. We assume uniformly charged hydrogen like plasma with $Z = 1$, density $n_e = Zn_i = n_p = 10^{28}/m^3$, and $\omega/\omega_p = 5$ giving the laser wavelength $\lambda \approx 66$ nm as considered in Ref.[8]. Corresponding critical density is $n_c \approx 2.52 \times 10^{23}\text{cm}^{-3}$.

2. Kinetic models of ν_{ei}

Assuming the Maxwellian velocity distribution for electrons and stationary ions, the quantum mechanical expression [8] of energy absorption rate $\dot{\varepsilon} = \langle e n_e \boldsymbol{v}_e \cdot \boldsymbol{E}\rangle$ is given by

$$\dot{\varepsilon} = \frac{8\sqrt{2\pi}Z^2 n_e n_i}{v_e^3}\omega^2 \sum_{n=1}^{\infty} n^2 \int_{k=0}^{\infty} \frac{dk \exp[-\frac{1}{2}(n\omega/kv_e)^2 - \frac{1}{2}(k/k_B)^2]}{k^3|\varepsilon_{RPA}(k,n\omega)|^2} \frac{\sinh(n\omega/k_B v_e)}{(n\omega/k_B v_e)} \int_0^1 J_n^2(kr_0 z)\,dz, \quad (1)$$

which relates $\nu_{ei} = 4\pi(\omega^2/\omega_p^2)\dot{\varepsilon}/\langle \boldsymbol{E}^2\rangle$. Here $\varepsilon_{RPA}(k,n\omega)$ is the quantum mechanical dielectric function (which is a complex quantity) in the random phase [8, 14] approximation (RPA) due to Lindhard [15], v_e is the electron velocity, $k_B = 2v_e/\hbar$ (note $\hbar = 1$ in a.u.) is the inverse of the DeBroglie wavelength λ_B, n is the order of the Bessel function $J_n(\mu)$, $r_0 = E_0/\omega^2$ is the excursion of a free electron in the laser field, and k signifies the inverse of the distance of the electron from the colliding ion. The integrals and the summation (in Eq.(1)) are computed numerically with upper limits of $n = n_{max}$, and $k = k_{max}$. Conventionally $k_{max} = v_{th}^2/Z$ [6, 8] is chosen. $\varepsilon_{RPA}(k,\omega)$ can be obtained [14] from

$$\varepsilon_{RPA}(k,\omega) = 1 + (\omega_p/kv_e)^2\left[(f(x-q/2) - f(x+q/2))/2q\right], \quad (2)$$

where $f(x) = \pi^{-1/2}\int_{-\infty}^{\infty} dt \exp(-t^2)/(t-x)$ is the plasma dispersion function, $x = \omega/\sqrt{2}kv_e$, and $q/2 = k/\sqrt{2}k_B$. $f(x)$ is related to the the Dawson's integral $D(x) = \exp(-x^2)\int_0^x dt \exp(t^2)$ as $f(x) = -2D(x) + \mathrm{i}\sqrt{\pi}\exp(-x^2)$. Calculation of $\varepsilon_{RPA} = \varepsilon_1 + \mathrm{i}\varepsilon_2$, leads to write real and imaginary parts as $\varepsilon_1(k,\omega) = 1 + \frac{1}{\sqrt{2}}(\omega_p/\omega)^2(\omega/kv_e)^2 G_2(k,\omega)$, $\varepsilon_2(k,\omega) = \sqrt{\frac{\pi}{2}}(\omega_p/\omega)^2(\omega/kv_e)^3 G_1(k,\omega)$, where $G_1(k,\omega) = \exp\left[-((\omega/kv_e)^2 + (k/k_B)^2)/2\right]\frac{\sinh(\omega/k_B v_e)}{(\omega/k_B v_e)}$, $G_2(k,\omega) = (k_B/k)\left[D_+(k,\omega) - D_-(k,\omega)\right]$ and $D_\pm(k,\omega) = D([\omega/kv_e \pm k/k_B]/\sqrt{2}$. For $k_B \to \infty$ (for $\hbar \to 0$) classical expressions of G_1, G_2, ε_1, ε_2 are obtained as G_1^{CL}, G_2^{CL}, ε_1^{CL}, ε_2^{CL} respectively. Corresponding classical $\varepsilon_{RPA}^{CL} = \varepsilon_1^{CL} + \mathrm{i}\varepsilon_2^{CL}$ gives

$$\dot{\varepsilon}^{CL} = \frac{8\sqrt{2\pi}Z^2 n_e n_i}{v_e^3}\omega^2 \sum_{n=1}^{\infty} n^2 \int_{k=0}^{\infty} \frac{dk}{k^3|\varepsilon_{RPA}^{CL}(k,n\omega)|^2} \exp\left[-(n\omega/kv_e)^2/2\right] \int_0^1 J_n^2(kr_0 z)\,dz. \quad (3)$$

Assuming $|\varepsilon_{RPA}^{CL}|^2 \approx 1$, the Dawson-Oberman model [6] of ν_{ei} can be deduced from Eq.(3) as

$$\dot{\varepsilon}_{Dawson}^{CL} = 8\sqrt{2\pi}Z^2 n_e n_i v_e^{-3}\omega^2 \sum_{n=1}^{\infty} n^2 \int_{k=0}^{\infty} k^{-3}\exp\left[-(n\omega/kv_e)^2/2\right] dk \int_0^1 J_n^2(kr_0 z)\,dz. \quad (4)$$

We use Eqs.(1), (3), (4) to obtain ν_{ei}.

3. Results for ν_{ei}

For numerical calculations of ν_{ei} we choose $n_{\max} = 20$, and $k_{\max}$ is discretized up to $m_{\max} = 100$ distinct values such that $k_{\max} = m_{\max}\Delta k$, and $k = m\Delta k$. Above parameters are kept unchanged unless mentioned explicitly. Special attention is given to the low temperature regime where anomalous variation is found to be pronounced.

Figure 1 shows normalized frequency $\nu_{\mathrm{ei}}/\omega_{\mathrm{p}}$ versus normalized velocity v_0/v_{th} using expressions (1), (3), (4) represented by "Bornath-Q", "Bornath-C", "Dawson-C" respectively with (a) $T_{\mathrm{e}} = 10$ eV and (b) $T_{\mathrm{e}} = 15$ eV. Here, conventional $k_{\max} = v_{\mathrm{th}}^2/Z$ has been used. For $v_0/v_{\mathrm{th}} > 1$, classical approximations ("Bornath-C" and "Dawson-C") match very well with the quantum result ("Bornath-Q"), but they under-estimate the frequency for $v_0/v_{\mathrm{th}} < 1$ at low temperature where quantum results are important. In all cases, however, the universally accepted feature is that $\nu_{\mathrm{ei}}/\omega_{\mathrm{p}}$ remains almost constant up to a value of $v_0/v_{\mathrm{th}} \approx 1$, and then it drops rapidly when $v_0/v_{\mathrm{th}} \gtrsim 1$ as in Ref.[8] (see Fig. 1 of Ref.[8]). The comparison between Figs.1(a), 1(b) shows that $\nu_{\mathrm{ei}}/\omega_{\mathrm{p}}$ increases from a value $\approx 2\times10^{-4}$ to $\approx 7\times10^{-3}$ for $v_0/v_{\mathrm{th}} < 1$ as T_{e} is increased from 10 eV to 15 eV. *This* anomalous increase of frequency with temperature is a new finding in this work within the existing models. To show it more clearly, Fig.2 depicts $\nu_{\mathrm{ei}}/\omega_{\mathrm{p}}$ versus T_{e} for a fixed intensity $\sim 10^{14}\,\mathrm{Wcm}^{-2}$. It shows that $\nu_{\mathrm{ei}}/\omega_{\mathrm{p}}$ increases monotonically with increasing T_{e} upto a maximum about $T_{\mathrm{e}} \approx 35$ eV (indicated by a shaded region), and then $\nu_{\mathrm{ei}}/\omega_{\mathrm{p}}$ decreases as T_{e} is increased beyond $T_{\mathrm{e}} \approx 35$ eV. Just after the maximum the rate of decrease of $\nu_{\mathrm{ei}}/\omega_{\mathrm{p}}$ is much slower than the conventional $\nu_{\mathrm{ei}}/\omega_{\mathrm{p}} \propto T_{\mathrm{e}}^{-3/2}$ fall, shown by a thin dashed line. Similar anomalous variation can also be shown for the quantum case [16] with DeBroglie wavelength λ_B and the corresponding cut-off $k_{\max} = 1/\lambda_B = 2v_{\mathrm{th}}$.

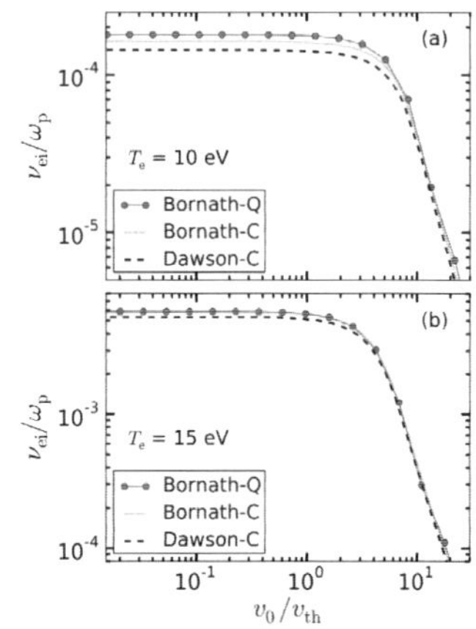

Figure 1. Frequency $\nu_{\mathrm{ei}}/\omega_{\mathrm{p}}$ versus v_0/v_{th} for (a) $T_{\mathrm{e}} = 10$ eV, and (b) $T_{\mathrm{e}} = 15$ eV, using $k_{\max} = v_{\mathrm{th}}^2/Z$, $Z = 1$, $n_{\mathrm{p}} = 10^{28}/m^3$, and $\omega/\omega_{\mathrm{p}} = 5$ giving $\lambda \approx 66$ nm.

So far, the effect of laser field strength has been disregarded in $k_{\max}$ (and in earlier works) which is not justified. $k_{\max}$ should depend on parameters of the laser field through the ponderomotive velocity v_0 in some form and the above conventional relation $k_{\max} = v_{\mathrm{th}}^2/Z$ should be modified [7, 11] depending on the total velocity $v = (v_0^2 + v_{\mathrm{th}}^2)^{1/2}$ of the electron. Replacing v_{th} by v in $k_{\max}$, and taking $k_{\max} = v^2/Z$ we show below that the conventional $I_0^{-3/2}$ decrease of ν_{ei} is also not obeyed in the low temperature and low intensity regime. The traditional quantum cut-off $k_{\max} = 2v_{\mathrm{th}}$ also does not depend on laser parameters and needs modification when electrons are driven by the laser field. For the quantum case, we use total velocity dependent $k_{\max} = k_B = 2v$ to study absorption of laser light in under-dense plasma in Sec.4.

4. Anomalous absorption of light wave in under-dense plasma

Anomalous increase of ν_{ei} with temperature or intensity occurs in the low temperature and low intensity regime which can be used to find absorption of laser light in an under-dense plasma slab. The fractional absorption [1] of light at a normal incidence can be written as $\alpha = 1 - I_{\mathrm{t}}/I_{\mathrm{inc}} = 1 - \exp(-2\kappa_i L)$, where $I_{\mathrm{inc}}, I_{\mathrm{t}}$ are the incident and the transmitted intensity of light, $\kappa_i = (n/n_{\mathrm{c}})\nu_{\mathrm{ei}}/v_{\mathrm{g}}$, $v_{\mathrm{g}} = c\sqrt{1-n/n_{\mathrm{c}}}$ is the group velocity of light, and L is the thickness of the plasma slab. For $\kappa_i \ll 1$, $\alpha \approx 2(n/n_{\mathrm{c}})\nu_{\mathrm{ei}}/v_{\mathrm{g}}$ holds which shows α should vary similar to ν_{ei} w.r.t. T_{e} and I_0 for a fixed plasma density and laser frequency. For illustration we assume $L = 200\lambda$ and keep other parameters as above. Figure 2 shows that anomalous frequency increase is possible if T_{e} is less than the value at which respective frequency maximum occurs.

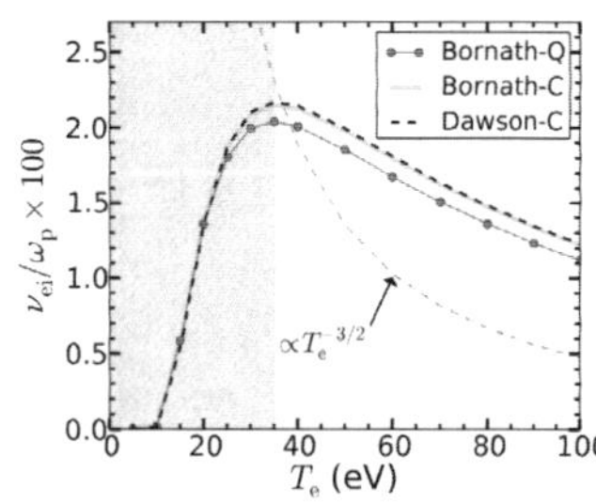

Figure 2. $\nu_{\mathrm{ei}}/\omega_{\mathrm{p}}$ vs T_{e} with $k_{\max} = v_{\mathrm{th}}^2$, $I_0 = 10^{14}\,\mathrm{Wcm}^{-2}$. Shaded region shows regime of anomalous increase of ν_{ei}. $T_{\mathrm{e}}^{-3/2}$ line shows conventional scaling. Other parameters are as in Fig.1.

Figure 3 shows variation of ν_{ei}/ω_p, and corresponding α against I_0 using (i) $T_e = 15$ eV with the classical cut-off $k_{max} = v^2/Z$ (in a,b), and (ii) $T_e = 5$ eV with the quantum cut-off $k_{max} = 2v$ (in c,d). In both cases we have now used total velocity dependent cut-offs. It is seen that ν_{ei} and α grow hand in hand up to a maximum value with increasing intensity up to $\approx 10^{16}\,\mathrm{Wcm}^{-2}$ (with the classical cut-off) and $\approx 5\times10^{15}\,\mathrm{Wcm}^{-2}$ (with the quantum cut-off), and after that they fall together. This anomalous growth of α was found in some experiments [1, 10] and an explanation was provided [11] for this fact with a classical scattering model [7] using total velocity dependent $\ln\Lambda(v)$. Here we find similar anomalous increase of α versus I_0 with more accurate kinetic models. But, discrepancy persists among these models which may be as large as 20% (for ν_{ei}) in the low intensity regime. When I_0 exceeds I_c corresponding to the maximum of ν_{ei} (or α), discrepancy among these models disappears.

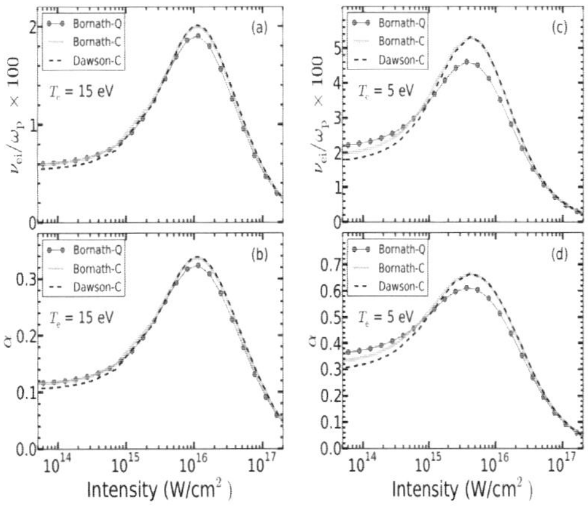

Figure 3. Variation of ν_{ei}/ω_p and α with intensity for $n_p = 10^{28}/m^3$ and $n_p = n_c/25$. For (a,b) $T_e =$ 15 eV classical cut-off $k_{max} = v^2/Z$ and for (c,d) $T_e =$ 5 eV quantum cut-off $k_{max} = 2v$ are used. Thickness of the plasma slab is $L = 200\lambda = 13.38\times10^{-6}$ m.

5. Summary

Electron-ion collision frequency (ν_{ei}) and corresponding fractional absorption (α) of laser pulses in an under-dense plasma has been re-examined with classical and quantum kinetic models of ν_{ei}. Conventionally, it is believed that ν_{ei} and α should decrease as $T_e^{-3/2}$ and $I_0^{-3/2}$ w.r.t. T_e and I_0 respectively. Contrarily, it is found to be true *only* in the high temperature (> 15 eV) and high intensity $> 5\times10^{15}\,\mathrm{Wcm}^{-2}$ regime. The conventional k_{max} only depends on the thermal velocity [6, 8], which is not justified. It must depend on the ponderomotive velocity $v_0 = E_0/\omega$. For a given $T_e < 15$ eV and a total velocity $v = (v_{th}^2 + v_0^2)^{1/2}$ dependent k_{max}, as the intensity is increased from a lower value $< 5\times10^{15}\,\mathrm{Wcm}^{-2}$, the frequency ν_{ei} is also found to increase monotonically up to a maximum value similar to ν_{ei} versus T_e at a constant I_0. The fractional absorption α which is almost proportional to ν_{ei}, follows the same anomalous variation w.r.t. T_e and I_0 as ν_{ei}. Thus, in both ways, we show growth of ν_{ei} and α with T_e and with I_0 up to a maximum using classical and quantum cut-off k_{max}. Anomalous increase of ν_{ei} with T_e and with I_0 were not pointed out earlier with kinetic models, because low temperature regime was not investigated in detail.

6. References

[1] Shalom Eliezer, *The Interaction of High-Power Lasers with Plasmas, (IOP Publishing, Bristol, 2002).*
[2] L. Schlessinger and J. Wright, Phys. Rev. A **20**, 1934 (1979).
[3] P. Hilse, M. Schlanges, Th. Bornath, and D. Kremp, Phys. Rev. E **71**, 056408 (2005).
[4] M. Moll, M. Schlanges, Th. Bornath, and V. P. Krainov, New Journal of Physics **14**, 065010 (2012).
[5] V. P. Silin, Sov. Phys. JETP **20**, 1510 (1965).
[6] C. D. Decker, W. B. Mori, J. M. Dawson, and T. Katsouleas, Phys. Plasmas **1**, 4043 (1994).
[7] P. Mulser, F. Cornolti, E. Bésuelle, and R. Schneider, Phys. Rev. E **63**, 016406 (2000).
[8] Th. Bornath, M. Schlanges, P. Hilse, and D. Kremp, Phys. Rev. E **64**, 026414 (2001).
[9] A. Brantov, W. Rozmus, R. Sydora, C. E. Capjack, V. Yu. Bychenkov et al., Phys. Plasmas **10**, 3385 (2003).
[10] D. Riley, L. A. Gizzi, A. J. Mackinnon, S. M. Viana, and O. Willi, Phys. Rev. E **48**, 4855 (1993).
[11] M. Kundu, Phys. Plasmas **21**, 013302 (2014).
[12] John Dawson and Carl Oberman, Phys. Fluids **5**, 517 (1962).
[13] S.H. Kim and H.E. Wilhelm, Phys. Fluids **25**, 668 (1982).
[14] A.V. Latyshev, A.A. Yushkanov, arXiv:1003.2531 [math-ph].
[15] J. Lindhard, Dan. Mat.-Fys. Medd. **28**, No. 8 (1954).
[16] M. Kundu, Phys. Rev. E **91**, 043102 (2015).

Journal of Physics: Conference Series **717** (2016) 012069 doi:10.1088/1742-6596/717/1/012069

Spectral line shapes of L-shell transitions in Ne-like iron

Roberto C. Mancini

Department of Physics
University of Nevada, Reno, USA

E-mail: rcman@unr.edu

Abstract. Photon-energy-resolved large-scale opacity calculations employ Stark broadened spectral line shapes only to account for the contribution of K-shell line transitions. Detailed ion broadening effects are not considered for L- and M-shell transitions. We present Stark broadening calculations for the line profiles of L-shell transitions linking ground state and singly excited states in Ne-like iron ions. These detailed line shapes have been computed in the standard Stark broadening theory approximation taking into account the effect of both static ions and dynamic electrons. The results show the importance of the ion's effect on the line broadening of several L-shell line transitions.

1. Introduction

The calculation of photon-energy-resolved opacities in plasmas depends on atomic physics, equation of state, and spectral line shapes. When the case of application involves L- and M-shell line transitions the calculation becomes challenging because of the very large number of transitions that have to be taken into account. In addition, if the density is sufficiently high so that the broadening of the spectral line shapes by the electric microfields of electrons and ions in the plasma is important, i.e. the Stark broadening effect, then the computation of the spectral line shapes themselves becomes an issue. Spectral line shapes are critical in plasma opacity calculations since their broadening, blending and degree of overlapping can change the opacity by significant amounts. In turn, this can impact the radiation transport and energetics of the plasma.

Currently, spectral line shapes in large scale opacity calculations involving L- and M-shell transitions are modelled as Voigt line profiles that account for natural, Doppler and an impact broadening contribution due to collisions with plasma electrons [1-5]. Detailed ion broadening effects are not considered. However, the splitting and shifting of atomic energy levels due to the Stark effect produced by the ion's microfield distribution can lead to the occurrence of forbidden (field-dependent) line transitions as well as significant changes to the broadening and shape of the line profiles. The key to overcome this limitation in the spectral line shapes of L- and M-shell transitions in large-scale opacity calculations is to develop both suitable theory approximations and fast algorithms that will permit the consideration of detailed ion broadening effects.

Furthermore, the discrepancies found between theory results and observations in iron opacity experiments performed at the Z Facility of Sandia National Laboratories [6] motivate, among other efforts, an examination of the spectral lines shapes of L-shell transitions in Ne-, F- and O-like iron ions. Here, we discuss the case of Stark-broadened spectral line shapes of Ne-like iron for the case of transitions linking the ground state and singly excited states.

2. Line profile calculations

The Stark-broadened spectral line shape calculations were performed with the multi-electron radiator line shape code MERL [7-9]. The physics model implemented in MERL computes the line profile within the scope of the standard Stark broadening theory approximation [10], i.e. assuming the effects of static ions and dynamic electrons. The electric microfield distribution function was computed with the APEX model and code [11], and the required atomic physics data (fine structure energy levels and reduced electric dipole matrix elements) were obtained using Cowan's atomic structure code [12].

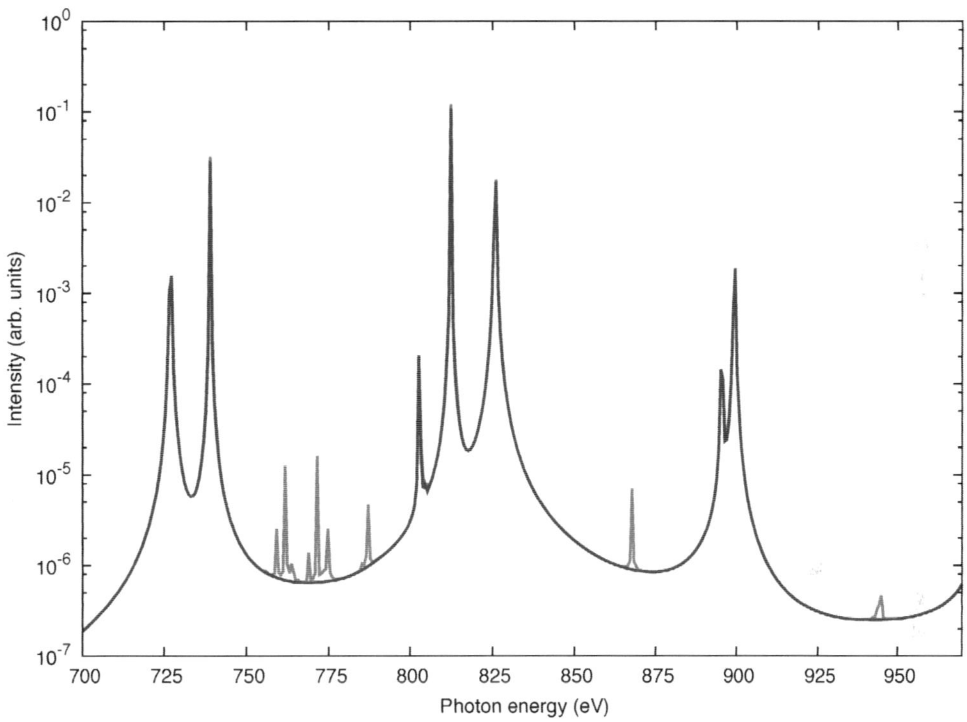

Figure 1. Stark-broadened Ne-like iron L-shell spectrum due to line transitions between the ground state and singly excited n=3 energy levels. Blue: result not including static ion's effect. Red: result including static ion's effect.

Figures 1 and 2 display the results of the line shape calculations done for an electron temperature T_e=195 eV and an electron density $N_e=1\times10^{22}$ cm^{-3}. The ion temperature was assumed to be equal to that of the electrons, and the plasma composition was 50/50 iron/magnesium. These plasma conditions are relevant to those of the iron opacity experiments performed at the Z Facility [6]. Fig. 1 shows the 700 eV to 970 eV photon energy range which has contributions from n=2-3 line transitions, while Fig. 2 shows the 970 eV to 1350 eV range which is dominated by contributions from n=2-4, n=2-5, n=2-6 and n=2-7 line transitions. The upper energy levels of the transitions were populated with a Boltzmann factor, i.e. an equilibrium distribution of population was assumed. This assumption can be easily removed since MERL can read in as part of the input data that defines a case of application, a non-equilibrium population distribution [8,13]. The ionization potential of Ne-like iron is 1263 eV. In both figures the effect of the static ions is emphasized by showing results not including and including the contribution from the plasma ions to the line shape calculation, i.e. performing the calculation for a zero value of the ion's microfield versus calculating the electron broadened line shape as a function of the ion's microfield value and then integrating it over the microfield distribution function. On the one hand, and for these plasma conditions, the effect of ion broadening in the n=2-3 line transitions is negligible and the Stark broadening is dominated by the collisions with the electrons (see Fig. 1). We note that the peaks between the main spectral features represent forbidden transitions that arise due to the ion's field mixing effect. On the other hand, Fig. 2 shows that the ion's effect is more important for line transitions arising from n=4, 5, 6 and 7. This is illustrated by both the appearance of forbidden components and the effect on the broadening and shape of the line profile of the integration over the ion's microfield distribution function.

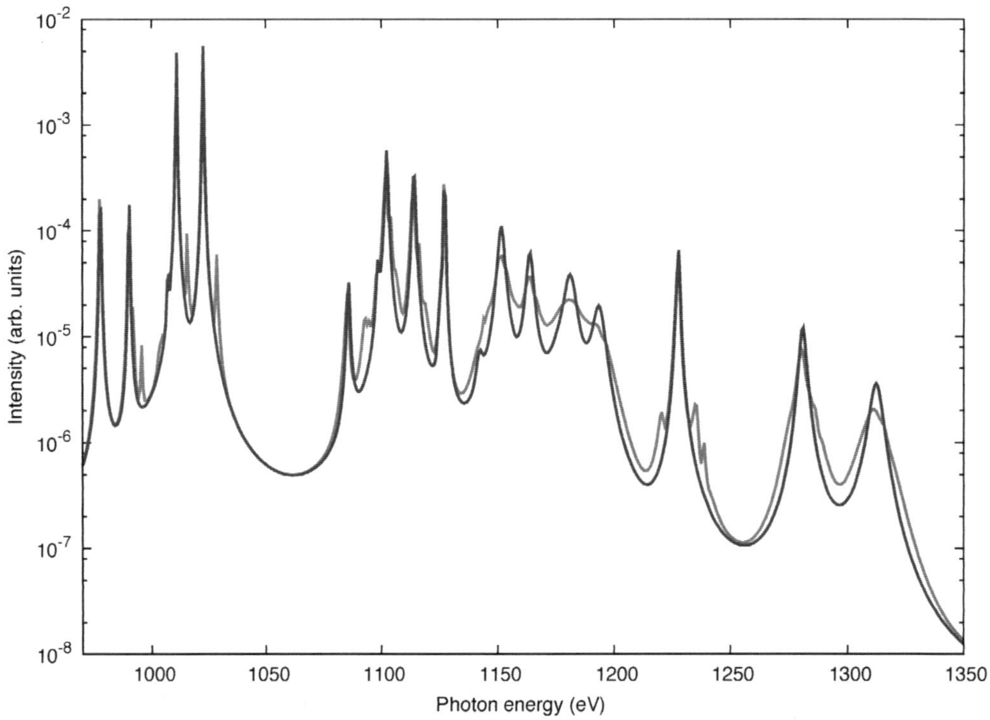

Figure 2. Ne-like iron L-shell spectrum due to line transitions between the ground state and singly excited n=4, 5, 6 and 7 energy levels. Blue: result not including static ion's effect. Red: result including static ion's effect.

3. Conclusions

In order to illustrate the effect of the ion's contribution to the Stark broadening of the L-shell spectrum of Ne-like iron for line transitions between the ground state and singly excited states, a detailed Stark broadening calculation has been performed for plasma conditions relevant to the iron opacity experiments performed at the Z Facility [6]. The ion's contribution impact the spectral line shape through the appearance of forbidden components and by changing the broadening and shape of the line profiles. Including these detailed line profiles in the analysis of the L-shell transitions observed in the iron opacity experiments at Z can further constrain the analysis of the data. To this end, work is in progress to use these line shapes in opacity calculations, consider additional electron density values and transitions involving doubly excited states, and perform line profile calculations for L-shell line transitions in F- and O-like iron ions.

This work was supported by a contract from Sandia National Laboratories.

References

[1] Iglesias C A and Rogers F 1996 *Astrophysical Journal* **464** 943.

[2] Mondet G, Blancard C, Cossé P and Faussurier G 2015 *Astrophysical Journal Supplement Series* **220** 2.

[3] Colgan J, Kilcrease D P, Magee N H, Sherrill M E, Abdallah, Jr. J, Hakel P, Fontes C J, Guzik J A and Mussack K A 2016 *Astrophysical Journal* (in press).

[4] Hansen S B, Bauche J, Bauche-Arnoult C and Gu M F 2007 High Energy Density Physics **3** 109.

[5] Pain J C, Gilleron F and Blenski T 2015 Laser and Particle Beams **33** 201.

[6] Bailey J E, Nagayama T, Loisel G P, Rochau G A, Blancard C, Colgan J, Cossé P, Faussurier G, Fontes C J, Gilleron F, Golovkin I, Hansen S B, Iglesias C A, Kildrease D P, MacFarlane J J, Mancini R C, Nahar S N, Orban C, Pain J C, Pradhan A K, Sherrill M and Wilson B G 2015 *Nature* **517** 56.

[7] Woltz L A and Hooper, Jr C F 1988 *Physical Review A* **38** 4766.

[8] Mancini R C, Kilcrease D P, Woltz L A and Hooper, Jr. C F 1991 *Computer Physics Communications* **63** 314.

[9] Mancini R C, Iglesias C A, Ferri S, Calisti A and Florido R 2013 *High Energy Density Physics* **9** 731.

[10] Griem H R *Line spectral broadening by plasmas* 1974 Academic Press (New York, London).

[11] Iglesias C A, DeWitt H E, Lebowitz J L, Mac Gowan D and Hubbard W H 1985 *Physical Review A* **31** 1698.

[12] Cowan R D *The theory of atomic structure and spectra* 1981 University of California Press (Berkeley).

[13] Wolts L A, Jacobs V L, Hooper, Jr. C F and Mancini R C 1991 Physical Review A **44** 1281.

Intense laser driven collision-less shock and ion acceleration in magnetized plasmas

K Mima[1], Q Jia[2], H B Cai[2], T Taguchi[3], H Nagatomo[4], J R Sanz[5] and J Honrubia[5]

[1] The Graduate School for the Creation of New Photonics Industries, Osaka, Japan,
[2] HEDPS, Center for Applied Physics and Technology, Peking University and Institute of Applied Physics and Computational Mathematics, Beijing, China
[3] Faculty of Engineering, Setunan University, Osaka, Japan
[4] Institute of Laser Engineering, Osaka University, Japan
[5] School of Aeronautics, University Politecnica de Madird

k.mima@gpi.ac.jp

Abstract. The generation of strong magnetic field with a laser driven coil has been demonstrated by many experiments. It is applicable to the magnetized fast ignition (MFI), the collision-less shock in the astrophysics and the ion shock acceleration. In this paper, the longitudinal magnetic field effect on the shock wave driven by the radiation pressure of an intense short pulse laser is investigated by theory and simulations. The transition of a laminar shock (electro static shock) to the turbulent shock (electromagnetic shock) occurs, when the external magnetic field is applied in near relativistic cut-off density plasmas. This transition leads to the enhancement of conversion of the laser energy into high energy ions. The enhancement of the conversion efficiency is important for the ion driven fast ignition and the laser driven neutron source. It is found that the total number of ions reflected by the shock increases by six time when the magnetic field is applied.

1. Introduction

The generation of strong magnetic field with a laser driven coil has been demonstrated by many experiments [1][2][3][4]. Recently, explored are their applications to the magnetized fast ignition (MFI) [5][6], and the magnetized collision-less shock in the astrophysics[7],[8]. In this paper, we investigate the external magnetic field effect on the collision-less shock driven by intense short pulse laser and ion acceleration in near relativistic cut-off density (NRCD) laser plasmas.

In the previous paper [9], it is found that magnetic field fluctuations excited by the Weibel instability grow nonlinearly after the saturation of the linear instability, when the electron cyclotron frequency of the external magnetic field is higher than 5% of the electron plasma frequency. In the simulation, it is found that the magnetic fluctuation energy with external B field is higher by 10 times than that without B. The fluctuating magnetic fields propagate with the shock and scatter ions at the shock front. As a result of the scattering, more ions are accelerated by the shock. In this paper, the shock wave generation by the ultra-intense short pulse (UISP) laser interactions with NRCD plasmas is studied by theory and simulation in the both cases without and with external B field. It has been found that ions are reflected and accelerated by the electrostatic field when there is no external B and the laser intensity is lower than 10^{20}W/cm^2 [10]. However, when the external magnetic field is applied, the magnetic fluctuation amplitude increases [9] and ions are strongly reflected and efficiently accelerated at the shock front.

The increase of total energy of accelerated ions is applicable to the ion driven fast ignition and the efficient laser driven neutron source.

2. Shock relations in magnetic fluctuations driven by the two stream and Weibel instabilities

Electromagnetic fluctuations in NRCD plasmas are high enough to cause strong scattering of electrons. So, electrons can be assumed in the local thermal equilibrium. Namely, the electron dynamics follows the thermodynamics. When the heat conduction is neglected and the adiabatic relation is assumed kept for electron fluid, then,

$$\frac{P_e}{P_{e0}} = (\frac{n}{n_0})^{\gamma_e},\ en_e\frac{d\phi}{dx} - \frac{d}{dx}P_e = 0, \text{ and } \frac{n_e}{n_{e0}} = (\frac{\gamma_e - 1}{\gamma_e}\bar{\phi} + 1)^{\frac{1}{\gamma_e - 1}}, \quad (1)$$

where P_e is the electron pressure, n_e is the electron density, γ_e is the adiabatic constant of electron, which is 4/3 in the highly relativistic case, $\bar{\phi} = en_{e0}\phi/P_{e0}$ is the normalized electrostatic potential, P_{e0} and n_{e0} are the up steam electron pressure and electron density. Here in Eq.(1), the electron inertia is neglected by assuming that the shock velocity is lower than the electron thermal velocity and the second equation of (1) were combined to obtain the last equation of (1).

When there is no external magnetic field, ions are not much scattered by magnetic fluctuations. So, the ion dynamics follows the Vlasov equation of non-magnetized plasmas which yields the ion distribution function, $F(x,p)$, with $n_i = \int F(x,p)dp$. In the moving frame of the shock, the distribution function satisfies, $p_x\frac{\partial F}{\partial x} - e\frac{\partial \phi}{\partial x}\frac{\partial F}{\partial p} = 0$, when the shock is stationary. Here, p_x is the ion momentum in x (which is the direction of the shock propagation) and ϕ is the electrostatic potential. The solution of the Vlasov equation is given by $F = F_-\left(\frac{p_x^2}{2M} + e\phi\right) + F_+\left(\frac{p_x^2}{2M} + e\phi\right)$. Here, F_+ and F_- are the distribution functions for ions moving from left to right and form right to left in Fig.1, respectively. For simplicity, we will consider the water bag model [11] in which the distribution is assumed constant in the area between two particle orbits as shown in Fig.1.

The distribution function in the up-stream ($x > x_s$) is given as follows.
$F_- = F_0\ (p_x < 0)$, when
$W_{max} - e\phi > \frac{p_x^2}{2M} > 0, or\ W_{min} - e\phi$,
where $\frac{F_0}{n_{io}} = \frac{1}{\sqrt{2M}(\sqrt{W_{max}} - \sqrt{W_{min}})}$. Note that n_{i0} is the incoming ion density from the up stream. In the down stream: Here, W_{max} *and* W_{min} are the maximum and the minimum ion energy at $x = \infty$, respectively. $F_+ = F_0\ (p_x > 0)$,
for $W_{min} - e\phi\ or\ 0 < \frac{p_x^2}{2M} < e(\phi_0 - \phi) + \Delta$,
where $e\phi_0 + \Delta$ is the maximum energy of accelerated ion. The distribution function in the down stream ($x < x_s$) is given as, $F_+ = F_0\ (p_x < 0)$ for $e(\phi_0 - \phi) < \frac{p_x^2}{2M} < W_{max} - e\phi$. When $n_i(\phi)$ is given as the function of ϕ by integrating the above distribution function, we get the equation for the electrostatic potential by combining the Poisson equation, Eqs.(1), and ion density, as follows.

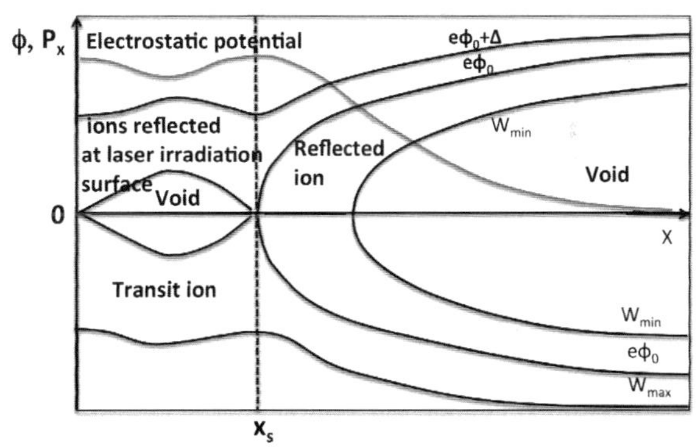

Fig.1 Ion phase space distribution and potential profile

$$\frac{d^2\bar{\phi}}{dx^2} = (\frac{\gamma_e - 1}{\gamma_e}\bar{\phi} + 1)^{\frac{1}{\gamma_e - 1}} - \bar{n_i}(\bar{\phi}) = -\frac{\partial U(\bar{\phi})}{\partial \bar{\phi}}, \quad (2)$$

where x is normalized by the Debye length: λ_D. Here, $U(\bar{\phi})$ is the Sagdeev potential. Note that $\bar{\phi}$ is assumed the coordinate of a virtual particle which is moving in the potential of $U(\bar{\phi})$. When a part of ions are reflected in front of the potential hump at x_s, then the $U(\bar{\phi})$ have different values in the region: $x > x_s$:up-stream and in the region: $x < x_s$:down stream. Actually, it is possible that in the up-stream: $x > x_s$, $U(\bar{\phi}) = 0$ at $\bar{\phi} = 0$, and $\bar{\phi}_0$ and in the down stream: $x < x_s$, $U(\bar{\phi}) = 0$ at $\bar{\phi} = \bar{\phi}_0$ and $\bar{\phi}_b < \bar{\phi}_0$. Here, $\bar{\phi}_b$ is the minimum of the potential ripple in the down stream. Namely, the equation (2) could have an asymmetric solution as shown in the Fig.1. This asymmetry is due to the ion reflection and the solution is called the electro static collision-less shock wave.

The potential and ion phase space distributions in the simulation of the intense short pulse laser interaction with NRCD plasmas as shown in the Fig.2.

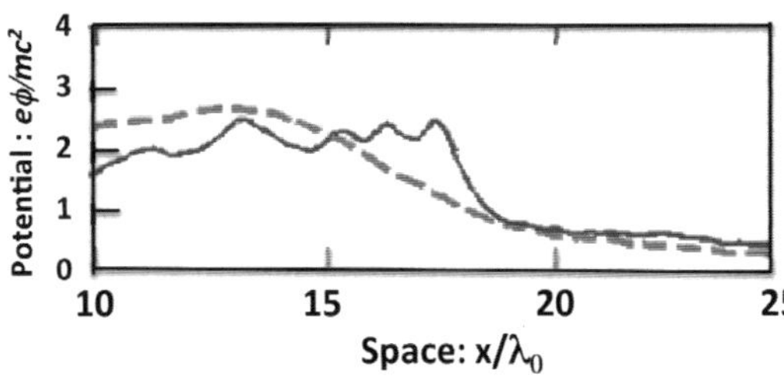

(A)

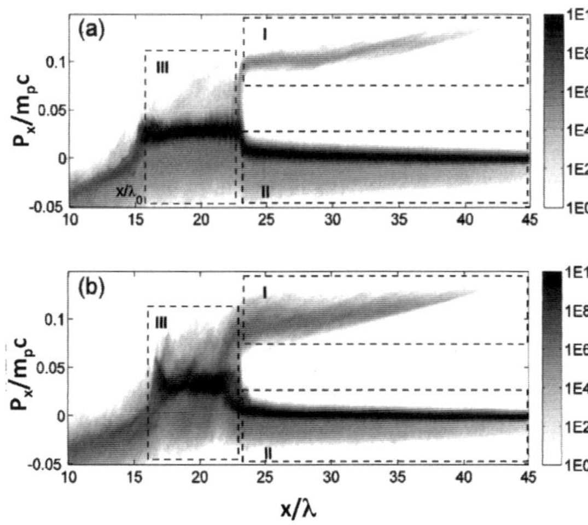

(B)

Fig.2 PIC simulations for the laser intensity of 10^{19} W/cm^2 and the plasma density of $7n_c$ with and without external magnetic field. A) Potential profiles, solid line: without B field, broken line: with external B field (4kT)), B) Ion phase space (x,P_x), (a) without B field and (b) with B field. I, II, and III in (B) indicate reflected ion, up-stream ion and down stream ion, respectively.

The phase space distribution of ions and potential profile in the simulation without external magnetic field as shown in Fig.2 B)-a) and b) agree with the theoretically prediction as shown in Fig.1. In the simulation without B, the reflected ion minimum velocity is $v = 0.045c$ in the shock wave frame which is equal to the ion velocity accelerated by the potential energy $e\phi_0 = 0.8$ MeV.

When the external magnetic field is applied, the magnetic field fluctuations are enhanced to scatter and reflect ions. In this case, the ion dynamics is described as a fluid. Then the ion stationary flows are described the following three conservation laws. Ion mass conservation: $\rho v = F($ constant$)$.
Ion momentum conservation: $\rho v^2 + P_i + P_e - \frac{4}{3}\mu \frac{d}{dx} v = \rho_0 v_0^2 + P_{i0} + P_{e0}$. Here, P_i and P_e are the ion and electron pressure. Energy conservation: $\frac{d}{dx}[\frac{\gamma_i}{\gamma_i - 1} P_i v] + \frac{d}{dx}\frac{\rho v^3}{2} = -\frac{Z}{M} e\rho v \frac{d\bar{\phi}}{dx} + \frac{4}{3}\mu \frac{d}{dx}(v \frac{dv}{dx})$.
Here, γ_i and μ represent the ion adiabatic constant and the ion viscosity. The electron density given by Eq.(1), the above conservation laws, and the charge neutrality condition yield the following equation for the flow velocity: v,

$$\left[\gamma_i P_{i0} - \frac{1}{2} F\big((\gamma_i + 1)v - (\gamma_i - 1)v_0\big)\right](v - v_0) + \gamma_e P_{e0} v[1 - (\frac{v_0}{v})^{\gamma_e} + \frac{\gamma_e(\gamma_i - 1)}{\gamma_i(\gamma_e - 1)}((\frac{v_0}{v})^{\gamma_e} - \frac{v_0}{v})] = -\frac{2}{3}\mu \frac{dv^2}{dx} \quad (3)$$

For $x \rightarrow \pm\infty$, the right hand side of the equation (3) is zero for a shock wave. Then, we obtain $v = v_0$ for the up stream ($x \rightarrow +\infty$) and $v = v_1$ in the down stream ($x \rightarrow -\infty$). The jump condition satisfies,

$$\frac{v_1}{v_0} = \frac{\gamma_i-1}{\gamma_i+1} + \frac{2\gamma_i}{\gamma_i+1}\frac{P_{i0}}{Fv_0} + \frac{2\gamma_e}{\gamma_i+1}\frac{P_{e0}v_1}{F\ (v_1-v_0)\ v_0}[1-(\frac{v_0}{v_1})^{\gamma_e} + \frac{\gamma_e(\gamma_i-1)}{\gamma_i(\gamma_e-1)}((\frac{v_0}{v_1})^{\gamma_e} - \frac{v_0}{v_1})] \ . \quad (4)$$

The meaning of Eq.(4) is clear when $\gamma = \gamma_e = \gamma_i$. Defining the Mach number by $M_S^2 = \frac{\rho_0 v_0^2}{\gamma(P_{e0}+P_{i0})}$, then, Eq.(4) is reduced to $\frac{v_1}{v_0} = \frac{\gamma-1}{\gamma+1} + \frac{2}{\gamma+1}\frac{1}{M_S^2}$ which is the Rankine-Hugoniot of a neutral fluid.

In the case of radiation pressure driven EM shock, we can assume that $P_{e0} \gg P_{i0}, \gamma_e = 4/3$ (in the relativistic limit of electron EOS) , and $\gamma_i = 2$ for the 2D simulations with longitudinal external magnetic field. So, the Eq.(4) is reduced to,

$$\frac{\rho_0}{\rho_1} = \frac{v_1}{v_0} = \frac{1}{3} + \frac{4}{9M_S^2}\frac{v_1}{v_1-v_0}\left(1 + (\frac{v_0}{v_1})^{\frac{4}{3}} - 2\frac{v_0}{v_1}\right) . \quad (5)$$

The Eq. (5) indicates that the shock compression ratio: ρ_1/ρ_0 is 3 in the strong shock limit (large Mach number) in the two dimensional space.

According to the simulation results of Fig.2, the potential jumps were about 0.8 MeV for both with and without B field and the up stream electron temperatures were about 0.5 MeV without B field and 0.3MeV with B field, respectively. The density jumps evaluated by Eq.(1) with those electron temperatures and the potential jumps are 2.0 without B field and 2.8 with B field, which are consistent with those obtained in the simulations. Furthermore, the Mach number with B field is calculated by Eq.(5) is 2.7 for the density jump of 2.8. The simulation of Fig.2-B(b) indicates the shock velocity is $0.06c$ which corresponds to the Mach number of 2.6 since the ion acoustic velocity is $0.018c$ near the shock front of the up stream. This reasonably agrees with the jump condition: Eq.(5)

3. Discussions and summary

The case (I): kinetic ion dynamics with adiabatic electron compression and the case (II): viscous ion flow with adiabatic electron compression are proposed to interpret the 2-D PIC simulation results on the shock wave driven by laser radiation pressure in the NRCD plasmas. We found that the case (I) corresponds to the collision-less shock wave of the non-magnetized ions and the case (II) corresponds to the magnetized ions, where the enhanced magnetic field fluctuations scatter ions strongly. Namely, the fluid model with finite viscosity is applicable for the ion dynamics. The scale length of the shock structure in the case (I) is the electron Debye length. On the other hand, the scale length of the case (II) is $\frac{\mu}{\rho v_0} \sim \frac{c_S^2}{v_0 \omega_{ci}} \sim \frac{\lambda_D \omega_{pi}}{\omega_{ci}} \sim 5\lambda_0$, which reasonably agrees with the broken line of the Fig.2 (A), where the magnetic fluctuation amplitude is about 20kT near the shock front.

Finally, as it is seen by comparing Fig.2-B(a) and (b), the number of accelerated ion is lager by about 5 times in the magnetized plasma. This means the coupling efficiency from laser to ion energy will be enhanced by 5 times and it will improve the performance of the ion driven fast ignition.

Reference

[1] H. Daido, F. Miki, K. Mima, et al., Phys.Rev. Lett., 56, No.8, 846(1986)
[2] H. Daido, K. Mima, F. Miki, et al., Jpn. J.Appl. Phys. 26, 1290 (1987)
[3] P.Y. Chang, G. Fiksel, M.Hohenberger, et al., Phys.Rev. Lett. *107*, 035006 (2011)
[4] J.J. Santos, M.Bailly-Grandvaux, L.Giuffrida, et al., New J. Phys. *17*, 083051 (2015)
[5] K. Mima. et al., Proceedings of IFSA(2013), Nara, Japan
[6] T.Taguchi, et al, Proceedings of IFSA(2013), and J.Honrubia, etal, Proceedings of IFSA(2013)
[7] Gargate, L, and Spitkovski, A. Astro. Phys. J.744, 67 (2012)
[8] Y. Kuramitsu, et al, Physical Review Letters, 106, 175002 (2011)
[9] F. Fiuza, et al., Phys. Rev. Lett. 109, 215001 (2012)
[10] Qing Jia, K.Mima, et al, Phys. Rev. E 91, 023107 (2014)
[11] See for instance, K.M. Robers and H.L.Berk, Phys. Rev. Letters 19, 297(1967)

Plasma structure and energy dependence in a magnetic thrust chamber system

T Morita[1], N. Yamamoto[1], R Kawashima[2], N Saito[2], M Edamoto[2], S Fujioka[3], Y Itadani[2], T Johzaki[4], S Miura[2], Y Mori[5], H Nishimura[3], A Sunahara[6], A Yogo[3] and H Nakashima[1]

1. Department of Energy Engineering Science, Faculty of Engineering Sciences, Kyushu University, 6-1 Kasuga-koen, Kasuga, Fukuoka 816-8580, Japan
2. Interdisciplinary Graduate School of Engineering Sciences, Kyushu University, 6-1 Kasuga-koen, Kasuga, Fukuoka 816-8580, Japan
3. Institute of Laser Engineering, Osaka University, 2-6 Yamada-oka, Suita, Osaka 565-0871, Japan
4. Graduate School of Engineering, Hiroshima University, Higashi-Hiroshima, Hiroshima 739-8527, Japan
5. The Graduate School for the Creation of New Photonics Industries, Hamamatsu, Shizuoka 431-1202 Japan
6. Institute for Laser Technology, Suita, Osaka 565-0871, Japan

E-mail: morita@aees.kyushu-u.ac.jp

Abstract. We demonstrate a magnetic thrust chamber system, in which an expanding plasma is controlled by an external magnetic field to produce a thrust. The plasma structure and energy dependences are discussed in terms of the drive laser energy and magnetic field strength. The density distribution from two different experiments show identical structure despite the laser energy is different by two order of magnitude when the ratio of magnetic field to plasma energy is more or less same. The experimental results indicate that this ratio is one of the essential factors to extrapolate the plasma dynamics for much larger energy such as inertial confinement fusion plasmas.

1. Introduction

The interaction between high-temperature plasmas and magnetic field is observed in various physical phenomena, for example, in laboratory: magnetic confinement fusion, fusion reactor wall, and laboratory experiments to study astrophysical phenomena. Furthermore, the magnetic field can convert the plasma thermal energy to kinetic energy to generate thrust for space propulsion systems. This is one of the strong candidates as the thrust system of laser fusion rocket (LFR), in which a high-temperature fusion plasma is controlled by a magnetic field. We call this specific system for LFR as a magnetic thrust chamber[1].

Previous numerical and theoretical works have shown that it produces both large impulse and high specific impulse which are required for manned interplanetary spaceflight[2, 3]. We have recently performed experiments with laser-produced ablation plasmas to demonstrate the magnetic thrust chamber systems: the direct measurement of an impulse with a thrust stand[4], a diamagnetic cavity formation[5], and observation of plasma deceleration which result in the thrust generation[6]. Though these previous experimental researches have shown the thrust

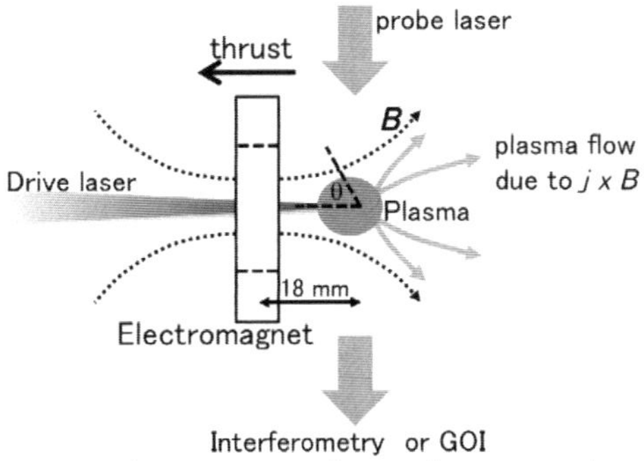

Figure 1. An experimental setup with an electromagnet and a 500 μm sphere CH target. The inner and outer diameter of the coil are 50 mm and 26 mm, respectively. The target is located at 13 mm from the coil surface. The drive laser irradiates the target to produce the plasma in an external magnetic field. The plasma is diagnosed with interferometry or GOI.

generation, energy dependence on the plasma structure formation and thrust generation have never been discussed. This energy dependence (energy scaling-law) is one of the important factors to design a future system LFR which utilize larger plasma kinetic energy resulting from a fusion reaction.

In this paper, we demonstrate a magnetic thrust chamber by using an electromagnet and two different laser systems. Despite the laser energies of two systems are different by two orders, the density distributions are remarkably similar when the ratio of magnetic field energy to plasma kinetic energy is similar.

2. Experiment

The experiment was performed with Extreme Ultra-Violet (EUV) database laser and Gekko-XII (GXII) laser at Institute of Laser Engineering, Osaka University. A plastic (CH) sphere target with the diameter of 500 μm was irradiated by a laser pulse (drive laser) to produce an expanding plasma as a model of an inertial confinement fusion. Figure. 1 shows the schematic view of a magnetic thrust chamber consisting of an electromagnet and CH sphere target. The drive laser irradiates the left-side of the target. The plasma initially expands leftwards and it interacts with an external magnetic field. Diamagnetic current induced in the plasma ($\boldsymbol{j}$) and the external magnetic field ($\boldsymbol{B}$) result in a $\boldsymbol{j} \times \boldsymbol{B}$ force acting on the plasma to produce a thrust on the magnet.

The experimental conditions are shown in Table 1. The laser energies are $\sim$6 J in 9 ns pulse width with EUV and $\sim$600 J in 1.3 ns with GXII with 500 μm focal spot diameter in FWHM. The plasmas were diagnosed with interferometry and self-emission gated optical imager (GOI) by using an intensified charge coupled device (ICCD) camera with a gate width of 10 ns. GOI measures the spatial distribution of self-emission intensity at a wavelength of 660 nm with the width of 10 nm in FWHM and interferometry gives electron density assuming that the plasma distributes axially symmetrically. In general, the emission comes from thermal bremsstrahlung and/or H-α emission at 656 nm, and strongly depends on the plasma density.

The electromagnet is a 96-turn (axially 8 turns and radially 12 turns) coil with inner and outer diameters of 26 mm and 50 mm, and with the thickness of 10 mm. The center of the coil is at $z = -18$ mm [$z = -18$ mm in Figs. 2(a) and 2(b)].

We estimate the energy conversion ratio as an absorption coefficient η_{ab} as a function of laser intensity I: $\eta_{ab}/\ln(1-\eta_{ab}) = 10^{11}\lambda\omega/cI$ [7], where λ is a characteristic length $\lambda = |(1/n_c)(dn_e/dx)|^{-1}$, n_e and n_c are electron and critical densities, ω is the frequency of the drive laser, and c is the speed of light. Both EUV and GXII experiments were performed with similar ratio of magnetic field energy E_B to plasma energy E_p, where $E_B = LI^2/2$, L is an indactance of the coil, I is the maximum current flowing in the coil, and $E_p = \eta_{ab}E_L$: $E_B/E_p \sim$ 1.4 for EUV and 1.2 for GXII experiments, where η_{ab} is 1.0 and 0.27, respectively.

Table 1. Experimental conditions

Parameters	EUV	GXII
Laser Energy E_L [J]	~ 6	~ 600
Pulse width (FWHM) [ns]	~ 9	~ 1.3
Wavelength [nm]	1064	1053
Focal spot diameter (FWHM) [μm]	500	500
Average laser intensity [Wcm^{-2}]	4.5×10^{11}	2.4×10^{14}
Magnetic field at the target [T]	0.2	1.1
Magnetic field Energy E_B [J]	8.2	190

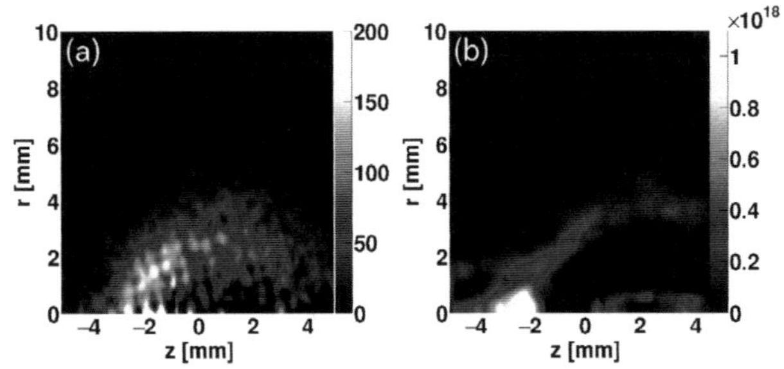

Figure 2. (a) The emission intensity from EUV experiment in arbitrary unit and (b) the electron density from GXII experiment in cm^{-3}. Both data are Abel-inverted assuming axial symmetry along z-axis.

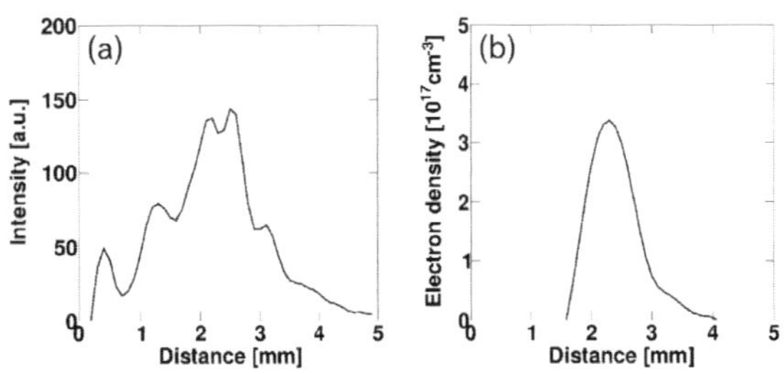

Figure 3. (a) The emission intensity and (b) electron density distribution at $\theta = 60°$ obtained from Figs. 2(a) and 2(b).

3. Results and discussions

Figures 2(a) and 2(b) show self-emission and electron density distributions obtained from EUV and GXII experiments, respectively at $t = 200$ ns. Experimental configuration is more or less axially symmetrical and we calculated the radial distributions with Abel inversion technique assuming the symmetry on z-axis. The drive laser irradiates the target at $z = 0$ from left-side and the plasma expands in $-z$ direction initially. The plasma stops expanding due to the interaction with the magnetic field later in time, and the plasma makes a shell-like conical structure balancing with the magnetic pressure. This conical shapes have been observed in numerical simulation by Nagamine *et al* [1] with much larger energy of 4 MJ as a model of a laser fusion plasma. Here, in the EUV experiment, the magnetic skin depth is $c/\omega_{pe} \sim 5.3$ μm, which is much smaller than typical plasma size of $\sim$4 mm. Moreover, the electron-ion collision frequency is $\nu_{ei} \sim 6.7 \times 10^{10}$ s^{-1} which is similar scale to the electron gyrofrequency $\omega_{ce} \sim 3.5 \times 10^{10}$ rad/s, and the magnetic diffusion coefficient $\eta = \nu_{ei}(c/\omega_{pe})^2 \sim 1.9$ m^{2}s^{-1} with typical plasma parameters in the EUV experiment: $T_e \sim 8$ eV, $n_e \sim 1 \times 10^{18}$ cm^{-3}, $A = 12$, $Z = 3$, and $B = 0.2$ T. The diffusion time of the magnetic field is $\tau \sim l^2/\eta \sim 8$ μs, where $l = 4$ mm is the typical plasma size. The diffusion time is much larger than plasma expansion time $\sim$200 ns, and therefore, the magnetic cavity should be formed in the expanding plasmas as observed in the previous experiment[5].

The line-out plots are shown, for example, in Figs. 3(a) and 3(b) taken at $\theta = 60°$. The origin corresponds to the initial target position $(z, r) = (0, 0)$ in Fig. 2. The peak positions are at $\sim$2.3 mm in both data.

Figure 4 shows the peak positions as a function of the angle θ derived from Figs. 2(a) (6 J)

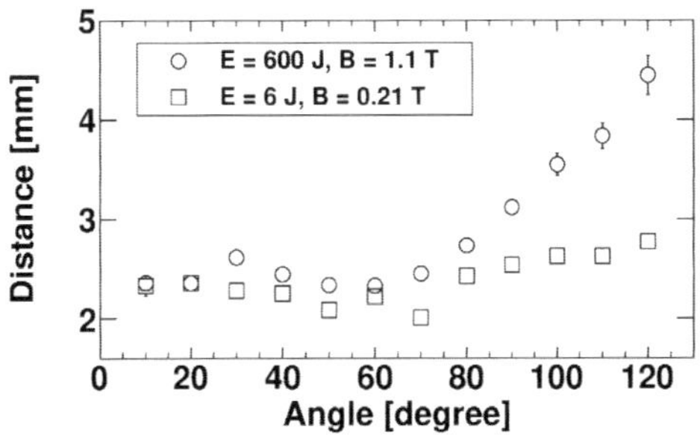

Figure 4. The distance between the initial target position and the peak intensity of the GOI data [Fig. 2(a) with 6 J laser energy] and the interferometry data [Fig. 2(b) with 600 J laser energy] as a function of the angle as shown in Fig. 1.

and 2(b) (600 J). Both results show similar tendency in the range $0 < \theta < 90°$, while the peak positions from 600 J case are further than those from 6 J case at $90° < \theta$. In small angle $\theta < 90°$, the plasma is initially produced by laser-irradiation, hence the plasma energy can be estimated by the drive laser energy and the absorption coefficient. In the range $\theta > 90°$, however, the plasmas does not expand directly from the initial position but flows from the region $z < 0$ along the magnetic field lines interacting with the field. The plasma drift velocity in EUV experiment [in Fig. 2(a)] is much smaller than the other. In this case, the plasma is easily trapped by the magnetic field, resulting in smaller structure in larger angle θ.

4. Summary

We have conducted two experiments to model a magnetic thrust chamber system with the drive laser energy of ~6 J and ~600 J with EUV and GXII laser systems. The density distributions are obtained with optical diagnostics and they show a shell-like conical structure, meaning the suppression of the plasma expansion by an external magnetic field. The structures obtained from both experiments show more or less same, under the condition of the similar ratio of the magnetic field energy to the plasma energy. This result indicates that the energy ratio is one of the essential factors in the magnetic thrust chamber system and it can help us to develop the energy scaling-law for future higher-energy experiments using fusion plasmas with such as National Ignition Facility and Laser Mega Joule. Nevertheless, there are still plenty of experimental work to be done to build an energy scaling-law in a magnetic thrust chamber: structure formation in different energy ratio, spherical irradiation, numerical analysis, and connection with thrust generation and thrust efficiency depending on the plasma and magnetic field energies.

Acknowledgments

The authors acknowledge the dedicated technical support by the staff at EUV and GXII laser facilities. This work was supported by the Japan Society for the Promotion of Science (JSPS) KAKENHI (Grant Numbers 25420852 and 15K18283) and by the joint research project of the Institute of Laser Engineering, Osaka University (2014B1-31 and 2014A1-31).

References

[1] Nagamine Y and Nakashima H 1999 *Fusion Science and Technology* **35** 62–70
[2] Hyde R, Jr L W and Nuckolls J 1972 *AIAA Paper* 72–1063
[3] Hyde R 1983 *Lawrence Berkeley Laboratory* UCRL-88857
[4] Maeno A, Yamamoto N, Nakashima H, Fujioka S, Johzaki T, Mori Y and Sunahara A 2011 *Applied Physics Letters* **99** 071501
[5] Yasunaga M, Maeno A, Yamamoto N, Nakashima H, Fujioka S, Sunahara A, Mori Y and Johzaki T 2012 *Transactions of the Japan Society for Aeronautical and Space Sciences, Aerospace Technology Japan* **10** Pb_109–Pb_114
[6] Kawashima R, Morita T, Yamamoto N, Saito N, Fujioka S, Nishimura H, Matsukuma H, Sunahara A, Mori Y, Johzaki T and Nakashima H 2016 *Plasma and Fusion Research* **11** 3406012 (2016)
[7] Chester A, Letokhov V and Martellucci S 1988 *Laser Science and Technology* (Springer)

Progress in the development of the MARBLE platform for studying thermonuclear burn in the presence of heterogeneous mix on OMEGA and the National Ignition Facility

T J Murphy, M R Douglas, J R Fincke, R E Olson, J A Cobble, B M Haines, C E Hamilton, M N Lee, J A Oertel, N A G Parra-Vasquez, R B Randolph, D W Schmidt, R C Shah, J M Smidt and I L Tregillis

Los Alamos National Laboratory, P. O. Box 1663, Los Alamos, New Mexico 87545 USA

E-mail: tjmurphy@lanl.gov

Abstract. Mix of ablator material into fuel of an ICF capsule adds non-burning material, diluting the fuel and reducing burn. The amount of the reduction is dependent in part on the morphology of the mix. A probability distribution function (PDF) burn model has been developed [6] that utilizes the average concentration of mixed materials as well as the variance in this quantity across cells provided by the BHR turbulent transport model [3] and its revisions [4] to describe the mix in terms of a PDF of concentrations of fuel and ablator material, and provides the burn rate in mixed material. Work is underway to develop the MARBLE ICF platform for use on the National Ignition Facility in experiments to quantify the influence of heterogeneous mix on fusion burn. This platform consists of a plastic (CH) capsule filled with a deuterated plastic foam (CD) with a density of a few tens of milligrams per cubic centimeter, with tritium gas filling the voids in the foam. This capsule will be driven using x-ray drive on NIF, and the resulting shocks will induce turbulent mix that will result in the mixing of deuterium from the foam with the tritium gas. In order to affect the morphology of the mix, engineered foams with voids of diameter up to 100 microns will be utilized. The degree of mix will be determined from the ratio of DT to DD neutron yield. As the mix increases, the yield from reactions between the deuterium of the CD foam with tritium from the gas will increase. The ratio of DT to DD neutrons will be compared to a variation of the PDF burn model that quantifies reactions from initially separated reactants.

1. Introduction

The success of inertial confinement fusion relies on the ability to implode a capsule of DT fuel in a way that prevents the introduction of capsule material into the fuel [1]. Mix of capsule material can reduce the performance of ICF capsules in a number of ways. Among these is the dilution of fusion fuel, resulting in lower number density of the reacting ions, and an overall reduction in nuclear yield.

The degree to which mix affects the yield is affected by the degree to which the mix is complete (Figure 1). For complete atomic mixing, the dilution is maximized. For incomplete

Journal of Physics: Conference Series **717** (2016) 012072 doi:10.1088/1742-6596/717/1/012072

mixing, regions of locally higher fuel density can occur and the performance degradation is reduced.

The effect of incomplete, heterogeneous mixing on thermonuclear yield can be described using a probability distribution function (PDF) [2] to model the concentration of DT fuel within a region of mixed fuel and capsule material. The PDF model uses parameters provided by turbulent mixing models that attempt to capture the important physics in a computationally efficient manner, allowing large-scale simulations of imploding capsules. One such model is the BHR model [3] and its revisions [4, 5]. Results from the BHR model are used in conjunction with the PDF burn model to describe concentration variations at the sub-grid level to determine the effect on burn [6].

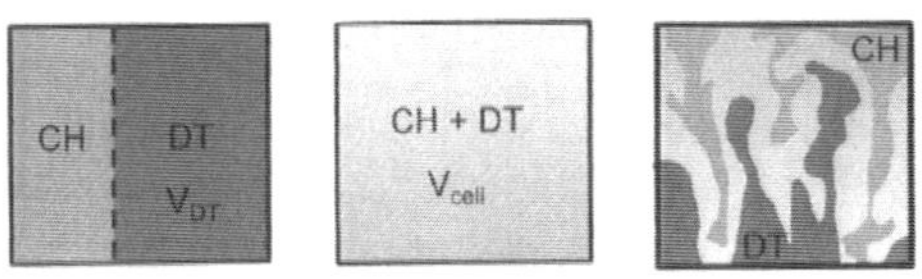

Figure 1. Computational cells in which the CH ablator and DT fuel are (left) completely separated, (center) completely atomically mixed, and (right) heterogeneously mixed with a range of concentrations.

2. Concept of the MARBLE experiment

Marble is an experimental campaign intended to quantify the effect of heterogeneous mix on fusion burn. Capsules filled with deuterated plastic foam and tritium gas will be imploded and the yield of 14-MeV DT neutrons will be measured to provide a measure of the degree to which the foam and gas mixed. In contrast to the case in which more complete mix of capsule material into fuel reduces the reactivity, increased mix of the components in this separated-reactant experiment increases the production of DT neutrons. To make the experiment less sensitive to shot-to-shot drive and capsule variations, the DT neutron yield will be normalized to the 2.45-MeV DD neutron yield to give the DT to DD yield ratio.

To vary the heterogeneity of the mix, the foam will be engineered to produce large "macropores" (Figure 2). The density of the interstitial foam will be set so that the average foam density in the capsule is known, and the gas pressure set so that the average concentration of deuterated plastic and tritium is known, and only the morphology of the mix changes. This morphology will be controlled by specifying the diameter of the macropores. Turbulent instability growth [7] with smaller characteristic length scales leads to faster development of atomic mix, so the smaller the macropore diameter, the more quickly and completely the components will mix to produce DT yield.

This design has a number of advantages over previous experiments:

- The amount and initial configuration of mix is defined (mix mass evolution removed)
- DT burn occurs near the center of the capsule where temperature gradients are less steep than at the ablator-fuel interface
- As in other separated reactant experiments [8], the DT yield is co-located with the mix

Simulations indicate that initial macropores with diameters of about 100 μm will result in significant heterogeneity at burn time, while macropores with diameters of about 10 μm will result in near fully atomically mixed components.

3. Status of the development of MARBLE

The ability to carry out the MARBLE campaign depends on developing the capability to fabricate the necessary targets. This includes the ability to create low-density deuterated foams,

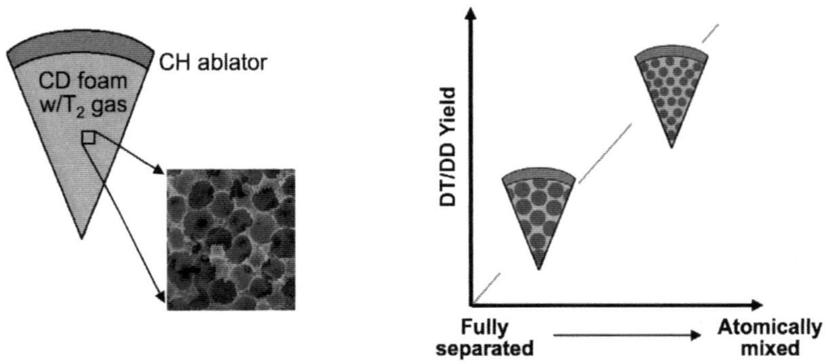

Figure 2. Pie diagram showing the capsules that will eventually be used in the MARBLE campaign. The ratio of DT to DD neutrons will increase as mix increases with smaller pore size.

to create foam with macropores of specified diameter, to encapsulate these foams into ICF targets, and to introduce tritium gas into the capsules. When in a close-pack structure, the macropores make up ~74% of the foam volume, and less if the ideal close-pack structure is not achieved. Techniques have been developed to machine foam spheres, characterize their macropore structure, and enclose them in capsules made from two hemispheres. Progress in the development of targets for MARBLE has been rapid and is described elsewhere [9].

Implosions of foam-filled capsules have been performed on the OMEGA laser [10] using direct drive, and on the National Ignition Facility [11] using the single-shock, exploding pusher platform [12]. These experiments were designed to test the ability of codes to correctly model the performance of foam-filled capsule implosions.

On OMEGA, capsules with 50%-deuterated foam and tritium gas fill were imploded and the DT neutron yields were measured. Due to the early state of target fabrication, many of the targets had nonuniformities in the thickness of the capsules. As a result, many of the capsules also exhibited a displacement of the imploded core from the center of the capsule in time-integrated x-ray images. When restricting the data set to those implosions with core displacements of less than 50 μm (Figure 3), the DT yield is approximately 30% of preshot calculations using the RAGE code [13]. The scaling with foam density and tritium fill pressure are seen to be in qualitative agreement with simulations.

On the National Ignition Facility, MARBLE capsules have been imploded using the indirect drive exploding pusher platform [12]. Two of the implosions were performed using the partially deuterated foam with no gas fill. Two more included a hydrogen fill gas at 1 and 3 mg/cm^3. The DD neutron yield from these experiments (Figure 4) ranges from 50 to 85% of that predicted in simulations, depending on the hydrogen gas density and the code [13, 14] used for the simulations. Bang times were within 100 ps of these simulations.

During FY 2016, two days of experiments will be performed on the OMEGA laser using capsules with engineered foams having pore sizes ranging from 30 to 90 microns in diameter. These will be the first implosion experiments utilizing the engineered foams, and will be the first test of the MARBLE concept.

NIF experiments will continue to develop the indirect drive platform, with experimental series to measure the yield from deuterium gas in CH foam-filled capsules, a series to establish the platform using fully deuterated foam with hydrogen gas fills, and then finally the introduction of tritium/hydrogen mixtures into the gas to confirm the DT to DD yield ratio from a fine-pore foam. Engineered foams with macro-pores will be utilized on NIF starting in fiscal year 2017.

Additional supporting experiments are planned for OMEGA. These include experiments to

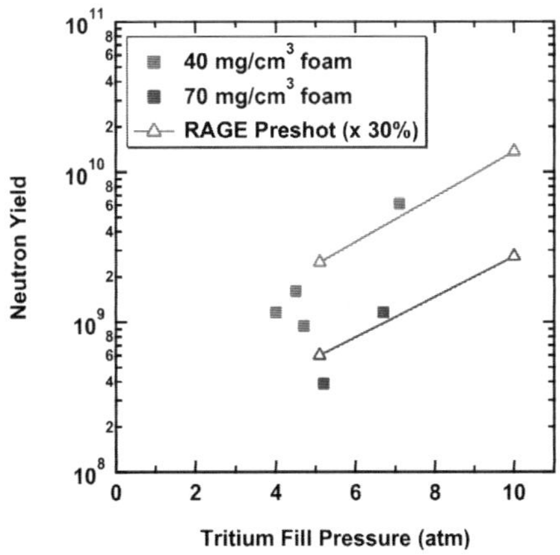

Figure 3. DT neutron yield from OMEGA NIF capsules filled with $CD_{0.5}H_{0.5}$ foam and 4 to 10 atm T_2 gas at room temperature. Yield measurement uncertainty is about 5%.

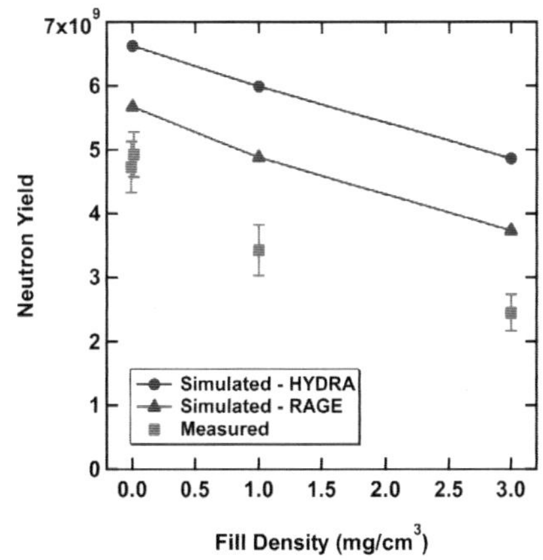

Figure 4. Simulated and measured DD neutron yield from NIF MARBLE capsules filled with $CD_{0.5}H_{0.5}$ foam and 0 to 3 mg/cm^3 H_2 gas.

visualize the interaction of a shock with a low-density void in higher density foam. These experiments will be used to validate the ability of codes to simulate the evolution of the macro-pores in the MARBLE experiments prior to burn time.

4. Conclusions

Development of a platform for studying the effect of heterogeneous mix on thermonuclear burn is underway. The ability to produce deuterated foam-filled capsules for OMEGA and NIF has been demonstrated. Simulations of OMEGA and NIF experiments are in good agreement with data. Initial experiments utilizing foams with macropores specified to control the heterogeneity of the mix will begin on OMEGA in 2016 and on NIF in 2017.

This work is supported by US DOE/NNSA, performed at LANL, operated by LANS LLC under contract DE-AC52-06NA25396.

References

[1] Smalyuk V A *et al*, 2014 *Phys. Plasmas* **21** 056301
[2] Delichatsios M A and Mathews M K 1989 *Fire Safety Science — Proc. of 2nd Intl. Symp.* p 149
[3] Besnard D, Harlow F H, Rauenzahn R M and Zemach C 1992 "Turbulence Transport Equations for Variable-Density Turbulence and Their Relationship to Two-Field Models," Los Alamos National Laboratory Report No. LA-UR-12303 (unpublished)
[4] Stalsberg-Zarling K and Gore R 2011 "The BHR-2 Turbulence Model: Incompressible Isotropic Decay, Rayleigh-Taylor, Kelvin-Helmholtz, and Homogeneous Variable Density Turbulence," Los Alamos National Laboratory Report No. LA-UR-11-04773 (unpublished)
[5] Banerjee A, Gore R A and Andrews M J 2010 *Phys. Rev.* E **82** 046309
[6] Fincke J R, unpublished
[7] Mikaelian K O 1989 *Physica D* **36** 343
[8] Smalyuk V A *et al*, 2014 *Phys. Rev. Lett.* **112**, 025002
[9] Randolph R B *et al*, "Process Development and Micro-Machining of Marble Foam Cored Rexolite Hemi-Shell Ablator Capsules," submitted to *Fus. Sci. Tech.*
[10] Boehly T R *et al.*, 1997 *Opt. Comm.* **133** 493
[11] Moses E I 2008, *Fusion Sci. Tech.* **54**, 361
[12] Le Pape S *et al.*, 2014 *Phys. Rev. Lett.* **112** 225002
[13] Gittings M *et al.*, 2008 *Computational Science & Discovery* **1** 015005
[14] Marinak M M *et al*, 1996 *Phys. Plasmas* **3** 2070

Amorphous nanostructuralization in HOPG by 10^{14} W cm^{-2} laser

Yasuhiko NISHIMURA[1,2], Yoneyoshi KITAGAWA[1], Yoshitaka MORI[1], Katsuhiro ISHII[1], Ryohei HANAYAMA[1], Hirozumi AZUMA[3], Tatsumi HIOKI[4], Teppei NISHI[5], Tomoyoshi MOTOHIRO[4,5], Osamu KOMEDA[6], Takashi SEKINE[7], Nakahiro SATO[7], Takashi KURITA[7], Toshiyuki KAWASHIMA[7], Hirofumi KAN[7], Atsushi SUNAHARA[8], Yasuhiko SENTOKU[9], and Eisuke MIURA[10]

[1] The Graduate School for the Creation of New Photonics Industries, 1955-1 Kurematsu-cho, Nishi-ku, Hamamatsu, Shizuoka 431-1202, Japan
[2] Toyota Technical Development Corporation, 1-21 Imae, Hanamoto-cho, Toyota, Aichi 470-0334 Japan
[3] Aichi Synchrotron Radiation Center, 250-3 Minamiyamaguchi-cho, Seto, Aichi, 489-0965 Japan
[4] Green Mobility Collaborative Research Center, Nagoya University, Furo-cho, Chikusa-ku, Nagoya, Aichi, 464-8601, Japan
[5] Toyota Central Research and Development Laboratories, Inc., 41-1 Yokomichi, Nagakute, Aichi 480-1192, Japan
[6] Toyota Motor Corporation, 1200 Mishuku, Susono, Shizuoka 410-1193, Japan
[7] Hamamatsu Photonics K.K., 1820 Kurematsu-cho, Nishi-ku, Hamamatsu, Shizuoka 431-1202, Japan
[8] Institute for Laser Technology, 1-8-4 Utsubo-honmachi, Nishi-ku, Osaka 550-0004, Japan
[9] Department of Physics, University of Nevada, Reno 1664 N Virginia Street, Reno, NV 89557, USA
[10] The National Institute of Advanced Industrial Science and Technology, 1-1-1 Umezono, Tsukuba, Ibaraki 305-8568, Japan

E-mail: `yasuhiko2460@gpi.ac.jp / yasuhiko.nishimura@mail.toyota-td.jp`

Abstract. This reports provide an amorphous nanostructuralization technique on the surface modification in Highly Oriented Pyrolytic Graphite (HOPG) by using a femtosecond laser. We showed, for the first time, that the surface of HOPG is changed to the amorphous nanostructuralization graphite by using a femtosecond laser-driven compression technique. Our results also suggest that the HOPG surface is changed until the deeper area from the surface by the laser-driven shock wave. A single shot of a femtosecond laser beam ($1.27 \sim 1.33 \times 10^{14}$ W cm^{-2} in intensity, with 2 mm-diameter, and 110 fs in pulse width) is irradiated under the vacuum ambience onto a 2 mm-thick of HOPG. The calculated impact pressures on a sample was $8.3 \sim 8.7$ GPa. Crystal structure in the HOPG were analyzed using a Raman spectroscopy and an X-ray diffraction, those analyzing depth from the surface were 50 nm and 350 μm, respectively.

1. Introduction

The recently study of materials surface modification is performed by using a femtosecond laser-driven compression techniques [1–5]. In this paper, we demonstrate an amorphous nanostructuralization in highly oriented pyrolytic graphite (HOPG) by laser irradiation with $10^{14}\,\mathrm{W\,cm^{-2}}$.

Tsujino et al., reported that the laser of 130 fs, 5.5 mJ was irradiated to the silicon and they observed the β-Sn crystal grains of a few tens nm [1]. Cuq-Lelandais et al., reported that the intensity of $0.6\,\mathrm{PW\,cm^{-2}}$ (300 fs in pulse duration and 30 J in energy with 4 mm-diam. on the sample) laser was irradiated to aluminum and they demonstrated a spallation mechanism and propagation of laser-driven compression in metal [2]. Matsuda et al., reported the nanocrystallization of iron in depth of $2\,\mu$m from the surface by intense-laser multiple irradiation of 130 fs, 0.2 mJ [3, 4]. Sano et al., also reported that a hexagonal crystal diamond was synthesized in the HOPG by laser-driven compression with $2\times10^{15}\,\mathrm{W\,cm^{-2}}$ [5]. Those reports have described a phase transition, spallation mechanism, propagation of laser-driven compression, nanocrystallization of metal, and synthesis, but they did not describe a nanostructuralization formation in depth direction of graphite in detail.

We are studying the surface modification of ceramics by using the ultra-intense laser-driven compression technique. We revealed the multilayered polycrystallization in single crystal yttria-stabilized zirconia by counter-irradiation with a single shot of ultra-intense laser ($10^{17}\,\mathrm{W\,cm^{-2}}$) in our previous research [6]. Here, we provide that the HOPG is changed to nanostructuralization in depth direction by using a laser-driven compression technique in a single shot of femtosecond laser irradiation in $10^{14}\,\mathrm{W\,cm^{-2}}$. We report an amorphous nanostructuralization in depth $350\,\mu$m from the surface by a Raman spectra and an X-ray diffraction (XRD) analysis results.

2. Experimental procedure

This experiments have been performed on the graduate school for the creation of new photonics industries's "7 TW" facility [7–11]. Its laser beam is focused to $10\,\mu$m-diameter at the focal position by using a off-axial parabolic mirror. In order to give a $10^{14}\,\mathrm{W\,cm^{-2}}$ of irradiation intensity on the HOPG surface, it was placed at 5 mm away from the focal position along the laser-incidence axis. An irradiation laser beam diameter of the surface was 2 mm in this layout. The irradiation laser beam on the HOPG were $430\sim450$ mJ, 110 fs in pulse width and $0.82\,\mu$m in wavelength. The shape of HOPG was 10 mm $\times$ 10 mm square of 2 mm-thick (Furuchi Chemical Corp.). The ambient pressure was $\sim10^{-3}$ Pa.

When a laser irradiates on the HOPG surface, the surface is pushed by the high pressure driven by the ponderomotive force of the laser [12, 13]. We can estimate the pressure P by;

$$P = (1+\eta)\,\frac{I_L}{c}, \tag{1}$$

where η is the reflectively of the laser at the surface, which is nearly 1 for the pulse duration. I_L and c are, respectively, the laser intensity and the speed of light. The impact pressure given to the HOPG in this experiment is $8.3\sim8.7$ GPa, using a Formula (1).

We analyzed the irradiated HOPG surface by using the following techniques. The laser Raman spectroscope (NRS-5100, JASCO of Aichi Center for Industry and Science Technology) analyzed the structure of the molecular state, and the XRD equipment (BL5S2 beam line of Aichi Synchrotron Radiation Center) analyzed the crystalline. The beam of laser Raman spectroscope (532 nm in wavelength, $1\,\mu$m-diam. in spot size) was illuminated on the HOPG and line-scanned with $10\,\mu$m steps of 2 mm length in irradiated area, and skin depth of laser beam was 50 nm. The X-ray beam of BL5S2 irradiation angle to the HOPG was $5.000\pm0.004^\circ$ in a grazing incidence, beam size was 0.6×0.5 mm, energy was 12.4 keV, observation angle was separated with 5 step in $30\sim70$ degrees, irradiation time was 3 minute per step, and the skin depth of the X-ray at this time was $350\,\mu$m.

3. Results and discussions

The laser irradiation trace and laser Raman line scanning image shows in the Figs.1 (a) and (b). The irradiation trace seems to discolored a little bit, and it is formed wavy pattern that is dependent on the influence of the energy distribution in the irradiation laser beam. As shown in Fig.1 (a), the irradiation pattern inside the encircled reflects square beam pattern. The laser Raman image also shows that the distribution of irradiated area peaks are changed to depending on the irradiation trace. And irradiated and nonirradiated area have different peak.

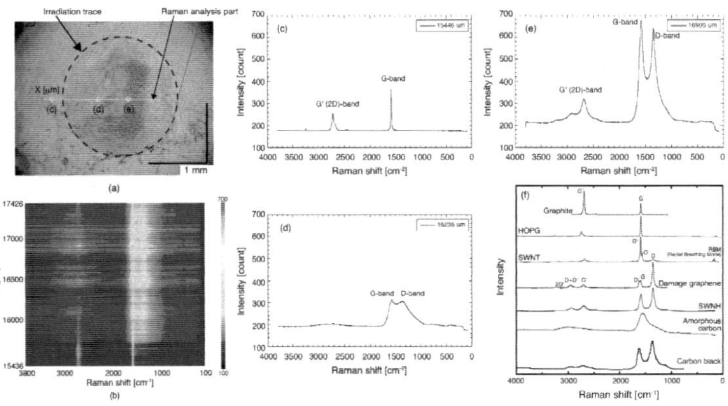

Figure 1. Observation results of the Raman spectrum from the irradiated HOPG. (a) irradiation trace. Inside of dashed black line is an irradiation trace. Green line is a scanning place of (b). (b) laser Raman line scanning image. (c), (d), and (e) are Raman spectrum in analysis point of (a), respectively, those position X are 15446 mm, 16236 mm, and 16906 mm. (f) Raman spectrum from variety carbon materials.

We performed a detailed analysis to selecting three analysis-points from the irradiation trace. Those analysis points are shown in (c), (d), and (e) of Fig.1 (a). We discussed in comparison with the Raman spectra of various carbon materials shown in the Fig.1 (f). Nonirradiated area of Fig.1 (c) agree with a Raman spectra of HOPG, certainly. On the other hand, the irradiation areas of Figs.1 (d) and (e) are appeared D-band peak, and these spectrum are very different with Fig.1 (c). Figures 1 (d) and (e) was similar to the amorphous carbon and the carbon black (nano-powders) having a nanostructure, respectively. Those discussions also suggest that the molecular structure in irradiated area is changed to the amorphous nanostructuralization carbon in depth of 50 nm from the surface by 10^{14} W cm^{-2} femtosecond laser irradiation.

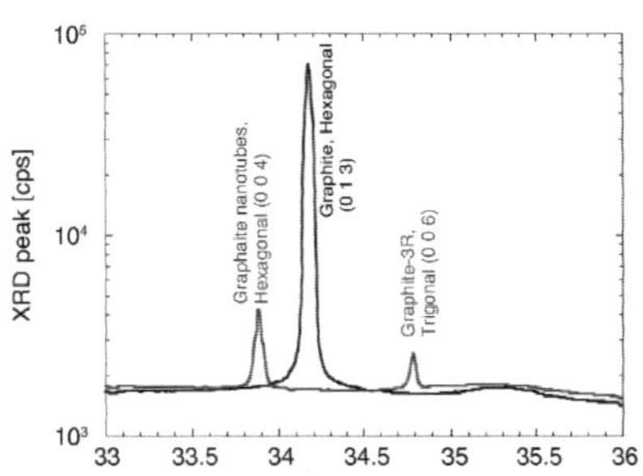

Figure 2. Observation result of the X-ray diffraction profiles. Black line is a nonirradiated HOPG, and blue line is an irradiated HOPG. The different crystalline is observed in an irradiated HOPG.

We next analyzed the inside structure of irradiated HOPG by the XRD analysis. The XRD profile from an irradiated and a nonirradiated HOPG area shows in the Fig.2. The strong XRD peak from nonirradiated HOPG (Black line) is disappeared by laser irradiation shown in irradiated HOPG (blue line) of Fig.2, and the two new peaks appear on it. Those phenomenon also are suggested which occurred between with 350 μm in depth from the surface. Since the shift amount is too large, we considered that the original peak is not shifted or separated by the residual stress of laser irradiation like a blue line peaks. Then, we analyzed an identification of these diffraction peaks. The peak on the left side of blue line in the Fig.2 agreed with a nanotubes (004), and the right side agreed with a graphite-3R (006), respectively.

Here, we hypothesized why our results agreed with the nano-structures peak in Raman and XRD analysis. The HOPG surface is ablated with 10^{14} W cm^{-2} femtosecond laser. Its surface is additional heated to the radial direction with radiant heat of generated plasma, and at the same time, to the depth direction with the ablation driven shock wave is propagated, which pressure is $8.3 \sim 8.7$ GPa. The surface is heated with pulse duration and its is quenched. Under the irradiation area, it is presumed to have broken to nanosized in the same as our previous studies [6]. Especially, the HOPG having a sheet structure is torn to pieces in a depth direction by the shock wave. Then, those pieces are changed to amorphous-like nanostructuralization which edge are rounded to individually by residual stress in quenching process.

Accordingly, we considered that the results of Raman and XRD analysis agreed with the nanostructuralization peak of carbon. Although it is possible to explain our results by this hypothesis, we are thinking that only these discussion is insufficient to reveal this mechanism. We now are performing an additional experiment and detailed analysis, and will reveal this mechanism in a next step.

4. Conclusions

We demonstrated a single shot of the irradiation in 2 mm-thick of Highly Oriented Pyrolytic Graphite (HOPG) with the laser intensity in $1.27 \sim 1.33 \times 10^{14}$ W cm^{-2}. Crystal structure in the HOPG were analyzed using a Raman spectroscopy and an X-ray diffraction, those analyzing depth from the surface were 50 nm and 350 μm, respectively. We showed, for the first time, that the surface of HOPG is changed to the amorphous nanostructuralization graphite by the femtosecond laser-driven compression technique. Our results also suggested that the HOPG surface is changed until the deeper area from the surface by the laser-driven shock wave.

Acknowledgements

We are deeply grateful to Dr Takeuchi Y and the KURE-1 laser technical crew of Hamamatsu Photonics K. K. for their exceptional support during these experiments. We acknowledge Dr Nakamura N, Mr Kondo T, Mr Kuzuya T, Mr Tsuji S, and Mr Okijima J of Toyota Motor Corporation, Mr Fujine M of Aichi Steel Corp., and Dr Kajino T of Toyota Central Research and Development Laboratories, Inc., for valuable discussions of analysis support. We acknowledge Mr Mizuno J, Mr Tanahashi N, and Dr Fukuda N of Aichi Center for Industry and Science Technology, for valuable disucussions of laser Raman measurement. We acknowledge Mr Nakanishi Y of Aichi Science and Technology Foundation for valuable discussions of X-ray diffraction measurement. Thanks to Mrs Suita S of The Graduate School for the Creation of New Photonics Industries for experiment preparation support.

References

[1] Tsujino M et al 2012 Journal of Japan Laser Processing Society **19**-1 54-59
[2] Cuq-Lelandais J P et al 2009 J. Phys. D: Appl. Phys. **42** 065402
[3] Matsuda T, Sano T, Arakawa K, Sakata O, Tajiri H and Hirose A 2014 Appl. Phys. Express **7** 122704
[4] Matsuda T, Sano T, Arakawa K and Hirose A 2014 Appl. Phys. Lett. **105** 021902
[5] Sano T et al 2009 J. Phys.:Conf. Ser. **165** 012019
[6] Nishimura Y et al Y 2015 J Phys. D: Appl. Phys. **48** 325305
[7] Mori Y et al 2013 Nucl. Fusion **53**, 073011
[8] Mori Y, Fukumochi S, hama Y, Sentoku Y and Kitagawa Y 2007 Int. J. Mod. Phys. B **21**, 572
[9] Mori Y and Kitagawa Y 2012 Phys. Plasmas **19**, 053106
[10] Mori Y and Kitagawa Y 2013 Appl. Phys. B Lasers and Optics **110**, 57-64
[11] Sekine T et al 2010 Opt Express **18**, 13927-13934
[12] Wilks S C, Kruer W L, Tabak M and Langdon A B 1992 Phys. Rev. Lett. **69**, 1383
[13] Sentoku Y, Kruer W, Matsuoka M and Pukhov A 2006 Fusion Sci. Technol. **49**, 278

Journal of Physics: Conference Series **717** (2016) 012074 doi:10.1088/1742-6596/717/1/012074

Stark effect modeling in the detailed opacity code SCO-RCG

J-C Pain[1], F Gilleron[1] and D Gilles[2]

[1] CEA, DAM, DIF, F-91297 Arpajon, France

[2] CEA, DSM, IRFU, F-91191 Gif-sur-Yvette, France

E-mail: jean-christophe.pain@cea.fr

Abstract. The broadening of lines by Stark effect is an important tool for inferring electron density and temperature in plasmas. Stark-effect calculations often rely on atomic data (transition rates, energy levels,...) not always exhaustive and/or valid for isolated atoms. We present a recent development in the detailed opacity code SCO-RCG for K-shell spectroscopy (hydrogen- and helium-like ions). This approach is adapted from the work of Gilles and Peyrusse. Neglecting non-diagonal terms in dipolar and collision operators, the line profile is expressed as a sum of Voigt functions associated to the Stark components. The formalism relies on the use of parabolic coordinates within SO(4) symmetry. The relativistic fine-structure of Lyman lines is included by diagonalizing the hamiltonian matrix associated to quantum states having the same principal quantum number n. The resulting code enables one to investigate plasma environment effects, the impact of the microfield distribution, the decoupling between electron and ion temperatures and the role of satellite lines (such as Li-like $1sn\ell n'\ell' - 1s^2n\ell$, Be-like, etc.). Comparisons with simpler and widely-used semi-empirical models are presented.

1. Introduction and outline of the method

In hot dense plasmas encountered for instance in inertial confinement fusion (ICF), the line broadening resulting from Stark effect can be used as a diagnostics of electronic temperature T_e, density n_e and ionic temperature T_i. This represents a challenging task, since from astrophysical dilute plasmas to ultra-dense nuclear fuel in ICF, the density varies by twenty orders of magnitude. The capability of the detailed opacity code SCO-RCG [1, 2] was recently extended to K-shell spectroscopy (hydrogen- and helium-like ions), following an approach proposed by Gilles and Peyrusse [3]. Ions and electrons are treated respectively in the quasi-static and impact approximations and the line profile reads

$$\phi(\nu) \propto \frac{1}{\pi} \int \mathrm{Re}\left[\mathrm{Tr}\{\hat{d}.\hat{X}^{-1}\}\right] W(F)dF, \tag{1}$$

where $\hat{X} = 2i\pi(\nu+\nu_1) - i\hat{H}(F)/\hbar - \hat{\Lambda}_c$, ν_1 being the frequency of the lower state and $\hat{H}(F) = \hat{H}_0 - \hat{d}.F$ the hamiltonian of the ion in the presence of an electric field F following the normalized distribution $W(F)$. $\hat{H}_0$ is the hamiltonian without electric field while $\hat{d}$ and $\hat{\Lambda}_c$ represent respectively the dipole and collision operators. The trace (Tr) runs over the various states of the upper level. If $\Delta\nu_D$ is the Doppler width and w_k the weight of the kth Stark component, neglecting non-diagonal terms in dipolar and collision operators, the line profile can be written as a sum of Voigt (V) functions (parametrized as in Ref. [4]):

$$\phi(\nu) = \frac{1}{\sqrt{\pi}}\frac{1}{\Delta\nu_D}\int_0^\infty W(F)dF\left[\sum_k w_k(F)V(x_k, y_k)\right] \; ; \; x_k = \frac{\nu-\nu_0-c_k(F)}{\Delta\nu_D} \; ; \; y_k = \frac{\langle k|\hat{\Lambda}_c|k\rangle}{2\pi\Delta\nu_D}, \tag{2}$$

where ν_0 is the frequency of the line without external field and

$$c_k(F) = \langle k| - \hat{d}.F|k\rangle \; ; \; \hat{\Lambda}_c = \frac{4\pi}{3}n_e\left(\frac{e}{\hbar}\right)^2 \hat{d}.\hat{d}\left(\frac{2m}{\pi k_B T_e}\right)^{1/2}\ln\left(\frac{\lambda_{DH}Z}{n^2 a_0}\right), \tag{3}$$

λ_{DH} being the Debye-Hückel length.

2. Hydrogen-like ions

Stark effect for hydrogenic ions can be calculated in parabolic coordinates using the basis states $|nqm_\ell\rangle$, where $q = n_1 - n_2$, n_1 and n_2 being the so-called parabolic quantum numbers, related by $n_1 + n_2 + |m_\ell| + 1 = n$, $-\ell \leq m_\ell \leq \ell$ being the magnetic orbital quantum number. The perturbation $\hat{d}$ is diagonal in this basis and a 2nd-order development gives

$$\langle nqm_\ell| - \hat{d}.F|nqm_\ell\rangle = \frac{3}{2}\frac{ea_0}{Z}nqF - \frac{1}{16}\frac{e^2a_0^2}{(2Ry)}\left(\frac{n}{Z}\right)^4\left(17n^2 - 3q^2 - 9m_\ell^2 + 19\right)F^2. \tag{4}$$

However, the fine-structure hamiltonian $\hat{H}_0$ is diagonal in the subset of states $|n\ell sjm_j\rangle$. In order to diagonalize the total hamiltonian $\hat{H}$ in such a basis, the Stark matrix element is

$$\begin{aligned}\langle n\ell sjm_j| - \hat{d}.F|n\ell' sj'm_j\rangle = & \sum_{m_s=-1/2}^{1/2}\sum_{q=-(n-1-|m_\ell|)}^{n-1-|m_\ell|,2}(-1)^{\ell+\ell'-1+3m_j-m_s-q-n}[\ell,\ell',j,j']^{1/2} \\ & \times\begin{pmatrix}\ell & s & j\\ m_\ell & m_s & -m_j\end{pmatrix}\begin{pmatrix}\ell' & s & j'\\ m_\ell & m_s & -m_j\end{pmatrix} \\ & \times\begin{pmatrix}\frac{n-1}{2} & \frac{n-1}{2} & \ell\\ \frac{m_\ell-q}{2} & \frac{m_\ell+q}{2} & -m_\ell\end{pmatrix}\begin{pmatrix}\frac{n-1}{2} & \frac{n-1}{2} & \ell'\\ \frac{m_\ell-q}{2} & \frac{m_\ell+q}{2} & -m_\ell\end{pmatrix} \\ & \times\langle nqm_\ell| - \hat{d}.F|nqm_\ell\rangle,\end{aligned} \tag{5}$$

with $s = 1/2$, $m_\ell + m_s = m_j$ and $[x] = 2x + 1$. Fig. 1 displays a comparison between our previous semi-empirical modeling (Refs. [5, 6]) and the present work in the case of Ar XVIII Ly$_\alpha$ line. Fig. 2 shows Mg XII Ly$_\beta$ profile in three different conditions.

3. Helium-like ions

We consider the transitions $1sn\ell\ ^1P - 1s^2$, $n \geq 2$. For $n \geq 5$, the perturbation due to field F is much larger than the separation between terms, the levels are quasi-hydrogenic and He lines are modeled as Ly-like lines (with $Z \rightarrow Z - 1$). For $n < 5$, singlet-triplet mixing is neglected and two-electron wavefunctions $\Psi(1, 2)$ of singlet states are built as [9]:

$$\Psi(1,2) = \frac{1}{\sqrt{2}}\left[\psi_{100}^Z(1)\psi_{nlm_\ell}^{Z-1}(2) \pm \psi_{nlm_\ell}^{Z-1}(1)\psi_{100}^Z(2)\right] \times \chi_{a(+),s(-)} \tag{6}$$

where ψ_{100}^Z is the wavefunction of the fundamental state of an hydrogenic ion of charge Z and $u_{nlm_\ell}^{Z-1}$ the wavefunction of the excited state nlm_ℓ of an hydrogenic ion of charge $Z - 1$. $\chi_{a,s}$ are antisymmetric (resp. symmmetric) spin functions. Such an approximation is valid

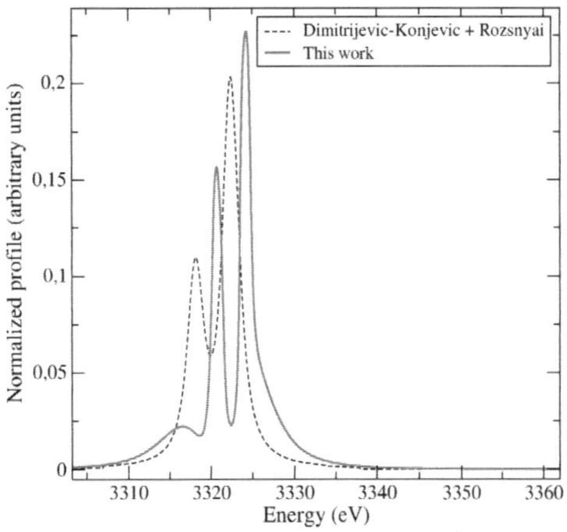

Figure 1. Ly$_\alpha$ line for an Ar plasma at T_e=T_i=700 eV and ρ=3.98 g/cm^3 (example chosen in Ref. [7]).

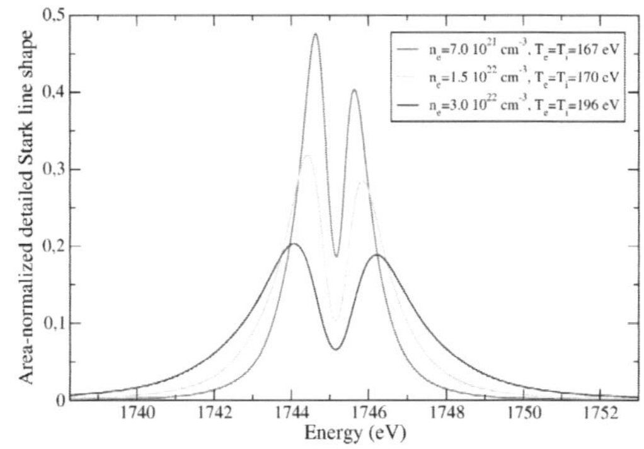

Figure 2. Ly$_\beta$ line for a Mg plasma in three different conditions (example chosen in Ref. [8]).

for highly charged ions when the electron-nucleus interaction overcomes the Coulomb electron-electron repulsion. The hamiltonian $\hat{H}_0 - e(z_1 + z_2)F$ is diagonalized in the sub-space of states $|1s; n\ell m_\ell; S\rangle$ with S=0 for singlet states and S=1 for triplet states. For He$_\alpha$, the resonance line ($1s2p\ ^1P - 1s^2$) requires the energies of terms $1s2s\ ^1S$ and $1s2p\ ^1P$ and the intercombination line ($1s2p\ ^3P - 1s^2$) the energies of terms $1s2s\ ^3S$ and $1s2p\ ^3P$.

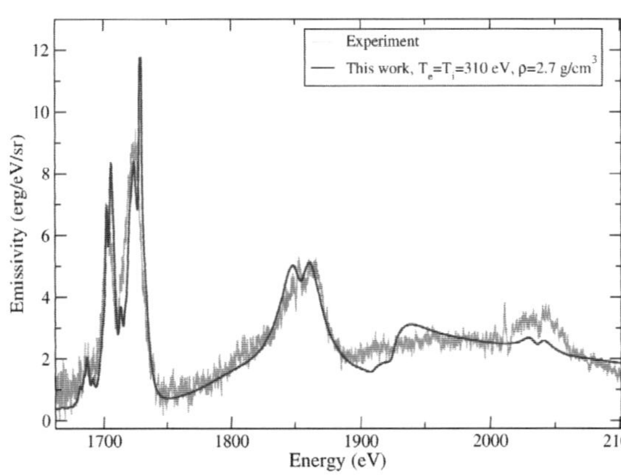

Figure 3. Measured emission of aluminum "buried layers" heated by an ultra-short laser [10] (emissive volume: 400 μm^2 × 0.5 μm, duration: 3 ps) compared to SCO-RCG prediction.

4. Interpretation of a "buried-layer" experiment on aluminum

Fig. 3 shows our interpretation of the recently measured emission of aluminum micro-targets buried in plastic ("buried layers") and heated by an ultra-short laser [10]. Fig. 5 displays the He$_\beta$ profile with different models for the microfield distribution function $W(F)$ (see Fig. 4): Holtsmark model, neglecting ionic correlations and electron screening (see Ref. [11] and references therein), Mayer and "nearest neighbour" (NN) distributions, valid for strongly

Journal of Physics: Conference Series **717** (2016) 012074 doi:10.1088/1742-6596/717/1/012074

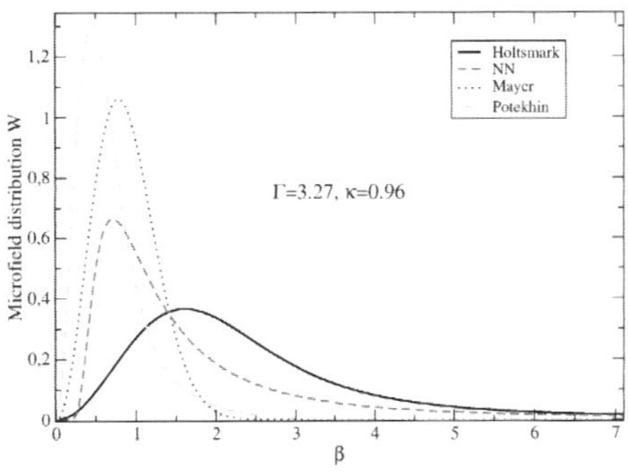

Figure 4. Different microfield distributions for an Al plasma at T_e=T_i=310 eV and ρ=2.7 g/cm^3 (units of $\beta = F/F_0$, where $F_0 = Z^*e/r_{ws}^2$).

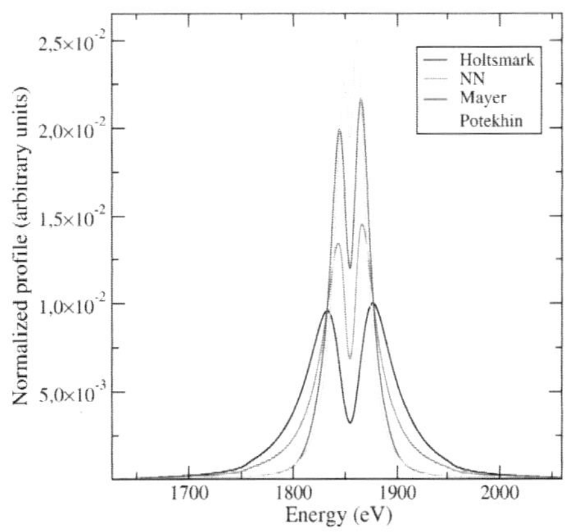

Figure 5. He$_\beta$ line for the different microfield distributions of Fig. 4.

Model / FWHM	Ly$_\alpha$	Ly$_\beta$	He$_\alpha$	He$_\beta$
This work	1.80	27.03	2.46	32.84
Dimitrijevic-Konjevic	2.19	53.73	2.07	61.92

Table 1. FWHM of Ly$_{\alpha,\beta}$ and He$_{\alpha,\beta}$ lines for an Al plasma at T_e=T_i=310 eV and ρ=2.7 g/cm^3.

coupled plasmas, and a combination of APEX (Adjustable Parameter EXponential) method with Monte Carlo simulations proposed by Potekhin et al. [11] and parametrized by ionic coupling $\Gamma = (Z^*e)^2 / (r_{ws}k_BT_i)$ and electron degeneracy $\kappa = r_{ws}/\lambda_{TF}$ constants, r_{ws} being the Wigner-Seitz radius and λ_{TF} the Thomas-Fermi screening length. Table 1 shows the values of the full width at half maximum (FWHM) of Ly$_{\alpha,\beta}$ and He$_{\alpha,\beta}$ lines with the new and previous (Refs. [5, 6]) modelings of Stark effect in SCO-RCG.

In the future, we plan to investigate the importance of autoionizing states $1s2\ell2\ell'$ and $1s2\ell3\ell'$ of He$_\beta$ (in the present work we only took into account $2\ell2\ell'$) and to include the line $1s3d\ ^1D_2$ - $1s^2\ ^1S_0$ induced by the field (mixing states $1s3d\ ^1D_2$ and $1s3p\ ^1P_1$) as well as the lines $1s3d\ ^3D_2$ - $1s^2\ ^1S_0$ and $1s3s\ ^3S_1$ - $1s^2\ ^1S_0$. We also started to study the Stark-Zeeman splitting.

Acknowledgments
We would like to thank V. Dervieux, B. Loupias and P. Renaudin for the experimental spectrum.

References
[1] Pain J-C, Gilleron F and Blenski T 2015 *Laser. Part. Beams* **33** 201
[2] Pain J-C and Gilleron F 2015 *High Energy Density Phys.* **15** 30
[3] Gilles D and Peyrusse O 1995 *J. Quant. Spectrosc. Radiat. Transfer* **53** 647
[4] Humlíček J 1979 *J. Quant. Spectrosc. Radiat. Transfer* **21** 309
[5] Dimitrijević M S and Konjević N 1980 *J. Quant. Spectrosc. Radiat. Transfer* **24** 451
[6] Rozsnyai B F 1977, *J. Quant. Spectrosc. Radiat. Transfer* **17** 77
[7] Mancini R C, Iglesias C A, Ferri S, Calisti A and Florido R 2013 *High Energy Density Phys.* **9** 731
[8] Nagayama T et al. 2014 *Phys. Plasmas* **21** 056502
[9] Bethe H A and Salpeter E E 1957 *Quantum Mechanics of one- and two- electron atoms* (Berlin: Springer)
[10] Dervieux V et al. 2015 *High Energy Density Phys.* **16** 12
[11] Potekhin A Y, Chabrier G and Gilles D 2002, *Phys. Rev.* E **65** 036412

Journal of Physics: Conference Series **717** (2016) 012075 doi:10.1088/1742-6596/717/1/012075

Behaviour of rippled shocks from ablatively-driven Richtmyer-Meshkov in metals accounting for strength

S Opie[1], S Gautam[1], E Fortin[1], J Lynch[1], P Peralta[1,a], and E Loomis[2]

[1]School for Engineering of Matter, Transport and Energy, Arizona State University, Tempe, AZ 85287, USA

[2]Plasma Physics (P-24), Los Alamos National Laboratory, Los Alamos, NM 87544, USA

[a]email: pperalta@asu.edu

Abstract. While numerous continuum material strength and phase transformation models have been proposed to capture their complex dependences on intensive properties and deformation history, few experimental methods are available to validate these models particularly in the large pressure and strain rate regime typical of strong shock and ramp dynamic loading. In the experiments and simulations we present, a rippled shock is created by laser-ablation of a periodic surface perturbation on a metal target. The strength of the shock can be tuned to access phase transitions in metals such as iron or simply to study high-pressure strength in isomorphic materials such as copper. Simulations, with models calibrated and validated to the experiments, show that the evolution of the amplitude of imprinted perturbations on the back surface by the rippled shock is strongly affected by strength and phase transformation kinetics. Increased strength has a smoothing effect on the perturbed shock front profile resulting in smaller perturbations on the free surface. In iron, faster phase transformations kinetics had a similar effect as increased strength, leading to smoother pressure contours inside the samples and smaller amplitudes of free surface perturbations in our simulations.

1. Introduction

Hydrodynamic instabilities are a dominant feature of many High-Energy-Density (HED) systems [1]. The growth of these instabilities depends on material phase and intrinsic fields that perturb the hydrodynamics away from an ideal fluid flow [2-4]. Many materials can retain significant resistance to shear deformation at the large strain rates developed under large transient pressure conditions [5]. This strength is known to decrease Rayleigh-Taylor (RT) and Richtmyer-Meshkov (RM) instability growth rates relative to those under ideal hydrodynamic conditions [3-5]. Little is known, however, about the effects of material strength on these instabilities as phase boundaries (solid-solid, solid-liquid, etc.) are approached. Specifically, the behaviour of shock and release waves undergo sharp changes near these boundaries [6], suggesting that significant changes to the growth rate of instabilities may occur at perturbed solid-state interfaces. It is also likely that material strength could mitigate instability growth in inertial confinement fusion (ICF) capsules using metal ablators [7]. Here we describe a novel experimental setup for validating and calibrating material models in loading regimes typical of hydrodynamic instabilities in solids. We study two materials, copper to explore material strength sensitivity, while iron is used to study phase transformation effects from the α to ε transition [6].

2. Experimental Details

Shock experiments were conducted via laser ablation of a sample's free surface as shown in Figure 1a. Pressure was controlled by varying the laser intensity, which determined material strength/phase regimes the sample experienced during loading. Laser pulses lasted for about ~ 7 ns and produced pressures from 10 to 30 GPa. Two sample geometries were used. Flat samples with an approximate 100 μm thickness were tested first and used to correlate a known laser intensity pulse and resulting velocity history at the opposite free surface, to a pressure boundary condition (BC) via simulation. The laser BC to pressure BC calibration was performed with a range of laser energies. All free surface velocities were recorded with a Line VISAR (Velocity Interferometer System for Any Reflector).

Another set of samples had the ablated surface perturbed with a square wave, as shown in Figure 1a, by a photolithography process [8]. Samples had wavelength (λ), amplitude (η), and thickness (t), of approximately 150 μm, 20 μm, and 100 μm, respectively. The perturbations produced a rippled shock front that imprinted the initially flat free surface with a perturbation at breakout. This, in turn, resulted in an RM instability. Out-of-plane displacement measurements were recorded at the free surface with Transient Imaging Displacement Interferometry (TIDI) [9], as seen in Figure 1b. Only two framing cameras were used to obtain two dynamic images per sample. Free surface velocity was monitored with Line VISAR (Figure 2a). The use of TIDI was of interest in this experiment because it provided an alternative for capturing the RM instability evolution. Other instability studies [3] have used radiography, which is not as sensitive to the small displacements seen here. All experiments were carried out at the TRIDENT Laser Facility at Los Alamos National Laboratory (LANL).

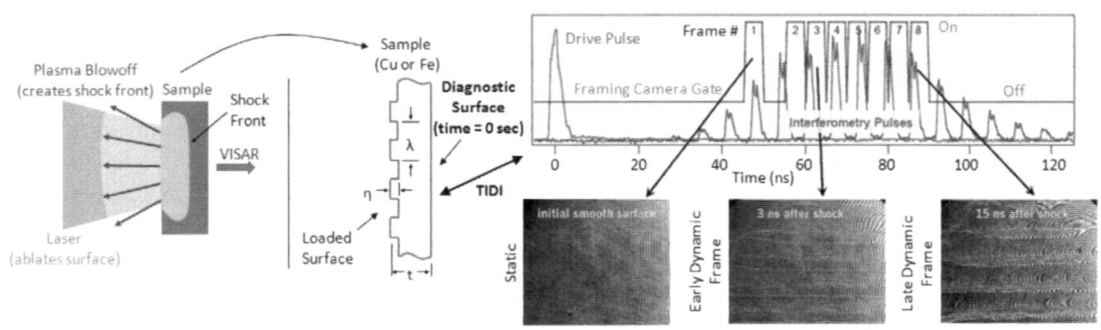

(a) (b)

Figure 1. Schematic of experimental setup and diagnostics. (a) Laser ablation and sample configuration. (b) Pulse train of laser illumination to capture raw TIDI data with framing cameras.

3. Experimental Results

Figure 2 shows typical experimental results obtained with the VISAR and TIDI diagnostics. Figures 2a and 2b (raw VISAR images) are interferograms of a line on the diagnostic surface (x-axis) with time (y-axis). A vertical line out of post-processed Line VISAR interferograms produces a velocity history (Figure 2c) of a particular point on the diagnostic surface. The velocity histories of flat samples (Figures 2b and 2c) have the usual form seen in laser ablation or plate impact experiments, i.e., the surface accelerates and decelerates as a whole with the shock front and release wave respectively. The free surface of the perturbed samples, however, evolves from an initially flat surface (Figure 2d) to a perturbed surface (Figures 2e and 2f) with a wavelength equal to the etched perturbations that were laser loaded. The perturbations evolve higher harmonics with time due to interactions with the release wave. Note that Figures 2d and 2e are post-processed displacement plots from raw static and dynamic images (interferograms) similar to those in Figure 1b.

4. Simulation Results and Discussion

Two dimensional, plane strain simulations were performed to assess sensitivity of the behavior to strength and phase transformation properties. Calculations were carried out in ABAQUS/Explicit^{TM}, via a user material subroutine (VUMAT) with a model similar to that used in [10]. An equation of state (EOS) [11] was used for the volumetric response and a Preston-Tonks-Wallace (PTW) model for the deviatoric response [12]. In iron, the phase transformation kinetics followed the phenomenological model shown in [13]. Separate EOS and strength parameters were used for each phase.

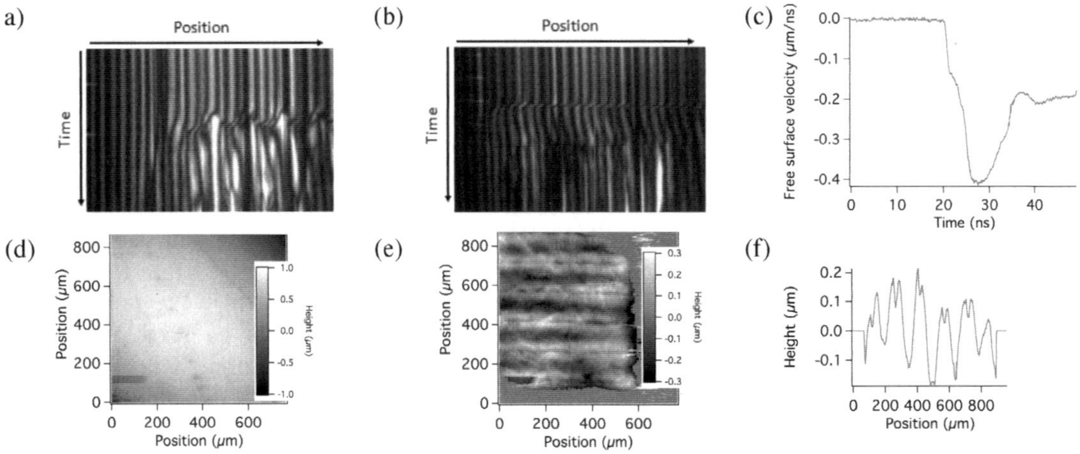

Figure 2. Typical experimental data. (a) and (b) raw Line VISAR of perturbed and flat Cu samples respectively. (c) Line out of flat sample Line VISAR data. (d) Pre-shot TIDI data of perturbed sample. (e) Dynamic TIDI data of perturbed sample. (f) Line out of dynamic TIDI data.

Copper and iron unitless PTW parameters were initially taken from literature [12,14] and adjusted to match the shock front profiles from the flat sample VISAR records. For both phases of iron the s_0, s_{inf}, y_0, y_{inf}, and y_1 parameters [12] were multiplied by 1.75 relative to [14]. This translated the thermal activation region up along the vertical flow stress axis to better match the elastic precursor seen here. In copper, y_0, y_{inf}, and y_2 were changed from 1e-4, 1e-4, and 0.575 to 1e-3, 1e-5, and 0.250, respectively, which increased the initial yield stress and kept the saturation stress unchanged. Strain rate sensitivity and phonon drag characteristics were the same for both materials as in [12] and [14].

Phase kinetic parameters for iron were initially set at ν = 1e5 s^{-1} and B = 3375 J/kg, where ν represents a reference transformation rate and B is akin to an energy barrier [13]. These values reproduced the data in [6] and references therein well. However, for our relatively thin samples changing B to 688 J/kg (or ν to ~1e7 s^{-1}) reproduced velocimetry data better, suggesting a very fast transformation (~1 ns), defined as a 95% transformation at the large driving forces used here [15]. More work is needed to understand the discrepancy in B, i.e., a better kinetics model may be needed.

Tables 1 and 2 show peak-to-peak (P-P) amplitudes obtained from perturbed sample TIDI data and from simulations using the parameters calibrated to flat sample VISAR data. These baseline parameters were used to look at the perturbation growth sensitivity to in material parameters. For copper, the strain rate sensitivity was increased (~2x) and decreased (~0.5x) by likewise changes in the $\varkappa$ and β PTW values [12] relative to the baseline, while other PTW values were changed to retain similar quasistatic strength across all simulations. As shown elsewhere [3] strength parameters were found to have a significant effect, simulations showed that the perturbed shock front profile traveling through the sample was less perturbed in samples with increased strength. In iron simulations parameters controlling phase kinetics were varied. Faster kinetics led to more uniform pressure contours, having a similar result on transient free surface ripple formation as increasing the strength of the material. All simulations showed smaller permanent deformations, relative to the transient deformations, after the shock release wave reached the free surface. Simulations showed that with a longer pressure (laser) pulse the transient perturbations at the free surface would have grown at a quasi-linear rate typical of small amplitude RM instabilities [2]. Figures 3 and 4 illustrate the simulation results, in Figure 4 the 'slow' kinetics had ν = 1.00e5 s^{-1} and B = 3375 J/kg.

Table 1. Cu (s25288, 19J) amplitudes, experimental and baseline simulation results.

Time[a] (ns)	Experimental Amplitude P-P (microns)	Simulation Amplitude P-P (microns)
6.2	0.20 +/- 0.05	0.18
12.7	0.65 +/- 0.05	0.62

a. Time since shock breakout.

Table 2. Fe (s25293, 55J) amplitudes, experimental and baseline simulation results.

Time[a] (ns)	Experimental Amplitude P-P (microns)	Simulation Amplitude P-P (microns)
9.1	0.48 +/- 0.07	0.50
28.2	0.15 +/- 0.05	0.17

a. Time since elastic wave breakout.

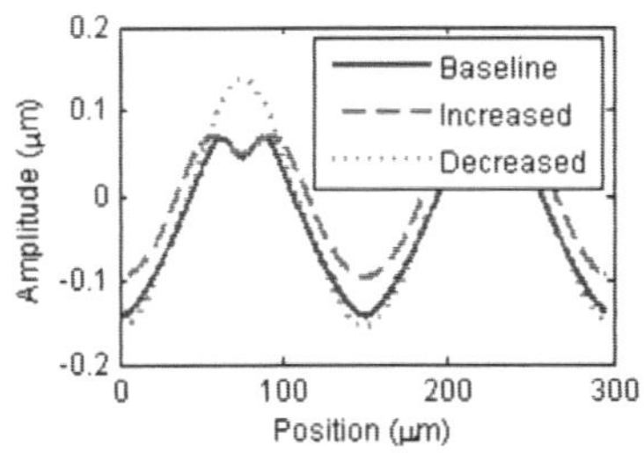

Figure 3. Copper free surface simulations results at 6.2 ns since shock breakout.

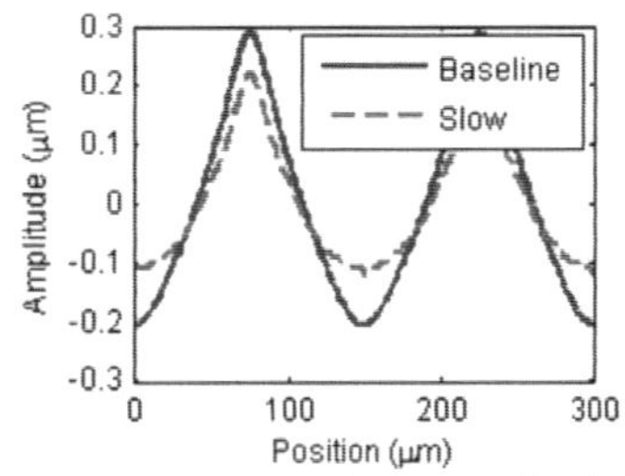

Figure 4. Iron free surface amplitude simulation results at 9.1 ns since elastic wave breakout.

5. Conclusions

Anticipation and control of hydrodynamic instabilities is critical in many applications, particularly ICF design. To accurately predict instability evolution a comprehensive material model is needed that accounts for the wide variation in strength and phases that evolve with the loading conditions. We have described a novel experimental setup that provides data to validate material models/parameters in high strain rate and pressure regimes typical of hydrodynamic instabilities in solids. Preliminary experiments and simulations show that experimental results of this setup are sensitive to material properties controlling strength and phase kinetics, and are useful for supplying additional experimental data for model validation in addition to typical velocity measurements, e.g., Line and Point VISAR.

References

[1] Zhou C and Betti R 2007 *Phys. Plasmsa* **14** 072703
[2] Atzeni S And Meyer-ter-Vehn J 2004 *The Physics of Inertial Fusion* (Clarendon Oxford)
[3] Park H et al. 2010 *Phys. Rev. Lett.* **104** 135504.
[4] Ortega A et al. 2015 *J. Mech. Phys. Solids* **76** p.291
[5] Barton N and Rhee M 2013 *J. App. Phys.* **114** 123507
[6] Boettger J and Wallace D 1997 *Phy. Rev. B* **55** p.2840
[7] Loomis E et. al. 2010 *Phys. Plasmas* **17** 056308
[8] Peralta P et. al. 2015 *Phil. Mag. Lett.* **95** p.67
[9] Greenfield S et. al. 2000 *Laser-Induced Damage in Optical Mat.* **4065** p.557
[10] Zuo Q, Harstad E, Addessio F and Greeff C 2006 *Mod. Sim. Mat. Sci. Eng.* **14** p.1465
[11] Heuze O 2006 *AIP Conf. Proc.* **845** p.212
[12] Preston D, Tonks D and Wallace D 2003 *J. App. Phy.* **93** p.211
[13] Greeff C, Rigg P, Knudson M, Hixson R and Gray G 2003 *Bull. of Amer. Phy. Soc.* **48.4** p.56
[14] Belof J et. al. 2012 *AIP Conf. Proc.* **1426** p.1521
[15] Jensen B, Gray G and Hixson R 2009 *J App. Phy.* **105** 103502

Acknowledgments
This work was supported by the US Department of Energy, Office of Fusion Energy Science under grant # DE-SC0008683. Access to the TRIDENT facility at LANL is gratefully acknowledged as well.

Journal of Physics: Conference Series **717** (2016) 012076 doi:10.1088/1742-6596/717/1/012076

Proposed pushered single shell capsule design for the investigation of mid/high Z mix on the NIF

Ryan Sacks, Robert Tipton, and Frank Graziani

Lawrence Livermore National Laboratory, 7000 East Ave, Livermore CA 94550

E-mail: sacks2@llnl.gov

Abstract. The CD Mix campaign has given a detailed explination of the mix mechanics in the current ignition capsule designs by investigating the relationship between material mixing, shell-fuel interfaces, and the change in thermonuclear yield given a deuterated layer in the capsule. Alternative ignition scenarios include the use of double shell designs that incorporate high-Z material in the capsule. Simulations are conducted on a proposed capsule platform using the ARES code on a scaled capsule design using a partially reduced glass capsule design. This allows for the inclusion of deuterium on the inner surface of the pusher layer similar to the CD mix experiments. The presence of silicon dioxide allows for the investigation of the influence of higher Z material on the mixing characteristics.

1. Introduction

The original low-foot implosions used in the National Ignition Campaign suffered from degraded performance through the mixing of ablator material into the central hotspot. In order to combat this two different experimental paths have been followed in regard to mix, mitigation and measurement. This lead to the high-foot three shock implosions [1] to reduce mix and the CDMix experiments to measure the amount of mixing through by measuring the ratio between DT and TT neutron yields [2]. The purpose of this capsule design is to use the diagnostics developed by the CDMix campaign, and the hohlraum-capsule ratio from the two-shock campaign [3] to develop a design to investigate the mix of mid-Z and high-Z material. This has applications to double-shell target designs [4][5] which are an alternative ignition scenario, see figure 1 for a shell design.

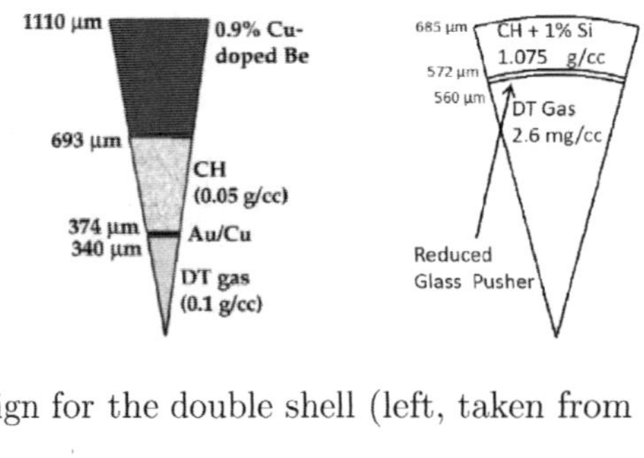

Figure 1. Capsule design for the double shell (left, taken from [4]) and the proposed pushered single shell (right).

The design of the capsule begins with attempting to match the outer radius of the two-shock target design (see figure 1) as the outer radius is matched to that of two shock while the thickness and radius of the pusher shell is detailed from General Atomics. The partially reduced glass portion of the capsule requires further elaboration. When producing glass capsules, General Atomics first sprays a mandrel with CH highly doped with silicon. The capsule is then fired to remove the mandrel and baked to drive the CH material out of the shell and replace it with oxygen, thus leaving only SiO_2 behind. For the proposed capsule this process is interrupted to allow some of the CH material to remain in the shell. This is done with the desire to replace the CH with CD in order to replicate the nuclear mix signal that was developed by the CDMix campaign. After this process was completed, General Atomics reported the total density of the shell as 1.8 g/cc with an approximate atom fraction break down of 23.5% Si, 47.5% O, 12.5% C, and 16.5% H atom fractions [6]. With this design, a radiation drive based on that used in the two-shock campaign is adjusted so that the velocity profile of the surrogate capsule is matched to that of a Mo-Al double shell (see figure 2) in an attempt to match instability growth. The simulations are a first approximation and the EoS for glass is used.

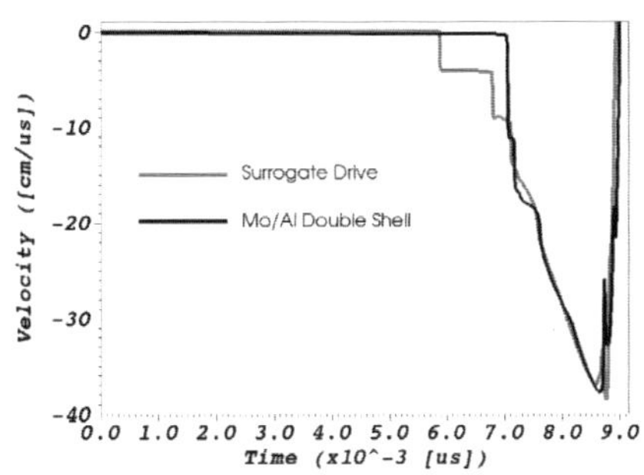

Figure 2. Comparison of velocity profiles for a Mo-Al double shell (black) and the proposed shell design with a surrogate drive (red). The velocity profiles match closely which is desired to promote similar instability growth.

2. Simulations and Results

The ARES [7] code it utilized to perform 1 and 2 D simulations of the capsule with a time dependent temperature source. The 1-D simulations were conducted using multi-group radiation physics and transit time matching at the interface boundaries between the ablator-pusher and pusher-fuel. Two sets of calculations were performed in 1-D, clean simulations with no mix physics, and mix simulations using the KL model where the mix settings were tuned to match two-shock experimental data. The metrics for all simulations are detailed in Table 1. The 1-D clean simulations show the primary shock propagates through the ablator material, is incident on the ablator-pusher, with the shock then incident on the pusher-gas boundary where it both transmits into the gas and is reflected back into the pusher. There is a second shock that catches up with the reflection in the pusher, merges and transmits into the gas where it meets the first shock as both travel towards capsule center (see figure 3). This also compresses the pusher portion of the target to high densities, the peak density of which corresponds well with the time of peak neutron production (see figure 4).

The temperature plot for the 1-D clean simulation shows that the core hotspot in the gas corresponds both to the maximum density as seen in figure 4 and with the time of peak neutron production (see figure 5). However clean simulations must be paired with mix simulations in order to give insight into how the capsule will perform. By examining the metrics in Table 1, it

Table 1. Capsule Performance Metrics

	$TotalNeutronYield$	$MaxIonTemperature$	$Bangtime$
1-D Clean	3.1×10^{15}	$5.3keV$	$8.90ns$
1-D Mix	1.6×10^{14}	$4.3keV$	$8.84ns$
2-D Clean	2.8×10^{15}	$5.2keV$	$8.90ns$

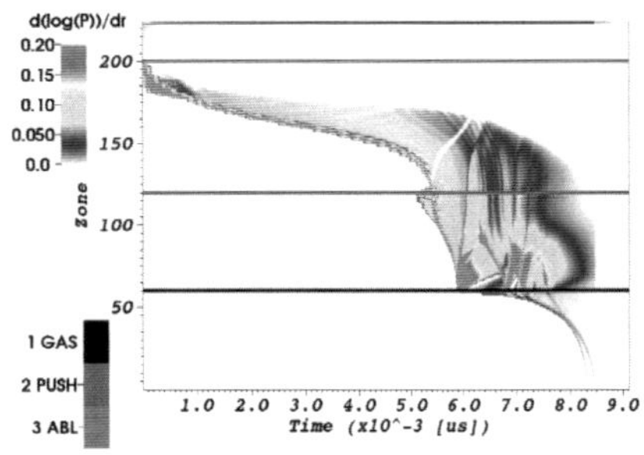

Figure 3. Shock plot for the clean 1-D simulation. The solid lines running horizontally represent, from top to bottom, the channel-ablator, ablator-pusher, and pusher-gas interfaces.

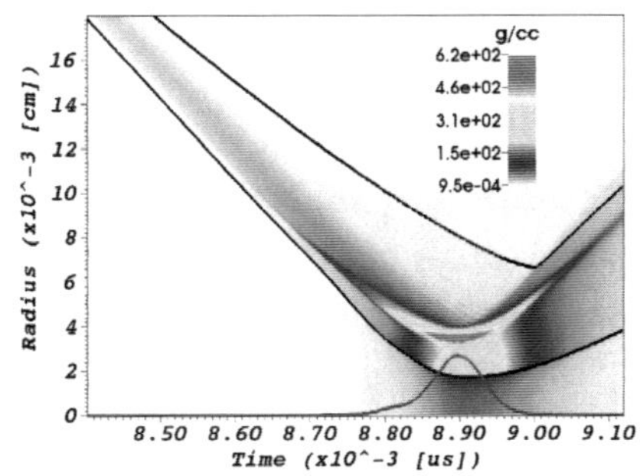

Figure 4. Density time history. The solid brown curve represents the neutron production vs time. The solid black lines represent the Lagrangian material boundaries.

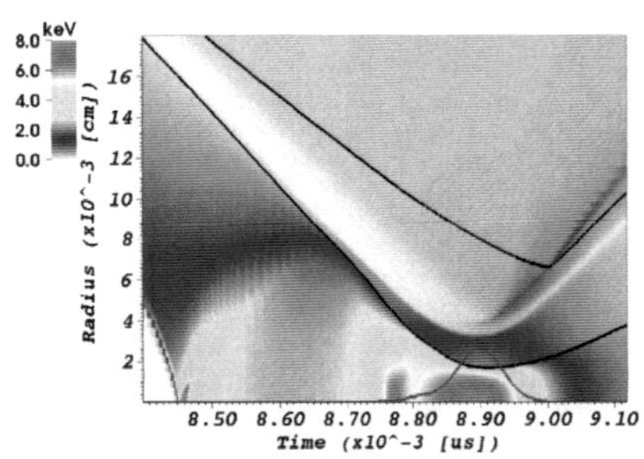

Figure 5. Temperature time history of the proposed shell design. The solid brown curve represents the neutron production vs time. The solid black lines represent the Lagrangian material boundaries.

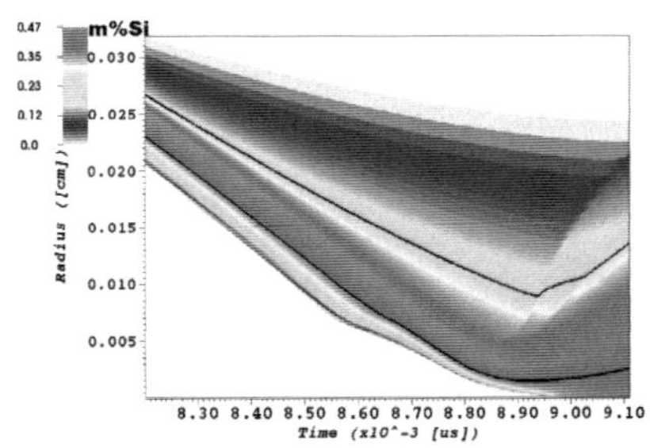

Figure 6. Silicon mass fraction time history. It is shown that there is a healthy amount of silicon that reaches into the capsule center, explaining the lower capsule metrics and slightly faster bangtime.

is seen that the YOC is 5.2% while the ion temperature drops by approximately 20%. This is understood by plotting the silicon mass fraction vs time in the capsule which shows the silicon reaching the center of the target, truncating the burn and depressing the temperature (see figure 6). These results however are still encouraging as the ion temperature and neutron yields are still robust enough for diagnostics.

Preliminary work has been performed on 2-D simulations, and the results of the 2-D clean simulations are available in Table 1, figure 7, and figure 8. The metrics show close to 1-D performance, which is expected for a converged 2-D clean simulation, while the figures show preliminary mix results. Figure 7 shows the mixed temperature profile, which when compared to the silicon mass fraction present in the capsule in figure 8 shows that the maximium temperature is reached in the core where the silicon has not penetrated. This gives the desired platform, as the silicon is migrating from the pusher region into the gas region, which in turn affects the hotspot size, shape, and neutron performance, as can be seen by the 5.2% YOC in the 1-D simulations.

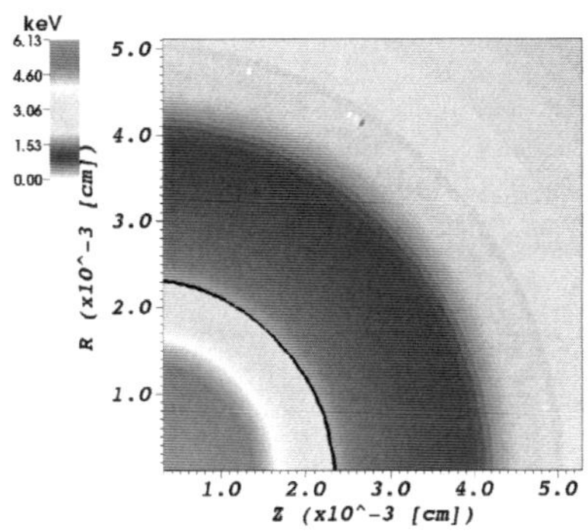

Figure 7. Temperature at capsule center at 8.82 ns for the 2D mix simulation. The solid black line represents the original Lagrangian zone boundary between the pusher and gas.

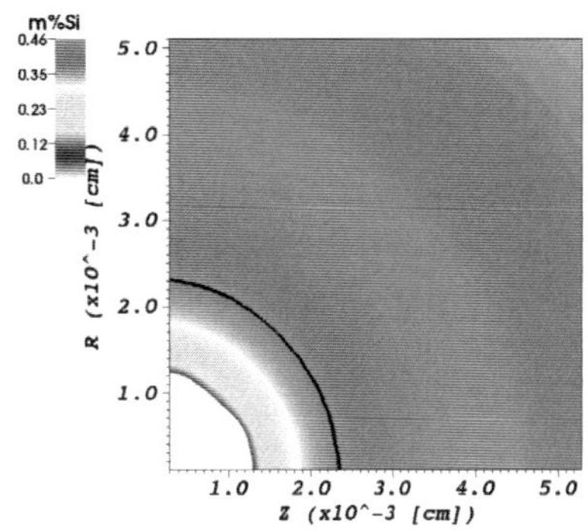

Figure 8. Silicon mass fraction in the capsule for the 2D mix simulation at 8.82 ns. The solid black line represents the original Lagrangian zone boundary between the pusher and gas.

3. Conclusions and Future Work

The result of this study is that the new proposed design has the neutron yields and temperatures required for a mix platform. The material developed by General Atomics allows for the investigation of higher Z mix while retaining the ability from the CDMix experiments to utilize a nuclear diagnostic to quantify mix. However further 2-D mix simulations are required along with more robust equations of state for the partially reduced glass material for more accurate simulation results. Additionally the use of other mix models, such as KLa, KLab, LES, and diffusion models should be used to simulate mix into the hotspot.

Acknowledgments

This work was performed under the auspices of the U.S. Department of Energy by Lawrence Livermore National Laboratory under Contract DE-AC52-07NA27344. Information management release number is LLNL-PROC-679613.

References

[1] O.A. Hurricane, D.A. Callahan et al., Phys. Plasmas 21, 056314 (2014).
[2] D.T. Casey, V.A. Smalyuk et al., Phys. Plasmas 21, 092705 (2014).
[3] Communications with S.A. MacLaren.
[4] P.A. Amendt, J.D. Colvin et al., Phys. Rev. Lett. 9, 2221 (2002).
[5] H.F. Robey, P.A. Amendt et al., Phys. Rev. Lett. 103, 145003 (2009).
[6] Communications with M. Hoppe and A. Nikroo of General Atomics.
[7] R.M. Darlington, Ph.D. Thesis U.C. Davis, UCRL-LR-136160

Efficient fast electron generation in an interaction of Intense, ultrashort laser with metal nanoparticle coated dielectric target

Deep Sarkar, Sheroy Tata, Moniruzzaman Shaikh, Amit D. Lad, Amitava Adak, Subhrangsu Sarkar , Pushan Ayyub and G. Ravindra Kumar

Tata Institute of Fundamental Research, Mumbai, India.

Email: deep.sarkar@tifr.res.in

Abstract. Hot electron generation in intense laser-matter interaction studies is a topic of great interest due in significant part to its applications in fast ignitor scheme in Inertial Confinement Fusion (ICF). We measure the hot electron energy spectrum from Ag nanoparticle coated fused silica target (100 μm thick) interacting with an intense ($I{\sim}10^{18}W/cm^2$), short pulse ($\tau{\sim}$ 30× $10^{-15}s$) laser and compare the results with those of an uncoated fused silica. Enhancement in hot electron energy and hard x-ray yield is measured as a function of thickness of Ag nano-coating, varied from tens of nm to hundreds of nm. The hot electron temperatures and integrated x-ray yield are observed to be greater for subwavelength film thicknesses for the case of a p-polarized laser. Such results indicate that metal nanoparticle layers have an important role to play in the enhancement of laser-plasma coupling efficiency for short scale-length plasmas created in femtosecond laser interactions.

1. Introduction

An intense, ultrashort, laser pulse interacting with solid matter creates a solid density plasma via different ionization mechanisms [1] and couples with the plasma thereafter via collective processes engendering the generation of hot electron jets [2], ion acceleration [3], ultrashort X-ray pulses [4], high harmonics [5] and other kinds of radiation such as Cherenkov [6] and THz radiation [7]. Several standard diagnostics are employed to determine the efficiency of coupling between the laser light and the pre-ionized hot, dense matter. One such diagnostic is Electron Spectrometry (ESM) [8] which characterizes the energy spectra of the hot electrons. The distribution curve is a typical bi-Maxwellian [9] exhibiting two temperatures. The hot electron temperature T_{hot} may scale either as the square root [10] or one-third power [11] of the laser intensity depending on the dominant absorption mechanism.

Measurement of integrated X-ray yield (bremsstrahlung and $K_{\propto}$ emission) is another diagnostic of the laser-plasma coupling efficiency as higher yield in X-rays imply greater absorption which in turn implies hotter electrons. Structured targets [12] such as nanorod arrays [13], nanometric subwavelength gratings [14] and microstructures [15] are known to enhance the conversion efficiency of laser energy into x-rays and hot electrons. Such an enhancement is attributed to the local electric field enhancement of several orders of magnitude at the surface structures due to the "lightning rod effect" [16]. Consequently, multifold enhancement in X-ray yield has been observed for such structured samples [12-15].

In our experiment, we have investigated the effect of sputter deposited layers of Ag nanoparticles with subwavelength thickness (~ 30-100 nm) on the laser energy conversion efficiency by measuring the hot electron spectrum by ESM and X-ray yield as a function of film thickness. The hot electron temperatures and x-ray yields for different film thicknesses have been compared with that of a plane, uncoated fused silica target of thickness $100\mu m$ which also serves as a substrate for the coated targets.

In the next section, we describe our experimental set-up in detail following which we present our findings and discuss their implications. A summary is presented in subsection 2.1.

2. Experimental setup

The experiment was performed with the 100 TW laser facility at TIFR (Ti: Sapphire, 800nm central wavelength) which delivers 25 fs pulses at a repetition rate of 10 Hz. The ASE contrast of the pulse was measured to be $\sim 10^{10}$. The laser beam is focused onto the target to a spot size of 10 μm (FWHM) using an (f/2.5) Off-axis parabolic mirror. ESMs were placed along target front, target rear and laser direction simultaneously to obtain information on the dominant heating mechanism. A NaI (Tl doped) detector coupled with a photomultiplier tube was placed at the target front at a distance of 90cm from the target surface. To reduce counts due to noise from cosmic particles, the detector was placed inside a thick lead cylinder and lead apertures were used to avoid saturation and pile-up due to excess X-ray flux. A schematic of the experimental setup is shown below (Fig.1).

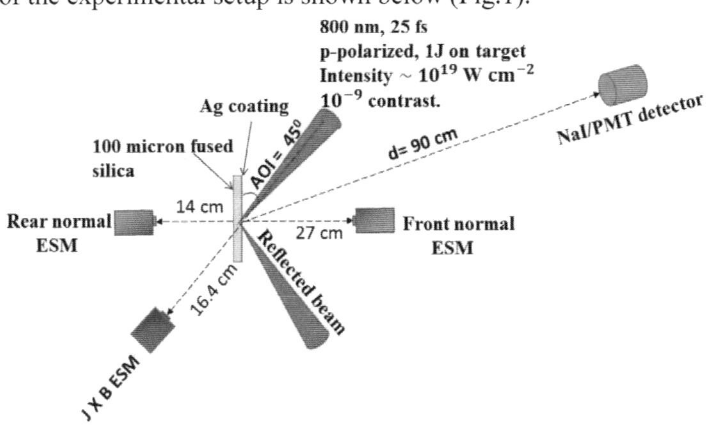

Fig. 1. **Schematic of experimental setup.**

As shown in Fig. 1, the laser beam was incident on the target sample at 45° angle of incidence, with its electric field polarized in the plane of incidence (p-polarized). The intensity on target was $\sim 10^{18} W/cm^2$. Both the electron energy spectra and the integrated X-ray yield were averaged over a total of around 30-40 shots for the coated samples and 70 shots for the plane FS sample.

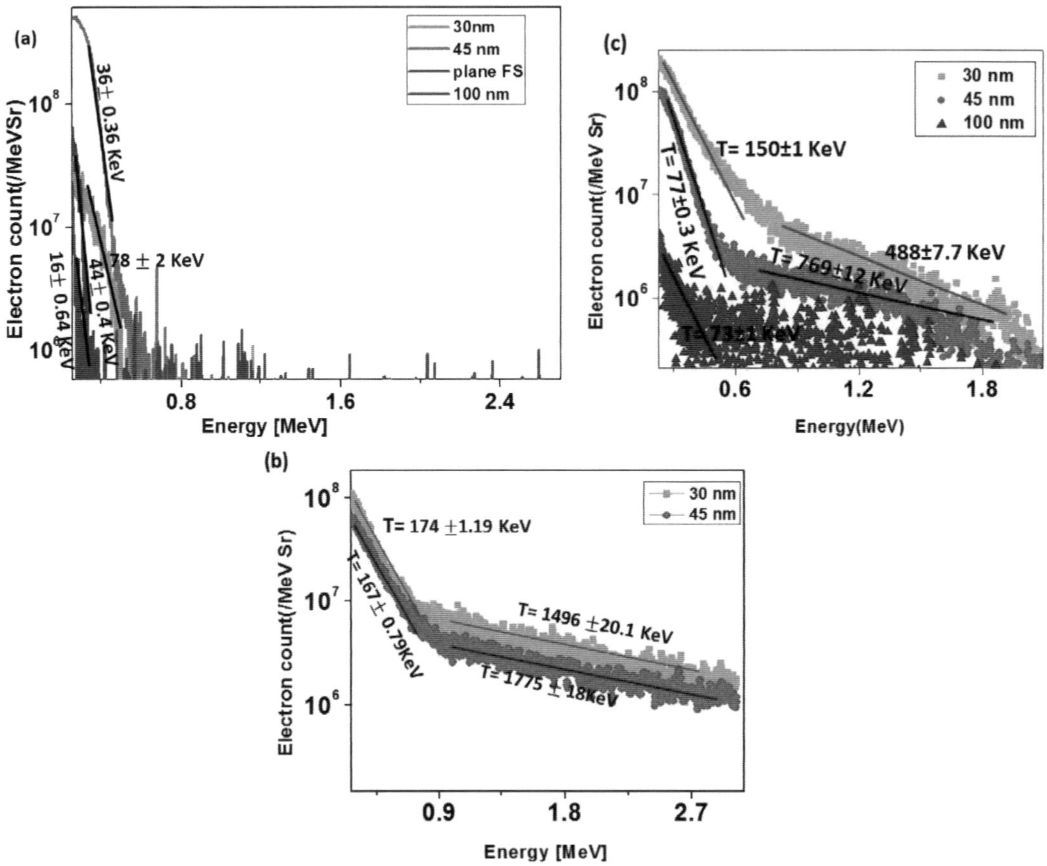

Fig. 2. (a) Energy spectrum along target front normal (b) Hot electron energy spectrum for 30 nm (orange) and 45 nm (green) coating thickness along J×B direction. (c) ESM energy spectrum along target rear normal (no trace obtained for plane FS).

Results. Fig. 2 shows the electron energy spectra for the coated and plane targets along the three directions. Along all three directions, target with film thicknesses around 30–45 nm yield electrons with highest temperatures. The temperatures are highest along the J×B direction, revealing J×B heating as the dominant heating mechanism for our case. The above trend clearly indicates a role of film thickness in optimizing the laser energy coupling efficiency with the plasma. The significantly lower electron temperatures in case of 100 nm film coated target, on the other hand, seems to imply the thickness dependence is weak or altogether absent for coating thicknesses above 100nm.

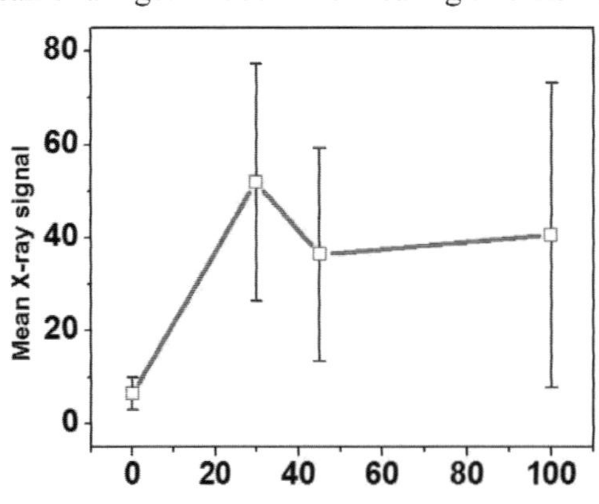

Fig. 3. Integrated x-ray yield as a function of coating thickness

The X-ray yield result (Fig.3) show a similar trend of enhancement for the nanocoated targets but the thickness dependence is not prominent due to large error bars in the yield values owing to statistical fluctuations in the X-ray signal. This may be due to the X-rays being generated both from the coating and the substrate interface.

2.1. Discussion
In our experiment, we have assumed absence of any laser instability mechanism (TPD, SRS) affecting the x-ray and hot electron generation, due to the extremely high contrast (~ 10^{10} ns contrast) of the laser pulse [17] and requirement of high threshold intensity for the SRS [18] to occur. Further, all ion motions and related instabilities (ion-acoustic waves, Langmuir decay instability) were ignored due to the extremely short time-scales of interaction involved.
In summary, we performed an experimental investigation of laser-plasma coupling efficiency in Ag nanoparticulate coatings for intense, femtosecond pulses by measuring the hot electron energy spectrum and time integrated X-ray yield and compared the results with that of a bare FS target. It was found that the electron temperatures were the highest for 30 and 45 nm film thickness with J×B heating as the dominant acceleration mechanism. The X-ray yield was higher for coated targets as compared to the plane one by about 5 orders of magnitude but the thickness dependence of the enhancement was unclear due to large statistical fluctuations in the signal. These results indicate the presence of a length scale besides the morphology of the coating which must be taken into account to control the generation of hot electrons in a high intensity laser-matter interaction. However, for an accurate physical interpretation of the underlying mechanism, simulations are required.

References
[1] P. Gibbon, *Short Pulse Laser Interactions with Matter An Introduction* (Imperial College Press, 2005) ; M. Protopapas *et al.*, Rep. Prog. Phys. **60** (1997).
[2] L. Gremillet *et al.*, Phys. Rev. Lett. **83,** 5015 (1999).
[3] M. Hegelich *et al.*, Phys. Rev. Lett. **89,** 085002 (2002).
[4] J.D. Kmetec *et al.*, Phys. Rev. Lett. **68,** 1527 (1992).
[5] A. Tarasevitch *et al.*, Phys. Rev. A **62,** 023816 (2000); M. Zepf *et al.*, **58,** R5253 (1998).
[6] F. Brandl *et al.*, Europhys. Lett., **61** (5), 232 (2003).
[7] H. Hamster *et al.*, Phys. Rev. Lett. **71,** 2725 (1993).
[8] K.A. Tanaka *et al.*, Rev. Sci. Instrum. **76,** 013507 (2005).
[9] Estabrook & Kruer, Phys. Rev. Lett. **40,** 42(1978).
[10] S.C. Wilks *et al.*, Phys. Rev. Lett., **69,** 1383 (1992).
[11] F. Beg *et al.*, Phys. Plasmas **4,** 447 (1997).
[12] P.P. Rajeev *et al.*, Phys. Rev. Lett. **90,** 115002 (2003).
[13] S. Mondal *et al.*, Phys. Rev. B **83,** 035408 (2011).
[14] S. Kahaly *et al.*, Phys. Rev. Lett. **101,** 145001 (2008).
[15] M. Krishnamurthy *et al.*, Optics Express **20,** 5754 (2012).
[16] P.P. Rajeev *et al.*, Opt. Lett. **29,** 2662-2664 (2004).
[17] F. Baffigi *et al.*, Phys. Plasmas **21,** 072108 (2014).
[18] H.A. Baldis *et al.*, Phys. Fluids **26,** 1364 (1983).

Magntohydrodynamic behavior of capacitor-coil target toward alternative inertial confinement fusion

T Sasaki[1], S Oyama[1], Y Sugimoto[1], K Takahashi[1], T Kikuchi[1], N Harada[1], H Nagatomo[2], S Fujioka[2], and A Sunahara[3]
[1]Nagaoka University of Technology, Nagaoka, Niigata, Japan
[2]Osaka University, Suita, Osaka, Japan
[3]Institute of Laser Technology, Osaka, Osaka, Japan

E-mail: sasakit@vos.nagaokaut.ac.jp

Abstract. To understand its magnetohydrodynamic behaviors and the electrical properties, we proposed to evaluate both experimental observations and numerical simulations. Electrical conductivity for nickel in warm dense matter (WDM) state has been measured with an exploding wire in a quasi-isochoric vessel. The result shows that the electrical conductivity for nickel in WDM is relatively high from the comparison of the electrical conductivities for several materials in WDM state. However, the skin effect in the capacitor-coil target will be neglected from the estimation. A two-dimensional magnetohydrodynamic simulation for the capacitor-coil target has been demonstrated. The results shows that the distribution of B-field in the capacitor-coil target depends on the electrical conductivity model.

1. Introduction

Critical to fast ignition [1] is the transport of the laser-generated fast electrons and their associated heating of compressed DT fuel. The coupling efficiency of laser energy to these fast electrons and the energy deposited in the fuel should be improved by use of the guiding method [2, 3]. The fast electrons are transported with the strong magnetic field, which is order of several teslas, generated by a capacitor-coil target [4]. The capacitor-coil target as shown in Fig. 1 consists of two disks with a single-tune coil. When kJ and ns laser irradiates on the disk, the super-thermal hot electrons emits from the disk. The electrons as a current generates the strong magnetic field in the single tune coil. The generated magnetic field using capacitor-coil target depends on the coil materials [5, 6]. The generated magnetic field should be considered the generation rate of super-thermal hot electrons and the skin effect as an electrical conductivity for ablated plasma at the generation of the magnetic field. To understand its magnetohydrodynamic behaviors and the electrical properties, we proposed to evaluate experimental observations and numerical simulations.

To understand its magnetohydrodynamic behaviors and the electrical properties, we proposed to evaluate both experimental observations and numerical simulation. Electrical conductivity for nickel in warm dense matter (WDM) state has been measured with an exploding wire in a quasi-isochoric vessel. To understand the magnetohydrodynamic behavior in the capacitor-coil target, a two-dimensional MHD simulation has been also demonstrated.

Journal of Physics: Conference Series **717** (2016) 012078 doi:10.1088/1742-6596/717/1/012078

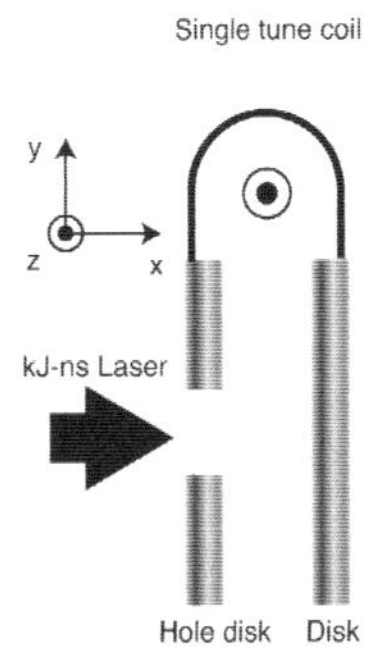

Figure 1. Image of capacitor-coil target.

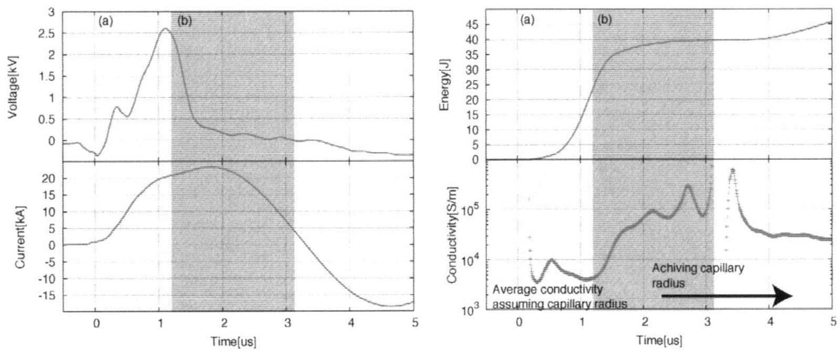

Figure 2. Time-evolution of exploding wire for nickel in the quasi-isochoric vessel. (1) voltage, (2) current, (3) input energy of the wire/plasma, and (4) electrical conductivity.

2. Evaluation of Electrical Conductivity using Exploding Wire Discharge with Isochoric Vessel

To observe the electrical conductivity in WDM, we have demonstrated an exploding wire discharge with quasi-isochoric vessel. The experimental setup of the exploding wire discharge with quasi-isochoric vessel as shown in Ref. [7]. A capacitor bank; C consists of $2 \times 1.89\mu$F low-inductance capacitors, which drive a wire explosion. The capacitor bank was charged up to 8 kV, and it was switched by a low-inductance triggered spark gap. The wire/plasma was confined by the glass tube having $\phi = 1$ mm in diameter with $l = 10$ mm in length. The wire diameter was selected to be 300 μm, considering the skin depth and the plasma density of the sample.

Figure 3 shows the typical voltage-current waveforms, input energy, and electrical conductivity of the sample. As shown in Fig. 3, the wire was exploded at 1.2 μs from the voltage-current waveform. It means that the region (a) in Fig. 3 corresponds to the wire-vapor condition. After exploded wire, the wire-plasma expands in the isochoric vessel. The wire/plasma expansion velocity is estimated to be 3×10^3 m/s from the optical emission of wire-plasma. The uniformity time of density distribution in the expanded wire plasma from exploding wire discharge in water is several hundreds ns, which depends on the sound velocity of the wire/plasma. Thus, the region (b) in Fig. 1 is the uniform wire/plasma state. After the region (b) in Fig. 3, the vessel was broken by the exploding wire/plasma. Observed electrical conductivity estimated by the plasma column size and voltage-current waveform is $10^4 \sim 10^5$ S/m. Since the optical emission from wire/plasma is black-body, the wire/plasma temperature is estimated to be 5000 K.

From the comparison of the previous experimental results as shown in Ref. [8, 9, 10, 11], the electrical conductivities of the other materials are estimated to be 10^4 S/m at the same density-temperature. It means that the electrical conductivity of nickel is higher than that of the other materials at the same plasma density. From the measured electrical conductivities of several materials in WDM state, the skin depth $\delta = \sqrt{1/\pi\mu_0 f \sigma}$, where f is the typical frequency as 200 MHz, is larger than the wire diameter of the capacitor coil.

3. Magnetohydrodynamics Simulation on Capacitor-coil Target

To understand the dependence of generated magnetic field on electrical conductivity distribution in matter, the magnetohydrodynamics (MHD) of single tune coil in the capacitor-coil target with two-dimensional cylindrical geometry has been calculated. The simulation is not incorporated the experimental results. The MHD behavior of capacitor-coil target is solved by conventional

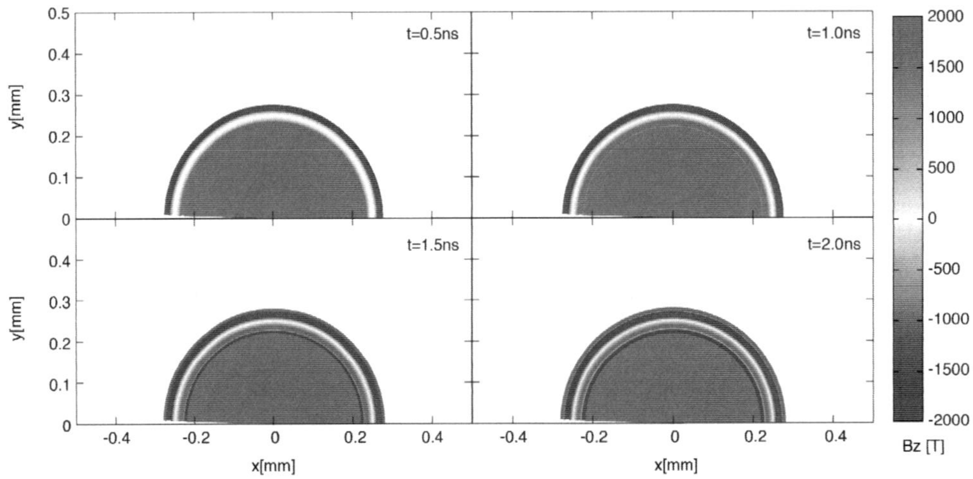

Figure 3. Time-evolutions of B_z-field profile in x-y plane, which is calculated using constant electrical conductivity for the wire.

MHD equations. The magnetic pressure in the equation of motion was neglected because the kinetic pressure of the plasma was higher than the magnetic pressure. The wire material, which is determine the equation of state and the transport properties, is aluminium. Coil and wire diameters of the capacitor-coil target at the initial condition were set to be 550 μm and 50 μm, respectively. The electrical conductivity in vacuum region is set to be 10^{-5} S/m.

The capacitor-coil target during the irradiation of intense laser behaves like a current source. The maximum interior magnetic flux density of capacitor-coil coil is approximately 1.5 kT. From the results, to simplify the current evolution $I(t)$, we have set sinusoidal wave form with 4 MA of maximum current and 500 MHz of frequency. The displacement current was also neglected since the propagation time of the electromagnetic wave from the inner wire surface to the center of the coil was estimated to be 1 ps. The typical timescale of magnetohydrodynamic motion is longer than that of the electromagnetic wave. The timescale in the MHD simulation is ensured.

Figure 3 shows the time-evolutions of B_z-field profile, which is calculated using constant electrical conductivity for the wire. The electrical conductivity of wire is set to be 10^5 S/m. The results shows that B_z-field in the coil is almost uniform until $t = 1$ ns. After $t = 1$ ns, the isolated reversal B_z-field is observed. However, the B_z-field at the center of coil is positive until $t = 2$ ns.

Figure 4 shows the time-evolutions of B_z-field profile, which is calculated using Lee-More model [12] for the wire. The results indicated that B_z-field in the coil using Lee-More model is rather small compared to that using constant conductivity model. Because the electrical conductivity of wire using Lee-More model is 10^6 S/m, the B-field diffuses into the coil. However, it is noted that the spatial-temporal distribution of B-field depends on the total current in the coil. Thus, the equivalent circuit model for capacitor-coil target or the total current into the coil will be included the simulation model.

4. Concluding Remarks

To understand the magnetohydrodynamic behavior and electrical properties of the laser-capacitor target we have evaluated by experimental observations and numerical simulations. Electrical conductivity for nickel in WDM state has been measured with the exploding wire in the quasi-isochoric vessel. The result shows that the electrical conductivity for nickel in WDM

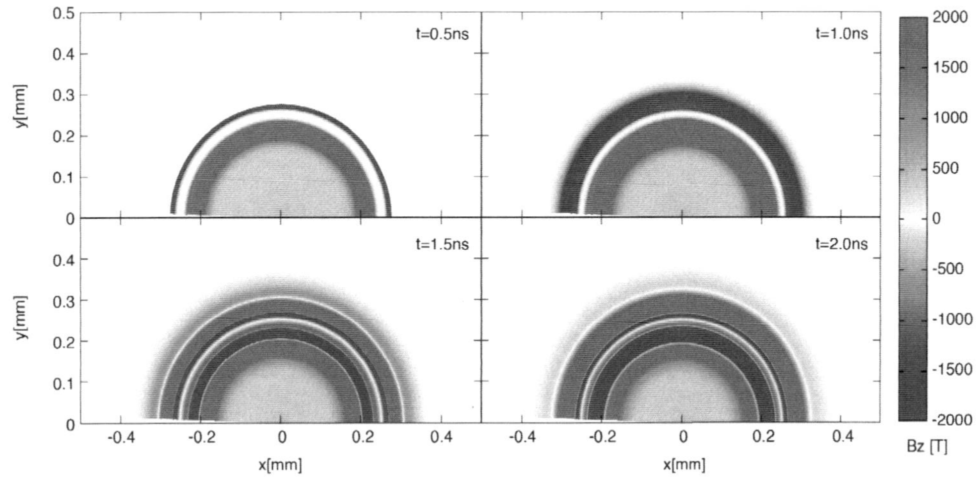

Figure 4. Time-evolutions of B_z-field profile in x-y plane, which is calculated using Lee-More model [12] for the wire.

is relatively high from the comparison of the electrical conductivities for several materials in WDM state. However, the skin effect in the capacitor-coil target will be neglected from the estimation. A two-dimensional magnetohydrodynamic simulation for capacitor-coil target has been demonstrated. The results shows that the distribution of B-field in the capacitor-coil target depends on the electrical conductivity model.

The equivalent circuit model for capacitor-coil target or the total current into the coil will be included the simulation model.

Acknowledgements

This work was partly supported by the joint research project of the Institute of Laser Engineering, Osaka University (under contract subject "2015B1-13").

References

[1] Kodama R, *et al* 2001 *Nature* **412** 798
[2] Kemp A J, Fiuza F, Debayle A, Johzaki T, Mori W, Patel P K, Sentoku Y and Silva L O 2014 *Nuclear Fusion* **54** 054002
[3] Robinson A P L, *et al* 2014 *Nuclear Fusion Fusion* **54** 054003
[4] Fujioka S, *et al* 2012 *Plasma Physics and Controlled Fusion* **54** 124042
[5] Fujioka S, *et al* 2013 *Scientific Reports* **3** 1170
[6] Santos J J, *et al* 2015 *New Journal of Physics* **17** 083051
[7] Amano Y, *et al* 2012 *Review of Scientific Instruments* **83** 085107
[8] Sasaki T, *et al* 2010 *Physics of Plasmas* **17** 084501
[9] Sheftman D and Krasik Y E 2010 *Physics of Plasmas* **17** 112702
[10] Sasaki T, *et al* 2006 *Laser and Particle Beams* **24** 371
[11] Krisch I and Kunze H J 1998 *Physical Review E* **58** 6557
[12] Lee Y T and More R M 1984 *Physics of Fluids* **27** 1273

Journal of Physics: Conference Series **717** (2016) 012079 doi:10.1088/1742-6596/717/1/012079

Short-pulse, compressed ion beams at the Neutralized Drift Compression Experiment

P A Seidl[1], J J Barnard[2], R C Davidson[3], A Friedman[2], E P Gilson[3], D Grote[2], Q Ji[1], I D Kaganovich[3], A Persaud[1], W L Waldron[1] and T Schenkel[1]

[1]Lawrence Berkeley National Laboratory, Berkeley, California, USA
[2]Lawrence Livermore National Laboratory, Livermore, California, USA
[3] Princeton Plasma Physics Laboratory, Princeton, New Jersey, USA

PASeidl@lbl.gov

Abstract. We have commenced experiments with intense short pulses of ion beams on the Neutralized Drift Compression Experiment (NDCX-II) at Lawrence Berkeley National Laboratory, with 1-mm beam spot size within 2.5 ns full-width at half maximum. The ion kinetic energy is 1.2 MeV. To enable the short pulse duration and mm-scale focal spot radius, the beam is neutralized in a 1.5-meter-long drift compression section following the last accelerator cell. A short-focal-length solenoid focuses the beam in the presence of the volumetric plasma that is near the target. In the accelerator, the line-charge density increases due to the velocity ramp imparted on the beam bunch. The scientific topics to be explored are warm dense matter, the dynamics of radiation damage in materials, and intense beam and beam-plasma physics including select topics of relevance to the development of heavy-ion drivers for inertial fusion energy. Below the transition to melting, the short beam pulses offer an opportunity to study the multi-scale dynamics of radiation-induced damage in materials with pump-probe experiments, and to stabilize novel metastable phases of materials when short-pulse heating is followed by rapid quenching. First experiments used a lithium ion source; a new plasma-based helium ion source shows much greater charge delivered to the target.

1. Introduction

Intense pulses of ions in the MeV range enable new studies of the properties of matter ranging from low intensity (negligible heating, but active collective effects due to proximate ion trajectories in time and space), to high intensity where the target may be heated to the few-eV range and beyond. By choosing the ion mass and kinetic energy to be near the Bragg peak, dE/dx is maximized and a thin target may be heated with high uniformity [1], thus enabling high-energy density physics (HEDP) experiments in the warm dense matter (WDM) regime. The Neutralized Drift Compression Experiment (NDCX-II) was designed with this motivation [2-4].

Reproducible ion pulses ($N>10^{11}$ /bunch), with bunch duration and spot size in the nanosecond and millimeter range, meet the requirements to explore the physics topics identified above. The formation of the bunches generally involves an accelerator beam with high perveance and low emittance, attractive for exploring basic beam physics of general interest, and relevant to the high-current, high-intensity ion beams needed for heavy-ion-driven inertial fusion energy [5].

Furthermore, short ion pulses at high intensity (but below melting) enable pump-probe experiments that explore the dynamics of radiation-induced defects in materials [6]. For high peak currents and short ion pulses, the response of the material to radiation may enter a non-linear regime due to the overlapping collision cascades initiated by the incident ions. These effects may be transient (no memory effect at a subsequent pulse) and the short, intense pulses of ions provide an opportunity to observe the time-resolved multi-scale dynamics of radiation-induced defects [6-8]. In addition, by measuring the ion range during the course of the ion pulse, the effects of defects and heating on range can be observed. The time-resolved information provides insight and constraints on models of defect formation and in the design of structural materials, for example, for fission and fusion reactors.

2. New helium ion source and opportunities for intense ion beam physics

The first target experiments used beam pulses of Li^+ accelerated to 1.2 MeV and focused to a beam radius, r ≅ 1mm and duration of 2 ns FWHM (peak current ~18 A/cm^2). These conditions were used to first commission the integrated accelerator components and then in target experiments, for example, to characterize dose rate effects on the ionoluminescence of yttrium aluminium perovskite (YAP). Single-shot YAP scintillator streak spectrometer results showed wavelength structures and motivate follow-up measurements with other materials while varying the focused intensity. These and other results are described in Ref. [9].

Recently, we have installed and used a new multicusp, multiple-aperture plasma ion source. The source can generate high purity ion beams of, for example, protons, helium, neon and argon. To date, we have used it solely for the generation of He^+ ions, where the source injects significantly greater charge than the lithium ion source. Furthermore, helium at about 1 MeV is nearly ideal for highly uniform volumetric energy deposition, because particles enter thin targets slightly above the Bragg peak energy and exit below it, leading to energy loss in the target, uniform within several percent.

The large extraction area is a novel feature of this new plasma source (38 cm^2). To control the plasma meniscus geometry over such a large area and maintain low emittance, the ion extraction gap is established between two parallel plates with many aligned holes, millimeter-diameter with a few-millimeter pitch. A filament driven plasma is formed for 1 millisecond. The plasma facing hole plate, which defines the ion-emission surface, is biased to nearly the same potential as the filament. One kilovolt ions are extracted between the parallel hole plates during the 1-μs high voltage pulse applied via voltage division to both the 3-mm gap and the ≈0.5 m injector column.

The average extracted current density over the 7-cm diameter emission plane is in the range 1-5 mA/cm^2, depending on the voltage, helium gas flow, and filament discharge settings, which are controlled stably and with high reproducibility. The beam is formed with low emittance ($\varepsilon_n < 2$ mm•mrad estimated from simulations) and with a manageable gas load into the downstream accelerating structure. Particle-in-cell simulations helped determine injector voltage settings so that the ion trajectories are similar to those of the previous lithium ion source in the accelerator. Operating parameters of the new ion source are summarized in Table 1, and a fuller description of the source design and characterization may be found in Ji et al. [10].

Table 1. Operating parameter ranges for the new helium ion source in NDCX-II.

I_{arc}	1-4 A	Flow rate	45 standard cc/min
$V_{filament\ bias}$	100 V	P_{source}	3.5×10^{-3} Torr
$P_{filament}$	5A x 50 V	$P_{injector}$	8×10^{-6} Torr
$V_{injector}$	135 kV	$P_{accelerator}$	$\leq 3 \times 10^{-6}$ Torr

An ion induction accelerator is capable of simultaneously accelerating and rapidly compressing beam pulses by adjusting the slope and amplitude of the voltage waveforms in each gap. In NDCX-II, this is accomplished with 12 compression and acceleration waveforms driven with peak voltages ranging from 15 kV to 200 kV and durations of 0.07-1 μs.

The first seven acceleration cells are driven by spark-gap switched, lumped element circuits tuned to produce the required cell voltage waveforms. These waveforms ("compression" waveforms because of their characteristic triangular shape) have peak voltages ranging from 20 kV to 50 kV. An essential design objective of the compression pulsers is to compress the bunch to <70 ns so that it can be further accelerated and bunched by the 200-kV Blumlein pulsers which drive the last five acceleration cells.

In the final drift section, the bunch has a head-to-tail velocity ramp that further compresses the beam by an order-of-magnitude. The space-charge forces are sufficiently high at this stage to require that dense plasma, generated prior to the passage of the beam pulse, neutralizes the beam self-field and to enables focusing and bunching of the beam to the millimeter and nanosecond range [11]. Since the coasting beam compresses with negligible space-charge repulsion, the spot size and duration are limited mainly by voltage waveform fidelity and the chromatic aberrations and emittance of the beam.

The accelerator provides a novel platform to extend the limits of intense beam and beam-plasma physics. For example, the ion-electron two-stream instability has been predicted [12], and it may be observed in NDCX-II by passing a nearly constant energy beam through a plasma. The manifestation would be a significant transverse defocusing and longitudinal bunching of the specially prepared beam. In some parameter regimes in an inertial fusion energy reactor chamber, this could be a critical issue for the transport of a much higher current ion beam in the presence of a significant number of background electrons. In NDCX-II, the effect is normally absent because of the imposed velocity ramp on the beam distribution.

Another interesting opportunity would be to demonstrate the collective focusing of an ion beam in a weak magnetic field [13]. The focusing occurs due to the formation of a strong radial self-electric field due to the rearrangement of the plasma electrons moving with the beam, and in response to a weak magnetic field. The magnetic field is established by the final solenoid near the end of the neutralized drift section. But, instead of using the full strength of the final solenoid (8 Tesla) to focus the beam, equivalent focusing may be achieve with only ~0.02 Tesla. One challenge is to limit the source of electrons to the plasma region upstream of the solenoid. If demonstrated to be practical, the focusing magnet strength would be greatly reduced with a corresponding reduction of the stray magnetic field at the nearby target plane – as preferred for some beam-target interaction experiments.

Non-neutral drift compression and beam pulse manipulation is of interest for future high intensity machines. For example, a nearly monoenergetic beam may be formed at the axial location where the significant velocity tilt is removed by the longitudinal space charge of the intense beam pulse, attractive for reducing chromatic effects when focusing to small spot size. The flexibility to independently trigger and shape the waveforms of each induction cell allows the formation of varied pulse shapes, such as foot—peak waveforms, as generally desired for indirectly driven fusion targets. Finally, double pulses of K^+ or Li^+ ions, separated by 0.05 - 0.5 μs, have been created from single triggers of the high voltage pulses in the injector and accelerator [6]. A single bunch is injected into the accelerator but was cleaved due to the shifted timings of the firing of the compression waveforms. Double pulses are of interest for example, in pump-probe experiments of radiation effects in matter.

3. Recent results and outlook

Using the new helium ion source, the transverse distribution of the focused beam at the target plane is characterized by a thin scintillator and a gated image-intensified CCD camera. In Fig. 1 the beam distribution is peaked with 80% of the integrated charge within a radius of 1.2 mm.

Previously, for the Li^+ beam, the charge in the highly compressed bunch was 1-2 nC. Simulations indicate that the focused charge is expected to ultimately reach over 30 nC per pulse while maintaining a small focal spot size and bunch duration, since much greater charge is now available from the new helium source. Toward that goal, we have measured 15 nC (9×10^{10} ions/pulse) in a few-mm focal spot with the He^+ beam (Fig. 2). A near-term objective is to achieve a pulse duration of 1-2 ns by model-derived tuning of the acceleration and bunching waveforms and the solenoid field strengths.

With the range of intensities and dose rates presently available, there is an opportunity to create novel states of matter with the NDCX-II beam, such as local nitrogen-vacancy center formation in nitrogen-implanted diamond by transient heating, followed by rapid quenching [14]. Annealing of specially prepared materials often takes place on a timescale of seconds to minutes, such as in the formation of paramagnetic materials [15]. With NDCX-II, we have an opportunity to locally excite materials with intense ion pulses on a ns timescale and stabilize novel phases by rapid quenching.

Intense ion beams from induction accelerators have complementary advantages (e.g., low energy spread, benign radiation environment) vs. beams derived from laser-plasma acceleration [16].

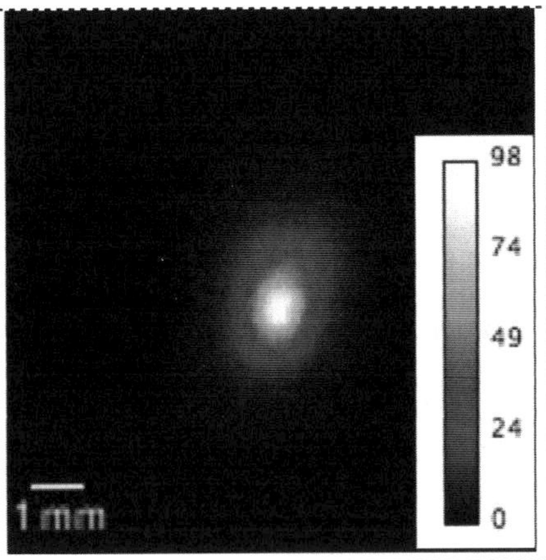

Figure 1. The transverse (x, y) distribution of the He^{+} beam is shown with a color calibration bar indicating the relative intensity. With the new helium source, NDCX-II can provide 1-MeV helium ion beams with an energy deposition greater than 0.7 joules/cm^2 (~40 A/cm^2).

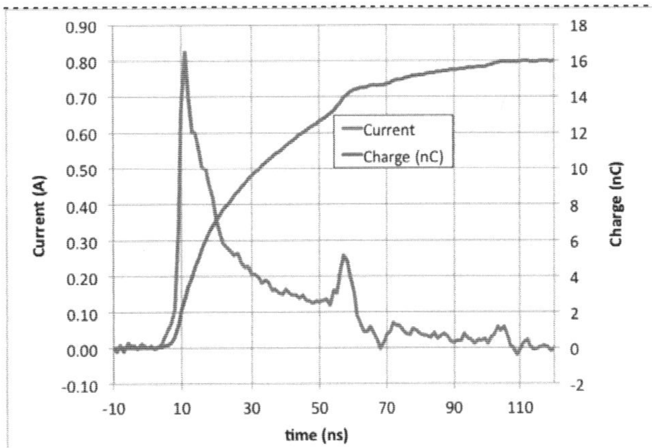

Figure 2. The full temporal distribution of the pulses is ~50 ns with a ~1 ns rise time, though still with a broader tail than desired. The integrated charge in the pulse is ≈15 nC. The pulse was measured with a fast Faraday collector isolated from the neutralizing plasma with hole plates with low geometric transparency.

Acknowledgments
This work was supported by the Office of Science of the US Department of Energy under contracts DE-AC02–05CH11231 (LBNL), DE-AC52-07NA27344 (LLNL) and DE-AC02-09CH11466 (PPPL).

References
[1] Grisham L R 2004 *Phys. Plasmas* **11** 5727
[2] Friedman A, et al. 2010 *Phys. Plasmas* **17** 056704
[3] Barnard J J, et al. 2014 *Nucl. Inst. Meth.* **A733** 45
[4] Waldron W L, et al. 2014 *Nucl. Inst. Meth.* **A733**
[5] Bangerter R O, Faltens A and Seidl P A 2013 *Rev. Accel. Sci. & Tech.* **6**, 85
[6] Schenkel T, Lidia, S M, Weis, C D, Waldron W L, Schwartz J, Minor A M, Hosemann P and Kwan J W 2013 *Nucl. Inst. Meth.* **B315** 350
[7] Persaud A, Barnard J J, Guo H, Hosemann P, Lidia S, Minor A M, Seidl P A and Schenkel T 2015 *Phys. Procedia* **66** 604
[8] Bai X M, et al. 2010 *Science* 327, 1631
[9] Seidl P A, et al. 2015 *Nucl. Inst. Meth.* **A800** 98–103
[10] Ji Q, et al. 2016 *Rev. Sci. Inst.* **87** 02B707
[11] Gilson E P, et al. 2013 *Nucl. Inst. Meth.* **A733** 75
[12] Tokluoglu E and Kaganovich I D 2015 *Phys. Plasmas* **22** 040701
[13] Dorf M, Davidson R C, Kaganovich I D, and Startsev E A 2012 *Phys. Plasmas* **19**, 056704
[14] Schwartz J, et al. 2014 *J. Appl. Phys.* **116**, 214107
[15] Bienfait A, et al. 2015 *Nature Nano.* doi:10.1038/nnano.2015.282
[16] Busold S, et al. 2015 *Scientific Reports* 5, 12459

Nuclear science research with dynamic high energy density plasmas at NIF

D A Shaughnessy[1], N Gharibyan[1], K J Moody[1], J D Despotopulos[1], P M Grant[1], C B Yeamans[1], L Berzak Hopkins[1], C J Cerjan[1], D H G Schneider[1] and S Faye[1,2]

[1]Lawrence Livermore National Laboratory, P.O. Box 808, Livermore, CA 94551, USA

[2]Department of Nuclear Engineering, University of California, Berkeley, CA 94720, USA

shaughnessy2@llnl.gov

Abstract. Nuclear reaction measurements are performed at the National Ignition Facility in a high energy density plasma environment by adding target materials to the outside of the hohlraum thermo-mechanical package on an indirect-drive exploding pusher shot. Materials are activated with 14.1-MeV neutrons and the post-shot debris is collected via the Solid Radiochemistry diagnostic, which consists of metal discs fielded 50 cm from target chamber center. The discs are removed post-shot and analyzed via radiation counting and mass spectrometry. Results from a shot using Nd and Tm foils as targets are presented, which indicate enhanced collection of the debris in the line of sight of a given collector. The capsule performance was not diminished due to the extra material. This provides a platform for future measurements of nuclear reaction data through the use of experimental packages mounted external to the hohlraum.

1. Introduction

The National Ignition Facility (NIF) is an Inertial Confinement Fusion (ICF) research facility located at Lawrence Livermore National Laboratory (LLNL). During a typical ICF shot, a plastic capsule filled with an equimolar mixture of deuterium and tritium (DT) is compressed to high densities and temperatures by high energy lasers [1,2,3]. The resulting fusion between the two hydrogen isotopes produces alpha particles and a significant yield of neutrons with a narrow energy distribution centered around 14.1 MeV. For a given implosion, increased capsule compression results in a higher fuel areal density (ρr), which produces an increased number of neutrons that have scattered off residual DT fuel to lower energies before escaping the capsule. Both the 14.1 MeV and lower-energy scattered neutrons interact with materials in proximity to the capsule, resulting in nuclear activation reactions and induced radioactivity [4]. By placing specific materials in and surrounding the capsule assembly, nuclear reaction rates can be determined by collecting the resulting debris and measuring the amount of reaction product produced. Through the use of the suite of nuclear diagnostics at NIF [4-12], the corresponding neutron spectrum is determined, resulting in the evaluation of nuclear activation cross sections. This method has applications in investigating nuclear reactions in the high energy density plasmas produced during ICF shots, where plasma screening may affect neutron- and charged-particle reaction rates compared to equivalent accelerator-based measurements [13].

Based on early results measuring gold isotope ratios produced via neutron interactions with the gold hohlraum [4], adding materials to the outside of the thermo-mechanical package (TMP) that surrounds the target and hohlraum assembly should provide enough target atoms for measuring cross sections from neutron activation reactions. We will discuss the first experiment designed to evaluate how addition of materials to the TMP affects capsule performance and the distribution of solid debris resulting from the interaction of the lasers with the capsule assembly. If a significant quantity of reaction products can be retrieved from the NIF chamber, this platform could be used on subsequent DT shots for measuring nuclear reaction cross sections in support of national security missions or fundamental nuclear science.

2. Experimental configuration

2.1. Shot implementation

NIF shot N141130-001 was an indirect-drive exploding pusher (IDEP) [14] capsule. A 1075-μm outer radius CH plastic capsule with a wall thickness of 120 μm was filled with a 50:50 mix of DT gas at a nominal density of 7.5 mg/cc. The capsule was mounted in a 5.75-mm diameter gold hohlraum filled with He at 16 torr. The laser delivered 910 kJ of 351-nm light in a 4.5-ns pulse with a peak power of 330 TW. Neutron yield was 5.8×10^{14} at a measured ion temperature of 5.1 keV.

2.2. Modified thermo-mechanical package

The NIF hohlraum was surrounded by an aluminum (Al) TMP, was used to hold the assembly in place in the target chamber. The TMP features a diagnostic band around the center that has windows for x-ray diagnostics. A standard TMP was used with the addition of 0.5-mm thick foils of thulium (Tm) and neodymium (Nd) metals (>99% pure) fixed to the outside around the circumference above and below the diagnostic band (Figure 1). The rare earth foils were coated with several nm of Al to prevent oxidation and flaking of the material during handling. The foils were configured such that the Nd foils were facing the Diagnostic Instrument Manipulator (DIM) located on the equator of the NIF target chamber at coordinates (90,78), and the Tm was facing the DIM at (90,315). The total mass of added material was 70.8 mg of Nd and 98.8 mg of Tm.

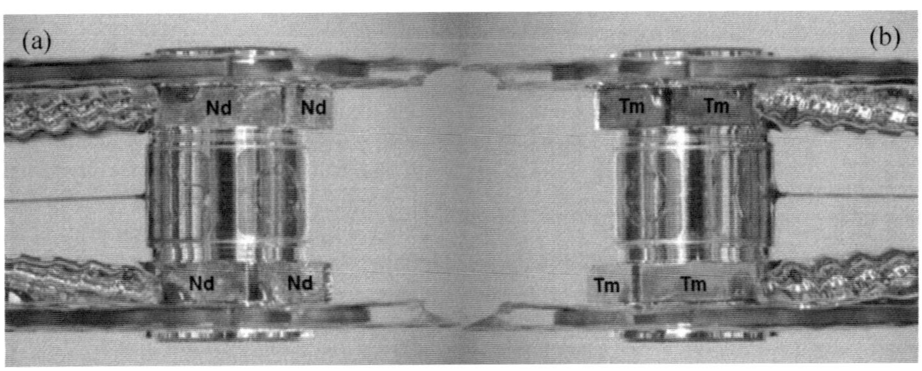

Figure 1. Modified TMP for shot N141130-001 with (a) 0.5-mm Nd foils (70.8 mg total) facing the DIM at (90,78) and (b) 0.5-mm Tm foils (98.8 mg total) facing the DIM at (90,315).

2.3. Solid radiochemistry diagnostics

Post-shot capsule debris is collected using the Solid Radiochemistry (SRC) diagnostic [4], which consists of 5-cm diameter, 1-mm thick metal discs that are housed in a cap attached to the snout of a

DIM via a quick-release bracket. Up to four SRC collectors can be fielded on each of the three DIM locations, two above the snout and two below, at a distance of 50 cm from target chamber center (TCC). This gives three individual lines of sight for the collectors at coordinates (90,78) and (90,315) (collectors facing the target assembly) and at (0,0) (collectors facing the north laser entrance hole on the top of the hohlraum). After a shot, the DIMs are retracted from the NIF chamber and the quick-release brackets are removed. The SRC collectors are then disassembled from the housing and delivered to the LLNL Nuclear Counting Facility (NCF) for gamma radiation counting. For this shot, all 12 of the SRC collectors were vanadium (V) foils.

After radiation counting, the collectors were treated with 9M HCl / conc. HNO_3 acid mixture to remove the resulting debris along with the first ~10 μm (~78 mg) of the surface of the V collectors. In addition, unused collectors and collectors previously exposed to the inside of the NIF chamber were processed as blanks. The lanthanides of interest (Nd and Tm) were separated from the excess V through cation-exchange chromatography and recovered in 1% high-purity HCl. These samples were then analyzed via ICP-MS to determine the absolute amounts of Nd and Tm collected. The (n,2n) activation products from the collected debris (^{147}Nd and ^{168}Tm) were used as internal tracers for determining the radiochemical yields from leaching, separation and recovery. The uncertainties in the reported collection yields include counting statistics from gamma spectroscopy (1σ), a 2% absolute efficiency of the HPGe detectors, the standard deviation of five individual measurements from mass spectrometry (2σ) and reported uncertainties in nuclear data.

3. Results

The active collection surface of any given SRC collector (12.566 cm^2) corresponds to 0.04% of the total 4π solid angle. Each of the individual 12 collectors was processed as a separate sample in order to determine the distribution of debris in the chamber. Table 1 gives the amounts of materials recovered for Nd and Tm in terms of the fraction of the total ingoing mass that was collected. The DIM samples from (0,0) collected almost none of the TMP materials and it appears that under these experimental conditions the polar positioner had no sensitivity to collection of the target assembly.

Table 1. Percent of the initial Nd and Tm mass collected on the SRC collectors as a function of DIM position.

TMP material	DIM (0,0)	DIM (90,78)	DIM (90,315)
Nd	$(3.04\pm0.51)\times10^{-5}$ %	0.692 ± 0.066 %	$(1.31\pm0.17)\times10^{-4}$ %
Tm	$(1.22\pm0.39)\times10^{-5}$ %	$(3.17\pm0.75)\times10^{-5}$ %	0.471 ± 0.011 %

Results from the equatorial DIMs suggest that the SRC collectors are most efficient at collecting material in their direct line of sight. The Nd foils were placed on the side of the TMP facing DIM (90,78), where the bulk of the Nd was observed. Likewise, Tm was measured only on the collectors on DIM (90,315), which were facing the Tm foils. In addition, the fraction of material collected on each of the equatorial lines of sight was typically greater than the geometric 0.04% solid-angle collection area. Therefore, the debris collection was enhanced in the equatorial direction over what has been observed in shots that did not include extra mass on the TMP.

The nuclear reaction products identified in the gamma spectra of the debris collectors included ^{169}Tm(n,2n)^{168}Tm ($t_{1/2}$=93.1±0.2 d), ^{142}Nd(n,2n)^{141}Nd ($t_{1/2}$=2.49±0.03 h), ^{148}Nd(n,2n)^{147}Nd ($t_{1/2}$=10.98±0.01 d) and ^{150}Nd(n,2n)^{149}Nd ($t_{1/2}$=1.728±0.001 h) [15]. The production cross sections for these reactions are 1.5-2 b [16]. The IDEP platform has very little capsule compression (downscattered ratio (DSR) < 0.2%), so the majority of the neutrons interacting with the rare-earth foils were 14.1 MeV, with very few lower energy neutrons present in the neutron spectrum.

4. Discussion and future work

Based on these results, it was concluded that material was collected from the TMP with efficiency greater than what was expected from solid angle alone (0.04%) for collectors in the same line of sight as the material on the outside of the TMP. In addition, NIF shot performance was unaffected by the additional mass. These results suggest that target materials can be added to the TMP to perform nuclear data measurements in "ride-along" fashion. As long as the DT yield of the capsule is high enough to produce sufficient reaction products for collection and analysis ($\geq 10^6$), a wide variety of materials could be used to field nuclear data measurements on any DT shot.

The IDEP platform offers an advantage in that all of the neutrons can be considered to have an energy of 14.1 MeV. Fielding similar experiments on a high-compression capsule will result in neutrons with a mixture of energies, 14.1 MeV as well as lower energy neutrons from down-scatter off unreacted DT fuel, other fusion reactions (i.e., DD and TT) and breakup reactions. Reaction products will therefore be produced both by activation at 14 MeV and via moderated neutron capture, which is most sensitive to lower energy neutrons, but does have a non-negligible cross section at 14 MeV as well. Selecting materials that are monoisotopic, or where the (n,2n) production rates are well known, will distinguish between products produced from a mixture of high and low neutron energies. Ultimately, this technique will be used for measuring unknown nuclear reaction data through the use of external target materials added to the hohlraum.

Acknowledgments

The authors thank the NIF Engineering and Operations staff, in particular the Target Diagnostic Factory and the NIF Radiation Control technicians. They also thank the staff of the LLNL Nuclear Counting Facility, Brian Yoxall for target engineering and Rachel Lindvall and Greg Klunder for performing the mass spectrometric analyses. This work was performed under the auspices of the U.S. Department of Energy by Lawrence Livermore National Laboratory under Contract DE-AC52-07NA27344. This work was funded by the Laboratory Directed Research and Development Program at LLNL under project tracking code 13-ERD-036.

References

[1] Nuckolls J, Wood L, Thiessen A and Zimmermann G 1972 *Nature (London)* **239** 139
[2] Nuckolls J H 1982 *Phys. Today* **35** 24
[3] Moses E I, Boyd R N, Remington B A, Keane C J and Al-Ayat R 2009 *Phys. Plasmas* **16** 041006
[4] Shaughnessy D A *et al.* 2014 *Rev. Sci. Instrum.* **85** 063508
[5] Shaughnessy D A, Velsko C A, Jedlovec D R, Yeamans C B, Moody K J, Tereshatov E, Stoeffl W and Riddle A 2012 *Rev. Sci. Instrum.* **83** 10D917
[6] Bleuel D L *et al.* 2012 *Rev. Sci. Instrum.* **83** 10D313
[7] Frenje J A *et al.* 2010 *Phys. Plasmas* **17** 056311
[8] Casey D T *et al.* 2011 *Rev. Sci. Instum.* **82**, 073502
[9] Casey D T, Frenje J A, Gatu Johnson M, Seguin F H and Li C K 2012 *Rev. Sci. Instrum.* **83** 10D912
[10] Gatu Johnson M, Frenje J A, Casey D T, Li C K and Seguim F H 2012 *Rev. Sci. Instrum.* **83** 10D308
[11] Glebov V Yu *et al.* 2010 *Rev. Sci. Instrum.* **81** 10D325
[12] Lerche R A *et al.* 2010 *Rev. Sci. Instrum.* **81** 10D319
[13] Boyd R N, Bernstein L A and Brune C 2009 *Phys. Today* **62** 60
[14] Le Pape S *et al.* 2014 *Phys. Rev. Lett.* **112** 225002
[15] Firestone R B, Shirley V S, Baglin C M, Chu S F and Zipkin J 1996 *Table of Isotopes* 8th ed. (New York: John Wiley & Sons, Inc.)
[16] Chadwick M B *et al.* 2011 *Nucl. Data Sheets* **112** 2887

Multi-scale modelling for HEDP experiments on Orion

N J Sircombe, M G Ramsay, S J Hughes and D J Hoarty

AWE, Aldermaston, Reading, Berkshire, RG7 4PR, UK
nathan.sircombe@awe.co.uk

Abstract. The Orion laser at AWE couples high energy long-pulse lasers with high intensity short-pulses, allowing material to be compressed beyond solid density and heated isochorically. This experimental capability has been demonstrated as a platform for conducting High Energy Density Physics material properties experiments. A clear understanding of the physics in experiments at this scale, combined with a robust, flexible and predictive modelling capability, is an important step towards more complex experimental platforms and ICF schemes which rely on high power lasers to achieve ignition. These experiments present a significant modelling challenge, the system is characterised by hydrodynamic effects over nanoseconds, driven by long-pulse lasers or the pre-pulse of the petawatt beams, and fast electron generation, transport, and heating effects over picoseconds, driven by short-pulse high intensity lasers. We describe the approach taken at AWE; to integrate a number of codes which capture the detailed physics for each spatial and temporal scale. Simulations of the heating of buried aluminium microdot targets are discussed and we consider the role such tools can play in understanding the impact of changes to the laser parameters, such as frequency and pre-pulse, as well as understanding effects which are difficult to observe experimentally.

1. Orion and the need for integrated modelling

Orion is a purpose built laser facility for High Energy Density (HEDP) experiments at the Atomic Weapons Establishment (AWE) in the UK[1][2], consisting of ten long-pulse beams (optimized to ~1 ns pulse duration, each providing 500 J at a wavelength of 351 nm) and two short-pulse beams (each ~500J in 0.5 ps at 1054 nm). Orion also has the option for frequency doubling one of the short-pulse beams to enhance the pulse temporal contrast[3], which we demonstrate here has a significant impact on the coupling of short-pulse energy into the target. One of the primary uses of Orion has been in the study of high-density, high-temperature plasma properties such as opacity and equation of state; the first data shots from the facility studied the material properties of dense plasmas created by a combination of short-pulse laser heating and compression by laser driven shocks[4]. Aluminum samples buried in plastic or diamond foils, compressed to densities ranging from 1 to 10 g/cc were heated to electron temperatures between 500 and 700 eV, and the plasma conditions diagnosed by K-shell emission spectroscopy.

Modelling these experimental campaigns calls for a multi-scale approach. The laser–plasma interaction (LPI) requires a kinetic description of the plasma and a mesh capable of resolving the dispersion of EM waves. The dense material of the bulk target, away from the LPI region, is dominated by collisional physics where the transport of hot particles generated by LPI is governed by their slowing and stopping in the dense material and their need to draw a return current. On longer timescales, the hydrodynamic response of the target to the compression beam (initially) and the pressure generated from isochoric heating (at late times) must be handled with a radiation hydrodynamic model. The approach taken at AWE has been to integrate a number of codes which capture the detailed physics for each spatial and temporal scale[5]: the PIC code EPOCH[6][1] for modelling the short-pulse LPI; the Monte-Carlo hybrid code THOR for modelling fast-particle transport through dense plasma; and the radiation hydrodynamics code CORVUS, which captures long-pulse physics, long timescale hydrodynamic effects and radiation transport.

2. First and second harmonic irradiation of diamond and plastic

We consider both diamond and plastic targets, with buried aluminium microdots, of the type used in [4]. These are compressed with 300J of third harmonic (i.e. 351 nm) light from one of Orion's long-pulse beams and heated with one of the short-pulse beams. The inter-beam timing is such that the shock driven into the target by the long-pulse has not broken out by the time the heating pulse arrives. This timing is somewhat conservative, we do not aim to heat at the point of peak density which occurs as the shock arrives at the material sample as this would not be practical, or reproducible, in experiment. A short-pulse pre-pulse, based on Orion measurements[2], is applied to the front of the target. The density and temperature profiles are mapped to EPOCH where the main short-pulse interaction is modelled. Considering first 500 J at 1ω in a 27.4 μm FWHM Gaussian spot with a 700 fs FWHM supergaussian pulse incident on a diamond target.

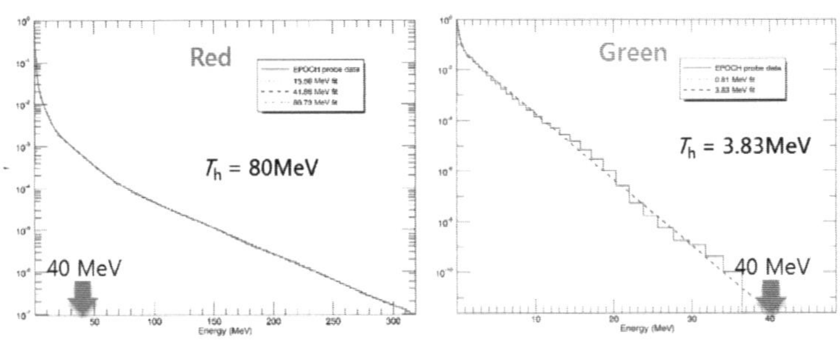

Figure 1. EPOCH generated energy distributions for the interaction of 500J of 1ω (left) and 100J of in 2ω (right) light with the pre-plasma generated by the laser pre-pulse. The distribution is not characterized by a single temperature.

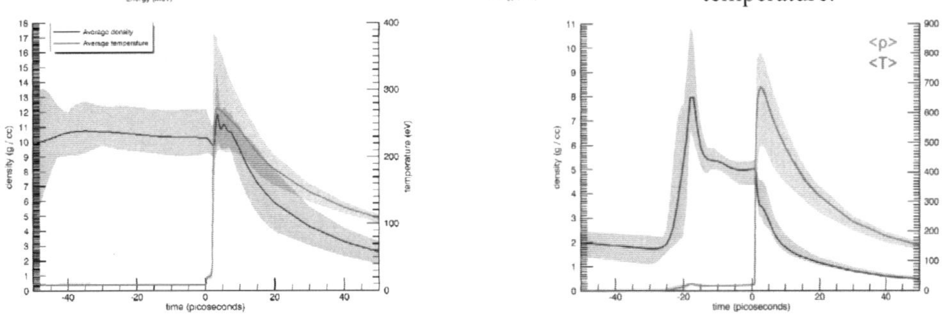

Figure 2. Density and temperature history for an aluminium microdot target. Left: a diamond target compressed with a third harmonic long-pulse and heated with 500 J of 1ω light. Right: a plastic target similarly compressed but heated with 100J of 2ω light.

Figure 1 shows the modelled time-integrated energy spectrum of the hot electrons generated by the short-pulse LPI. Simulations also indicate a high divergence (~50% of the hot electrons fall within a half-angle of 49°), it is worth noting that the high divergence of the hot electron source could lead to the generation of an x-ray spot larger than the laser spot or the effective hot electron source. Figure 2a shows the density and temperature history for the aluminium microdot. Shock compression of the sample drives it to densities in excess of 10 g/cc, but the average temperature of the aluminium peaks at just 250 eV.

The use of 2ω short-pulses during the material properties campaigns on Orion, and its predecessor HELEN, has proven very effective at heating buried layer and buried microdot target. The low pre-pulse, a side-effect of the frequency doubling process, generates less pre-plasma accordingly. Combined

[1] Part funded by the UK EPSRC grants EP/G054950/1, EP/G056803/1, EP/G055165/1 and EP/ M022463/1

with the higher critical density for green light, this helps to couple the incident beam to the target plasma at a higher density, resulting in a colder fast electron spectrum which is better able to deposit its energy in the target over the spatial scales of interest. The conversion from 1ω to 2ω comes at a cost in total energy, with approximately 100 J delivered to target. We consider a 0.2 μm thick aluminium microdot mounted in 10 μm of plastic. CORVUS modelling shows the lower pre-pulse does reduce the level of pre-plasma. Short-pulse LPI modelling in EPOCH shows that the hot electron spectrum is much colder in the 2ω case (see Figure 1). Figure 2b shows the temperature and density history of the microdot. A peak density in excess of 10 g/cc is achieved, however the average density in the layer when the heating pulse arrives is closer to 5 g/cc. The density of the sample begins to fall as the layer heats and expands, the lower density plastic proving a less effective tamper than the diamond which provides a higher pressure to resist the layer expansion. Adjusting the timing to ensure that the short-pulse heating occurs on the rising edge of the shock could help mitigate this effect, but as discussed above the shock timing is chosen to reflect experimental practicalities. A peak (spatially averaged) temperature of 600 eV is achieved at a density of ~4.5 g/cc. These results are consistent with those seen in experiments [4]

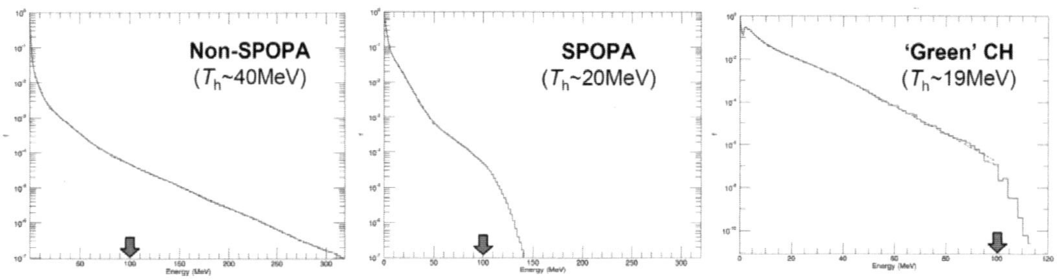

Figure 3. Hot electron spectra from EPOCH for interaction of a 500J, 1 ω, 700 fs short-pulse beam with pre-plasma conditions modelled in CORVUS. Moving from contrast levels of ~10^7 (left) to high contrast (~10^{10}) operation using the SPOPA front end (centre) removes the high energy tail of the electron distribution. Adopting pre-plasma conditions equivalent to those experienced with 2ω operation (a contrast of ~10^{14}) has a relatively small impact on the hot electron spectrum (right).

The poor coupling between the heating beam and the material sample observed (both in experiments and in modelling) in the case of 1ω light, relative to 2ω, may be due to the high energy tail in the electron distribution. The highest energy electrons are generated from the interaction with the diffuse pre-plasma generated by the laser pre-pulse. Improvements to the Orion front-end can now deliver significantly higher 1ω contrast – modelling suggests that this has a significant impact on the energy spectrum of hot electrons generated (see Figure 3). However, there may not be much scope for improvement through further increases in contrast. In all the cases discussed here, the improvements to the pre-plasma conditions and the cooling of the electron spectrum do not make a significant impact on the coupling to the material sample. While the effect of changes to the pre-plasma in these cases has brought dramatic changes to tail of the distribution, the bulk remains largely unaffected. Combined with the experimental and modelling results for 2ω, this suggests that planned uplifts in Orion's 2ω capabilities may be the best route to further improvements in target heating.

The results summarized in Figure 2 take max, min and spatially averaged values for density and temperature over the material sample at each point in time. From these time histories, it is clear that there are variations over the sample observed in both quantities. These variations appear to be the result of transverse gradients in the plasma conditions of the layer (see Figure 4). Until recently, it has not been possible to observe gradients experimentally, but the recent development of a spatial-imaging spectrometer on Orion has produced results which appear to indicate the presence of spatial temperature gradients in targets with a buried scandium microdot, in the case where the laser focus is smaller than the microdot. While defocused laser spots are often employed, resulting in a spot radius comparable or greater than the microdot, these effects merit further study - in particular it will be important to understand what impact, if any, the presence of transverse gradients could have on the plasma conditions inferred from both the streaked and time integrated spectroscopy – detailed post-processing of CORVUS simulations with detailed atomic physics codes will be required.

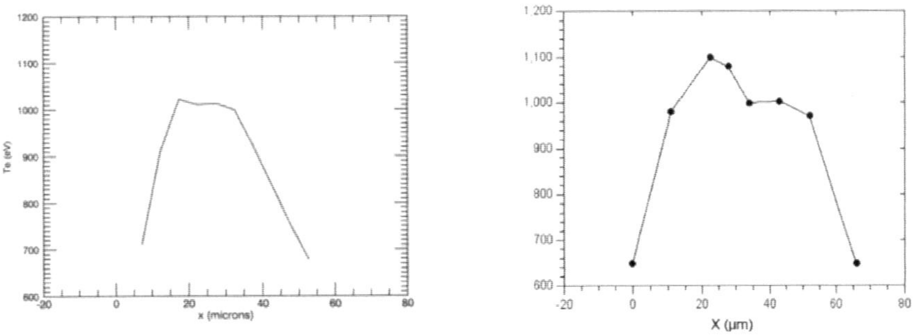

Figure 4. Transverse temperature gradients in short-pulse heating of thin material samples with a radius greater than the laser spot size. Left: profile from a simulation of a buried aluminium layer showing peak temperature in the layer. Right: profile implied by the time integrated emission spectra from a scandium microdot recorded by the spatial imaging spectrometer. Results are indicative of the presence of transverse temperature gradients both in simulations and experiment which warrants further study.

3. Discussion

Since becoming operational, Orion has enabled AWE to conduct an extensive series of HEDP material properties experiments to study plasmas at high temperatures and densities. These campaigns have utilized Orion's most notable capabilities – its combination of short and long-pulse beamlines, and ability to deliver second harmonic high intensity pulses. These experiments place challenging requirements on modelling – as any campaign attempting to marry such a range of spatial scales, temporal scales, densities and energies would. Here we have outlined AWE's integrated modelling approach which builds on in-house radiation hydrodynamics capability and collaborative work with the broader community on PIC modelling. The results discussed here are consistent with the early Orion experiments, in particular the observed variation between heating with 1 and 2ω light, which has proven difficult to reproduce in simulations to date. In future advances like the spatial imaging spectrometer, improvements to the fidelity of the diagnostics and the use of a broader range of material samples, which helps to provide a more reliable 'thermometer', will facilitate direct comparison between the modelling and experiment. At the same time, short-pulse capability campaigns on Orion will offer a means to study short-pulse LPI in detail to help build confidence in the modelling approach.

Developing and exercising a modelling capability to address these challenges is essential for understanding the physics, for example diagnosing the impact of effects which are difficult to diagnose experimentally, and facilitating the development of more ambitious platforms in future. Orion-scale HEDP modelling and experiments offers an opportunity to: study material properties under extreme conditions; interrogate fast electron properties, and study the impact of changes to laser and plasma conditions; and provide an essential route to building confidence in the multiscale modelling capability required for many alternative ignition schemes.

References

[1] A Randewich and C Danson. High Power Laser Science and Engineering, **2**, e40 (2014)
[2] N Hopps *et al.* Plasma Phys. Control. Fusion, **57**, 064002 (2015)
[3] D Hillier *et al.* Applied Optics, **52**, 4258, (2013), D Hillier *et al.* Applied Optics, **53**, 6938, (2014)
[4] D J Hoarty *et al.* HEDP, **9**, 661 (2013)
[5] N J Sircombe, S J Hughes and M G Ramsay, New J. Phys. **15** 025025 (2013)
[6] T D Arber *et al.* Plasma Phys. Control. Fusion. **57**, 113001, (2015)

Equations of State for Ablator Materials in Inertial Confinement Fusion Simulations

P A Sterne, L X Benedict, S Hamel, A A Correa, J L Milovich, M M Marinak, P M Celliers and D E Fratanduono

Lawrence Livermore National Laboratory, Livermore, CA 94550, USA
Email: sterne1@llnl.gov

Abstract. We discuss the development of the tabular equation of state (EOS) models for ablator materials in current use at Lawrence Livermore National Laboratory in simulations of inertial confinement fusion (ICF) experiments at the National Ignition Facility. We illustrate the methods with a review of current models for ablator materials and discuss some of the challenges in performing hydrocode simulations with high-fidelity multiphase models. We stress the importance of experimental data, as well as the utility of ab initio electronic structure calculations, in regions where data is not currently available. We illustrate why Hugoniot data alone is not sufficient to constrain the EOS models. These cases illustrate the importance of experimental EOS data in multi-megabar regimes, and the vital role they play in the development and validation of EOS models for ICF simulations.

1. Introduction

The equation of state (EOS) of the ablator material controls many aspects of inertial confinement fusion (ICF) performance. The EOS strongly affects the ablation front pressure and temperature, controls the development of shocks at the ablation front, and determines the timing as the initial and subsequent shocks and release waves propagate through the ablator material. The EOS also controls the impedance match at the ablator-fuel interface [1], and so affects energy transfer via hydrodynamic processes into the fuel. Inertial Confinement Fusion (ICF) simulations at Lawrence Livermore National Laboratory (LLNL) use tabular EOS models to describe the EOS properties of the ablators and fuel. In this paper, we briefly describe some of the approaches used to construct these EOS models and highlight some ongoing work on selected ablator materials.

2. EOS Construction

Two different approaches are used to construct tabular EOS models for ICF ablators. The first is an extended version of the QEOS method [2] in which a global EOS is constructed using a simplified match between the solid and liquid phases that neglects latent heats. These models extend to high temperatures using the Purgatorio atom-in-jellium model [3] for the electron-thermal contributions. This includes atomic shell structure effects, which tends to give a better overall description of the ablation properties and resultant shock timings than the previously-used Thomas-Fermi approach.

The second approach uses an explicit multiphase EOS construction method. This is more complicated and time consuming than the simpler QEOS-based approach, since the resulting EOS depends sensitively on free-energy minimization across several phases. This approach, which is capable of capturing the complicated two-phase structure and latent heats of transition between solid and liquid phases, has been used for the high-density carbon (HDC) EOS [4]. We will discuss some of the challenges in using these high-fidelity EOS models in hydrocode simulations.

EOS models are constructed by fitting experimental and theoretical data using physics-based models that extend over wide ranges of density and temperature. The EOS is constructed to fit the experimental data where available; ab initio calculations using quantum molecular dynamics (QMD) and path-integral Monte Carlo (PIMC) methods [5] are also used in the absence of experimental data.

3. Glow Discharge Polymer (GDP)

GDP is a carbon-hydrogen polymer with an approximate composition $CH_{1.36}$ [6]. The recommended EOS model, leos 5400, is constructed to fit a variety of data, most notably Hugoniot data [6], and "keyhole" data that measures the shock velocity change at the GDP-deuterium interface [1,7]. The Hugoniot data can be fit non-uniquely in a variety of ways, since trade-offs can be made between cold curve variations and thermal contributions, such as the anharmonic Grueneisen gamma term. The keyhole data provides a constraint on the off-Hugoniot behavior of the material, which narrows the possible trade-off choices and so constrains the EOS. Figure 1 shows the measured shock velocities in GDP and the resulting shock in deuterium, using leos 1014 for the deuterium EOS [8]. The two GDP EOS models 5310 and 5400 have similar Hugoniots (not shown), but the resulting deuterium shock velocities are very different, due to their differing off-Hugoniot behavior. This illustrates the importance of including data in addition to the Hugoniot data when constructing EOS models.

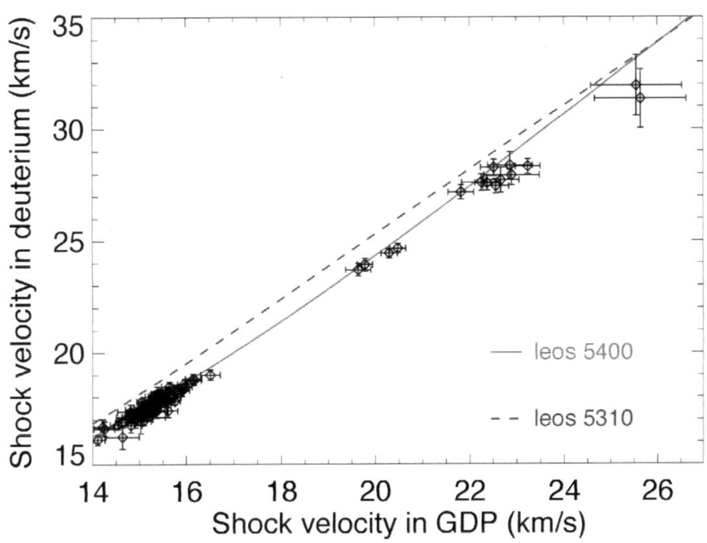

Figure 1. Measured shock velocities in GDP and resulting shock velocity in deuterium. Crosses are experimental data with error bars [9]. Lines are calculated using EOS models leos 5400 and 5310.

4. Beryllium

The currently recommended model for beryllium is XEOS 40. This is a QEOS model using Purgatorio-based electron-thermal contributions at high temperature. A multiphase model was also developed [10]; it is similar to XEOS 40, over much of its range, but has a more approximate extension to the Purgatorio-based higher temperature limit. The solid-solid phase transitions in Be do not appear to have a large effect on the EOS, so the benefit of the multiphase approach appears limited in this case, which justifies our current recommendation of the QEOS-based model.

In order to assess the accuracy of this EOS, we performed QMD simulations covering the Hugoniot pressure range from 3 to 7 Mbar using the VASP code [11]. We also estimated systematic uncertainties by performing calculations at one density with both local density and generalized-gradient [12] exchange-correlation functionals, and by comparing 2- and 4-electron pseudopotentials. The differences between these different approximations were small, and on the same order as the difference between XEOS 40 and QMD results [13]. This suggests that the current EOS agrees fairly well with QMD simulations, but a slight adjustment may be needed. Future EOS modifications are planned, pending upcoming Hugoniot measurements in this pressure range on the Sandia Z machine.

5. Boron Carbide

Recent measurements on boron carbide provided Hugoniot data in the previously-unmeasured 3-8 Mbar region [14]. A new EOS model, leos 2121 was constructed to fit this data. In addition to Hugoniot data, the experimental data was analyzed to extract Eulerian sound speeds [15], which was found to disagree with the new EOS model. By simultaneously adjusting the cold curve and the Grueneisen gamma model, a new EOS model, leos 2122, was constructed to fit both sets of data, as shown in Fig. 2. This is another example of how it is not sufficient to constrain the EOS model with Hugoniot data alone. Additional data, in the form of sound speeds or release data, can help to resolve ambiguities in the decomposition of the EOS into cold and thermal terms.

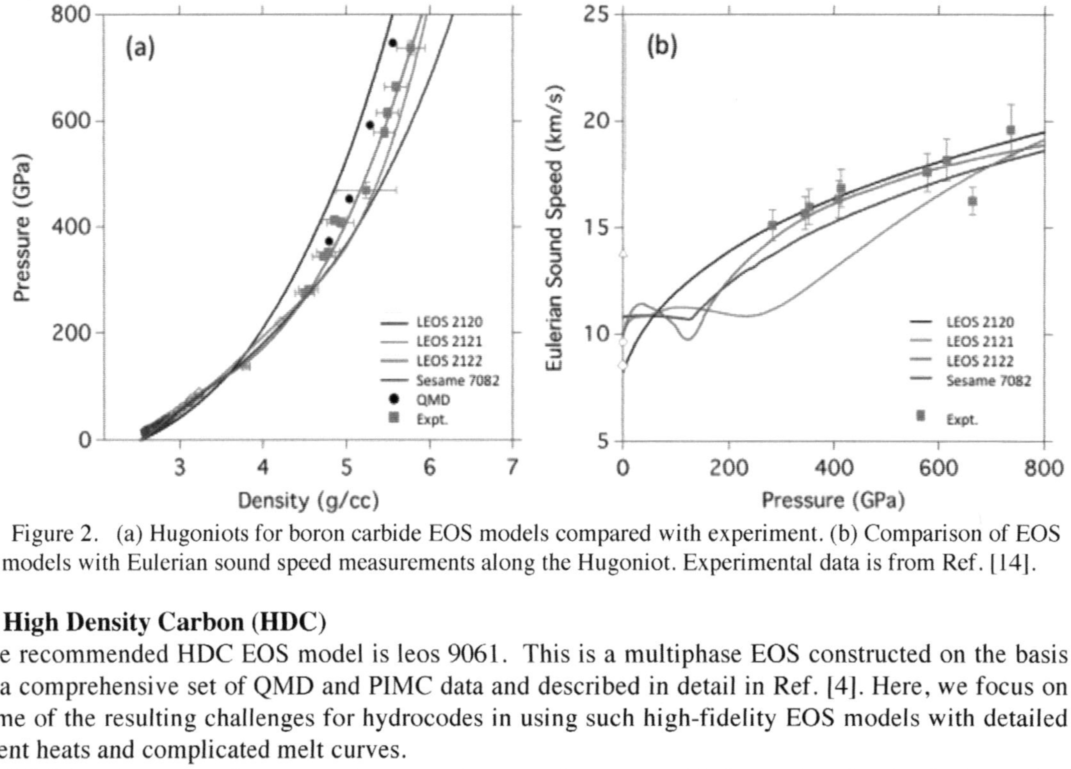

Figure 2. (a) Hugoniots for boron carbide EOS models compared with experiment. (b) Comparison of EOS models with Eulerian sound speed measurements along the Hugoniot. Experimental data is from Ref. [14].

6. High Density Carbon (HDC)

The recommended HDC EOS model is leos 9061. This is a multiphase EOS constructed on the basis of a comprehensive set of QMD and PIMC data and described in detail in Ref. [4]. Here, we focus on some of the resulting challenges for hydrocodes in using such high-fidelity EOS models with detailed latent heats and complicated melt curves.

HDC has the unusual property that its melt curve decreases with increasing density in the region where the Hugoniot crosses the melt curve. The EOS in this region features two solid phases and the melt phase, which results in a complicated set of isotherms for the ion-thermal energy, shown in Fig. 3. In hydrocodes, the derivatives of the energy with respect to temperature and density are used to estimate the change in density and temperature for a zone during a time-step. For this EOS, the derivatives near the melt region vary strongly with both density and temperature, which can result in anomalously large projected density or temperature updates. These, in turn, cause apparent noise in the calculations due to the creation of spurious shocks and releases caused by these unphysical density/temperature updates. In order to mitigate this effect in 1D implosions in the Hydra [16] code, it was necessary to place upper limits on the allowed changes in density and temperature at each time step. In subsequent work, changes have also been made to iterate on the projected update so that the change is not solely dependent on the prediction from the initial derivatives, but is corrected by the actual EOS at the proposed density-temperature value. These steps greatly reduced the noise in 1D simulations. Work is ongoing to address similar features in 2D simulations.

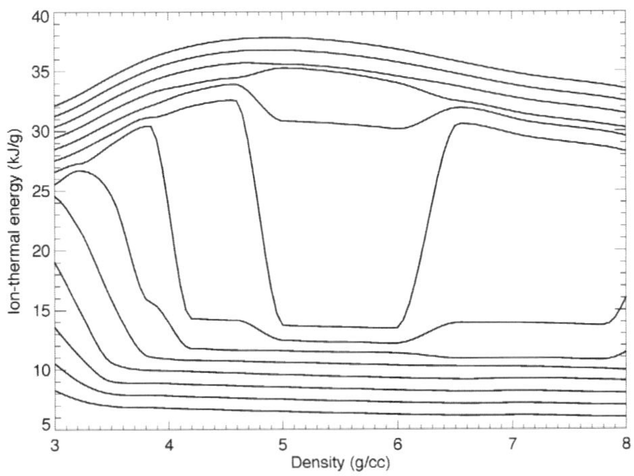

Figure 3. Ion-thermal energy isotherms for leos 9061 from 4000 K to 10000 K in 500 K steps.

7. Conclusions

The examples presented here have illustrated some of the challenges in constructing and using tabular EOS models, and some of the approaches that have been used to address these challenges. We stressed the importance of using electron-thermal models such as the Purgatorio model that includes shell-structure effects in order to produce accurate ablation-front behavior. We illustrated the importance of experimental data in the relevant density-temperature region, both Hugoniot data and additional data in form of sound-speed or release data at material interfaces, in order to resolve ambiguities between cold-curve and thermal EOS contributions. We indicated how ab initio QMD calculations can complement experimental data. We illustrated some of the problems in representing the physics associated with phase transitions in hydrocodes, and our initial steps to resolve these problems. The equation of state remains an important component in modeling ICF implosions, and improvements in EOS models based on additional experimental data and better theoretical models will lead to more accurate EOS representations and improved hydrodynamic modeling in ICF simulations.

8. Acknowledgments

This work was performed under the auspices of the U.S. Department of Energy by Lawrence Livermore National Laboratory under contract DE-AC52-07NA27344.

References

[1] Robey H F et al 2012 *Physics of Plasmas* **19** 042706
[2] More R M et al 1988 *Phys. Fluids* **31** 3059
[3] Wilson B et al 2006 *Journal of Quantitative Spectroscopy and Radiative Transfer* **99** 658
[4] Benedict L X et al 2014 *Phys. Rev. B* **89** 224109
[5] Pollock E L and Ceperley D M 1984 *Phys. Rev. B* **30** 2555
[6] Barrios M A et al 2012 *J. Appl. Phys.* **111** 093515
[7] Hamel S et al 2012 *Phys. Rev. B* **86** 094113
[8] Kerley G I 2003 *Sandia National Laboratories Report* SAND2003-3613
[9] This set of 94 data points is an update of subsets that were previously published in Refs. [1,7]
[10] Benedict L X et al 2009 *Phys. Rev. B* **79** 064106
[11] Kresse G and Furthmüller J 1996 *Phys. Rev. B* **54** 11169
[12] Perdew J P et al 1996 *Phys. Rev. Lett.* **77** 3865
[13] Sterne P A and Hamel S 2015 unpublished
[14] Fratanduono D E et al, in preparation
[15] Fratanduono D E et al 2014 *J. Appl. Phys.* **116** 033517
[16] Marinak M M et al 2001 *Physics of Plasmas* **8** 2275

Experimental and numerical studies on plasma behavior flowing across perpendicular magnetic field

T Takezaki, K Takahashi, T Sasaki, T Kikuchi, and N Harada

Nagaoka University of Technology, Nagaoka, Niigata, Japan

E-mail: `ttakezaki@stn.nagaokaut.ac.jp`

Abstract. To understand particle acceleration mechanisms in a collisionless shock, we have investigated the behaviors of a one-dimensional fast plasma flow in a perpendicular magnetic field by experimental and numerical simulations in a laboratory scale experiment. The velocity of the plasma flow generated by a taper-cone-shaped plasma focus device has varied by the gradient of the perpendicular magnetic field. The plasma flow has accelerated by applying the magnetic field with the negative gradient. To clarify the behavior of the plasma flow in the perpendicular magnetic field, numerical simulations based on an electromagnetic hybrid particle-in-cell (PIC) method have been carried out. These results indicate that the magnetic field gradient affects the plasma flow velocity.

1. Introduction

To understand astrophysical phenomena, laboratory scale experiments with well-defined behavior are required for an experimental evaluation of several numerical model studies [1–8]. A collisionless shock plays a key role in generation of high-energy particles in outer space. The particle acceleration mechanisms in the collisionless shock have been considered the Fermi acceleration, which works diffusive shock acceleration due to electromagnetic turbulence, and the shock drift or surfing accelerations obtained by the upstream magnetic field in a magnetized shock [9–13]. For the experimental evaluation of these acceleration mechanisms, Drake [3] has estimated the required conditions, which depend on the hypersonic plasma flow and the magnetic field, to generate the collisionless shock in laboratory scale experiments.

In order to generate the hypersonic plasma flow in laboratory scale experiments, a method by using laser ablations or pulsed-power discharges has been demonstrated [14–19]. To simplify the plasma behavior and the distribution of applied magnetic field, we have developed a taper-cone-shaped plasma focus device by using the pulsed-power discharge [20]. The device has generated a one-dimensional fast plasma flow with well-defined behavior for understanding astrophysical phenomena.

In this study, to understand the particle acceleration mechanisms in the collisionless shock, the experimental and the numerical simulations in the laboratory scale experiment have been demonstrated. To evaluate the plasma flow as a function of magnetic field distributions, we have investigated behaviors of the plasma flow generated by the taper-cone-shaped plasma focus device in two types of perpendicular magnetic field distributions. The numerical simulations based on an electromagnetic hybrid particle-in-cell (PIC) method have been carried out for understanding the effect of the magnetic field gradient on the plasma flow.

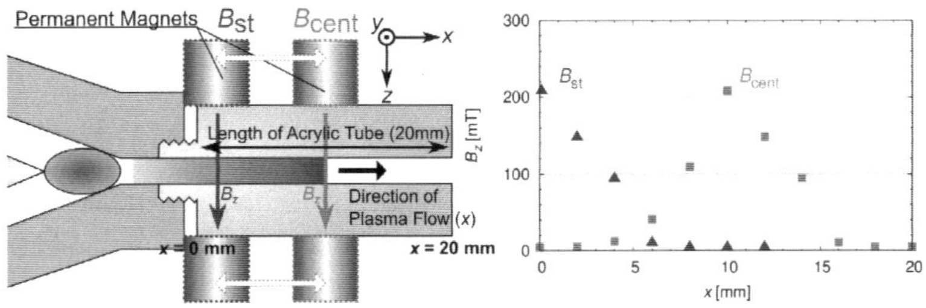

Figure 1. Experimental setup and magnetic field distributions in the acrylic tube.

2. Behaviors of plasma flow generated by taper-cone-shaped plasma focus device in perpendicular magnetic field

The taper-cone-shaped plasma focus device with an acrylic tube produces the quasi-one-dimensional hypersonic plasma flow [20]. The length of the acrylic tube is $L = 20$ mm. The velocity of the plasma flow using helium gas discharge is about 10km/s which the Mach number corresponds to $M_a \sim 10$. To investigate the behavior of the plasma flow in perpendicular magnetic field, permanent magnets ($\o = 5$ mm and $h = 5$ mm) were set on the acrylic tube, as shown in Fig. 1. The direction of the applied magnetic field is z perpendicular to the propagation direction x of the plasma flow. To evaluate the plasma flow as a function of magnetic field distributions, we have changed the position of the permanent magnets. The magnets were located at 10 mm from the end of the electrode called as B_{cent}. On the other hand, the magnets were located at 0 mm called as B_{st}. The difference of magnetic field gradients affects the propagation of the plasma flow.

Figure 2 shows streak images of the plasma flow in the acrylic tube with each magnetic field distribution. The initial gas pressure in the chamber is 2.0 Pa with B_{cent} shown in Fig. 2(a), and 0.5 Pa with B_{st} shown in Fig. 2(b). As shown in Fig. 2(a), the velocity of the plasma flow with B_{cent} decreased from 14 km/s to 10 km/s. By contrast, the velocity of the plasma flow with B_{st} increased from 12 km/s to 14 km/s shown in Fig. 2(b). These results indicate that the magnetic field gradient has affected the variation of the plasma flow velocity.

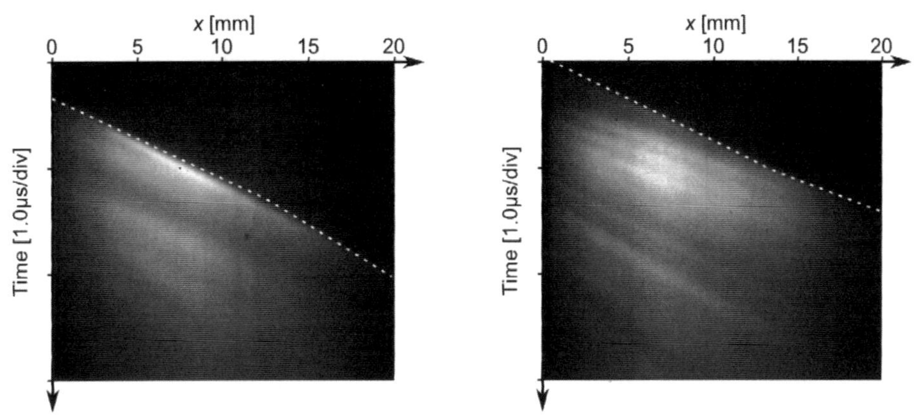

(a)With B_{cent} (b)With B_{st}

Figure 2. Streak images of the plasma flow in the acrylic tube. (a)With the magnetic field distribution B_{cent} at 2.0 Pa, (b)with the magnetic field distribution B_{st} at 0.5 Pa.

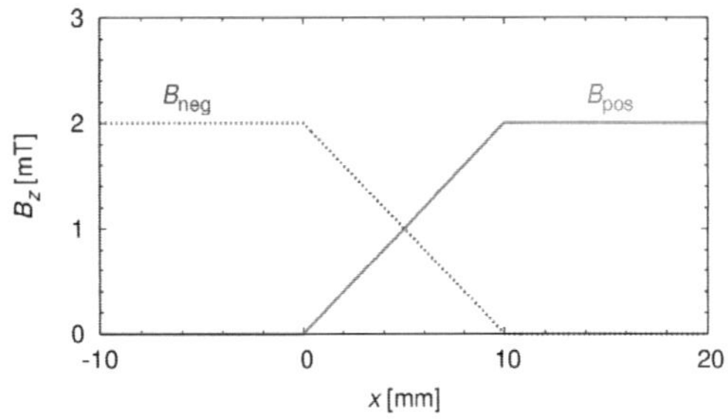

Figure 3. Magnetic field distribution B_z for the initial condition of the numerical simulations. To simplify the model of the magnetic field gradient, the positive gradient magnetic field B_{pos} and the negative gradient magnetic field B_{neg} are applied.

3. Numerical simulations based on electromagnetic hybrid PIC

In order to understand the behavior of the fast plasma flow generated by the taper-cone-shaped plasma focus device in the perpendicular magnetic field, we have carried out four-dimensional space (x, v_x, v_y, v_z) numerical simulations based on an electromagnetic hybrid PIC method [21, 22] with the simple modeling of the magnetic field gradient as shown in Fig. 3. In the initial condition of the numerical simulation, the thermodynamic equilibrium is assumed due to the plasma focus. The ions having the thermodynamic equilibrium, which the average velocity is 10 km/s, are set from -10 mm to 0 mm on the axis x. These ions are assumed fully single ionized condition. The ion temperature $T_{\rm i}$ and the ion number density $n_{\rm i}$ are respectively estimated to be 9000 K and 9.5×10^{18} m^{-3} by using Rankine-Hugoniot relation with strong shock condition for the initial gas pressure $P_0 = 0.01$ Pa, the room temperature $T_0 = 300$ K, and the Much number $M_{\rm a} = 10$. The region from 0 mm to 20 mm indicates the region of the acrylic tube. The initial shock front is set at $x = 0$ mm. The collision between the ions and the neutral particles forehand the shock front is neglected because of the Knudsen number $K_{\rm n} = \lambda_{\rm in}/L \sim 50 > 1$, where $\lambda_{\rm in}$ is the mean free path between the ion-neutral collision, and L (= 20 mm) is the system scale length. The mean free path $\lambda_{\rm in}$ is estimated as $\lambda_{\rm in} = 1/4\sigma_{\rm He}n_0$, where $\sigma_{\rm He}$ $(= 10^{-19}$ m$^2)$ is the cross section of helium for electron [23], n_0 is the number density of the neutral particles forehand the shock. To simplify the model of the magnetic field gradient, we applied the positive gradient magnetic field called as $B_{\rm pos}$ and the negative gradient magnetic field called as $B_{\rm neg}$, as shown in Fig. 3.

Figure 4 shows the time evolution of the ion number density $n_{\rm i}$ (upper graph) and the time evolution of the shock front (lower graph). The time evolutions of the density differ depending on the magnetic field distributions. However, the discontinuous surface of the density has been formed in both magnetic field distributions because the magnetic pressure affects the plasma flow. To compare with the experimental results, we followed the discontinuous surface as the shock front of the plasma flow such as the streak images. The dotted lines, as shown in the lower graphs of Fig. 4, indicate the initial velocity of the plasma flow. The initial velocity of the plasma flow has been faster than the initial average velocity of the ions, because the faster ions in the initial Maxwell distribution propagate at first and form the discontinuous surface. The plasma flow with $B_{\rm pos}$ has decelerated. On the other hand, the plasma flow with $B_{\rm neg}$ has accelerated. These results indicate that the acceleration and deceleration of the plasma flow have varied by the direction of the magnetic field gradient.

4. Conclusion

To understand particle acceleration mechanisms in the collisionless shock, we have investigated the behavior of the plasma flow across the perpendicular magnetic field by the experimental and the numerical simulations in the laboratory scale experiment. The velocity of the plasma flow generated by the taper-cone-shaped plasma focus device has varied by the gradient of the perpendicular magnetic field. The plasma flow has accelerated by applying the magnetic field with the negative gradient. The numerical simulation results show that the acceleration (or

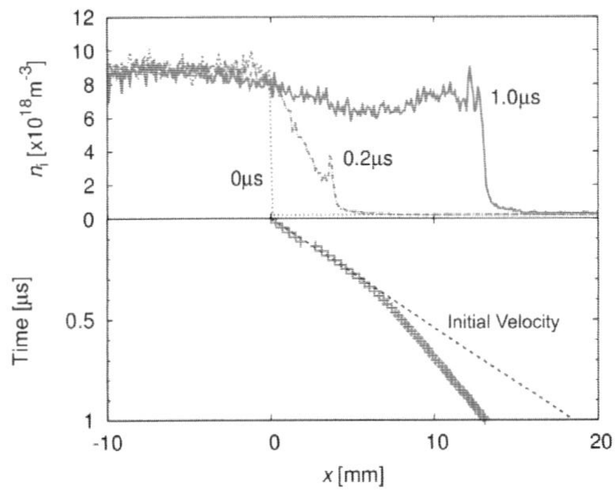

(a)With B_{pos}

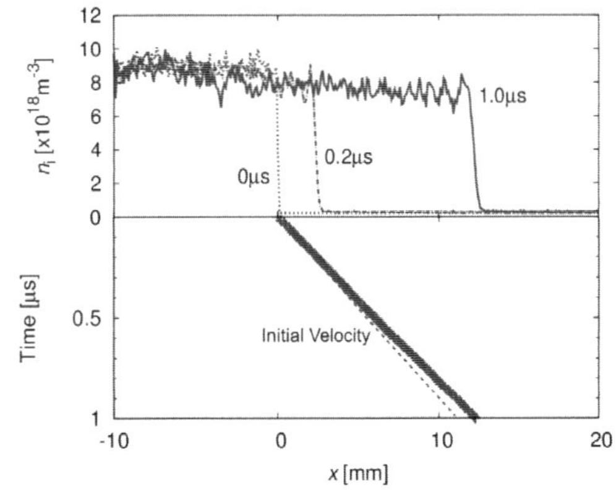

(b)With B_{neg}

Figure 4. Time evolutions of the ion number density n_{i} (upper graph) and the shock front (lower graph). The initial shock front has been set at $x = 0$ mm. (a)With the positive gradient magnetic field B_{pos}, (b)with the negative gradient magnetic field B_{neg}. The dotted lines in the lower graphs indicate the initial velocity of the plasma flow.

deceleration) of the plasma flow has depended on the direction of the magnetic field gradient. The magnetic field gradient induces a charge separation by grad-B drift. The plasma flow accelerates by the electric field induced by the grad-B drift with the negative gradient.

To understand the fast plasma flow as a function of several magnetic field structures, we will experimentally evaluate the ion energy distribution function and its plasma parameters.

Acknowledgments

This work is partly supported by the NIFS Collaboration Research Program (NIFS14KLEH042).

References

[1] Leroy M *et al.* 1982 *Journal of Geophysical Research: Space Physics* **87** 5081–5094
[2] Ryutov D *et al.* 1999 *The Astrophysical Journal* **518** 821
[3] Drake R 2000 *Physics of Plasmas* **7** 4690–4698
[4] Quest K B 1986 *Journal of Geophysical Research: Space Physics* **91** 8805–8815
[5] Kato T N and Takebe H 2008 *The Astrophysical Journal Letters* **681** L93
[6] Gargaté L and Spitkovsky A 2012 *The Astrophysical Journal* **744** 67
[7] Stockem A *et al.* 2014 *Scientific Reports* **4**
[8] Bulanov S *et al.* 2015 *Plasma Physics Reports* **41** 1–51
[9] Bell A 1978 *Monthly Notices of the Royal Astronomical Society* **182** 147–156
[10] Blandford R D and Ostriker J P 1978 *The Astrophysical Journal* **221** L29–L32
[11] Decker R and Vlahos L 1985 *Journal of Geophysical Research: Space Physics* **90** 47–56
[12] Begelman M C and Kirk J G 1990 *The Astrophysical Journal* **353** 66–80
[13] Hoshino M and Shimada N 2002 *The Astrophysical Journal* **572** 880
[14] Bell A *et al.* 1988 *Physical Review A* **38** 1363
[15] Courtois C *et al.* 2004 *Physics of Plasmas* **11** 3386–3393
[16] Kondo K *et al.* 2006 *Review of Scientific Instruments* **77** 036104
[17] Morita T *et al.* 2010 *Journal of Physics: Conference Series* **244** 042010
[18] Kuramitsu Y *et al.* 2011 *Physical review letters* **106** 175002
[19] Adachi K *et al.* 2015 *IEEJ Trans. FM* **135** 366–372
[20] Sasaki T *et al.* 2014 *J. Phys. Soc. Proc* **1** 015096
[21] Winske D *et al.* 2003 *Space Plasma Simulation* (Springer) pp 136–165
[22] Tsuno S *et al.* 2014 *IEEE Trans. Plasma Sci.* **42** 3732–3741
[23] Mitchner M and Kruger C H 1973 *Partially ionized gases* (Wiley New York)

Journal of Physics: Conference Series **717** (2016) 012084 doi:10.1088/1742-6596/717/1/012084

LMJ & PETAL Status and first experiments

J-L Miquel

CEA/DAM Île de France, Bruyères le Châtel, 91297 Arpajon Cedex, France

jean-luc.miquel@cea.fr

Abstract. The laser Megajoule (LMJ) facility is designed to provide the experimental capabilities to study High Energy Density Physics (HEDP). The LMJ is part of the Simulation Program, which combines improvement of physics models, high performance numerical simulation, and experimental validation. Since the operational commissioning of the LMJ, in October 2014, several experimental campaigns have been achieved. They have demonstrated the good performances of LMJ and demonstrated its aptitudes to perform experiments for the Simulation Program. The PETAL project consists in the addition of one short-pulse (ps) ultra-high-power, high-energy beam (kJ) to the LMJ facility. The first high energy test shots in the compressor stage of PETAL, performed in May 2015, have reached a power of 1.2 PW. PETAL will offer a combination of a very high intensity multi-petawatt beam, synchronized with the nanosecond beams of the LMJ. This combination will expand the LMJ experimental field in HEDP. LMJ-PETAL is open to the academic communities; the first experiments are planned in 2017.

1. Introduction

The laser Megajoule (LMJ) facility, developed by The Commissariat à l'Energie Atomique et aux Energies Alternatives (CEA), is designed to provide the experimental capabilities to study High Energy Density Physics (HEDP). LMJ is part of the Simulation Program which is a keystone of guarantee of French nuclear deterrence. It combines improvement of physics models, high performance numerical simulation, and experimental validation. The heart of this Program is the Simulation Standard which couples Reference data, Physical models and Numerical methods. It will be regularly upgraded according to experimental results obtained on LMJ.

LMJ offers unique capabilities for the Simulation Program. When completed, the 176 beams of the facility will deliver a total energy of 1.4 MJ of 0.35 μm (3ω) light and a maximum power of 400 TW. A large panel of experiments will be done to study physical processes at temperatures from 100 eV to 100 keV and pressures from 1 Mbar to 100 Gbar. These experiments will validate the advanced theoretical models of the Simulation Program, and provide accurate data (equation of state, atomic and nuclear data). Among these experiments, Ignition is the most exciting challenge since ICF experiments define the most stringent specifications.

The operational commissioning of the LMJ, with the first bundle (eight beams), was declared in October 2014. Since this date, several experimental campaigns have been achieved, in order to qualify the LMJ experimental capabilities, and to validate radiative hydrodynamics simulations. They have demonstrated the LMJ aptitudes to perform experiments for the Simulation Program.

2. Program overview

The CEA is developing a thematic approach for the experimental program on LMJ; this program is coupled to a progressive power and energy increase of the facility. LMJ will enhance its capacities in the next years with the completion of other bundles and a full set of diagnostics. The program is based on a gradual pathway with dedicated experiments. Eight experimental topics have been identified for the Simulation program and summarized in table 1.

Table 1. Experimental topics of the Simulation program.

Topic	Mechanisms to be addressed	Physics to be controlled
Hohlraum energetics	Laser plasma interaction, X-ray conversion	Radiation flux
Fundamental data's	Equation of state, Opacities	Matter's behavior under high pressure and temperature
Radiation transport	X-ray absorption, losses, reemission	Energy transport
Implosion hydrodynamics	Implosion velocity, Shock tuning	Compression
Hydrodynamic instabilities	Instabilities growth, turbulence	Mixing
Fusion studies	Thermodynamic conditions, initiation of fusion reactions	Ignition conditions
Ignition	Study of different kind of ignition targets	DT burning
Applications	Coupling of ignition target with another target	Complex powerful system

The carrying out of these experimental themes depends on the experimental capabilities of the facility and particularly the laser configuration (number of beams) and the availability of diagnostics. Therefore a diagnostics development plan has been developed, it is summarized in the table 2.

This program allows to draw a robust roadmap for ignition: during the next years LMJ will increase its operational capabilities with the completion of new beams, providing more energy, and the implementation of new diagnostics. This period of time will be useful for the handling of the facility, including training of the operational crew and control of performances (precisions, reproducibility). This phase will also be exploited to tackle some of the experimental topics of the Simulation program:

- Radiation transport taking into account losses, and energy balance in hohlraums in order to control the radiative drive as a function of time,
- Fundamental data with the obtaining of equation of state and opacities of materials relevant for the Simulation program
- Hohlraum energetics with the qualification of smoothing, and control of crossed beam energy transfer, in He/H_2 gas-filled hohlraums,
- Implosion hydrodynamics in planar and then convergent geometry,
- Hydrodynamic instabilities from linear growth to turbulence, in planar and convergent geometry,
- And, at last, Fusion studies with drive adjustment in order to manage a more than 350 km/s implosion velocity with the full LMJ.

After LMJ completion, some topics as Radiation transport and Fundamental data will be continued at higher energy, and some other as Implosion hydrodynamics and Hydrodynamic instabilities will be focused on ignition target, using cryogenic equipment, in order to control mix in the hot spot. Then experiments will be dedicated to the Ignition topic to obtain a significant gain, and the target design will be improved to increase the gain. The last topic, Applications, dedicated to the control of powerful system will couple an ignition target to another type of target to obtain, for instance, Fundamental data's in a very high temperature and pressure domain.

All the experiments performed during the phase of power increase and after completion of LMJ will contribute to improve the simulation standards.

Table 2. Diagnostics development plan. Some diagnostics (number 6,7 and 10) are developed in the framework of PETAL+ project (see section 5.3).

	Type	Performances : spectral range, resolution, field of view	Date of delivery
1	Gated hard X-ray Imager 1	12 images, 0.5 – 15 keV, 30 μm resol./ 3 mm fov	2014
2	Drive diagnostic DMX : Broad-band X-ray Spectrometer Grating spectrometer Laser Entrance Hole (LEH) imager	 0.03 – 20 keV 1 – 5 keV 0.5 - 2 keV, 100 μm res./2 mm fov)	2015
3	Gated hard X-ray Imager 2	12 images, 0.5 – 15 keV, 150 μm res./ 15 mm fov	2015
4	Streaked soft X-ray Imager	2 images, 0.05 – 1.5 keV, 30 μm / 5 mm fov	2016
5	Broad-band X-ray Spectr. (miniDMX)	0.03 – 4 keV	2016
6	*Hard X-ray Spectrometer (PETAL+)*	*15 – 100 keV*	*2016*
7	*Electron Spectrometer (PETAL+)*	*5 – 150 MeV*	*2016*
8	EOS pack	2 VISARs, 1D & 2D SBO, reflectivity	2017
9	Streaked hard X-ray Imager	4 images, 0.5 – 10 keV, 50 μm/5 mm fov	2017
10	*Ion Spectrom. and Imager (PETAL+)*	*0,1 – 200 MeV*	*2017*
11	Polar X-ray Imager 1 (LEH)		2017
12	Full Aperture Backscatter System	Raman & Brillouin range	2018
13	Near Backscatter Imager	Raman & Brillouin range	2018
14	Polar X-ray Imager 2 (LEH)		2018
15	Enhanced resol. hard X-ray Imager	5 μm resolution	2019
16	High resolution X-ray Spectrometer		2019
17	Neutron pack		2019
18	Gated soft X-ray Imager		2020
19	High resol. X-ray Spectro-Imager		2020

3. LMJ present status

A description of the LMJ facility can be found in [1, 2], and a schematic view in figure 1. In the laser bays, the framework and equipment of the four laser bays are completed, the optical and electronic components are now being installed. The first bundle is completed, activated and used since 2014, four other bundles are being mounted and will be activated in 2016 & 2017. In the switchyard each bundle is divided into two quads of four beams which are directed to the upper and lower hemispheres of the chamber using six transport mirrors per beam. In the target bay the beams are converted to 3ω and focused using gratings in the SCF (System for frequency Conversion and Focusing).

The first high energy test shots at 1ω and then at 3ω on calibrated calorimeter were performed in August and September 2014 and gave a good spatial uniformity [2]. Alignment and synchronization of the 8 beams at the chamber center was achieved with a dedicated diagnostic: the "Nanojoule active target" at the center of the target chamber, which uses a CCD for alignment and measurement of focal spot at low energy, and silicon photodiode for synchronization of the beams.

In the Target bay, the first diagnostics manipulator (SID) and associated equipment have been qualified mid-2014. LMJ diagnostics are complex systems coupling several measurements; at present, 3 systems are activated (see table 2): two X-ray Imaging systems and one drive diagnostic system (DMX). The first X-ray imager has been used to qualify the pointing accuracy, which is between 19 and 75 μm (6 shots) for a specification of less than 100 μm.

According to these good results, the LMJ has demonstrated that it meets all specifications required to begin experiments. The official commissioning was declared by Prime Minister Manuel Valls on October 23rd 2014.

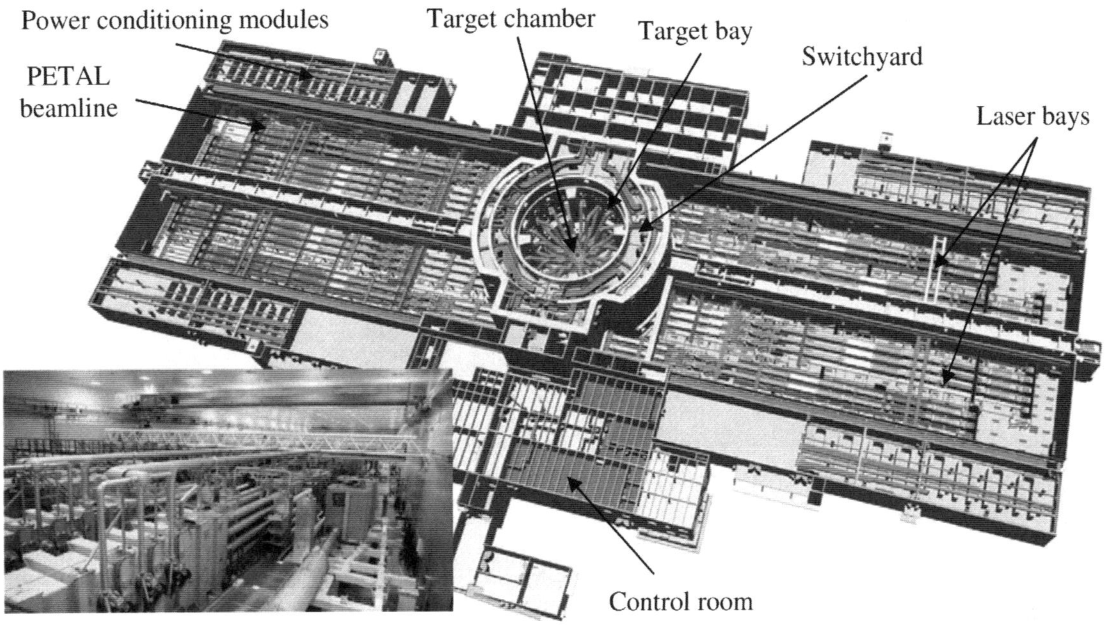

Figure 1. Schematic view of the Laser MegaJoule facility and picture of one of the four laser bays with 7 LMJ bundles and PETAL beamline on the right which occupies the place of one LMJ bundle.

4. LMJ experiments

The first LMJ Campaign took place in October 2014. The objective was to demonstrate LMJ capabilities to perform experiments for the Simulation program. The experiment was dedicated to radiation transport, and particularly to the dynamics of slot closure, in well-controlled material, due to the radiative flux produced by a gold hohlraum. Closure dynamics was diagnosed by auto-radiography, explicitly the hard x-ray produced by the impact of beams on the hohlraum wall was used as a back-lighter. Several materials have been studied: Ta_2O_5 aerogel and Gold samples. Nice results were obtained from the first shot with Ta_2O_5 aerogel sample with a 200 μm thickness and 100 and 75 μm slot widths. Details of the phenomena were well predicted by the simulations as the late phenomena of the closure dynamics showing a denser plasma zone in the middle of the slot due to plasma collision (image on the right in figure 2).

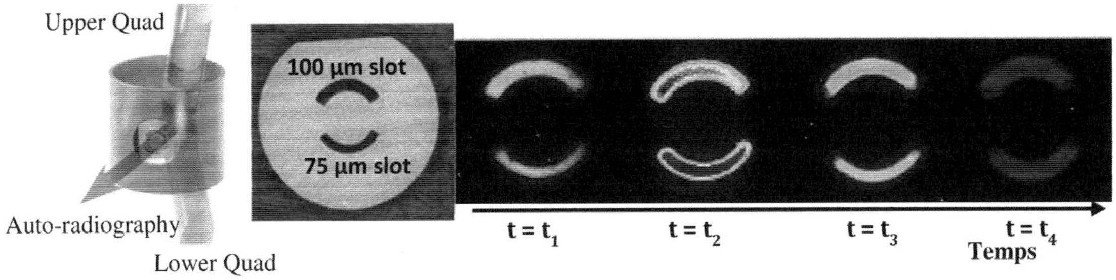

Figure 2. Schematic of the slot closure experiment, and example of data obtained on a Ta_2O_5 aerogel.

A second campaign was performed in May 2015 and dedicated to asymmetrically driven implosion with the objective of qualification of the radiographic capabilities of LMJ. The implosion of capsule in hohlraum with LEH shield heated by the first quad was diagnosed by Ti or Sc back-lighter produced by the other quad used with an optimized pulse (1 ns pre-pulse followed by a 2 ns pulse). The target presented several complexities, as the two diagnostic holes in a 3D geometry for radiographic axis, and 3D plastic plugs (50 μm wall thickness, 500 μm diameter) to avoid diagnostic holes closing.

Figure 3 shows an example of the nice results obtained during this campaign. The egg shape of the capsule, predicted by simulation, is due to the anisotropic distribution of the drive around the capsule.

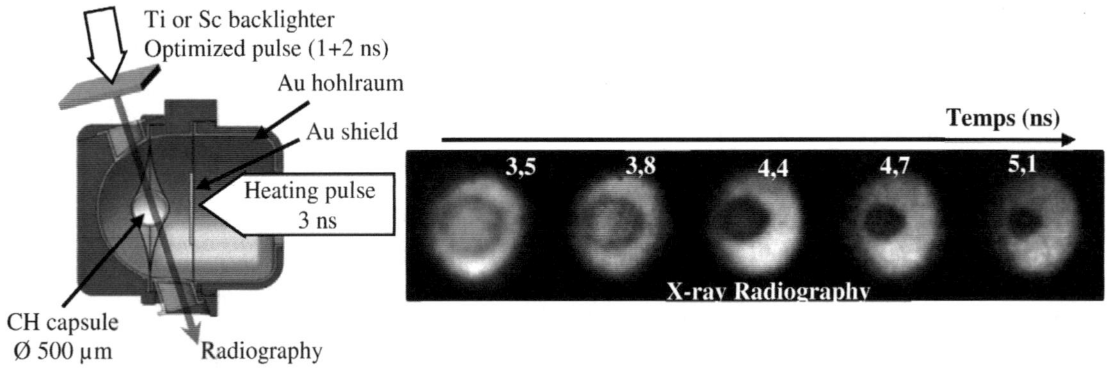

Figure 3. Schematic of the asymmetrically driven implosion experiment, and example of data obtained with a Ti back-lighter.

The program will continue with :

- the pursuit of Asymmetrical implosion with a capsule positioning at different distances from the shield, and a modification of capsule thickness to control the implosion symmetry ;
- Hydrodynamic instabilities with the characterization of the plasma jet produced by a local defect in a planar geometry ;
- Radiation transport with the pursuit of slot closure with quantitative measurement of losses, Radiative balance in hohlraums and Rosseland opacities;
- Hohlraum energetics with X-ray conversion and Characterization of magnetic fields;
- EoS of reference materials (Quartz, Al, Diamond) at the beginning …

These experiments require new diagnostics that will be delivered in the next years : see table 2.

5. PETAL status

The PETAL project consists in the addition of one short-pulse (ps) ultra-high-power, high-energy beam (kJ) to the LMJ facility. PETAL will offer a combination of a very high intensity multi-petawatt beam, synchronized with the nanosecond beams of the LMJ. This combination will expand the LMJ experimental field in HEDP. A description PETAL can be found in [1]

The first high energy test shots in the compressor stage of PETAL were performed in May 2015; they demonstrated the PW capabilities of PETAL with a 1.2 PW power shot (840 J energy and 700 fs duration) [3]. This PW power has then been brought to the LMJ target chamber center in December 2015, and a test shot coupling LMJ and PETAL has been performed at the same date.

PETAL energy is today limited by damage threshold of optics after compression. PETAL gratings have been optimized and a damage threshold above 4J/cm² (cross section) has been obtained [4]. Nevertheless test on transport mirror revealed a 2.4 J/cm² damage threshold at 500 fs; this corresponds to a maximal energy in the beam of ~1 kJ, taking into account beam modulations.

The PETAL performance before entering the compressor will be improved in the next year with an upgraded spatial uniformity and a better filling of sub-aperture of the beam. With these improvements a 2 PW beam is easily reachable. Moreover a better characterization will be obtained with the activation of all diagnostics (pulse contrast, focal spot, …). Concerning mirror damage threshold, new technologies are needed. Several ways of improvement have been identified with new coating materials, improvement of coating processes, new design of layers, and adjustment of layers thickness. Another matter is the electromagnetic pulse (EMP) generation; before performing experiments, EMP mitigation is required in order to protect LMJ equipment. Several concepts have been identified and are under test [5].

The PETAL+ Project is dedicated to the first PETAL plasma diagnostics [6]. It is funded by the

French National Agency for Research (ANR), managed by the University of Bordeaux and developed by CEA. It includes manipulators derived from LMJ SID (for diagnostics compatibility); two modules of electron spectrometer using a magnetic field for 5 - 150 MeV electrons; a hard X-ray spectrometer using a transmission crystals (Quartz and LiF) for 15 – 100 keV X-ray; and a charged particles diagnostic made of proton spectrometer and imaging system (for proton-radiography) for 100 keV-200 MeV proton, coupled to an electron spectrometer for 100 keV – 150 MeV electrons using two Thomson parabolas.

Experiments combining LMJ and PETAL will start in 2016, giving the possibility to address a new physics.

6. Access of the Academic Community to LMJ-PETAL

The CEA-DAM has promoted for several decades national and international collaborations. Between 2005 and 2014, access to the LIL facility has been given to the scientific communities. With the LMJ and PETAL facilities, the CEA-DAM is once again in a position to welcome national and international teams. The academic access to LMJ-PETAL and the selection of the proposals for experiments is done through the Institute Laser & Plasmas (ILP) with the help of the PETAL international Scientific Advisory Committee. The LMJ-PETAL User guide provides the necessary technical references to researchers for the writing of Letter of Intent of experimental proposals to be performed on LMJ-PETAL. Regularly updated version of this LMJ-PETAL User guide is available on LMJ website at http://www-lmj.cea.fr/en/ForUsers.htm.

The first experimental configuration (end 2016) will include 4 LMJ quads and the PETAL beam in the equatorial plane of the target chamber. For the first call for experiments (2017-2018), 4 experiments have been selected from 16 proposals:

- Study of the interplay between B field and heat transport in ICF conditions (Magnetic reconnection), proposed by Dr R. Smets (LPP, Ecole Polytechnique, France).
- Amplification of magnetic fields in radiative plasmas - Magnetogenesis and turbulence in galaxy, proposed by Prof. G. Gregori (Department of Physics, University of Oxford, UK).
- Strong Shock generation by laser plasma interaction in presence or not of laser smoothing (SSD) in the context of shock ignition studies, proposed by Dr. S. Baton (LULI, Ecole Polytechnique, France), and Dr. X. Ribeyre (CELIA, France).
- Interacting radiative shock: an opportunity to study astrophysical objects in the Laboratory, proposed by Dr. M. Koenig (LULI , Ecole Polytechnique, France).

The next call for proposals will be announced in 2016 for experiments to be performed in 2019.

Acknowledgements

The author would like to thank the numerous contributors to this paper who cannot be completely cited.

The PETAL project is being performed under the auspices of the Conseil Régional d'Aquitaine, of the French Ministry of Research and of the European Union and with the scientific support of ILP. The development of PETAL diagnostics takes place within the Equipex PETAL+ funded by the French National Agency for Research (ANR) and coordinated by the University of Bordeaux.

References

[1] Miquel J L, Batani D and Blanchot N 2014, *Review of Laser Engineering* **42** 131-6
A. Casner *et al.* 2015, *High Energy Density Physics* **17** 2-11
[2] Vivini P *et al.* 2015, *High Power Lasers Fusion Research III* : *Proceedings SPIE* **9345** 934503
[3] Blanchot N *et al., these proceedings*
[4] Neauport J *et al.* 2007, *Opt. Express* **15** 12508-22
[5] Bardon M *et al., these proceedings*
[6] Batani D *et al.* 2014 *Phys. Scr.* **161** 014016-22
Ducret J E *et al.* 2013 Nucl. Instrum. Methods Phys. Res. Sect. A **720** 141-43

Journal of Physics: Conference Series **717** (2016) 012085 doi:10.1088/1742-6596/717/1/012085

Status and progress of the National Ignition Facility as ICF and HED user facility

B M Van Wonterghem, R L Kauffman, D W Larson and M C Herrmann

Lawrence Livermore National Laboratory, P.O. Box 808, Livermore, CA, USA

vanwonterghem1@llnl.gov

Abstract. Since its completion in 2009, the National Ignition Facility has been operated in support of NNSA's Stockpile Stewardship mission, providing unique experimental data in the high energy density regime. We will describe the progress made by the National Ignition facility in the user office and management, facility capabilities, target diagnostics and diagnostics development. We will also discuss the results of a major effort to increase the shot rate on NIF. An extensive set of projects, developed in conjunction with the HED community and drawing on best practices at other facilities, improved shot rate by over 80% and recently enabled us to deliver 356 target experiments in FY15 in support of the users. Through an updated experimental set-up and review process, computer controlled set-up of the laser and diagnostics and disciplined operations, NIF also continued to deliver experimental reliability, precision and repeatability. New and complex platforms are introduced with a high success rate. Finally we discuss how new capabilities and further efficiency improvements will enable the successful execution of ICF and HED experimental programs required to support the quest for Ignition and the broader Science Based Stockpile Stewardship mission

1. Introduction

The National Ignition Facility is the world's largest laser facility for Inertial Confinement Fusion (ICF) and High Energy Density (HED) science research. NIF is a 192 beam Nd-glass laser system that has demonstrated operation producing 1.8 MJ of 0.35 ∝m wavelength light with more than 500 TW of peak power.[1] Completed in 2009, initial experiments were done through 2012 as part of the National Ignition Campaign (NIC) conducting ignition and supporting experiments using indirect drive to implode cryogenic layered DT ignition CH targets.[2] NIF supports four main programs as part of its role in the NNSA's Stockpile Stewardship program: ICF, HED, Discovery Science (DS), and other National Security Applications (NSA). NIF is increasing its shot rate to provide increased opportunity for all its users. It continues to enhance its capabilities for improving efficiency and providing expanded capabilities for its users.

NIF continues making progress toward its ignition goal. Implosion experiments done using laser pulses producing high adiabat and lower instabilities growth implosions have resulted in implosions with significant heating of the DT fuel due to energy redeposition from the fusion produced alpha

Journal of Physics: Conference Series **717** (2016) 012085 doi:10.1088/1742-6596/717/1/012085

particles.[3] Symmetry control and laser-plasma instabilities in these gas-filled hohlraums remain an issue.[4] The support tents that hold the capsule in place were found to provide a seed for hydrodynamic instabilities also distorting the imploding fuel.[5] Experiments are presently investigating hohlraum dynamics to improve symmetry control and laser plasma instabilities. New methods for supporting the capsule are being developed to reduce the effect on implosions. NIF is also doing experiments to broaden its ICF studies using alternative ablators such as high density carbon (HDC) [6] and Be [7] and investigating alternative implosion geometries such as double shell targets and foam lined capsules. A program investigating direct drive implosions on NIF has also begun.

NIF has significantly expanded its efforts in HED science. A major effort is studying materials at high pressures. It has performed experiments measuring the equation of state of carbon and iron to 5 terapascals relevant to planetary formation.[8] Platforms have been developed to measure x-ray diffraction from crystals dynamically compressed to high pressures. A platform is being developed to measure the strength of materials at high pressures by measuring the reduction in growth of hydrodynamic instabilities due to strength. The long pulses and high energy of NIF allows it to study hydrodynamic instabilities in unique regimes. NIF is performing experiments studying complex hydro in shear flow into nonlinearity. NIF has developed the capability to delay beams by tens to one hundred nanoseconds to produce x-ray backlighters to enable these experiments. Experiments have also begun to measure nuclear cross sections using NIF implosions. The relatively pure 14 MeV neutron source and low background offer the possibility of making high fidelity cross section measurements, NIF is also being developed as a low debris, high power x-ray source for x-ray effects testing.

2. Improved NIF Operations

NIF has continually worked to improve its operational efficiency. The number of NIF target shots by quarter is shown in figure 1. NIF performed 356 target experiments in FY15,far exceed the stretch goal of 300 laid out in the 120-day study. The same study had a stretch goal of 400 target shots in FY16. The four main programs utilizing the NIF have all benefited from the increase in shot rate. This is consistent with the goal of NIF to be a user facility serving a broad base of users. NIF has established a NIF user group and has an annual meeting of the users to discuss issues and plan for future development of the facility. [9]

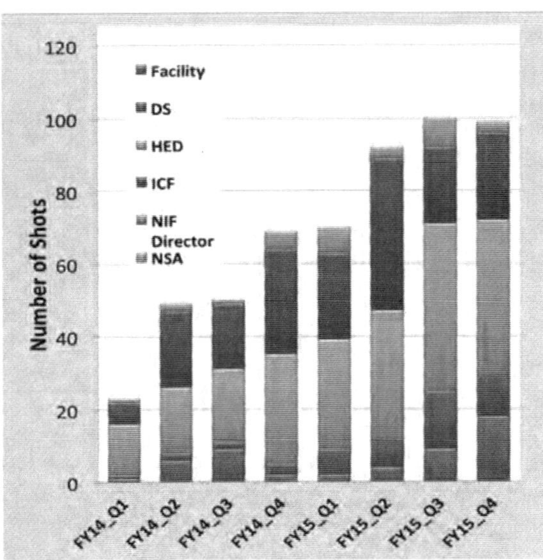

Figure 1, NIF shots by quarter. Actual shots through Q3 of FY15. Planned shots for Q4 of FY15.

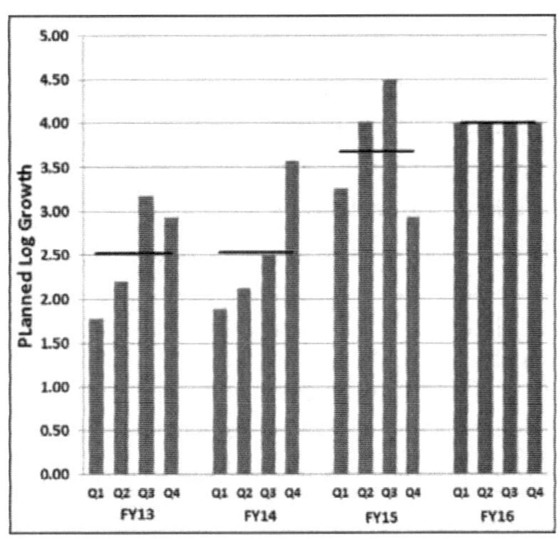

Figure 2. NIF final optics usage

Journal of Physics: Conference Series **717** (2016) 012085 doi:10.1088/1742-6596/717/1/012085

A significant improvement in the shot rate has been due to operational efficiencies. A number of workshops were held in 2013 with operations managers from other large facilities and a number of potential improvements were identified. NIF has adopted a fixed schedule for target shots and maintenance. The operations crew has examined all of its procedures streamlining many of them making it more efficient. A number of hardware changes have been engineered to improve efficiency. The NIF User Office, working with NIF Data Systems has also improved its user interfaces, improving and streamlining the tools needed for experimental set up. NIF continues these efforts as part of its continuing improvement program.

To meet its user requirements, NIF routinely operates above the 3ω laser fluence damage threshold for its final optics. Part of managing operational efficiency is managing optics use. NIF optics usage is shown in figure 2 in terms of log growth of damage sites that is a measure of optics damage. Currently, the facility replaces ~40 3ω optics per week. A recycle loop has been developed to mitigate and repair damaged optics, and lower the cost of operating above the damage threshold. NIF has an active program for improving damage thresholds. This includes improving finishing capabilities and managing optics damage. As these technologies are introduced into production operations at higher damage growth is possible, as seen in figure 2, which enables users to request higher energy for experiments without significant additional cost, improving operational efficiency.

Improvements are also being made in efficiencies for target fabrication to meet the demands of an increasing shot rate. Most of the indirect drive target components are manufactured at General Atomics while assembly and metrology are done at LLNL. In recent years target fabrication has moved to a clean room environment to improve target quality and yield. Improved efficiencies are being realized by increasing automation. Examples of these are shown in figures 3 and 4. Figure 3 shows a photograph of an automated proofing station that can process twice as many targets as the old proofing station. An automated robotic hohlraum insertion and alignment station is shown in figure 4. These streamlining projects are expected to increase target fabrication capacity by 30-50%.

Target chamber hardware is being added to improve shot capabilities and efficiency. The Advanced Target Laser Alignment System (ATLAS) is being commissioned for faster and more accurate target and diagnostic alignment. The system consists of a laser and sensors that can monitor inside the targets chamber the position of fiducials placed on diagnostics and other hardware for accurate positioning.

Figure 3. Automated proofing station

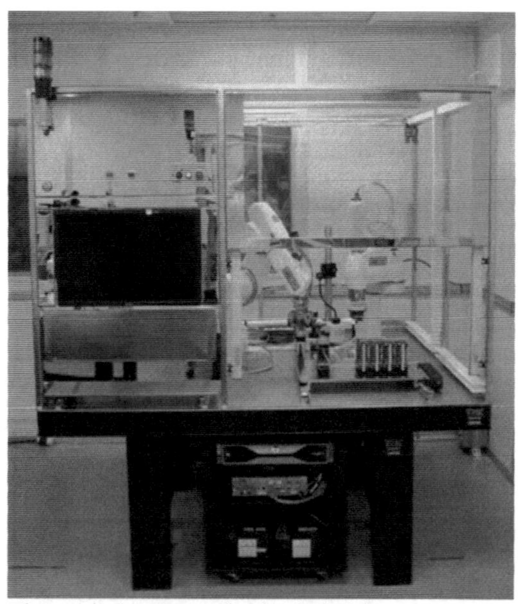

Figure 4. Automated robotic hohlraum insertion and alignment

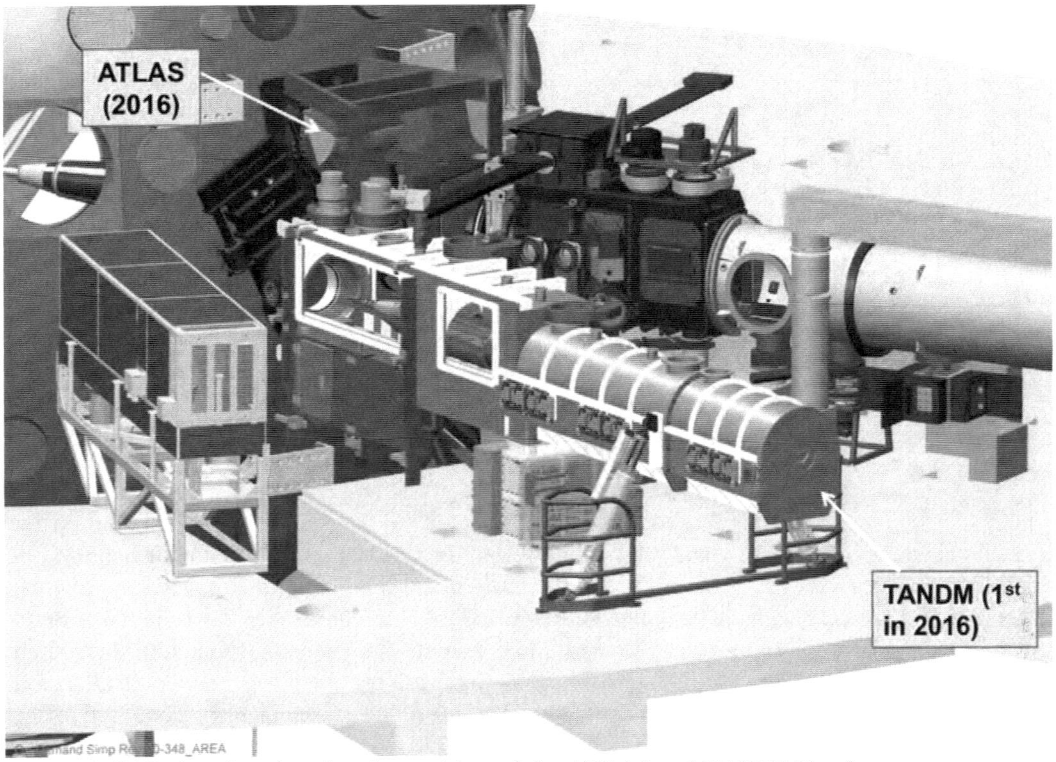

Figure 5, CAD drawing showing the position of the ATLAS and TANDM hardware

The system will replace the Opposed Port Alignment System (OPAS) presently being used for DIM instrument alignment freeing up valuable real estate at the equator of the target chamber. The system has already demonstrated that it can align diagnostics on the Polar DIM that in the past has been difficult since it did not have an OPAS. A new manipulator called the Target and Diagnostic Manipulator (TANDM) is being added. TANDM will be able to field DIM based diagnostics and serve as a warm target positioner. This addition will provide another view of the target for diagnostics and flexibility for target fielding.

3. New Diagnostic Capabilities

NIF continues to add to its impressive diagnostic capabilities. The Advanced Radiographic Capability (ARC) for high energy backlighting was commissioned in 2015. New x-ray imaging and x-ray spectroscopy capabilities have recently been added. New implosion diagnostics are being implemented to provide near polar views of the implosion.

ARC adds short pulse capability to the NIF beams using chirped pulse amplification. A chirped pulse is generated in the NIF ARC preamplifier and stretched for amplification. Large gratings in the target area recompress the beam that will be focused near chamber center on targets to produce high energy backlighting for radiography of high areal density targets such as ICF capsules or high-Z hydrodynamics experiments. Initially two NIF beams will be converted to short pulse with each beam segmented into two beamlets. Initial operation is planned for each beamlet to operate at 0.9 kJ in a 30 ps pulse for intensities greater than 10^{17} W/cm^2. Shorter pulse lengths and higher energies will be commissioned in the future.

New X-ray imaging and x-ray spectroscopy capabilities are being developed. The Dilation Imager for X rays at Ignition (DIXI) [10] is a new x-ray imager developed at General Atomics that can image the imploding core of ICF targets with 10 ps resolution, more than five times faster than previous systems. DIXI converts an x-ray image to an electron image at a photocathode. The electron image is

Figure 6. DIXI mounted on the NIF target chamber

accelerated using a ramped voltage to time dilate the image, transported by an imaging magnetic field to a gated detector. DIXI mounted on the NIF chamber is shown in figure 6. DIXI has already shown implosion features not observed in standard gated images due to the poorer time resolution. DIXI also has an advantage that it is well shielded since it is outside of the target chamber allowing it to operate on yield shots. Time resolution has been added to one of the Static X-ray Imagers to time resolve the x rays produced by the lasers hitting the hohlraum wall. A two-frame Single Line of Sight (SLOS) camera developed at Sandia National Laboratories has been retrofitted to the lower SXI producing two-dimensional images with 2 ns resolution. Four frame and eight frame versions of the camera are planned for a number of other applications.

Several x-ray spectrometers are being developed for NIF experiments. The NIF X-ray Spectrometer (NXS), designed and built by LLE at the University of Rochester, mates to an x-ray streak camera for recording time resolved spectra.[11] It can be fielded with ten different elliptically bent crystal configurations to cover the spectral range from 2 keV to 18 keV. The spectrometer has been used to measure x-ray line intensities of tracers to characterize hohlraum conditions. A time integrated x-ray spectrometer, VIRGIL, is being fielded to measure the x-ray spectrum from 1.5 keV to 6 keV along the Dante line of sight. The spectrometer will be used to better characterize the x-ray spectrum in the band for better fidelity measurements of x-ray preheat. Additional spectrometers planned include a high resolution spectrometer to measure line width of Kr K-shell radiation for characterizing imploded cores and a high resolution spectrometer for opacity experiments.

Experiments and calculations indicate that ICF implosions may have significant anisotropies. Three-dimensional simulations show distorted cores due to drive asymmetries and target features. Nuclear activation detectors (NAD) placed around the chamber measure an anisotropy attributed to variations in the shell areal density.[12] Low mode asymmetry from hohlraum symmetry has been measured along two orthogonal lines of sight, but most other measurements have a limited number of views. New lines of sight for other nuclear diagnostics are being developed to obtain a more complete understanding of the implosion. Neutron time of flight (nToF) spectrometers are used to measure the DT and DD neutron yields and ion temperature and to measure the DT down scatter ratio.[13] NIF currently has three lines of sight, two nearly orthogonal near the equator and one near the bottom of the chamber or south pole. A fourth line of sight is being added to the north pole with the detector on the roof. This will have two detectors with

Figure 7. The NXS being fielded on a DIM at NIF

nearly opposing views to detect if there is bulk motion of the fuel at stagnation. Presently there is only one line of sight at the equator for the Neutron Imaging System (NIS).[14] A second line of sight near the top of the chamber is being developed. A third line of sight orthogonal to these two views is being considered. An additional line of sight for DIXI near the top of the chamber is also being planned.

4. Summary

NIF continues to develop as a unique experimental facility for HED research. It is improving both is operations efficiency through operational improvements and equipment and expanding its shot rate to accommodate the requests of its large user base. All NIF programs saw their shots and capabilities increase: ICF, HED, the Discovery Science programs and National Security Applications. NIF is adding new target experimental capabilities and platforms benefiting. The ARC laser capability was completed in 2015 providing short pulse capability on four beamlets for high energy backlighting and short pulse physics research. NIF continues to add target diagnostics capabilities incorporating state of the art techniques, consistent with the National Diagnostics Plan. The facility provides new and exciting physics data supporting scientific progress in all areas of the HED science and NNSA's Stockpile Stewardship Program.

Acknowledgments

This work performed under the auspices of the U.S. Department of Energy by Lawrence Livermore National Laboratory under Contract DE-AC52-07NA27344.

References

[1] Haynam C A, et al. 2007 *Applied Optics* **46** 3276-303
[2] Edwards M J, et al. 2013 *Physics of Plasmas* **20** 070501
[3] Hurricane O A, et al. 2014 *Nature* **506** 761-4
[4] Hurricane O A, et al. 2014 *Physics of Plasmas* **21** 056314
[5] Tommasini R, et al. 2015 *Physics of Plasmas* **22** 056315
[6] Meezan N B, et al. 2015 *Physics of Plasmas* **22** 062703
[7] Simakov A N, et al. 2014 *Physics of Plasmas* **21** 022701
[8] Smith R F 2014 *Nature* **511** 330
[9] See https://lasers.llnl.gov/for-users
[10] Nagel S R, 2012 *Rev. Sci. Instrum.* **83** 10E11
[11] Perez F, 2014 *Rev. Sci. Instrum.* **85** 11D613
[12] Yeamans C B *Rev. Sci. Instrum.* **83** 10D315
[13] Glebov V Yu *Rev. Sci. Instrum.* **83** 10D309
[14] Merrill F E, *Rev. Sci. Instrum.* **83** 10D317

Journal of Physics: Conference Series **717** (2016) 012086 doi:10.1088/1742-6596/717/1/012086

The Nexawatt: A Strategy for Exawatt Peak Power Lasers Based on NIF and NIF-like Beam Lines

C P J Barty

Lawrence Livermore National Laboratory
Livermore, CA 94551

barty1@llnl.gov

Abstract. An exawatt-scale, short-pulse amplification architecture based upon a novel pulse compressor arrangement and amplification of long-duration chirped beam pulses is described. This architecture is capable of extracting the full, stored energy of a NIF or NIF-like beam line and in doing so produce from one beam line a near-diffraction-limited, laser pulse whose peak power would exceed 0.2 EW. The architecture is well suited to either low-f-number focusing or to multi-beam, dipole focusing concepts that in principle enable focused intensities in the range of 10^{26} W/cm^2 or 5 orders of magnitude beyond that possible from present PW systems based on chirped pulse amplification.

1. Introduction

Modern, inertial confinement fusion lasers based on Nd:glass have amplification bandwidths that are capable of supporting sub-picosecond pulses. With the implementation of chirped pulse amplification (CPA), it is possible for beam lines at the National Ignition Facility (NIF) at the Lawrence Livermore National Laboratory (LLNL), the Laser Mega-Joule (LMJ) facility in Bordeaux, France, the LFEX laser at the Institute for Laser Engineering in Osaka, Japan and the Omega EP facility at the Laboratory for Laser Energetics in Rochester, New York to create petawatt peak power laser pulses of nominally 1-kJ energy [1]. While these system are at the forefront of present high-energy, high-peak power capabilities, they utilize only a small fraction (typically less than 10%) of the potential stored energy of the underlying Nd:glass laser amplification system. A single beam line at the NIF, for example, has a stored energy in excess of 25 kJ [2]. Scaling of existing PW systems to higher peak power and high energy per beam is limited by intensity dependent damage of final optics, lack of sufficient pre-amplifier bandwidth to support shorter duration pulses and insufficient stretched pulse durations required to avoid the onset of b-integral-dependent, non-linear damage of the main amplifiers and transport optics.

The NIF exawatt or "Nexawatt" concept circumvents traditional CPA limitations via a combination of chirped pulse and chirped beam [3] amplification, a novel pulse compressor arrangement capable of compressing 20-ns duration pulses and an increase in the final beam area via splitting beams before final compression. When combined with wide-bandwidth pre-amplifiers [4] these techniques can in principle produce near-diffraction limited pulses with peak powers in excess of 200 PW and pulse durations of order 100 fs from a single amplification beam line. It should be noted that amplification to the 25 kJ level of 100 fs pulse bandwidths is possible without having to change the main Nd:phosphate amplifier glass but does require significant modification of the pre-amplification stages to include both optical parametric amplification and Nd:silicate amplifiers. These modifications enable the creation of joule-level, seed pulses whose spectrum is shifted and wider than the natural FWHM gain bandwidth of kJ-class Nd:phosphate amplifiers. Saturated amplification of such seed pulses to > 20 kJ output produces a net bandwidth capable of supporting sub-100-fs pulses. For the remainder of this manuscript we posit the existence of appropriate pre-amplification to enable 100 fs pulse production. [4]

The Nexawatt strategy also avoids final optic damage by increasing the beam area by splitting the amplified beam prior to the final compressor optics and then coherently recombining the beams prior to focusing. This approach is compatible with a variety of final focusing arrangements including low-f-number focusing and dipole focusing [5]. As such a Nexawatt beam line can potentially produce peak intensities in excess of 10^{26} W/cm^2 or more than 5 orders of magnitude beyond the existing state of the art for existing PW laser systems.

2. Nexawatt Architecture

Existing high-energy, petawatt lasers such as the ARC PW at LLNL [6] are based upon CPA in large-aperture, Nd:phosphate glass amplifiers and utilize meter-scale, high-damage-threshold, multilayer-dielectric gratings to compress the chirped pulses to their final, ps pulse duration. The exact size of the gratings in a typical four grating compressor such as that shown in Figure 1 is set by the combination of the projected area of the input beam on the grating at use angle and the duration of the chirped pulse that is to be compressed. For 1780 gr/mm gratings used in ARC, the highest efficiency and damage threshold is obtained at a use angle ~76.5 degrees. [6] As illustrated in Figure 1, at this use angle a chirped pulse with a duration of 3.34 ns and a full beam aperture of 37 cm x 37 cm would require a second (and third) grating of dimensions 37 cm x 200 cm. While the maximum aperture of the existing gratings used in the ARC system is currently 40 cm x 95 cm, the tools used to fabricate these gratings are in principle capable of producing a 200 cm wide optic via stitched lithographic exposure. For the remainder of this manuscript we will posit the existence of 200 cm wide gratings. In Figure 1, it is evident that the beam area in the dispersed dimension has been increased from 37 cm to 46 cm at the position between gratings 2 and 3. At this position one has a partially compressed pulse that has spectral content that is distributed both in time and in space. The center portion of the chirped-beam pulse has a pulse duration equal to half of the input pulse duration while the edges of the

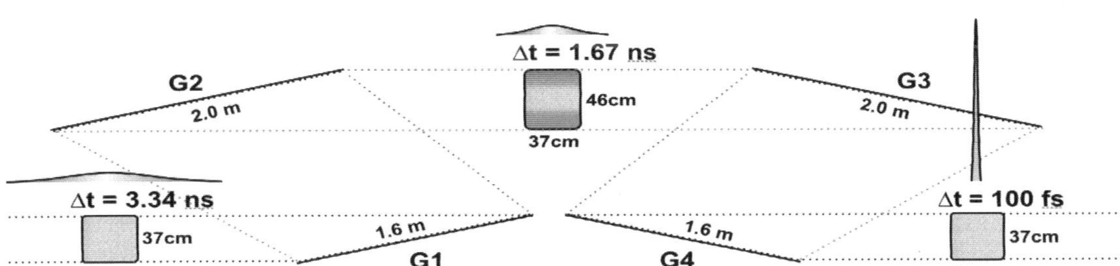

Figure 1. Standard parallel grating pulse compressor with 3.34 ns input chirped pulse duration and NIF beam dimensions

beam pulse do not contain the full spectral content of the input beam and are somewhat longer in duration. The second parallel grating pair (gratings G3 & G4) removes the spatial chirp from the beam and creates the remaining delay required to fully compress the pulse to its ultimate pulse duration of 100 fs. As is evident from examination of Figure 1, only 20% of the area of gratings G2 and G3 actually contributes to creating the delay required for pulse compression. The remaining 80% of the grating aperture is simply required because of the size of the input beam.

Because of b-integral limits [2], a 3.34 ns pulse can only be safely amplified in a NIF amplification chain to approximately 4.2 kJ prior to compression. To achieve higher energy, one must utilize a

Journal of Physics: Conference Series **717** (2016) 012086 doi:10.1088/1742-6596/717/1/012086

longer duration chirped pulse. In principle it is possible to operate within the b-integral limits of NIF and produce a 25 kJ output pulse from a single beam line if the amplified pulse duration is 20 ns. It is certainly possible to create a chirped pulse of 20 ns duration but compression with a 4-grating compressor such as that in Figure 1 would require a grating aperture for gratings G2 and G3 of nearly 4.5 meters!

The Nexawatt amplification architecture and its six grating pulse compressor is illustrated in Figure 2. In this architecture the output from the pulse compressor is a beam with an area equal to half of the standard NIF beam, i.e. 18.5 cm x 37 cm while the input to the amplification chain is a chirped beam pulse with an area equal to the standard NIF beam, i.e. 37 cm x 37 cm. The duration of the chirped beam pulse is 20 ns and therefore can be amplified to ~25 kJ safely. In the compressor the first two gratings invert the spatial chirp and produce a delay such that the pulse after G2 is 11.7 ns. This 8.3 ns FWHM delay can be determined geometrically given the dispersion illustrated in Fig. 1 and the increased separation and beam area of gratings. The second two gratings further compress the pulse to 3.4 ns and create a chirped beam pulse with the same area as the input beam pulse but with the spatial chirp identical to the input beam-pulse. The final two gratings remove the beam chirp and provide the final 3.34 ns of delay required to produce an ~100 fs compressed pulse. It should be noted that this architecture only requires two 2-meter wide gratings yet is capable of compressing a 20 ns pulse.

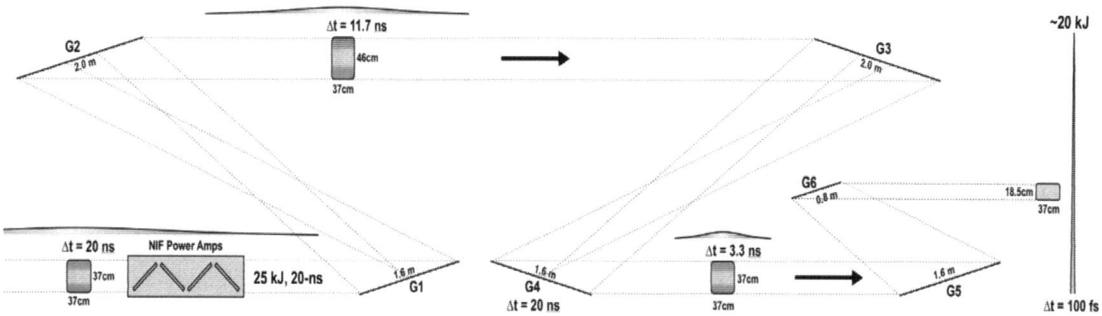

Figure 2. Nexawatt amplification and compression architecture for 0.2 EW pulse production

Gratings G1, G2, G3 and G4 operate within the long pulse damage limits of MLD gratings [6]. Of course the final grating (G6) illustrated in Figure 2 would be incapable of handling a 20-kJ, 100-fs pulse because of its limited 18.5 cm by 37 cm beam area. To alleviate this problem, the Nexawatt concept splits the beam after the fourth grating (G4) and before the fifth grating (G5) into N identical "beamlets". As illustrated in Figure 3, it is possible to split the beam with an appropriate beam splitter arrangement so that each new beamlet is not only identical in energy but also in dispersion or chirp. The number of beamlets required for safe operation is determined by the damage threshold of the final grating and focusing optics.[6] Using the existing damage characteristics for ARC optics [6] and assuming a total combined output of 20-kJ in 100-fs, approximately 20 beamlets are required for safe operation. Combining the individually compressed pulses to form a coherent beam is possible with "cooperative-target", beam-phasing architectures that have been previously investigated for beam phasing of ARC sub-apertures [7]. Such techniques can also be used to correct for beam spatial distortions and in principle to produce near-diffraction limited output. The multi-beamlet, final-focusing geometry of the Nexawatt concept is highly flexible and compatible with either low f-number focusing as illustrated in Figure 3 or with higher-intensity, dipole focusing concepts. [5] Dipole focusing could in principle produce intensities of greater than 10^{26} W/cm^2 from the 0.2 EW Nexawatt output of a single NIF beam line. It should be noted that only gratings G5 and G6 need be enclosed in a vacuum chamber as the pulse duration is long (3.4 ns) prior to grating G5.

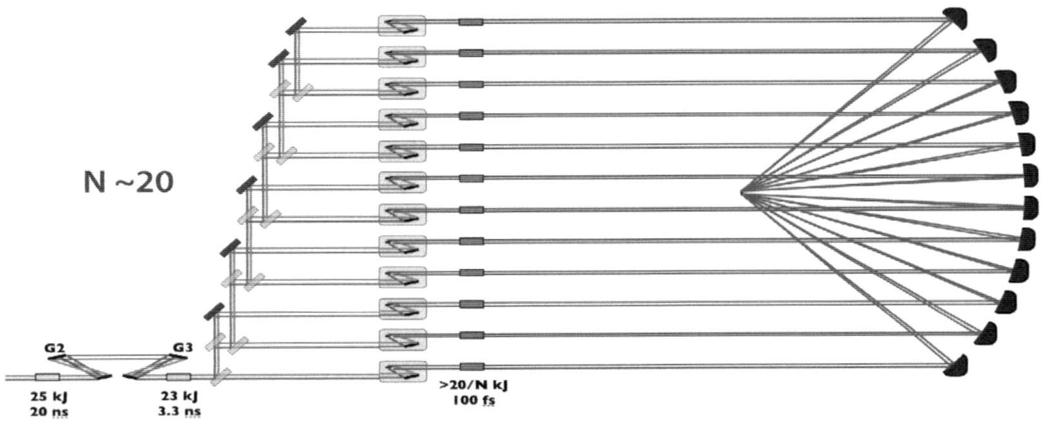

Figure 3. Nexawatt beam splitting and final focusing concept

3. Conclusions

The Nexawatt concept enables full extraction of the stored energy from large-scale, Nd:glass laser amplifiers and the generation of exawatt-scale, 100-ps pulses. It increases short pulse amplification efficiency from <10% to nearly 100%. The concept is based on demonstrated designs and damage thresholds for final optics, is compatible with existing tools for fabrication of compressor gratings and requires only two, 2-meter compressor gratings. The concept operates within the established pulse-width dependent amplification capabilities of NIF and NIF-like amplifiers and unlike other exawatt concepts that aim to combine individually amplified beams, the Nexawatt concept phases "identical" beams after amplification and is compatible with existing beam phasing technologies. The multi-beamlet final output of the Nexawatt is compatible with dipole focusing and theoretically could enable extension of ultrahigh intensity laser science by > 5 orders of magnitude.

This work was performed under the auspices of the U.S. Department of Energy by Lawrence Livermore National Laboratory under Contract DE-AC52-07NA27344

References

[1] Zuegel J D, Borneis S, Barty C P J, Legarrec B, Danson C, Miyanaga N, Rambo P K, Le Blanc C, Kessler T J, Schmid A W, Waxer L J, Kelly J H, Kruschwitz B, Jungquist R, Moses E, Britten J, Jovanovic I, Dawson J and Blancot N 2006, *Fusion Science & Tech* **49**, 453-482

[2] Haynam C A, Wegner P J, et al. 2007, *Applied Optics* **46**, 16, 3276-3303

[3] Barty C P J 2004 Optical Chirped Beam Amplification & Propagation, *US Patent #6,804,045*

[4] Gaul E, Martinez M, Blakeney J, Jochmann A, Ringuette M, Hammond D, Borger T, Escamilla R, Douglas S, Henderson W, Dyer G, Erlandson A, Cross R, Caird J, Ebbers C, and Ditmire T 2010, *Applied Optics*, **49**, 9, 1676-1681

[5] Gonoskov A, Bashinov A, Gonoskov I, Harvey C, Ilderton A, Kim A, Markland M, Mourou G and Sergeev A 2014, *Physical Review Letters*, **113**, 014801

[6] Barty C P J, Key M, Britten J, Beach R, Beer G, Brown C, Bryan S, Caird J, Carlson T, Crane J, Dawson, J, Erlandson A, Fittinghoff D, Hermann M, Hoaglan C, Iyer A, Jones L, Jovanovic I, Komashko A, Landen O, Liao Z, Molander W, Mitchell S, Moses E, Nielsen N, Nguyen H, Nissen J, Payne S, Pennington D, Risinger L, Rushford M, Skulina K, Spaeth M, Stuart B, Tietbohl G and Wattelier B 2004 *Nuclear Fusion* **44**, 266-275

[7] Baker K L, Homoelle D, Utterback E, Stappaerts E, Siders C W and Barty C P J 2009, *Optics Express* **17**, 19 16696

Journal of Physics: Conference Series **717** (2016) 012087 doi:10.1088/1742-6596/717/1/012087

Design of a north pole Neutron Time-of-Flight (NTOF) system at NIF

J A Caggiano, F Barbosa, T J Clancy, M J Eckart, G Grim, E P Hartouni, R Hatarik, H Khater, A Lee, M Sampson, D B Sayre, C Yeamans and M Yeoman

Lawrence Livermore National Laboratory, Livermore, CA 94550

E-mail: caggiano1@llnl.gov

Abstract. A north pole NTOF system for neutron spectroscopy is being implemented at the NIF. The design is centered around a fast scintillator with low mass housing fielded 21.6m from target chamber center at $\theta=18°, \phi=304°$. The line-of-sight (LOS) features a primary port collimator, two secondary collimators in the intervening concrete floors, and a beam dump with a backscatter shield. Because the detector is being fielded on the roof of the NIF building, diagnostic options such as optical and electrical attenuation are remotely controlled, saving setup time and increasing shot rate. The expected performance of the diagnostic is excellent with high sensitivity to both high-energy reaction-in-flight neutrons as well as lower energy down-scattered neutrons.

1. Introduction

For several years now, observed asymmetries of the NIF cryogenic DT layered implosions have been discussed at great length because of their impact on the performance (yield) of these experiments. Recently, a fully 3D simulation, with a significant P1 asymmetry imposed upon it, shows drastic anisotropies in ion temperature, down-scattered neutron (DSN) ratios, and neutron activation as well as DD and DT neutron peak shapes, consistent with observed data[1]. The changes in peak shapes can be so prominent that a new ad hoc parameterization, that includes elements of skew and kurtosis, was created to better describe the emerging neutron spectrum in velocity space [2] and is currently being incorporated into the NTOF data analysis[3]. Concurrently, the flange neutron activation diagnostic (FNAD) has shown that the cryogenic layered implosions exhibit a systematic collection of fuel mass at the north pole[4], but no other neutron detector is located there to confirm this observation.

As a result, the NIF is deploying a spectroscopic NTOF detector atop the target chamber at NIF (θ=18°). The detector package is identical to the other three LOS of spectroscopic neutron detectors (θ–ϕ = 90°-174°, 116°-316°, and 151°-56°) that are based on a low-mass (low scatter background) "JacBlac" housing and an octagonally-shaped, bibenzyl organic scintillator 10cm across by 2.5cm thick [5]. The LOS penetrates the target chamber, two floors, the target bay ceiling, and terminates in a hut on the roof of the NIF building.

2. System Overview

The system was designed to produce maximum sensitivity to desired neutrons while minimizing scattered and background radiation. The detector is placed 21.6m away from target chamber center

(TCC) and will occupy the 18°-304° LOS on the NIF. A port collimator is located between 6 and 7 meters from TCC. The LOS penetrates two floors, one steel, the other concrete, and the NIF ceiling before intercepting the detector and terminating in the beam dump. **Figure 1** shows the engineering rendering/drawing of the LOS, and **Figure 2** shows the simplified MCNP6 [6] model. The modeling activities drive the design requirements and ultimately the layout of the LOS elements.

Figure 1. LOS layout. Primary neutron beam is the red line and terminates on the detector.

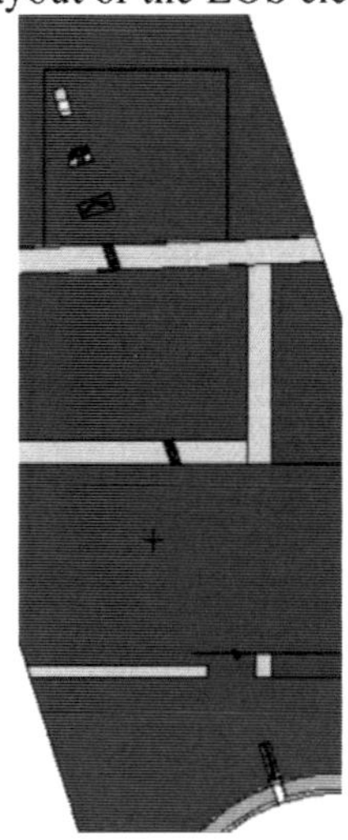

Figure 2. Two-dimensional slice from MCNP6 model of Fig. 1.

3. Collimation

The port collimator defines the beam. The port collimator is 33” long (constrained by facility layout) and consists of borated HDPE with a copper sleeve insert. The sleeve has a cone-shaped aperture that matches the divergence of the beam to minimize intra-collimator neutron scattering and gamma-ray production. The collimator endplate is made of tungsten to reduce the flux of the 477 keV gamma-rays at the scintillator created by low-energy neutron capture in the boron in the borated HDPE.

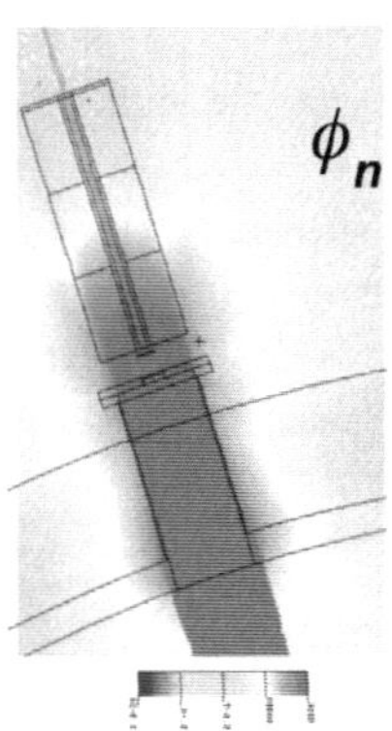

Figure 3. Port collimator simulation showing neutron flux and emergence of beam at downstream end.

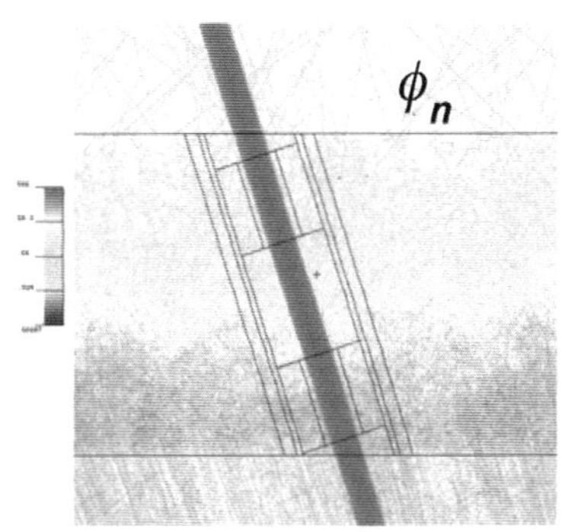

Figure 4. Floor collimator simulation showing a beam of neutrons that has been formed by passing through the port collimator.

Simulations of the LOS with a directed neutron beam demonstrate the efficacy of the collimator (**Figures 3, 4**). The flux of neutrons and gamma rays through the sides of the collimator is attenuated by more than a 1000 times, more than that provided by the target chamber aluminum and concrete.

4. Detector and recording

The detector is based on the JacBlac design [5] and is designed to optimize the signal to background. The detector assembly is mounted on a light-weight frame 1 meter above the floor to minimize scatter background.

Several features of the detector mount are unique to this north pole NTOF installation. The neutral density filters are automatically insertable, mounted on a wheel that is remotely controlled. Programmable electronic attenuators are also used. Tektronix DPO7254 scopes will record the photodetector (Photek photomultiplier and photodiode tubes) signals after approximately 10m of LMR600 cable. The system is similar to other fielded NTOF systems described elsewhere [7].

5. Beam dump and backscatter shield

A beam dump and backscatter shield combination stops the beam while shielding the PMTs from backscattered radiation from the dump. The beam dump is constructed of 18" of steel and 12" of concrete in the beam direction, and is a 12" square in cross section. The backscatter shield is constructed of 8" of lead plates and a truncated cone of concrete .

The efficacy of the backscatter shield can be judged by looking at the shadow cast upon the PMTs. **Figure 5** shows the gamma ray flux from a MCNP6 simulation where neutrons are striking the beam dump. The PMTs clearly reside in the shadows of the backscatter shield. The shield casts a "shadow" on the PMTs and the background reduction within the shadow is approximately 4x (**Figure 6**).

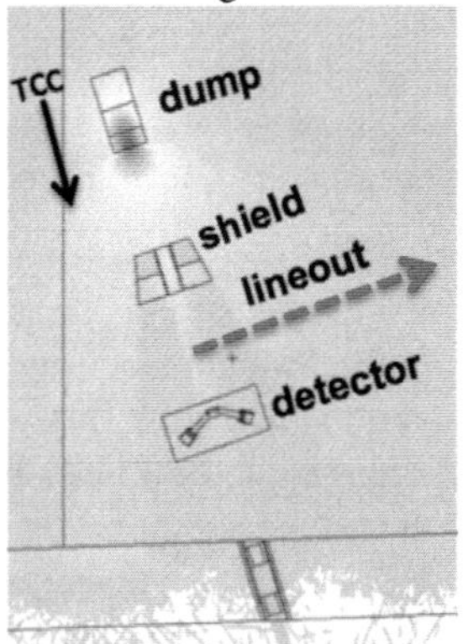

Figure 5. Gamma ray flux showing shadowing of PMTs from backscattered radiation from beam dump.

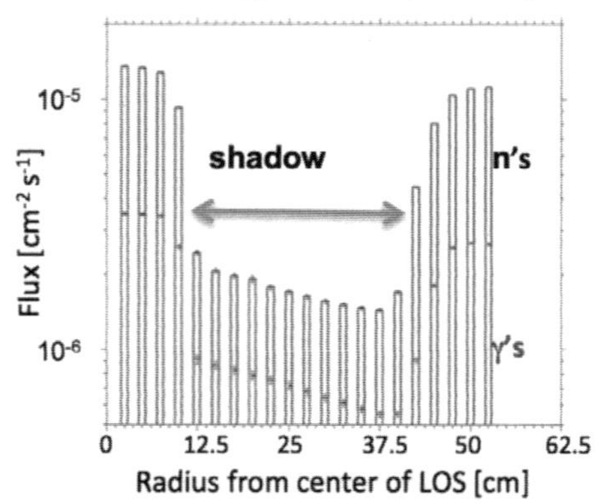

Figure 6. Flux vs radius from center of LOS behind the backscatter shield (lineout of Fig. 5). The depth of the "shadow" cast by the shield is about 4x for both neutrons and γ-rays.

6. Expected Performance

The NP NTOF LOS is designed using MCNP6 simulations to keep scattered radiation to levels lower than true diagnostic signal levels. **Figure 7** shows a simulation of the temporal flux of neutrons and gamma-rays arriving at the detector. The simulation starts DT neutrons (E=14.029 MeV) at $t=0$. The bumps in the gamma-ray flux (red) are created when neutrons strike objects in the LOS. For a cryogenic layered DT shot, both high and low energy neutrons carry diagnostic value. The down-scattered neutrons in the range of 10-12 MeV are a measure of the compression/convergence of the capsule and the subsequent ρR of the fuel over the burn duration. Note that the background radiation in that time window from 455-498ns is lower by at least an order of magnitude. The high energy neutrons arriving before the primary DT neutrons are "reaction-in-flight" (RIF) neutrons, coming from up-scattered D and T fusing with stationary fuel atoms creating higher energy neutrons up to 32 MeV. Those neutrons are hard to measure due to the low levels (typically ~10^{-5} of the primary neutrons) and are a very sensitive measure of cold fuel areal density and mix [8]. Figure 7 shows that the background radiation should not be a significant issue for measuring the RIF neutrons. Even accounting for possible errors of up to 2.5x in the simulation, the RIF window (from 283 to 408 ns) provides a high signal to background ratio.

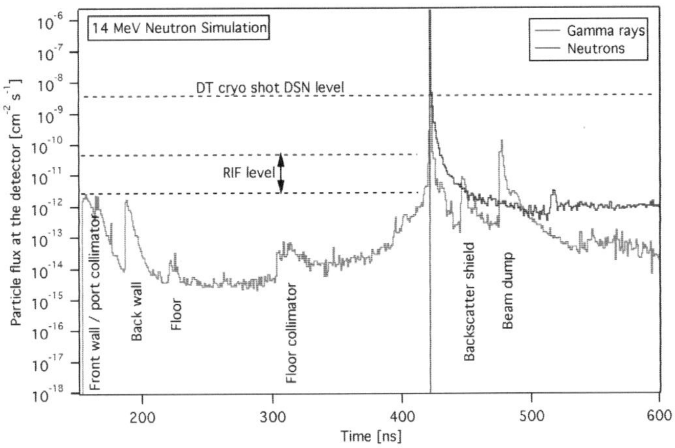

Figure 7. Neutron (blue) and gamma-ray (red) flux at the detector.

7. Summary

A north pole NTOF LOS is being built and will provide high fidelity neutron spectral measurements from NIF implosions. The LOS has been carefully designed to minimize background radiation from interfering with sensitive diagnostic measurements such as DSN and RIF, while maximizing sensitivity to the neutron signal of interest.

8. References

[1] B. Spears *et al.*, Phys. Plasmas **22** 056317 (2014); D. Munro, private communication.
[2] D. Munro, submitted to Nucl. Fus. (2015).
[3] R. Hatarik *et al.* in preparation (2015).
[4] D. Bleuel *et al.*, Rev. Sci. Inst. 83, 10D313 (2012).
[5] J.A. Caggiano *et al.* (2015) to be submitted to Rev. Sci. Inst. October 2015.
[6] J.T. Goorley, *et al.*, "Initial MCNP6 Release Overview", Nucl. Technol., 180, 298-315 (2012).
[7] T.J. Clancy *et al.*, Proc. SPIE 9211, "Target Diagnostics Physics and Engineering for Inertial Confinement Fusion III", 92110A (2014).
[8] A.C. Hayes *et al.* LA-UR-09-04487, "Reaction-in-Flight Neutrons as a Signature for Shell Mixing in NIF capsules" (2009).

Acknowledgements:

This work was performed under the auspices of the U.S. Department of Energy by the Lawrence Livermore National Laboratory under Contract DE-AC52-07NA27344.

Journal of Physics: Conference Series **717** (2016) 012088 doi:10.1088/1742-6596/717/1/012088

Performance measurements on NIF beamlines for future experiments to support polar direct drive

J K Crane[1], B Kruschwitz[2], S T Yang[1], M Bowers[1], D Browning[1], T Budge[1], D Canning[2], J Chou[1], A Consentino[2], J M Di Nicola[1], S Dixit[1], C Dorrer[2], G Erbert[1], R Hackel[1], J Heebner[1], E Hill[2], M Johnston[1], J Kelly[2], J Kwiatkowski[2], M Shaw[1], L Smith[1], P Wegner[1] and J Zuegel[2]

[1]Lawrence Livermore National Laboratory, Livermore, CA, U.S.A.
[2]Laboratory for Laser Energetics, Rochester, NY, U.S.A.
crane1@llnl.gov

Abstract. We are studying the implementation of polar direct drive (PDD) ignition experiments on the National Ignition Facility (NIF) laser system. Part of this preparation involves testing the performance of the NIF laser system over a broader span of center wavelengths, 3.6 nm, where the laser currently operates and that gain models describe. The temporal shape for the PDD pulses consists of a drive pulse preceded by three lower power "picket pulses". These picket pulses require a multi-FM sinusoidal phase modulation format with a bandwidth of ~ 200 GHz and a more dispersive grating in the preamplifier module (PAM) for smoothing-by-spectral-dispersion (SSD). In this paper we discuss recent measurements of gain on the NIF laser system over this broader wavelength range. We measured FM-to-AM conversion over the 3.6 nm wavelength range. The possibility of pinhole closure due to the larger bandwidth and dispersion associated with multi-FM SSD was studied at LLE on the OMEGA EP laser.

1. Introduction

Laser-driven, inertial confinement fusion requires uniform illumination of a spherical target by x-rays in indirect drive or multiple, high power laser beams in direct drive [1]. The National Ignition Facility[2,3] was designed for indirect-drive fusion where the laser beams are focused into the top and bottom of a cylindrical hohlraum. The hohlraum forms a high temperature x-ray oven that produces the necessary rapid ablative compression of the spherical fuel target [1]. The current cylindrical arrangement of NIF beams can still be used for spherical illumination in polar direct drive (PDD) [4]. PDD requires several modifications to the NIF laser to satisfy the higher degree of illumination uniformity necessary for direct drive. The focus of the laser beams on the spherical fuel target requires a higher degree of smoothing to minimize the speckle present due to spatial coherence. The smoothing of the laser beam on the target surface is accomplished in two ways: rapid dithering of the focal spot in one dimension to smear the speckle pattern[5] produced by the phase plates and splitting each beam into orthogonal polarizations that do not interfere at the target surface. To further increase laser coupling into the target the interacting laser beams are operated at two wavelengths separated by 1.2 nm at the 3rd-harmonic, 351 nm. The large wavelength separation reduces cross-beam energy transfer (CBET) from plasma coupling between laser beams, another source of nonuniformity in the target illumination [4].

PDD will use a special temporal pulse format that consists of three short "picket" pulses that precede a main pulse that is similar to a NIF drive pulse [3]. The picket pulses are phase modulated by the following three rf frequencies, f_n, and modulation depths, $m(f_n)$: f_1=21.165 GHz, $m(f_1)$=0.45 radians; f_2=22.837 GHz, $m(f_2)$=1.04 radians; f_3=31.881 GHz, $m(f_3)$=2.07 radians, producing a bandwidth of ~200 GHz. [6,7] These frequencies were determined by modeling the target surface illumination and picking frequencies that did not generate resonances in the surface spatial frequencies that couple to the target [4]. The main drive pulse will be phase modulated at the normal NIF FM frequencies, 3 GHz for stimulated Brillouin scattering suppression (SBS) and 17 GHz for SSD with m(3GHz)=5.13 radians and m(17 GHz)=1.32 radians.

Journal of Physics: Conference Series **717** (2016) 012088 doi:10.1088/1742-6596/717/1/012088

The frequency modulation is converted to angular dispersion in the far-field by a diffraction grating in the preamplifier module, PAM. For future PDD smoothing experiments we added a 1700 line/mm grating in one NIF PAM to produce ~4X the angular dispersion of a 1050 line/mm NIF grating. The line density of the grating yields an angular dispersion for the modulated pulse that is just within the smallest spatial filter pinhole in the NIF laser chain, 150-μrad radius for the full-size NIF beam.

2. Results

2.1. Energetics

To mitigate the effect of CBET, PDD requires the wavelengths of the upper and lower hemisphere NIF beamlines to be separated by 3.6 nm at 1053 nm. This wavelength offset exceeds the current NIF 1ω accessible wavelength tuning range of 2 nm (i.e. from 1052 to 1054 nm). To determine the injection energy to the main laser (ML) amplifier required to produce the 16 kJ output energy at these wavelengths, we ran Virtual BeamLine (VBL)[8] simulations of the NIF main laser as a function of injection energy at the Input Sensor Package (ISP) and wavelength. The main laser (ML) amplifiers use combinations of Hoya LHG-8 and Schott LG-770 laser glass, which have gain spectra that fall off differently on either side of gain peak. The gain available at the extreme wavelengths will be constrained by the beamlines with the extreme combinations of laser glasses.

We conducted a series of NIF shots from 4 bundles (i.e. 32 beamlines) at the two chosen offset wavelengths: 1051.5 nm and 1055.0 nm. The initial shot energy was set to 3kJ to minimize amplifier saturation. After each shot, the ML gain in each beamline is computed using the injected energy and loss through the ML. Since slab amplifiers in each beamline are comprised of different mixtures of Hoya LHG-8 and Schott LG-770 glass, regression analysis of gains from the 32 beam-lines with their different slab amplifier composition was carried out to extract the small signal gain of each type of glass separately. Briefly, the ML gain from each beam-line is cast into the following additive form by taking the natural logarithm of the measured gain: $\ln(G) = N_H \times \ln(G_H) + N_S \times \ln(G_S)$,
where G is the measured ML gain, $N_{H,S}$ are the number of Hoya and Schott slabs in the beamline respectively, and $G_{H,S}$ are the single pass slab gain of the Hoya and Schott glass slab, respectively.

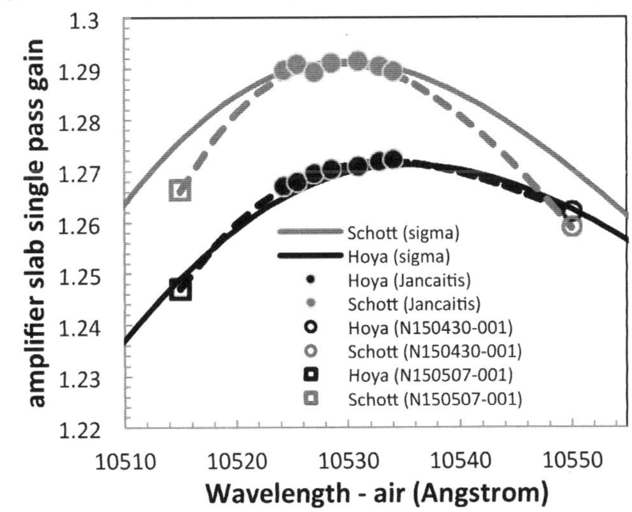

Figure 1. Normalized single pass Hoya and Schott amplifier slab gain as a function of wavelength.

Regression is performed over the 32 beamlines that participated in the shot. The result of the regression analysis is shown in Figure 1, where open circles and squares denote the extracted small-signal gain at the extreme wavelengths. The filled-in circles are small-signal gain measured previously

Journal of Physics: Conference Series **717** (2016) 012088 doi:10.1088/1742-6596/717/1/012088

[9]. Also included in the plot is the gain calculated using emission cross-section values that were measured off-line [10]. These are shown as solid lines. Finally, dashed lines are 3rd order polynomial fits to the measured gains, including gains measured near the peak as well as the ones at the wavelength extremes. As shown in Figure 1, the measured gain, the calculated gain and the polynomial fits all agree well for Hoya glass even at the extreme wavelengths. For Schott glass, however, the measured gain agrees well at the peak but falls faster than the calculation in the wings. We performed a second series of shots on these same 32 NIF beamlines and demonstrated the capability to reach 16 kJ at 1051.5 and 1055 nm.

2.2. Pinhole closure measurements.

With the large spectral dispersion and wide modulation bandwidth inherent in the multi-FM SSD system, the threat of pinhole closure must be considered. Irradiance of the pinhole edge, which increases with both the spectral bandwidth and the dispersion, can generate plasma that expands into the center of the pinhole, causing dangerous modulation of the forward-going beam at the tail end of the pulse [11]. To evaluate the threat of pinhole closure, tests were performed on a beamline of the OMEGA EP laser, which features a very similar architecture to a NIF beamline [12].

No sign of pinhole closure was observed on OMEGA EP up to the maximum 1ω power of 0.6 TW, which is equivalent to 0.3 TW/beam at 3ω, the expected picket power level for Polar Direct Drive picket shots on NIF. Because the OMEGA EP spatial filter, at f/51, is significantly faster than the f/80 NIF transport spatial filter, the pinhole irradiance on these shots was approximately 2.5 times higher than expected on NIF, and hence this was a conservative test. Irradiance near the edge of the pinhole, as measured by a Schlieren diagnostic, was shown to be significantly below the level required to close our pinhole, according to reference [11]. Figure 2(a) shows the measured irradiance distribution in the output pinhole of OMEGA EP for the 0.6-TW shot, with the pinhole edge indicated by a white circle. Figure 2(b) shows a lineout in the dispersion axis from this shot, with the intensity scaled to the NIF transport spatial filter, and with data from a shot with no dispersed bandwidth shown for reference. The irradiance required to close the pinhole at the end of a 10-ns temporally flat pulse, calculated from reference [8] and scaled to the NIF geometry, is 485 GW/cm^2 and is indicated by the green line. For ideal alignment into the output pinhole, the measured intensity is approximately 100 times below this closure threshold. Alternatively, the figure shows that pinhole irradiance will be below the closure level over a large pointing tolerance of ±50 μrad. Hence, pinhole closure is not a significant threat for the proposed multi-FM SSD system at the power levels being considered for NIF.

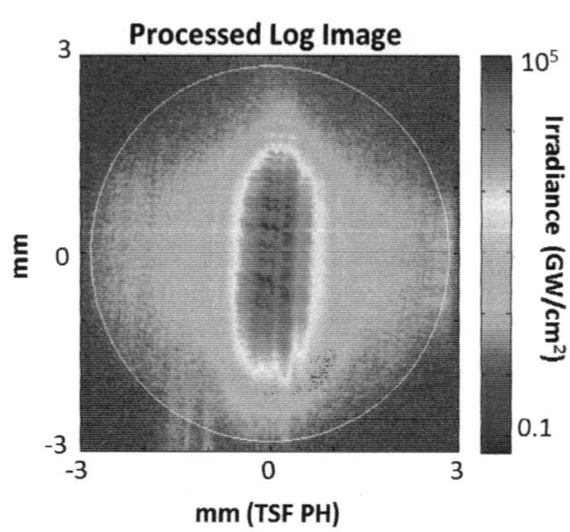

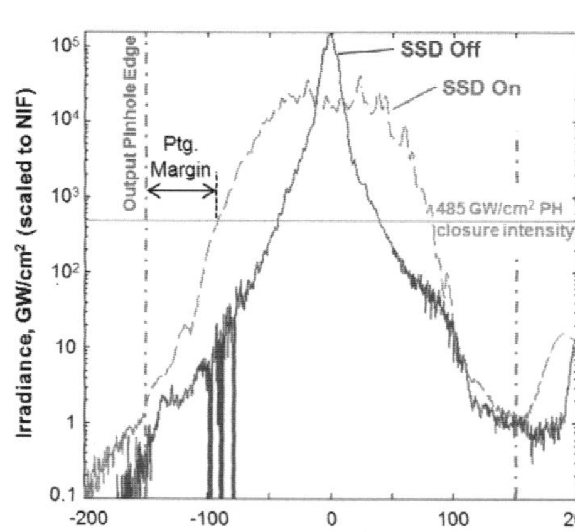

Figure 2. Far-field irradiance measurements from 0.6-TW shots at 1ω, obtained from experiments with multi-FM SSD beam propagation on OMEGA EP: (a) log-scale irradiance distribution in the transport spatial filter (TSF) output pinhole as measured using a Schlieren

configuration of the Far-Field Imaging diagnostic, with the white circle indicating the edges of the 150-μrad half-angle pinhole; (b) lineout of the irradiance in the dispersion direction, scaled to the NIF transport spatial filter geometry, for the multi-FM SSD system (red) and low-bandwidth system (blue), and with the pinhole edges and pinhole irradiance required for closure of a 10-ns pulse both indicated.

3. Summary and future work

The results presented in this paper demonstrate that we can operate the NIF beamlines at the wavelengths and energy required for future PDD experiments. Measurements on Omega EP at the final filter pinhole edges show that we are a factor of ~100 below the irradiance for damage. Recently we installed the multi-FM chassis in the NIF MOR and a 1700 line/mm grating in a NIF PAM and operated a quad of NIF beamlines at the equivalent power level, modulation bandwidth, and angular dispersion necessary to demonstrate the high degree of smoothing required for future PDD experiments. The first of these smoothing experiments will be performed in January 2016 [13].

4. References

[1] Lindl J 1995 *Phys. of Plasmas* **2**, 3933
[2] Miller G H *et al* 2004 *Opt. Eng.* **43**, 2841.
[3] Haynam C A *et al* 2007 *Appl. Opt.* **46**, 3276
[4] Collins T J B *et al* 2012 *Physics of Plasmas* **19**, 056308
[5] Rothenberg J E *et al* 1999 *Solid State Lasers for Application to Inertial Confinement Fusion* **3492**, 51
[6] Kruschwitz B E *et al* 2013 *High Power Lasers for Fusion Research II* **8602**, 86020E-1 (Bellingham, WA SPIE)
[7] Dorrer C *et al* 2013 IEEE *J. Sel. Topics in Quant. Electr.* **19** 3500112
[8] Sacks R A *et al* 2015 *High Power Lasers for Fusion Research III* **9345**, 93450M (Bellingham, WA SPIE)
[9] K. Jancaitis 2013 *Internal LLNL memorandum.*
[10] R. Page 2006 *Internal LLNL memorandum.*
[11] Murray J E *et al* 2000 *Appl. Opt.* **39**, 1405
[12] Kelly J H *et al* 2013 *High Power Lasers for Fusion Research II* **8602**, 86020D-1 (Bellingham, WA SPIE)
[13] Hohenberger M *et al* 2015 *Physics of Plasmas* **22**, 056308

Acknowledgement

This work was performed under the auspices of the U.S. Department of Energy by the Lawrence Livermore National Laboratory under contract DE-AC52-07NA27344, and at the University of Rochester's Laboratory for Laser Energetics under Award Number DE-NA0001944.

The preliminary design of the optical Thomson scattering diagnostic for the National Ignition Facility

P Datte[1], J S Ross[1], D Froula[2], J Galbraith[1], S Glenzer[5], B Hatch[1], J Kilkenny[1], O Landen[1], A M Manuel[1], W Molander[1], D Montgomery[3], J Moody[1], G Swadling[1], J Weaver[4], G Vergel de Dios[1] and M Vitalich[1]

[1]Lawrence Livermore National Laboratory, Livermore, California, USA
[2]Laboratory for Laser Energetics, University of Rochester, Rochester, New York, USA
[3]Los Alamos National Laboratory, Los Alamos, New Mexico, USA
[4]Plasma Physics Division, Naval Research Laboratory, Washington DC, USA
[5]SLAC National Accelerator Laboratory, Menlo Park, California, USA
LLNL-CONF-680048

datte1@llnl.gov

Abstract. The National Ignition Facility (NIF) is a 192 laser beam facility designed to support the Stockpile Stewardship, High Energy Density and Inertial Confinement Fusion programs. We report on the preliminary design of an Optical Thomson Scattering (OTS) diagnostic that has the potential to transform the community's understanding of NIF hohlraum physics by providing first principle, local, time-resolved measurements of under-dense plasma conditions. The system design allows operation with different probe laser wavelengths by manual selection of the appropriate beamsplitter and gratings before the shot. A deep-UV probe beam (λ_0 between 185-215 nm) will optimally collect Thomson scattered light from plasma densities of 5×10^{20} electrons/cm^3 while a 3ω probe will optimally collect Thomson scattered light from plasma densities of 1×10^{19} electrons/cm^3. We report the phase I design of a two phase design strategy. Phase I includes the OTS recording system to measure background levels at NIF and phase II will include the integration of a probe laser.

1. Introduction

A Diagnostic Instrument Manipulator (DIM) based optical Thomson scattering diagnostic (OTS) is being designed for operation at the National Ignition Facility (NIF) to characterize under-dense plasmas [1]. The diagnostic is being built to characterize plasma conditions in ICF hohlraums[2,3]. This information is critical to understand crossed-beam energy transfer, hohlraum wall motion and wall-hohlraum/gas-fill mix, all of which can greatly impact symmetry control and implosion performance. Typical gas-filled hohlraums have densities of 1×10^{20} electrons/cm^3 to 1×10^{21} electrons/cm^3 and temperatures of 2.5 to 5 keV. With a planned scattering angle of ~11 degrees for these conditions we are expecting a collective scattering signal on the order of a hundred nano-joules into the diagnostic for a 10J probe at 5ω. In this regime, scattering from the blue shifted electron plasma wave (EPW) feature will allow us to characterize both the electron temperature and density. The ion acoustic feature (IAW) feature will be used to measure the plasma flow velocity and the ion temperature. We expect modest gradients for our scattering volumes which are ~50 microns in length. The system is designed to measure density to within 20% and the temperature to 25%. This accuracy is primarily limited by the expected hohlraum background, which is driving the choice of the probe beam wavelength to 5ω. The system is focused on the blue shifted EPW resonance due to the lower expected background at these wavelengths. One peak should be sufficient for an accruement fit and measurement [4]. The OTS diagnostic will be inserted into the target chamber by a DIM which requires it to have a compact design, see figure 1. The baseline concept of the system includes a blast window, an unobscured collection telescope, transport and focusing optics, two crossed Czerny-Turner spectrometers and a shared optical streak camera photocathode, located inside of an airbox. The collection telescope is an off-axis Schwarzschild design that relays a 50 µm spot (Thomson volume) to the entrance of the spectrometers. One high resolution (~0.6 meter) spectrometer will measure the ion acoustic wave feature (e.g. 206-214 nm for a 5ω probe laser) and one low resolution (~0.15 meter) spectrometer will measure a tunable 50 nm band to measure the electron plasma wave feature. The EPW spectrometer range will be tunable from 150-400 nm. The outputs of the spectrometers are relayed to a gated optical streak camera with selectable sweep windows between 3 and 40 ns. The inherent optical path delay between the two spectrometer systems allow the output signals to be superimposed on the same photocathode separated by approximately 5 ns and having a spatial extent of 20mm. We report on the current design status and operational performance at NIF.

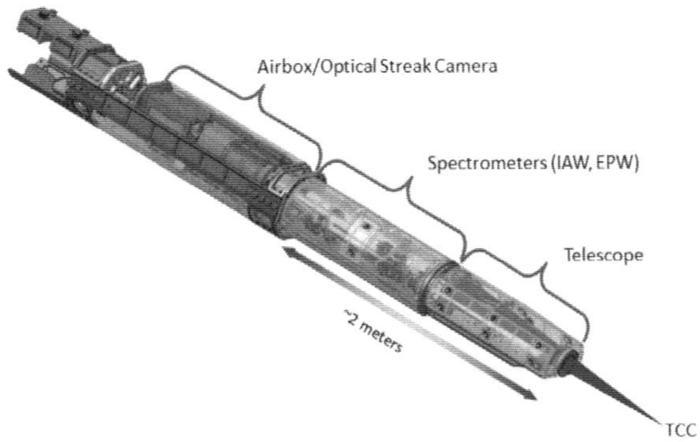

Figure 1. The Diagnostic Instrument Manipulator (DIM) for optical Thomson scattering, which includes the telescope, spectrometer group and the airbox optical streak camera assembly.

2. Optical Design

2.1 Telescope Collection System

The optical design is separated into two major sections, the telescope and the spectrometers. Following the spectrometers is the airbox assembly containing the optical streak camera.

The telescope consists of an off-axis Schwarzschild design of 9.8° allowing an unobstructed view of the targets. The f/8.3 collection optics is the fastest system possible while still maintaining the temporal resolution requirement of 200 ps set by the relationship between the grating illumination of the IAW spectrometer and the telescope focal length. An off-axis parabolic mirror focuses the collimated light onto a pinhole that defines the entrance aperture for the two spectrometers. The system magnification from target chamber center to the streak camera photocathode is 2.0 for the IAW and 2.7 for the EPW. Preceding the telescope is a nominal (100 mm diameter x 8 mm thick) MgF_2 blast window which blocks particulate debris emanating from the laser target interaction during the shot.

2.2 Spectrometer System

The spectrometer group contains two Czerny-Turner type spectrometers that disperse the Thomson scattered light into two distinct bands, the IAW band and the EPW band. The spectrometers share the same 135 µm diameter input aperture pin hole (50 µm in the target image plane) and disperses the bands following a beam splitter that separates the light into two wavelength groups. The deep UV band 150-200 nm is reflected off the front surface and the near UV band 206-214 nm is passed through the splitter. Table 1 is a list of the associated design values for the optical system.

Table 1. Spectrometer design specifications for the optical Thomson scattering system.

Specification	IAW	EPW
Spectrometer size (meter)	0.58	0.14
Spatial extent of recording window (mm)	20	20
Desired bandwidth (nm)	4[a]	50
Wavelength band (nm)	206-214	150-200
Spectrometer resolution ($\delta\lambda/\lambda$)	0.0001	0.01
Nominal grating spacing (gr/mm)	2400	1200
Grating order	2	1
Dispersion at photocathode (nm/mm)	0.2293	4.437
Time resolution (ps)	200	200

[a] The 4 nm band is smaller than the full band. The grating will be adjusted to record any 4 nm band in the overall specified wavelength band.

An unfolded optical model describing the key parameters of the optical Thomson scattering collection system is shown in figure 2.

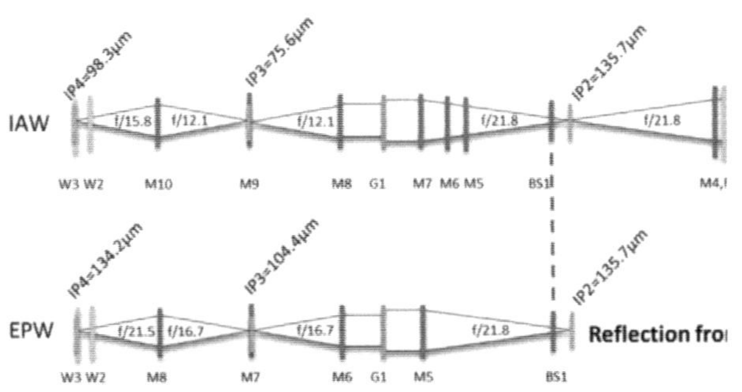

Figure 2. The unfolded optical layout of the NIF DIM based optical Thomson scattering system is described using optical component callouts following the direction of light propagation from right to left. M describes mirrors, F describes filters, BS describes beam splitters, G describes gratings, and W describes windows. The labeled f/#s are based on the beam diameter at the primary mirror (M1) (D_1 =154.3 mm). The IAW magnification is optimized for the resolution element of the streak camera recording system measurement to be a nominal ~100 μm.

The (0.14 meter) EPW spectrometer is a low resolution design with expected resolution $\delta\lambda/\lambda$ = 0.00331 and operated in first order. The output of the spectrometer is passed to an intermediate image plane that allows for wavelength masking as required. The image plane is then relayed to the input window of the streak tube for recording.

The (0.58 meter) IAW is the high resolution spectrometer design with expected resolution $\delta\lambda/\lambda$ = 0.00011 and is operated in second order. The image size on the grating also sets the temporal resolution limit of 200 ps [5]. The input light is processed from the upstream beam splitter in transmission, dispersed and then directed to an intermediate image plane where the probe wavelength line can be masked. This intermediate image plane is re-imaged onto the photocathode of the streak tube. There is an inherent delay between the two spectrometers of approximately 5 ns where the EPW is recorded first and the IAW is recording second on the same recording window. The IAW second order grating configuration was selected based on the availability of gratings with the desired grove density and blaze angle that matched the requirements for this system.

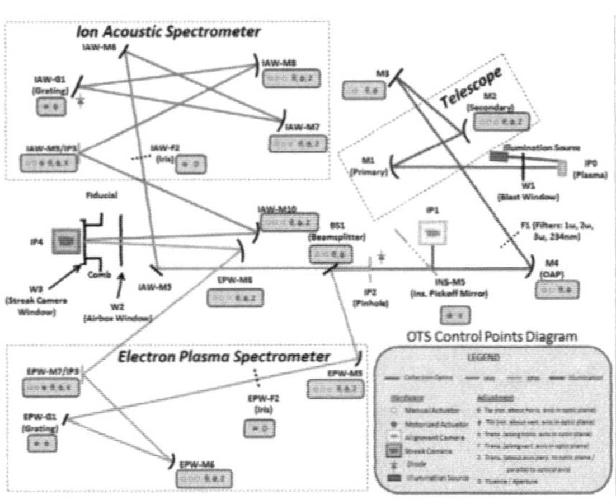

Figure 3. The folded flat model of the NIF DIM based optical Thomson optical detector system. The configuration includes the telescope, two spectrometers and optical streak camera in an airbox.

Figure 3 above describes the complete collection system. There are several undesirable sources of light that are managed in the telescope and spectrometer sections of the instrument. In the telescope section, the debris window, telescope optics and relay optics are coated to reject 1ω, and 2ω laser light. The strongest attenuation occurs at the blast shield to prevent high 1ω fluence at the secondary mirror where the light is converging before collimation. Also in the telescope are filters operating in transmission that adds additional attenuation to the 1ω, 2ω, 3ω and 3/2ω before the light enters into the pin-hole aperture of the spectrometer. The light that enters into the spectrometer is separated into two bands at the beam splitter. Light outside the two bands is further attenuated by the splitter before the light hits this grating. The light that hits the grating is dispersed and light out of band is not seen by the photocathode. Undesirable light that is inside the measurement band, is blocked by an adjustable masked the tacks independently from the grating orientation. The mask is located at the intermediate image plane (IP3) and is adjustable in the wavelength plane. This mask is designed to block the probe laser wavelength and other in band undesirable scattering features.

3. Recording System

The recording system for the Thomson scattering is an optical streak camera utilizing a Photonis, Inc. P510 sealed streak tube with an S20 photocathode deposited on a CaF_2 window. The data from each spectrometer is recorded in the same sweep window with the IAW spectral content arriving approximately 5 ns later than the EPW spectral content. Simulated response from a NIF hohlraum target has been generated using the expected system throughput, quantum efficiency and estimated back ground levels. The synthetic data described in figure 4 shows the temporal separation between the EPW and IAW spectrometers and their nominal wavelength band.

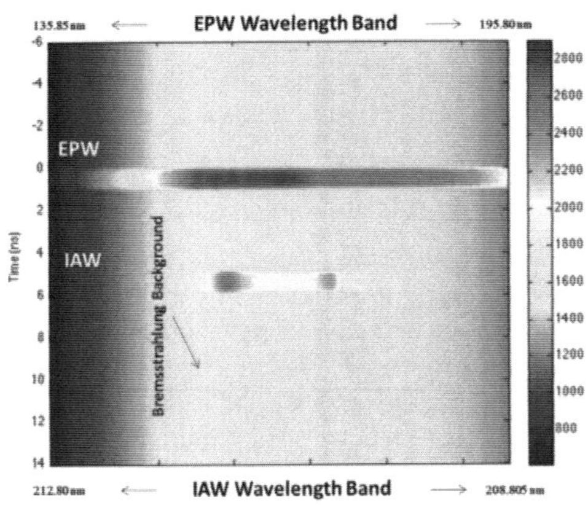

Figure 4. Simulated recorded optical Thomson scattering data as expected on the 30 ns recording window. The data includes quantum efficiency, background levels, the expected noise levels and expected system throughput.

5. Summary

An optical Thomson scattering system is being designed to measure the IAW and EPW features of various plasma environments generated at the NIF. The recording system resides inside a NIF DIM and can be operated in the equator or the polar locations. The system will operate in the deep-UV and provide temporal information during the plasma evolution. The system will be implemented in two phases, where phase I will be used to measure the NIF background environment in the deep UV and make Thomson scattering measurements using a 3ω probe. Phase II implementation shall include a UV laser operating at the 5th laser harmonic.

6. References

[1] Glenzer S, et al., *Phys Rev Lett* 79, pp 1277–1280 (1997).
[2] Froula D H, Ross J S, Divol, L and Glenzer, S, *Review of Scientific Instruments* 77, pp 10E522 (1-7) (2006).
[3] Miller G, Moses E I, and Wuest C R, *Opt. Eng* 43, pp 2841–2853 (004).
[4] Ross J S, Divol L, Froula D H and Glenzer S H, *Journal of Instrumentation*, Vol 6, pp P08004 (1-8), August 2011.
[5] Visco A, Drake R P, Froula D H, Glenzer S H and Pollock P P, 2008 *Review of Scientific Instruments* 79 pp 10F454 (1-4).

Acknowledgements

*This work performed under the auspices of the U.S. Department of Energy by Lawrence Livermore National Laboratory under Contract DE-AC52-07NA27344.

A scalable high-energy diode-pumped solid state laser for laser-plasma interaction science and applications

M De Vido, K Ertel, P D Mason, S Banerjee, P J Phillips, T J Butcher, J M Smith, W Shaikh, C Hernandez-Gomes, R J S Greenhalgh and J L Collier

Central Laser Facility, STFC Rutherford Appleton Laboratory, Didcot, OX11 0QX, United Kingdom

mariastefania.de-vido@stfc.ac.uk

Abstract Laser systems efficiently generating nanosecond pules at kJ energy levels and at multi-Hz repetition rates are required in order to translate laser-plasma interactions into practical applications. We have developed a scalable, actively-cooled diode-pumped solid state laser amplifier design based on a multi-slab ceramic Yb:YAG architecture called DiPOLE (Diode-Pumped Optical Laser for Experiments) capable of meeting such requirements. We demonstrated 10.8 J, 10 Hz operation at 1030 nm using a scaled-down prototype, reaching an optical-to-optical efficiency of 22.5%. Preliminary results from a larger scale version, delivering 100 J pulse energy at 10 Hz, are also presented.

1. Introduction

The ongoing strong effort to improve the understanding of laser-plasma interactions and to convert this knowledge into viable real-world applications has given a great impetus to the development of high-energy laser sources. These lasers have so far relied on flashlamp-pumped amplifier technology, which severely limits the repetition rate and the electrical-to-optical efficiency. This results in a limitation on the amount of science which can be carried out and prevents the translation of laser-plasma interactions into practical applications. An innovative approach, based on diode-pumped solid-state laser (DPSSL) technology, allows overcoming these limitations and is currently under development within the framework of the DiPOLE project at STFC's Central Laser Facility. DPSSL systems can be either used directly to compress matter to extreme densities or to drive high-energy femtosecond laser chains for the generation of laser-driven secondary sources of particles (electrons, protons) or photons (from THz to gamma), making them suitable for a wide range of laser-plasma applications such as inertial confinement fusion, intense X-rays generation and particle acceleration research. In this paper, results from a cryogenically-cooled, multi-slab DPSSL amplifier, delivering 10J pulses at 10 Hz repetition rate, will be presented along with initial results from a 100 J, 10 Hz system under development.

2. DiPOLE: 10 J, 10 Hz prototype amplifier

The DiPOLE amplifier concept, already extensively described in previous publications [1, 2], is based on a diode-pumped, cryogenically cooled, multi-slab ceramic Yb:YAG scalable architecture designed for amplifying 1030 nm pulses with a time duration between 2 ns and 10 ns to kJ-level energies at multi-Hz repetition rates. A scaled-down prototype, designed to produce 10 J pulses at 10 Hz repetition rate with a target optical-to-optical efficiency of 25%, has been built to test the viability of DiPOLE concept and is used as a test-bed for further developments [3].

2.1. Energy and beam quality

As shown in figure 1, at 10 Hz repetition rate and at a temperature of 140 K, the system amplified 10ns pulses to an energy of 10.8 J for a pump pulse energy of 48 J, corresponding to an optical-to-optical conversion efficiency of 22.5% and an electrical-to-optical efficiency of 12.8%. To the best of our knowledge, this is the highest energy level achieved by a cryogenically-cooled Yb-doped laser and the highest efficiency achieved by a multi-J DPSSL system. However, while the specification on output energy was fulfilled, the design target of 25% optical-to-optical efficiency has not been met yet. Design modifications have been put in place to reach higher efficiency values on future systems.

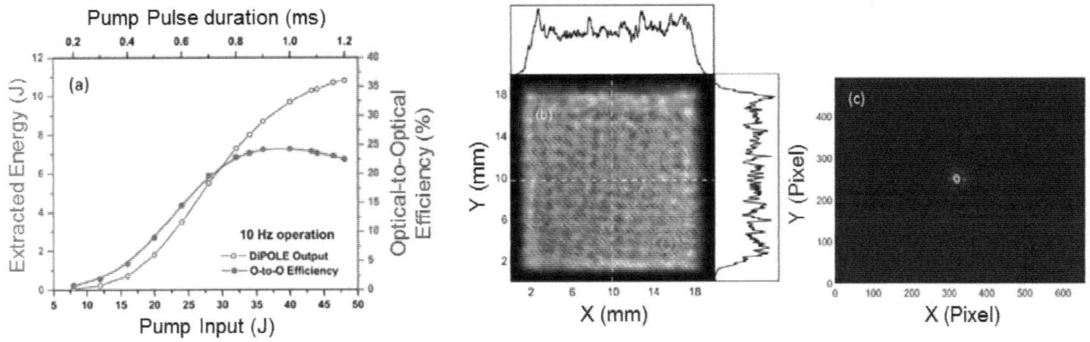

Figure 1. (a) DiPOLE performance at 10 Hz, 140 K: output energy, optical-to-optical efficiency of the system. Near-field (b) and far-field (c) profiles measured at 10 Hz and 10.8 J.

Figure 1 also shows the near-field (b) and the far-field (c) beam profiles. Comparison with a predicted diffraction limited spot-size showed that the far-field spot is 2 and 1.9 times diffraction limited in the x and y-axes respectively. Inhomogeneity on the near-field profile is partly due to diffraction effects caused by optical filters used in the beam diagnostic system.

2.2. Long-term stability tests

In order to validate the long-term operation of the DiPOLE amplifier at the design baseline fluence of 2 J/cm^2 chosen for a 100 J upgrade, the system was operated at 10 Hz repetition rate and at a pulse energy of 7 J for more than 50 hours (figure 2). Data was collected in runs lasting from 4 to 6 hours. Shot-to-shot energy stability over this period was 0.85% rms.

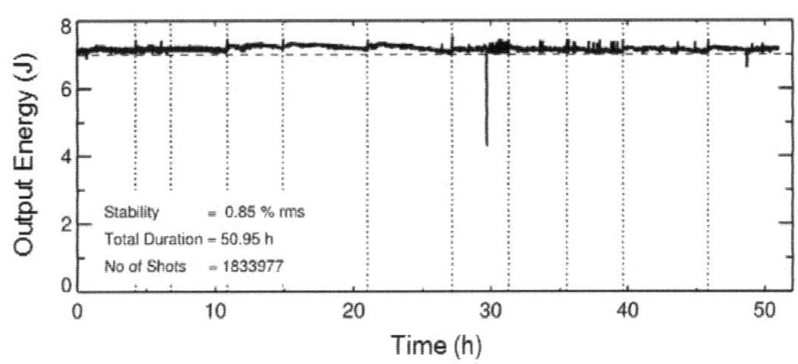

Figure 2. Output energy during long-term operation of the DiPOLE amplifier at 7 J, 10 Hz. Vertical lines denote start and finish of individual runs.

Long-term and short-term beam pointing stability plots are shown in figure 3(a) and 3(b), respectively. While short-term beam pointing stability was affected by random fluctuations characterised by an amplitude smaller that the diffraction limit of the system, a drift in the beam pointing was observed during long-term operation. This drift could be easily compensated by active beam stabilisation systems, a measure which will be implemented both in DiPOLE and in new systems.

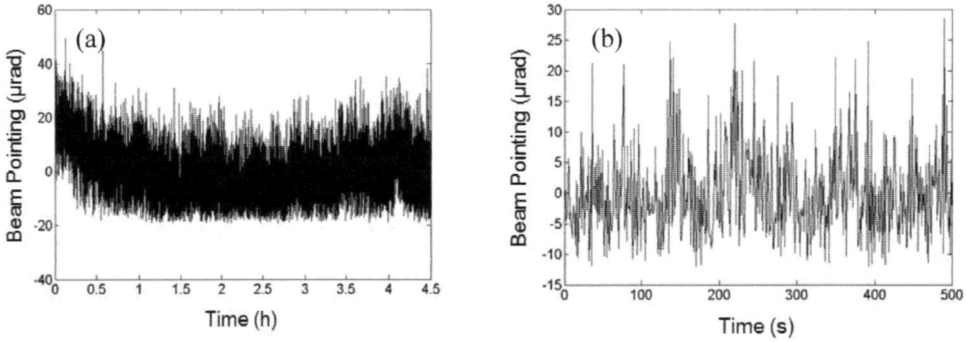

Figure 3. Long-term (a) and short-term (b) beam pointing stability at 10 Hz, 7 J operation.

2.3. *Second harmonic generation results*

Second harmonic generation (SHG) tests were carried out to assess the suitability of the system for pumping femtosecond laser chains. An LBO crystal in Type-I configuration was employed. As shown in figure 4(a), 82% conversion efficiency was achieved at 10 Hz, 6.25 J operation.

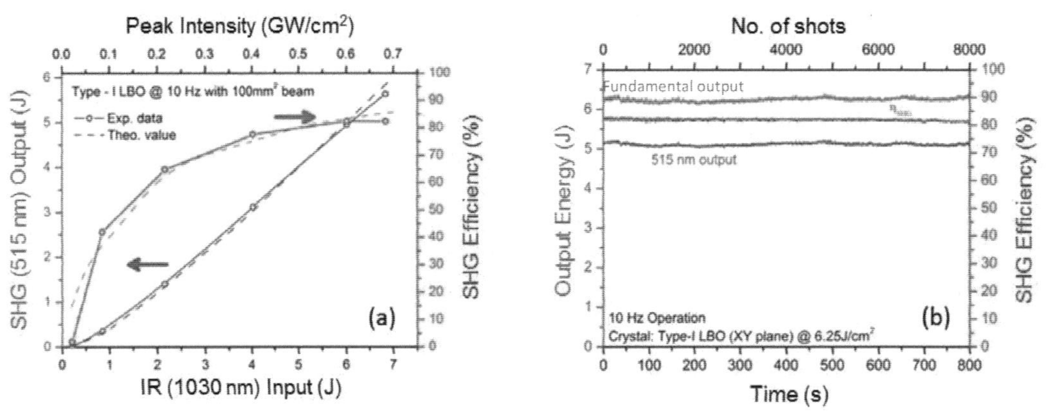

Figure 4. (a) Energy of second harmonic beam and conversion efficiency. (b) Long-term operation.

The system was operated for 8000 shots without any damage to the LBO crystal (Figure 4(b)). The energy stability for the second harmonic was 1.4% rms.

3. DiPOLE100: 100 J, 10 Hz upgrade

Results shown above have confirmed the validity of the DiPOLE concept, giving confidence for energy up-scaling. A new system, under construction at the CLF for the HiLASE project in the Czech Republic [4] and described in previous publications [5], is designed to amplify 1030 nm pulses, with super-Gaussian (order ≥ 8) square spatial profile and arbitrarily shaped temporal profile between 2 and 10 ns, to an energy of 100 J with a repetition rate up to 10 Hz. The architecture is similar to DiPOLE with front-end, 10 J stage and additional 100 J stage.

Journal of Physics: Conference Series **717** (2016) 012090 doi:10.1088/1742-6596/717/1/012090

3.1. Initial results

First amplification tests with the 10 J stage were carried out at a repetition rate of 1 Hz and a seed energy of 33 mJ from the front-end. Figure 5(a) shows output energy as a function of pump pulse duration. The maximum energy obtained was 9.1 J with a pump energy of 33 J. This equates to an optical-to-optical efficiency of 27.5%, a significant improvement over DiPOLE, achieved through a higher peak pump power and reduced passive losses.

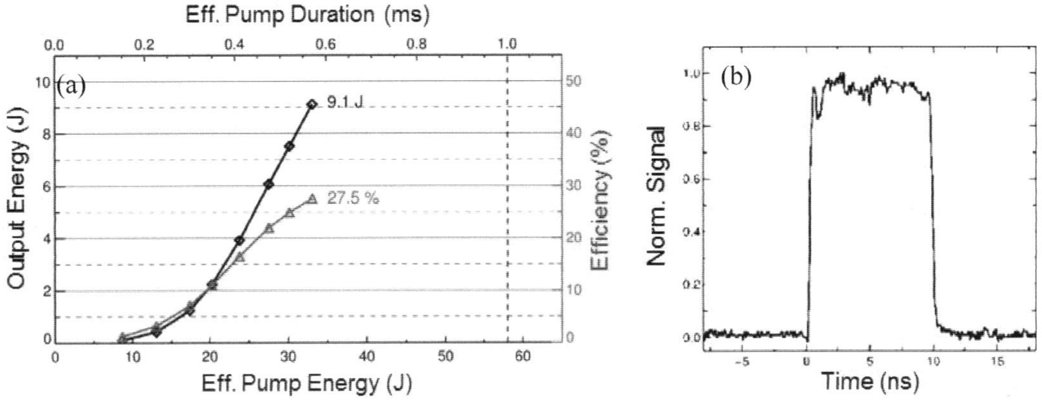

Figure 5. (a) Output energy and optical-to-optical efficiency of the 10 J amplifier operated at 1 Hz repetition rate at different pump pulse energies. (b) Temporal pulse profile at 9.1 J.

4. Conclusions

The validity of DiPOLE concept has been demonstrated using a scaled-down 10 J, 10 Hz prototype. A 100 J, 10 Hz system is under construction at STFC's Central Laser Facility. While the architecture of the new system is similar to the scaled-down prototype, great effort has been devoted to improve the performance and the reliability of the system. Initial results from the 10 J stage confirmed a relevant increase in optical-to-optical efficiency compared to the prototype. Moreover, while the output energy was limited to 9 J to minimise risk of laser-induced damage to the gain medium, the system has the potential to produce higher energies since more than 58 J pump energy is available. Advanced processing techniques for ceramic Yb:YAG are currently being investigated to increase gain media resilience to intense laser irradiation for reliable operation at higher fluence levels [6]. Initial results described in this paper confirm the effectiveness of amplifier architecture design improvements and indicate that demonstration of the 100 J target specification will be achieved in the near future.

References

[1] Ertel K, Banerjee S, Mason P D, Phillips P J, Siebold M, Hernandez-Gomez C and Collier J L 2011 *Optics express* **19**.27, 26610-26

[2] Mason P D *et al* 2015 *Applied Optics* **54**.13, 4227-38

[3] Banerjee S, Ertel K, Mason P D, Phillips P J, De Vido M, Smith J M, Butcher T J, Hernandez Gomez C, Greenhalgh R J S and Collier J L 2015 *Optics Express* **23**.15 19542-51

[4] Divoky M *et al* 2014 *High Power Laser Science and Engineering* **2**: e14

[5] Mason P *et al* 2015 *SPIE Optics + Optoelectronics* International Society for Optics and Photonics 951302-951302-9

[6] De Vido M *et al* 2014 *SPIE Laser Damage* International Society for Optics and Photonics 92371M-92371M-8

Journal of Physics: Conference Series **717** (2016) 012091 doi:10.1088/1742-6596/717/1/012091

Development of Enhanced, Permanently-Installed, Neutron Activation Diagnostic Hardware for NIF

E R Edwards[1], D R Jedlovec[2], J A Carrera[2], and C B Yeamans[2]

[1]University of California, Berkeley, Berkeley, CA, USA
[2]Lawrence Livermore National Laboratory, Livermore, CA, USA

E-mail: edwards76@llnl.gov

Abstract. Neutron activation diagnostics are baseline neutron yield and flux measurement instruments at the National Ignition Facility. Up to 19 activation samples are distributed around the target chamber. Currently the samples must be removed to be counted, creating a 1-2 week data turn-around time and considerable labor costs. An improved system consisting of a commercially available $LaBr_3$(Ce) scintillator and Power over Ethernet electronics is under development. A machined zirconium-702 cap over the detector is the activation medium to measure the ^{90}Zr(n,2n)^{89}Zr reaction. The detectors are located at the current neutron activation diagnostic sites and monitored remotely. Because they collect data in real time yield values are returned within a few hours after a NIF shot.

1. Introduction

Neutron Activation Diagnostics (NADs) are used at the National Ignition Facility (NIF) to measure the neutron yield by activating a material through neutron reactions and measuring the γ-rays from the subsequent decay [1]. 19 Flange-NADs (FNADs) are distributed around the NIF target chamber to measure the spatial distribution of the yield. The FNADs are Zr pucks that measure the neutron flux at each location using the reaction ^{90}Zr(n,2n)^{89}Zr. The ^{89}Zr has a 78.41 hr half life [2] and emits a 909 keV γ-ray. This reaction has a 12.1 MeV threshold and therefore is sensitive only to 14.1 MeV primary DT neutrons. Currently they must be removed from the NIF target bay after each shot for counting, which is both time and labor intensive.

Two $LaBr_3$(Ce) scintillation detectors (Figure 1) were placed in the NIF target chamber at two of the current FNAD locations (Figure 2) [3]. Each detector receives its power from a Power over Ethernet (PoE) cable, which also transfers the data to the control room. The data is collected in list mode, which means that each event is recorded as a channel (energy) with a timestamp. Instead of a Zr puck, a Zr cap is placed over the crystal. The first detector has been in the target bay for 18 high-yield and many low-yield shots as of November 2015 and shows no signs of performance degradation from radiation damage. A yield measurement is returned in less than 8 hours.

2. ^{89}Zr ground state

Histograms are made by summing the number of events in each channel within a set time range (Figure 3). The time bin is chosen depending on the half life of the isotope of interest (30 minutes for ^{89}Zr). Many of these histograms are created and stacked together to show the spectrum as a function of time (Figure 5a). Energy and time are the axes and the colors represent the number of counts in each energy/time bin. Since all of the original data remains unchanged, shorter or longer time intervals can be used in the post-processing to highlight certain features of the spectrum.

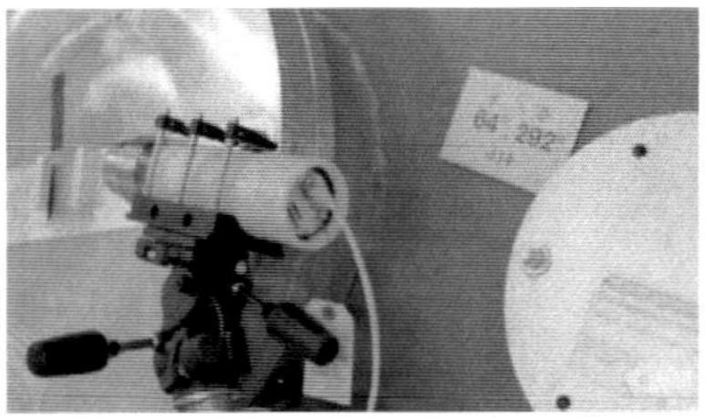

Figure 1. Lanthanum bromide detector at an FNAD port [4].

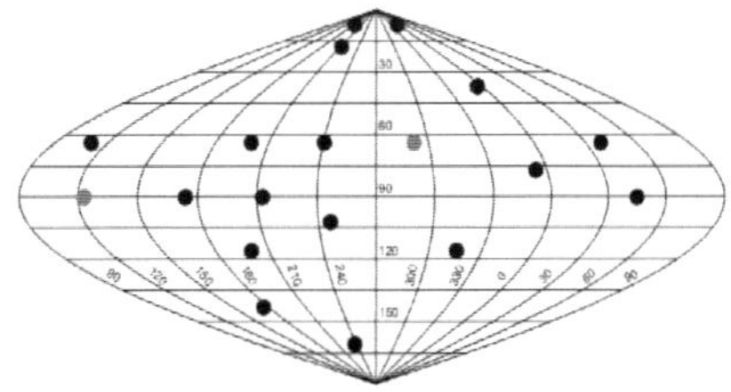

Figure 2. Current locations of the 19 Flange-NADs. The two prototype locations are shown in red.

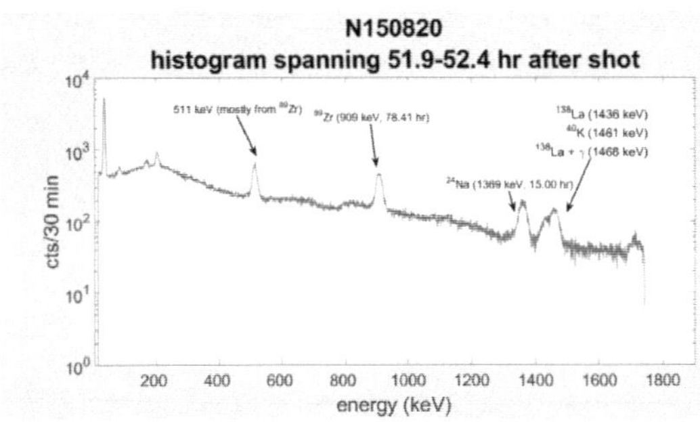

Figure 3. A histogram spanning a 30-minute interval starting 51.9 hr after N150820-003. Many of these histograms are stacked in time to form Figure 5a.

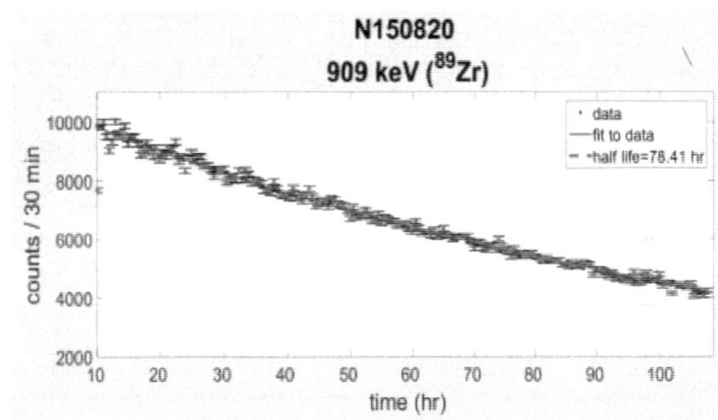

Figure 4. An energy range can be selected for a region surrounding a particular photopeak. Then the counts in the peak at each time interval are plotted as a function of time, and fit with an exponential. The resulting equation can be used to extrapolate the activity at shot time, which is used to find the yield.

The isotopes are identified by the energy and half lives of the photopeaks. To find the half life, an energy range surrounding the 909 keV peak is selected and the peak is fit. The peak area is plotted in time and an exponential decay is fit (Figure 4). The known ^{89}Zr half life of 78.41 hr is used, so the fit value is only a check. To calculate the neutron yield, the decay curve of ^{89}Zr is extrapolated to time zero using time bins of 30 minutes. Since the calibration is linear (Eq. 1 in [1]), the equation of the line can be used to convert counts to yield (Figure 8). The real-time system has shown that it is a viable replacement for a single FNAD position.

3. ^{89m}Zr: Neutron Isomeric Ratio Detector (NIRD)

Neutrons can also activate ^{90}Zr to form the metastable state ^{89m}Zr, which decays primarily by a 588 keV γ-ray with a half life of 4.18 minutes (Figures 9 and 10). It is not possible to observe it with the current FNAD system because the ^{89m}Zr has completely decayed by the time the FNAD pucks are recovered. It is sometimes observed with the $LaBr_3$ detectors. After each shot the activation of materials in the target chamber and electrons from β^- decays from activated La and Br within the detector create a high background that dominates the spectrum until it decays. After higher-yield ($>10^{15}$) shots the background is so high that none of the 588 keV γ-ray is visible. By the time the background has decayed, ^{89m}Zr has also completely decayed. However, part of the ^{89m}Zr decay is visible after medium-yield ($1\text{-}5*10^{14}$) shots (Figure 6) and when the detector is placed far (28 m) from the neutron source in a shielded alcove (Figure 7). Because the cross section for producing ^{89}Zr and ^{89m}Zr depends on neutron energy, the observed ^{89m}Zr/^{89}Zr ratio could be used to determine the bulk motion of the neutron source [5]. However, the high background prevents a precise measurement of ^{89m}Zr at this time.

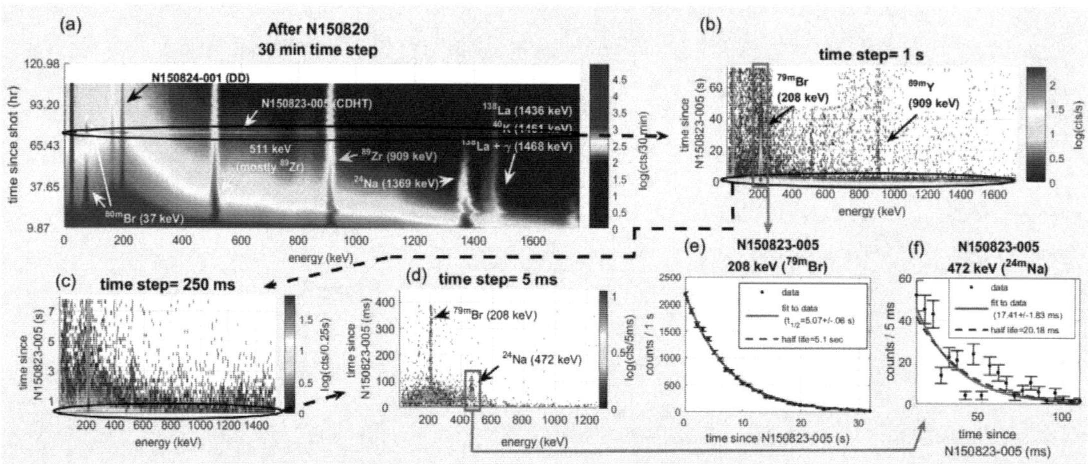

Figure 5. The detector continued running many days after N150820-003 and observed shots N150823-005 and N150824-001. Since the data is stored as a list of times and channels, it can be processed using varying time bins. Here the time bins were used to focus on N150823-005. By taking smaller and smaller time steps, different isotopes are observed (5a-5d). In 5a, the post-shot background has decayed enough so the natural radioactivity of ^{138}La in the detector is visible. 5e and 5f show the decays from ^{89m}Br and ^{24m}Na that can be observed.

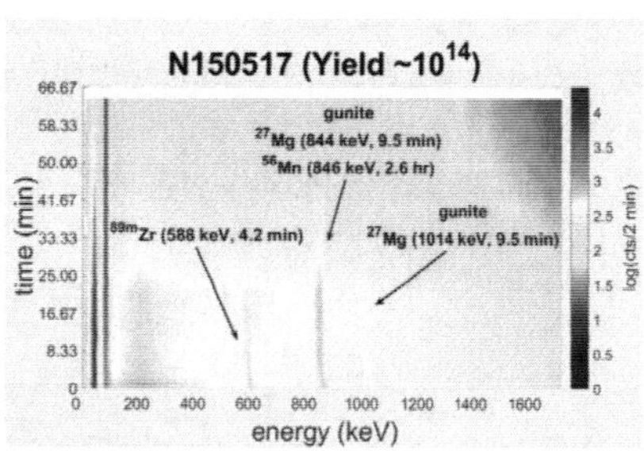

Figure 6. Example of ^{89m}Zr visible after a medium-yield shot, with detector on an FNAD port.

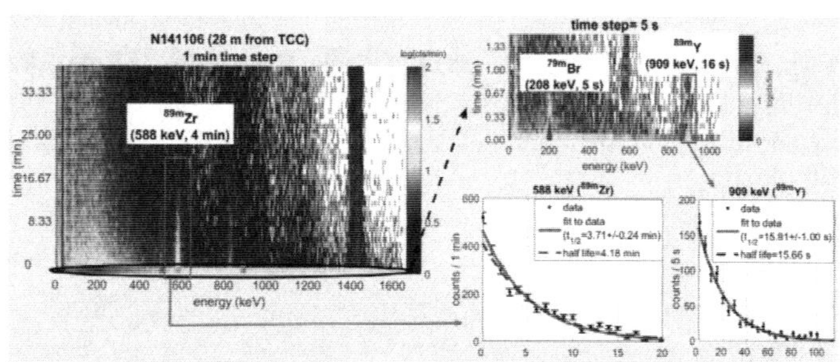

Figure 7. The background in the neutron alcove at 28 m and behind a shield wall is much lower, making it possible to see short-lived isotopes.

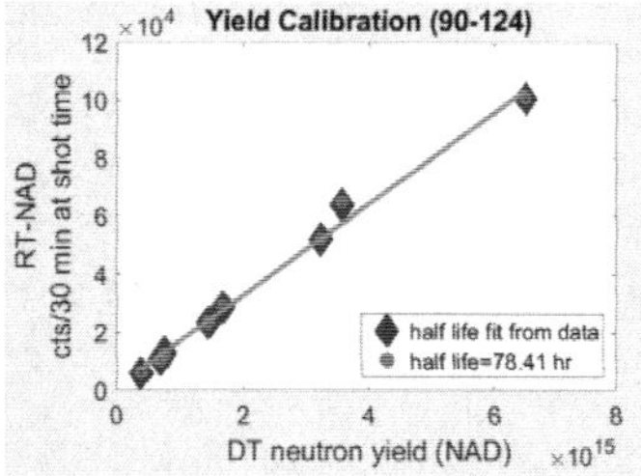

Figure 8. RT-NAD response calibration to yield.

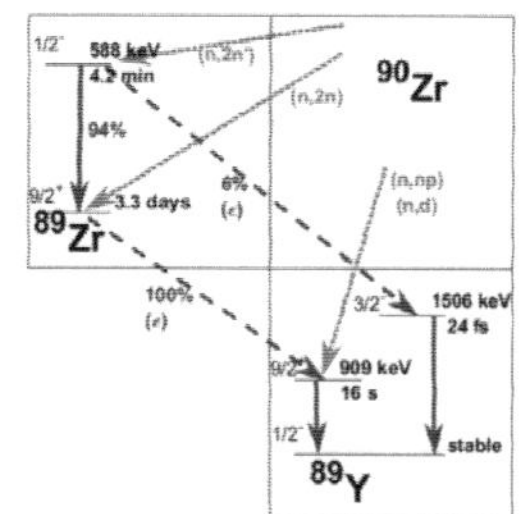

Figure 9. Reactions and decays in Zr and Y.

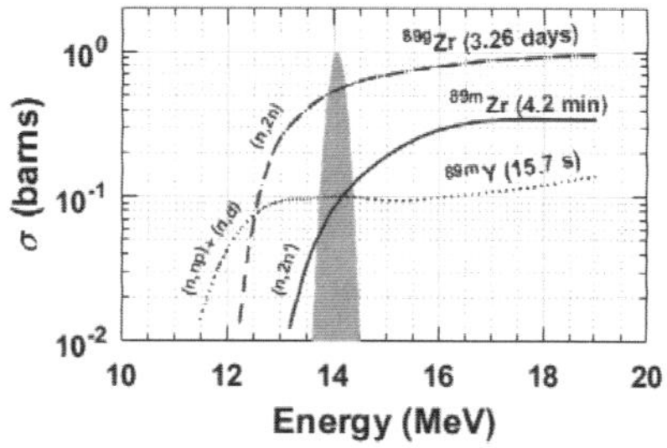

Figure 10. Neutron reactions on ^{90}Zr with DT neutron spectrum for reference. Cross sections from [6]-[10].

4. Short-lived isotopes: ^{89m}Y, ^{24m}Na, ^{79m}Br

Short-lived isotopes are observed when the detector is turned on before they have completely decayed, which for the shortest ones means the detector has to run through the shot. Furthermore, the background level must be low enough to see the signal. So far the short-lived isotopes have not been analysed in depth but they have the potential to be used to resolve the low-energy neutron spectrum. For example, the threshold for the ^{27}Al(n,α)^{24}Na reaction is ~4.5 MeV but ^{79}Br(n,n')^{79m}Br is only ~210 keV. Reactions with low enough thresholds can also be used on DD shots to estimate the yield. A shot with mid-10^{14} yield or higher causes a recoverable upset, and the detector cannot be restarted in time to catch the shortest-lived isotopes. Lower yield shots usually do not upset the detector.

The 909 keV γ-ray associated with ^{89}Zr decay is the relaxation of ^{89m}Y, which has a 16 s half life. Compared to the decay time of ^{89}Zr the half life of ^{89m}Y is negligible. The decay of ^{89m}Y produced directly by ^{90}Zr(n,np)^{89m}Y or ^{90}Zr(n,d)^{89m}Y (Figures 9 and 10) can be observed when it is produced in a short time (Figures 5b and 7).

One example of running through shots occurs after N150820-003 (Y_n=7*10^{14}). The detectors were restarted after the DT shot but ran through N150823-005 (3 days later), which had a DT yield of ~10^{12}, and N150824-001 (4 days later, Y_{DD}=5.2*10^{10}). N150823-005 was an opportunity to observe very short-lived isotopes since the activation background was low. ^{24m}Na ($t_{1/2}$=20 ms) is produced by the (n,α) reaction on ^{27}Al.

^{79}Br is 51% of the natural Br that makes up the scintillator crystal. Inelastic neutron scatter reactions with thresholds above 210 keV create ^{79m}Br with a 5.1 s half life. This reaction has been observed for low yield shots and with the detector 28 m from TCC. Both ^{24m}Na and ^{79m}Br are good examples of the ability to see isotopes with extremely short half lives. N150823-005 also demonstrates the usefulness of being able to run constantly and to analyze the data any time post-shot.

5. Conclusions

The primary goal of developing a more efficient and functional alternative to FNADs is achieved with RT-NAD. A linear calibration between DT neutron yield and RT-NAD response has been demonstrated. Furthermore, very short-lived isomers are observed under certain conditions and have the potential to become additional diagnostics. There is a greater choice in material since the half life does not have the limitation of the current FNADs. Operation is more efficient since they do not have to be removed to be counted, and a yield value is returned within hours instead of days. In the future the cap can be replaced with a different material to study other neutron reactions: for example, to measure DD and TT yields.

References

[1] Bleuel D *et al* 2012 *Rev. Sci. Instrum.* **83** 10D313
[2] Singh B 2013 *Nuc. Data Sheets* **114** 1-208
[3] Jedlovec D, Edwards E, Carrera J and Yeamans C 2015 *Proc. SPIE 9591, Target Diagnostics Physics and Engineering for Inertial Confinement Fusion IV* vol 9591 ed J Koch and G Grim (San Diego: SPIE)
[4] Crane R 2015 *CNN* http://www.cnn.com/videos/tech/2015/10/02/orig-zp-nuclear-fusion-power-pioneers.cnn/video/playlists/pioneers-orig/
[5] Yeamans C, Bleuel D and Bernstein L 2012 *Rev. Sci. Instrum.* **83** 10D315
[6] Semkova V, Bauge E, Plompen A and Smith D 2010 *Nuc. Phys.* A **832** 149–69
[7] Chadwick M *et al* 2011 *ENDF/B-VII.1*
[8] Kanda Y 1972 *Nuc. Phys.* A **185** 177-195
[9] Abboud A, Decowski P, Grochulski W, Marcinkowski A, Piotrowski J, Siwek K and Wilhelmi Z 1969 *Nuc. Phys.* A **139** 42-56
[10] Pavlik A, Winkler G, Vonach H, Paulsen A and Liskien H 1982 *J. Phys.* G **8** 1283-1299

Journal of Physics: Conference Series **717** (2016) 012092 doi:10.1088/1742-6596/717/1/012092

Improved pulse contrast on the Texas Petawatt Laser

E Gaul, T Toncian, M Martinez, J Gordon, M Spinks, G Dyer, N Truong, C Wagner, G Tiwari, M E Donovan, T Ditmire and B M Hegelich

Center for High Energy Density Science, C1510, The University of Texas at Austin, Austin, Texas 78712, USA

gaul@physics.utexas.edu

Abstract. We have completed a pulse contrast upgrade on the Texas Petawatt Laser. This improvement enables the use of thin and reduced mass targets for ion acceleration, and reduces pre-plasma effects on all experiments. The new design starts with two BBO-based OPCPA stages pumped by an optically synchronized 8-ps laser. These stages amplify slightly chirped few ps pulses by six orders of magnitude. Next there are two LBO-based OPCPA stages that are pumped by 4 ns pulses. With much less gain than before, parametric fluorescence has been reduced by about three orders of magnitude. Prior to the upgrade, lenses caused pencil beam prepulses. Since tilting or wedging lenses was not a viable option, we replaced all lenses in the glass amplifiers with off-axis parabolic mirrors. There are still weak prepulses that we attribute to surface scattering. We eliminated thin transmissive optics to avoid post pulses that would result in prepulses by nonlinear (B-integral) conversion. This required us to reduce from eight to four passes in the 64-mm glass amplifier and to add a two-pass 25-mm "booster amplifier." As a final upgrade we added an Acousto-Optic Programmable Dispersive-Filter (AOPDF) to improve higher order dispersion and steepen the rising edge of the compressed pulse.

1. Motivation for improving contrast

The Texas Petawatt Laser [1] is a university research laser used for a variety of experiments that require ultrahigh electromagnetic fields, extreme laser intensities, and extreme temperatures and pressures. The laser facility has two target chambers: Target Chamber 1 (TC1) that uses fast focus optics (f/3 and soon f/1 off-axis parabolic mirrors) to achieve maximum intensities, generally for solid targets, and Target Chamber 2 (TC2) that has a slower focus f/40 mirror most often used for larger volume interactions with gas targets. Both target chambers have been very productive as high energy particle sources. TC1 produced the brightest ultrashort pulsed neutron source yet measured ($>10^{18}$ n/cm^2/s in a sub 50-ps pulse) [2]; the highest measured positron-to-electron ratio pair creation in a solid (~50% in a Pt rod) [3]; and high energy (~45 MeV pre-upgrade) proton yields. TC2 has yielded 2×10^7 fusion neutrons from deuterium clusters [4]; high charge (nC), low divergence (sub-mrad), quasi-monoenergetic high energy (2-3 GeV) laser wakefield accelerated electrons [5]; high brightness betatron x-rays [6]; and $\sim10^4$ muons from wakefield accelerated electrons hitting a tungsten converter.

All of the results mentioned could potentially benefit from higher laser main-pulse to prepulse contrast, and in fact ion acceleration has already shown improvement in initial experiments after the upgrade. For laser interactions with solid targets, sufficiently intense laser prepulses create preformed surface plasma which can affect the nature of the laser-target interaction. In the case of thin or mass-limited targets, which are known to produce the highest ion energies [7], the target may not even survive the prepulse. For protons and neutrons produced by cluster fusion, the yield can be reduced by a prepulse destroying the clusters. For wakefield acceleration, the impact of a prepulse on the gas cell is unclear and difficult to account for theoretically.

In 2014-2015, the Defense Advanced Research Project Agency funded a year-long Texas Petawatt Laser upgrade to advance the science and technology of neutron generation for radiography and other applications. With better contrast, the expectation was that ions of laser breakout afterburner ion acceleration [8] at higher energies and fluence create more neutrons from a knock-on target.

2. Pre-upgrade prepulse measurement and analysis

Prior to the upgrade, the Texas Petawatt Laser's chirped, broadband, 1054-nm seed was being amplified via three stages of optical parametric chirped pulse amplification (OPCPA) in BBO and YCOB, follow by 8-pass amplification in a 64 mm silicate Nd:glass rod, followed by 4-pass amplification in a 315 mm phosphate Nd:glass slab amplifier. The laser had typically been operated between 100 J and 180 J pulse energies, with pulse duration FWHM of 150-170 fs.

Parametric fluorescence from the parametric amplifiers showed up in the region ±4 ns from peak signal (Figure 1) as a pedestal, consistent with having a 4-ns OPA pump pulse. Depending on the amount of saturation and gain in a given stage, the amplitude of the prepulse pedestal could vary, and data to that effect is shown later for the post-upgrade performance. It is important to realize that the OPCPA fluorescence pedestal is not chirped and will not compress—thus the intensity contrast is four orders of magnitude better than the energy contrast seen in the photodiode measurement.

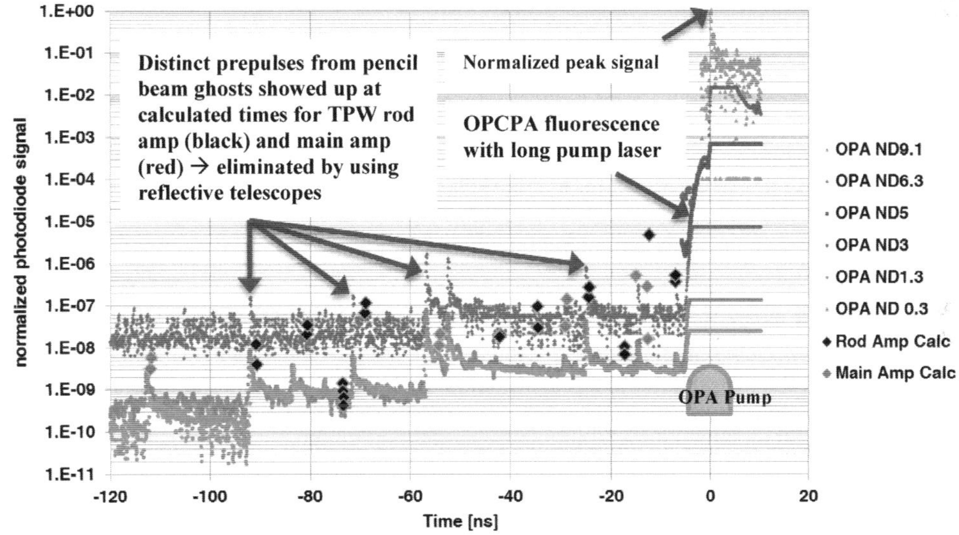

Figure 1. Laser pulse energy versus time as measured by a photodiode placed after the target chamber focus. This measurement was taken with various neutral density filters to get the needed dynamic range. Calculated prepulse arrival time and energy (diamonds) is consistent with the measurement.

The photodiode measurements of Figure 1 show discrete prepulses between the OPCPA fluorescence pedestal (-4 ns) and about -112 ns. These include the prepulses that come from multiple reflections in optics that produce ghost foci and pencil beams [9]. We have calculated the expected arrival time and amplitude of pencil beam prepulses, and can with good accuracy attribute the measured prepulses to a distinct set of reflections. Since the pencil beams are not relay imaged to the target, they will not focus at the target and will have better intensity contrast than the measured energy contrast.

Discrete prepulses can also arise from diffuse surface scattering at just the right location to propagate forward along the beam path; this prepulse source is uncorrected and we believe it to be the source of remaining prepulses. Unlike the main pulse, such a prepulse would fill the aperture of the downstream spatial filters, and thus would have better (higher) intensity contrast than energy contrast.

3. Nanosecond prepulse suppression

To eliminate lenses as a source of prepulses, we replaced all of the lenses beyond the OPCPA stages with off-axis parabolic (OAP) mirrors (Figure 2). Tilting or wedging lenses is not possible in a broadband short pulse system because of unacceptable chromatic aberration and wavefront errors. Removing the relay lens in the 315-mm power amplifier presented some geometrical challenges. We ended up using a 45° OAP reflecting vertically onto a fold mirror in a periscope configuration, which sent the laser pulse into a raised set of horizontal amplifier heads. The vacuum window leading into the main amplifier was optimized in tilt angle and thickness to avoid secondary reflections. We carefully designed mounts as the mirrors are less forgiving of pointing errors than lenses.

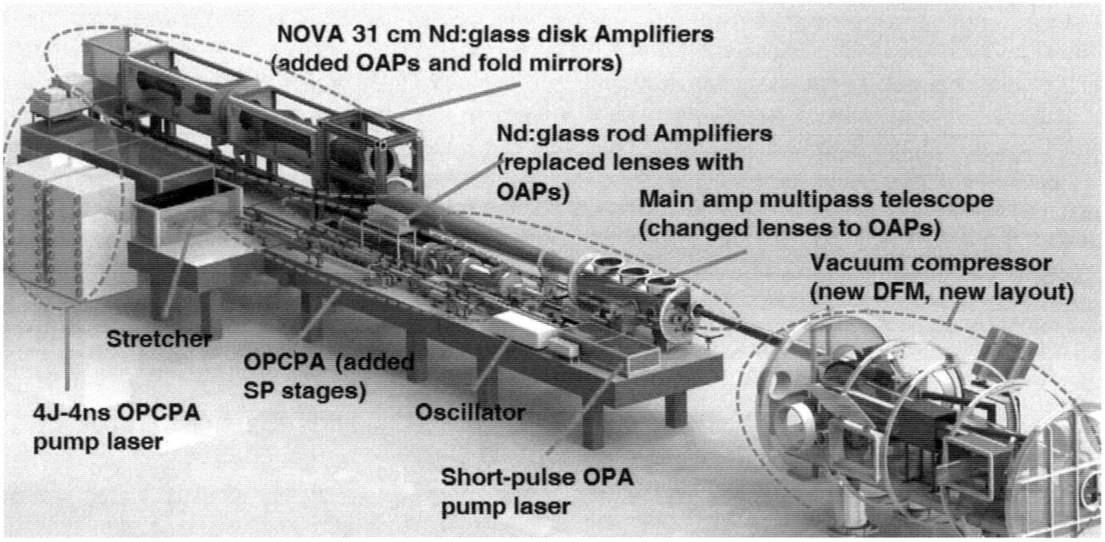

Figure 2. Rendering of the Texas Petawatt Laser upgrades.

4. Sub-nanosecond prepulse pedestal suppression

From -4ns until peak pulse energy we have at issue the parametric fluorescence pedestal and higher order (typically 4th order) uncompensated dispersion. In past experiments amplified spontaneous emission from both glass amplifiers had no effect on 400 nm thick targets and seemed negligible.

We addressed the fluorescence pedestal by implementing two "short-pulse OPCPA" stages using design principles described in [10]. We achieve a total gain of 10^6 in two BBO crystals each pumped by 5mJ, 8ps, 527nm pulses from an optically synchronized Nd:YLF laser (Ekspla). The amplified 8-ps pulse is then stretched to 4 ns and amplified in two LBO-based OPCPA stages. The net result is that the first two OPCPA stages only contribute to a pedestal within 8 ps of the main pulse arrival, and with a stronger seed, the 4-ns OPCPA gain and fluorescence are reduced by three orders of magnitude.

To correct higher order dispersion and steepen the rising edge of the pulse, we added an AOPDF (Dazzler from Fastlite) in the laser chain acting both as a picosecond stretcher and to pre-compensate for the residual spectral phase errors after the final pulse compression. The oscillator pulse is stretched by the AOPDF to approximately 2 ps by adding 60,000 fs^2 dispersion to the pulse. The fine-tuning of the compression at the end of the laser chain is done employing a closed loop between the AOPDF and a self-referring spectral interferometer (Wizzler-1030 from Fastlite) resolving the spectral phase. To minimize postpulses and nonlinear conversion to prepulses we eliminated all waveplates, minimized all double reflections in transmissive optics and added a booster rod amplifier to compensate the gain.

5. Contrast measurements on the upgraded Texas Petawatt Laser

Photodiode measurements show prepulses at -85 ns and -38.5 ns, which correspond to the round trip times in the main amplifier and rod amplifier, respectively (Figure 3). We believe these prepulses to be due to scattering into the spatial filter pinholes at angles that propagate through the system.

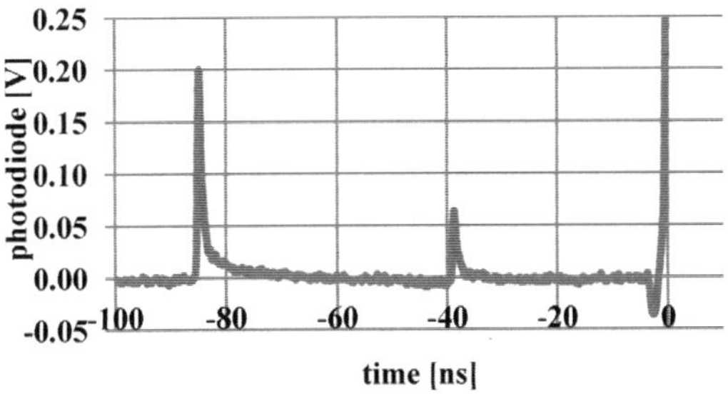

Figure 3. Measured pulse energy at target chamber focus from the upgraded laser as a function of time. The energy contrast is 3 x 10^8 for the -85 ns pulse. The noise floor is better than 10^{11}, and no other prepulses are evident. The estimated intensity contrast is 10^2-10^3 times better than the energy contrast due to different glass dispersion and lack of focusability.

Intensity contrast has been measured with a third-order autocorrelator. Figure 4 shows how the parametric florescence pedestal is affected by various levels of OPA saturation, and it shows, not unexpectedly, that the pedestal depends of how hard the crystals are pumped. This set of data was taken prior to optimizing the spectral phase of the laser pulse. Figure 5 shows the final contrast after optimization of the pulse duration, resulting in pedestal contrast of 5×10^8 at hundreds of ps delay, about three orders of magnitude in contrast better that prior to the upgrade.

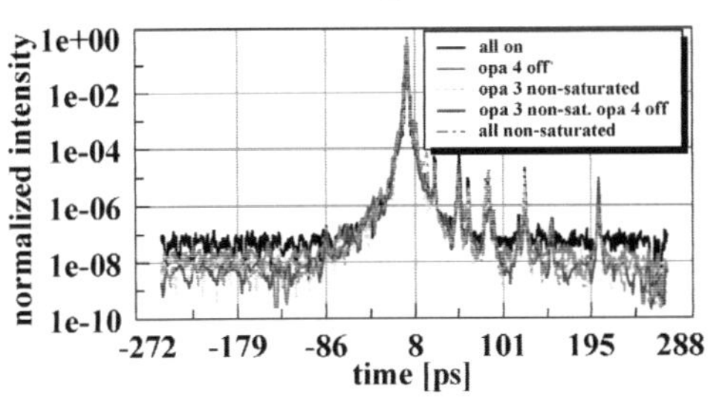

Figure 4. Long pulse OPA saturation affects the pedestal contrast.

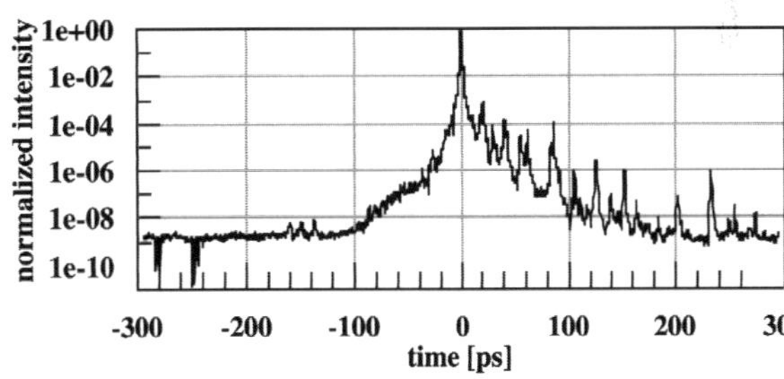

Figure 5. Picosecond intensity contrast is $\sim5\times10^8$. The small prepulses and the post pulses come from the measurement device.

We have added a second deformable mirror which has greatly improved the focus (typical Strehl ratio > 0.6) and improved operations. In June 2016 we will add an f/1 OAP, which we expect to take us to intensities over 10^{22} W/cm^2 and make the prepulse improvement even more relevant.

Acknowledgements
This work was supported by Defense Advanced Research Project Agency Contract 12-63-PULSE-FP014, and by National Nuclear Security Administration Cooperative Agreement DE-NA0002008.

References

[1] Gaul E *et al* 2010 *Appl.Opt.* **49** 1676 doi: 10.1364/AO.49.001676
[2] Pomerantz I *et al* 2014 *Phys. Rev. Lett.* **113** 184801
[3] Liang E *et al* 2015 *Sci. Rep.* **5** 13968 doi: 10.1038/srep 13968
[4] Bang W *et al* 2013 *Phys. Rev.* E **87** 023106; Bang W *et al* 2013 *Phys. Plas.* **20** 093104
[5] Wang X *et al* 2013 *Nat. Commun.* **4** doi: 10.1038/ncomms2988
[6] Wang X *et al* 2014 *CLEO* Paper FM1B.2 doi:10.1364/CLEO_QELS.2014.FM1B.2
[7] Hegelich BM *et al* 2013 *New J. Phys* **15** 085015
[8] Roth M *et al* 2013 *Phys. Rev. Lett.* **110** 044802; Jung D *et al* 2013 *Phys. Plas.* **20** 056706
[9] Murray J *et al* 1995 *Proc. SPIE* 2633 608–14 doi:10.1117/12.228267
[10] Dorrer C *et al* 2008 *CLEO/QELS* Paper JTuA45

A demonstration of ultra-high time resolution with a pulse-dilation photo-multiplier

J D Hares[1], A K L Dymoke-Bradshaw[1], T J Hilsabeck[2], J D Kilkenny[2], D Morris[2], C J Horsfield[3], S G Gales[3], J Milnes[4], H W Herrmann[5] and C McFee[6]

[1]Kentech Instruments Ltd., Wallingford, United Kingdom
[2]General Atomics, San Diego, California, USA
[3]Atomic Weapons Establishement, Aldermaston, United Kingdom
[4]Photek Ltd, St Leonards on Sea, United Kingdom
[5]Los Alamos National Laboratory, Los Alamos, NM, USA
[6]Sydor Instruments LLC, Rochester, New York, USA

E-mail: `jdh@kentech.co.uk`

Abstract. A novel microchannel plate (MCP) intensified high-speed photo-multiplier tube making use of pulse-dilation[1] has been tested. A ramped photo-cathode voltage followed by a relatively long drift region results in a transit time which is dependent on the photo-electron birth time. This leads to temporal magnification or dilation, so providing an enhancement in time resolution of the optical signal with respect to the electrical signal at the output anode. By this means a time resolution on the order of picoseconds may be realized with a substantially slower oscilloscope. The photo-electron signal is guided from a photo-cathode to an MCP by an axial magnetic field and a short input record length is stretched by a factor up to 40X to yield significantly improved time resolution at the photo-cathode. Results of the first measurements are presented.

1. Motivation

Diagnosing the evolution of plasma conditions during inertial confinement fusion reactions is challenging due to the short duration of the burn. At the National Ignition Facility (NIF), implosion burn widths are typically 150 ps and are expected to drop as performance improves. For adequate characterization of the dynamics near stagnation, detectors with temporal response on the order of 10 ps are required. There are several diagnostic techniques which record optical signals derived from the imploding core plasmas. These optical signals can be recorded with traditional photo-multiplier tubes but temporal resolution is limited to around 100 ps. Streak cameras can provide the required 10 ps response time, but they suffer from poor signal-to-noise performance in the harsh radiation environment surrounding the NIF implosion.

2. Electron Pulse-dilation

The pulse-dilation technique is a method for slowing down an electronic signal using velocity dispersion of electrons in a vacuum drift tube. The technique was first demonstrated by Prosser[1] in a device that was intended to slow down electrical signals so they could be recorded acurately using low bandwidth oscilloscopes. The electron velocity dispersion was created by

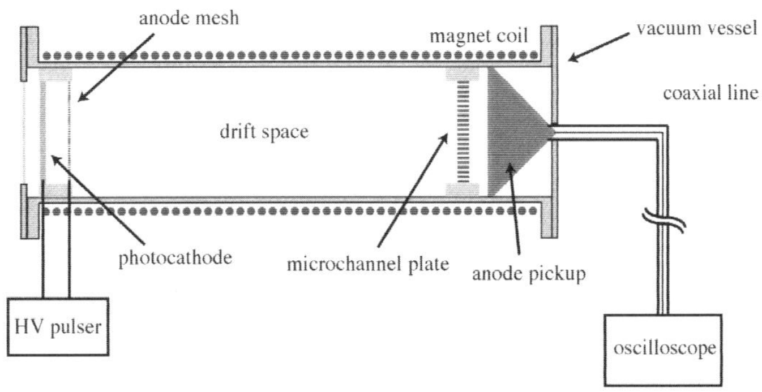

Figure 1. Schematic layout of electron pulse-dilation enhanced photomultiplier tube. Electrons emitted from the photocathode are accelerated by a ramped electric field and separate in time as they drift toward the collection anode.

ramping the potential difference between the electron emitting thermionic cathode and an anode mesh. More recently, the electron dispersion tube was coupled to a photo-cathode and a gated MCP detector to achieve 5 ps gating of 2D images[2] and an x-ray imager for the NIF was developed based upon this work known as DIXI[3]. Here, we employ pulse-dilation to create a photo-multiplier tube with reduced requirements on the bandwidth of the collection anode and recording oscilloscope.

Figure 1 shows a basic layout of the device which consists of a vacuum vessel containing a transmission photocathode, anode mesh, field free drift region, microchannel plate and collection anode. The photocathode - anode mesh gap is excited by a high voltage pulser which produces the birth time dependent velocity of the photoelectrons. The microchannel plate is biased DC and provides gain for the photoelectron signal. The microchannel plate output is collected near the end of the tube and coupled to a transmission line output. An energized coil wound around the vacuum vessel produces a weak magnetic field which guides the electrons down the tube.

3. Photoelectron Accelerating Field

For a reasonably compact device, the time varying potential applied to the anode-photocathode gap must have a ramp rate of 1 kV/ns or greater. The spatial gradient associated with these short duration ramps induces a discrepancy in the transit times of photoelectrons born at the same time but at different locations on the photocathode. This effect will signficantly limit the minimum acheivable temporal resolution unless it is corrected. One solution for mitigating this effect is to fashion the anode-photocathode gap into a microstrip transmission line structure and launch traveling waves from opposing directions along the line. For linearly varying waves, the spatial gradients exactly cancel and the transit time discrepancy for simultaneous photoelectrons vanishes.

Unfortunately, a linear potential ramp does not produce a uniform temporal magnification of the signal and accurate reconstruction of the input signal requires precise knowledge of the relative timing between the photocathode pulse and the output signal. For example, a linear ramp which produces a temporal magnification varying between 40 and 100 over a 100 ps input interval requires timing control of 5 ps to achieve a reconstructed signal error of less than 10%. To lessen this strict timing requirement, a shaped photocathode pulse can be driven which results in a uniform temporal magnification over the recording interval. In this case, the input signal can

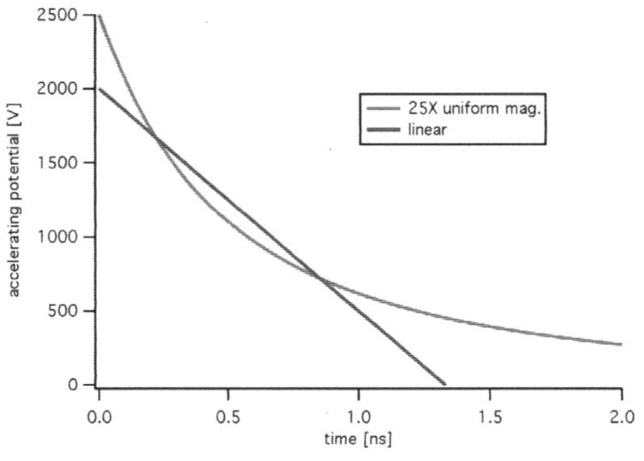

Figure 2. Ramp potentials applied to the anode-cathode gap. To acheive uniform temporal magnification, a shaped ramp is needed. In this experiment, a linear ramp was used which results in the temporal magnification increasing during the measurement.

be constructed by simply contracting the time axis of the output signal. However, the shaped pulse will have a spatial gradient which limits the temporal resolution to some extent. For 40X magnification, effective temporal spreads of 1, 4 and 9 ps are encountered for simultaneous photoelectrons born 10, 20 and 30 mm apart from one another. Therefore, a 10 mm diameter active area device with uniform temporal magnification and a 10 ps temporal resolution goal is achievable. A comparison of a linear and a shaped ramp is given in Figure 2.

4. Experimental Demonstration

The DIXI x-ray framing camera was modified in order to demonstrate an electron pulse-dilation photomultiplier tube. The phosphor window and CCD camera at the back end of DIXI were replaced with a simple 10 mm diameter anode which was coupled to a coaxial transmission line. A bias Tee was used to maintain the anode potential above that of the microchannel plate output in order to collect the electron signal. An input signal was derived from an EKSPLA PL2251A mode locked Nd:YAG laser operating at its fourth harmonic. A single 80 ps FWHM 266 nm laser pulse was divided with a beam splitter and each pulse was directed on to the photocathode. The pulses were spaced temporally by 160 ps by adjusting the beam paths and this temporal spacing was verified using DIXI in streak camera mode. Figure 3 shows the output signal measured on an Agilent DSO9104A 1 GHz bandwidth oscilloscope. Two pulses are clearly shown in the data with a relative temporal spacing of approximately 4 ns which demonstrates a temporal magnification of 25X. The pulse shapes in Figure 3 differ in height and width due to the non-uniform temporal magnification caused by the linear DIXI photocathode ramp.

5. Future Devices

Efforts are underway to produce a visible light sensitive device which is capable of 10 ps temporal resolution over a recording interval of 1 ns. This new tube has a 35 cm drift length and its S25 photocathode has an active diameter of 10 mm. The accelerating potential ramp is generated by a programmable waveform avalanche pulser which can be tuned to produce a uniform temporal magnification. The stretched electron signal exits the tube through a high speed anode output feed with 100 ps response time. Unlike an imaging system, only a modest magnetic field is required to prevent the drifting electrons from hitting the wall. Therefore, a DC solenoid is used which allows for repetitive operation.

A model for a 40 cm visible light sensitive pulse-dilation photo-multiplier tube is shown in

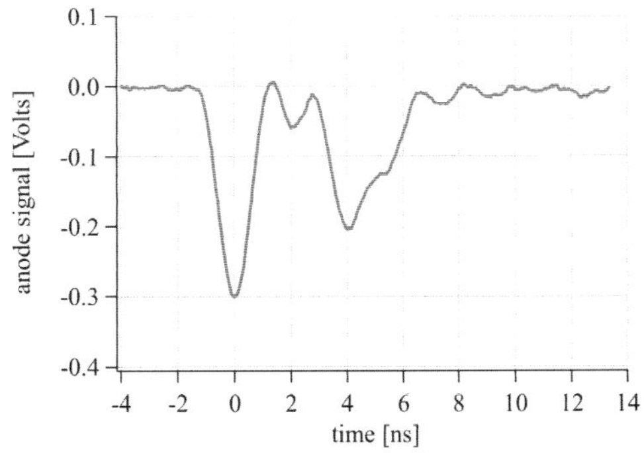

Figure 3. Oscilloscope trace clearly showing two independent pulses separated by 4 ns. The injected light pulses were spaced by 160 ps and the temporal magnification is 25.

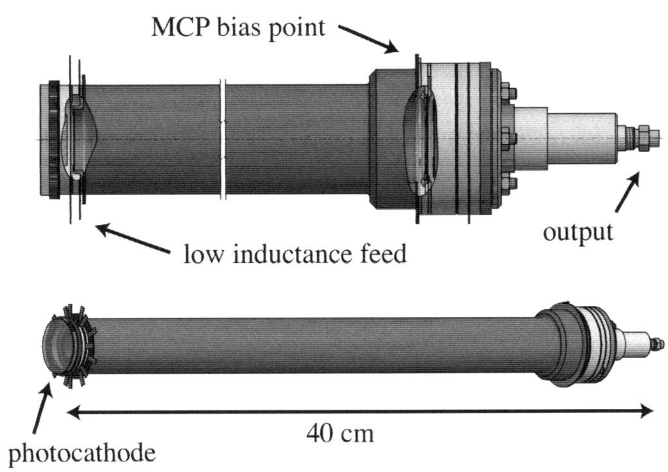

Figure 4. Visible light sensitive pulse-dilation photo-multiplier tube. The 40 cm tube length provides 10 ps resolution over a 1 ns recording interval.

Figure 4. The main tube is stainless steel which is welded at its ends to Kovar flanges and the various tube sections are joined together with brazed ceramic insulators. We have confirmed that the disturbance to the magnetic field from Kovar joints is not a problem. Screening is used to prevent the photocathode fast ramp from coupling to the output end of the tube. The device illustrated in Figure 4 is a prototype that will be used to demonstrate the potential of high speed pulse-dilation light sensing. Future devices based on this design will be constructed for specific applications. It is envisioned that pulse-dilation photo-multiplier tubes will provide a single channel alternative to streak cameras for light detection at the picosecond time scale.

References

[1] Prosser R D 1976 *J. Phys.* E **9** 57
[2] Hilsabeck T J *et al* 2010 *Rev. Sci. Instrum.* **81** 10E317
[3] Nagel S R *et al* 2012 *Rev. Sci. Instrum.* **83** 10E116

Journal of Physics: Conference Series **717** (2016) 012094 doi:10.1088/1742-6596/717/1/012094

Laboratory experiments on the formation and recoil jet transport of aerosol by laser ablation

Yoshi Hirooka[1], Kazuo A. Tanaka[2], Keisuke Imamura[2] and Katsuya Okazaki[2]

[1] National Institute for Fusion Science, and the Graduate University for Advanced Studies, 322-6 Oroshi, Toki, Gifu 509-5292.
[2] Osaka University, 2-1 Yamada-oka, Suita, Osaka565-087.

E-mail: hirooka.yoshihiko@nifs.ac.jp

Abstract. In a high-repetition rate inertial fusion reactor, the first wall will be subjected to repeated ablation along with pellet implosions, which then leads to the formation of aerosol to scatter and/or deflect laser beams for the subsequent implosion, affecting the overall reactor performance. Proposed in the present work is a method of in-situ directed transport of aerosol particles by the use of laser ablation-induced jet recoil momenta. Lithium and carbon are used as the primary ablation targets, the former of which is known to form aerosol in the form of droplet, and the latter of which tends to form carbon nanotubes. Laboratory-scale experiments have been conducted to irradiate airborne aerosol particles with high-intensity laser to produce ablation-induced jet. Data have indicated a change in aerosol flow direction, but only in the case of lithium.

1. Introduction

Along with repeated pellet implosions, the interior surface structures of the target chamber of a high-repetition rate inertial fusion reactor are exposed to short pulses of high-energy x-ray, unburned DT-fuel particles, He-ash and pellet debris. As a result, surface materials are subjected to ablation, emitting particles in the plasma state. Ablated plasma particles will either be re-deposited elsewhere on the wall or collide with each other, perhaps in the centre-of-symmetry region of the chamber volume. Colliding ablation plasma particles can lead to the formation of molecular clusters to grow into aerosol, possibly floating thereafter, which can then deteriorate the subsequent implosion performance via laser scattering, etc. Clearly, this points to a need for the in-situ removal or displacement of aerosol from the position of ignition. Proposed in the present work is a method of directed transport of aerosol particles by the use of laser ablation-induced jet recoil momenta.

2. Experimental

Employed in our previous studies is a laboratory-scale laser ablation experiment setup, referred to as LEAF-CAP for the Laboratory Experiments for Aerosol Formation by Colliding Ablation Plumes [1]. Used for the primary ablation process in the LEAF-CAP setup is a 1 Joule YAG laser beam (3rd harmonic), which is first split into two equal power beams, both line-focused down to 1cm x 0.01cm, irradiating two arc-shaped targets at 10Hz, each 6ns long, so that the energy density ranges from 3 to

10 J/cm^2/pulse. These two targets are set in vacuum, positioned in such a way that ablation plasma plumes will collide with each other in the center-of-arc (CoA) region, as shown in figure 1.

In our recent work, colliding ablation plumes from graphite have been found to generate aerosol in the form of carbon allotropes such as carbon nano-tubes [2], whereas those from metals tend to form aerosol in the form of droplets [3]. In both of these cases, aerosol particles have been observed to flow in the compound velocity direction of the colliding plumes, apparently cancelling the momenta in the counter directions and to be deposited eventually on the sampling plate.

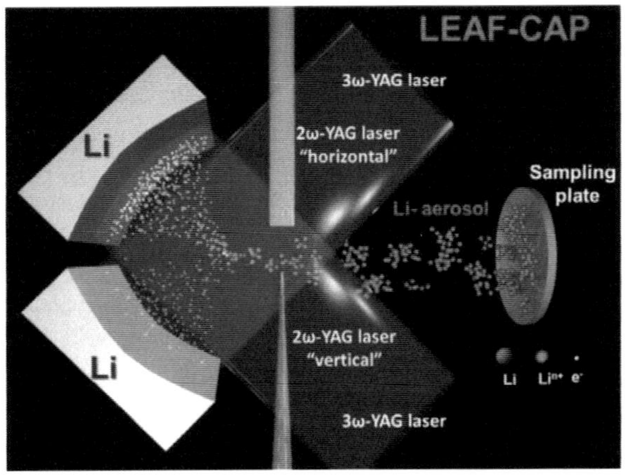

Figure 1 An artist rendering of the lithium aerosol formation process in the LEAF-CAP setup [1].

In the present work, the LEAF-CAP setup has been modified with an additional 250m Joule YAG laser beam (2nd harmonic), line-focused in the two directions, i.e. horizontal and vertical, to intersect the flow of aerosol at around 5mm away from the CoA-region at 10Hz, each 8ns long, as illustrated in figure 1. The energy density in this laser beam is set typically at 3.3J/cm^2/pulse. Also, a quartz crystal film thickness monitor is employed, set at the position of the sampling plate, so that the arrival rate of aerosol can be monitored in real-time during the double-action laser ablation process. One can then expect to see a decrease in molecular arrival rate when the redirection of aerosol occurs as a result of the ablation by 2ω-YAG laser irradiation. In parallel with aerosol arrival rate monitoring, visible spectroscopic measurements are conducted, viewing the region where the aerosol flow is intersected by the 2ω-YAG laser beam. In a separate experimental setup, where the film thickness monitor is removed, ion charge collector measurements have also been conducted.

3. Results and discussion

3.1. Visible spectroscopy measurements

Shown in figure 2 is the streaking light observed by an ICCD camera along with the vertically line-focused 2ω-YAG laser beam, intersecting a carbon aerosol flow generated from two carbon targets. Visible spectroscopic analysis has been done on this streaking light, and the data are shown in figure 3. Recognize the Swan band is observed, before 2ω-YAG laser irradiation, indicating the formation of C_2 molecules although it is partially covered (520nm~540nm) by the filter. As can be seen in figure 3, with 2ω-YAG laser irradiation, the swan band spectra are superimposed with line emissions from C^o, C^{+1} and C^{2+} which are believed to be generated by the decomposition of carbon aerosol.

Unlike carbon, lithium does not appear to exhibit any observable streak lines with the irradiation of 2ω-YAG laser beam, line-focused in either direction. As shown in figure 4, however, visible spectroscopy data indicate a significant increase in Li-I line intensities, all of which are related to neutral species, under identical conditions to those employed in the case of carbon.

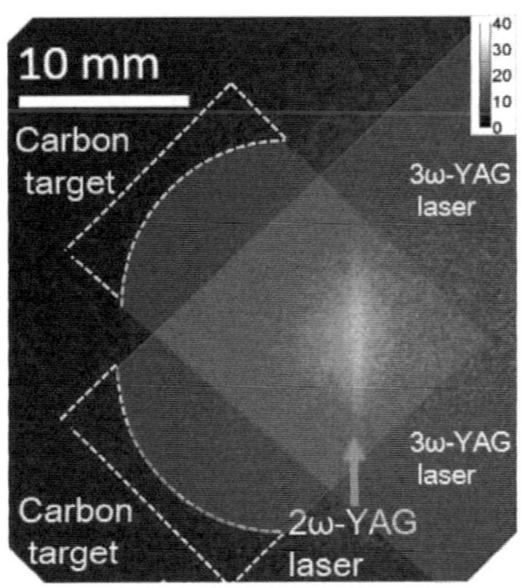

Figure 2 The streak of visible lights observed along with the vertically line-focused 2ω-YAG laser beam irradiation at an energy density of 3.3 J/cm^2/pulse.

3.2. Ion charge collector measurements

In addition to visible spectroscopy, a charge collector has also been used to detect the effect of the irradiation of 2ω-YAG laser. This charge collector is composed of an ion collector covered with a mesh and an isolated chassis. During its operation, the ion collector is DC-biased to -10V and the mesh and chassis are at the same floating voltage. Generally, it is known for this type of ion charge collectors that a sharp increase in signal is observed after a temporal indication of a negative current due to electrons arriving at the collector [4].

Shown in figure 5 are the results from ion current measurements on lithium ablation plasmas with and without 2ω-YAG laser irradiation, both indicating the general trend described above. With a few micro second of delay, a sharp increase in ion current has been observed, presumably due to an increase in plasma density by the 2ω-YAG laser irradiation which, however, is not indicated by the visible spectroscopy data shown in figure 4. From these data, the plasma (ion) density has been estimated to be of the order of 10^{10} 1/cm^3 at the position where the charge collector is placed.

3.3. Aerosol arrival rate measurements

In the case of carbon, no clear sign of ablation-induced jet effects has been identified in the aerosol arrival rate monitoring data. This may be due to the decomposition of carbon aerosol being rather isotropic or due to the insufficient energy density of 2ω-YAG laser irradiation. Consistently with this, in our previous work, carbon has been found to be most difficult to generate high-density ablation plasmas, compared with the other materials examined [2].

In the case of lithium, as shown in figures 6-(a) and (b), noticeable changes in molecular arrival rate have reproducibly been seen with the irradiation of 2ω-YAG laser beam, even as there is intrinsic fluctuation and/or noise in the arrival rate monitoring signal. In the case of vertically line-focused irradiation, the energy density of 2ω-YAG laser is particularly raised to 12J/cm^2/pulse so as to enhance the effect of ablation-induced jet.

Two possibilities have been investigated to interpret these observations: (1) First, the radiation pressure, induced by the 2ω-YAG laser irradiation is evaluated, using the model for laser driven ablative implosions [5]. The result indicates that the 2ω-YAG laser used in the present work is not strong enough to redirect the flow of aerosol.

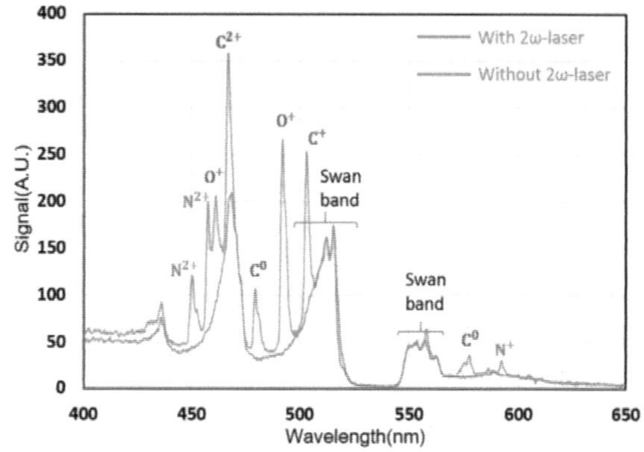

Figure 3 Visible spectroscopy of the carbon aerosol with and without 2ω-YAG laser beam irradiation. Data with 2ω-YAG laser beam irradiation are taken from the streak light shown in figure 2.

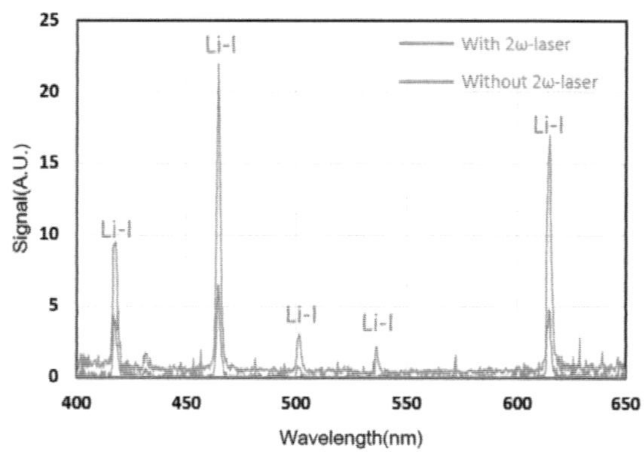

Figure 4 Visible spectroscopy of the lithium aerosol with and without 2ω-YAG laser beam irradiation.

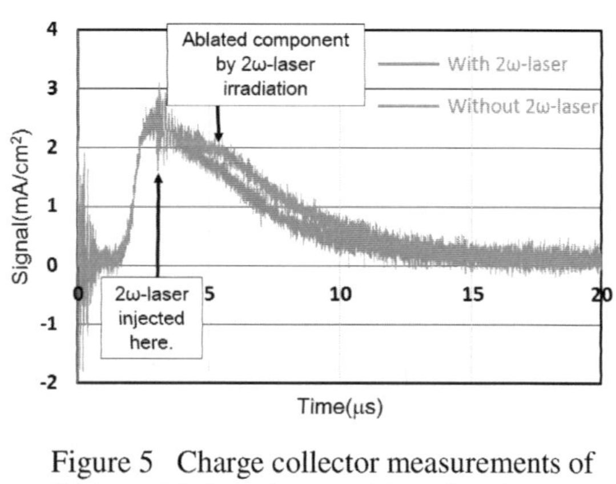

Figure 5 Charge collector measurements of lithium ablation plasma with and without 2ω-YAG laser beam irradiation.

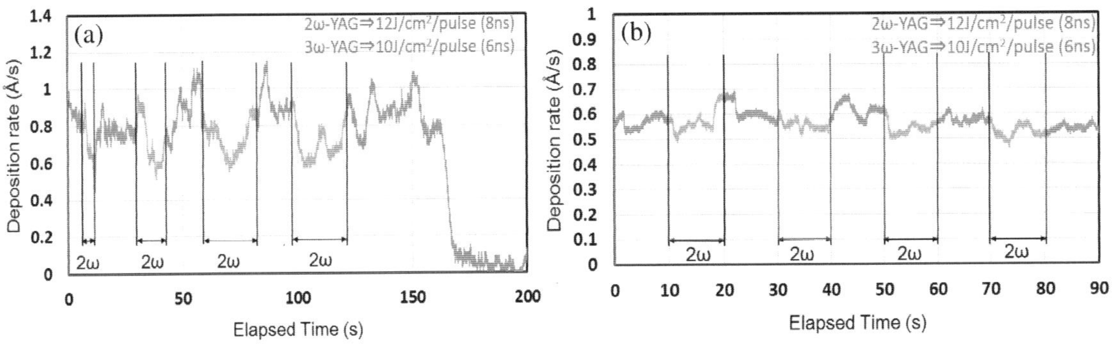

Figure 6 Lithium aerosol arrival rate data, measured by a quartz crystal film thickness monitor, with and without 2ω-YAG laser beam irradiation: (a) vertically line-focused and (b) horizontally line-focused. Irradiation periods of 2ω-YAG laser are indicated by the yellow-highlighted zones.

(2) Next, the effect of laser ablation-induced jet recoil is evaluated, assuming that lithium aerosol particles are spherical and for simplicity, only the hemisphere irradiated by 2ω-YAG laser is heated up although, in actuality, the entire body temperature increases. The velocity of aerosol powered by ablation-induced jet is found to be inversely proportional to the diameter. Importantly, the velocity must be below a certain level to be detected by the film thickness monitor with a sensor diameter of only 8mm. This simple ablation-induced jet model then predicts that only the aerosol particles with diameters larger than ~7μm could arrive at the sensor. As can be seen in figure 7, however, the number density of lithium aerosol particles satisfying this diameter requirement is not large, which is believed to have led to the difficulty in arrival rate measurements.

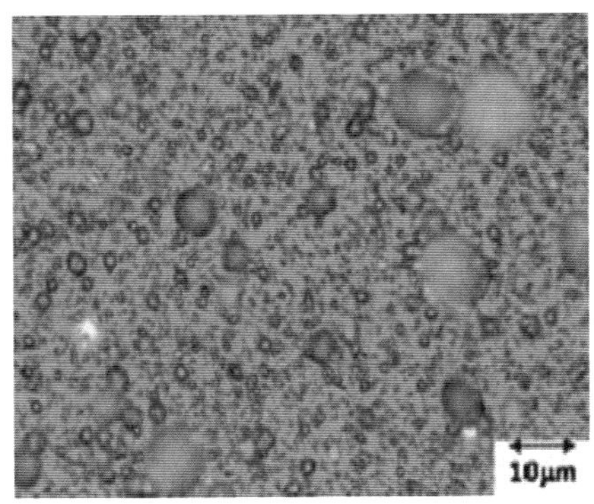

Figure 7 Lithium aerosol deposited on a pyrex glass sampling plate placed in the quartz crystal film thickness monitor [3].

4. Conclusion

In the present work, the possibility of applications of ablation jet momenta has been addressed for the directed transport of aerosol particles that are predicted to be formed in the pellet implosion region in a high-repetition rate inertial confinement reactor.

A series of laboratory-scale experiments have been conducted in which two consecutive laser ablation processes take place, the first one of which generates aerosol particles via laser ablation, and the second of which is intended to provide recoil jet momenta to re-direct the flow of airborne particles. Carbon and lithium are chosen as the primary ablation target materials.

Data from these experiments have indicated a change in aerosol flow direction when the secondary laser ablation takes place, but only in the case of lithium. Clearly, further effort is necessary to understand more details on the dynamics of aerosol transport and its redirection. Nonetheless, the present work has opened the door to a new area of research, contributing to the achievement of inertial fusion energy, or at least so we believe...

References

[1] Y. Hirooka et al., J. Phys. Conf. Ser. **244**(2010)032033(4pp).
[2] Y. Hirooka et al., Nucl. Fusion **54**(2014)022003(6pp)
[3] Yoshi Hirooka et al., Fusion Eng. Des. **87**(2012)760-1764.
[4] Israel Pelah, Phys. Lett. **59**(1976)348-350.
[5] Claire E. Max et al., Phys. Fluids **23**(1980)1620-1645.

Reactor for boron fusion with picosecond ultrahigh power laser pulses and ultrahigh magnetic field trapping

G H Miley, H Hora and G Kirchhoff

[1]Department of Nuclear, Plasma & Radiological Engineering, University of Illinois, Urbana IL, USA,
[2]Deptartment of Theoretical Physics, University of New South Wales, Sydney 2052, Australia
[3]UJK Management GmbH, Poing, Germany
*h.hora@unsw.edu.au

Abstract. Compared with the deuterium tritium (DT) fusion, the environmentally clean fusion of protons with ^{11}B is extremely difficult. When instead of nanosecond laser pulses for thermal-ablating driven ignition, picosecond pulses are used, a drastic change by nonlinearity results in ultrahigh acceleration of plasma blocks. This radically changes to economic boron fusion by a measured new avalanche ignition.

1. Introduction

In search for the aim of energy production by the nuclear fusion of light nuclei, an alternative may be available by using laser pulses of picosecond (ps) duration and powers in the range above dozens of petawatts (PW) up to terawatts (TW). The essential difference to the highly developed laser fusion by using huge lasers with nanosecond laser pulses of more than MJ energy [1] now with the achievements of "high foot" operation [2] is that the non-thermal forces of laser pulses are used in combination with recently achieved ultrahigh magnetic fields for trapping the reacting plasma. In addition use is done by the avalanche reaction of fusing light hydrogen (protons) with the boron isotope ^{11}B (HB11) with negligible problems of radioactive pollution [3][4] and direct conversion of the of the energy of the generated alpha particles into electric power [5][6].

This option was developed from the physics of the basically nonlinear processes opened by the interaction of lasers with plasmas nearly from the very beginning with the special turning point [7] of generating the >PW powers by Chirped Pulse Amplification CPA [8]. The ultrahigh acceleration was calculated since 1978 ([9]: Figs. 10.17 a&b) by direct conversion of laser energy into mechanical motion of macroscopic plasma bocks by nonlinear (ponderomotive) forces and first measured [10] in full agreement with the theory [11]. The necessary picosecond interaction time for igniting uncompresed solid density deuterium-tritium (DT) fusion fuel [12] could be shown to be possible for the otherwise extremely difficult clean fusion of HB11 [13][14]. The need of the ultrahigh magnetic fields [15] and of the avalanche reaction of HB11 [3][4] was the final step for this development.

In discussing this option [3], Crandall [6] mentioned that an enormous further work would be necessary to achieve this goal. The following views may summarize some of the tasks.

2. Laser development

The building of lasers of higher powers is exciting for all kinds of applications and sub-picosecond lasers have achieved 7.2 PW pulse powers [16]. New laser architectures based on existing high energy PW technologies [8] for production of 200 PW laser pulses from single beam lines of existing NIF and NIF-like lasers have been recently defined [17]. Laser pulses of 200 PW are on the way by using the existing NIF laser. NIF-like lasers can be reduced in size by many orders of magnitude after short laser-diode-pumped neodymium glass laser amplifiers are ready for use.

For the ps-block ignition of boron fusion, very high quality laser pulses with extreme contrast are necessary as seen from the repetition [18] of Sauerbrey's [10] ultrahigh acceleration in agreement with the theory. This technology can be studied immedietely with table top laser experiments for generating plasma blocks with several 100 MeV ion energy having space charge neutral ion densities million times higher than from accelerators. This is needed also for other applications as e.g. for hadron therapy of cancer [19].

Plasma mirror techniques can evaluate to what degree the contrast is necessary where it can be explored what initial density profiles of the irradiated targets should be used for optimized conditions of the nonlinear force driven dielectric explosion [4][9]. It had been shown numerically (see Figure 1 of [3]) that the use of a bi-Rayleigh initial density profile for a very low reflection of the laser light has the additional advantage that the plasma blocks for acceleration can have an increased thickness for the dielectric explosion. Evaluations with modified density variations of the target can strongly increase the energy of the ions in the accelerated plasma bocks [20]. Numerical studies for providing optimized targets e.g. as foam with a varying density profile can experimentally confirm the expected optimization for HB11 fusion.

3. Ultrahigh magnetic fields and the avalanche reaction

The measured ultrahigh magnetic fields of 4.5 kilotesla for about nanosecond by laser pulses with kJ energy of nanosecond duration [15] can be used to trap a cylindric plasma from solid density HB11. The generated fluid of alpha particles from the 0.2mm thin target expands slowly in radial direction (see [21]: Fig. 3) but with many1000km/s along the cylinder axis when trapped by 10 kT fields even when computing very pessimistically used binary fusion reactions, well with high gains of ignition [21]. When including the avalanche reaction and the subsequent expansion into a solid density mantle of 1mm radius around the initial 0.2mm cylinder, the block ignition by a ps laser pulse of 30kJ energy should produce more than GJ energy in alpha particles. Experimental studies of this generation of ultrahigh magnetic fields should be studied specifically to measure the trapping of the co-axial, initially solid HB11 fuel when compressed by the magnetic field in comparison to preceding and further generalized computations with cylindrical rotation [21][22][23][24].

Independent from these steps, the measurement of 10^9 HB11 reactions per steradian [25] at laser plane geometry reaction per laser irradiation on very special targets was perfectly repeated [26]. This is the basic measurement for concluding the avalanche process [3][4]. Further studies under varying conditions should well be performed to further confirm these very exceptionally increased HB11 reaction gains.

Then it may be interesting to combine these measurements with cylindrical HB11 targets in presence of the ultrahigh magnetic fields with measuring the generation of alpha particles. It can be expected that the reaction rate may be further increased above the values reached numerically [21] when the fact of the gyration of the alphas are added on top of the avalanche mechanism.

For further confirming the theory of the avalanche process, the elastic ion collisions in the broad energy range of 600keV energy will be further evaluated. This is a state of a highly non-equilibrium plasma. A number of results for describing these conditions are at hand in support of the avalanche reactions [27] for the centre of mass distribution in the reacting plasma. This can be combined with the results of multi-fluid computations where the generated fluid of alpha particles [21] resulted in the ignition process even for the pessimistic HB11 binary reactions without avalanche. This led to realize that the reacting plasma was in an extreme non-equilibrium state, and it was possible to use the elastic collisions of the alphas with protons and boron nuclei to reach the high fusion gains.

These results of the elastic nuclear interaction for the protons and boron nuclei at the very broad energy distribution with a maximum of ion interaction energies at 600 keV is exceptional for HB11 due to the fact that there is the very broad maximum of the energy dependence of the fusion reaction cross sections. HB11 has - in contrast to all other fusion reactions - a cross section very broadly elevated by a factor of more than ten for reproducing [27] the measured strong increase of the HB11 reaction gains [25].

4. Scheme for designing of a HB11 reactor

The numerical result of generation of more than one gigajoule energy (270 kWh) of alpha particles from a solid HB11 cylinder of 1 cm length and 1mm radius in a co-axial 10 kilotesla magnetic field during one ns by irradiation of one ps laser pulse of 30 kJ at one side at 10^{20} W/cm^2 [4][21] should convert at least the main part of the generated alpha particle energy into electric power by being slowed down electrostatically in a field of 1.4 megavolt (Figure 1) [4][5].

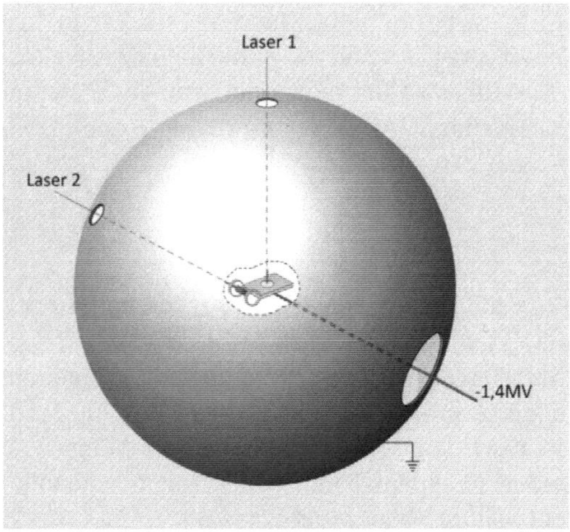

Figure 1. Scheme of a HB11 fusion reactor without any radioactive radiation problem is based on non-thermal plasma block ignition by nonlinear forces using a 30kJ-picosecond laser pulse 2. The central reaction unit (Figure 2 of [3]) located in the center of the reactor sphere is electric charged to the level of -1.4 million volts against the wall of the sphere such that the alpha particles (helium nuclei) produce more than a gigajoule energy, of which a small part is needed for the operation of the laser pulses [3][4][5].

The reaction in the center of the sphere with producing more than GJ energy generates a mechanical shock to the spherical container of the reactor. This is to be compared with the shock of chemical explosives. The generated momentum of the alpha particles compared with a chemical explosion depends on the square root of the energy of the particles which is for the nearly 10-Million times higher nuclear than that of chemical energy. The mechanical shock of the alpha particles to the reactor wall is then about 3000 times lower than for a comparable chemical explosive. In the case of the reactor of Figure 1, this corresponds then to a number of several grams of TNT.

It can be ascertained that the boron fusion reaction occurring only in the volume of few cubic millimetres and during the time of less than a few nanoseconds is limited to the controlled generation of fusion energy only and cannot be used for large scale nuclear reactions.

The technology for handling the loading of the central HB11 reaction unit into the sphere in Figure 1 within vacuum and on an electric potential of -1.4 megavolts, especially when operating with one reaction per second, needs a special design of the equipment to be used, but this should be on a level of present day's achievements. Special attention may have to be given to move the unit through gates from zero potential to the high potential in the interior of the reactor to be located at the right hand side of Figure 1.

If the reactor has to run with one shot per second, the direct current electrical energy in the pulse has to be converted according to the methods of the High Voltage Direct Current HVDC techniques in three phase ac current. This technology is completely developed for commercial use of power transmission over large distances [28][29][30] such that the generated power at the one Hertz operation corresponds to a 1000 Megawatt power station.

A very preliminary first estimation after subtracting the costs for the central reaction unit in the center of the sphere of Figure 1 (which is destroyed at each shot) and of the costs for the HB11 fuel and after subtracting the costs of the investment and that of the operation of the power station, the profit may arrive at up to $300Million per year of generated electric power grid.

References:

[1]Crandall, D. IFSA Seattle 2015 Conf. Abstracts p. 45
[2]Hinkel, D. E., Herzack-Hopkind, L.F. et al. IFSA Seattle 2015 Conf. Abstracts p. 163
[3]Hora, H., et al. IFSA Seattle 2015 Conf. Abstracts p. 409
[4]Hora, H., G. Korn, L. Giuffrida, D. Margarone, A. Picciotto, J.Krasa, K. Jungwirth, J. Ullschmied, P. Lalousis, S. Eliezer, G.H. Miley, S. Moustaizis and G. Mourou, 2015 Laser and particle Beams: 33, No. 4/2015, "First View" 14 July 2015. Cambridge University Press DOI: 10.1017/S026.03461.5000634
[5]Hora, H.and Kirchhoff, G., 2014 (23MAR) Method for generating electrical energy by laser-based nuclear fusion and laser fusion reactor. Internat. Patent Disclosure WIOP PCT (01 OCT 2015) WO 205/144190 A1
[6]Crandall, D. 2015 Discussion at [3]
[7]Strickland, D. and Mourou G., 1985 Opt. Commun. 56, 219
[8]Mourou, G., C.P.J. Barty & M.D. Perry. 1998 Ultrahigh-Intensity Lasers. Physics Today 51 No.1, 22; Barty C.P.J., M. Key, J. Britten, R. Beach, G. Beer, C. Brown, S. Bryan, J. Caird, T. Carlson, J. Crane, J. Dawson, A. C. Erlandson, D. Fittinghoff, M. Hermann, C. Hoaglan, A. Iyer, L. Jones, II, I. Jovanovic, A. Komashko, O. Landen, Z. Liao, W. Molander, S. Mitchell, E. Moses, N. Nielsen, H. H. Nguyen, J. Nissen, S. Payne, D. Pennington, L. Risinger, M. Rushford, K. Skulina, M. Spaeth, B. Stuart, G. Tietbohl, and B. Wattellier, 2004 Nuclear Fusion **44**, S266-275
[9]Hora, H., 1981 *Physics of Laser Driven Plasmas* Wiley, New York
[10]Sauerbrey, R. (1996) Acceleration of femtosecond laser produced plasmas. Physics of Plasmas 3, 4712-4716.
[11]Hora, H., J. Badziak, M.N. Read, Y.-T. Li, T.-J. Liang, H. Liu, Z.-M. Sheng, J. Zhang, F. Osman, G.H. Miley, W. Zhang, X. He, H. Peng, F. Osman, S. Glowacz, S. Jablonski, S. Wolowski, Z. Skladanowsli, K. Jungwirth, K. Rohlena & J. Ullschmied 2007 Phys. Plasmas 14, 072701
[12]Chu, M.S. 1972 Phys. Fluids 15, 412.
[13]Hora, H., 2009 *Laser and Particle Beams* 27, 207-222
[14]Hora, H., G.H. Miley, M. Ghorannviss, H. Malekynia, N. Azizi and X-T. He 2010 Energy and Environment Science 3, 479.
[15]Fujioka S., Zhe Z., et al. (2013) Scientif. Rep. 3, 1170-1176.
[16]Li, Ru Xin, X. Liang, Z. Gan, H. Lu, L. Yu, D. Yin, Y. Leng, X. Lu, C. Wang & Z. Xu (2015) Progress toward a 10PW ultra intense Laser Facility: Demonstration of 5 PW High gain Large Aperture Ti:sapphire Amplifier. 4th Advanced Lasers and Photon Sources (ALPS'15), Yokohama, Yapan, 22-24 April.
[17]Barty, C.P.J., 2015, The Nexawatt: Strategies for Exawatt peak power laser based on NIF and NIF-like beam lines. IFSA Seattle Conf. 2015, Book of Abstracts, p. 459.
[18]Földes, I., J.S. Bakos, K. Gal, Y. Juhasz, M. A. Kedves, G. Koscis, S. Syatmari, G. Verex. 2000, Laser Physics 10, 264.
[19]Banati, R., H. Hora, P. Lalousis, S. Moustaizis. Jour. Intense Pulsed Lasers & Application in Adv. Physics 4, No. 1, 11-16 (2014).
[20]Yazdani, E., R.Sadighi-Bonabi, et al. 2014 Journal of Applied Physics 116, 103302
[21]Lalousis, P., Hora and S. Moustaizis, 2014 Laser and Particle Beams 32, 409.
[22] Lalousis, P., H. Hora, S. Eliezer, J.-M. Martinez-Val, Moustaizis, G.H. Miley & G. Mourou 2013 Phys. Letters A377, 885
[23]Hora, H., P. Lalousis, S. Moustaizis 2014 Laser and Particle Beams 32, 63
[24]Lalousis, P., S. Moustaizis, H. Hora and G.H. Miley, 2015 Journal of Fusion Energy 34, 62
[25]Picciotto, A., D. Margarone, A. et al. 2014 Phys. Rev. X 4, 031030.
[26]Guiffirda, L., D. Margarone and G. Korn, (Discussions 8 Sept. 2015)
[27]Eliezer S., H. Hora, G. Korn, N. Nissim and J.M. Martinez-Val 2015 Nonequilibrium elastic nuclear collision model for avalanche generation at boron fusion, Phys. Lett. A (submitted)
[28]Gotham, D.J and Heydt G.T.1998 IEEE Transaction on Power Systems 13, 60
[29]Kanngiesser, K.W., D.H. Huang & H. Lips 1994 High-voltage direct current transmission-Systems and Planning. Siemens Monographs Munich, EV HA 7.
[30]Breuer, W., D. Povh, D. Retzmann, Ch. Urbanke and M. Weinhold 2007 Prospect of Smart Grid Technologies for a Sustainable and Secure Power Supply. 20th World Energy Congress, 11-15 November, Rome

Journal of Physics: Conference Series **717** (2016) 012096 doi:10.1088/1742-6596/717/1/012096

Modeling of energy transfer between two crossing smoothed laser beams in a plasma with flow profile

A. Colaitis[1], S. Hüller[2], V. T. Tikhonchuk[1], D. Pesme[2], G. Duchateau[1] and A. Porzio[2,3]

[1] Centre Lasers Intenses et Applications, Université de Bordeaux CNRS CEA, Talence, France
[2] Centre de Physique Thorique, CNRS Ecole Polytechnique, Palaiseau, France
[3] LAGA, Institut Galilée, Université Paris 13 CNRS, Villetaneuse, France

E-mail: huellercpht.polytechnique.fr

Abstract. We study the crossed beam energy transfer (CBET) between laser fields generated by optical smoothing methods. The energy transfer, as well as the angular distribution of the outgoing light fields are investigated for two incident smoothed laser beams in a plasma with a flow gradient, allowing for resonant transfer close to the sonic point. Simulations with the code HARMONY based on time-dependent paraxial light propagation are compared to simulations using a new approach based on paraxial complex geometrical optics (PCGO). Both approaches show good agreement for the average energy transfer past a short transient period, which is a promising result for the use of the PCGO method as a module within a hydrodynamics code to efficiently compute CBET in mm-scale plasma configurations. Statistical aspects related to role of laser speckles in CBET are considered via an ensemble of different phase plate realizations.

1. Introduction

For both Direct Drive and Indirect Drive concepts of laser fusion, a reliable modeling of laser propagation is of crucial importance. Such a modeling has to take into account the potential energy exchange between laser beams, and between groups of laser beams.

For this purpose we show here results of a comparison between two approaches to model crossed beam energy transfer (CBET), namely (i) a standard modeling of laser plasma interaction using a paraxial wave solver coupled to non linear fluid code for the plasma motion and (ii) a new approach that is based on paraxial complex geometrical optics (PCGO) and which can be implemented in radiation hydrodynamics codes. In this approach the path of multiple beamlets is integrated by a set of model equations [1, 2] for each time step of the hydro code.

In particular we investigate the crossing of laser beams, whose fields have been generated by optical smoothing methods as random (or kinoform) phase plates (RPP, KPP). The energy transfer considered in this comparison corresponds hence to a realistic situation that may appear in laser fusion configurations where wide beams with speckle structure overlap in a plasma profile [3]-[8]. We concentrate to one of the most likely configurations where resonant exchange between beams of equal laser frequency occurs, namely in a plasma with a flow gradient in the vicinity of sonic point. We compare the amplification of the beam that receives transfer from its counterpart between both models and analyze the outgoing laser fields behind the crossing zone with respect to their angular distribution and direction, and we investigate the statistical role of speckles.

2. Comparison: paraxial complex geometrical optics vs. paraxial propagation

We have carried out simulations in a two-dimensional (2D) box with 4000 wavelengths in length and 1800 wavelengths in width. Two laser beams are incident from the left-hand-side boundary, as shown in Fig. 1, with an angle of $\pm 10°$, hence the angle θ=20° between both beams. Beam 1 with the wave vector $\mathbf{k}_1$ and frequency ω_1 is denoted as the *pump* beam, beam 2 with $\mathbf{k}_2$ and ω_2 as the *probe* beam. As *probe to pump* ratio we denote the ratio between the average intensity (taken over the beam width) of both beams, $1{:}I_1/I_2$. Simulations with ratios 1:1, 1:8, and 1:64 are discussed here. The overlap of both beams forms a rhombus in the central part of the simulation box in which the plasma density is homogeneous along the common beam axis $\mathbf{x} \parallel (\mathbf{k}_1 + \mathbf{k}_2)$. In the direction of the difference wave vector $\mathbf{k}_1 - \mathbf{k}_2$ the electron density follows a parabolic shape $n_e(y) = n_0 \max\{0,\ 1 - [(y - 900\lambda)/1580\lambda]^2\}$ and has a flow with gradient as $V_p(y) = c_s(y - 657\lambda)/200\lambda$, with $n_0 = 0.1 n_c$ standing for the maximum electron density,c_s the ion sound velocity, and λ for the vacuum wave length. The critical density n_c and the wavelength λ correspond to beam 1. In our simulations beam 2 has the same frequency as beam 1, $\omega_1 = \omega_2$. The resonance conditions for crossed beam energy transfer correspond to stimulated Brillouin scattering: the matching conditions for frequency and wave vector between the ponderomotively driven ion acoustic density perturbations and the two overlapping laser waves, $\omega_s - \mathbf{k}_s \cdot \mathbf{V_p} = \omega_1 - \omega_2$ and $\mathbf{k}_s = \mathbf{k}_1 - \mathbf{k}_2$, are fulfilled for a flow profile with velocities close to the sound speed c_s inside the rhombus of the crossing beam, as chosen in our simulations.

We have computed the energy transfer with two approaches, namely a time-dependent paraxial code, coupled to fluid equations for the plasma fluid, with the code HARMONY [9, 10] and with the paraxial geometrical optics approach, based on a stationary CBET model [2, 11, 12]. The amplification in power, $\mathcal{T} = P_{\text{probe,out}}/P_{\text{probe,in}}$, of the *probe* beam at the right-hand-side boundary ("out") with respect to its incident power ("in") is shown in Fig. 2 for three different *probe:pump* ratios and as a function of the overlapped flux of the incident beams in their focal planes, $I_{14}\lambda^2$ $[10^{14} W cm^{-2} \mu m^2]$. The values from HARMONY have been taken after a transient period, see Fig. 2, lasting a few oscillation periods of $2\pi/\omega_s = \lambda/(2c_s \sin\theta/2)$. In the range of $I_{14}\lambda^2$ shown here the agreement between both models is very good. The values obtained are also in relatively good agreement with the model developed by McKinstrie et al [12]. Assuming that the SBS gain is determined by inhomogeneous plasma flow with $L_v = |V_p/\nabla V_p|$, given by $G \simeq 3.5(L_v/200\lambda)(n_e/0.1n_c)I_{14}\lambda^2/T_e(\text{keV})$, at θ =20°, the amplification in the probe beam

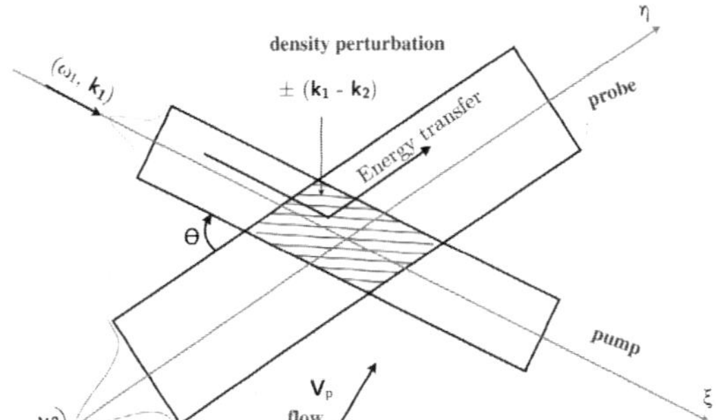

Figure 1. Scheme of two crossing laser beams in a plasma profile with flow. Both the pump and the probe are smoothed laser beam. Ponderomotively driven density perturbations arise where both beams overlap.

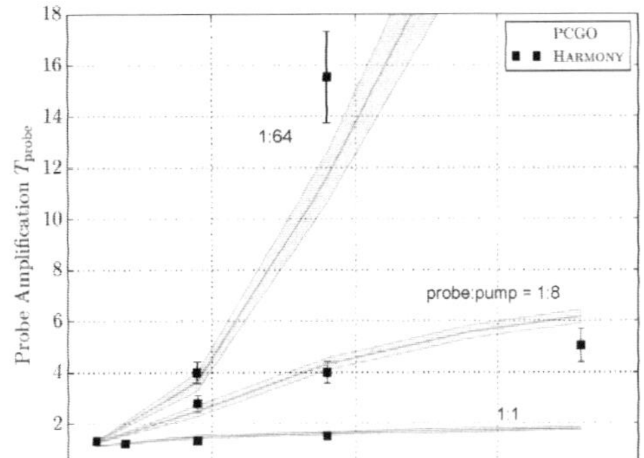

Figure 2. Probe amplification computed from simulations with the PCGO model (green curve) and with HARMONY (black) as a function of the overlapped beam flux $I_{14}\lambda^2$.

power is delimited by $\mathcal{T} = (\mathcal{R}G)^{-1}\log[1+e^{G}(e^{\mathcal{R}G}-1)]$ with $\mathcal{R}$ as the *probe:pump* ratio (see [12]). The latter eventually confirms the CBET algorithm chosen in the PCGO approach (see [2, 11]).

Both approaches use smoothed laser beams. Beam speckle patterns are generated by random numbers and therefore realization dependent. For the HARMONY simulations the beam shapes correspond to RPP speckle pattern in 2D while the beams in the PCGO mimic a speckle pattern shape by superposition of numerous (here: 100) beamlets, having usually a greater radius than laser speckles defined by the focusing f-number (here f =7). Error bars indicate the standard deviation around the ensemble average due to the spread from different (here: 16) realizations.

Also the angular distributions of the outgoing fields have been compared, where we also find good agreement between both approaches. Figures 3 and 4 show that due to *pump* depletion, which transfers gradually its flux to the *probe* beam, the angular distribution changes with increasing probe amplification $\mathcal{T}$. Figure 3 shows both the distribution of 5 different RPP realizations and for a 'regular' beams without speckles, being closer to the PCGO results in Fig. 4 (also shown here for 'regular' beams). While the overall tendency is the same, the differences illustrate the role of speckles. Ensemble average 'center of mass' values in angle $\langle\vartheta\rangle$ of the distributions are computed from 16 realizations, listed together with the standard deviation in Table 1. The angular dispersion σ_ϑ from the ensemble average due to the speckles is still dominated by the angular width of the smoothed beams, which is more pronounced for RPP than for PCGO beams where the width depends on both the number and the size of beamlets chosen. It is remarkable that the center of mass of the *pump* beam is systematically shifted due to depletion, which gives rise to additional angular dispersion. The statistical standard deviation $\sigma_{\mathcal{T}}$ in $\mathcal{T}$ observed in the HARMONY simulations, Figs. 2 and 5, is proportional to the probe amplification itself, so that the relative deviation $\sigma_{\mathcal{T}}/\mathcal{T}$ is stable for a fixed f-number. Figure 5 illustrates the importance of the f-number: for small speckles with f =5, the relative dispersion is reduced to <10% with respect to 13% for larger speckles with f =7. Note that $\sigma_{\mathcal{T}}/\mathcal{T}$ is overestimated in 2D simulations compared to 3D geometry. However, the greater angular spread in speckle beams compared to regular beams is a an important feature both in 2D and 3D. For successive beam crossings, the importance of speckles to both the energy transfer and the beam direction may increase, depending on the number of encounters n and the strength

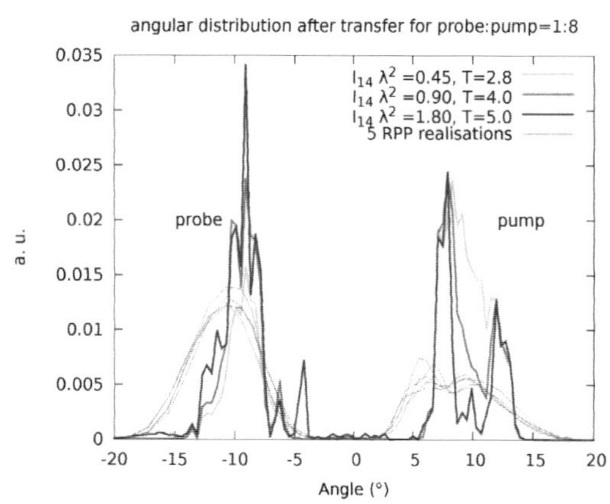

Figure 3. From HARMONY: angular distribution of transmitted *pump* and *probe* beams, with 'regular' beams without speckles, for $I_{14}\lambda^2$ =0.45 (green), 0.9 (red), and 1.8 (blue). Blue thin lines: smoothed distributions corresponding to 5 RPP realizations at $I_{14}\lambda^2$ =1.8.

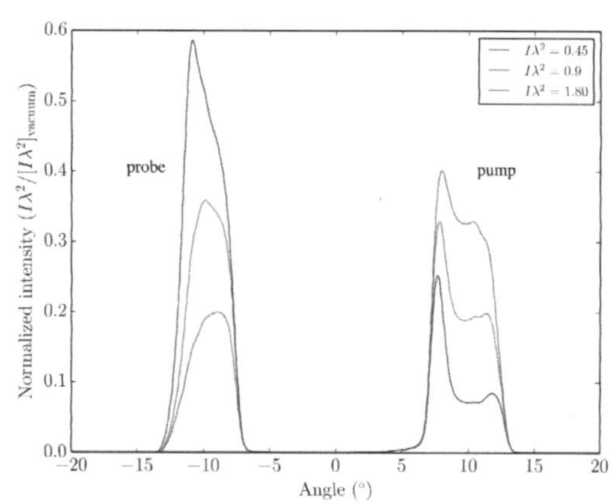

Figure 4. From PCGO: angular distribution of *pump* and *probe* beams, with 'regular' beams without pseudo-speckles [1], for $I_{14}\lambda^2$ =0.45 (green), 0.9 (red), and 1.8 (blue), corresponding as in Fig. 3 to probe amplification values $\mathcal{T}$ = 2.8, 4, and 5, respectively.

		probe		pump	
$I_{14}\lambda^2$	$\mathcal{T}$	$\langle\vartheta\rangle$	σ_ϑ	$\langle\vartheta\rangle$	σ_ϑ
0.45	2.8	-9.7°	±1.7°	9.0°	±1.9°
0.9	4.0	-10.0°	±2.2°	8.7°	±2.3°
1.8	5.0	-10.8°	±2.5°	8.6°	±3.0°

Table 1. Expectation values $\langle\vartheta\rangle$ and standard deviation σ_θ in angle of the transmitted *pump* and *probe* beams, for indicated values of overlapped laser flux $I_{14}\lambda^2$, and the corresponding probe amplification $\mathcal{T}$, computed from 16 different realizations. The angles of the incident light are $\vartheta_{\rm pump} = -\vartheta_{\rm probe} = \theta/2 = 10°$, each with an angular width $\pm(1/2)\tan^{-1}(1/2f) \simeq \pm 2°$ for $f = 7$.

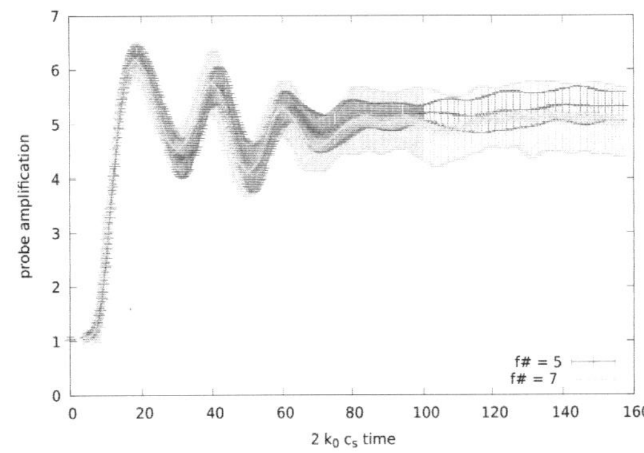

Figure 5: Time history of probe amplification $\mathcal{T}(\pm\sigma_T)$ of 16 RPP realizations with HARMONY, for f-numbers f =5(red) and f=7(green); *probe:pump* = 1:8, at $I_{14}\lambda^2$ =1.8.

of the transfer. Assuming independence of the successive beam crossing, the variance σ_T^2 will add up over all encounters, $\sum_{i=1}^{n}\sigma_{T,i}^2$. By disregarding decrease of $\sigma_{T,i}$ (e.g. pump depletion), we can estimate that the deviation due to speckle contributions increases with $\sqrt{n}$.

3. Conclusions

The comparison between simulations with HARMONY and the paraxial complex geometrical optics approach shows good agreement concerning the energy transfer, and the angular distribution of the transmitted fields. The PCGO approach is computationally less expensive and can be implemented in radiation hydro codes in a similar manner as standard ray tracing methods, so that PCGO can be efficiently used for mm-scale simulations for ICF design. We find that CBET from smoothed laser beams is mainly governed by the parameters characterizing the average beam. Depletion of the *pump* beam that transfers energy to the *probe* beam leads to a modification of the angular distribution, and in particular to an angular deviation of the *pump* beam. The contribution to CBET due to speckles decreases with the number of speckles involved in the crossing volume, which depends on the focusing f-number.

Acknowledgments

This work was carried out within the framework of the EUROfusion Consortium. It has received funding from the Euratom research and training programme 2014-2018 under grant agreement No 633053, and from the *Agence Nationale de Recherche*, project no. ANR-12-BS04-0006. The views and opinions expressed herein do not necessarily reflect those of the European Commission.

References

[1] Colaitis A et al 2015 Phys. Rev. E **91** 013102
[2] Colaitis A 2015 Ph D thesis Université de Bordeaux, http://www.theses.fr/en/2015BORD0253
[3] Kirkwood R K et al 1996 Phys. Rev. Lett. **76** 2065
[4] Kruer W L et al 1996 Phys. Plasmas **3** 382
[5] Michel P et al 2009 Phys. Rev. Lett. **102** 025004 and Phys. Plasmas **16** 042702
[6] Igumenshchev I V et al 2010 Phys. Plasmas **17** 122708
[7] Igumenshchev I V et al 2012 Phys. Plasmas **19** 056314
[8] Myatt J F et al 2014 Phys. Plasmas **21** 055501
[9] Hüller S et al 2006 Phys. Plasmas **13** 22703
[10] Masson-Laborde P E et al 2014 Phys. Plasmas **21** 032703
[11] Colaitis A et al 2016 Phys. Plasmas **23** 032118, http://dx.doi.org/10.1063/1.4944496
[12] McKinstrie C J et al 1996 Phys. Plasmas **3** 2686

Comparative study on cost evaluation and network visualization of particle accelerator components for heavy ion inertial fusion

A Inoue[1], K Takahashi[1], T Sasaki[1], T Kikuchi[1], Nob Harada[1] and J J Barnard[2]

[1] Nagaoka University of Technology, Nagaoka, Niigata, Japan
[2] Lawrence Livermore National Laboratory, Livermore, California, USA

E-mail: akie_inoue@stn.nagaokaut.ac.jp

Abstract. By visualizing accelerator system components in heavy ion inertial fusion, the connection between the components becomes clear. We clarify an influential component on the entire cost by the relation of node connections due to the visualization result. Since a low cost component affects a high cost component, not only the cost estimation but also the relation between the components is considerable and important issue. A cost estimation result changing with an induction core cost indicates no influences in the rate of details.

1. Introduction

Inertial confinement fusion (ICF) is a method for realizing a fusion energy system. Several energy drivers such as laser, pulsed-power discharge induced X-ray, and charged particle beams are candidates for an ICF system. Heavy ion inertial fusion (HIF) is one of the methods for ICF driven by intense heavy ion beams. To accelerate heavy ions, we must consider the space charge effect, because of the high-current beams are required in the HIF system. For this reason, as methods to accelerate the heavy ions, various accelerator concepts are proposed as the HIF driver [1]. However, all the methods require a huge particle accelerator. Moir [2] has estimated the cost of HIF system "HYLIFE-II", and compared the costs in the cases for lasers and heavy ion beams as the energy driver.

Figure 1 shows the typical system with a linear accelerator (linac) [3]. In this system, each component is connected to the linac. In the case of recirculator as shown in Fig.2, several accelerator components such as an injector, induction units, bending magnets, injection and extraction kicker systems, and so on, are included in the system. After the acceleration of the heavy ions by using the small circular accelerator, the heavy ions transit to next circular accelerator (medium energy ring), and are further accelerated. Figures 3 and 4 indicate the connection between the components at each system. However, these costs and connections are difficult to design the HIF system because of the unclear and complex connections of the each component.

In this study, we researched the cost evaluation and the visualization of system network. To visualize the HIF systems and the cost weights proposed by Meier et al [4] and Newton et al [5,6], we have used Gephi [7], which is software to visualize the network connections. The cost of the HIF system has been suggested various estimations including the details. However, the relation between the components in HIF system is unclear. To understand the relations, we considered a correlation diagram of HIF system using Gephi.

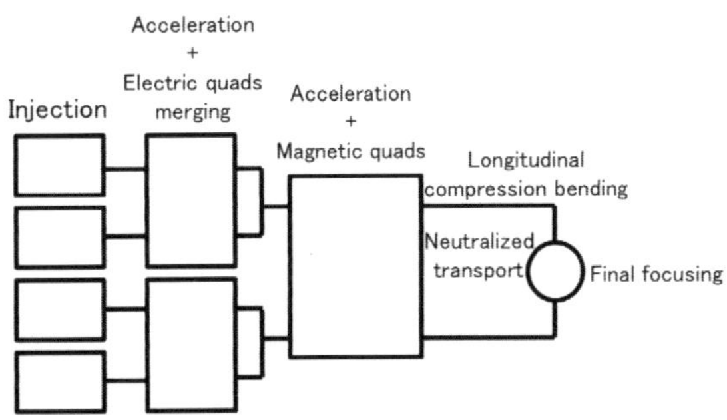

Figure 1. HIF system with linac [3]

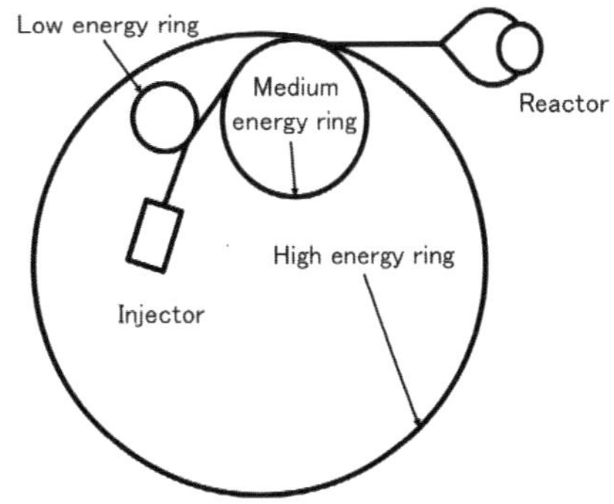

Figure 2. HIF system with recirculator [5]

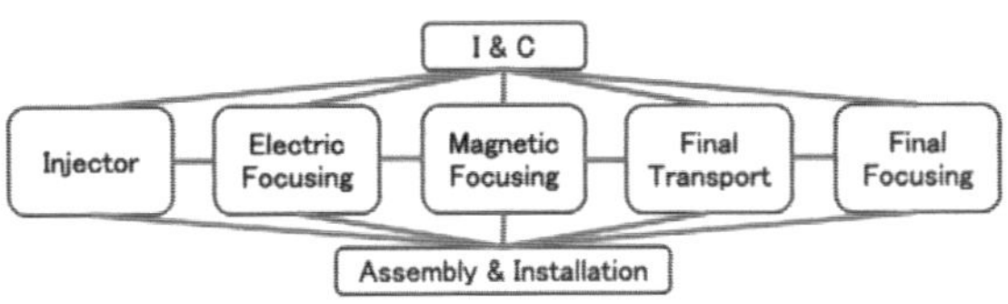

Figure 3. Component model of linac system

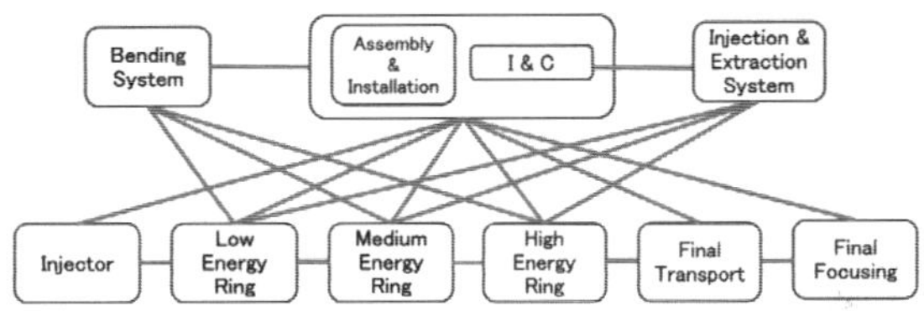

Figure 4. Component model of recirculator system

2. Classification of cost and components

Table 1. Meier's system (Linac) cost

Component		Cost [M$]	Rate [%]
Operation Cost	Assembly and Installation	322.0	23.0
	I&C	112.0	8.0
Total Accelerating System Cost	Electrostatic Focusing	140.0	10.0
	Magnetic Focusing	672.0	48.0

Table 2. Newton's system (Recirculator) cost

Component		Cost [M$]	Rate [%]
Operation Cost	Administrative	38.5	7.7
	Controls	18.3	3.7
	Engineering	38.5	7.7
	Facilities	54.4	10.9
	Installation	38.5	7.7
Total Accelerating System Cost	Alignment system	30.6	6.1
	Focus magnet system	44.7	8.9
	Prime power system	45.0	9.0
	Vacuum system	52.0	10.4
	Accelerating system	67.1	13.4

We show the details of "Operation cost" and "Total Accelerating System Cost" in Tables 1 and 2. Tables 1 and 2 show the cost details of Operation and Total Accelerating System at each system. Because each HIF accelerator system has different components for the above cost estimations, we categorized the individual element costs as "Operation cost" and "Total Accelerating System Cost." The cost and the rate (percentage of the individual element cost to the total cost) were cited from Refs.[4] and [5,6] in the tables.

3. Visualization of cost and connection between components for accelerator system

3.1. Visualization of connection between components of accelerator

Figure 5 shows the visualization of both HIF systems, according to the cost estimation summarized in the previous section and the component models as shown in Figs.3 and 4. The circle size shows the cost, and the circle color shows the number of in-degree, which is the incident number of arrows into the node. In the higher value of in-degree, the color becomes thick. It implies that a starting point node (component) of an arrow affects a terminal node of the arrow.

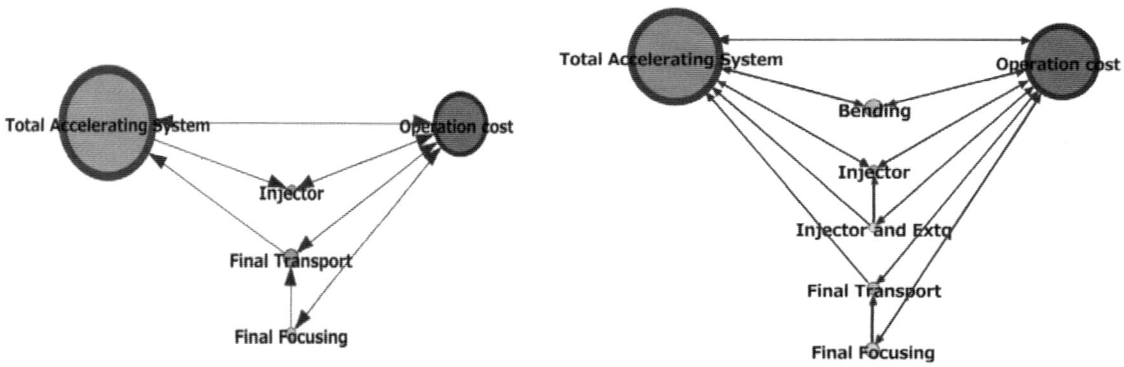

(a) Meier's system (Linac) (b) Newton's system (Recirculator)

Figure 5. Visualization of connection between accelerator components for linac and recirculator systems.

We assume that the HIF accelerator system is designed by the requirements from the fuel pellet side. The Total Accelerating System is the highest in all the costs, and the Operation cost is the second component. The number of connected nodes in the Total Accelerating System and the Operation Cost is almost same.

In the Meier's system model, the number of nodes connecting with the Total Accelerating System is three, and the number of nodes connecting with the Operation cost is four. On the other hand, the number of nodes connecting with the Total Accelerating System is five, and the number of nodes connecting with the Operation cost is five in the Newton's system model. Because, the Newton's system model has the components for the Bending and the Injection and Extraction due to a recirculator configuration, while the Meier's system model has no component for the Bending and the Injection and Extraction, which the system consists of a linac. As shown in Fig.5(a), in the Meier's system model, the cost of the Final Transport component is small. However the Final Transport affects the Total Accelerating System, which is the highest cost component. It is predicted that the modifications of the Final Transport design give the changes of the Total Accelerating System. As a result, a cheap component such as the Final Transport may control an expensive component.

In the case of Newton's system model, as shown in Fig.5(b), the bending and the Injection and Extraction components are extra equipment in comparisons with the linac system. The Injection and Extraction component affects directly the Total Accelerating System and the Injector components. For this reason, the Injection and Extraction component affects indirectly the Total Accelerating System via the Injector. Consequently, although the cost of the Injection and Extraction is cheap, the influence is not low impact on the entire system design and cost.

In both the systems, the Total Accelerating System and the Operation costs are high, and have large number of nodes connecting to the other components. As a result, both the components are critical in the HIF system. However, as mentioned above, a node with the low cost such as the Final Transport affects the high cost component such as the Total Accelerating System. Consequently, not only the cost estimation but also the relation between the components is considerable issue.

3.2. System cost estimation for induction core cost

Figure 6 shows the total cost and the details at each system. The two graphs at the left side show the original costs in Refs. [4] and [5,6]. The Total Accelerating System and Operation cost are accounted for the majority of the total cost. The two graphs at the right side show the details of costs after changing the cost of the induction core. The cost of induction core in the Newton's system model was 5 \$/kg. On the other hand, the cost was estimated as 20 ~ 150 \$/kg [8]. For this reason, we also estimated the total system cost when the cost of the induction core quadrupled. In comparison with the estimations, the total cost increases, however it was found that the rate of cost details is similar.

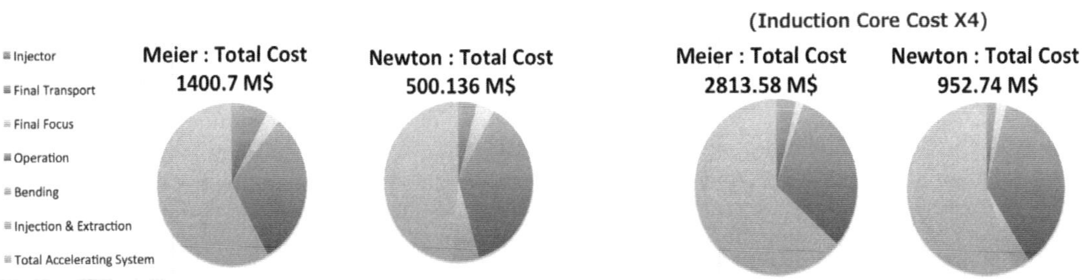

Figure 6. Comparison of cost as a function of induction core cost at each system model

4. Conclusion

By visualizing the components of accelerator system in the HIF using Gephi, the connection between the components became clear. Because it was difficult to understand from the only cost estimation, the component network visualization has a possibility to find the unclear fact.We clarified an influential component on the entire cost by the in-degree of the node. Because the Operation cost is connected to many nodes, it is considered particularly important element to determine the total cost. In addition, element number of system is also important. The Newton's system has more components than that for the Meier's system. It suggested that each component is closely related to other components. Not only the cost estimation but also the relation between the components was considerable and important issue. The induction core cost was main for the HIF accelerator system. The cost estimation changing with the induction core cost indicates no influences in the rate of the details. Considerations of any other costs in the above, such as social cost, public relation cost, and ethical education cost, are in our near future work for the economics in the HIF system.

References

[1] Barnard J J *et al* 1998 *Nucl. Instrum. Methods Phys.* Res. **A415** p 218
[2] Moir R W 1998 *Proc. ICENES-9* p 12
[3] Bangerter R O 1993 *Il Nuovo Cimento* **106A** 1445
[4] Meier W R *et al* 1998 *Nucl. Instrum. Methods Phys.* Res. **A415** p 249
[5] Barnard J J *et al* 1993 *Phys. Fluids* **B5** p 2698
[6] Newton M A *et al* 1991 *Proc. Particle Accelerator Conf.* p 2595
[7] Bastian M et al 2009 "*Gephi: an open source software for exploring and manipulating networks*" *International AAAI Conference on Weblogs and Social Media.*
[8] Molvic A W and Faltens A 2002 *Phys. Rev. ST Accel. Beams* **5** 080401

Journal of Physics: Conference Series **717** (2016) 012098 doi:10.1088/1742-6596/717/1/012098

Mechanical design of experimental apparatus for FIREX cryo-target cooling

A Iwamoto[1], T Norimatsu[2], M Nakai[2], H Sakagami[1], S Fujioka[2], H Shiraga[2] and H Azechi[2]

[1]National Institute for Fusion Science, Toki, Gifu, Japan
[2]Institute of Laser Engineering, Osaka University, Suita, Osaka, Japan

E-mail: iwamoto.akifumi@LHD.nifs.ac.jp

Abstract. Mechanical design of an experimental apparatus for FIREX cryo-target cooling is described. Gaseous helium (GHe) sealing system at a cryogenic environment is an important issue for laser fusion experiments. The dedicated loading system was designed for a metal gasket. We take U-TIGHTSEAL® (Usui Kokusai Sangyo Kaisha. Ltd.) with an indium plated copper jacket as an example. According to its specification, a linear load of 110 N/m along its circumference is the optimum compression; however a lower load would still maintain helium (He) leak below the required level. Its sealing performance was investigated systematically. Our system demanded 27 N/mm of the load to keep He leak tightness in a cryogenic environment. Once leak tightness was obtained, it could be reduced to 9.5 N/mm.

1. Introduction

Cryogenic target layering of First Ignition Realization Experiment (FIREX) has been studied. A target is cooled by ambient GHe at low pressure with precise temperature control. To make sure of He leak tightness, mass-produced metal gaskets, such as Helicoflex delta®(Garlock) and U-TIGHTSEAL® (Usui Kokusai Sangyo Kaisha. Ltd.), have been applied for vacuum sealing in a dedicated apparatus [1], and their validity has been confirmed in a cryogenic environment. According to their instructions, the optimum linear loads to metal gaskets are more than a hundred and several tens N/mm along its circumference. Therefore, mechanical structures must be rigid for their applications. To date, we have employed a standard configuration with a pair of flanges tightened by bolts.

For laser fusion experiments, 10's of ms before laser irradiation, a cryogenic target must be exposed to target chamber space. Based on the ILE cryogenic target system which was produced for central ignition experiments [2], its target container is separated to the upper and bottom parts. It must be He leak tight for target cooling. Instead of the above-mentioned standard configuration, rigid structures sustain pushing a pair of flanges against each other, and indium, which is good at a cryogenic seal [3], is used as a sealing material. Its disadvantage is that expertise on indium sealing is required. For the routine of laser experiments, stable sealing performance is demanded.

One solution would be metal gasket seal. To incorporate its stable sealing performance within a cryogenic system for laser fusion experiments, not the optimum but a required load to keep He leak tightness is substantial for system design. We designed and constructed a preliminary seal and loading system for cryogenic use. Onset compression loads of GHe leak are researched systematically. The required load on a metal gasket and related system design are discussed.

2. Mechanical design of apparatus

2.1. Design of cryogenic seal with metal gasket

Metal gaskets possess stable seal performance when an industrial product is manufactured to high tolerances. Figure 1 represents an example of a metal gasket. It consists of a spring and jacket. The elasticity of the spring and the plasticity of the jacket make the metal gasket a high performance seal. Figure 2 shows a characteristic curve. There is a hysteresis effect because of permanent deformation of the spring and jacket. The leak rate will have dropped below the required level when a load reaches F0 on the compression curve. Applications of the optimum compression load F2 are recommended according to instructions. Decompressing the load below F1 results in the leak rate exceeding the required level again. The compression load between F0 and F2 is the design point of our cryogenic sealing system.

A metal gasket loading system and target holder with a flange is shown in Figure 3. A screw type loading mechanism with a load cell can control the load to a metal gasket precisely. A spring section makes the stress to a metal gasket uniform. To reduce heat in-leak to cryogenic parts, Glass Fiber Reinforced Plastic (GFRP) is used in the bottom part of the loading system. The target holder is inserted into a target can as shown in figure 4. The counterpart of a pair of flanges is equipped on the target can with a groove. A metal gasket U-TIGHTSEAL® with an inner diameter of 81 mm and a height of 3.1mm (see figure1) is employed. We select an indium plated copper jacket gasket for cryogenic use. The indium coating must ensure He leak tightness. The announced optimum linear compression load is 110 N/mm along its circumference; requiring a total compression of 29.0 kN on the gasket. Highly rigid structures would be required to be proof against a load of this magnitude. To combine structural design with high cooling performance, a metal gasket should be used at a low compression load range. Its possibility is investigated in the next chapter.

Figure 1. An example of metal gasket, U-TIGHTSEAL®. It is a gasket with an indium plated copper jacket.

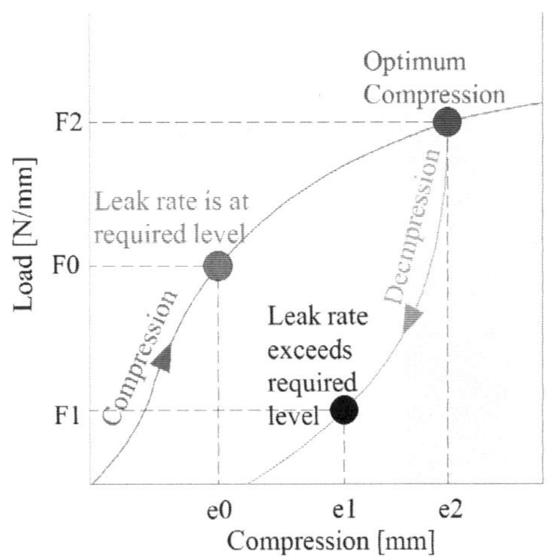

Figure 2. A characteristic curve of metal gasket. After a cycle of compression and decompression, permanent deformation of a spring and jacket remains.

2.2. Cooling system with reduced vibration

For system cooling, a 4 K Gifford-McMahon (GM) cryocooler (Sumitomo Heavy Industries, Ltd.) is employed. It becomes a vibration source. For target observations, a vibration free system is ideal. The target can is connected to the cryocooler using flexible thermal conductive links and is supported by a different structure with vibration control from the cryocooler. Vibration reduction has been demonstrated, and cooling performance has been reported [4].

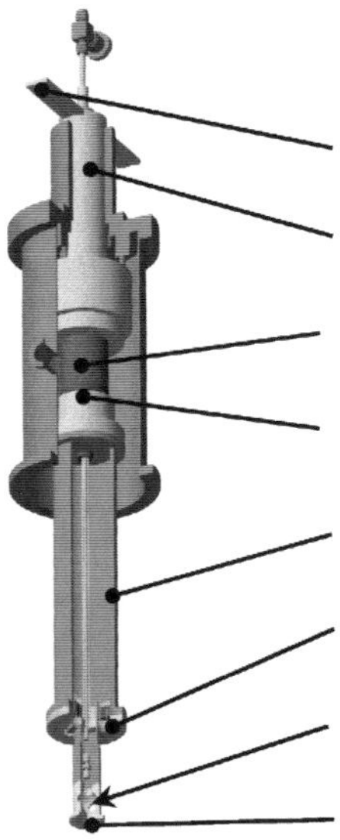

Figure 3. Gasket loading system.

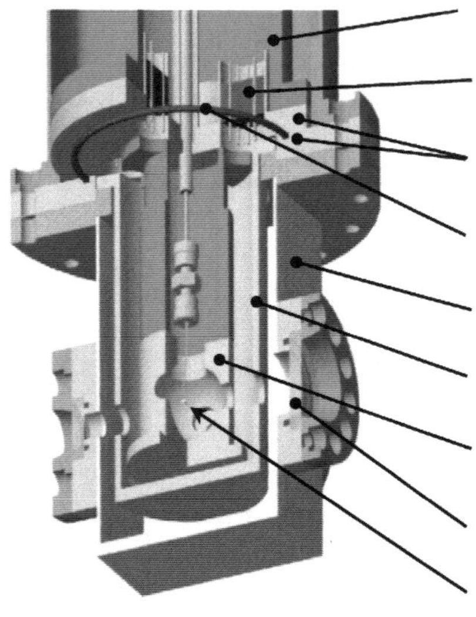

Figure 4. Target can. The target holder is inserted in it.

3. Experiments

A required compression load on the metal gasket is researched. Figure 5 shows the diagram of He leak tests. The gasket was compressed in the range of 8.6 N/mm to 27 N/mm. GHe was filled in the target can at ~1 kPa. Its sealing performance was evaluated using a He leak detector (HELEN A-221M-LD, Anelva Corporation) which can measure a leak in the range of 10^{-5} to 10^{-12} Pa m^3/s. The threshold intrinsic He leak rate was decided to be 1.0 x 10^{-11} Pa m^3/s. The measurements were systematically conducted at room temperature and in a cryogenic environment.

Initial compression loads of 14 N/mm and 27 N/mm were tested. At first, a He leak rate with a load of 13.7 N/mm was measured at room temperature. As filling the target can with GHe at 1.07 kPa, no leak was detected under the condition with 3.0 x 10^{-11} Pa m^3/s of a background leak rate. Then the system was cooled down to ~10 K. The GHe pressure was decreased to 49 Pa and the load was settled down to 14.3 N/mm. At this point, a leak rate was 2.9 x 10^{-11} Pa m^3/s. As GHe was added into the target can up to 1.00 kPa, the leak rate increased to 7.6 x 10^{-11} Pa m^3/s. A load of 14.3 N/mm is not large enough to reach the allowable leak rate level. In spite of the leak, a 200 L/s class turbo molecular pump could keep an insulation vacuum. After the measurements, the system was warmed to room temperature and the used metal gasket was replaced with a new one. A He leak rate with a 26.7 N/mm load was measured at pressure of 1.03 kPa in the target can. No leak was detected under the condition with 1.5 x 10^{-12} Pa m^3/s of a background leak rate. Then the system was cooled down to 7.4 K. The GHe pressure was decreased to 48 Pa and the load was settled down to 26.9 N/mm. While the pressure added to 1.02 kPa, the leak rate was constant at 1.2 x 10^{-12} Pa m^3/s. Eventually, no leak was detected. While decompressing the gasket by ~0.96 N/mm, leak rates were measured. The system could keep He leak tightness down to 9.46 N/mm, and then an intrinsic He leak of 8.1 x 10^{-11} Pa m^3/s was detected at 8.61 N/mm. When it was recompressed to 14.5 N/mm, He leak tightness was recovered. After the successful first cool down, the system was warmed up to room temperature and the load was released. Using the same metal gasket, the second cycle was started. A load of 26.6 N/mm could keep He leak tightness against 1.13 kPa GHe pressure at room temperature again. A back ground leak rate was 6.9 x 10^{-11} Pa m^3/s. The system was cooled down to 7.4 K. While the GHe pressure was increased from 49 Pa to 1.02 kPa, the leak rate was stable at 5.6 x 10^{-11} Pa m^3/s. While decompressing the gasket by ~1.9 N/mm, leak rates were measured. When the loads were decreased to 17.3 N/mm and then 15.3 N/mm, measured rates monotonically increased to 1.0 x 10^{-10} Pa m^3/s and 2.1 x 10^{-10} Pa m^3/s, respectively. When the gasket was recompressed again to 24.0 N/mm, He leak tightness was recovered.

After the second cool down, the system was warmed to room temperature and the compression was unloaded. In the third cycle, a load of 26.5 N/mm could keep He leak tightness against 1.27 kPa GHe pressure at room temperature. After cooling down to 7.3 K, no leak was detected. A back ground leak rate was 1.9 x 10^{-11} Pa m^3/s. Decompression to 19.2 N/mm allowed an intrinsic He leak of 1.1 x 10^{-9} Pa m^3/s, and then recompression to 26.9 N/mm could keep He leak tightness again. Figure 6 shows intrinsic He leak rate versus linear compression load in a cryogenic environment. Because of the hysteresis effect of metal gasket, the onset load of He leak in the decompressing process becomes high by cooling cycles. Judging from the experimental results, a compression load of 27 N/mm was large enough to keep He leak tightness, and furthermore the metal gasket has the potential for multi-time use. We would combine high cooling performance with metal gasket sealing for a cryogenic system.

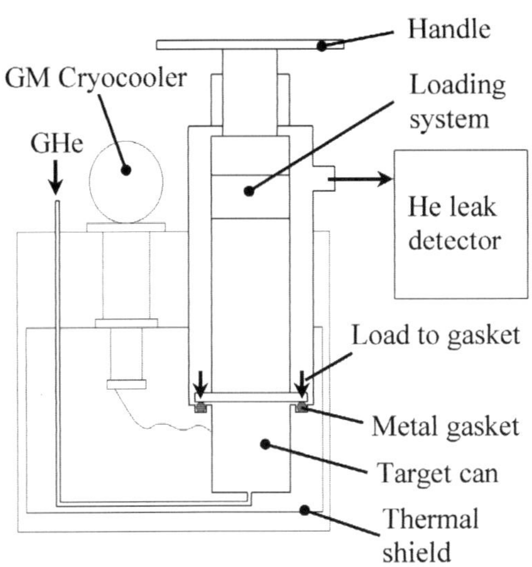

Figure 5. Diagram of He leak test.

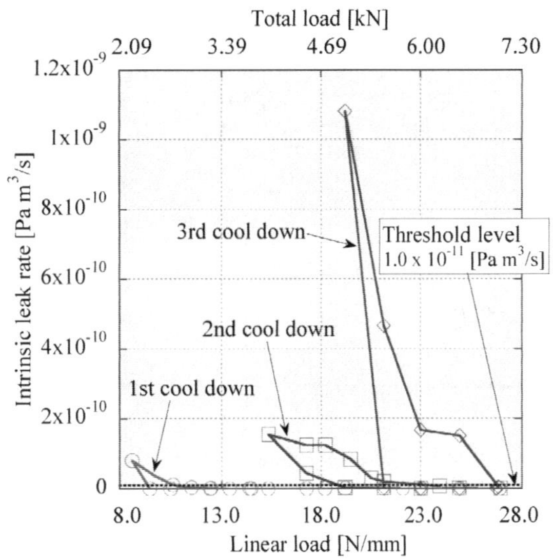

Figure 6. Intrinsic He leak rate versus linear compression load at 7.3 - 7.4 K in the case of the initial load of 27 N/mm.

4. Summary

For the mechanical design of a cryo-target apparatus, the sealing performance of U-TIGHTSEAL® with an indium plated copper jacket was systematically tested. Despite the optimum compression linear load of 110 N/mm, the required load is found to be 27 N/mm for our system, and therefore high rigid structure is not demanded. Furthermore it possesses the potential for multi time use. High cooling performance would be combined with a metal gasket sealing system.

References

[1] Iwamoto A, Maekawa R, Mito T, Okamoto M, Motojima O, Sugito S, Okada K, Nakai M, Norimatsu T and Nagai K 2006 Cool-down performance of the apparatus for the cryogenic target of the FIREX project *Fusion Eng. Des.* **81** 1647-52

[2] Norimatsu T, Katayama H, Mano T, Takagi M, Kodama R, Tanaka K A, Kato Y, Yamanaka T Nakai S, Nishino Y, Nakai M, Yamanaka C 1988 Fabrication of a cryogenic target for inertial confinment fusion experiments *J. Vac. Sci. Technol.* **A 6** 3144-47

[3] Iwamoto A, Maekawa R, Mito T, 2001 Kapitza conductance of an oxidized copper surface in saturated He II *Cryogenics* **41** 367-371

[4] Iwamoto A, Norimatsu T, Nakai M, Sakagami H, Shiraga H, Azechi H, 2016 Cool-down performance of the new apparatus for fuel layering demonstrations of FIREX targets *to be published in J. Phys.: Conf. Ser.*

An accelerator facility for WDM, HEDP, and HIF investigations in Nazarbayev University

M. Kaikanov[1,4], K Baigarin[1], A Tikhonov[1], A Urazbayev[1], J W Kwan[1], E Henestroza[1], G Remnev[2], B Shubin[2], A Stepanov[2], V Shamanin[2] and W L Waldron[3]

[1]National Laboratory Astana, Astana, Kazakhstan
[2]Tomsk Polytechnic University, Tomsk, Russia
[3] Lawrence Berkeley National Laboratory, Berkeley, USA

[4]marat.kaikanov@nu.edu.kz

Abstract. Nazarbayev University (NU) in Astana, Kazakhstan, is planning to build a new multi-MV, ~10 to several hundred GW/cm^2 ion accelerator facility which will be used in studies of material properties at extreme conditions relevant to ion-beam-driven inertial fusion energy, and other applications. Two design options have been considered. The first option is a 1.2 MV induction linac similar to the NDCX-II at LBNL, but with modifications, capable of heating a 1 mm spot size thin targets to a few eV temperature. The second option is a 2 - 3 MV, ~200 kA, single-gap-diode proton accelerator powered by an inductive voltage adder. The high current proton beam can be focused to ~1 cm spot size to obtain power densities of several hundred GW/cm^2, capable of heating thick targets to temperatures of tens of eV. In both cases, a common requirement to achieving high beam intensity on target and pulse length compression is to utilize beam neutralization at the final stage of beam focusing. Initial experiments on pulsed ion beam neutralization have been carried out on a 0.3 MV, 1.5 GW single-gap ion accelerator at Tomsk Polytechnic University with the goal of creating a plasma region in front of a target at densities exceeding ~10^{12} cm^{-3}.

1. Introduction

Nazarbayev University in Astana, Kazakhstan, is planning to build a new multi-MV, ~10 to several hundred GW/cm^2 ion accelerator research facility within a few years. The facility will be used in studies of material properties at extreme conditions, for hydrodynamic experiments relevant to ion-beam-driven inertial fusion energy, and other beam applications. This new accelerator facility, which is named Nazarbayev University Research Accelerator (NURA), is conceived to follow one of the two design options described below.

The first and more mature option is a 1.2 MV induction linac with an architecture similar to the Neutralized Drift Compression Experiment (NDCX-II) at LBNL [1, 2], but with two important changes. The second option being considered is a 2-3 MV, ~200 kA, single-gap-diode proton accelerator based on an inductive voltage adder. The high current proton beam pulse of ~ 50 ns is focused to ~1 cm spot size to obtain power densities of several hundred GW/cm^2, in order to heat thick targets to temperatures of tens of eV. In comparison to the linac, the ion beam generated by a high-power ion diode typically has higher current but lower brightness (ion temperature of >10 eV instead of 0.1 - 0.5 eV for the linac). The longer pulse length due to the high power diode accelerator is more favorable for studying High Energy Density Physics (HEDP) hydrodynamics at the tens of nanoseconds time scale and at tens of eV temperature. An innovative solution to shorten the pulse length is needed in order to use a diode accelerator for (sub-ns) isochoric heating of materials at Warm Dense Matter conditions.

In both cases, a common requirement to achieving high beam intensity on target and pulse length compression is to utilize beam neutralization at the final stage of beam focusing. For example, NDCX-II makes use of a neutralized drift section to compress the beam pulse length to sub-ns. The required neutralizing plasma density near the target is n $\sim 10^{13}$ cm^{-3}. Due to the much higher beam current, the single-gap ion diode may require even higher plasma densities to neutralize the beam charge.

Initial experiments on pulsed ion beam neutralization has been carried out on a 0.3 MV, 1.5 GW single-gap ion accelerator at Tomsk Polytechnic University [4]. This accelerator can deliver hundreds of amperes of proton beam with kinetic energy of a few hundred keV. The goal is to create a plasma region in front of a target, at densities exceeding n $\sim 10^{12}$ cm^{-3}, for studying beam neutralization and compression. Progress of this experiment and the NURA facility design is reported in this paper.

2. NURA linear induction accelerator

The initial concept for NURA was to use the NDCX-II technology [1, 2] to build an accelerator for WDM studies. A modified design was completed in 2014 for NURA with changes to optimize cost and performance [3]. The main differences from the original NDCX-II design was that NURA will use helium ions (instead of lithium) and electrostatic quadrupole beam transport (instead of solenoids).

Through neutralized drift compression, the 70 nC beam charge at the target will have < 1 ns pulse duration, which will be focused to a beam spot size about 1 mm to reach power densities of ~10 GW/cm^2. This short beam pulse can isochorically heat a thin target to a temperature > 1 eV before hydro-expansion takes place, thus suitable for warm dense matter (WDM) studies.

In the next section we describe an alternate option for building NURA based on a Single-Gap accelerator (High-Power Diode) design. In this case, a higher beam power can be delivered but over a longer pulse length, thus allowing hydrodynamic motion of the heated target material.

3. NURA single-gap diode accelerator powered by inductive voltage adder

While building a NURA induction linac can be viewed as following the approach of Heavy Ion Fusion (HIF), an alternate option for NURA is to follow the approach of Light Ion Fusion (LIF) which uses magnetically insulated single-gap diode accelerator. It has already been demonstrated that light ion beams can deliver up to several TW/cm^2 to heat targets to hundreds of eVs.

Both the HIF and LIF programs had gone through several decades of hardware and beam physics development. Although HIF is currently the leading concept for ion-beam driven inertial fusion, the light ion diode machine can still be a competitive and cost effective tool in providing high power ion beam pulses for HEDP studies, as long as it is not being evaluated as an inertial fusion driver.

The pulse length from a single-gap diode is typically in the range of several tens of ns (without invoking longitudinal beam compression). In this time scale, the heating process is not isochoric. For example, the sound speed of the gold plasma is about 2.5 km/s. For a few micrometer thick targets, the time scale for hydro motion is about few ns. Thus the single-gap accelerator is not suitable for WDM studies because target heating would not be isochoric. A NURA facility based on the light ion diode accelerator can be useful for experiments to study hydrodynamics of high energy density materials such as acoustics, phase-transitions, plasma formation, implosion, and radiation. Furthermore, the same machine can be used to generate high power electron beams for hydrodynamics and radiation studies.

3.1. Design parameters for a NURA single-gap diode

Several single-gap accelerators using pulsed-power technology were built during the 1980's and 1990's by various LIF programs, including KALIF, ETIGO, PROTO, GAMBLE, [5, 6, 7, 8] etc. For example, KALIF was based on a pulsed-forming line that can deliver 50 ns, 1.7 MV, 600 kA beam pulses. The generated proton beam could produce up to 1 TW/cm^2 of heating power on target. On the other hand, KALIF-HELIA [9] has a pulsed-power generator based on an Inductive Voltage Adder (IVA) which produces 40 ns, 6 MV, 400 kA beam pulses (2.4 TW on target). The IVA structure provides an upgrade path to higher voltages and higher power.

Figure 1 shows a conceptual drawing of the NURA Single-Gap Accelerator based the KALIF-HELIA approach. It will deliver 30 ns, 2 MV, 500 kA proton beam pulses. If the beam can be focused down to a 1 cm diameter spot size, the power density on target will be 1 TW/cm^2 and the material can be heated to ~10 eV temperature. The voltage, current, pulse length, and repetition rate are not considered major difficulties in such a design. We believe that the most challenging task is generating a good quality beam from a carefully designed ion source, and a low aberration focusing system, such that the ion beam can be focused to a spot size small enough to meet the power density requirement.

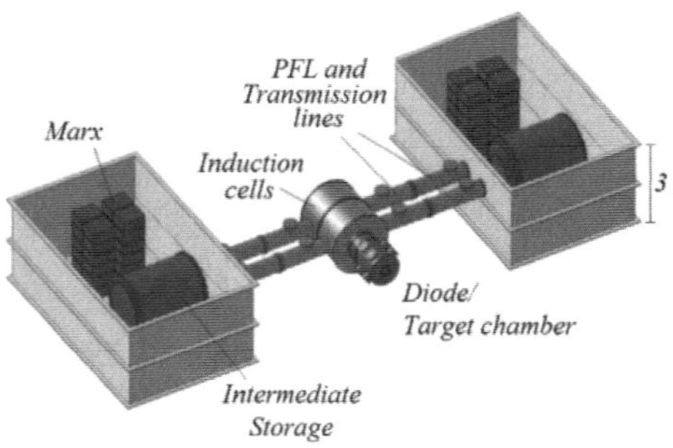

Fig. 1. NURA single-gap diode accelerator.

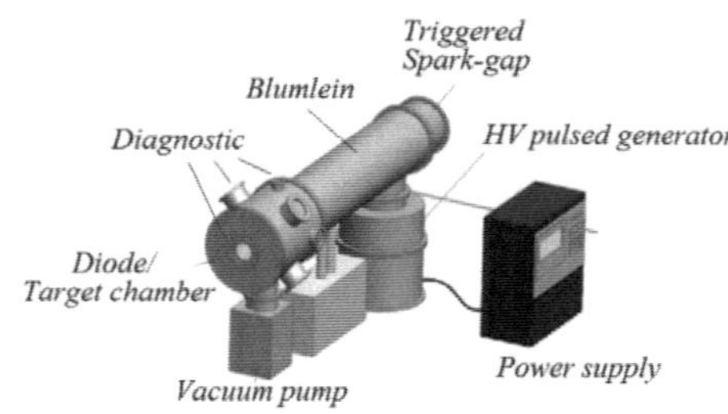

Fig. 2. A single-gap diode accelerator test stand.

3.2. A single-gap accelerator test stand is being constructed at NU
At present, NU, in collaboration with Tomsk Polytechnic University, is building a scaled version of the NURA single-gap diode accelerator. This pilot project is expected to be completed in one year. The test stand, as shown in Fig. 2, is designed to produce 10 kA of protons at 600 keV with a 30 ns pulse length. Beam power on target will be fractions of GW/cm^2 depending on the focal spot size (<5 cm). The test stand will be used to study beam transport and neutralization, for technical development of the NURA components including target diagnostics, and for other high intensity charged particle beam applications such as material modifications.

4. Neutralized beam transport experiment
A recent experiment was carried out at Tomsk Polytechnic University using the TEMP-4 single-gap accelerator to examine the effect of beam neutralization by a plasma. The accelerator produces 10 kA of proton beam in a conical shape with a kinetic energy of 370 keV and a pulse length of <100 ns [6]. Figure 3 shows the experimental layout.

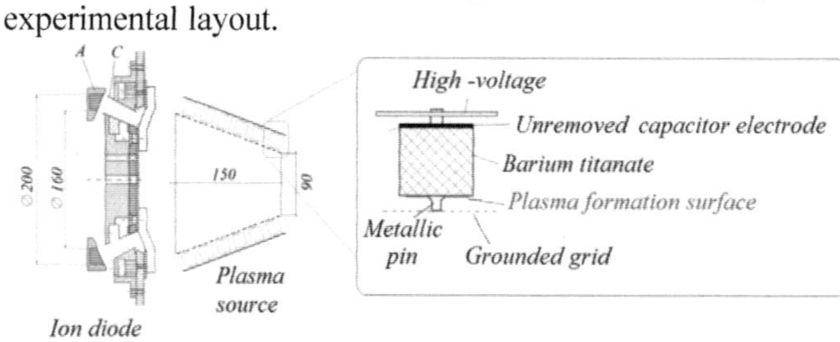

Fig. 3. Experiment layout of the Neutralized Transport Experiment.

In the standard configuration, the beam is neutralized by a background air pressure of $5*10^{-4}$ -10^{-3} Torr. In this experiment, the diode chamber was pumped down to below 10^{-5} Torr to minimize the presence of residual gas and a plasma was injected into the beam path before the beam arrival.

Figure 4 shows the conical ferroelectric plasma source that was built to generate plasma by flashover discharge. The plasma source is a hollow cone made of an array of barium titanate ceramic capacitors with relative permittivity ε = 900 - 1000. High voltage (17 kV, 4 μs) is applied to the outer electrode and a grid on the inner surface connects all capacitors to ground. Flashover discharge, starting at the triple points of metal-ceramic-vacuum contact, is formed on the ceramic surface and the generated plasma flows through the grounded grid to fill the inner region of the conical plasma source.

Outer high-voltage electrode   Plasma formation region

Fig. 4. Conical plasma source.

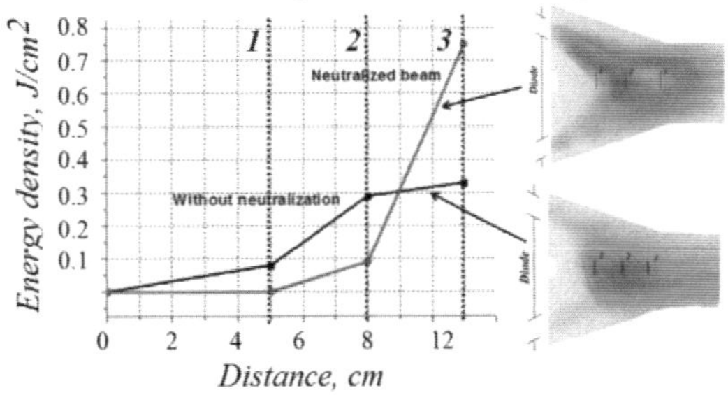

Fig. 5. Result of the Neutralized Transport Experiment.

By using a biased Faraday cup we estimated the plasma density to be > 10^{10} cm^{-3} near the grounded grid of the plasma source, where the beam is expected to be transported. A 1.5 cm diameter copper target was used a as calorimeter to measure the total energy of the ion beam at different distances from the cathode (5, 8, and 11 cm). The results showed that the ion beam had been neutralized, thus affecting the beam focus. The energy density of the ion beam at the focal spot increased by a factor of 2, with corresponding radiosensitive film confirming similar results as shown in Fig. 5.

5. Summary and conclusion

We have considered two options for NU to enter the field of High Energy Density Physics. An NDCX-type accelerator facility for WDM experiments, or a single-gap diode type accelerator that is more suitable for ICF-relevant hydrodynamic experiments. The second option can also be used for other high power beam pulse applications including electron beams. Within one year, a single-gap accelerator test-stand will be available for experiments at NU. In the mean time, we have started conducting beam neutralization experiments using the ion gun at Tomsk Polytechnic University.

6. Acknowledgements

The authors would like to thank H.J. Bluhm, P. Ni, A. Ratti, P. Seidl, J. Barnard, B. Cauble, D. Hoffmann, I. Kaganovich, R. Sazonov, K. Nielsen, and I. Smith for valuable discussions and inputs. Their participation has provided encouragement to advancing the NURA project.

References

[1] P.Seidl, et al., "Short-Pulse Ion Beams at Berkeley Lab", this conference
[2] W.L. Waldron, et al. Nucl. Inst. Methods Phys. Res., vol. A733, 226 (2014).
[3] K. Baigarin, et al., Proc. 20th International Symposium on Heavy-ion Inertial Fusion, Lanzhou, China, August 11-15, 154 (2014).
[4] E.G. Furman et al. Technical Physics, Vol. 52, No. 5, 621 (2007)
[5] H. Bluhm, et al, Proc. 5th IEEE Int. Pulsed Power Conf., Arlington, VA, p 114 (1985)
[6] G. E. Remnev, et al., Surf. Coat. Technol. 114, 206 (1999).
[7] D. Johnson, et al, Journal of Applied Physics 53, 4579 (1982)
[8] G. Cooperstein, et al, Proc. 3rd Int. Conf. on High-Power Electron and Ion Beam Research & Technology, p 567 (1979)
[9] P. Hoppe, et al, Proc. 12th Int. Conf. on High-Power Particle Beams, BEAMS '98, p. 218 (1998)

Journal of Physics: Conference Series **717** (2016) 012100 doi:10.1088/1742-6596/717/1/012100

Numerical analysis of applied magnetic field dependence in Malmberg-Penning Trap for compact simulator of energy driver in heavy ion fusion

T Sato[1], Y Park[2], Y Soga[2], K Takahashi[1], T Sasaki[1], T Kikuchi[1], and Nob.Harada[1]

[1] Nagaoka University of Technology, Nagaoka, Niigata, Japan
[2] Kanazawa University, Kanazawa, Ishikawa, Japan

tomohiro_sato@stn.nagaokaut.ac.jp

Abstract. To simulate a pulse compression process of space charge dominated beams in heavy ion fusion, we have demonstrated a multi-particle numerical simulation as an equivalent beam using the Malmberg-Penning trap device. The results show that both transverse and longitudinal velocities as a function of external magnetic field strength are increasing during the longitudinal compression. The influence of space-charge effect, which is related to the external magnetic field, was observed as the increase of high velocity particles at the weak external magnetic field.

1. Introduction

An energy driver development is a key issue for inertial fusion energy. In inertial confinement fusion (ICF) driven by heavy ion beams (HIB), several system configurations have been considered [1-4]. These systems at the final stage of the particle accelerator system require the high-current HIB [4]. To obtain the high-current HIB, a beam bunch should be compressed more than 10 times in the longitudinal direction [5-10]. By the pulse compression in the longitudinal direction, HIB undergoes a transition from an emittance-dominated regime to a space-charge-dominated regime. Because the required parameters for HIB in ICF are far from that of the conventional beams, the beam behaviors, which is in the condition dominated by the space-charge effect, should be researched well for the energy driver development.

Instead of the HIB, the equivalent beam dynamics can be simulated with a compact experimental apparatus using the electron beam, which is in the condition dominated by the space-charge effect. For this reason, the beam dynamics simulation in the pulse compression of the HIB has been demonstrated using the Malmberg-Penning trap device [11]. Simulating beam parameters in a compact experimental apparatus using the electron beam are evaluated by the tune depression, which is an index for space charge strength in the beam bunch. In this index, the space charge strength is evaluated as the dependence on external magnetic field strength and number density on the space-charge effect [12].

In this study, we have numerically investigated the dependence on external magnetic field for the beam dynamics in the Malmberg-Penning trap device [13] as the compact simulator of HIB. For this purpose, the multi-particle numerical simulation based on particle-particle method is performed the compression process in different external magnetic field conditions, and we analyzed real and phase space distributions.

2. Malmberg-Penning trap device

A Malmberg-Penning trap device [13] is a confinement device of electron cloud, and can investigate non-neutral plasmas on controlling the density distribution. The parameters of this device are maximum external magnetic field of 0.1 T along the axial direction, and the voltage of well shaped potential barriers is -100 V at both axial ends. The electron motions are limited for the radial direction by the uniform magnetic field applied in the axial direction. The electron trapped in the axial direction by the well shaped potential with negative barriers at both ends.

The experimental simulation of the pulse compression using the Malmberg-Penning trap device is performed by manipulating the barrier voltage to reduce the space between the potential barriers.

3. Numerical simulation model

Numerical simulations are performed under the experimental conditions as longitudinal compression process of HIB using the Malmberg-Penning trap device. In the numerical simulations, ten times compression from the initial condition, in which the compression ratio is 10, was carried out using the multi-particle simulation code [14], which represents a particle-particle model [15]. The particle position $\mathbf{x}_i$ for i th super particle is calculated by

$$\frac{d}{dt}\mathbf{x}_i = \mathbf{v}_i \tag{1}$$

where the particle velocity $\mathbf{v}_i$ for i th super particle is temporally changed by the external magnetic fields and the interaction with j th super particle. The particle velocity is calculated by Newton-Lorentz equation of motion

$$m\frac{d}{dt}\mathbf{v}_i = \frac{e^2}{4\pi\varepsilon_0}\sum_{j\neq i}^{N}\frac{\mathbf{x}_i - \mathbf{x}_j}{(|\mathbf{x}_i - \mathbf{x}_j|^2 + s^2)^{3/2}} - e\mathbf{v}_i \times \mathbf{B}_{ext} \tag{2}$$

where m is the mass of electron, e is the elementary charge, ε_0 is the permittivity in vacuum, $\mathbf{B}_{ext} \equiv (0,0,B_z)$ is the external magnetic flux density, s is the softening parameter, respectively.

At the initial condition, the electron number was set to be 1.6×10^8 with thermodynamic equilibrium distribution as 1 eV. The electrons were converted to 10000 super-particles that maintaining the mass-to-charge ratio and arranged uniformly in cylindrical space with 10 mm in radius and 180 mm in axial length. The boundary conditions at the both axial ends are set to be reflection boundary for the particles. The compression time and the external magnetic flux density are 100 ns and 0.01 T and 0.1 T, respectively.

4. Calculation results and Discussions

Figure 1 shows the charge density distribution in transverse real space for each external magnetic field density of 0.01 T and 0.1 T. The color map shows the normalized charge density logarithmic scale by the maximum value at each time. In comparison to the different strengths of the external magnetic field, the particles are confined by the potential barriers and the external magnetic field during the longitudinal compression in both cases. The results also show that the charge density distributions in both the strengths of external magnetic field are similar. From the initial condition, the Larrmor radii estimated by the thermal velocity of 1 eV in the external magnetic field of 0.01 T and 0.1 T are 0.42 mm and 0.042 mm, respectively. In the case of weak external magnetic field as 0.01 T, the some particles deviate from the beam radius during the compression.

Figure 2 shows the normalized speed distribution in transverse space for each external magnetic field. At the initial condition, the normalized speed distribution in transverse fits with Maxwell distribution. As shown in Fig. 2, particles with higher velocity increase as the deviation from the Maxwell distribution for each external magnetic field. The rate of higher velocity particles in external magnetic field of 0.01 T is higher than that in external magnetic density of 0.1 T. These results mean that the transverse velocity of the particles increase the longitudinal-transverse coupling of momentum due to the Coulomb interactions during the longitudinal compression.

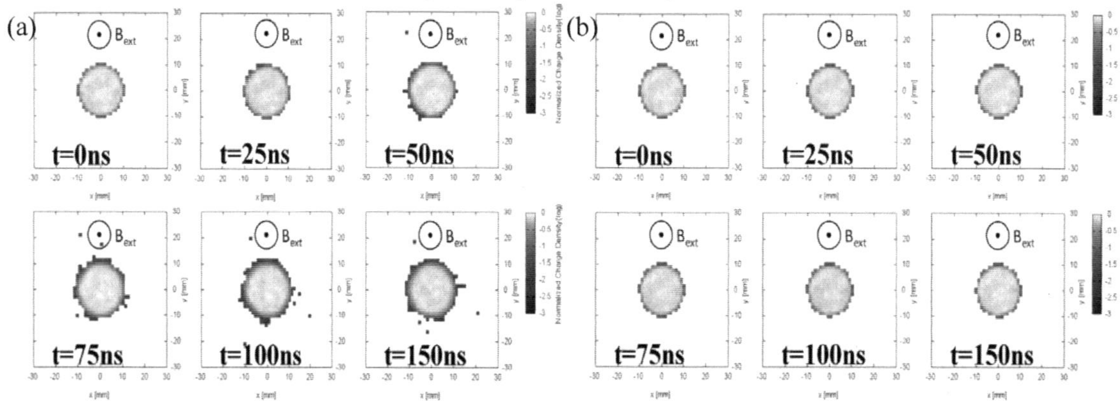

Figure 1. Normalized charge density distribution in transverse real space at each external magnetic flux density for (a) 0.01 T and (b) 0.1 T.

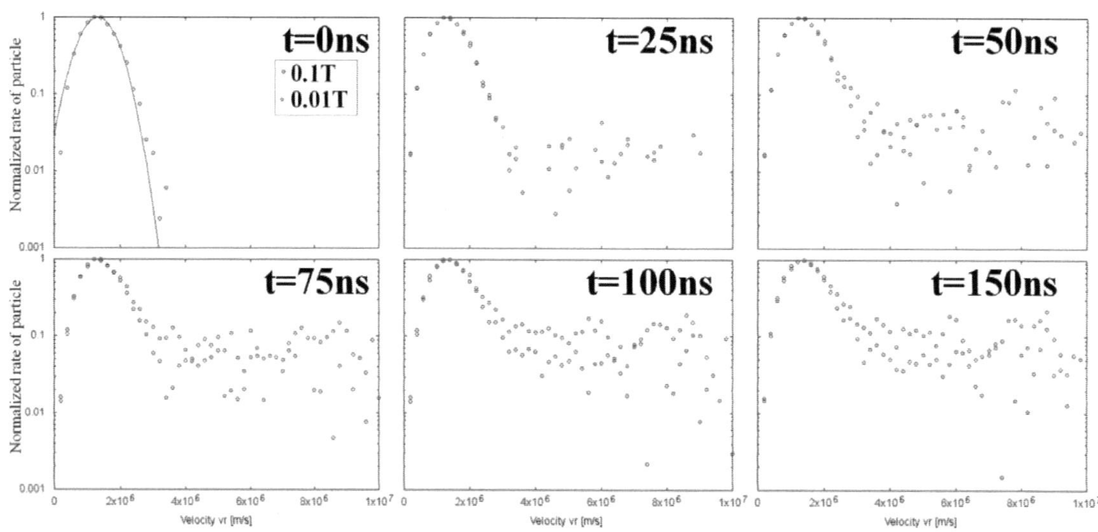

Figure 2. Normalized speed distributions in transverse direction at each external magnetic field.

Figure 3 shows the charge density distributions in the longitudinal phase space at the initial condition and at 150 ns for each external magnetic field density of 0.01 T and 0.1 T. The color map shows the normalized charge density logarithmic scale by the maximum value at each time. As shown in Fig. 3, the longitudinal velocity spread for both cases increase during the longitudinal compression. It should be noted that the longitudinal velocity distribution under the weak external magnetic field shows a larger spread in comparison to that under the strong external magnetic field. Therefore, the longitudinal velocity spread during the longitudinal compression depends on the external magnetic field.

From these results, both the transverse and longitudinal velocities increase during longitudinal compression for each external magnetic field. The Larmor radii estimated by the thermal velocity of 1 eV at each external magnetic field are considered to be negligible smaller than the beam radius. In other word, thermodynamic particle behavior is confined by the external magnetic field. However, the particles having high velocity have large Larmor radius in weak external magnetic field of 0.01 T. The some particles that deviate from the beam radius in the external magnetic field of 0.01 T would have

Journal of Physics: Conference Series **717** (2016) 012100 doi:10.1088/1742-6596/717/1/012100

resulted by the Coulomb interaction during the longitudinal compression. This result suggests that the influence of space-charge effect depends on external magnetic field, as estimated by the tune depression, which is an index for space charge strength in the beam bunch [3].

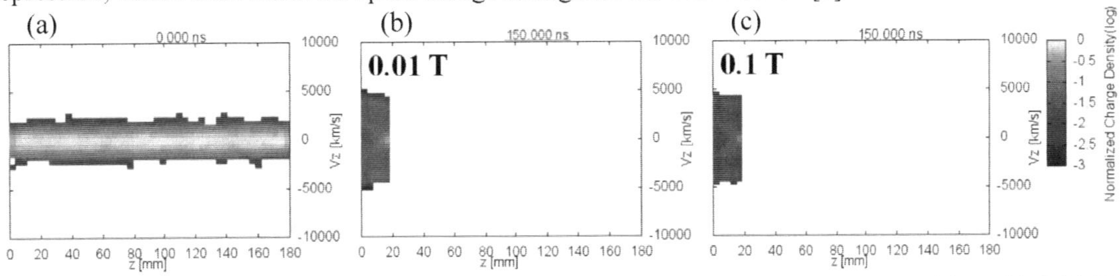

Figure 3. Normalized charge density in longitudinal phase space, (a) at initial condition, (b) 0.01 T, and (c) 0.1 T, after 250 ns in longitudinal pulse compression.

5. Conclusion

To simulate the pulse compression of space charge dominated beams for ICF driven by HIB, we have demonstrated the multi-particle numerical simulations as the equivalent beam dynamics using the Malmberg-Penning trap device. From the results of multi-particle numerical simulation for the compression process with different external magnetic fields, both the transverse and longitudinal velocities increase during longitudinal compression for each external magnetic field. In the weak external magnetic field as 0.01 T, some particles deviate from the beam radius during the longitudinal compression. The longitudinal velocity spread in the case of weak external magnetic field expands from the comparison of that in the case of strong external magnetic field. These mean that the space-charge effect influence depends on the external magnetic field.

In conclusion, the beam dynamics in Malmberg-Penning trap device has the influence of space-charge effect that depends on the external magnetic field. When thermodynamic particle behavior is confined by the external magnetic field, the influence of space-charge effect becomes strong in the case of weak external magnetic field.

References

[1] Bangerter R O 1993 *il Nuovo Cimento* **106A** 1445
[2] Prior C R 1998 *Proc. EPAC1998* 323
[3] Yu S S *et al* 2003 *Fusion Science Tech.* **44** 266
[4] Barnard J J *at el* 1998 *Nucl. Instrum. Methods Phys. Res.* A **415** 218
[5] Kim C H *et al* 1986 *AIP Conf. Proc.* **152** 264
[6] Ho D D M *et al* 1991 *Part. Accelerators* 35 15
[7] Lee E P *et al* 2002 *Laser Part. Beams* **20** 581
[8] De Hoon M J L *et al* 2003 *Phys. Plasmas* **10** 855
[9] Qin H et al *2004 Phys. Rev. STAB* **5** 034401
[10] Sharp W M *et al* 2005 *Nucl Instrum Method Phys Res* A **544** 398
[11] Park Y *et al* 2013 *NIFC PROC* **93** 84
[12] Kikuchi T et al *in this conference*
[13] Soga Y *et al* 2006 *Phys. Plasmas* **13** 052105
[14] Kikuchi T *et al* 2015 *Progress in Nuclear Energy* 82 pp126-129
[15] Hockney R W and Eastwood J W 1988 *Computer Simulation Using Particles* (Bristol and Philadelphia: IOP publishing)

Journal of Physics: Conference Series 717 (2016) 012101 doi:10.1088/1742-6596/717/1/012101

Numerical analysis corresponding with experiment in compact beam simulator for heavy ion inertial fusion driver

T Kikuchi[1], Y Sakai[2], T Komori[1], T Sato[1], J Hasegawa[2], K Horioka[2], K Takahashi[1], T Sasaki[1] and Nob. Harada[1]

[1] Nagaoka University of Technology, Nagaoka, Niigata, Japan
[2] Tokyo Institute of Technology, Yokohama, Kanagawa, Japan

E-mail: tkikuchi@vos.nagaokaut.ac.jp

Abstract. Tune depression in a compact beam equipment is estimated, and numerical simulation results are compared with an experimental one for the compact beam simulator in a driver of heavy ion inertial fusion. The numerical simulation with multi-particle tracking is carried out, corresponding to the experimental condition, and the result is discussed with the experimental one. It is expected that the numerical simulation developed in this paper is useful tool to investigate the beam dynamics in the experiment with the compact beam simulator.

1. Introduction

In heavy ion inertial fusion (HIF), which is inertial confinement fusion driven by heavy ion beams, space charge dominated beam physics is critical research topic [1]. Because the beam parameters are far from those of the heavy ion beam produced by the existing particle accelerator complex [2]. The heavy ion accelerator is undoubtedly needed for HIF, however the large size is a problem issue for the HIF research and development. For this reason, compact beam equipments with electrons were proposed and developed [3, 4, 5]. Not only the experimental approaches but also the numerical analyses with multi-particle tracking were carried out according to the experimental condition [6, 7, 8].

In particular, longitudinal pulse compression is critical manipulation in the final stage of HIF driver system. The beam current is increased drastically with the extreme pulse compression. The beam parameters are changed dynamically during the pulse compression, and achieve from an emittance-dominated regime to a space-charge-dominated regime [1, 2].

In this study, the tune depression in the compact beam equipment is estimated, and the calculation results are compared with the experimental result. The multi-particle tracking based on particle-particle method is carried out, and the numerical result is discussed with the experimental one.

2. Simulating beam parameter in estimation from tune depression

The tune depression k/k_0, which is an index for space charge strength in the beam bunch, is given by (Eq.(4.99) in Ref.[9])

$$k/k_0 = \sqrt{1+u^2} - u\,, \tag{1}$$

where u is the dimensionless parameter (Eq.(4.92) in Ref.[9])

$$u = \frac{K}{2\,k_0\,\varepsilon}\,. \tag{2}$$

Here K is the generalized perveance (Eq.(4.24) in Ref.[9])

$$K = \frac{\omega_{\mathrm{p}}^2\,a^2}{2\,\beta^2 c^2}\,, \tag{3}$$

and $k_0 = |qB|\,/\,(2\,m\,c\,\beta\,\gamma)$ is the wave number (Eq.(4.86b) in Ref.[9]) and ε is the emittance (Eq.(4.72) in Ref.[9]), which is approximated as temperature in this paper assuming with a thermal equilibrium condition,

$$\varepsilon = \frac{2\,a}{\beta\,c}\left(\frac{k_{\mathrm{B}}\,T_{\mathrm{e}}}{\gamma\,m}\right)^{1/2}, \tag{4}$$

where a is the beam radius, β is the velocity divided by speed of light c, q is the charge of particle, B is the applied magnetic flux density, m is the mass of particle, γ is the relativistic factor, k_{B} is the Boltzmann constant, T_{e} is the temperature, respectively. The relativistic plasma frequency is defined by $\omega_{\mathrm{p}}^2 = q^2\,n/(\epsilon_0\gamma^3 m)$ (Eq.(4.2) in Ref.[9]), where n is the number density and ϵ_0 is the permittivity in a vacuum (the mass m is replaced by the relativistic mass $\gamma\,m$, and since the electric repulsion is reduced by magnetic attraction, the potential due to the self fields is replaced by $1/\gamma^2$. Finally, the nonrelativistic plasma frequency is replaced with the relativistic one by adding $1/\gamma^3$.)

Consequently, Eq.(2) is rewritten by

$$u = \left|\frac{q}{B}\right| \frac{n\,a}{4\,\epsilon_0\gamma^2}\left(\frac{\gamma\,m}{k_{\mathrm{B}}\,T_{\mathrm{e}}}\right)^{1/2}. \tag{5}$$

By using Eq.(1) with Eq.(5), the tune depression is estimated as shown in Fig. 1. In the case of this experiment, $a = 0.5$ mm, $q = -1.6022\times10^{-19}$ C, $\gamma = 1$, $T_{\mathrm{e}} = 0.1$ eV, $m = 9.1094\times10^{-31}$ kg, and $B = 11, 22, 33$ mT are assumed, respectively. The electron beam currents are -265 μA for 11 mT, -278 μA for 22 mT, and -248 μA for 33 mT, as the initial condition in 100 ns pulse duration with the flat-top profile. After the pulse compression, the beam current was increased to be about 20 times in comparison with the initial one [3, 4].

For the final pulse compression in HIF driver, the tune depression is estimated in the range from 0.9 (before pulse compression) to 0.16 (after pulse compression) [10]. In comparisons with Fig. 1, it is expected that the compact beam simulator covers the space charge strength during the pulse compression in the final stage of HIF driver.

3. Numerical simulation corresponding to experimental condition

Figure 2 shows the computational box for the numerical simulation corresponding to the experimental condition [3, 4]. In the compact beam equipment, the electrons are emitted from the thermal cathode [3, 4], and are transported through the solenoid transport line after the modulation gap. At the gap, the longitudinal velocity of electrons is modulated with the voltage produced by induction modulators for the drift compression. The behavior of electron beam is

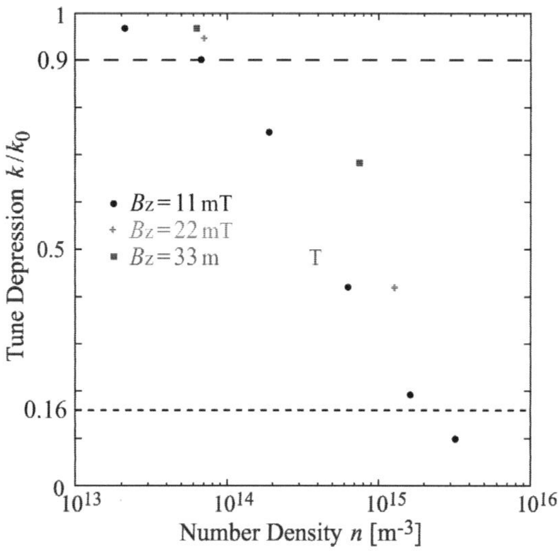

Figure 1. Tune depression as a function of number density at typical experimental parameters [3, 4]. Black circles •, red crosses +, and blue squares ■ indicate the values at each strength of magnetic flux density.

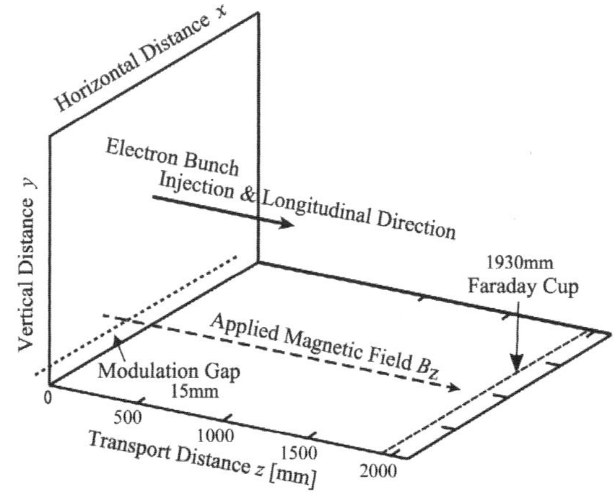

Figure 2. Calculation condition corresponding with experiment in compact beam simulator [3, 4].

assumed with non-relativistic regime, because the kinetic energy of electrons is 2.8 keV at the maximum.

According to the above conditions, we carried out the multi-particle tracking simulations to confirm the beam dynamics in the compact beam simulator. Molecular Dynamics (MD) simulation [11], which represent a so-called Particle–Particle (PP) method [12], is applied to the calculation of particle dynamics in three-dimensional space [8]. The time-dependent particle position $\mathbf{x}_i \equiv \{x_i(t), y_i(t), z_i(t)\}$ for i th super particle is obtained by

$$d\mathbf{x}_i/dt = \mathbf{v}_i, \tag{6}$$

where the particle velocity $\mathbf{v}_i$ for i th super particle is temporally changed by the external applied electric and magnetic fields and the interaction with j th super particle, and is calculated by

$$m_i \frac{d\mathbf{v}_i}{dt} = q_i \mathbf{E}_{\text{ext}} + \frac{q_i q_j}{4\pi\,\epsilon_0} \sum_{j \neq i} \frac{\mathbf{x}_i - \mathbf{x}_j}{\{(\mathbf{x}_i - \mathbf{x}_j)^2 + s^2\}^{3/2}} + q_i \mathbf{v}_i \times \mathbf{B}_{\text{ext}}, \tag{7}$$

where m_i is the mass of i th super particle, q_i is the charge of i th super particle, ϵ_0 is the permittivity in vacuum, $\mathbf{E}_{\text{ext}}$ and $\mathbf{B}_{\text{ext}}$ are the external applied electric field and the external applied magnetic flux density, respectively. Here s is the softening parameter [13], and $s = 2\,r_e$ is applied in this study, where r_e is the classical electron radius. The applied electric field $\mathbf{E}_{\text{ext}}$ is produced by the modulation voltage, and is only given in the modulation gap placed at the entrance of transport line.

Figure 3 shows the beam current waveform at $z = 1.93$ m after the modulation gap. Figure 4 shows the compression ratio, which is defined by the ratio of the beam current at each time to the initial one, at $z = 1.93$ m after the modulation gap. In the experiment with the compact beam simulator, the beam current was measured by using the Faraday cup [3, 4]. The initial beam current is -265 μA, and the applied magnetic flux density is 11 mT in the solenoid transport line. The beam current was compressed as highest as 25 times in comparison to the initial one as shown in Fig. 4. As shown in Figs. 3 and 4, the numerical results reproduce the experimental one. Consequently, it is implied that the numerical simulation developed in this paper is useful to discuss the beam behavior in the compact beam equipment.

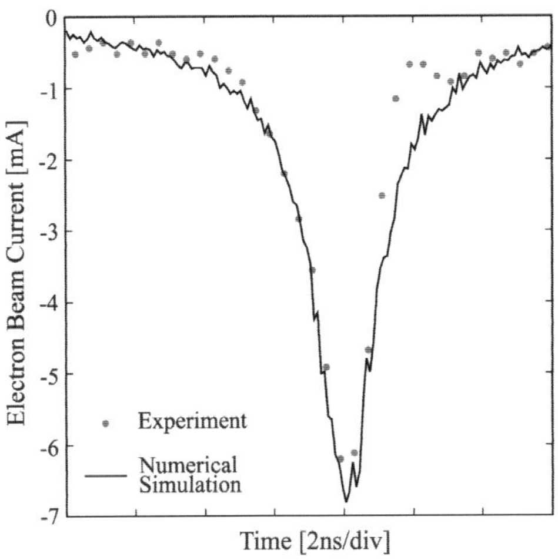

Figure 3. Beam current waveform at transport distance of 1.93 m after modulation gap. Red-full circles indicate experimental results, and solid line shows numerical result.

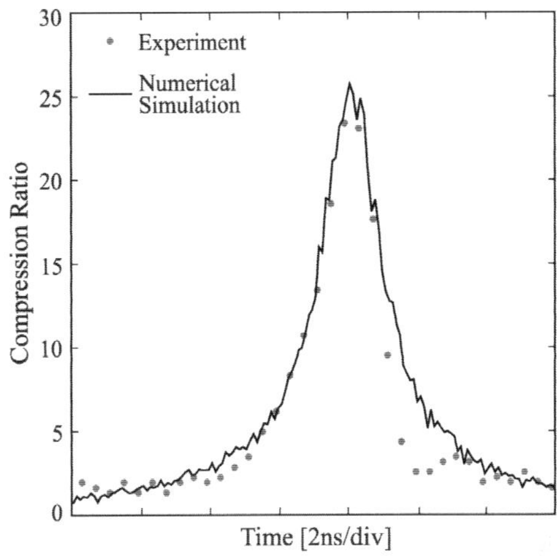

Figure 4. Compression ratio at transport distance of 1.93 m after modulation gap. Red-full circles indicate experimental results, and solid line shows numerical result.

4. Conclusion

The tune depression in the compact beam equipment was estimated, and the calculation results were compared with the experimental result in the compact beam simulator for the final stage of HIF driver system. The numerical simulation with the multi-particle tracking was carried out, and the calculation result reproduced the experimental result. It was expected that the numerical simulation developed in this paper is useful tool to investigate the beam behavior in the compact beam equipment. The HIF system is designed to clear the beam behavior by using the approaches with the compact beam equipment and the numerical simulation.

References

[1] Horioka K *et al* 2009 *Nucl. Instrum. Methods Phys. Res.* **A 606** 1
[2] Kikuchi T *et al* 2007 *Nucl. Instrum. Methods Phys. Res.* **A 577** 103
[3] Nakayama A *et al* 2013 *EPJ Web Conf.* **59** 09005
[4] Sakai Y *et al* 2014 *Nucl. Instrum. Methods Phys. Res.* **A 733** 70
[5] Park Y *et al* 2013 *NIFS PROC* **93** 84
[6] Kikuchi T *et al* 2013 *EPJ Web Conf.* **59** 09004
[7] Sato T *et al in this conference*
[8] Kikuchi T *et al* 2015 *Prog. Nucl. Energy* **82** 126
[9] Reiser M 1994 *Theory and Design of Charged Particle Beams* (New York: Wiley)
[10] Kikuchi T *et al* 2004 *Phys. Rev. ST-AB* **7** 034201
[11] Blaum K and Herfurth F 2008 *Trapped Charged Particles and Fundamental Interactions* (Berlin Heidelberg: Springer)
[12] Hockney R W and Eastwood J W 1988 *Computer Simulation Using Particles* (Bristol and Philadelphia: IOP publishing)
[13] Yamamoto K *et al* 2006 *Phys. Plasmas* **13** 012106

Energy distribution of fast electrons accelerated by high intensity laser pulse depending on laser pulse duration

Sadaoki Kojima[1], Yasunobu Arikawa[1], Alessio Morace[1], Masayasu Hata[1], Hideo Nagatomo[1], Tetsuo Ozaki[2], Shohei Sakata[1], Seung Ho Lee[1], Kazuki Matsuo[1], King Fai Farley Law[1], Shota Tosaki[1], Akifumi Yogo[1], Tomoyuki Johzaki[3], Atsushi Sunahara[4], Hitoshi Sakagami[2], Mitsuo Nakai[1], Hiroaki Nishimura[1], Hiroyuki Shiraga[1], Shinsuke Fujioka[1] and Hiroshi Azechi[1]

[1]Institute of Laser Engineering, Osaka University, 2-6 Yamada-oka, Suita Osaka, 565-0871, Japan
[2]National Institute for Fusion Science, 322-6, Oroshi, Toki, Gifu, 509-5292, Japan
[3]Hiroshima Univ., 1-4-1, Kagamiyama, Higashi-Hiroshima, 739-8527, Japan
[4]Institute for Laser Technology, 2-6 Yamada-oka, Suita Osaka, 565-0871, Japan

E-mail: kojima-s@ile.osaka-u.ac.jp

Abstract. The dependence of high-energy electron generation on the pulse duration of a high intensity LFEX laser was experimentally investigated. The LFEX laser (λ = 1.054 and intensity = $2.5 - 3 \times 10^{18}$ W/cm^2) pulses were focused on a 1 mm^3 gold cubic block after reducing the intensities of the foot pulse and pedestal by using a plasma mirror. The full width at half maximum (FWHM) duration of the intense laser pulse could be set to either 1.2 ps or 4 ps by temporally stacking four beams of the LFEX laser, for which the slope temperature of the high-energy electron distribution was 0.7 MeV and 1.4 MeV, respectively. The slope temperature increment cannot be explained without considering pulse duration effects on fast electron generation.

1. Introduction

In the fast-ignition (FI) scheme, the energy of the ignitor laser beams is converted into fast electrons, and the fast electrons heat a compressed fuel core to the ignition temperature. Laser power on the order of petawatts is required to induce fusion ignition using the FI scheme [1].

A realistic ignitor laser pulse must have a relatively long pulse duration ($\approx$ 20 ps) compared to that of standard petawatt laser systems ($\leq$ 10 ps) to keep the peak intensity of the ignitor beam within the acceptable level (10^{21} W/cm^2) [2]. This is because a higher-intensity laser generates higher-energy electrons that deposit less energy in the fuel core. When the pulse duration is much longer than 1 ps, the fast-electron generation can be affected significantly by plasma expansion that takes place over the duration of the ignitor laser pulse. A recent computational study indicated that a longer ignitor laser pulse produces a plasma with a longer density scale length, and consequently more energetic electrons are produced in the longer under-dense region of the

Journal of Physics: Conference Series **717** (2016) 012102 doi:10.1088/1742-6596/717/1/012102

plasma [3]. In this study, we investigate the dependence of the fast-electron energy distribution on the pulse duration of a high-intensity LFEX laser under pre-plasma-free conditions.

2. Reduction of preformed plasma by using plasma mirror

The experiment was carried out using the LFEX laser system at the Institute of Laser Engineering at Osaka University. The LFEX system consists of four beams. Each beam was compressed by two sets of double gratings, and they were focused onto the target by a common f/10 large off-axis parabolic mirror. Each LFEX laser pulse ($\lambda = 1.054\mu m$) is able to deliver > 500 J in a 1.2 ps duration. The on-target FWHM spot size was 70 μm, and 30% of the laser energy was contained in the spot. At the rear side of the LFEX system, the intensity contrast ratio of the LFEX laser was 10^{-9} at 2 ns before the main pulse, as measured by a 20 ps rise-time photodiode connected to a 20-GHz oscilloscope. The detailed shape of the LFEX pulse including the preceding pulses was measured at the front end of the LFEX system with a third-order correlation . A two-dimensional (2D) radiation hydrodynamic simulation (PINOCO-2D) [4] predicted the plasma scale length just before the main pulse irradiation to be 7 μm. This contrast is not small enough to experimentally investigate the effect of the pulse duration on fast-electron generation under pre-plasma-free conditions. (When laser interaction with overdense plasma occurs for a pulse duration of 4 ps and focal intensity of 3×10^{18} W/cm^2, the plasma scale length is predicted by particle-in-cell calculations to be about 10 μm. The initial plasma scale length should be an order of magnitude smaller than that.

A plasma mirror (PM) was installed in the LFEX laser system to significantly reduce the scale length of the preformed plasma prior to the arrival of the main pulse. The PM is an anti-reflection coated spherical concave glass substrate with a reflectivity of < 0.5%. Using the PM considerably improved the contrast ratio of the LFEX pulse by a factor of 10^2. Due to the geometric constraints of the LFEX laser system, the PM was placed behind the target. The LFEX beams were focused 3 mm above the target and defocused on the PM. In this configuration, 99.5% of the foot pulse and pedestal passes through the PM and only 0.5% is reflected and focused by the PM. Once a plasma was produced on the PM surface by the leading edge of the main pulse, the reflectivity of the PM increased to > 50%. The reduction of the preformed plasma produced by the preceding pulses was confirmed using optical interferogram diagnostics. A 1 mJ laser pulse with a wavelength of 630 nm and a pulse duration of 20 ps was used to measure the density profile for the preformed plasma. Plasma expansion was observed 1.3 ns before the arrival of the main pulse without the PM, while the target retained its sharp boundary 140 ps before the arrival of the main pulse at the PM. Figure 1 shows interferogram images of a preformed plasma measured with and without the PM at the same laser intensity. Calculations of the scale length for the preformed plasma carried out using the PINOCO code agree well with the experimentally measured values. The PINOCO calculations indicate that the density scale length for the preformed plasma is 1.5 μm with the PM. Details of the analysis will be reported elsewhere.

3. Dependence of fast electron energy distribution on pulse duration

Each pulse of the LFEX laser has a FWHM duration of 1.2 ps and a peak intensity of 2.5×10^{18} W/cm^2. The LFEX laser pulses are temporally stacked to create several pulse shapes of various pulse duration and intensity, as shown in Fig. 2. These pulses were focused by the PM onto a 1 mm^3 cubic gold block. Energy distributions of the fast electrons were measured with a vacuum electron spectrometer located at 20.9° from the incident axis of the LFEX beams. The slope temperature of the fast-electron energy distribution was 0.7 MeV in Case A and 1.4 MeV in Case B. In Case B, the slope temperature was two times higher than that in Case A, even though the peak intensities in Cases A and B are almost same. This increase in slope

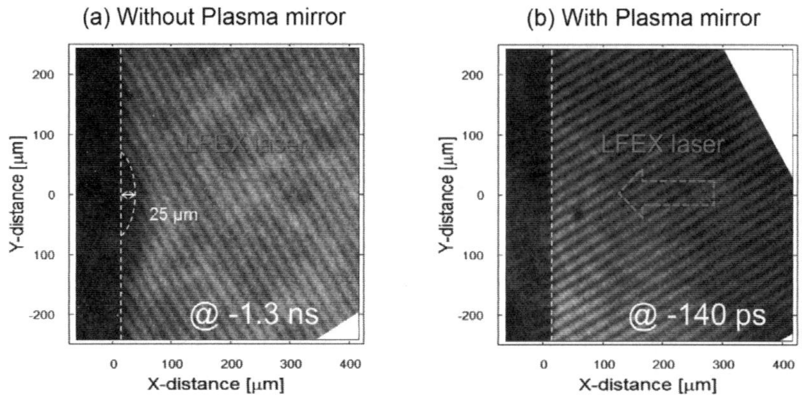

Figure 1. (Color online) Interferograms of preformed plasma on the target with (right) and without (left) the PM at the same laser intensity. The preformed plasma was observed using optical interferogram diagnostics with a 630 nm probe pulse of 21 ps duration. When using the PM, a preformed plasma was not created 140 ps before the main peak, while a preformed plasma was observed 1.3 ns before the main peak for the original LFEX.

temperature is inconsistent with conventional energy scaling [5][6][7], which does not take into account the effects of pulse duration.

We attempted to reproduce the experimental results using a two-dimensional particle-in-cell code (PICLS2D) [8]. The gold block target was simulated using a 20 μm planar plasma with an electron number density of $40n_{\rm cr}$. The initial surface of the plasma has a preformed plasma with a scale length of 1 μm from $0.1n_{\rm cr}$ to $40n_{\rm cr}$, where $n_{\rm cr}$ is the critical electron density. The intensity and pulse duration used in the calculation were the same as those in the experiment. The PICLS2D calculation shows that energetic electrons are generated in a long-scale-length, under-dense plasma produced by the main laser pulse during the main pulse duration. For a 1.2 ps pulse, the energetic electrons were primarily generated at the peak intensity of the laser pulse, while for a 4.0 ps pulse, energetic electrons were mainly accelerated by the latter part of the longer main pulse ($>$ 3 ps).

4. Conclusion

We have experimentally investigated the dependence of the fast-electron energy distribution in FI on the LFEX laser pulse duration. Preformed-plasma-free conditions were achieved by installing a PM in the LFEX laser system. The slope temperatures for the fast electrons accelerated by 2.5×10^{18} W/cm^2 were 0.7 MeV and 1.4 MeV for FWHM pulse durations of 1.2 ps (Case A) and 4.0 ps (Case B), respectively. The temperature increment cannot be explained without considering the effects of pulse duration on the fast electron generation. The calculations carried out using the PICLS2D code show that the plasma expansion driven by the main laser pulse during the main pulse duration results in the generation of a high-energy tail in the fast-electron distribution. Future calculations must consider interference between the beams in both the time and space domains. Because the 1.054 μm beams overlap each other with an incident angle of 2.86° with respect to the target normal in the experiment, the calculated separation between interference fringes is 10 μm ($<$ laser focal spot) under these conditions[9]. The interference fringes create perturbations on the critical surface of an expanding plasma due to the spatial nonuniformity of the focus pattern intensity. The density profile of the expanded plasma may be significantly affected by the hydrodynamic instabilities initiated by the perturbation. Details of this analysis will be published elsewhere in the near future.

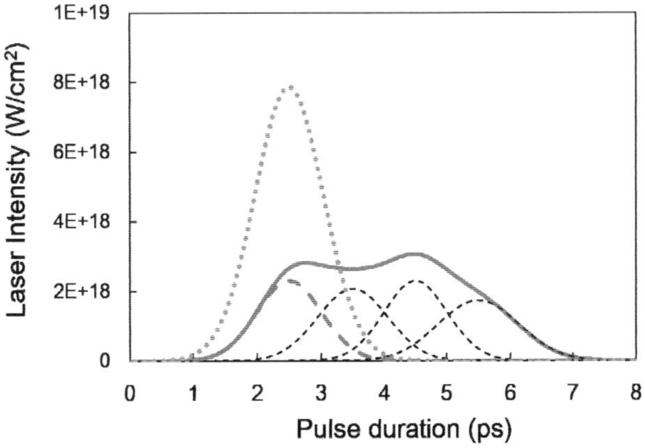

Figure 2. (Color online): LFEX laser pulses temporally stacked to generate various pulse shapes. Case A (red broken line): 1.2 ps single pulse with peak intensity of 2.5×10^{18} W/cm^2; Case B (blue solid line): 1.2 ps trained four pulse with pulse intensity of 3×10^{18} W/cm^2; Case C (green dotted line): 1.2 ps temporal overlap four pulse with pulse intensity of 7.6×10^{18} W/cm^2.

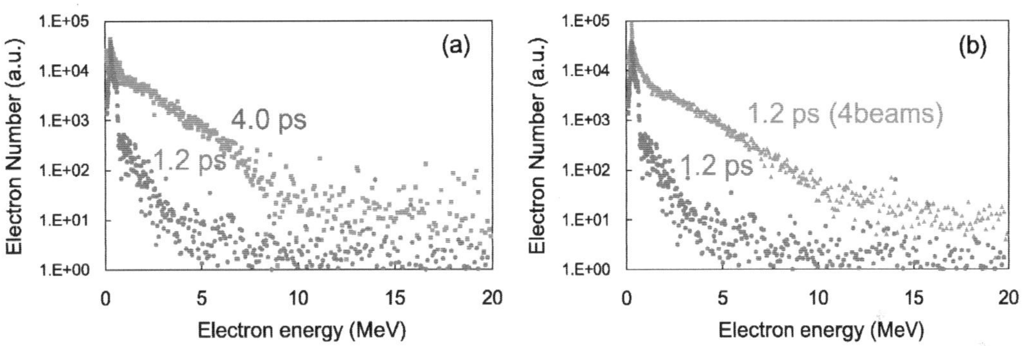

Figure 3. (Color online) (a) Electron energy distributions for the same laser irradiation intensity but different pulse durations (Case A vs. Case B). (b) Electron energy distributions for the same pulse duration but different laser irradiation intensities (Case A vs. Case C).

Acknowledgments

This work was partly supported by the Japan Society for the Promotion of Science under Grants-in-Aid for Scientific Research (A) No. 24244095, (A) No. 26246043, Young Scientists (A) Nos. 24684044 and 24686103, (B) Nos. 70724326 and 15K17798 Challenging Exploratory Research No. 25630419, and JSPS Fellows No.14J06592. Additional support was received from the auspices of the Japanese Ministry of Education, Culture, Sports, Science and Technology (MEXT) project on "Promotion of Relativistic Nuclear Physics with Ultra-intense laser", a NIFS collaboration research program (NIFSKUGK057, NIFSKUGK070 and NIFS14KNSS054), and a ILE Osaka University collaboration research program.

References

[1] S. Atzeni. Physics of Plasmas **6**, 8 (1999)
[2] A.J. Kemp, F. Fiuza, A. Debayle et al. Nuclear Fusion **54**, 054002 (2014)
[3] A. J. Kemp, L. Divol. Physical Review Letters **109**, 195005 (2012)
[4] H. Nagatomo, T. Johzaki, T. nakamura et al. Physics of Plasmas **14**, 056303 (2007)
[5] S. C. Wilks, W. L. Kruer, M. Tabak et al. Physical Review Letters **69**, 9 (1992)

[6] F. N. Beg, A. R. Bell, A. E. Dangor et al. Physics of Plasmas **4**, 447 (1997)
[7] M. G. Haines, M. S. Wei, F. N. Beg et al. Physical Review Letters **102**, 045008 (2009)
[8] Y. Sentoku, A. J. Kemp et al. Journal of computational Physics **227**, 6846-6861 (2008)
[9] H. Sakagami, T. Johzakai, A. Sunahara et al. Proc. 39th EPS Conf. on Plasma Physics P2.122 (2012)

Journal of Physics: Conference Series **717** (2016) 012103 doi:10.1088/1742-6596/717/1/012103

High dynamic range multi-channel cross-correlator for single-shot temporal contrast measurement

A Kon, M Nishiuchi, H Kiriyama, K Ogura, M Mori, H Sakaki, M Kando and K Kondo

Quantum Beam Science Directorate, Japan Atomic Energy Agency,
8-1 Umemi-dai, Kizugawa, Kyoto 619-0215, Japan

E-mail: kon.akira@jaea.go.jp

Abstract. We have developed a multi-channel cross-correlator for high dynamic range ($>10^{10}$), single-shot temporal contrast measurements. The correlator utilizes a third-order cross-correlation technique and has a reference channel, to be normalized by the measured peak intensity, and four independent optical delay lines. The measurement results of the shot-to-shot temporal contrast clearly show the intensity fluctuations of short pre-pulses at -4.5 ps and -26 ps before main pulse.

1. Introduction

The temporal contrast of an ultra-high intensity laser, which is defined as the ratio between the peak and pre-pulse intensities, is a crucial parameter for laser–plasma interaction experiments because of the influence of the pre-formed plasma before the arrival of the main pulse on the laser-plasma interaction. This interaction has been strongly indicated in experiments of particle acceleration [1] [2]; thus, it is important to measure the temporal contrast of a high-power laser pulse. Recently, several methods have been developed for the measurement of temporal contrast, using instruments such as FROG [3], Wizzler [4], and SPIDER [5]. These spectral techniques can measure a precise pulse duration and temporal contrast within 10's of ps over a dynamic range of 10^6. Moreover, a scanning cross-correlator has been widely used for temporal contrast measurements with a wide temporal range (>100 ps) and high dynamic range ($>10^{8\text{-}10}$) [6] [7]. However, because of the low repetition rate of the high-power laser system, the intensity fluctuations due to energy, temporal and spatial properties of the pump laser and/or seed laser, and so on, a shot-to-shot contrast measurement is required. For this reason, several techniques that enable a single-shot measurement have been developed; for example, the cross-correlator based on a pulse replicator [8], fiber-array [9], and time-frequency coding [10]. These and the above-mentioned spectral techniques can typically achieve the single-shot measurement with picosecond time resolution and a time window of up to ~30 ps.

We have developed a multi-channel cross-correlator (MCCC) [11] that can carry out the shot-to-shot contrast measurement with a maximum dynamic range of $>10^{10}$ and the maximum time window up to > 200 ps. The MCCC is based on a third-order cross-correlation technique, which is typically used as the delay-scanning correlator. The MCCC has four independent delay lines and channels that can measure the temporal contrast with a single-shot operation. In comparison with the commercial instrument and previous studies, one advantage of using the MCCC is that it has a high dynamic range of $>10^{10}$ owing to the independent channels and a high fluence of the incident laser. Secondly, the MCCC has a reference channel that can measure the shot-to-shot peak intensity [7]. Because the pre-pulse intensity is normalized by the peak intensity, it can identify the real fluctuations of the pre-pulse intensity.

In this paper, we report a single-shot measurement of the temporal contrast using the MCCC system. We demonstrate the maximum high dynamic range and present the results of the delay-scanning temporal contrast, which are compared to the results of a commercial cross-correlator (SEQUOIA by Amplitude Technologies). Moreover, we demonstrate the shot-to-shot temporal contrast, clearly showing intensity fluctuations of the short pre-pulses at -4.5 ps and -26 ps.

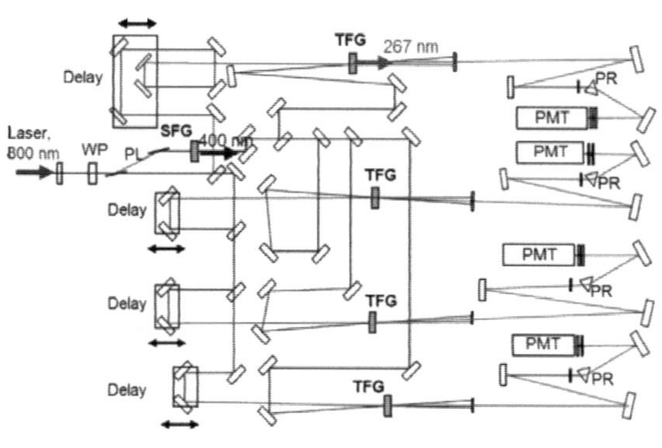

Figure 1. Setup of a multi-channel cross-correlator (MCCC). ND: neutral density filter, PL: polarizer, WP: half-wave plate, SFG: second-frequency generation crystal, TFG: third-frequency generation crystal, PR: prism, PMT: photomultiplier tube.

2. Experimental method

We have demonstrated a single-shot temporal contrast measurement using the MCCC, based on the third-order cross-correlation technique [7]. The cross-correlation technique is widely used for the estimation of the temporal profile. In this technique, the third harmonic pulse is generated by mixing the first and the second harmonic pulses in a nonlinear crystal. The temporal profile of the pulse can be measured by varying the delay time τ . In order to achieve the single-shot measurement using the MCCC, the third harmonic signal is normalized by the peak intensity $()0 = \tau$ with a reference channel in every shot. Note that this peak intensity cannot be expressed as the absolute value. In order to decide the real peak intensity, it is necessary to measure the laser energy and the pulse duration, the focal spot size on target with an extra instrument.

In the experiment, along with the MCCC, the JLITE-X laser system at KPSI (JAEA) [12] is used. Figure 1 shows the setup of the MCCC. The input laser pulse, with a pulse energy of < 8 mJ, diameter of 4 mm, pulse duration of 50 fs, and central wavelength of 800 nm, is split by a half-wave plate and two polarizers. The frequency of the s-polarized pulse is doubled by a 0.5-mm-thick type-I BBO crystal (a diameter of 5mm). The MCCC setup separates the s-polarized and p-polarized pulses into four channels and delivers them to each third-frequency generation crystals through a delay line. The energy branching ratio for the channels is 1-ch: 2-ch: 3-ch: 4-ch = 9: 0.1: 0.45: 0.45. Each channel has an optical delay (the minimum time step being >13 fs) from -950 ps to -150 ps (1-ch), and from -450 ps to +80 ps (2-4-ch). The fundamental and the second harmonic pulses are crossed on a 0.2-mm-thick type-I BBO crystal for the third harmonic generation. A photomultiplier tube (PMT, R759, Hamamatsu Inc.) measures the intensity of the third-order frequency pulse, which is generated in the crystal, after the pulse passes through a slit, 2–4 dichroic mirrors, a fused silica prism, neutral density (ND) filters, and a band pass filter with a center wavelength of 266 nm. The PMTs are calibrated with the combination of photodetectors (Thorlabs, DET210) and ND filters. Additionally, SEQUOIA is also used to cross-calibrate the MCCC in the temporal contrast measurement experiment. The incident laser energy into SEQUOIA is found to be less than 3 mJ/cm^2.

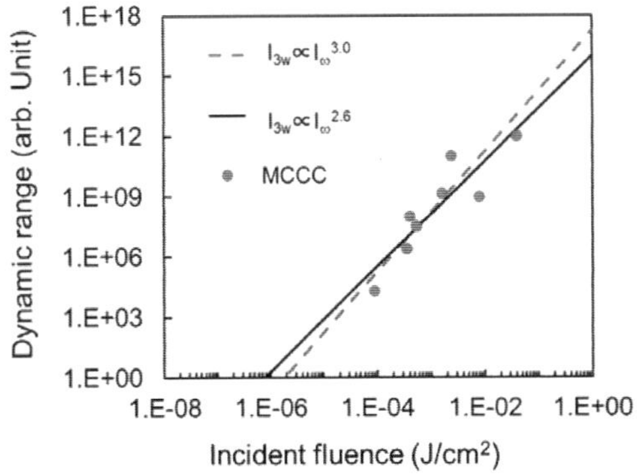

Figure 2. The dynamic range versus the incident fluence of the MCCC (solid curve) and SEQUOIA (solid triangle). The solid line and the dotted line show the power law fitting of the data with the MCCC curve ($I_{3\omega} \propto I_{\omega}^{2.6}$) and $I_{3\omega} \propto I_{\omega}^{3}$.

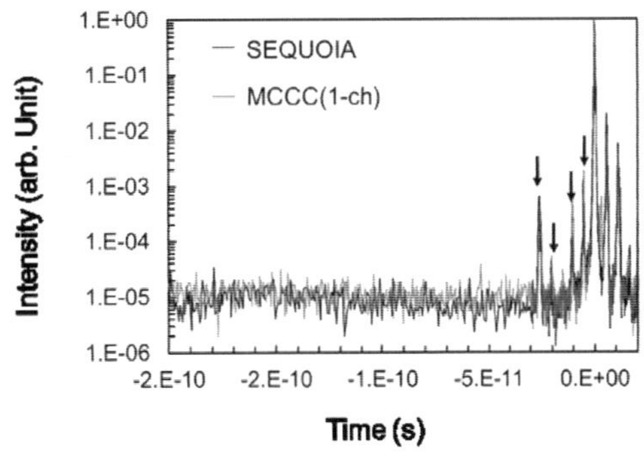

Figure 3. Result of the delay-scanning temporal contrast with SEQUOIA and the MCCC (1-ch).

3. Experimental results and discussion

The dynamic range measurement results with the MCCC by changing the optical density of the filters are shown in Fig. 2. The dynamic range increases to $\sim 10^{12}$ with the increase of the fluence of the incident laser. The dynamic range is defined as the ratio between the signal of the third harmonic and noise. The third harmonic intensity $I_{3\omega}$ is proportional to the cube of the fundamental harmonic intensity I_{ω} in the third-order correlator [7]. The slight difference between the fitting result and the cubic law in Fig. 2 may result from the phase mismatch, group velocity mismatch, or energy loss in the optical devices.

The delay-scanning temporal contrast was measured, and the results are compared with those obtained with the commercial cross-correlator. Fig. 3 shows the temporal contrast measured by 1-ch of the MCCC and SEQUOIA. Each point shows the average of 5 shot number/time at 8 mJ/cm². The scanning results of 1-ch are in good agreement with SEQUOIA data. The intensity (at $\sim 10^{-5}$), from -200 ps to -30 ps, mainly results from the amplified spontaneous emission of a regenerative amplifier [14]. The short pre-pulses in the range of -30 ps to -4 ps (the arrows in Fig.3) are due to the slight cavity misalignment of the regenerative amplifier.

In the single-shot temporal contrast measurement, we measured the intensity fluctuations of a short pre-pulse. The peaks at -4.5 ps and -26 ps are attributed to the delays of 4-ch and 1-ch, respectively. The pre-pulse intensity and the time delay of each channel is precisely decided from the measurement results of the scanning contrast ratio for each channel. The other channel is used for the detection of the noise level, when the second harmonic pulse is dumped before the crystal for the third harmonic generation. The fluctuations at a given time measured in 500 shots are shown in Fig.4. The pre-pulse intensity is normalized by peak intensity in every shot. The RMS of the pre-pulse at -4.5 ps and -26 ps are 51% and 39%, respectively. The fluctuation in a short pre-pulse is caused by variations in the main pulse, such as variation of the laser energy or the temporal and spatial profiles. For example, in the high gain amplification of the regenerative amplifier (a gain of approximately 10^{6}), taking into account the small energy fluctuation of 1.5% (RMS) with the pump energy (Surelite-10, Continuum), the fluctuation of the output energy can be estimated to be ~24% by using a small signal gain equation [15]. This estimation is not completely consistent with the experimental results; it is possible that this inconsistency is a result of the fluctuation of pump energy. To determine the origin of the intensity fluctuation, it is necessary to measure the energy, and the temporal and the spatial properties of the pump lasers and seed laser simultaneously.

4. Conclusion

We have developed and tested a multi-channel cross-correlator for the single-shot measurement of the temporal contrast with a high dynamic range (>10^{10}). In the delay-scanning temporal contrast measurement, a good agreement between the results produced by both the MCCC and the commercial cross-correlator was observed. In the single-shot measurement, the shot-to-shot pre-pulse intensity at -4.5 ps and -26 ps was observed in 500 shots. The difference of the two intensity fluctuations is likely due to the origin of the pre-pulses. However, to further investigate the source of the shot-to-shot fluctuations observed in this experiment, more detailed measurements of the physical parameters of the laser are necessary. Moreover, by using the information of the pre-pulses shot by shot obtained with the MCCC, hydrodynamic simulation enables us to characterize the plasma density profile and temperature caused by the pre-pulse.

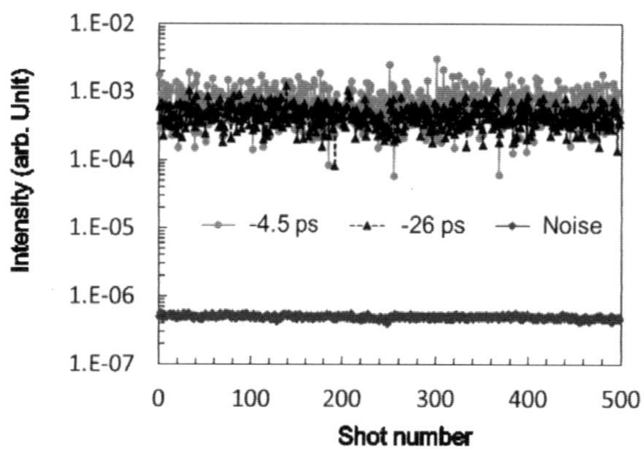

Figure. 4 The shot-to-shot fluctuation of pre-pulse intensity at -4.5 ps (circle) and -26 ps (triangle) in 500 laser shots. The noise level of the PMT is shown in blue.

Acknowledgments

We acknowledge the expert support of J-KAREN operator team at the Japan Atomic Energy Agency and JSPS Core-to-Core Program on International Alliance for Material Science in Extreme States with High Power Laser and XFEL.

References

[1] Baton S D, et al. 2008 Phys. Plasmas 15 042706
[2] Batani D, J et al. 2010 New J. Phys. 12 045018
[3] Yellampalle B, Kim K and Taylor A J 2008 Opt. Lett. 33 2854
[4] Moulet A, Grabielle S, Cornaggia C, Forget N and Oksenhendler T 2010 Opt. Lett. 35 3856–8
[5] Dorrer C, De Beauvoir B, Le Blanc C, Rousseau J P, Ranc S, Rousseau P, Chambaret J P and Salin F 2000 Appl. Phys. B Lasers Opt. 70 1644–6
[6] Luan S, Hutchinson M H R, Smith R A. and Zhou F 1993 Meas. Sci. Technol. 4 1426–9
[7] Divall E J and Ross I N 2004 Opt. Lett. 29 2273–5
[8] Dorrer C, Bromage J and Zuegel J D 2008 Opt. Express 16 13534–44
[9] Wang Y, et al. 2014 Sci. Rep. 4 3818
[10] Collier J, C Hernandez-Gomez, R Allott, C Danson, and Hall A, 2001 Laser Part. Beams 19,231
[11] Kon A, Nishiuchi M, Kiriyama H, Ogura K, Mori M, Sakaki H, Kando M and Kondo K, "High dynamic range cross-correlator for shot-to-shot measurement of temporal contrast", Jpn. J. Appl. Phys. (to be submitted).
[12] Mori M, et al.2008 IEEE Trans. Plasma Sci. 36 1872–7
[13] Itatani J, Faure J, Nantel M, Mourou G and Watanabe S 1998 Opt. Commun. 148 70–4
[14] Koechner W, 1996 Solid-State Laser Engineering (Springer, Berlin, Heidelberg 1996)

Design and operation of the multiple-pulse driver line on the OMEGA laser

T Z Kosc[1], J H Kelly, E M Hill and L J Waxer

Laboratory for Laser Energetics, University of Rochester,
250 East River Road, Rochester, NY 14623-1299

tkos@lle.rochester.edu

Abstract. The multiple-pulse driver line (MPD) provides on-shot co-propagation of two separately generated pulse shapes in all 60 OMEGA beams at the Laboratory for Laser Energetics. Smoothing by spectral dispersion (SSD), which increases the laser bandwidth, can be applied to either one of the two pulse shapes, thereby enabling dynamic bandwidth reduction. The design of the MPD required careful consideration of beam combination as well as the minimum pulse separation for two pulses generated by two separate seed sources. A new combined-pulse–shape diagnostic needed to be designed and installed after the last SSD grating. The capability of MPD to reduce dynamic bandwidth has been used on a series of campaigns on OMEGA and the performance data are presented.

1. Introduction

The 60-beam OMEGA laser at the Laboratory for Laser Energetics (LLE) is used for basic research on inertial confinement fusion. A laser pulse with a maximum duration of 4 ns and a peak power of 0.5 TW propagates through seven stages of amplification after which frequency-conversion crystals convert the IR beams (1053 nm) to the UV (351 nm). Two-dimensional smoothing by spectral dispersion (SSD) applies bandwidth to the laser beam to improve far-field irradiance [1]. Additional beam smoothing is obtained by distributed polarization rotators [2].

Researchers at LLE are attempting to demonstrate hydro-equivalent ignition—an implosion that, if scaled upward, would produce ignition. Cross-beam energy transfer (CBET) [3] has been identified as a mechanism that reduces drive to the implosion. CBET reduction is required to demonstrate hydro-equivalence. The multiple-pulse driver line (MPD) project supports the CBET mitigation initiative by providing on-shot co-propagation of two separate pulse shapes in all 60 OMEGA beams. The two co-propagating pulse shapes would typically be (1) a series of 100-ps "picket" pulses followed by (2) a longer square or shaped "drive" pulse. MPD makes it possible to reduce dynamic bandwidth, where the bandwidth is applied only to the initial picket portion of a pulse shape. Dynamic bandwidth reduction increases energy in the drive because the absence of bandwidth results in more-efficient frequency conversion and an increase in the drive-pulse energy. The possibility of propagating different spatial profiles for each pulse shape has been included, which, in conjunction with a special phase plate, enables on-shot focal-spot-size changes or discrete-step focal "zooming."

2. Multiple-pulse driver-line design

A new regenerative amplifier (regen), Phoenix (PHX), has been activated to provide the additional pulse for the MPD project. Since the PHX beam co-propagates with the SSD beam, its beam

[1] Author to whom any correspondence should be addressed.

characteristics, such as spatial profile, size, and collimation, must match the existing SSD metrics. A compensating grating is added to the PHX path to introduce temporal shear, because after combination, the PHX beam propagates though the final SSD grating, which removes the temporal shear of the previous SSD gratings. In figure 1, the picket pulse (A) is generated with the SSD regen and bandwidth is added. The drive pulse (B) is generated using the PHX regen. Both pulses are combined using a 90/10 mirror and co-propagated to the driver line. Regen synchronization is achieved using the time-multiplexed pulse-shaping system, which feeds multiple front ends with a single arbitrary waveform generator and integrated front-end system [4]. Regen-to-regen jitter is of the order of 10 ps.

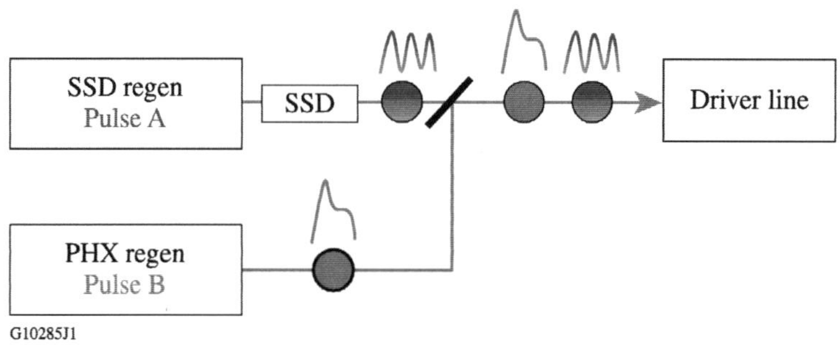

Figure 1. Block diagram of the multiple-pulse driver-line concept. Pulse A from the SSD regen has a picket-pulse shape and bandwidth is added to its spectrum. It is combined with pulse B, produced by the PHX regen, which has the drive-pulse shape. SSD: smoothing by spectral dispersion.

Most requirements for the combined pulse shape, such as the maximum pulse length, peak power, and intensity, are defined by standard OMEGA laser specifications [5]. The specification of the temporal separation τ of the co-propagating pulses, defined as the time between the peak of the final picket (pulse A) and the equivalent power point on the drive (pulse B), is the main requirement to be determined [figure 2(a)]. Conservatively, it is assumed that the picket and drive pulses are coherent. Even when SSD is applied, there is a finite probability of interference. Calculations are based on a 100-ps/decade rise and fall times for picket and drive in the IR. For a given temporal separation, the IR picket and drive fields are added and the square of the sum gives IR intensities. Frequency-conversion simulations [6] are used to determine the UV intensity. The temporal separation between the picket and drive pulses is reduced until the picket UV intensity exceeds the maximum-allowable intensity of 0.9 GW/cm^2 [figure 2(b)]. As the picket approaches the drive pulse, the contrast between the two features decreases. The picket intensity increases due to interference between the constituent pulses.

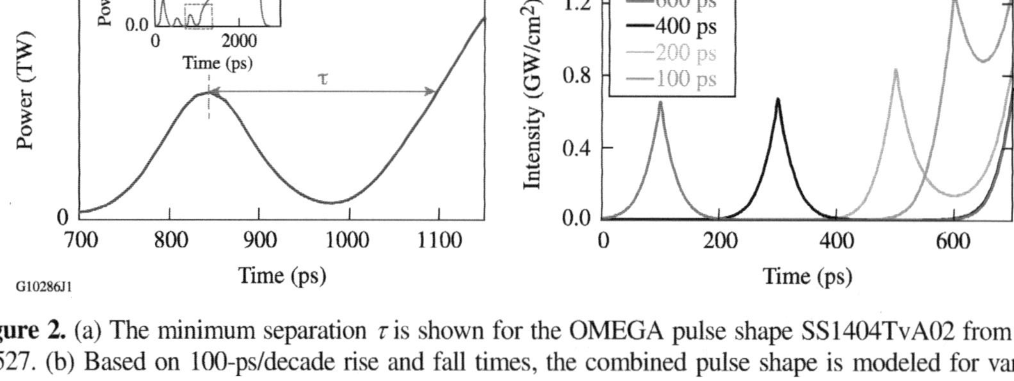

Figure 2. (a) The minimum separation τ is shown for the OMEGA pulse shape SS1404TvA02 from shot 71527. (b) Based on 100-ps/decade rise and fall times, the combined pulse shape is modeled for various separations τ between the picket (100 ps, 10-J UV) and the equivalent drive intensity.

Journal of Physics: Conference Series **717** (2016) 012104 doi:10.1088/1742-6596/717/1/012104

Modeling shows that final picket-energy variations will exceed ±10% for picket-drive separations below 300 ps (independent of picket intensity). This picket-energy variation of the final picket is in addition to the standard ±10% shot-to-shot picket energy fluctuation specification that current OMEGA performance requires. The specification for τ is set to 250 ps.

3. Driver-line diagnostic upgrade

In preparation for shot operations, it is necessary to know whether the propagating pulse meets the safety requirements of the laser system. Predicting the full MPD pulse shape from two separate measurements of the regenerative amplifier outputs prior to beam combination is deemed insufficient, so a driver-line diagnostic upgrade (DDU) was designed to measure the combined pulse shape.

The DDU design is based on the requirement that both pre-shot and on-shot pulse-shape predictions be acquired [4]. The pre-shot measurement is performed in the far field. There is little energy in the pre-shot mode, so the beam is coupled into a fiber launcher with a microscope objective. The beam is easily focused into a diffraction-limited spot on a fiber launcher as long as the SSD bandwidth is not applied. Once SSD is applied, however, the signal cannot be measured well because the spectral components spread out well beyond the 5-μm fiber core. The on-shot measurement has sufficient energy to be conducted in the near field, and data can be acquired when SSD is on or off. Additional filtration is used as necessary. The fiber tip is carefully placed at an image plane to minimize unwanted amplitude modulation when SSD is on.

4. MPD performance and capabilities

Successful activation of MPD requires that both SSD and PHX drivers match the standard baseline driver performance characterized by several metrics: square-pulse distortion, pulse-shape production, picket stability, output polarization, collimation, spatial profile, and system output energy. All required metrics were met. Picket stability and well-matched spatial profiles were demonstrated.

Picket-energy stability was measured using the new DDU combined-pulse-shape diagnostic [figure 3(a)]. When the entire pulse was generated from one driver, the stability of the first and second pickets was the same. In the MPD configuration, picket stability decreased as the picket energy increased, and the energy stability of the second picket was worse by ~2%. These results demonstrated the increased picket-power fluctuations resulting from interference between the picket-pulse and drive-pulse electric fields. Figure 3(b) shows the outstanding match between the SSD and PHX spatial profiles at a near-field location immediately following the first three-way beam split in the laser system. The PHX profile matched the SSD profile to within 3% at the 1/e2 contour. Well-matched spatial profiles are key to efficient energy extraction on OMEGA.

Picket-pulse shapes have been successfully propagated in all 60 beams of the OMEGA laser in the MPD configuration with SSD bandwidth turned both on and off. For both SSD bandwidth conditions, the picket-energy stability in the MPD configuration remained <10% rms (root mean square). On-target UV energy delivered with dynamic bandwidth reduction met the expected ~8% energy

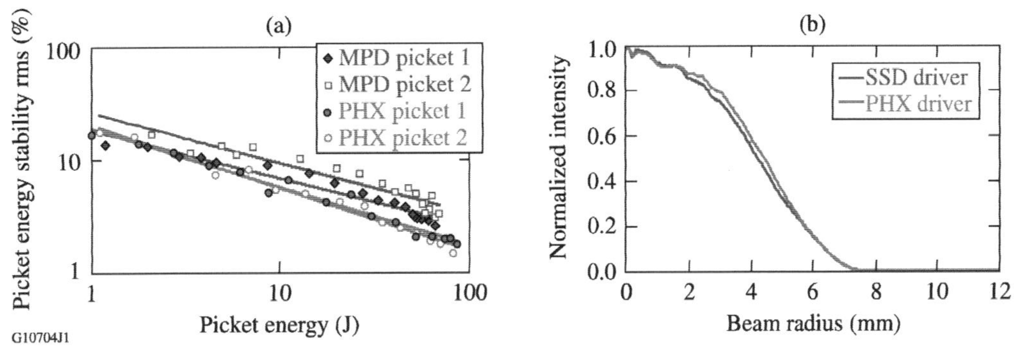

Figure 3. (a) Energy stability (% rms of a 32-point average) of both pickets was the same for the PHX driver. Stability of the second picket decreased for co-propagating beams (τ ~350 ps). (b) The azimuthal average of the near-field spatial profile in Leg 2 of the A-stage split.

increase. The effect of optical interference between pulses manifests itself in the jitter between the picket and the drive pulses, which decreases as τ increases (figure 4). Jitter is reduced when SSD bandwidth is turned on. All pulse shape features maintain their integrity at the limit of $\tau = 250$ ps.

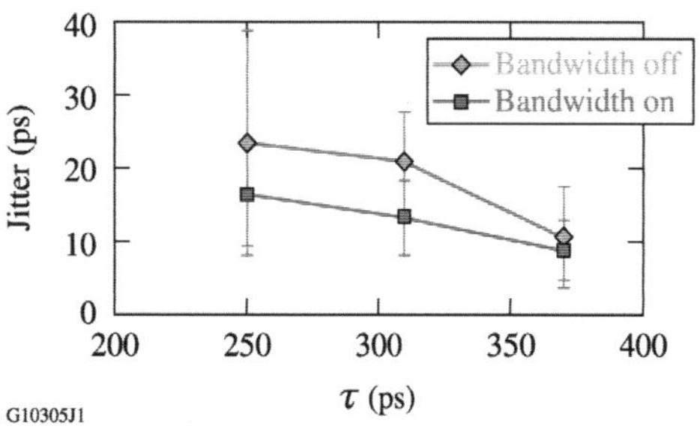

Figure 4. A series of three triple-picket pulse shapes with a decreasing τ were investigated to verify that the pulse shape's integrity was maintained at $\tau = 250$ ps.

The future capability of discrete-step focal zooming has also been provided, whereby various spatial profiles will be propagated in the MPD. Focal zooming requires a reduced drive focal-spot size relative to that of the pickets. Specifically designed apodizers, placed equidistant from the beam combiner in both beam paths, can reshape pulses from the SSD and the PHX driver into a smaller sub-aperture beam and an annulus, respectively. Unique distributed phase plates are also required to achieve the reduced focal-spot size. Configuration challenges include achieving and maintaining (1) the energy ratio between pulses and (2) a safe temporal- and spatial-pulse overlap.

5. Summary

The multiple-pulse driver line is designed to co-propagate two beams that can vary in pulse shape, bandwidth, and spatial profile in all 60 OMEGA beams. The driver-line diagnostic upgrade measures the combined pulse shapes after the final grating in the SSD system, and provides the first opportunity to gauge the effect of SSD in the driver line before any beam splitting and additional amplification. The reactivated SSD and the new PHX drivers meet the required baseline performance. Two beams having different pulse shapes and bandwidth conditions are co-propagated, and successful activation of the multiple-pulse driver line on the OMEGA laser is demonstrated. The multiple-pulse driver line is a key project in support of CBET mitigation and demonstration of hydro-equivalent ignition on LLE's OMEGA Laser System.

Acknowledgment

This material is based upon work supported by the Department of Energy National Nuclear Security Administration under Award Number DE-NA0001944, the University of Rochester, and the New York State Energy Research and Development Authority. The support of DOE does not constitute an endorsement by DOE of the views expressed in this article.

References

[1] Skupsky S, Short R W, Kessler T, Craxton R S, Letzring S and Soures J M 1989 *J. Appl. Phys.* 66 3456
[2] Boehly T R, Smalyuk V A, Meyerhofer D D, Knauer J P, Bradley D K, Craxton R S, Guardalben M J, Skupsky S and Kessler T J 1999 *J. Appl. Phys.* 85 3444
[3] Froula D H, Kessler T J, Igumenshchev I V, Betti R, Goncharov V N, Huang H, Hu S X, Hill E, Kelly J H, Meyerhofer D D *et al.* 2013 *Phys. Plasmas* 20 082704
[4] Dorrer C, Bittle W A, Cuffney R, Spilatro M, Hill E M, Kelly J H, Kosc T Z, and Zuegel J D presented at CLEO 2015, San Jose, CA, 10–15 May 2015
[5] Kosc T Z, Kelly J H, Hill E M, Dorrer C, Waxer L J, and Donaldson W R 2015 *Proc. SPIE* 9345 Paper 93450H
[6] Craxton R S 1980 *Opt. Commun.* 34 474

Journal of Physics: Conference Series **717** (2016) 012105 doi:10.1088/1742-6596/717/1/012105

Picosecond laser-driven terahertz radiation from large scale preplasmas of solid targets

G Q Liao[1], Y T Li[1,4], C Li[1], L N Su[1], Y Zheng[1], M Liu[1], J Dunn[2], J Nilsen[2], J Hunter[2], W M Wang[1,4], Z M Sheng[5,3,4] and J Zhang[3,4]

[1] Beijing National Laboratory for Condensed Matter Physics, Institute of Physics, Chinese Academy of Sciences, Beijing 100190, China

[2] Lawrence Livermore National Laboratory, 7000 East Avenue, Livermore, CA 94551, USA

[3] Key Laboratory for Laser Plasmas (MoE) and Department of Physics and Astronomy, Shanghai Jiao Tong University, Shanghai 200240, China

[4] Collaborative Innovation Center of IFSA (CICIFSA), Shanghai Jiao Tong University, Shanghai 200240, China

[5] SUPA, Department of Physics, University of Strathclyde, Glasgow G4 0NG, UK

E-mail: ytli@iphy.ac.cn

Abstract. The terahertz (THz) radiation from the front of solid targets with a large-scale preplasma irradiated by relativistic picosecond laser pulses has been studied. The THz radiation measured at the specular direction nonlinearly increases with laser energy and an optimal plasma density scalelength is observed. Particle-in-cell simulations indicate that the radiation can be attributed to the model of mode conversion. While the THz radiation near the target normal direction is saturated with laser energy and plasma scalelength. Unlike the radiation in the specular direction, the transient current formed at the plasma-vacuum interface could be responsible for the radiation near the target normal.

1. Introduction

The generation of high-power terahertz (THz) radiation sources is one of the biggest challenges to nonlinear THz photonics, in which regime new frontier applications in many fields will be opened up [1]. Various THz sources based on different mechanisms have been developed over last decades. Recently the THz generation from laser-produced plasmas has attracted much attention [2-5] since the plasma is free of breakdown restrictions. The THz pulses with energies of >100 μJ/pulse have been demonstrated from the rear side of foils irradiated by relativistic femtosecond (fs) laser pulses [6]. The THz generation critically depends on the laser and plasma conditions. For the THz generation from laser-solid interactions reported so far, most experiments are performed with fs-lasers and plasmas with a short density gradient. The generation mechanisms are attributed to transient electron currents.

To exploit new THz sources and get a more comprehensive understanding on the THz generation mechanisms, here we report on the THz radiation by the relativistic picosecond (ps) laser pulses interacting with large-scale plasmas in front of solid targets. We have studied the THz radiation as a function of the laser energy and plasma density scalelength. The different THz characteristics observed at different directions imply different generation mechanisms. The plasma wave-based and transient current-based models are used to explain the THz radiation.

2. Results and discussions

The experiment was performed using the COMET laser system at the Lawrence Livermore National Laboratory. The experimental layout was similar to that described in Ref. [5]. A *p*-polarized 1053 nm, 0.5 ps laser pulse with a contrast ratio of $\sim 10^{-5}$ was focused onto a 1 mm thick, 10 mm × 20 mm copper (Cu) slab at an incidence angle of 62.5°. The laser focal spot diameter was ~7 μm, and the corresponding intensity was $\sim 5\times 10^{17}$W/cm^2 for 1 J laser energy. The THz radiation was measured at 62.5° (specular), 0° (normal) and −17.5° with respect to the target normal (minus means the backward direction relative to the laser propagation direction). In the experiment, we adjusted the preplasma profile by introducing a prepulse in advance of the main pulse and varying the timing between the prepulse and main pulse. The plasma density scalelength was inferred from the Abel inversion of interferograms measured by a Normaski interferometer.

Figure 1 shows the THz radiation at different directions as a function of the laser energy. With the increase of laser energy, the radiation increases first. However, when the laser energy is >3.5 J, the radiation at 0° and -17.5° starts to get saturated, and even decreases slowly with the laser energy. While the THz radiation at 62.5° is still increased. Fitting the data for the laser energies >3.5 J with a power law gives a power exponent ~3±0.5, as shown with the red dashed curves in figure 1. There is no evidence of saturation for the THz radiation at 62.5°. Regarding the relative strength, the THz radiation at 0° is the lowest. With increasing of laser energy, the THz radiation at 62.5° will surpass those at other directions. At the laser energy of ~9 J, the THz radiation at 62.5° is ~10 times higher than that at -17.5°.

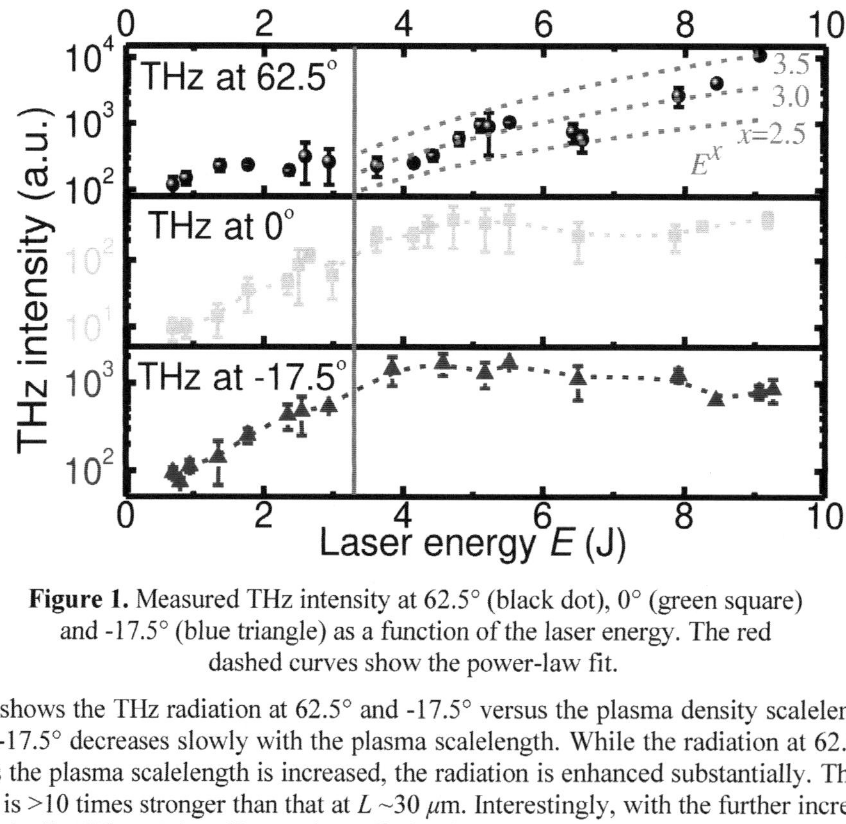

Figure 1. Measured THz intensity at 62.5° (black dot), 0° (green square) and -17.5° (blue triangle) as a function of the laser energy. The red dashed curves show the power-law fit.

Figure 2 shows the THz radiation at 62.5° and -17.5° versus the plasma density scalelength L. The radiation at -17.5° decreases slowly with the plasma scalelength. While the radiation at 62.5° is much different. As the plasma scalelength is increased, the radiation is enhanced substantially. The radiation at L ~45 μm is >10 times stronger than that at L ~30 μm. Interestingly, with the further increase of L, it drops dramatically. When L is ~60 μm, the radiation becomes comparable to that at L ~30 μm. The ratio of the radiation intensity at L ~60 μm to that at L ~30 μm is ~1.2. It is indicated that there exists an optimal L for the THz radiation at 62.5°.

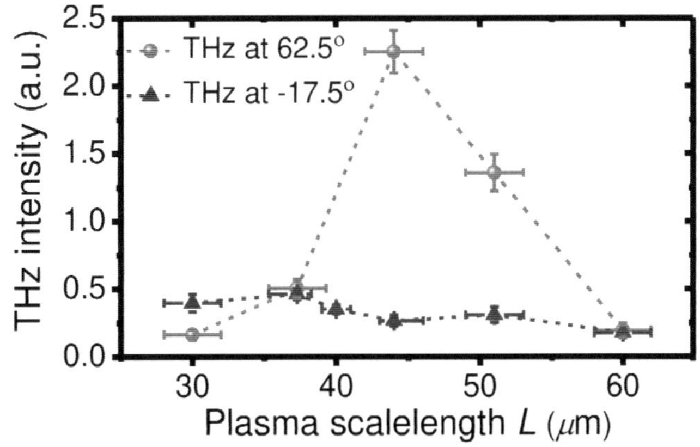

Figure 2. Measured dependence of THz radiation at 62.5° (red dot) and -17.5° (blue triangle) on the preplasmas density scalelength L.

Additionally, we measured the polarization of THz radiation by inserting a wire-grid polarizer in the detection path. The THz radiation at 62.5° behaves like p-polarized. Nevertheless, the THz intensity detected at 0° and -17.5° does not change much when varying the angle of the polarizer, indicating that it could be circularly or radially polarized.

Given the THz radiation at 0° and -17.5° has similar features, while the radiation at 62.5° is rather different, there could be two generation mechanisms, among which one is responsible for the THz radiation at 62.5° while the other for the radiation at 0° and -17.5°.

The model of linear mode conversion [7] can well explain the THz radiation at 62.5°. The model predicts there should be an optimal plasma configuration for the most efficient THz generation, which is in good agreement with the experimental. The laser pulse is too long (0.5 ps) to excite strong high-frequency laser wakefields directly. Nevertheless, in this case where the intense laser pulses propagating in large-scale plasmas, stimulated Raman scattering (SRS) will occur [8] and excite large-amplitude plasma waves. Besides, plasma waves could also be induced by the self-modulation instability (SMI) [9]. In the plasma with an increasing density, a part of the excited plasma waves could be converted into the electromagnetic waves through mode conversion. The mode-converted electromagnetic radiation emits mainly in the specular direction of the laser incidence. Hence the model is mainly valid for the THz radiation at 62.5° in our case. We have performed particle-in-cell (PIC) simulations. In the simulations, strong p-polarized multi-cycle THz radiation is observed to be emitted in the specular direction. The THz radiation as a function of the plasma scalelength L are simulated. We find that the dependence of the THz radiation on L is similar to the experimental. A similar optimal L is also observed in the simulations. In the non-relativistic case, according to the model of linear mode conversion, the THz energy is predicted to scale with the square of the laser intensity since the laser wakefield is approximately proportional to the laser intensity [7]. While in our present relativistic case, the instability growth rates of both SRS and SMI increase with the laser intensity, and subsequently the THz energy will increase with the laser intensity faster than the square law. This can explain the observed dependence of the THz radiation at 62.5° on the plasma scalelength and laser energy. More detailed discussions can be found in Ref. [5].

The transverse transient current induced by the lateral ponderomotive force in the underdense region may explain the radiation at 0° and −17.5°. Our previous two-dimensional PIC simulations show that, no matter the laser pulses irradiate the inhomogeneous underdense plasmas either normally or obliquely, a single-cycle THz pulse is observed to be emitted from front of the plasma slab in both cases [10]. The velocity of the hot electrons lateral transport current is usually much lower than the light velocity in vacuum [11]. And hence the generated radiation mainly locates near the target normal. Considering the two-dimensional radial distribution of the lateral current [12] in the case with large-scale plasmas, the induced radiation would be radially-polarized. This agrees with our polarization measurements for the radiation at 0° and −17.5°.

In previous experiments [13], the x-ray emission generated by hot electrons is also observed to be saturated with the laser energy and preplasmas. The similar features between the THz radiation and lateral electron transport imply that the THz radiation at 0° and −17.5° is generated by the lateral electron transient current. We have also observed similar saturation in the fs laser-driven THz radiation at a small laser incidence angle [14].

3. Conclusions

We have studied the THz radiation from relativistic ps laser interactions with large-scale plasmas in front of solid targets. The characteristics of THz radiation are observed to be dependent on the observation angle, which implies that there could be more than one generation mechanism. By using PIC simulations, we attribute the THz radiation in the specular direction mainly to the mode conversion of SRS-excited plasma waves, and the THz radiation near the target normal to the transient current at the plasma-vacuum interface. The integrated different types of THz radiation from laser-solid interactions may enable diverse applications with different requirements.

Acknowledgements

This work is supported by the National Basic Research Program of China (Grant Nos. 2013CBA01501), the National Nature Science Foundation of China (Grant Nos. 11520101003, 11135012, 11375262, 11421064, 11129503 and 11375261).

References

[1] Tonouchi M 2007 *Nat. Photonics* **1** 97
[2] Leemans W P *et al* 2003 *Phys. Rev. Lett.* **91** 074802
[3] Xie X, Dai J and Zhang X C 2006 *Phys. Rev. Lett.* **96** 075005
[4] Li Y T *et al* 2012 *Appl. Phys. Lett.* **100** 254101
[5] Liao G Q *et al* 2015 *Phys. Rev. Lett.* **114** 255001
[6] Gopal A *et al* 2013 *Phys. Rev. Lett.* **111** 074802
[7] Sheng Z M, Mima K, Zhang J and Sanuki H 2005 *Phys. Rev. Lett.* **94** 095003
[8] Kruer W L 1988 *The Physics of Laser Plasma Interactions* (New York, Addison-Wesley)
[9] Krall J, Ting A, Esarey E and Sprangle P 1993 *Phys. Rev. E* **48** 2157
[10] Dong X G, Sheng Z M, Wu H C, Wang W M and Zhang J 2009 *Phys. Rev.* E **79** 046411
[11] Bowes B T, Langhoff H, Downer M C, Wilcox M, Hou B, Nees J and Mourou G 2006 *Opt. Lett.* **31** 116
[12] Lin X X *et al* 2012 *Laser Part. Beams* **30** 39
[13] Reich Ch, Uschmann I, Ewald F, Düsterer S, Lübcke A, Schwoerer H, Sauerbrey R, FöRster E and Gibbon P 2003 *Phys. Rev.* E **68** 056408
[14] Liao G Q *et al* 2016 *Phys. Plasmas* **23** 013104

Journal of Physics: Conference Series **717** (2016) 012106 doi:10.1088/1742-6596/717/1/012106

Overview of LMJ alignment to target chamber center and very first results

M Luttmann[1], V Denis[1], P Gendeau[1], C Lanternier[1], P Baudon[1], Y Schiano[1], L Espert[1], F Seguineau[1], O Henry[1], D Raffestin[1], E Compain[2], L Delage[2], C Lecot[2] and N Chapron[3]

[1] *Commissariat à l'Énergie Atomique,* CEA, CESTA, F-33116 Le Barp, France
[2] *Bertin Technologies groupe CNIM, 155 Rue Louis Armand 13290 Aix en Provence, France*
[3] *AREVA TA Chantier LMJ, Bât. 692 F-33116 Le Barp Cedex, France*

E-mail : Michel.luttmann@cea.fr

Abstract: The laser Megajoule (LMJ) is a high energy laser facility (176 UV beam lines) designed for High Energy Density Physics. The end of the year 2014 was devoted to the progressive commissioning of the LMJ to achieve the first experimental campaign with 8 beams. This paper describes the alignment system developed in order to focus the laser beams and point the plasma diagnostics on the target. Alignment performances have been validated at very low power (1J front end shots) on an active target supporting a CCD sensor. Additional high power pointing shots have been performed to complete the characterization. We have demonstrated beam to target positioning accuracy < 52µm rms which is the LMJ requirement.

1. INTRODUCTION

The Laser MégaJoule (or LMJ), is a high power laser facility being constructed at the CEA Cesta site near Bordeaux (France) to study high energy density physics [1]. For the most demanding experiences the alignment requirements are 50 µm rms for the laser quad (groups of 4 beams) and 15µm rms for the target [2]. Plasma diagnostics are located all around the target chamber to measure and follow the physical phenomena occurring during an experiment. Some of these diagnostics need to be aligned up to better than 10% of their optical field to minimize field loss due to misalignment. In order to shorten the chain of dimensions and to reduce the number of sensors, the laser beams and the plasma diagnostics are aligned with a single sensor named the Common Reference.

2. LMJ ALIGNMENT SYSTEMS AROUND THE TARGET BAY

A simplified diagram of one LMJ beam line is shown Figure 1, together with the main components involved in laser, target and Plasma diagnostics alignment. A front end and a preamplifier module (PAM) generate a 39mm x 38mm beam at 1ω (λ=1053nm). This initial pulse is injected in the main amplifier where it is amplified to more than 15 kJ @ 3ns. Then the beams propagate in the experience bay where a series of 6 mirrors directs the beams into the Final Optic Assemblies (FOA). The FOA is composed with two KDP crystals (tripling the frequency) sandwiched between a 3ω focusing grating and a 1ω grating [3]. A continuous phase plate (CPP) is placed at the input of the FOA to produce a top hat focal spot. The whole FOA can be moved to place the focal spot at the desired aim point [4]. FOA alignment principle is based on a double autocollimation at 3ω and at 1ω on the CPP [2]. 3ω autocollimation is measured on the Common Reference whereas 1ω autocollimation is measured by the Alignment sensor (AS) located at the end of the amplification section just after the last spatial filter hole. Six tele-microscopes are positioned around the chamber allowing the target alignment. The tele-microscopes operate in pairs and are placed face to face relatively to the target chamber center. This system called SOPAC works with green light (532 nm) [5] [6]. The origin of the shot coordinate system is given by the Common Reference. Plasma diagnostic aim point is materialized by the intersection of two focusing green laser pointers (532nm) attached to their end. In the common reference plane, the distance between the two spots gives the plasma diagnostic focalization whereas the mean position of the spots provides the pointing information.

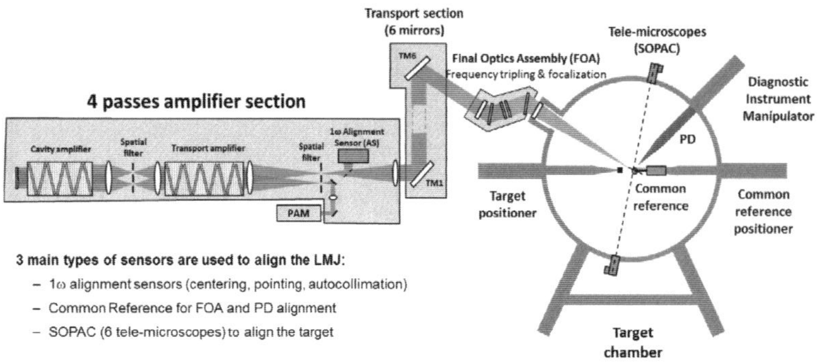

Figure 1 : simplified diagram for one LMJ beam line showing the main equipments involved during the alignment phase prior to the shot.

3. RESULTS OBTAINED DURING FIRST BUNDLE COMMISSIONING

The end of the year 2014 was devoted to the progressive commissioning of the LMJ to achieve the first experimental campaign with 8 beams [7]. Concerning alignment, the commissioning was performed in 4 steps:

- Laser beam alignment on the common reference
- Target alignment with the SOPAC
- Beam to target alignment at low energy (1J rod shot) on the Nanojoule Active target (NAT) which is a target equipped with a CCD sensor
- Quad to target alignment via high energy pointing shots

3.1 Laser beam alignment on the common reference

As mentioned earlier the laser beams are aligned via a double autocollimation process on the FOA continuous phase plates. The alignment sequence starts with the 3ω autocollimation on the common reference. Each phase plate has two actuators with sub μrad accuracy which are used to superimpose the 4 FOA spots on the common reference. The whole FOA structure is moved with two rotations (pointing) and one translation (focus) to locate the FOA focal spot at the desired point on the common reference. The loop convergence tolerance is 1.4 μrad which correspond to 11 μm on the target. Once the FOA is aligned, we can move to the 1ω autocollimation. Adjustment is done using the TM6 (the last transport mirror, as shown in Figure 1) tilt actuators. The loop convergence tolerance is 2.8 μrad which correspond to 22 μm on the target. At this point the laser beams should be aligned on the common reference. This can be checked by firing rod shots on the common reference to get the beam spot locations. Measurements have shown pointing errors exceeding 100 μm (see black square points in Figure 2). Fortunately these errors are constant with time. They can therefore be taken into account via a simple offset correction during the 3ω autocollimation process. After offset corrections, one can see on Figure 2 that beams pointing error is 39μm rms which is below the required 50 μm rms.

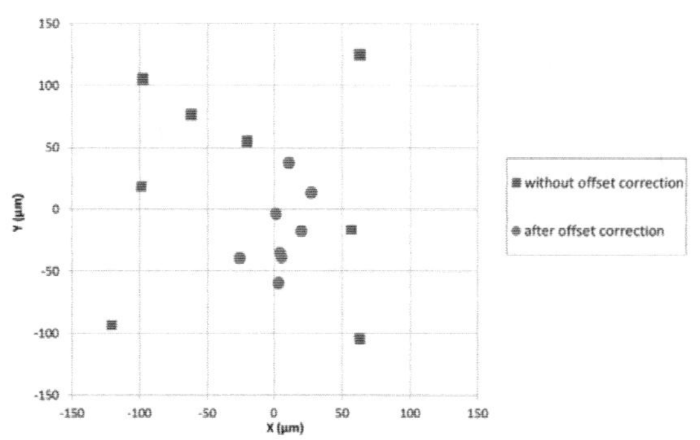

Figure 2 : Beam pointing errors measured on the common reference with rod shots.

3.2 Target alignment with the SOPAC

The SOPAC is made of 6 identical viewers (or tele-microscopes) placed face to face around the target chamber. They define 3 lines of sight. Four viewers are placed orthogonally at the chamber equator while the last two viewers are placed near the poles (16°). Each tele-microscope is equipped with 4 episcopic light sources and one less intense source for the shadow-viewing mode. Shadow viewing is the nominal mode for cryogenic target alignment since it substantially reduces the light flux reaching the target. Each tele-microscope has a 40 x 60 mm field of view. It has been optimized to have a point spread function limited by the diffraction and to exhibit less than 2µm rms total distortion in a ϕ 26 mm field. In site measurements have also shown that they were very stable (<1 µm/ hour) [6].

The LMJ target can be aligned in two operation modes:

- The first operation mode is manual. It is based on a specific GUI, with which an operator try to superimpose real-time target images with dedicated alignment reticles (see Figure 3). Alignment accuracy is strongly dependents on target geometry and target contour quality. Target orientation tuning is also difficult for the operator. This mode is therefore mainly used to pre-aligned the target and possibly to check target position before the shot.
- The second alignment mode is semi-automatic. It is based on image processing and mathematical alignment algorithms. Spheres have been added around the target to facilitate the target alignment process. The SOPAC precisely locates the spheres in each of the six tele-microscope image plan via image processing. Thanks to target metrology we know the position of target interest points relative to the 4 spheres and the sphere positions relative to each other. This target model is projected by the SOPAC on the 6 image plans. An alignment measurement consists then in minimizing the distance between the model and the experimental sphere positions. The system is strongly overdetermined since we have 2x4x6 = 48 equations for 6 degrees of freedom. The alignment performances of this method are impressive : loop tolerance performance < 2µm and 10 µrad per axe, converge in typically 2 iterations (less than 5 minutes), reconstruction error between the target model (4 spheres) and the SOPAC measurements is typically < 10 µm rms. This gives an idea of the coherence of the different data.

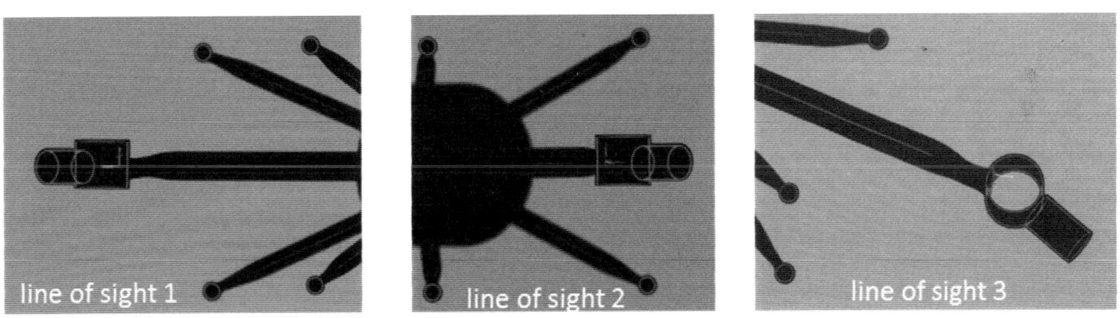

Figure 3 : image of a target on 3 SOPAC viewers. Alignment reticles used for manual alignment are drawn in white. The 4 spheres added around the target are used for semi-automatic alignment.

3.3 Test of beam to target alignment on the nanojoule active target

A specific target has been designed to measure directly shape and position of the focal spots on the target. This target is called NAT (Nanojoule Active Target) and is composed of a CCD sensor surrounded by small spheres. Their positions are accurately measured relative to the CCD pixel matrix. NAT is held by the target positioner and is aligned with the same process than a real target. It is a reusable target and sees all the alignment defects except those induced by high energy shots. NAT is also useful to validate plasma diagnostics alignment by locating the two green pointers focal spot on the CCD matrix. An example of measurements is shown on Figure 4. Measurements performed on the NAT have shown that de beam to target requirements (50µm rms for the quad + 15 µm for the target i.e. 52 µm rms) where fulfilled on low energy shots.

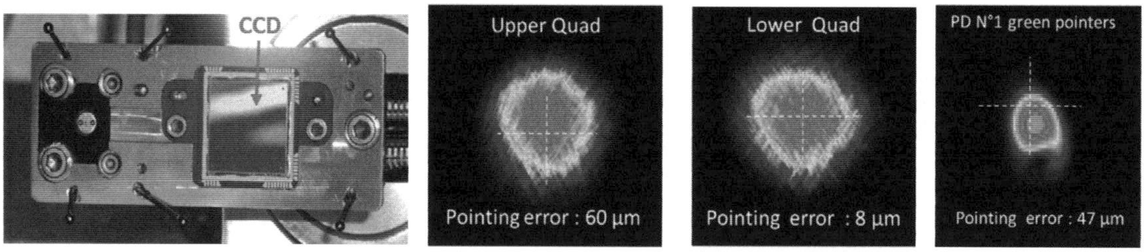

Figure 4 : Nanojoule Active Target and pointing measurement examples.

3.4 Test of beam to target alignment via pointing shots

The last step concerning alignment commissioning is to check the alignment performances on power shots. This is performed on dedicated pointing shots. Pointing target consists of a thin gold plate on which are arranged CH patterns. The center of the target is given by the middle of the hypotenuse of three small CH discs (see Figure 5). X-Ray emission is recorded on a time resolved hard X-ray imager with a spatial and temporal resolution of respectively 30 µm and 110 ps [8][9]. Six pointing targets have been shot during the first LMJ experimental campaign. All of them have been done with SSD beam smoothing and phase plates producing a 940µm focal spots. Various laser energies (from 1 kJ to 10 kJ), pulse durations (from 0.7 ns to 3 ns) and configurations (upper quad, lower quad and both quad) were used. Figure 5 shows two images recorded during pointing shots. Alignment errors are given in the target plan. Due to the beam to target angle of incidence a correction must be applied to get the alignment error in the plan perpendicular to the quad direction. After correction, the quad to target alignment error measured on the 6 shots is 48µm rms. This meets the 52 µm required for the LMJ.

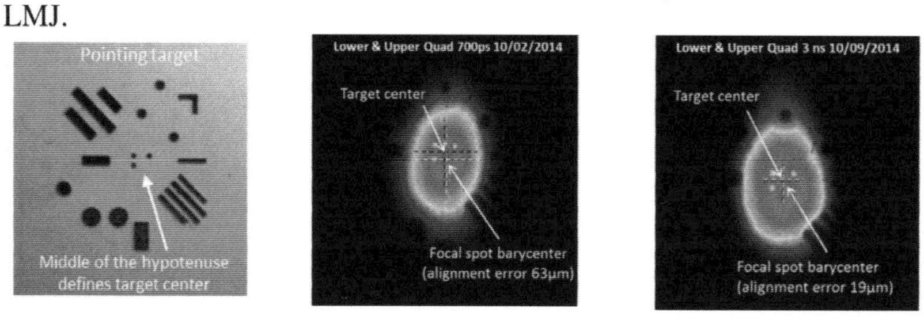

Figure 5 : pointing target and pointing shots images recorded on the time resolved X-ray imager.

References

[1] *LMJ-Petal User Guide available* at http://www-lmj.cea.fr

[2] Luttmann M et al 2011 Laser megajoule alignment to target center *Proc. SPIE **7916** High Power Lasers for Fusion Research 79160*

[3] Adolf A et al 1999 Final optics design for the megajoule laser *Proc. SPIE **3492** Third International Conference on Solid State Lasers for Application to Inertial Confinement Fusion p 44–50*

[4] Hugget A et al 2007 New design of the laser megajoule final optics assembly *Proc. SPIE **6665***

[5] Geitzholz M et al 2006 Review of laser megajoule target area: design and processes *J. Phys. IV France **133** p 631-636*

[6] Compain E et al 2015 Very high stability systems: LMJ target alignment system and MTG imager test setup *to be published SPIE JENA conference 2015*

[7] Vivini P et al 2015 The LMJ: overview of recent advancements and very first experiments *Proc. SPIE9345, High Power Lasers for Fusion Research III, **934503***

[8] R. Rosch et al 2011 Premiers diagnostics X pour le LMJ *UVX 2010 p 145-15.*

[9] Trosseille C et al 2014 Overview of the ARGOS X-ray framing camera for Laser MegaJoule *Review of Scientific Instruments **85**, 11D620*

Continuous distributed phase-plate advances for high-energy laser systems

J A Marozas, T J B Collins, J D Zuegel, P W McKenty, D Cao, S Fochs and P B Radha

Laboratory for Laser Energetics, University of Rochester,
250 East River Road, Rochester, NY 14623-1299, USA

jimijam@lle.rochester.edu

Abstract. The distributed phase plate (DPP) design code *Zhizhoo'* has been used to design full-aperture, continuous near-field transmission optics for a wide variety of high-fidelity focal-spot shapes for high-energy laser systems: OMEGA EP, Dynamic Compression Sector (DCS), and the National Ignition Facility (NIF). The envelope shape, or profile, of the focal spot affects the hydrodynamics of directly driven targets in these laser systems. Controlling the envelope shape to a high degree of fidelity impacts the quality of the ablatively driven implosions. The code *Zhizhoo'* not only produces DPP's with great control of the envelope shape, but also spectral and gradient control as well as robustness from near-field phase aberrations. The focal-spot shapes can take on almost any profile from symmetric to irregular patterns and with high fidelity relative to the objective function over many decades of intensity. The control over the near-field phase spectrum and phase gradients offer greater manufacturability of the full-aperture continuous surface-relief pattern. The flexibility and speed of the DPP design code *Zhizhoo'* will be demonstrated by showing the wide variety of successful designs that have been made and those that are in progress.

1. Introduction

The symmetric-direct-drive (SDD) and polar-direct-drive (PDD) configurations utilized in inertial confinement fusion [1,2] (ICF) driven by high-powered lasers require target illumination that conforms to the design shape or objective with a high degree of fidelity. Nonuniformity in the lower spherical-harmonic ℓ modes can have a significant impact on ICF-target performance since these modes imprint for the longest period of time and are the most difficult to smooth.

Continuous phase plates are used in SDD and PDD ICF applications because they offer control of the far-field intensity envelope in the presence of typical laser system phase aberrations. The resultant time-averaged, far-field spot intensity has a well-controlled shape. The goal is to design phase-dislocation-free continuous phase plates that produce a speckled far field whose envelope and spectrum are controlled, unaffected by system aberrations and speckle that can be smoothed.

The following report describes a novel DPP design process that achieves higher fidelity to the design objectives relative to existing methods. The novel DPP design code is called *Zhizhoo'* and is capable of producing a continuous phase-dislocation–free DPP with low near-field modulation that achieves <1 to 2% weighted σ_{rms} error of the far-field spot shape in a few minutes on a multicored personal computer with optional GPU accelerations.

[1] Author to whom any correspondence should be addressed.

The versatility of the *Zhizhoo'* design technique is evident in its ability to craft far-field envelopes from simple super-Gaussian to rather arbitrary shapes [3]. The phase-plate design techniques presented here can be applied to phase plates with or without constraining the far-field power spectrum to lower spectral power in the long-wavelength band. The ability of this technique to calculate phase-dislocation-free continuous phase plates is closely linked to maintaining a correlation with the speckle pattern and minimizing the phase gradient [3]. Various phase-plate designs will be presented for a few high-power laser systems that highlight the various capabilities of *Zhizhoo'*.

2. *Zhizhoo'* DPP design tool

The MATLAB-based tool *Zhizhoo'* [3,4] crafts continuous DPP's; the salient features are outlined here:

(a) Employs feedback loop: Unlike other methods currently in use, *Zhizhoo'* employs a novel feedback technique as a fundamental tool to generate DPP profiles with tight control of the resultant far-field spot shape and phase plate; e.g., far-field shape, arbitrary azimuthal and radial variations, DPP phase gradient, DPP phase spectral control, and phase anomaly-free designs [3]. The algorithm employs a highly modified Fienup-type algorithm as part of the whole feedback loop [5,6]. The overall novel technique is unique in its approach and it is very fast due to the feedback (that distinguishes *Zhizhoo'* as it hastens convergence via augmentation) and the FFTW-based methods [7]. In addition, a robust phase-unwrapping algorithm is employed that solves Poisson's equation in the least squares sense (algorithm adapted from Ref. [8]).

(b) Designs far-field envelopes from simple super-Gaussian to rather arbitrary shapes: Simple or exotic far-field envelope shapes are effortlessly handled with *Zhizhoo'*. Wide design objectives and/or steep profiles will require correspondingly higher surface or phase gradients in the DPP. *Zhizhoo'* can maintain envelope control even down to the ~1% σ_{rms} level.

(c) Uses an optimal filter: An important aspect of the *Zhizhoo'* feedback-loop is the Wiener or optimal filter [3]. The Wiener filter employs the well-known speckle statistics from Goodman [9,10] to model the speckle "noise" to create an optimal filter that accurately extracts the true envelope shape.

3. *Zhizhoo'* intermediate NIF polar-direct-drive distributed phase-plate designs

The NIF PDD asymmetric far-field spot design objective is an ideal candidate to test the shape control capabilities of *Zhizhoo'*. The NIF PDD asymmetric spot shape is a composite spot consisting of a primary super-Gaussian plus an offset secondary ellipse that is modulated by an offset aperturing function referred to as "spot-masking apodization" (SMA). The asymmetric far-field spot objective for NIF PDD cannot be considered an ellipse nor can it be accurately represented as a distorted ellipse. The 43 × 43-cm sq-aperture intermediate NIF PDD design for one of the equatorial spots is shown figure 1(a) and the resultant speckled spot in figure 1(b). The effect of SMA is clearly observed in figure 1(b), where the over-the-horizon portion of the spot is occluded.

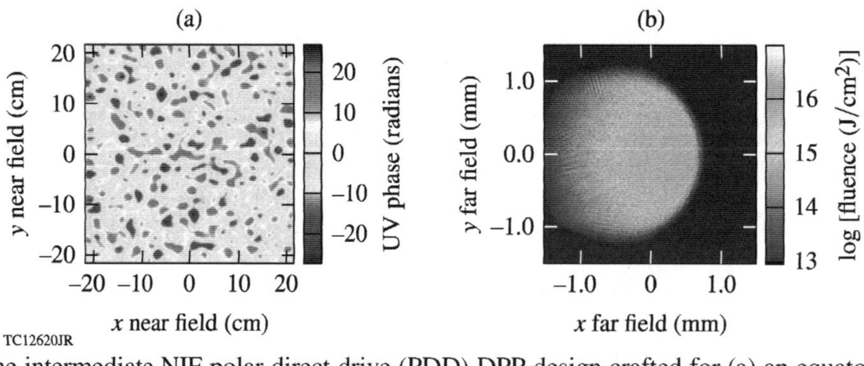

Figure 1. The intermediate NIF polar-direct-drive (PDD) DPP design crafted for (a) an equatorial beam profile and (b) the resultant speckled. The speckled image on the log scale demonstrates the remarkable speckle rejection and smooth profile at low intensity not obtainable using other methods. Note that the design objective function and the extracted envelope are nearly indistinguishable at <1% rms (root mean square) error.

It is crucial to the success of NIF PDD experiments that the DPP design prepared for the manufacturing process be as close as possible to the design objective. Otherwise, the far-field spot integrity severely degrades in the presence of both manufacturing phase error (MPE) and near-field wavefront error (WFE). A DPP design that initially has the highest integrity level will remain more intact, relative to an insufficient design. NIF's WFE was measured and imposed upon the DPP's for a worst-case analysis via *DRACO* hydrodynamic simulations. The strongest NIF WFE was a weaker aberration than a 25-μm-rms MPE that set the acceptable MPE tolerance to 25-μm rms.

During the NIF PDD (intermediate and ignition-scale) DPP design process, a potential manufacturing problem surfaced. The issue was the result of a combination of interferometric measurements and the machine's internal phase-unwrapping algorithms. The resulting unwrapped phase would produce areas of phase dropouts and occasionally large regions of π discontinuities. However, the phase-unwrapping procedure incorporated within *Zhizhoo'* is designed to be immune to areas of noise and regions of π discontinuities. It was demonstrated that the phase-unwrapping algorithm was more than capable of removing and correcting the corrupted phase data from the instrument[11]. Utilizing the phase-unwrapping algorithm from *Zhizhoo'* is a cost-effective remedy to procuring expensive interferometers. The algorithm is able to correct the phase errors from the intermediate energy-scale up to the ignition-scale designs.

4. Steep-profile, low-ripple flattop round spots

Low-ripple, flat-topped spots with steep profiles are additional design objectives compatible with the *Zhizhoo'* DPP design method. Traditionally, DPP's have difficulty designing for low-ripple, flattopped spots because the designs tended to ring as the spot shape rolls off to zero. In contrast, *Zhizhoo'*-crafted DPP's tend not to suffer the same fate because of the feedback control with augmentation of the designprofile.

The OMEGA EP laser required a redesign for its 1.8-mm-wide spot because of damage that the turning mirror suffered from high-level modulation caused by a retroreflection back through the focusing lens. The close proximity of the turning mirror posed a design challenge for *Zhizhoo'* by mandating wavelength control of the DPP's feature size. The design for the far-field envelope demanded a large flat area with a fast roll-off. The low-ripple (2.5%) resultant extracted envelope is shown in figure 2(a). The equivalent free-space back-propagation was determined to be 6m, which drove the DPP design to use large feature sizes to minimize near-field modulations; see figure 2(b). The larger feature sizes had the side effect of driving up the peak-to-peak phase depth of the DPP because of the smaller bandwidth distribution of the phase that also increases local phase gradients.

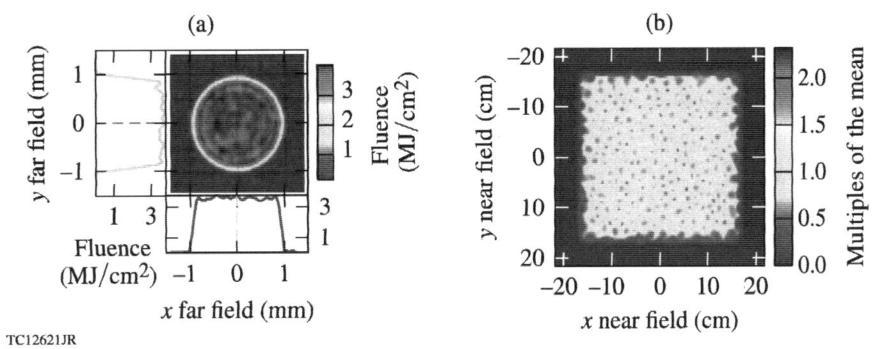

Figure 2. (a) The OMEGA EP low-ripple, 1.8-mm-wide far-field extracted envelope. (b) The resulting near-field low-level modulation from a retroreflection is indicative of the large DPP feature sizes.

The Dynamic Compression Sector (DCS) laser also required a low-ripple, flattopped spot, but with two additional attributes: decreased mid-range wavelength control (high pass) and flexible spot-shape control via dispersion control. The high-pass DPP design procedure was similar to the method reported in Ref. [3] and successfully reduced the power in the long- to mid-wavelength modes, even in the presence of predicted DCS WFE [see figure 3]. The DCS DPP design provides a trade-off among several smoothing attributes, including spot shape and intensity on target by adjusting a differential grating that changes the dispersion experienced by the 1-D, mFM SSD system [12].

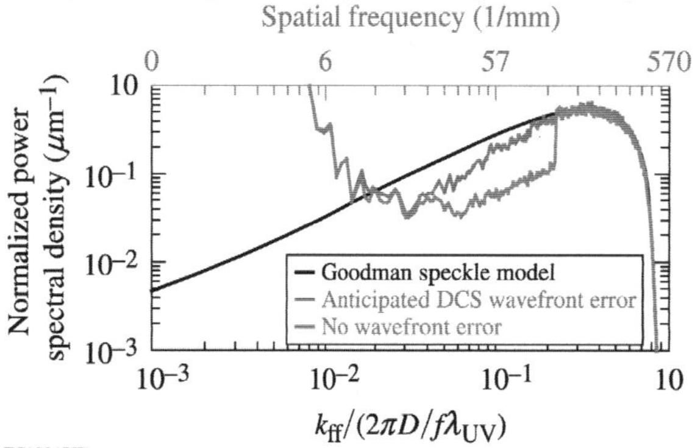

Figure 3. The far-field speckled spot spectrum for the DCS Laser System. The Goodman speckle model is shown as a reference to indicate the ability of the high-pass DPP design to modify the far-field spot's spectrum (blue trace). In the presence of predicted DCS laser WFE, the high-pass DPP design still maintains a decreased spectrum over the spectral band (red trace). DCS: Dynamic Compression Sector; WFE: wavefront error.

5. Conclusion

The continuous phase-plate design code *Zhizhoo'* is capable of crafting DPP's for a variety of high-power laser systems, each having different design constraints. *Zhizhoo'* designs continuous DPP's with simple envelope shapes or exotic shapes with asymmetry. The code *Zhizhoo'* crafts DPP's with a high degree of fidelity to the design objective. A higher-fidelity DPP design results in a more-faithful representation of the desired objective function when the DPP is subjected to WFE and MPE. The flexibility of the *Zhizhoo'* design code makes it easy to create multiple designs, even when the design requirements change, because *Zhizhoo'* can respond in a short period of time or produce multiple realizations to improve beam-overlap nonuniformity reduction.

Acknowledgment

This material is based upon work supported by the Department of Energy National Nuclear Security Administration under Award Number DE-NA0001944, the University of Rochester, and the New York State Energy Research and Development Authority. The support of DOE does not constitute an endorsement by DOE of the views expressed in this article.

References

[1] Verdon C P 1993 *Bull. Am. Phys. Soc.* **38** 2010
[2] Bodner S E, Colombant D G, Gardner J H *et al.* 1998 *Phys. Plasmas* **5** 1901
[3] Marozas J A 2007 *J. Opt. Soc. Am. A* **24** 74
[4] Inspired by Anishinaabe words *Zhizhoo'*, meaning "spreads someone using something," and *zhizhoo'an*, meaning "spreads something using something," just as the phase plate spreads the diffraction-limited, far-field spot to match the far-field objective. Resource: Nichols J D and Nyholm E 1995 *A Concise Dictionary of Minnesota Ojibwe* (Minneapolis: University of Minnesota Press)
[5] Gerchberg R W and Saxton W O 1972 *Optik* **35** 237
[6] Fienup J R 1982 *Appl. Opt.* **21** 2758
[7] Frigo M and Johnson S G 2005 *Proc. IEEE* **93** 216
[8] Ghiglia D C and Romero L A 1994 *J. Opt. Soc. Am. A* **11** 107
[9] Goodman J W 1988 *Introduction to Fourier Optics* (New York: McGraw-Hill)
[10] Goodman J W 1984 in *Laser Speckle and Related Phenomena* vol 9 Dainty J C ed Topics in Applied Physics (Berlin: Springer-Verlag) chapter 2 pp 9–75
[11] LLNL kindly fabricated *Zhizhoo'* crafted DPP prototypes and measured them using their interferometers which provided the data for this analysis.
[12] Marozas J A, Zuegel J D and Collins T J B 2007 *Bull. Am. Phys. Soc.* **52** 145; 2008 **53** 249; 2009 **54** 306; 2010 **55** 294

Journal of Physics: Conference Series **717** (2016) 012108 doi:10.1088/1742-6596/717/1/012108

Debris and shrapnel assessments for National Ignition Facility targets and diagnostics

N D Masters, A Fisher, D Kalantar, J Stölken, C Smith, R Vignes, S Burns, T Doeppner, A Kritcher and H-S Park
Lawrence Livermore National Laboratory, Livermore, CA 94550, USA

E-mail: masters6@llnl.gov

Abstract. High-energy laser experiments at the National Ignition Facility (NIF) can create debris and shrapnel capable of damaging laser optics and diagnostic instruments. The size, composition and location of target components and sacrificial shielding (e.g., disposable debris shields, or diagnostic filters) and the protection they provide is constrained by many factors, including: chamber and diagnostic geometries, experimental goals and material considerations. An assessment of the generation, nature and velocity of shrapnel and debris and their potential threats is necessary prior to fielding targets or diagnostics. These assessments may influence target and shielding design, filter configurations and diagnostic selection. This paper will outline the approach used to manage the debris and shrapnel risk associated with NIF targets and diagnostics and present some aspects of two such cases: the Material Strength Rayleigh-Taylor campaign and the Mono Angle Crystal Spectrometer (MACS).

1. Introduction

High-energy laser experiments at the National Ignition Facility (NIF) can create debris and shrapnel capable of damaging laser optics and diagnostic instruments [1, 2]. An important aspect of NIF operations, particularly with increased shot rates, is the identification and management of risks to the facility posed by debris and shrapnel. The NIF Debris & Shrapnel Working Group is charged with assessing potential risks, identifying the generation, nature and velocity of shrapnel and debris and the effect they may have on laser optics and target diagnostics and recommending strategies for managing these risks.

The formal approach to debris and shrapnel risk management is captured by the Risk Management Flowchart in figure 1. Hazards are largely due to two sources: 1. Diffuse, consisting mainly of so-called debris wind loads and 2. Local Effects, including shrapnel effects and x-ray ablation. Identification of these hazards for a given experiment may involve previous experience, analytic models, or detailed numerical simulation. The consequences of the identified hazards—potential damage to laser optics and target diagnostics—are then evaluated in terms of severity and probability. These are the risks. The appropriate risk reduction strategy—Avoidance, Mitigation, or Transference—can then be selected. Avoidance modifies the design (target, diagnostic, etc.) to minimize the potential risk. Mitigation applies controls to minimize the impact of risks (required minimum filtration, restrictions on diagnostic reuse, debris tests, inspections, etc.), or Transference (transferring risk to the facility to be accepted). The following case studies present examples of the risk assessment, analysis and management strategies outlined in figure 1.

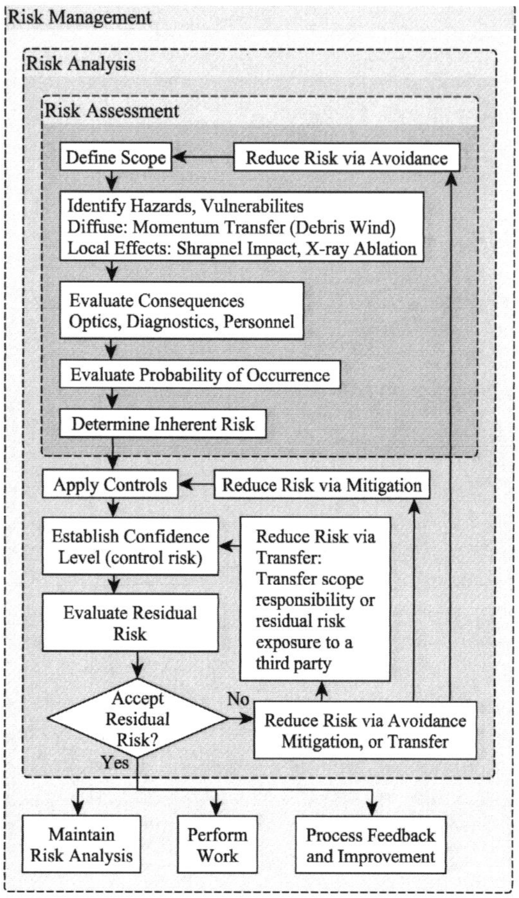

Figure 1: Flowchart of NIF Debris and Shrapnel Risk Management Approach.

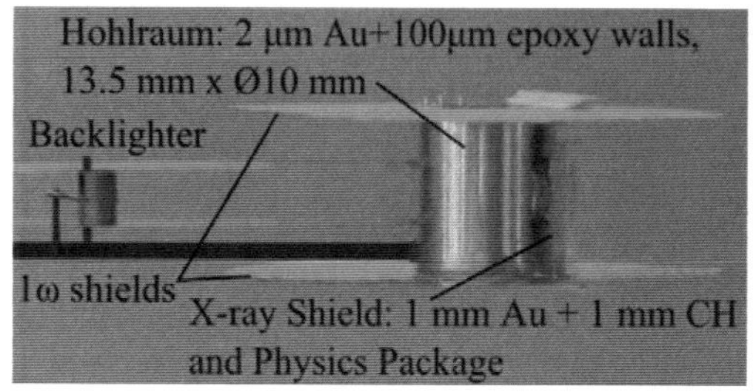

Figure 2: Final Material Strength Rayleigh-Taylor Target design.

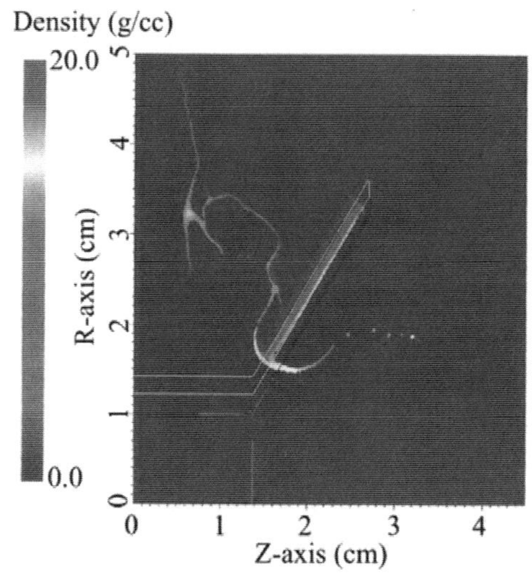

Figure 3: density plot from axisymmetric simulation of large shield MatStrTaRT at 5.5 μs (z-axis is horizontal and overlay shows initial material boundaries).

2. Case Studies

2.1. Material Strength Tantalum Rayleigh-Taylor (MatStrTaRT)

The MatStrTaRT target, shown in figure 2, is a warm, thin-walled, gold-epoxy target 13 mm long with inner and outer diameters of 10 mm and 10.224 mm, respectively (2 μm Au+110 μm CH walls). An 800 kJ laser pulse indirectly drives a planar ablator and tamper to produce Rayleigh-Taylor instabilities in a rippled sample. An x-ray backligher produces face-on 2D point-projection radiographs of the instability growth and reference samples on image plates in the High Energy Imaging Diagnostic (HEIDI) [3]. These measurements and used to develop material strength models at high pressures and strain rates [4, 5, 6]. HEIDI has a multilayer shielded enclosure (aluminum, tungsten, and hevimet) to reduce fluorescence and high-energy x-ray background. The open hevimet aperture (nominally ⊘10.75 mm) is positioned 62.5 mm from TCC.

Background x-ray emissions from the hohlraum and LEH are blocked from entering the HEIDI aperture by a flat 1 mm thick gold shield attached to—and covering the full extent of—the side of the hohlraum. The physics package is mounted in an aperture in this shield. LEH emissions are blocked by extending the shield above and below the LEH at 30° to avoid drive beam interference (see figure 2). Additional dimpled unconverted light shields minimize specular reflections of 1ω light from the target that could damage laser components. The initial design intended to shield LEH and plume emissions by extending the shields ≈7.5 mm above and below the LEH. Hydrocode simulations of this large shield design predicted significant solid and molten debris directed towards optics with sufficient size and velocity to damage several layers of NIF optics (see figure 3). For example, solid fragments up to 6.7 mg were identified with velocities of ≈ 2 km/s—sufficient to penetrate more than 18 mm of borosilicate glass (a 3.3 mm DDS is predicted to be penetrated by 6.7 mg fragments at velocities greater than ≈250 m/s, or alternatively by fragments at 2 km/s with masses greater than ≈150 μg). In terms of Risk Acceptance, superficial damage to the 3.3 mm disposable debris shields (DDS) may be acceptable, but full penetration and any predicted damage to the 10 mm thick Grated Debris Shields (GDS) or other optics is not.

Replacing the 10.75 mm diameter circular HEIDI aperture with a rectangular 6.5 mm × 10.75 mm aperture (see figure 4) allowed the shield extension to be reduced to 1.1 mm, above and below the LEH. Simulation of the new configuration predicted near full melt of the shield with little directed towards the optics, thus avoiding the risk of the initial design. As such the residual risks—significant x-ray and debris load on the primary diagnostic and only a nominal debris risk to the laser optics—could be transferred and accepted by the facility:

Post shot inspections from many MatStrTaRT shots have found no optics damage resulting from these shots. The HEIDI diagnostic [3] does collect significant debris and shows cratering of the tungsten alloy aperture as shown in the pre- and post-shot images (figures 4 and 5, respectively) of the nosecap from shot N130923. The strip of gold stuck to the aperture is approximately 7 mm × 1 mm.

2.2. Mono Angle Crystal Spectrometer (MACS)

The MACS diagnostic has been developed for x-ray Thompson scattering spectroscopy of matter and uses a curved Highly-Oriented Pyrolytic Graphite (HOPG) crystal to focus the x-ray scattering signal towards one strip of a NIF Gated X-ray Detector (GXD) [7]. Although the direct line-of-sight to the detector is blocked (see figure 6), the fragile HOPG crystals could be damaged by ablation or debris without x-ray filtration. X-ray ablation simulations found that monolithic or stacked polycarbonate main filter configurations, constrained by a maximum apparent thickness of ≈900 μm (for acceptable signal attenuation), could suffer spall failures and thus would not remove the risk. The addition of a thin tilted prefilter has been identified as a successful risk avoidance strategy; decoupling the x-ray and debris loads by allowing the blowoff from x-ray ablation of the prefilter to be directed away from subsequent filters [8]. A 25 μm kapton prefilter was selected (based on predicted ablation depth of 7 μm). The 700 μm main filter is sized to provide a total apparent thickness of ≤ 820 μm (the apparent thickness varies due to the tilting of the shields). The main filter covers a 32 mm × 17 mm aperture and remains subject to significant debris wind loading, despite x-ray decoupling. Five 1 mm wide support aluminum bars—machined as part of the main filter mount—span the height of the aperture to reduce the unsupported span of the main filter. Shadows cast by the support bars reduce the effective width of the curved HOPG crystal but are aligned parallel to the dispersive direction of the crystals and therefore do not interrupt the measured spectra. During a shot, the prefilter is expected to only move a small distance by the time fast moving plasma from the target reaches it. This may explain the patterns of burned and protected regions on the main filter, visible in some post-shot inspections (see figure 7) These inspections have found that this configuration has performed well. However, reuse of the MACS without replacing the prefilter, e.g, for sequential shots, should be avoided to mitigate the risk this would pose.

3. Conclusions

This paper has described some of the influence debris and shrapnel assessments have on the design of NIF targets and diagnostics. Careful assessment of the risks posed by any new target or diagnostic allows the hazards to be avoided or mitigated while facilitating the scientific goals and limiting the risk transferred to the facility. The risk reduction steps taken in the development of the MatStrTaRT target and MACS spectrometer demonstrate Avoidance, Mitigation and Transference strategies.

Figure 4: HEIDI nosecap before shot.

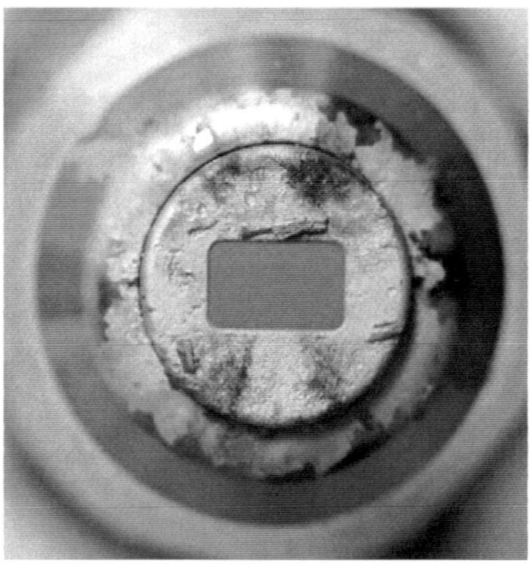

Figure 5: Significant debris deposition following shot, including large gold fragments.

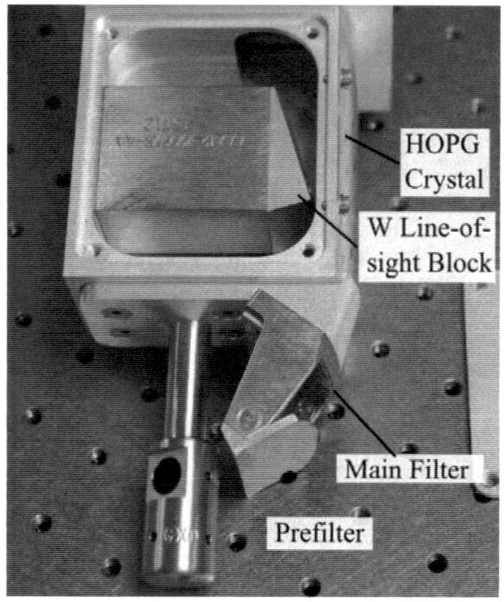

Figure 6: MACS Spectrometer with side panel removed showing line-of-sight block, location of crystals and pre- and main filters.

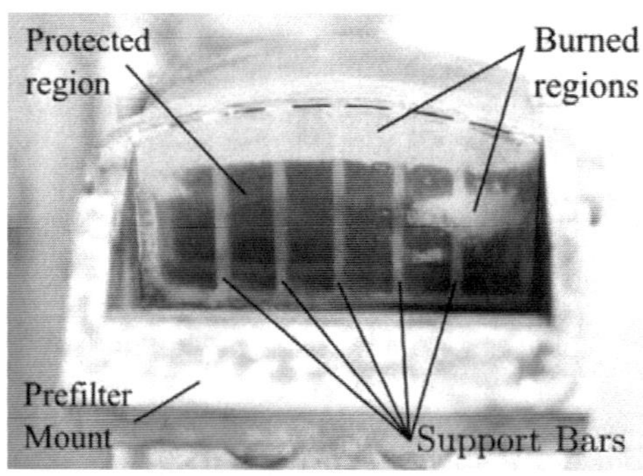

Figure 7: Post-shot inspection of the main filter showing absense of prefilter, main filter support bars and protected and burned regions on the main filter.

This work performed under the auspices of the U.S. Department of Energy by Lawrence Livermore National Laboratory under Contract DE-AC52-07NA27344.

References

[1] Eder D, Fisher A, Koniges A and Masters N 2013 *Nuc. Fusion* **53** 113037
[2] Eder D C, Koniges A E, Landen O L, Masters N D, Fisher A C, Jones O S, Suratwala T I and Suter L J 2008 *J. Phys.: Conf. Series* **112** 032023
[3] Ahmed M F, Mcnaney J M, Vignes R M, Smith C A, Masters N, Bailey C and Petre R B 2014 *Proc. SPIE. 9211, Target Diagnostics Physics and Engineering for Inertial Confinement Fusion III, 92110F*
[4] Park H S *et al.* 2016 These proceedings
[5] Park H S *et al.* 2015 *Phys. Rev. Lett.* **114**
[6] Prisbrey S T *et al.* 2012 *Phys. Plasmas* **19** 056311
[7] Döppner T *et al.* 2014 *Rev. Sci. Instrum.* **85** 11D617
[8] Fournier K B *et al.* 2010 *Rev. Sci. Instrum.* **81** 075113

Journal of Physics: Conference Series **717** (2016) 012109 doi:10.1088/1742-6596/717/1/012109

Gamma Ray Measurements at OMEGA with the Newest Gas Cherenkov Detector "GCD-3"

A M McEvoy[1,7], H W Herrmann[1], Y Kim[1], A B Zylstra[1], C S Young[1], V E Fatherley[1], F E Lopez[1], J A Oertel[1], T J Sedillo[1], T N Archuleta[1], R J Aragonez[1], R M Malone[2], C J Horsfield[3], M Rubery[3], S Gales[3], A Leatherland[3], W Stoeffl[4], M Gatu Johnson[5], W T Shmayda[6] and S H Batha[1]

[1] Los Alamos National Laboratory, Los Alamos, NM 87545, USA
[2] National Security Technologies, 182 Eastgate Dr, Los Alamos, NM 87544, USA
[3] Atomic Weapons Establishment, Aldermaston, Reading RGR 4PR, United Kingdom
[4] Lawrence Livermore National Laboratory, 7000 East Ave, Livermore, CA 94550, USA
[5] MIT Plasma Science and Fusion Center, 167 Albany St #204, Cambridge, MA 02139, USA
[6] Laboratory for Laser Energetics, 250 E River Rd, Rochester, NY 14623, USA
[7] Author to whom correspondence should be addressed.

Contact email: ammcevoy@lanl.gov

Abstract. Initial results from the newest Gas Cherenkov Detector (GCD-3) are reported demonstrating improved performance over previous GCD iterations. Increased shielding and lengthening of the Cherenkov photon optical path have resulted in a diminished precursor signal with increased temporal separation between the precursor and the primary DT Cherenkov signal. Design changes resulted in a measured GCD-3 sensitivity comparable to GCD-1 at identical 100 psia CO_2 operation. All metal gasket seals and pressure vessel certification to 400 psia operation allow for a GCD-3 lower Cherenkov threshold of 1.8 MeV using the fluorinated gas C_2F_6 as compared to the 6.3 MeV lower limit of GCD-1 and GCD-2. Calibration data will be used to benchmark GEANT4 and ACCEPT detector models. The GCD-3 acts as a prototype for the Super GCD being fielded at the National Ignition Facility (NIF) as part of the National Diagnostics Plan and will be installed at NIF in early 2016.

1. Introduction

The newest Gas Cherenkov Detector (GCD-3) [1,2,3] was initially fielded in mid-2014 at the Laboratory for Laser Energetics Omega laser facility and has since provided gamma ray measurements for a variety of experimental campaigns. The GCD-3 incorporates significant improvements over the earlier GCD-1 and GCD-2 models and acts as a prototype for the "Super GCD", one of eight "Transformative Diagnostics" to be installed at the National Ignition Facility (NIF) as part of the National Diagnostic Plan [4]. A diagram of the GCD-3 and the principles of detector operation is shown in figure 1.

Gamma rays produced during Inertial Confinement Fusion (ICF) implosions provide important information about fusion burn and the imploded capsule conditions. Typical implosions utilizing deuterium-tritium (DT) fuel generate high energy (up to 16.75 MeV) DT fusion gamma rays that provide a diagnostic signature directly related to the temporal shape and duration of the fusion burn profile. In addition, primary DT neutron interactions with capsule materials generate gamma rays that contain information about the areal density (ρR) of the ablator near the time of peak fusion burn.

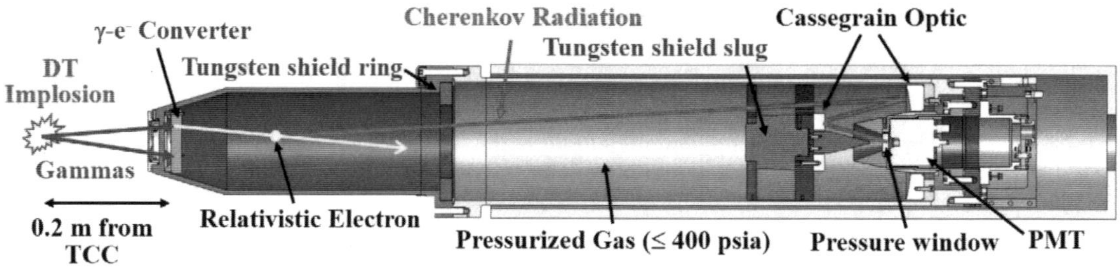

Figure 1. The GCD-3 converts gamma rays into Compton electrons, which generate Cherenkov photons inside of the pressurized gas cell. The Cherenkov photons are then focused onto a photomultiplier tube (PMT) using a Cassegrain optic.

2. GCD-3 Design Improvements

2.1. Optical Path Lengthening and Enhanced Shielding

Several improvements to the GCD-3 design were made to mitigate the impact of the "precursor" signal, which is caused by gamma rays that penetrate or bypass the tungsten shielding and interact directly with the pressure window or photomultiplier tube (PMT). First, the distance between the front of the detector and the primary Cassegrain reflector optic was increased by approximately 4 cm such that the time between the precursor and the Cherenkov signal was increased by ~0.25 ns. This additional time difference ensures that the tail end of the precursor signal does not overlap with the rising edge of the Cherenkov signal. As shown in figure 2, the GCD-1 precursor overlaps significantly with the primary gamma peak whereas the temporal separation between the GCD-3 precursor and the primary signal is large enough that the precursor signal completely returns to baseline prior to primary signal onset. The impact of the precursor can be very important for measurements where the primary gamma signals are of the same strength or are weaker than the precursor signal (e.g. at high Cherenkov energy threshold).

GEANT4 simulations indicated that a ring of hevimet tungsten-alloy (92.5% W, 5.25% Ni, 2.25% Fe chosen for its high density and machinability) placed just behind the front flange, as shown in figure 1, would reduce the amplitude of the precursor signal. The hevimet ring shields against diverging gamma rays that might strike the walls of the vessel behind the front flange and subsequently scatter into a trajectory that bypasses the central hevimet slug that shields the PMT from direct-shine gammas. While the hevimet ring reduces the number of Cherenkov photons that pass the front flange, the GCD-3 utilizes a larger diameter gamma-to-electron converter, larger gas volume and improved optical collection such that the detector sensitivity (integrated primary signal) is only ~15% lower than for GCD-1, as

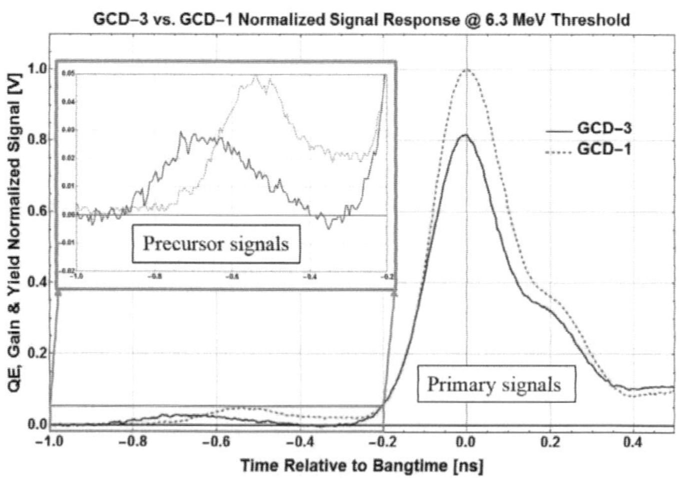

Figure 2. Yield normalized GCD-1 and GCD-3 signals, corrected for differences in PMT quantum efficiency and gain, recorded from Omega shot 77952. The detectors had identical gas fills (100 psia CO_2) and were located the same distance from TCC (20 cm). The GCD-1 (red, dashed line) precursor overlaps with the primary gamma signal while the GCD-3 (blue, solid line) precursor signal completely returns to baseline before the primary signal begins.

illustrated by the relative DT yield normalized signals from Omega shot 77952 shown in figure 2. The larger GCD-3 signal at around 0.4 ns is simply due to differences between the PMTs' unique impulse response shapes in the tail region.

2.2. Metal seals allowing for fluorinated gases at 400 psia

The GCD-1 and GCD-2 detectors utilize rubber O-rings and are pressure rated to an upper limit of 100 psia using non-fluorinated gases. Fluorinated gases such as SF_6 and C_2F_6 have a larger index of refraction than CO_2 at equivalent pressures and are therefore desirable for achieving low Cherenkov threshold energies. However, fluorinated gases can dissociate and release atomic fluorine, which poses a significant corrosive threat to the tritium recovery system at Omega. Stringent leak rate requirements were imposed on the GCD-3 to ensure that fluorine corrosion would not occur, which necessitated the use of all metal seals in place of rubber O-rings. Stainless steel knife-edge flanges were explosion bonded to the aluminum pressure vessel and sealed using copper and/or aluminum gaskets.

A brazed sapphire window separates the pressurized gas volume from the PMT. Initial attempts to install the sapphire window resulted in fracturing of two windows due to heat from the welding operation warping the surrounding metal. Redesigning the window housing resulted in a successful installation of a third backup sapphire window, however this window had not been coated with anti-reflective (AR) coating and the resulting light transmission through the window was predicted to be reduced by about 14%. The sensitivities of the GCD-3 and GCD-1 would be roughly comparable had an AR coated window been used. The GCD-3 has been routinely operated at 400 psia with CO_2 and C_2F_6, which corresponds to Cherenkov energy thresholds of 2.9 MeV and 1.8 MeV respectively.

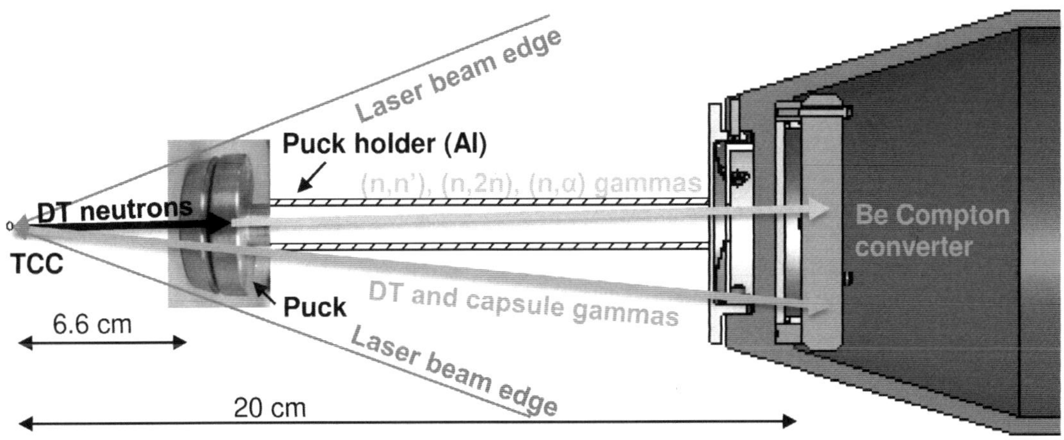

Figure 3. Neutron induced gamma ray studies may be performed using pucks affixed to the front of the GCD-3 (not to scale).

2.3. Neutron induced gamma ray studies using "pucks" mounted on GCD-3

A target, or "puck", holder consisting of a slender hollow aluminum tube may be affixed to the front of the GCD-3 for neutron induced gamma ray studies on the puck material chosen. 14 MeV DT neutrons interacting with the puck material produce gamma rays that may generate a unique GCD-3 signal if the resulting gamma ray energy is above the Cherenkov threshold energy. This puck signal is separated in time from the primary DT signal by about 1 ns when placed 6.6 cm from target chamber center (TCC), which is determined by the time of flight difference between the prompt (DT and capsule) gamma rays and the DT neutrons. A schematic of the puck assembly on the front of the GCD-3 is shown in figure 3.

Techniques for determining the ρR of the unablated plastic or high density carbon (HDC) shell mass at bangtime using previous gas Cherenkov diagnostics have been reported by Hoffman [5,6] and Cerjan [7]. A graphite puck of known density mounted on the front of GCD-3 allows for an in situ calibration of the detector response to 4.44 MeV $^{12}C(n,n')$ inelastic scatter gamma rays. The puck is also mounted at other positions around TCC in order to identify the angular distribution of the gamma rays produced, as described previously by Hoffman [8]. A similar glass (SiO_2) puck calibration, complete with angular scan, has been performed to extend the capability of the GCD-3 to determining ablator ρR for glass capsule implosions, with results to be published.

3. Future GCD-3 development

In August 2015 the GCD-3 was absolutely calibrated at the High Intensity Gamma-Ray Source (HIGS) at Duke University. The detector response was determined over a wide range of Cherenkov thresholds for different gases by exposing the GCD-3 to well-characterized quasi-mono-energetic gamma rays (Gaussian energy distribution with $\Delta E_\gamma / E_\gamma$~5% FWHM) produced by the HIGS free electron laser with gamma ray fluxes between 10^7 and 10^8 gammas/second. Several GCD-3 computational models are being developed using the Monte Carlo codes GEANT4 and ACCEPT, which will be benchmarked against the HIGS data as done previously with the GCD-1 and the Gamma Reaction History (GRH) diagnostics [9].

The GCD-3 will be installed into a 3.9 meter diagnostic well at NIF in early 2016 in preparation for the eventual construction and installation of the "Super GCD". A specially designed carrier support tube and mounting system has been designed to insert the GCD-3 to various depths within the diagnostic well in order to optimize sensitivity and reduce signal background due to laser plasma interaction (LPI) x-rays. At 3.9 m from TCC, the GCD-3 will be the most sensitive gamma-ray based Reaction History and ρR detector fielded at NIF and will enable a wide range of additional gamma-ray studies in support of ICF and discovery science campaigns.

4. Summary

Initial experimental results demonstrate improved GCD-3 performance over earlier GCD designs. Lengthening the Cherenkov photon optical path resulted in a larger temporal separation between the precursor and primary gamma-ray signals. The precursor signal magnitude was also significantly reduced due to improved tungsten shielding. The GCD-3 sensitivity was observed to be ~15% lower than GCD-1 while fielded simultaneously at 100 psia CO_2 on Omega DT implosions, largely due to the use of a back-up sapphire window that lacked AR coating. Metal seals on the GCD-3 allow for 400 psia operation with fluorinated gases and a lower Cherenkov energy threshold limit of 1.8 MeV using C_2F_6. Calibration data obtained at HIGS along with carbon and glass puck signal data will be used to benchmark GEANT4 and ACCEPT detector models that are currently being developed. The GCD-3 will be installed, acting as a prototype for the Super GCD, in a 3.9 meter diagnostic well at NIF in early 2016 to provide high sensitivity ρR measurements, reaction history and additional gamma ray measurements in support of the National Diagnostics Plan.

References

[1] Caldwell S, *et al.* 2003 *Rev. Sci. Instrum.* **74**, 1837
[2] Mack J, *et al.* 2003 *Nucl. Instrum. Methods Phys. Res. A* **513**, pp 566-572
[3] Herrmann H, *et al.* 2014 *Rev. Sci. Instrum.* **85**, 11E124
[4] Kilkenny J, *et al.* 2015 The National Diagnostic Plan *10th National Diagnostic Working Group Meeting* (Oct. 6, 2015 Los Alamos, NM)
[5] Hoffman N, *et al.* 2010 *Rev. Sci. Instrum.* **81**, 10D332
[6] Hoffman N, *et al.* 2013 *Physics of Plasmas* **20**, 042705
[7] Cerjan C, *et al.* 2015 *Physics of Plasmas* **22**, 032710
[8] Hoffman N, *et al.* 2013 *EPJ Web of Conferences* **59**, 13019
[9] Rubery M, *et al.* 2013 *Rev. Sci. Instrum.* **84**, 073504

Radiation emission from ultra-relativistic plasma electrons in short-intense laser light interactions

R Ondarza-Rovira[1] **and TJM Boyd**[2]
[1] Instituto Nacional de Investigaciones Nucleares, A.P. 18-1027, México 11801, Distrito Federal, Mexico
[2] Centre for Theoretical Physics, University of Essex, Wivenhoe Park, Colchester, CO4 3SQ, United Kingdom

E-mail: ricardo.ondarza@inin.gob.mx

Abstract. Intense femtosecond laser light incident on overcritical density plasmas has shown to emit a prolific number of high-order harmonics of the driver frequency, with spectra characterised by power-law decays. When the laser pulse is p-polarised, plasma effects do modify the harmonic spectrum, weakening the so-called universal decay index $p = 8/3$ to $5/3$. In this work appeal is made to a single particle radiation model in support of the predictions from particle-in-cell (PIC) simulations. Using these, we further show that the emission radiated by electrons –those that are relativistically accelerated inside the plasma, after being expelled into vacuum, the so-called Brunel electrons– is characterised not only by the plasma line but also by ultraviolet harmonic orders characterised by the $5/3$ decay index.

1. Introduction

Solid density plasma targets irradiated by highly intense femtosecond laser pulses have over recent decades proved to be a substantial source of high harmonic generation. Derived from the increasing levels in laser intensity, nowadays attainable in laboratory experiments, a corresponding increase in the spectral range beyond a thousand harmonics can be observed [1]. The scaling of the power efficiency η_m, with $m \gg 1$ being the harmonic order, was first shown empirically to scale as $\eta_m \sim 0.9\left(I_{20}\,\lambda_L^2\right)^2 (m/10)^{-p}$, [2], where I_{20} denotes light intensity normalized to 10^{20} W cm^{-2}, λ_L the wavelength expressed in microns, and $p = 5$. In particular, this scaling showed that the spectral decay index p decreased as I increased from weakly relativistic values ($a_0 \lesssim 1$) to moderately relativistic levels, typically $a_0 \sim 5$, where $a_0 \sim 8.54\left(I_{20}\lambda_L^2\right)^{1/2}$ is the normalized quiver momentum. For increasing values of I_{20} it was found that the decay index p weakens from 5.5 at $a_0 \sim 0.6$ to 3.4 at $a_0 \sim 2.7$ [3].

As laser intensities approached the ultra-relativistic range (UR) the spectral decay index p weakened further in the power-law decays $P_m \sim m^{-p}$. Baeva *et al.* (BGP) [4] formulated a model based on the assumption that the interaction physics remains invariant for constant values of a similarity parameter defined as $S = n_{e0}/n_c a_0$, where n_{e0}/n_c is the ratio of the initial electron plasma density to the critical density. This model predicted a decay index $p = 8/3$, claimed to be *universal*, over the entire harmonic spectral range and characterised by a cut-off. However, only a few experimental insights of the $p = 8/3$ decay have been found in spectra obtained from high-intensity laser-plasma interactions [5].

Boyd and Ondarza-Rovira (BOR) [6] identified a transition from the BGP index to one where $p = 5/3$ across a range of parameter combinations for $1 \lesssim S \lesssim 5$, showing that the deviation in the spectra from the 8/3 decay reflected details of the interaction physics in which emission at the plasma frequency ω_p and its harmonics distort the spectrum of laser harmonics. Plasma waves are excited by bunches of UR electrons, generated by Brunel absorption that couple strongly to the radiation field at the plasma surface. In this paper, we appeal to a hybrid particle-PIC code model and apply this to further explore the spectrum and its decay characteristics, together with that observed from the Larmor radiation of Brunel electrons.

2. Single particle model and PIC simulations

The model used in this paper was due originally to Weatherall [7], to provide insight into the generation of radiation from sources as diverse as relativistic electron beam devices in the laboratory to galactic jets. The original model provided a representation of highly localised intense electrostatic fields by a soliton-like structure. In the case of laser-plasma interactions, the Brunel electrons –those that are first pulled out of the plasma by the electric field of the laser pulse and then relativistically accelerated back into the plasma during the consecutive half optical cycle (Figs. 1a and 1b)– bunch to excite plasma waves, that in turn couple to the radiation field in the strong density gradients at the plasma surface. The bunching of Brunel electrons provides the driver for the generation of Langmuir waves. Figure 2a shows the density contours of the Langmuir field that increases in a confined region inside the plasma, and Fig. 2b the corresponding correlation of electron trajectories and beam crossings that generate the electrostatic field [8].

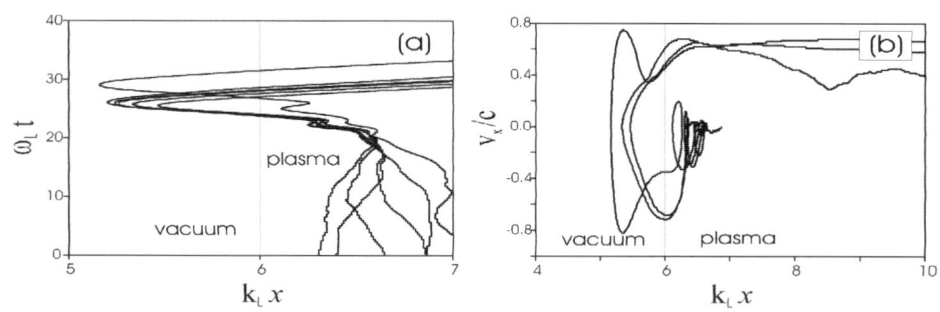

Figure 1: Brunel electrons followed when first expelled from the plasma and then reinjected back through the plasma surface: (a) particle trajectories and (b) normalized velocities.

These localised intense electrostatic fields are represented by a soliton-like structure $\boldsymbol{E}_{Lang} = -\frac{2}{3}\,\pi\, e\, n_e\, D\, \exp\left(-i\,\omega_p\, t\right)\, \hat{\mathbf{p}}$, where D is the scale length and the dipole moment is oriented in the direction $\hat{\mathbf{p}}$. In this context, relativistic electrons serve as the agent driving beam-resonant Langmuir waves. We have chosen n_e and D, tailored to conform to values measured in our PIC simulations.

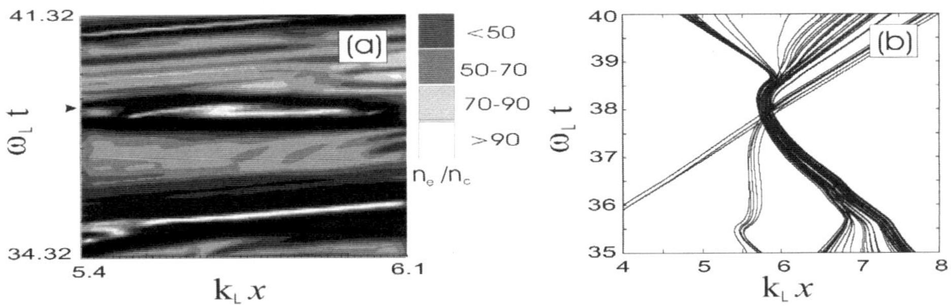

Figure 2: (a) Electron density contours inside the plasma where a high electrostatic field E_{Lang} is generated, and (b) the correlation of this field with Brunel trajectories and beam crossings.

Weatherall computed the emission spectrum from E_{Lang}, using the relativistic Larmor formula under the assumption that the soliton was decoupled from the electron beam, so that electrons followed an unperturbed linear trajectory in passing through its centre. Summing individual particle radiation over the electrons constituting the beam results in intrinsically weak emission, on account of a lack of coherence. However, Benford and Weatherall [9] noted that allowing for a spectrum of density correlations led to a plasma condensate spectrum characterised by a power-law decay, with decay index $p = 5/3$ across a range $\omega_p \lesssim \omega \lesssim \omega_c$; the cut-off frequency $\omega_c = 2\gamma_b^2\, c/D$ where γ_b is the relativistic beam gamma and c the speed of light in vacuum. Our approach differs in that we adopt E_{Lang} as a source field in the relativistic Lorentz equation, that allowed us to compute the radiation from an electron subject simultaneously to the electromagnetic laser fields and the electrostatic Langmuir field, this aligned along the pulse propagation. Whereas Weatherall appealed to plasma turbulence to provide beam coherence, in our case coherence results from relativistic electrons driven by the laser pulse itself.

In Figs. 3 a-b we present a comparison of two spectra generated from a PIC simulation and from the model here outlined, for the parameter combination $a_0 = 60, n_{e0}/n_c = 95$.

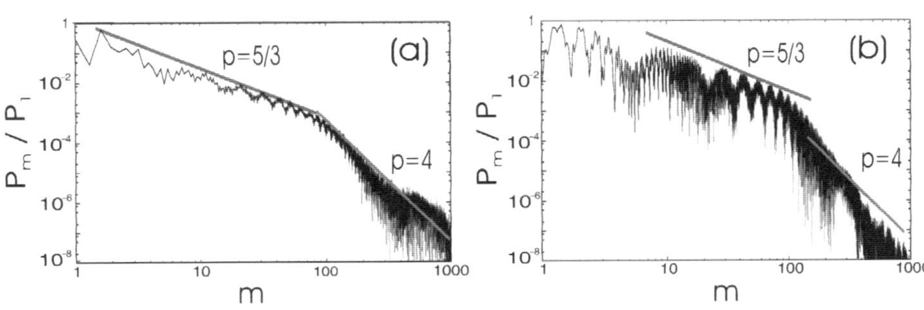

Figure 3: (a) PIC and (b) single particle emission for $a_0 = 60, n_e/n_c = 95$ and $\tau_p = 3$ osc showing transitions from regions characterised by $p = 5/3$ to $p = 4$. Plasma modulation effects and emission at ω_p are discerned.

3. Radiation from Brunel electrons

Once we observed the similarities of the radiation spectra from both the particle model described above and from PIC simulations, we further explored the radiated emission from those selected Brunel electrons, previously tracked from numerical runs. We computed the emission

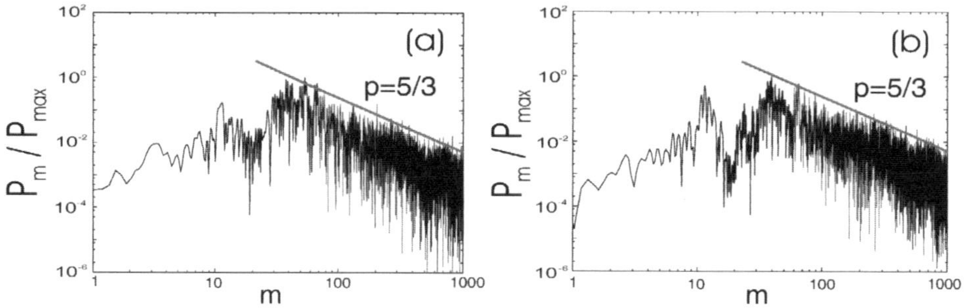

Figure 4: (a) and (b) Larmor radiation calculated from two different accelerated Brunel plasma electrons, as those shown in Fig. 1a, for the laser-plasma parameters as in Fig. 3. The harmonic spectra are characterised by a 5/3 power decay index over high harmonic orders, preceded by strong emission at the plasma line, $m_p = \omega_p/\omega_L$.

from the Larmor formula of a moving charged particle, to perceive that the contribution to the reflected field from the plasma electrons had in fact two significant characteristics: a robust emission at the plasma line and emission at high harmonic orders governed by a 5/3 power decay index. Figures 4 a-b show the harmonic spectra from two arbitrarily chosen Brunel plasma electrons.

4. Conclusions

In this work we have shown from PIC simulations that the radiation emission observed from the interaction of UR short pulses of laser light with dense plasma targets are in agreement with the single particle radiation model proposed, based on Brunel electrons perturbed by a strong localised electrostatic field that is generated inside the plasma by the action of the laser field. Furthermore, the Larmor radiation from Brunel electrons is consistent with both the harmonic spectrum reflected from the plasma, observed from simulations, and from that obtained from the single particle model.

Acknowledgements

The authors acknowledge support from CONACyT under Contract No. 167185.

References

[1] Burnett NH *et al* 1977 *Appl. Phys. Lett.* **31** 172; Carman RL *et al* 1981 *Phys. Rev.* A **24** 2649; Dromey B *et al* 2006 *Nat. Phys.* **2** 456; Kim IJ *et al* 2012 *Nat. Commun.* **3** 1231
[2] Gibbon P 1996 *Phys. Rev. Lett.* **76** 50
[3] Norreys PA 1996 *et al Phys. Rev. Lett.* **76** 1832
[4] Baeva T *et al* 2006 *Phys. Rev.* E **74** 046404
[5] Dromey B *et al* 2007 *Phys. Rev. Lett.* **99** 085001
[6] Boyd TJM and Ondarza-Rovira R 2008 *Phys. Rev. Lett.* **101** 125004; 2008 *J. Phys.* C **112** 022107
[7] Weatherall JC 1988 *Phys. Rev. Lett.* **60** 1302; Weatherall JC and Benford G 1991 *Astrophys. J.* **378** 543
[8] Boyd TJM and Ondarza-Rovira R 2010 *Phys. Lett.* A **374** 1517
[9] Benford G and Weatherall JC 1992 *Phys. Fluids* B **4** 4111

Journal of Physics: Conference Series **717** (2016) 012111 doi:10.1088/1742-6596/717/1/012111

The influence of magnetised electron transport on thermal self-focusing and channelling of nanosecond laser beams

Martin Read [1,2], **Robert Kingham** [2] **and John Bissell** [3]

[1] York Plasma Institute, University of York, Heslington, York, YO10 5DQ, UK
[2] Blackett Laboratory, Imperial College London, London, SW7 2AZ, UK
[3] Department of Physics, University of Bath, Claverton Down, Bath, BA2 7AY, UK

E-mail: martin.read@york.ac.uk

Abstract. The propagation of a nanosecond IR laser pulse through an under-dense $(0.01 - 0.1\,\mathrm{n_{cr}})$ magnetised laser-plasma is considered. The interplay between magnetised transport, B-field evolution and plasma hydrodynamics in the presence of a dynamically evolving beam are investigated by means of a paraxial wave solving module coupled to CTC, a 2D MHD code including Braginskii electron transport and IMPACT, a 2D implicit Vlasov-Fokker-Planck (VFP) code with magnetic fields. Magnetic fields have previously been shown to improve density channel formation for plasma waveguides however fluid simulations presented here indicate that Nernst advection can result in the rapid cavitation of magnetic field in the laser-heated region resulting in beam defocusing. Kinetic simulations indicate that strong non-local transport is present leading to the fluid code overestimating heat-flow and magnetic field advection and resulting in the recovery of beam channelling for the conditions considered.

1. Introduction

In recent years there has been an increasing use of applied magnetic fields in long-pulse (ns) laser-plasma experiments. Magnetic fields can be utilised for the purpose of easing the path to ignition in ICF (e.g., improving laser-plasma coupling in hohlraum targets [1] and increasing fusion yield in direct-drive experiments [2]), for their benefits to plasma wave-guide formation [3] and for the study of fundamental plasma physics.

Electron transport, magnetic field evolution and beam propagation are intricately linked in a magnetised plasma and under appropriate conditions, a range of magnetised transport phenomena are present. These transport effects can change the thermal and hydrodynamic evolution of the plasma and via the refractive index $\eta \propto (n_e/n_{cr})^{1/2}$, may affect laser propagation. Additionally, magnetic fields can both lead to instabilities and modify temperature scale-lengths, determining whether the local approximation is fulfilled which has ramifications on the validity of fluid codes. For these reasons, a complete model of electron transport in the presence of B-fields under such conditions is vital to the simulation and interpretation of laser-plasma experiments. This work investigates the interplay between laser propagation and plasma evolution accounting for a full range of magnetised transport phenomena.

2. Physical Background

Electron transport in the presence of a magnetic field is governed by the Braginskii transport equations [4] for Ohms law and heat-flow $\mathbf{q}$, given in equations (1) and (2), where $\underline{\underline{\alpha}}^c$, $\underline{\underline{\beta}}^c$ and $\underline{\underline{\kappa}}^c$ are the normalised resistivity, thermoelectric and thermal conductivity tensors and τ is the electron-ion collision time. Furthermore $\mathbf{C}$ is the plasma fluid velocity and $\mathbf{j}$, n_e, T_e and P_e give the electron current density, number density, temperature and pressure respectively.

$$en_e\left(\mathbf{E}+\mathbf{C}\times\mathbf{B}\right) = -\nabla P_e + \mathbf{j}\times\mathbf{B} + \frac{m_e}{e\tau}\underline{\underline{\alpha}}^c\cdot\mathbf{j} - n_e\underline{\underline{\beta}}^c\cdot\nabla T_e \tag{1}$$

$$\mathbf{q} = -\frac{n_e T_e \tau}{m_e}\underline{\underline{\kappa}}^c\cdot\nabla T_e - \underline{\underline{\beta}}^c\cdot\mathbf{j}\frac{T_e}{e} \tag{2}$$

The electric field $\mathbf{E}$ determined by Ohm's law is combined with Faradays law yielding an induction equation describing the evolution of the magnetic field $\mathbf{B}$. Of particular interest is the $\beta_\wedge$ component of the thermoelectric term $\left(\underline{\underline{\beta}}^c\cdot\nabla T_e\right)$ perpendicular to both the magnetic field and temperature gradient which gives rise to the Nernst effect – an advection of magnetic fields along with heat-flow down temperature gradients. In previous simulations [5], Nernst field advection was shown to have a dramatic effect on magnetic field evolution, leading to changes in the density and thermal profiles and causing the re-emergence of non-local transport in systems which were previously localised. Due to the intricate coupling between magnetised electron transport, hydrodynamics and a dynamically evolving laser, a numerical approach to modelling the problem is taken.

3. Code Development and Setup

The interaction of a laser with an under-dense magnetised plasma is modelled using two codes: CTC (Classical Transport Code) [6], a 2D MHD code which utilises Braginskii electron transport coupled to equations for continuity, momentum and energy, and IMPACT [7], an implicit 2D Vlasov-Fokker-Planck (VFP) code with magnetic fields which inherently incorporates transport. The codes share the same geometry – 2D slab Cartesian (x-y) with a perpendicular B-field component (B_z). CTC, benefits from a fast execution time and the means to enable/disable terms in the transport equations. IMPACT is computationally expensive but it is well known that the kinetic treatment, which accounts for non-local effects and non-Maxwellian distributions, is often required to correctly account for heat-flux in problems related to plasma ablation and ICF conditions. The laser is included in both codes using a newly developed module based on the paraxial wave treatment given by Sentis [8,9]. The laser is coupled to the plasma codes via inverse Bremsstrahlung heating and the plasma refractive index. The addition of the paraxial module allows for the modelling of laser-pulse dynamics within an evolving plasma.

For the results presented here, beam propagation over $\sim 1\,\text{ns}$ was modelled through a $\sim 1\,\text{mm}^2$ domain. The initial plasma temperature and density were given by $T_{e0} = 20\,\text{eV}$ and $n_{e0} = 1.5\times10^{19}\,\text{cm}^{-3}$ and applied fields ranged from $0-6\,\text{T}$. The input laser had intensity $I_L = 3.9\times10^{14}\,\text{Wcm}^{-2}$, wavelength $\lambda_L = 1054\,\text{nm}$ and spot-size $\phi_{fwhm} = 10\,\mu m$. Simulations were performed primarily using CTC to explore parameter space and the changes caused by the Nernst effect. IMPACT was used to look for non-local and kinetic effects. Parameter space exploration was limited by the computational power required to simulate macroscopic volumes of plasma over nanosecond timescales using a kinetic code and by the magnetothermal [6] instability – an instability resulting from the interplay between Nernst advection and Righi-Leduc heat-flow – which leads to unstable temperature and field profiles.

4. Results and Discussion

The CTC simulations showed significant changes in laser focusing dynamics over a 2 mm distance when Nernst advection was enabled/disabled as shown in figure 1 where the initial magnetic field

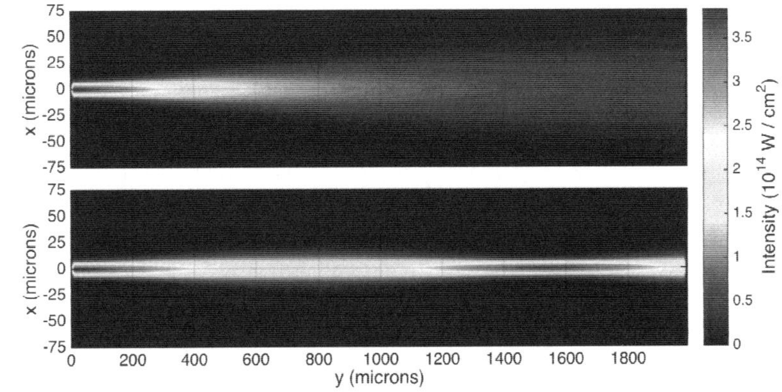

Figure 1. Changes to the intensity profile after ~ 350 ps. When the Nernst effect is enabled (upper), the laser defocuses whereas when it is disabled (lower), the beam demonstrates self-focusing due to plasma channel formation.

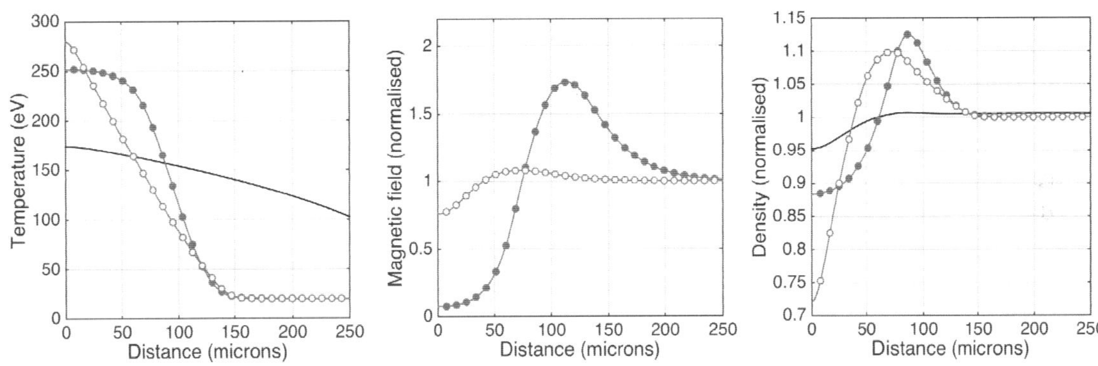

Figure 2. Temperature, density and B-field lineouts 1 mm into the domain with (closed circles) and without (open circles) Nernst advection. Unmagnetised case (solid line) shown for reference.

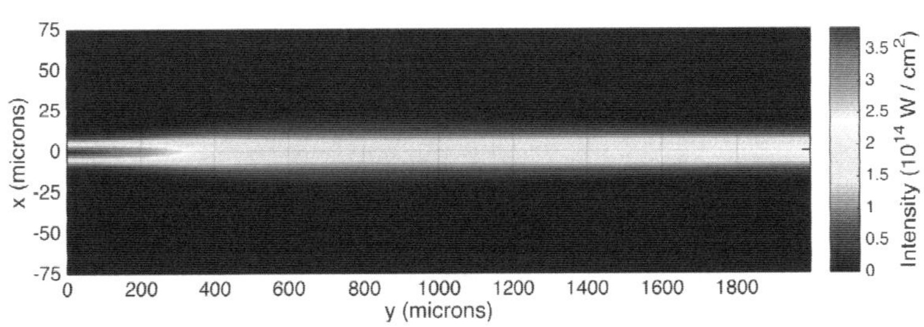

Figure 3. The intensity profiles from IMPACT (which includes Nernst advection) after $\sim$ 350 ps exhibit beam self-focusing behaviour.

was $B = 6\,\mathrm{T}$. Accounting for Nernst advection, after several hundred ps the laser defocuses as it propagates, whereas without Nernst advection, the beam continuously focuses and defocuses.

This is explained by observing the density, temperature and magnetic field profiles, transverse lineouts (1 mm into the domain) of which are shown in figure 2. Nernst advection (closed circles) leads to a rapid evacuation of 90% of the field from the central heated region after 350 ps compared with only $\sim 25\%$ when it is not accounted for (open circles). The reduction in field leads to the central region becoming un-magnetised, giving unsuppressed heat-flow and a wide, flat temperature profile. This affects the thermal pressure, resulting in a significantly modified density profile – shallow and broad when the Nernst effect is enabled and narrow and deep when it is disabled. In the broad profile, the beam can defocus. In the narrow profile, the beam is continuously channelled.

Figure 3 shows the corresponding intensity profile from the VFP code (IMPACT) which automatically includes Nernst advection. For these parameters, the kinetic approach results in beam channelling over the length of the domain unlike the defocusing behaviour shown in the fluid simulations including Nernst advection (Figure 1 (upper)). The change in focusing is likely due to non-local effects. Under these conditions, the transport in the heated plasma centre is non-local resulting in a decreased heat-flow in the regions of steepest temperature gradient. As shown in figure 4, CTC overestimates both the heat-flow out of the laser-heated region and the Nernst advection, resulting in a faster-than-expected cavitation of the B-field. To correctly match the non-local behaviour shown in the VFP code, CTC should make use of a flux limiter for the heat-flow and for the field advection. Choice of the correct limiter for Nernst field advection is the subject of ongoing work [10, 11].

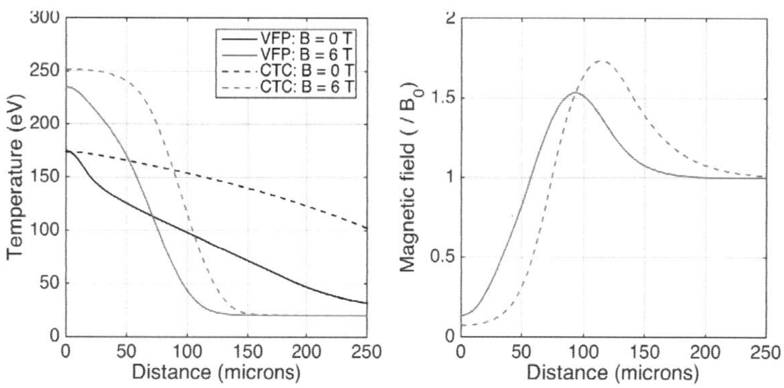

Figure 4. Temperature and B-field lineouts from CTC (dashed lines) and IMPACT (solid lines) 1 mm into the domain after ~350 ps

5. Conclusions

Nanosecond pulse propagation through under-dense magnetised plasmas has been investigated computationally. Laser-defocusing resulting from the feedback of Nernst advected B-field onto the density profile have been observed using the fluid code CTC coupled to a paraxial wave solving module. Kinetic simulations using the VFP code IMPACT show that under these particular conditions, beam defocusing does not arise as readily and beam channelling is maintained. Previous work with VFP simulations [5] has shown significant changes to B-field and density in kinetic simulations due to the Nernst effect and therefore it is likely that non-local transport has lead to a shift in the threshold beyond which defocusing occurs at these parameters.

Acknowledgments

The author would like to acknowledge the funding of the UK EPSRC (DTG No. EP/P504694/1) and the HPC facilities provided by the Imperial College High Performance Computing Service.

References

[1] Montgomery D S, Albright B J, Barnak D H, Chang P Y *et al.* 2015 *Phys. Plasmas* **22**
[2] Chang P Y, Fiksel G, Hohenberger M, Knauer J P *et al.* 2011 *Phys. Rev. Lett.* **107**
[3] Froula D H, Divol L, Davis P, Palastro J P *et al.* 2009 *Plas. Phys. Cont. Fus.* **51**
[4] Braginskii S I 1965 *Reviews of Plasma Physics* **1**
[5] Ridgers C P, Kingham R J and Thomas A G R 2008 *Phys. Rev. Lett.* **100**
[6] Bissell J J, Ridgers C P and Kingham R J 2010 *Phys. Rev. Lett.* **105**
[7] Kingham R J and Bell A R 2004 *J. Comp. Phys.* **194**
[8] Sentis R 2005 *ESAIM: Mathematical Modelling and Numerical Analysis* **39**
[9] Ballereau P H, Casanova M, Duboc F, Dureau D *et al.* 2007 *J. Sci. Comp.* **33**
[10] Bissell J J 2013 *Central Laser Facility Annual Report 2012-13*
[11] Davies J R, Betti R, Chang P Y and Fiksel G 2015 *Phys. Plasmas* **22**

Development of 4.5 keV monochromatic X-ray radiography using the high-energy, picosecond LFEX laser

H Sawada[1], S Fujioka[2], T Hosoda[2], Z Zhang[2], Y Arikawa[2], H Nagatomo[2], H Nishimura[2], A Sunahara[3], W Theobald[4], P K Patel[5] and F N Beg[6]

[1]University of Nevada Reno, Reno, NV, 89577-0220, USA
[2]Institute of Laser Engineering, Osaka University, Suita, Osaka 565-0871, Japan
[3]Institute for Laser Technology, Nishi-ku, Osaka, 550-0004, Japan
[4]Laboratory for Laser Energetics, University of Rochester, Rochester, NY, 14627, USA
[5]Lawrence Livermore National Laboratory, Livermore, CA 94550, USA
[6]University of California San Diego, La Jolla, CA 92093, USA

Email: hsawada@unr.edu

Abstract. Development of a monochromatic x-ray imaging system using a high-energy short-pulse laser LFEX and a spherical crystal is reported. Irradiation of the intense short-pulse laser produces a flash of 4.51 keV Ti K-alpha x-ray while the spherically bent quartz crystal provides a narrow spectral bandwidth and high spatial resolution. This high spatiotemporal imaging technique was applied for recording 2-D monochromatic x-ray images of laser-driven Fast Ignition targets. The results show a sufficiently high spatial resolution to characterize the implosion core, suggesting that the core information extracted from the radiograph images can be used to benchmark a 2-D radiation-hydrodynamic code for accurate hydrodynamic modelling and optimization of FI fuel assembly in the asymmetrical implosion.

1. Introduction

Fast Ignition (FI) [1, 2] is one of the advanced laser fusion concepts in inertial confinement fusion (ICF) with a potentially higher fusion gain than conventional ICF schemes. It uses a two-step approach to achieve fusion ignition: 1) a deuterated plastic (CD) shell or a shell filled with fusion fuels (e.g., deuterium and tritium gas) is first compressed by long pulse, nanosecond lasers to form a high density fuel core and 2) the dense fuel core is rapidly heated by injecting an ultraintense, picoseconds laser to ignition temperature. Creation of high areal density cores ($\rho R > 0.3$ g/cm^2) is essential in ICF for self-sustaining fusion reactions by alpha heating. [3] Specifically in the FI concepts that use energetic charged particles (electrons, protons or ions) for core heating, the heating efficiency with those energetic particles increases with its areal density because of shorter stopping ranges in the core. In cone-guided FI [4, 5], a re-entrant metal cone is inserted in the spherical shell to maintain the clear path for ignition laser from expanding coronal plasma. Simulations of such core formation require a 2-D Radiation-hydrodynamics code. It is important to benchmark the simulation code with high quality experimental measurements. Key benchmarking parameters for the FI fuel assembly are the core areal density, core size and standoff distance between the core and the cone tip. The core information can be obtained through 2-D monochromatic x-ray radiograph measurements. Monochromatic x-ray radiography using a spherically bent crystal has been developed using nano-second long-pulse lasers [6, 7] and picosecond short-pulse lasers [8, 9]. In particular, 4.5 keV Ti Kα radiography has been employed for imaging cone-guided targets, however, clear radiograph images of compressed cores have not been measured at the photon energy.[8] Here, the development of 4.5 keV Ti Kα monochromatic x-ray imager using high-energy, short-pulse laser LFEX (Laser for Fast ignition Experiments [10]) and a spherically bent crystal for the Gekko XII chamber is reported.

2. Monochromatic K-alpha x-ray crystal imager for joint GXII-LFEX experiments

X-ray backlight imaging is a vital technique to obtain density-related information of an object in high energy density sciences. To record 2-D monochromatic x-ray images of a laser-imploded core, a spherically bent quartz crystal was chosen as an x-ray optic. Spherical crystals have advantages of high-throughput, narrow spectral bandwidth (monochromatic) and high magnification geometry.[11] The monochromaticity of the reflected x-rays is achieved by the Bragg condition of the crystal. The quartz crystal that was used in the experiment had 20 mm diameter, [2023] Miller indices, 2d spacing of 2.749 Å, and a radius of curvature of R = 260 mm. It was operated at 1.0° off normal incidence at the 1st order reflection of 4.51 keV Ti K-alpha x-ray with the magnification of 11.5. The spectral bandwidth of the crystal was calculated to be ~ 5 eV in the diagnostic geometry, which is much narrower than the Ti K-alpha line spectrum to form monochromatic x-ray images. The temporal resolution of the diagnostic is determined by the pulse duration of the LFEX short-pulse laser of 1.6 ps, enabling flash radiography of implosion cores. The spatial resolution was estimated to be ~ 18 μm from the measured radiograph images.

3. Experiment

A monochromatic radiograph experiment was conducted using Gekko XII (GXII) and LFEX laser system at Institute of Laser Engineering (ILE), Osaka University [10]. Figure 1 shows a schematic of the experiment. A 200 μm diameter CD sphere directly attached on the outside tip of a cone (diamond-like carbon, DLC or gold) was irradiated by 9 beams of the GXII, nano-second laser with 300 J per beam (total of ~ 2 kJ) in a 1.3 ns FWHM Gaussian pulse shape. Near the peak compression, 1 kJ, 1.6 ps LFEX laser irradiated on a 20 μm thick Ti foil positioned at 2 mm from the cone-sphere target to generate 4.51 keV Ti K-alpha x-ray. The LFEX beam was defocused to ~300 μm spot to generate spatially uniform backlighter emission in the foil. The intensity of the K-alpha x-rays were monitored with a Bragg crystal spectrometer. The x-ray transmitted through the imploded target was reflected by the spherically bent crystal and recorded with an imaging plate (IP) detector. The timing between the implosion laser and the short-pulse irradiation was varied around the peak compression based on the shell trajectory measurements with a time-resolved x-ray streak camera. The nominal timing jitter of

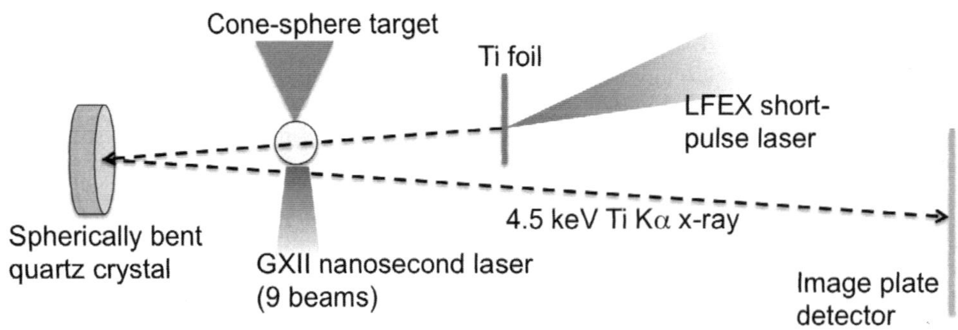

Figure 1 A schematic of the monochromatic K-alpha radiograph experiment. 9 GXII beams were irradiated on the CD sphere with a cone while the LFEX short-pulse was focused on a Ti backlighter foil.

the short-pulse laser was ~ a few hundred picoseconds in this experiment. The use of a solid CD sphere target allowed the radiograph measurements relatively insensitive to the mistiming of the backlighter beam. The initial areal density of the sphere was 0.02 g/cm^2 at the centre.

4. Results of the monochromatic x-ray radiographs

Figure 2 (a-c) shows measured monochromatic images with the crystal imager: short-pulse irradiated Ti foil, radiograph of undriven CD shell target and self-emission of a driven cone-sphere target. Direct short-pulse laser-foil interaction generated energetic electrons traversing through the entire foil, producing Ti Kα x-rays via collisions. The shape of the 1.0 mm by 1.0 mm Ti foil at projected angle and the defocused laser beam interaction region are clearly observed. The central x-ray emission in the foil was spatially uniform, which allows the interpolation of incident x-ray profile to be uniform in order to calculate 2-D transmission profiles. As shown in Figure 2 (b), x-ray attenuation by opaque objects such as the gold cone or a dense core was recorded on the backlighter emission as a shadow. The crystal imager works as monochromatic self-emission imaging system at the absence of the backlighter x-ray. Figure 2 (c) shows the monochromatic self-emission of a driven CD cone-sphere implosion. The emission signal was ~10 times lower than the backlighter x-ray.

Figure 2 (d) and (e) show a schematic of a cone-CD sphere target and a radiograph image of a driven target. The image was taken at 694 ps after the peak of 1.3 ns Gaussian pulse. The radiograph image shows the compressed CD target with its size of ~ 100 μm diameter and the gap between the core and cone tip. It also shows a deformation of the DLC cone tip due to the shock waves. The diagnostic spatial resolution of ~ 18 μm was estimated from the sharpness of the cone shadow. All images shown here suggest that the crystal imager developed has a sufficient spatial resolution to provide information necessary for characterization of FI targets. The temporal evolution of the implosion process can be recorded in a series of radiograph images by varying the timing between the driver and backlighter beams, which will be performed in future experiments.

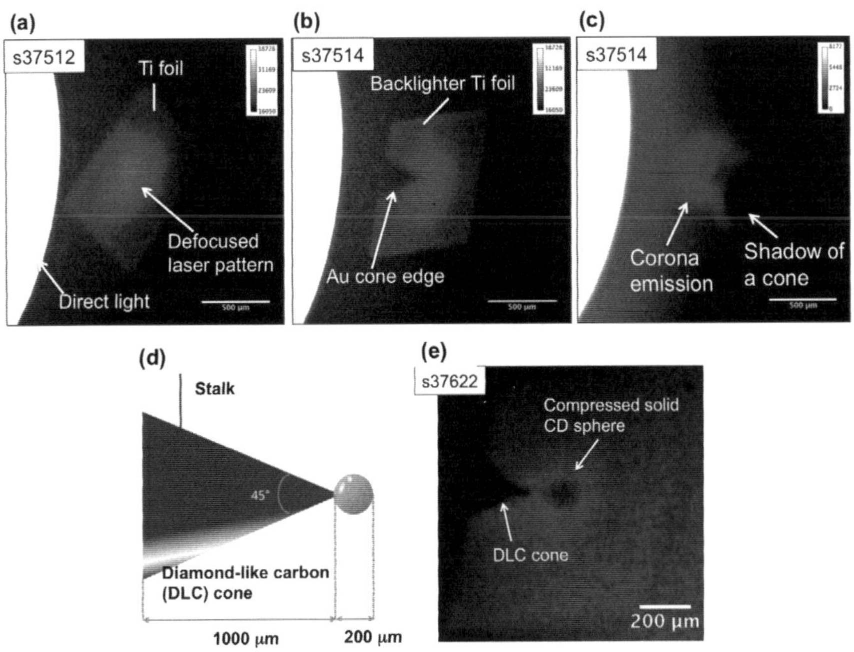

Figure 2 Recorded images of (a) monochromatic Kα emission of a Ti foil, (b) radiograph of undriven cone-CD shell and (c) self-emission of a driven cone-CD sphere target. (d) A schematic of a 200 μm diameter solid CD sphere attached on a diamond-like carbon (DLC) cone. (e) Radiograph of a compressed CD cone-sphere target at t=+694 ps after the peak of the Gaussian pulse.

5. Summary

A 4.51 keV monochromatic x-ray imager with a spherically bent quartz crystal has been developed for Gekko XII chamber using the high-power, short-pulse LFEX laser. High spatial and temporal resolutions of the diagnostic have been successfully demonstrated by recording 2-D monochromatic x-ray images of standalone foil, undriven and driven targets. The radiograph image of the driven cone-CD sphere target shows a smaller core size than the initial 200 μm diameter and a gap between the core and cone tip, suggesting that the spatial resolution of the crystal imager is adequate for quantitative characterization of the imploded target. In future experiments, a series of radiograph images will be taken to record a hydrodynamic evolution of implosion at various timings. The experimental data will be used to benchmark a 2-D radiation-hydrodynamic code for optimization of cone-guided FI targets.

6. Acknowledgement

The authors would like to thank the crews for the GXII and LFEX laser operation and the experimental group. This work was performed with the support and under the auspices of the NIFS Collaboration Research program (NIFS13KUGK072). H.S acknowledged the support by the UNR Provost office.

References

[1] Tabak M, Hammer J, Glinsky M E, Kruer W L, Wilks S C, Woodworth J, Campbell E M, Perry M D and Mason R J 1994, *Phys. Plasmas* **1** 1626
[2] Key M H 2007 Phys. Plasmas 14 055502
[3] Lindl J 1995 *Phys. Plasmas* **2** 3933
[4] Kodama R, Norreys P a, Mima K, Dangor a E, Evans R G, Fujita H, Kitagawa Y, Krushelnick K, Miyakoshi T, Miyanaga N, Norimatsu T, Rose S J, Shozaki T, Shigemori K, Sunahara A, Tampo M, Tanaka K a, Toyama Y, Yamanaka T and Zepf M 2001 *Nature* **412** 798–802
[5] Theobald W, Solodov A. A., Stoeckl C, Anderson K S, Betti R, Boehly T R, Craxton R S, Delettrez J A., Dorrer C, Frenje J A., Glebov V Y, Habara H, Tanaka K a., Knauer J P, Lauck R, Marshall F J, Marshall K L, Meyerhofer D D, Nilson P M, Patel P K, Chen H, Sangster T C, Seka W, Sinenian N, Ma T, Beg F N, Giraldez E and Stephens R B 2011 *Phys. Plasmas* **18** 056305
[6] Koch J A., Aglitskiy Y, Brown C, Cowan T, Freeman R, Hatchett S, Holland G, Key M, MacKinnon a., Seely J, Snavely R and Stephens R 2003 *Rev. Sci. Instrum.* **74** 2130
[7] Fujioka S, Fujiwara T, Tanabe M, Nishimura H, Nagatomo H, Ohira S, Inubushi Y, Shiraga H and Azechi H 2010 *Rev. Sci. Instrum.* **81** 22–5
[8] King J A, Akli K, Zhang B, Freeman R R, Key M H, Chen C D, Hatchett S P, Koch J A, MacKinnon A J, Patel P K, Snavely R, Town R P J, Borghesi M, Romagnani L, Zepf M, Cowan T, Habara H, Kodama R, Toyama Y, Karsch S, Lancaster K, Murphy C, Norreys P, Stephens R and Stoeckl C 2005 *Appl. Phys. Lett.* **86** 191501
[9] Stoeckl C, Fiksel G, Guy D, Mileham C, Nilson P M, Sangster T C, Shoup M J and Theobald W 2012 *Rev. Sci. Instrum.* **83** 033107
[10] Fujioka S. et al 2012 Plasma Phys. Control. Fusion 54 124042
[11] Koch J A, Landen O L, Barbee T W, Celliers P, Da Silva L B, Glendinning S G, Hammel B A, Kalantar D H, Brown C, Seely J, Bennett G R and Hsing W 1998 *Appl. Opt.* **37** 1784–95

Ablation plasma transport using multicusp magnetic field for laser ion source

K Takahashi, M Umezawa, T Uchino, K Ikegami, T Sasaki, T Kikuchi, and N Harada

Department of Electrical Engineering, Nagaoka University of Technology, Kamitomioka 1603-1, Nagaoka, Niigata, Japan

E-mail: kazumasa@vos.nagaokaut.ac.jp

Abstract. We propose a plasma guiding method using multicusp magnetic field to transport the ablation plasma keeping the density for developing laser ion sources. To investigate the effect of guiding using the magnetic field on the ablation plasma, we demonstrated the transport of the laser ablation plasma in the multicusp magnetic field. The magnetic field was formed with eight permanent magnets and arranged to limit the plasma expansion in the radial direction. We investigated the variation of the plasma ion current density and charge distribution during transport in the magnetic field. The results indicate that the plasma is confined in the radial direction during the transport in the multicusp magnetic field.

1. Introduction

Heavy ion inertial fusion (HIF) is considered as a candidate to realize power generation by nuclear fusion. The HIF requires heavy ion beams with high current to deposit sufficient energy to the fuel. For accelerating ion beams, several schemes have been considered using induction linac accelerator or recirculator[1]. For the induction acceleration scheme, ion sources, which can supply ion beams with current of 0.5 A and pulse duration of 20 μs, are required. The current is very huge compared to conventional heavy ion beams. To realize the beams, various types of ion sources are considered [2].

A laser ion source (LIS) is expected to be one of the candidates as the injector for providing the high current ion beam [3-5]. The LIS provides ion beams with a pulse due to using a pulsed laser and the ion current waveform indicates shifted-Maxwellian distributions [6]. The duration of the beam pulse can be varied by changing the plasma drifting distance between the surface of the laser target and the ion beam extraction point. However, the current density of the ion beam decreases significantly due to the three-dimensional expansion of the plasma [7]. To maintain the plasma density, applying confining magnetic fields have been investigated [8-10].

In addition, multibeamlet scheme has been considered as the method to supply ion beams with sufficient current for HIF [11, 12]. To extract uniform ion beams with large area, uniform plasma is required. To solve this problem, we proposed multicusp magnetic guiding using permanent magnets, which has been used for confinement of plasma for electron cyclotron resonance ion sources. To investigate the effect of guiding using magnetic field on the ablation plasma, we demonstrated the transport of the laser ablation plasma in the multicusp magnetic field.

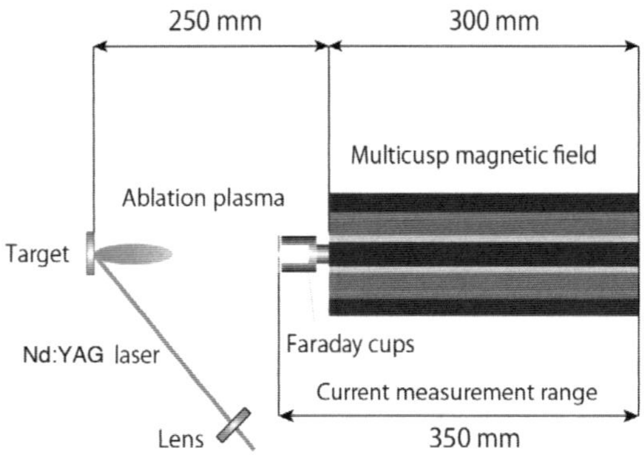

Figure 1. Schematic of the experimental setup.

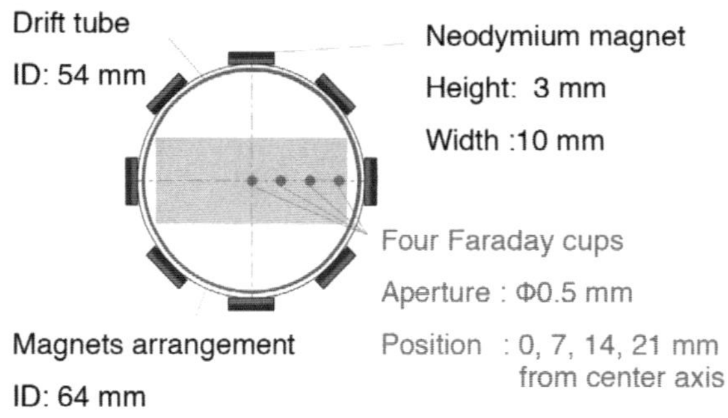

Figure 2. Cross section of the drift tube.

2. Experimental setup

Figure 1 shows the schematic of experimental setup. A Nd:YAG laser (wavelength: 532 nm) was focused on the copper target with the incident angle of 60 deg from the line perpendicular to the target surface. The pulse duration and energy of the laser are 16-18 ns and 402 mJ, respectively. The laser intensity was 2×10^8 W/cm^2 estimated by measuring the irradiated spot area of 0.12 cm^2. In the case of producing plasmas with the laser intensity of $\sim 10^8$ W/cm^2, singly charged ions are dominant in the plasma and the extracted ion beams showed normalized RMS emittance of 0.025π mm.mrad [13].

The magnetic field was formed with eight neodymium magnets, which have magnetic flux density of 0.2 T on the surface and arranged around a drift tube with inner diameter of 54 mm to limit the plasma expansion to radial direction as shown in figure 2. The magnetic field is placed at 250 mm from the laser target. The distribution of magnetic field between two magnets faced each other was measured using a Gauss meter. The plasma was transported over 300 mm in the magnetic field.

The plasma ion current was measured with Faraday cups biased at -30 V to return electrons. The Faraday cups have apertures of 0.5 mm at 0, 7, 14, and 21 mm from the center axis of the drift tube. The time of flight of the plasma can be obtained by the signal from a photodiode, which detects the scattered light generated by radiation of the laser.

3. Results and discussion

The ion current was measured as a function of the plasma drifting distance from the laser target. The measurements were performed three times at same position. Figure 3 indicates the ion current signal measured at 200 and 550 mm from the laser target on the center axis of the drift tube. Both signals have waveforms like shifted-Maxwellian distribution.

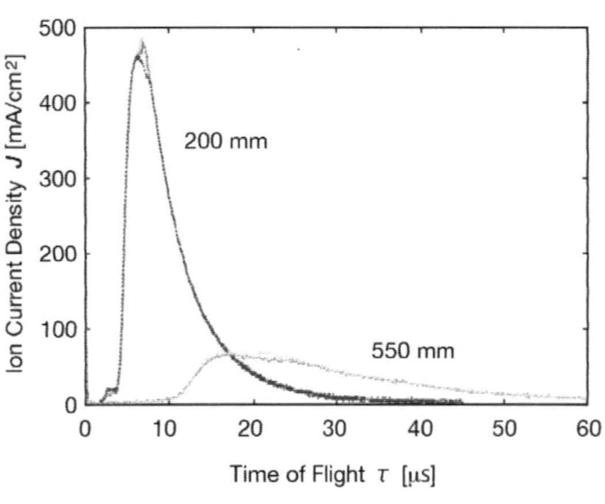

Figure 3. Ion current density J measured on the center axis of drift tube at L = 200 and 550 mm from the laser target. Each condition is performed three times.

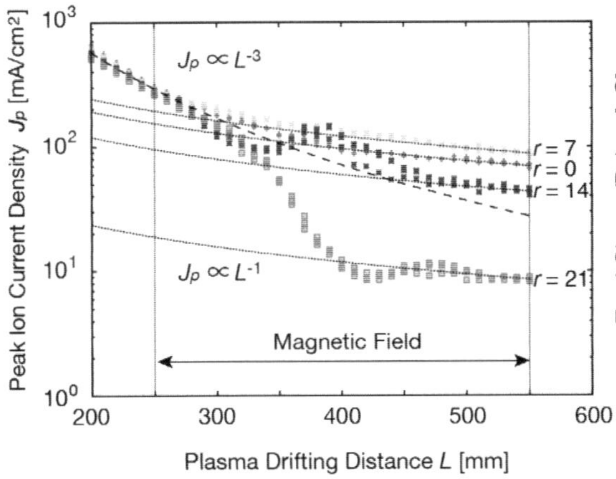

Figure 4. The variation of plasma ion current density at the peak of the waveform measured at $r = 0$ mm (red), 7 mm (green), 14 mm (blue), and 21 mm (pink) as a function of plasma drifting distance.

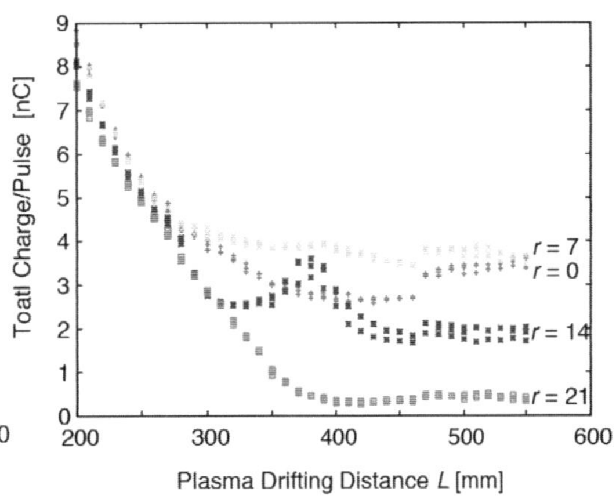

Figure 5. The variation of total charge/pulse obtained by integrating current waveform measured at r = 0 mm (red), 7 mm (green), 14 mm (blue), and 21 mm (pink) as a function of plasma drifting distance.

Figure 4 shows the variation of plasma ion current density at the peak of the waveform as a function of plasma drifting distance from the laser target L. The distribution of current density in radial direction is uniform without the magnetic field (L = 200 ~ 250 mm). In this region, the peak of the current density J_p decreases with the relation as $J_p \propto L^{-3}$ shown by the broken line. After injected into the magnetic field, current density increase compared to that without the magnetic field except for r = 21 mm. This implies that plasma ions move inside the drift tube from near wall where magnetic flux density is large. After going through transition region (L = 250 ~ 450 mm), current density decreasing corresponds to the dotted curves $J_p \propto L^{-1}$ ranging (L = 450 ~ 550 mm) at all radial positions.

Figure 5 shows the variation of total charge/pulse obtained by integrating ion current waveform. The variation corresponds to that of peak ion current density. The charges ranging (L = 450 ~ 550 mm) indicate almost constant. The results indicate that plasma expands only in longitudinal direction and laser ablation plasma can be confined in radial direction during transport in multicusp magnetic field.

Figure 6 indicates the distribution of the peak ion current density at L = 550 mm with magnetic flux density distribution. The plasma profile is almost consistent with the magnetic field. Although the plasma profile does not have flat top, it would be caused by the reason that the plasma profile near the center is on the way to be uniform. It is indicated that the variation of charge at r = 0 and 7 mm was not constant completely as shown in figure 5. To consider the effect of magnetic confinement, pressure of the plasma and magnetic field are estimated. The plasma pressure P can be derived using

$$P = (Z + 1)nkT. \quad (1)$$

Here Z is charge state of ions, k is the Boltzmann constant, T is temperature. Plasma ion density n is derived by the following equation,

$$J = Zenv, \quad (2)$$

where e is the elementary charge, v is velocity of the ions. Magnetic pressure P_B can be calculated as

$$P_B = B^2/2\mu_0 \,. \quad (3)$$

where B and μ_0 are magnetic flux density and magnetic permeability, respectively.

Assuming Z = 1 and adopting the peak current density J_p as J, estimated pressures based on the values in figure 6 are shown in figure 7. As shown in figure 7, magnetic pressure is much larger than

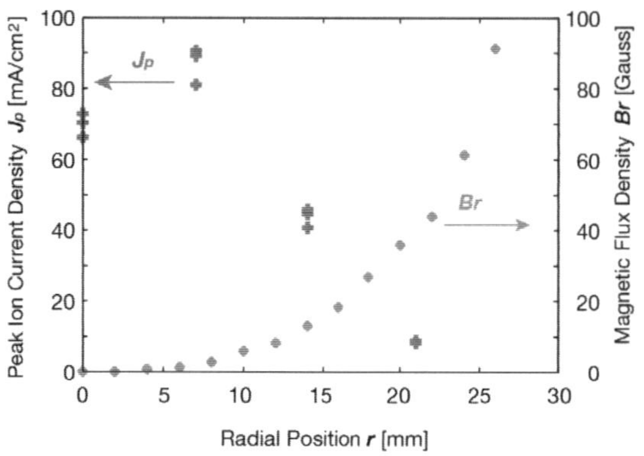

Figure 6. Distribution of peak ion current density in radial direction at L = 550 mm with magnetic flux density. Red and green plots indicate the current density and magnetic flux density, respectively.

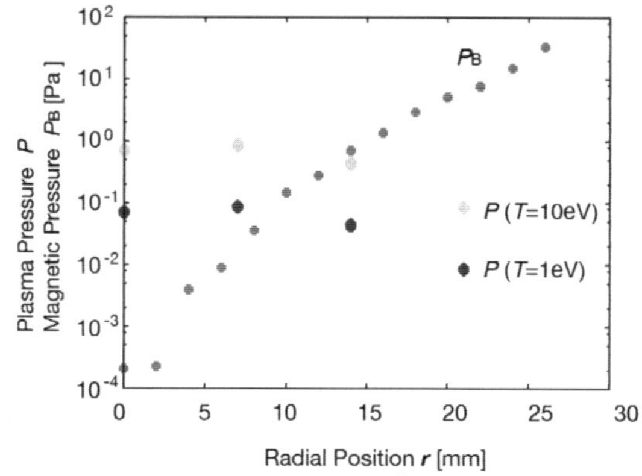

Figure 7. Comparison between plasma pressure P and magnetic pressure P_B at L = 550 mm. Red plots indicate magnetic pressure, green and blue plots show plasma pressure assuming temperature of 10 and 1 eV, respectively.

plasma pressure near wall of drift tube (r = 27 mm) and relatively uniform ion density is seen in the region of $P_B < P$.

Conclusion

To develop high current laser ion sources with large area, we demonstrated the transport of the laser ablation plasma in the multicusp magnetic field, which has much larger magnetic pressure than plasma pressure near the wall of drift tube. We investigated the variation of the plasma ion current density and charge distribution during transport in the magnetic field. The results indicate that the plasma is confined in the radial direction and expands in longitudinal direction during the transport in the multicusp magnetic field. In this study, the current density and pulse width of ~ 70 mA/cm^2 and ~ 20 μs pulse duration (FWHM) at 550 mm from the target have been achieved. Although the uniform density region was realized within the diameter of ~10 mmΦ, we need the uniform plasma with diameter of 30 mmΦ to get 0.5 A beam. Moreover, ions should be extracted as multi-beamlet from multi-apertures and larger uniform plasma will be required. Therefore, we will try to clarify the relation between the magnetic field and plasma profile in detail and get more uniform plasma in the radial direction.

References

[1] Bangerter R O, Faltens A and Seidl P A, 2013 *Rev. Accel. Sci. Tech.* **6** 85
[2] Kwan J W 2005 *IEEE Trans. Plasma Sci.* **33** 1901
[3] Barabash L Z *et al.* 1984 *Laser Part. Beams* **2** 49
[4] Hasegawa J *et al.* 2000 *Nucl. Instrum. Meth. Phys. Res.* B **161-163** 1104
[5] Okamura M *et al.* 2014 *Nucl. Instrum. Meth. Phys. Res.* A **733** 97
[6] Kelly R and Dreyfus R W 1988 *Surf. Sci.* **198** 263
[7] Baranov V Y *et al.* 1996 *Laser Part. Beams* **14** 347
[8] Okamura M *et al.* 2010 *Rev. Sci. Instrum.* **81** 02A510
[9] Takahashi K *et al.* 2013 *AIP Conf. Proc.* **1525** 241
[10] Takahashi K *et al.* 2013 *Plasma Fusion Res.* **8** 1206005-1
[11] Kwan J W *et al.* 2001 *Nucl. Instrum. Meth. Phys. Res.* A **464** 379
[12] Grote D P and Kwan J W 2003 *Phys. Rev. Special Topics Accel. Beams* **6** 014202
[13] Kanesue T, Tamura J, and Okamura M 2008 *Proc. EPAC08* (Genoa, Italy) 421

Journal of Physics: Conference Series **717** (2016) 012114 doi:10.1088/1742-6596/717/1/012114

Multiscale integral analysis of a HT leakage in a fusion nuclear power plant

M Velarde[1], J Fradera[2], J M Perlado[1], I Zamora[2], E Martínez-Saban[2], C Colomer[2] and P Briani[2]

[1] Institute of Nuclear Fusion–Technological University of Madrid, José Gutiérrez Abascal 2, 28006, Madrid Spain
[2] Idom Nuclear Services, Gran Vía Carlos III, 97 bajos, 08028, Barcelona, Spain

E-mail: marta.vmayol@upm.es

Abstract. The present work presents an example of the application of an integral methodology based on a multiscale analysis that covers the whole tritium cycle within a nuclear fusion power plant, from a micro scale, analyzing key components where tritium is leaked through permeation, to a macro scale, considering its atmospheric transport. A leakage from the Nuclear Power Plants, (NPP) primary to the secondary side of a heat exchanger (HEX) is considered for the present example. Both primary and secondary loop coolants are assumed to be He. Leakage is placed inside the HEX, leaking tritium in elementary tritium (HT) form to the secondary loop where it permeates through the piping structural material to the exterior. The Heating Ventilation and Air Conditioning (HVAC) system removes the leaked tritium towards the NPP exhaust. The HEX is modelled with system codes and coupled to Computational Fluid Dynamic (CFD) to account for tritium dispersion inside the nuclear power plants buildings and in site environment. Finally, tritium dispersion is calculated with an atmospheric transport code and a dosimetry analysis is carried out. Results show how the implemented methodology is capable of assessing the impact of tritium from the microscale to the atmospheric scale including the dosimetric aspect.

1. Introduction

The transfer of tritium from the IRC to the environment is conservatively assumed to occur when the IRC is operating with a single enclosure scheme through the tritium plant power conversion system (intermediate heat exchangers and helium blowers). Tritium loss is mainly caused by HT and T_2 permeation and simultaneous primary coolant leakage through the Heat Exchangers. Primary coolant chemistry appears to be the most effective way to control tritium permeation from the breeder into primary coolant and from primary coolant through Heat Exchangers by H_2 flux isotopic exchange or steel (EUROFER/INCOLOY) oxidation. A leakage at a NPP primary to secondary heat exchanger is considered for the present coupling example. Both primary and secondary loop coolants are assumed to be He. A leakage is set to happen inside the one of the HEX, leaking tritium in HT form to the secondary loop where it permeates through the piping structural material to the exterior. HVAC removes the leaked tritium towards the NPP exhaust. The HEX is modelled with system codes and coupled to CFD to account for tritium dispersion inside the NPP and site environment. Finally, tritium dispersion is calculated with an atmospheric transport code and a dosimetry analysis is carried out. Tritium environmental release rate limit are assumed to be ~ 1 g-T/y . This limit determines the key parameters for the tritium IRC (with HT and HTO as dominant molecules, and T_2, T_2O as marginal one). The mechanism of transfer to the environment is assumed to be through leakages or permeation at the exchange system of the power plant (primary to secondary loop). Hence, chemical process optimization of the primary system is a key factor that needs to be reanalyzed in terms of radiological impact. It is necessary to consider all the pathways of tritium from the reactor to the

Journal of Physics: Conference Series **717** (2016) 012114 doi:10.1088/1742-6596/717/1/012114

atmosphere and to model the transport phenomena adequately. Early and chronic doses, which have been evaluated for the Most Exposed Individual (MEI), have been calculated at particular distance bands from the release point.

2. Integral multiscale analysis example

1.1. Integral multiscale analysis. Primary to secondary leakage analysis

Different time and space scale calculation have to be carried out to perform an integral analysis of tritium dispersion from a leakage in a NPP. In the present work, an example for a HT leakage in a HEX is presented. Primary and secondary loop calculations are carried out with MELCOR[1] and GOTHIC[2] System Codes (SC) while larger scale calculation involving the NPP rooms and site are performed with ANSYS® Fluent® [3] CFD code. Atmospheric transport is calculated with UFOTRI Code [4]. Codes are coupled in order to achieve a best-estimate calculation with conservative parameters.

1.2. Tritium atmospheric and dosimetric analysis

For selecting the boundary conditions in the Tritium atmospheric dispersion at the macro scale, conservative parameters are used to define the input of the coupling with the data calculated with the CFD code. The methodology used for the calculation of the doses is probabilistic. The secondary phase in the tritium dispersion requires a coupling in the dispersion outputs and the tritium transport in all the body intake enclosures. In this phase, all mechanisms should be established where the deposition and incorporation in the environment can result in absorbed dose, including the oxidation of tritium in the case of HT.

3. HT leakage results

In the present work a leakage at a HEX is selected as an example to show the integral and multiscale analysis. As an example a 2.5 GW^{th} fusion power plant with three power loops is simulated with MELCOR and GOTHIC, coupled to ANSYS® Fluent® CFD code and atmospheric transport up to 1 km. The HEX primary pipe breaks and HT leaks to the secondary loop. Note that this is a hypothetical accident scenario designed to clearly show how different codes can be coupled to achieve an integral and multiscale analysis.

3.1 HEX MELCOR and GOTHIC results

MELCOR and GOTHIC have been used as system codes for the simplified example in order to analyze the behavior of a leakage at a HEX and to show its coupling to ANSYS® Fluent®. A small leakage, through a hole in a primary pipe, with a section of 7.5×10^{-5} m^2 is set to happen 1000 s after the system reaches the steady state. Results, shown in Fig. 1, present a leakage maximum mass flow rate of 0.82 kg/s . Tritium leakage is assumed to be only in form of HT and with a mass fraction of a 0.001. Note that permeation has not been calculated. The ratio of permeated tritium is set to be more than ten times that of a leakage at ambient conditions. The resulting GOTHIC data present a slightly larger (more conservative) leakage mass flow rate than that of MELCOR as shown in Fig. 1. Tritium in the secondary loop permeates to the environment. GOTHIC results are set as a time dependent BC at the considered room in the NPP. A simplified generic geometry for a fusion reactor site has been considered including a simple HVAC to the exhaust. A transient case has been run until the plume reaches its steady state and analyzed for surface concentrations (see. Fig. 2)

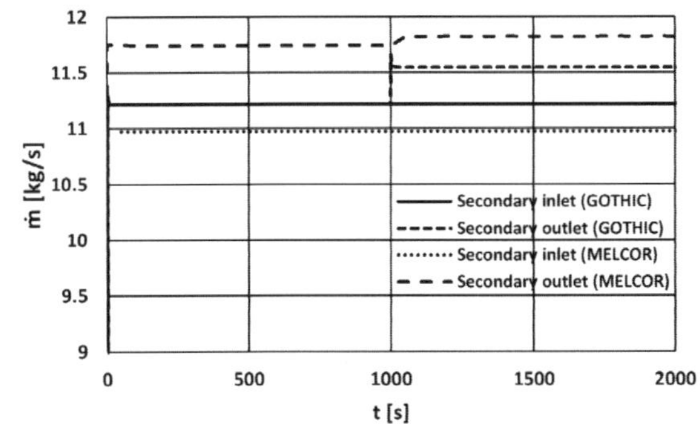

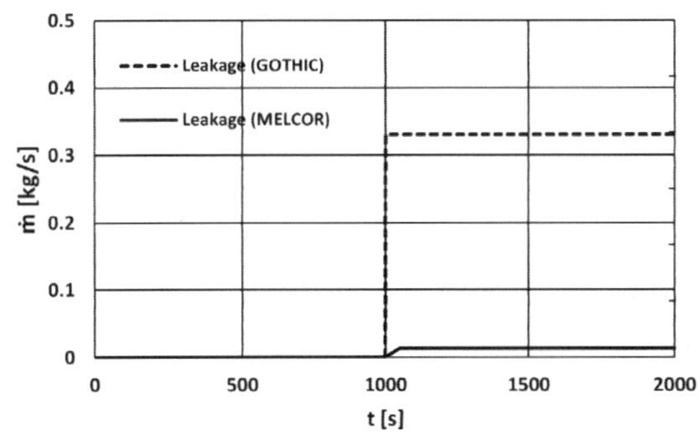

Fig. 1. MELCOR (at the left) and GOTHIC (at the right) HEX model secondary mass flow rate evolution.

Fig. 2. HT mass fraction on the site surface including buildings.

3.2 Atmospheric and Dosimetric analysis

The environmental conditions of the primary phase of tritium emission are factors that determine the intake concentrations by biological agents. Atmospheric processes have an important role mainly in the primary phase where tritium is not deposited in soil and plants in the environment. The accident takes placed at 30 m high stack and the rate of emission into the atmosphere is of 1.032 gram/s. The meteorological variables after the accident are also measured every 10 minutes of day 181 (the first of June). At that moment there is a contamination in the crops and therefore a dose by ingestion. The major chemical forms of tritium, HT and HTO, are strongly dependent on their velocity of deposition, **¡Error! No se encuentra el origen de la referencia.**. This parameter determines the concentration of tritium released to the atmosphere as HTO and the amount of tritium that penetrates into the lower layers of the soil below 5 cm from the surface, [5]. We have considered different releases, HT and HTO, for comparing the different chemical forms of tritium. But elemental tritium is observed. The doses resulting from inhalation are considered from two different perspectives: on one hand tritium enters the body from direct inhalation of the plume passage, and, on the other hand from (the plume) re-emission. For calculating the intake doses, the assumption made is that all of food consumed is produced locally and it is considered a compartment which is divided into an organic part and an inorganic part. The dose by ingestion of contaminated foods constitute the 97% of the contribution to doses in emissions to the atmosphere of elemental tritium when the crop is under development and the vegetal growth is ongoing. Concerning the contribution to the total dose due to inhalation, it has different components, one due inhalation by the direct pass of the tritium contaminated plume and the dose by re-emission of HT oxidized to HTO, from the surface of the soil. In the case of HT the dose due to tritium re-emission is a factor 3 higher than the dose due to inhalation directly from the tritium contaminated plume. This produce an increment of the atmospheric amount of HTO, therefore the dose due to inhalation of HTO constitute a 44.8% of the total dose.

Table 1. Most Exposed Individual, MEI, (Sv) values with downwind distance.

Distance (m)	HT (Sv)	HTO (Sv)
145	1.85E-04	3.88E-04
180	1.76E-04	3.60E-04
320	1.25E-04	2.54E-04

CONCLUSIONS

An integral and multiscale methodology application for Tritium leakages has been presented, showing that different techniques can be coupled to determine tritium transport at different time and space scales. All tritium transport phenomena have been assessed from a conservative point of view. A MELCOR and GOTHIC SC calculations together with a CFD analysis has been used to show how a detailed scale simulation can be coupled to atmospheric transport calculations. Plume and site concentrations can be analyzed at a micro scale giving valuable insight on the behavior of HT leakages. In any case the maxima dose overpasses the 1 mSv stipulated by the guidelines of regulation. This is due to the quantity of tritium deposited into the soil, which is lower than the atmospheric tritium in the first 200 meters of the accident where the tritium concentrations are always higher. The values of HT minor than the values of HTO are due to Dose Conversion Factor for a person older than 17 for HT is 4 orders of magnitude minor than HTO. The difference of doses for HT and HTO emission is due to the contribution to total doses by re-emission, which is important at short distance but not at large distances, because 95% of the HT deposited on the surface is returned to the air as HTO, increasing the atmospheric levels of this molecule. The HT contribution to MEI in the worst case (near to the point of emission) is a factor 3 lower than the correspondent HTO contribution.

ACKNOWLEDGMENT

EUROFUSION Enable Research ToIFE 2014

REFERENCES

[1] MELCOR, INL/EXT-07-12856, http://www5vip.inl.gov/technicalpublications/documents/3751088.pdf, 2014
[2] GOTHIC, NAI 8907-02 Rev 17 - GOTHIC Containment Analysis Package. User manual. Version 7.2a(QA)
[3] ANSYS® Fluent®, http://www.ansys.com/es_es/Productos/Flagship+Technology/ANSYS+Fluent, 2014
[4] W Raskob. UFOTRI: Program for Assessing the Off-Site Consequences from Accidental Tritium Releases. Report KfK-**4605**. (1990).
[5] W Raskob, M Velarde, J M Perlado. Tritium dose assessments with regulatory and advanced computer models for the potential european iter site vandellós (spain), *Fusion Science And Technology*., ISSN:1536-1055, vol. **48**/492-495, 2005
[6] M Velarde, L Sedano, J M Perlado. Dosimetric impact evaluation of primary coolant chemistry of the internal tritium breeding cycle of a fusion reactor DEMO, *Fusion Science and Technology* Vol. **54** Issue: 1, 122-126, 2008

TARDIS-C: A target diagnostic for measuring material structure at high pressure

R M Vignes, M F Ahmed, J H Eggert, A C Fisher, D H Kalantar, N D Masters, C A Smith and R F Smith

Lawrence Livermore National Laboratory, 7000 East Avenue, Livermore, CA, USA, 94550

E-mail: vignes2@llnl.gov

Abstract. A goal of the National Ignition Facility (NIF) at Lawrence Livermore National Laboratory is to better understand solid matter behavior at extreme conditions. Diagnostic tools such as the Target Diffraction In-Situ (TARDIS) have been designed to record data of solid material compressed to tens of Mbars over short time scales. NIF drive beams (~120 kJ) heat a carefully designed ablator to ramp compress the target to high pressure. A backlighter produces an x-ray source which is diffracted onto image plates through the compressed target. An unimpeded optical path allows Velocity Interferometer System for Any Reflector (VISAR) measurements to be recorded as the compression wave progresses through the target. To reduce the VISAR blast shield's exposure to debris and minimize contamination of the NIF chamber, a transparent barrier has been designed to contain debris within the TARDIS body.

1. Introduction

The National Ignition Facility (NIF) is fielding the Target Diffraction In-Situ (TARDIS) diagnostic to understand dynamic phase changes and structural transitions between state of matter at pressures and temperatures comparable to those found in "Super-Earths" (planets twenty times more massive than Earth). Ramp (shock-free) compression of the target material prevents material melting so that a solid matter state is retained while compressed to the extreme densities found in the center of planets. Experiments have been performed which ramp compressed diamond to five terapascals to obtain high pressure equation-of-state (EOS) data [1]. Future experiments include studying the EOS and high pressure melt curves of metals such as iron, and compressing diamond to break and reform atomic bonds from the SC1 state to the BC8 state to determine whether metastable BC8 carbon can exist at ambient conditions [2]. To prevent fast moving debris from these experiments from contaminating the NIF chamber or damaging diagnostics, a transparent barrier has been designed to block the VISAR aperture (Figure 1) while still allowing VISAR to record data, turning TARDIS into TARDIS-C.

2. Original TARDIS diagnostic

The TARDIS diagnostic is composed of two distinct components: the target assembly and the diagnostic assembly. Figure 1 contains both a cut-away and blow-out view of the device. The left of Figure 1 labels the components of the target assembly while the right labels the diagnostic assembly. The target assembly holds the target, provides a mount for the backlighter, shields the image plates from x-rays with the tantalum-10-tungsten (Ta10W) body, and is covered by a plastic-coated, aluminum dimpled shield to angularly disperse specular reflections of unconverted 1 and 2ω light to reduce their impact on laser optics. In the diagnostic assembly, the tungsten

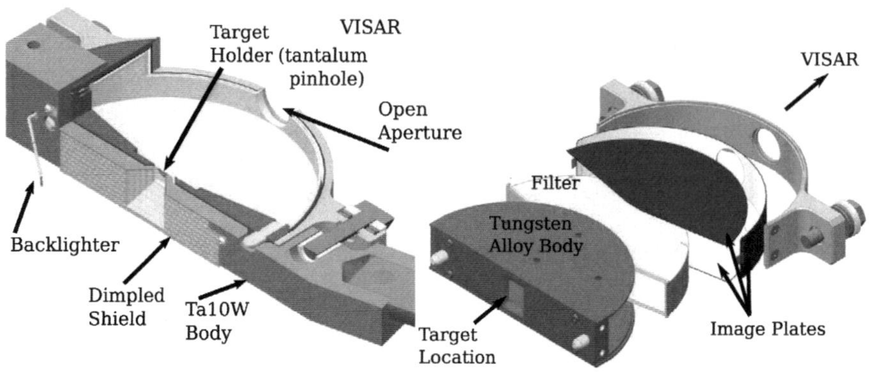

Figure 1. Two views of the TARDIS assembly, left labels the target assembly while the right labels the diagnostic assembly.

alloy body provides additional x-ray shielding for the reusable phosphorus image plates. Filters help to further reduce background signal from reaching the image plates. The image plates capture the sample's spectrography diffraction lines while the specimen is at peak compression. The backlighter, a thin foil material, emits a quasi-monochromatic He-α x-ray source when it is impinged upon by the lasers. The x-ray signal passes through the compressed sample material and the diffraction lines are then captured on the image plates. The diffraction lines contain information about the target material structure while under compression.

On the side of the diagnostic assembly opposite the target holder, an aperture provides a line of sight for VISAR measurements. VISAR records the target's free surface velocity, allowing the compressed state and drive pressure to be determined. The target material has an unimpeded path from the target holder through the image plate aperture and onto the VISAR blast shield. Further details of the diagnostic design, as well as alignment, operation, and post-shot handling are described in [3].

3. Debris threat and transparent barrier design considerations

While the NIF drive beams ($\sim$120 kJ laser pulse) vaporize or melt most of the ablator and target material, small pieces of unvaporized target debris and the TARDIS tantalum pinhole are driven toward the VISAR debris shield with velocities in the range of several kilometers per second. To prevent damage to the blast shield and to contain debris within the TARDIS body, an optically transparent barrier has been placed in the aperture to VISAR, as shown in Figure 2. To prevent altering the function of TARDIS, design of the barrier was highly constrained. The barrier's size and shape had to maximize the VISAR field of view without shadowing the image plates. External dimensions were limited by beam alignment fixturing. A single pane of protective transparent barrier would not suffice as the shrapnel impact shock would fracture the entire barrier and prevent a visual inspection from assessing whether debris had escaped the diagnostic enclosure. In developing a multi-pane barrier, care had to be taken to ensure the front panes would not cause debris to rebound and reenter the NIF chamber through the target holder entrance. Consideration also had to be given to the failed transparent material to ensure it did not create additional debris which could enter the NIF chamber.

4. TARDIS-C: TARDIS modified by a transparent debris "C"atcher

To design the transparent barrier, an estimate of the debris' mass and velocity distribution were obtained from radiation hydrocode (an LLNL code, ARES[4]) simulations of laser interaction

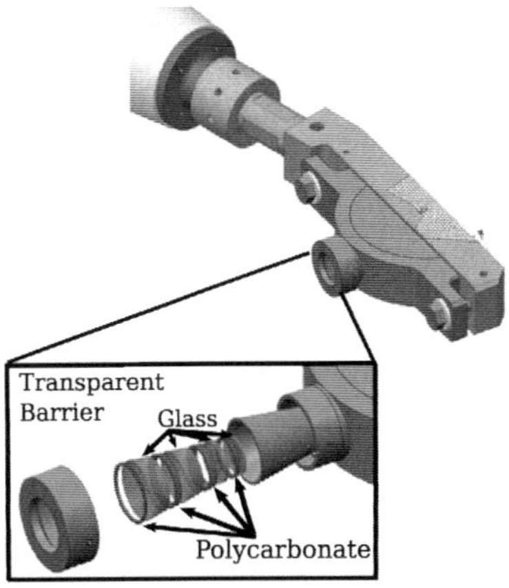

Figure 2. Transparent barrier modification to prevent contamination

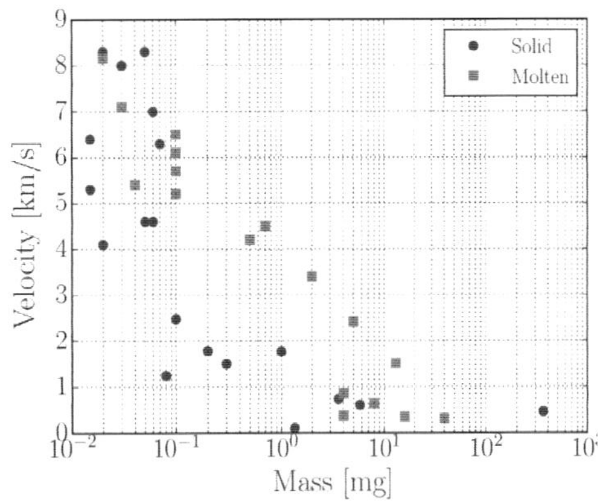

Figure 3. Debris field from target and pinhole impacting the transparent barrier

with the target and pinhole. The resulting derbis field is shown in Figure 3. The simulations predicted a mass of 468 mg impacting the barrier with a mass averaged velocity of 540 m/s. The debris includes material from both the pinhole and the target package. The molten material is traveling much quicker at 795 m/s but only accounts for 94 mg (20%) while the remaining mass, 374 mg, remains solid and travels with a mass averaged velocity of 475 m/s. A more conservative case, which consisted of all the laser energy being converted to kinetic energy to launch the entire intact pinhole (480 mg at 2,000 m/s) at the barrier, was also considered.

After many design iterations, the final barrier configuration, shown in the left of Figure 4, consisted of four panes of borosilicate float glass. In modeling the barrier, the work of [5, 6] were implemented in a constitutive response which modeled glass subjected to large strains, high strain rates, and high pressures. The constitutive model also accounts for material placement, such as whether the material is in the undamaged bulk, near a surface, or near damaged material. Polycarbonate was also considered for the barrier material, but when used alone as a barrier, it would shock melt and provided inadequate protection due to its relatively low density and melt temperature. When used in combination with borosilicate glass, the polycarbonate would be launched as a flyer plate which then impacted and heavily damaged the other barrier elements. Polycarbonate spacers were used to separate the glass panes, acting to isolate the rear glass panes from the initial debris impact shock and preventing damage of the rear panes. A comparison of the model predictions with experimental observations is shown in Figure 5. The simulations agreed well with the experimentally observed damage; both depict extensive damage to the front two panes, cracking and fracture of the third pane, and an intact, undamaged rear pane.

After much modeling effort, four design considerations which have the largest influence on barrier behavior were observed:

- Thin front glass panes – prevents debris rebounding back through target entrance and allows small, fast moving debris to be capture in the TARDIS body. Debris that does rebound off front glass has significantly reduced velocity
- Large air space between front glass panes – leaves room for debris to spall front panes instead of propagating damage to rear panes
- Thin, compliant spacer rings – minimizes load transmitted between glass panes, protecting rear panes from damage. Polycarbonate spacers minimize shock transmission to the glass
- Dense material is recommended for the barrier as more impact energy is absorbed in accelerating the heavier fragments

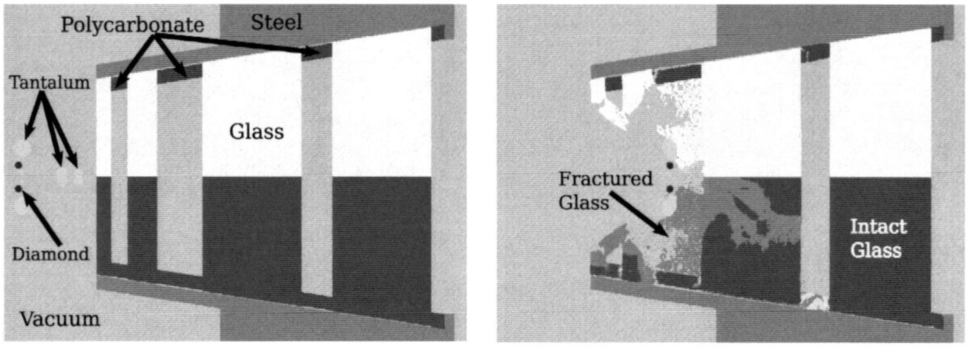

Figure 4. Transparent barrier, pre and post-impact. Due to design constraints, the barrier had to be less than 26.4 mm in length and have a radius of 11 mm at the VISAR exit and 7.16 mm at the target facing entrance. Top view in each figure shows the material while the bottom view depicts the damage (red→failed material).

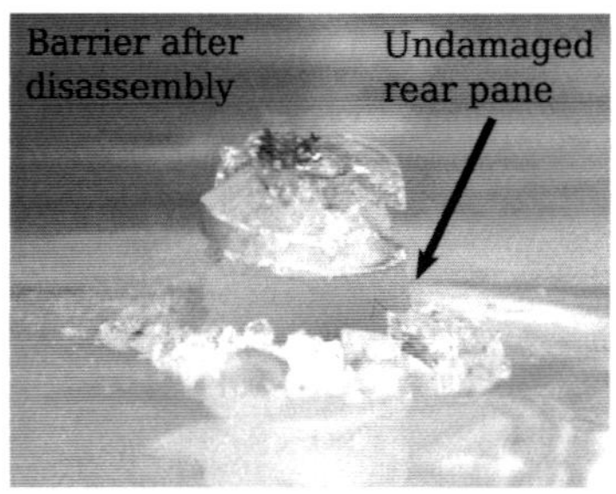

Figure 5. Transparent barrier recovered post-shot. Front two panes are completely damaged, third pane is in large fragments, while rear pane is undamaged, agreeing with model predictions.

5. Summary

A four pane borosilicate barrier is sufficent to prevent debris from contaminating the NIF chamber and damaging the VISAR shield. Thin front panes limit the rebounding of material while thick rear panes prevent material from escaping the barrier and allow a visual assessment to determine whether the barrier failed. Low density materials such as polycarbonate are not recommended for the barrier. If higher shot energies are needed, it may be possible to replace the rear glass panes with a stronger and higher density transparent material, such as sapphire.

Acknowledgments

This work was performed under the auspices of the U.S. Department of Energy by Lawrence Livermore National Laboratory under contract DE-AC52-07NA27344. LLNL-CONF-680336.

References

[1] Smith R F, et al 2014 *Nature.* **511**. 330-8.
[2] Sun J, Klug D D and Marto R 2009 *J. Chem. Phys.* **130**. 194512.
[3] Ahmed M F, House A, Smith R F, Ayers J, Lamb Z S and Swift D W 2013 *Proceedings of the SPIE 8850, Target Diagnostics, Physics and Engineering for ICF II.* (San Diego: Bellingham WA)
[4] B Division Code Group, Lawrence Livermore National Laboratory 2015. *ARES User Guide.*
[5] Johnson G R, Holmquist T J and Beissel S R 2003. *J. Appl. Phys.* **94(3)**. 1639-46.
[6] Holmquist T J and Johnson G R 2011 *J. Appl. Mech.* **78**. 051003.

Mounting, alignment and integration of large optics in China's high power laser

Hui Wang[1], **Zhao Xiong**[2], **Xiaodong Yuan**[2]

[1] Department of Mechanical Engineering, Tsinghua University, Beijing, 100084

[2] Research Center of Laser Fusion, China Academy of Engineering Physics, Mianyang, China, 621900

E-mail: xiong_022111@163.com

Abstract. SG-III, a high-power laser facility of China, is constructed to produce 0.18MJ energy for physical experiments under controlled laboratory conditions. Each laser beam requires the ability to align to a millimeter-sized target with a precision of 30 μm (RMS) and the single-beam energy will be up to 3.75 KJ. Arrayed along each beam-path, hundreds of optics must be positioned to stringent tolerances. Therefore, this paper introduces the approaches used by engineers to overcome the technical challenges on precise mounting, alignment and integration of large optics in china's high power laser facility.

1. Introduction

SG-III, the large high-power laser facility at China Academy of Engineering Physics, is constructed to create fusion conditions with controlled laboratory conditions [1]. Its 1053 nm output of 48 laser beams is frequency converted to produce 0.18 MJ energy and precisely focus onto a millimeter-sized target with a RMS deviation less than 30 μm. Therefore, arrayed along each beampath, 26 different types (reflecting mirrors, neodymium-doped laser glass, etc.), hundreds of large optics must be mounted and positioned to stringent tolerances, which put a huge technical challenge on the assembly and integration of optics for SG-III engineers.

2. The laser system of SG-III

The SG-III facility shares a similar schematic design with the National Ignition Facility (NIF) and Laser MégaJoule (LMJ) [2, 3], as shown in Fig.1. In each beampath, the laser pulse are generated, amplified, purified, transferred and finally, focused on one target. Hundreds of line-replaceable units (LRUs) are used as independent subsystems in each beampath, which represent a success of the philosophy of modularity [5]: large optics is packaged into LRU which can be installed by standardized joints in laser building to form laser beamline. The commissioning of the SG-III laser is quite unique as a consequence of the laser multi-pass architecture and the LRUs based mounting of optical components. All laser beampaths will be completely contained within the huge Laser and Target Area Building (LTAB), but the successful story in Optics Assembly Building (OAB) is critical to the SG-III engineering task. The role of the OAB is the assembly and alignment of Line replaceable units (LRUs) at rigorous cleanliness level (Class 100 clean-room). Driven primarily by ignition target performance including a specification of ±30μm RMS pointing at target center, great demands are placed on the precision alignment and transport of these optics. However, due to design of LRU, large optics module packages used for quick assembly, transportation, installation, and removal of the optics components and usually hold two or four separate optics, it precludes easy access to optical units after installation.

So, the optical components shall be assembled within the LRU structure with strict specifications of cleanliness, wavefront and off-line pre-alignment, in the OAB.

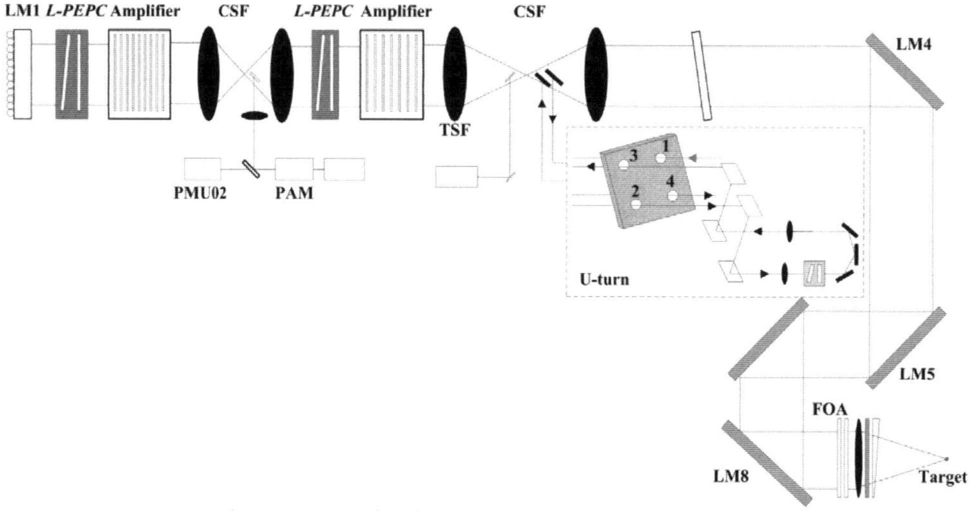

Fig.1 Schematic diagram of the SG-III laser beam

3. Mounting and alignment of large optics

In the Optics Assembly Building, optical components are mounted, fastened and pre-aligned within the LRU structure. To precisely hold the optics at design locations, tight LRU component positioning specifications are required – including mirrors (±3mm), lenses (±1mm) and pinholes/references (±1mm). A main challenge is how to keep the large optics stable at design positions even under ambient input excitations while with least stress/distortion of large aperture optics. This challenge comprises both design and manufacturing engineering which drives us to develop an integrated technical framework for performance evaluation and optimization (IPEO) on the large optics mounting under the full-life-cycle constrains. As shown in Fig.2, the assembly/mount of large optics is a multi-stages process and it must identify the various items to setup and validate the assembly/mount process while mechanics, optics and many others models relevant to the engineering feasibility and performance of LRU integration are used to support the development process.

Our practice on the switchyard transport mirrors [6] is a typical case, as the Fig.3 illustrates. The transport mirror plays an important role on laser beam's repositioning and pointing to the final target. And its surface requires unique wavefront specifications [7] because of the extremely high fluences and nonlinear processes at extreme laser power densities, for instance, roughness of surface less than 0.4 nm, RMS Gradient less than λ/75/cm, Peak-to-Valley less than λ/3, mount-induced surface aberration less than λ/4 while the laser wavelength (λ) is 632nm, and the deviation of mirror surface normal vector less than 50 microradians (μrad) when pre-alignment. So the first step in the performance evaluation and optimization process is to build the fundamental mechanics models relevant to mirror mounting and various factors such as gravity, tightening forces, components tolerances, shall be considered. Then, Opto-mechanical analysis and evaluation based on above models is implemented and the mirror surface performances, particularly, wavefront distortion, stress in mirror and pre-alignment within LRU, are kept within the required range by a closed-loop iterative process with the assembly/mounting parameters adjustment. In this process, both the methods of field optical measurements and 3D numerical modeling are proposed to predicate the surface distortion and alignments, etc. And others constraints including requirement for on-site equipment and human resources are also considered for determination on the feasibility of assembly operations. Finally, pilot test at the OAB will be executed to verify final results of mirror assembly/mount process (Fig.4). By now, with great efforts of several years, all the requirements for high cleanness, high accuracy and high efficiency have been met.

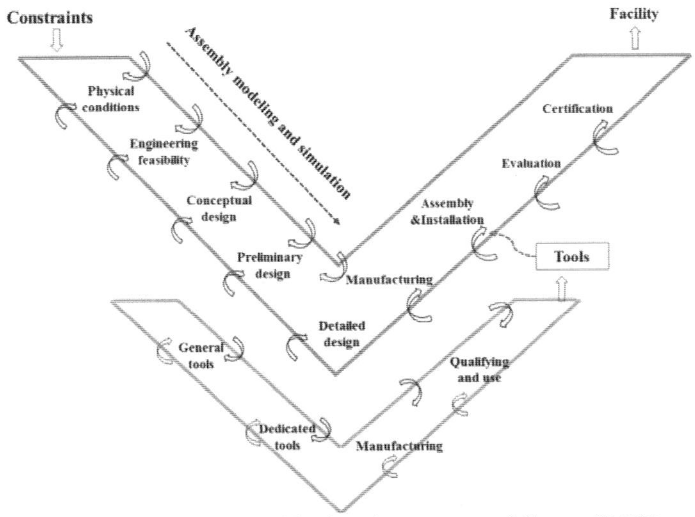

Fig.2 Integrated assembly development workflow of LRUs

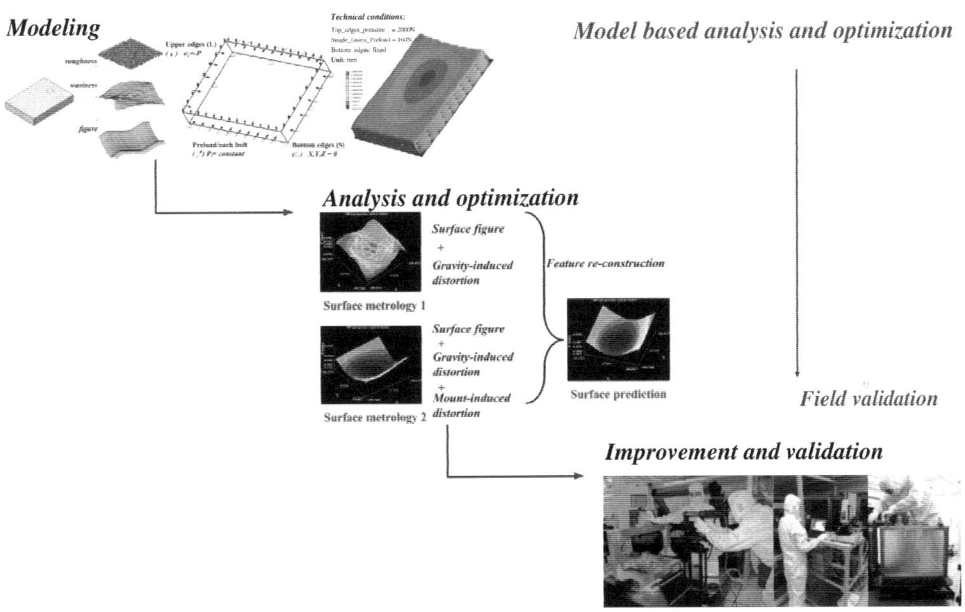

Fig.3 Assembly/mount performance evaluation and optimization of the transport mirror

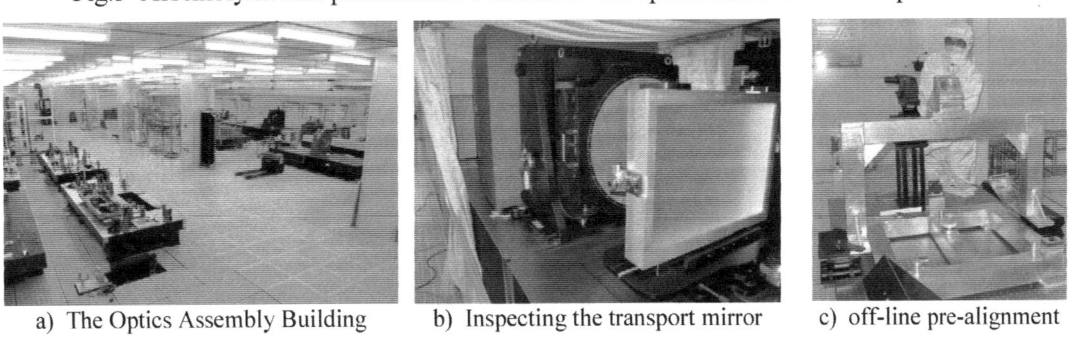

a) The Optics Assembly Building b) Inspecting the transport mirror c) off-line pre-alignment

Fig.4 Large optics assembly practices in OAB

After the LRU is assembled at OAB, the packaged LRU will be inserted by robotic transporters into kinematic mounts located on the supporting infrastructure in Laser Building and Target Area. Individual LRU's shall be positioned better than 1mm, and also for the center of large optics, better than 0.5% aperture, with state-of-the-art laser surveying. 3D virtual assembly environment and process simulation also are powerful tools before on-site assembly operation with a robotic alignment verification system, as shown in Fig.5, which help engineers correct thousands of minor operational problems and confirm the final operation solution. So, all the efforts on the integrated process development which incorporates both of engineering software and automatic equipment for large optics performance validation are crucial to the precise mounting and alignment of SG-III commissioning.

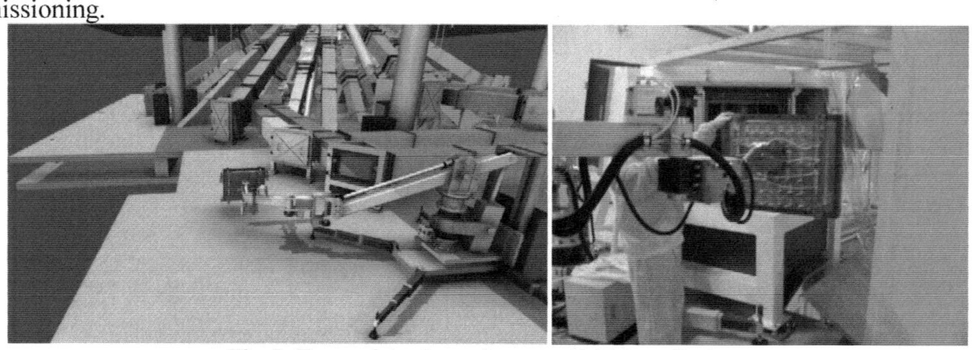

Fig.5 From virtual mock-up to field installation

4. Conclusions

The SG-III project has demonstrated the successful installation and commissioning of large optics on 48 beampaths, from the master oscillator to a 1mm sized target located at the center of the target chamber. And the laser pulse meets or exceeds the major requirements in terms of power, energy and pulse length producing the highest energy of 180 kJ at 3ω. Precise assembly, mounting and alignment of large optics are crucial to the success of the SG-III project. The integrated technical framework for the performance evaluation and optimization of large optics, demonstrates an important role on meeting the extreme requirements of SG-III installation and commissioning.

Acknowledgments

The authors gratefully acknowledge the support and contributions of our colleagues in the Research Center of Laser Fusion, China Academy of Engineering Physics. And this research is partially sponsored by National Natural Science Foundation of China (Grant 51575310).

References

[1] Wanguo Z et al. 2013 Development of laser technology in Research Center of Laser Fusion *High Power Laser and Particle Beams* **25** 3082

[2] English E et al. 1998 Optical system design of the National Ignition Facility *Proceedings of SPIE 3482: International Optical Design Conference* **3482** 726

[3] Michel L 1999 The French Megajoule Laser Project *Fusion Engineering and Design* **44** 43

[4] Zhiquan S 2000 Beam position stability analysis in the large solid state laser system *High Power Laser and Particle Beams* **12S** 167

[5] Larson D W 2004 NIF laser line-replaceable units (LRUs) *Proceedings of SPIE 5341: Optical Engineering at the Lawrence Livermore National Laboratory II, The National Ignition Facility* **5341** 127

[6] Hui W et al. 2015 Assembly error analysis for large aperture transport mirror in high power solid-state laser system *Acta Optica Sinica* **35** 0922004

[7] Hui W et al. 2015 Optomechanical analysis of the mounting performance of large laser transport mirrors Optical Engineering **54** 035107

9th International Conference on Inertial Fusion Sciences and Applications (IFSA 2015) IOP Publishing
Journal of Physics: Conference Series **717** (2016) 012117 doi:10.1088/1742-6596/717/1/012117

Nuclear Diagnostics at the National Ignition Facility, 2013-2015

C B Yeamans[1], W S Cassata[1], J A Church[1], D N Fittinghoff[1], M Gatu Johnson[4], N Gharibyan[1], R Határik[1], D B Sayre[1], H W Sio[4], R M Bionta[1], D L Bleuel[1], J A Caggiano[1], C J Cerjan[1], G W Cooper[6], M J Eckart[1], E R Edwards[5], S A Faye[5], C J Forrest[7], J A Frenje[4], V Yu Glebov[7], P M Grant[1], G P Grim[1], E P Hartouni[1], H W Herrmann[3], J D Kilkenny[2], J P Knauer[7], A J Mackinnon[1], F E Merrill[3], K J Moody[1], M J Moran[1], R D Petrasso[4], T W Phillips[1], H G Rinderknecht[1], D H G Schneider[1], S M Sepke[1], D A Shaughnessy[1], W Stoeffl[1], C A Velsko[1], P Volegov[3]

[1] Lawrence Livermore National Laboratory, Livermore, California, USA
[2] General Atomics, San Diego, California, USA
[3] Los Alamos National Laboratory, Los Alamos, New Mexico, USA
[4] Massachusetts Institute of Technology, Cambridge, Massachusetts, USA
[5] University of California, Berkeley, Berkeley, California, USA
[6] University of New Mexico, Albuquerque, New Mexico, USA
[7] University of Rochester, Laboratory for Laser Energetics,Rochester, New York,USA

Email: yeamans1@llnl.gov

Abstract. The National Ignition Facility (NIF) relies on a suite of nuclear diagnostics to measure the neutronic output of experiments. Neutron time-of-flight (NTOF) and neutron activation diagnostics (NAD) provide performance metrics of absolute neutron yield and neutron spectral content: spectral width and non-thermal content, from which implosion physical quantities of temperature and scattering mass are inferred. Spatially-distributed flange-mounted NADs (FNAD) measure, with nearly identical systematic uncertainties, primary DT neutron emission to infer a whole-sky neutron field. An automated FNAD system is being developed. A magnetic recoil spectrometer (MRS) shares few systematics with comparable NTOF and NAD devices, and as such is deployed for independent measurement of the primary neutronic quantities. The gas-Cherenkov Gamma Reaction History (GRH) instrument records four energy channels of time-resolved gamma emission to measure nuclear bang time and burn width, as well as to infer carbon areal density in experiments utilizing plastic or diamond capsules. A neutron imaging system (NIS) takes two images of the neutron source, typically gated to create coregistered 13-15 MeV primary and 6-12 MeV downscattered images. The radiochemical analysis of gaseous samples (RAGS) instrument pumps target chamber gas to a chemical reaction and fractionation system configured with gamma counters, allowing measurement of radionuclides with half-lives as short as 8 seconds. Solid radiochemistry collectors (SRC) with backing NAD foils collect target debris, where activated materials from the target assembly are used as indicators of neutron spectrum content, and also serve as the primary diagnostic for nuclear forensic science experiments. Particle time-of-flight (PTOF) measures compression-bang time using DT- or DD-neutrons, as well as shock bang-time using D^3He-protons for implosions with lower x-ray background. In concert, these diagnostics serve to measure the basic and advanced quantities required to understand NIF experimental results.

Journal of Physics: Conference Series **717** (2016) 012117 doi:10.1088/1742-6596/717/1/012117

1. NIF Nuclear Diagnostics

The National Ignition Facility (NIF) relies on a suite of nuclear diagnostics to measure the neutronic output of experiments in inertial confinement fusion, high energy density physics, and fundamental sciences. These instruments are essential to understanding complex stagnation physics. The NIF Conceptual Design Review provided the framework for laser diagnostics [1] and target diagnostics [2], including substantial conceptual design of nuclear diagnostics [3], well in advance of anticipated full-scale fusion experiments. After commissioning, the NIF Execution Plan allowed for refinement of design [4],[5]. Subsequent full-scale ignition experiments [6] utilized these and became the physics and experiential basis [7] for the development described herein. New simulations have better diagnostic output capability [8], so better relation can be made between measurements and underlying physical quantities. Substantial evolution in capability is required to understand high convergence implosions: plasma conditions, signatures of alpha heating, and artifacts of fundamentally-3D effects.

1.1. Neutron Time-of-Flight (NTOF)

NTOF instruments measure the capsule neutron spectrum from three unique directions, with a fourth under construction, by amplifying the light output of a bibenzyl scintillator [9], [10]. The physical quantities of plasma temperature and scattering mass are inferred from the measured neutron spectrum: width and non-thermal content. To cover the widest range of yields and neutron energies, both precision collimated [11] and robust radiation-hard designs [12] were implemented [13]. Precision and yield ranges are being substantially improved [14]. Improved analysis capabilities include absolute timing, advanced spectrum fitting [15], and spectral moment analysis [16].

1.2. Neutron Activation Diagnostics (NADs)

NADs provide the basic performance metric of absolute neutron yield [17],[18],[19]. Additionally, spatially-distributed flange-mounted NADs (FNADs) measure, with nearly identical systematic uncertainties, primary DT neutron emission to infer a whole-sky neutron field [20]. Real-time NADs, based on a lanthanum bromide scintillator and compact gamma spectrometer installed in the target bay, are being developed to provide immediate postshot readout of yield in a modular and expandable integrated system design [21]. This system will have additional physics capability to measure multiple DT neutron reactions [22] and other neutron source terms [23].

1.3. Magnetic Recoil Spectrometer (MRS)

The MRS shares few systematics with NTOF and NADs, and as such is deployed as an independent measurement of the primary neutronic quantities [24],[25]. A deuterated plastic foil converts neutrons to deuterons, and subsequent magnetic fields bend particles to CR39 plastic detectors. Particle tracks are developed by etching and are analyzed for number and size to unfold the source neutron spectrum. The MRS measures fuel and ablator areal density (ρR) of the implosion [26], as well as detecting signatures of non-3D effects, such as bulk neutron velocity [27].

1.4. Gamma Reaction History (GRH)

The gas-Cherenkov Gamma Reaction History (GRH) instrument [28],[29],[30] records four energy channels of time-resolved gamma emission to measure nuclear bang time and burn width [31], as well as infer carbon ρR [32] in experiments utilizing plastic or diamond capsules. GRH utilizes Cherenkov radiation emission from Compton-converted electrons to detect high-energy gamma rays of nuclear origin. Forward-fit of the signal [33],[34] provides burn-averaged observables, including total DT fusion yield [35], total ρR, ablator ρR, and fuel ρR [36],[37]. A Mach-Zehnder recording system [38] meets the need for measurement of ultra-fast optical pulses [39] over large physical areas [40] necessitated by the NIF physical expanse and complicated target bay geometry. A new instrument with lowest-energy threshold at 1.8 MeV and high sensitivity, currently known as Super GCD (or GCD-3 at OMEGA), is being developed for use at OMEGA and NIF [41]. The Gamma-to-Electron Magnetic Spectrometer (GEMS) [42],[43] is designed to measure prompt γ-ray spectra during high-yield deuterium-tritium implosions when implemented [44], and will be capable of detecting the diagnostic signatures of alpha heating [45].

1.5. Neutron Imaging System (NIS) and Coregistered Neutron and X-ray Imaging (CNXI)
NIS takes two highly-resolved images of the neutron source, typically gated to create coregistered primary (13-15 MeV) and downscattered (6-12 MeV) images [46],[47] after single-image reconstruction [48]. A second neutron imaging system is being built to enable multiple views and eventually full 3D neutron source reconstruction [49]. CNXI measures neutrons and x rays on a common line of sight using an image plate stack with interstitial n-p converter foils [50].

1.6. Radiochemical Analysis of Gaseous Samples (RAGS)
RAGS pumps target chamber gas to a chemical reaction and fractionation system designed with pulse-counting gamma spectrometers, allowing measurement of radionuclides with half-lives as short as 8 seconds [51],[52]. Operation of the system takes place automatically after a NIF shot to collect, concentrate, and analyze noble gas samples for radioactive isotopic composition with efficiencies of 80-100%. Additionally, stable samples may be retrieved for mass spectroscopy. Measurements of fission product yields support nuclear forensic science experiments [53]. Capsule tracer studies may also be performed, probing ablator/fuel mix [54] and plasma stopping power [55].

1.7. Solid Radiochemistry Collectors (SRC)
SRCs with backing NAD monitor foils collect target debris [56]. Activated gold from the hohlraum is collected, chemically leached from collectors [57], and the ratio of ^{198}Au (low-energy capture) to ^{196}Au (high-energy n,2n) is used as a spatially-averaged measure of fuel and ablator ρR [58],[59]. They also serve as the primary diagnostic for nuclear forensic science experiments [60],[61]. Additionally, the backing NADs are used as a diagnostic of primary DT neutrons, downscattered neutrons, and neutrons above the primary DT neutron energy created by reactions of upscattered D and T in flight [62],[63].

1.8. Particle Time-of-Flight (PTOF)
PTOF uses a CVD diamond detector to measure the compression bang time using DT- or DD-neutrons [64],[65]. It also measures shock bang time using D^{3}He-protons [66] and ρR evolution [67] in implosions with lower x-ray background. A shielded detector using a bending magnet allows interrogation of more meaningful surrogate reactions [68].

References
[1] Kilkenny J D *et al.* 1995 *Rev. Sci. Instrum.* **66** 288
[2] Leeper R J *et al.* 1997 *Rev. Sci. Instrum.* **68** 868
[3] Moran M J and Hall J 1997 *Rev. Sci. Instrum.* **68** 521
[4] Murphy T J *et al.* 2001 *Rev. Sci. Instrum.* **72** 773
[5] Glebov V Yu *et al.* 2006 *Rev. Sci. Instrum.* **77** 10E715
[6] Lindl J D *et al.* 2014 *Phys. Plasmas* **21** 129902
[7] Kilkenny J D *et al.* 2016 *Fus. Sci. Tech.* **69** dx.doi.org/10.13182/FST15-173
[8] Clark D S *et al.* 2015 *Phys. Plasmas* **22** 022703
[9] Határik R *et al.* 2012 *Rev. Sci. Instrum.* **83** 10D911
[10] Határik R *et al.* 2015 *J. Appl. Phys.* **118** 184502
[11] Glebov V Yu *et al.* 2010 *Rev. Sci. Instrum.* **81** 10D325
[12] Moran M J 2012 *Rev. Sci. Instrum.* **83** 10D312
[13] Murphy T J *et al.* 2001 *Rev. Sci. Instrum.* **72** 850
[14] Grim G P *et al.* these proceedings
[15] Határik R *et al.* these proceedings
[16] Munro D H *J. Nucl. Fus.* accepted
[17] Barnes C W, Murphy T J and Oertel J A 2001 *Rev. Sci. Instrum.* **72** 818

[18] Bleuel D L *et al.* 2012 *Rev. Sci. Instrum.* **83** 10D313
[19] Cooper G W *et al.* 2012 *Rev. Sci. Instrum.* **83** 10D918
[20] Bleuel D L *et al.* 2013 *EPJ Web of Conferences* **59** 13015
[21] Jedlovec D R, Edwards E R, Carrera J A and Yeamans C B 2015 *Proc. SPIE* **9591** 95910H
[22] Yeamans C B, Bleuel D L and Bernstein L A 2012 *Rev. Sci. Instrum.* **83** 10D315
[23] Edwards E R, Jedlovec D R, Carrera J A and Yeamans C B these proceedings
[24] Frenje J A *et al.* 2010 *Phys. Plasmas* **17** 056311
[25] Casey D T *et al.* 2013 *Rev. Sci. Instrum.* **84** 043506
[26] Gatu Johnson M *et al.* 2014 *Rev. Sci. Instrum.* **85**, 11E104
[27] Gatu Johnson M *et al.* 2013 *Phys. Plasmas* **20** 042707
[28] Moran M J 1999 *Rev. Sci. Instrum.* **70** 1226
[29] Malone R M *et al.* 2008 *Rev. Sci. Instrum.* **79** 10E532
[30] Hoffman N M *et al.* 2010 *Rev. Sci. Instrum.* **81** 10D332
[31] Herrmann H W *et al.* 2010 *Rev. Sci. Instrum.* **81** 10D333
[32] Hoffman N M *et al.* 2013 *Phys. Plasmas* **20** 042705
[33] Labaria G R *et al.* 2013 *Proc. SPIE* **8602** 86020C
[34] Sayre D B *et al.* 2012 *Rev. Sci. Instrum.* **83** 10D905
[35] Kim Y *et al.* 2012 *Phys. Plasmas* **19** 056313
[36] Hoffman N M *et al.* 2013 *EPJ Web of Conferences* **59** 13019
[37] Cerjan C *et al.* 2015 *Phys. Plasmas* **22** 032710
[38] Miller E K *et al.* 2012 *Rev. Sci. Instrum.* **83** 10D719
[39] Milnes J S *et al.* 2012 *Rev. Sci. Instrum.* **83** 10D301
[40] Milnes J S, Horsfield C J, Conneely T M and Howorth J 2014 *Rev. Sci. Instrum.* **85** 11E601
[41] Herrmann H W *et al.* 2014 *Rev. Sci. Instrum.* **85** 11E124
[42] Kim Y *et al.* 2012 *Rev. Sci. Instrum.* **83** 10D311
[43] Kim Y *et al.* 2014 *Rev. Sci. Instrum.* **85** 11E122
[44] Herrmann H W *et al.* these proceedings
[45] Church J A *et al.* these proceedings
[46] Sweezy J E *et al.* 2006 *Rev. Sci. Instrum.* **77** 10E722
[47] Merrill F E *et al.* 2012 *Rev. Sci. Instrum.* **83** 10D317
[48] Volegov P *et al.* 2014 *Rev. Sci. Instrum.* **85** 023508
[49] Fittinghoff D N *et al.* 2015 *Proc. SPIE* **9591** 95910E
[50] Merrill F E *et al.* 2014 *Rev. Sci. Instrum.* **85** 11E614
[51] Stoyer M A *et al.* 2012 *Rev. Sci. Instrum.* **83** 023505
[52] Shaughnessy D A *et al.* 2012 *Rev. Sci. Instrum.* **83** 10D917
[53] Cassata W S *et al. J. Radioanal. Nucl. Chem.* accepted
[54] Colvin J, Cerjan C, Hoffman R, Stoyer M and Amendt P 2008 *Phys. Plasmas* **15** 102704
[55] Hayes A C *et al.* these proceedings
[56] Gostic J M *et al.* 2012 *Rev. Sci. Instrum.* **83** 10D904
[57] Grant P M *et al.* 2015 *J. Radioanal. Nucl. Chem.*, v. **303**, 1851
[58] Shaughnessy D A *et al.* 2014 *Rev. Sci. Instrum.* **85** 063508
[59] Hagmann C *et al.* 2015 *Rev. Sci. Instrum.* **86** 076105
[60] Gharibyan N *et al.* 2015 *J. Radioanal. Nucl. Chem.* **303** 1335
[61] Shaughnessy D A *et al.* these proceedings
[62] Grim G P *et al.* 2014 *Proc. SPIE* **9211** 921108
[63] Hayes A C *et al.* 2015 *Phys. Plasmas* **22** 082703
[64] Rinderknecht H G *et al.* 2012 *Rev. Sci. Instrum.* **83** 10D902
[65] Schmid G J *et al.* 2003 *Rev. Sci. Instrum.* **74** 1828
[66] Sio H W *et al.* these proceedings
[67] Rinderknecht H G *et al.* these proceedings
[68] Rinderknecht H G *et al.* 2014 *Rev. Sci. Instrum.* **85** 11D901

Development of a WDM platform for charged-particle stopping experiments

AB Zylstra[1,2], JA Frenje[1], P E Grabowski[3], C K Li[1], G W Collins[4], P Fitzsimmons[5], S Glenzer[6], F Graziani[4], S B Hansen[7], S X Hu[8], M Gatu Johnson[1], P Keiter[9], H Reynolds[5], J R Rygg[4], F H Séguin[1] and R D Petrasso[1]

[1] Plasma Science and Fusion Center, Massachusetts Institute of Technology, Cambridge, MA 02139; [2] Los Alamos National Laboratory, Los Alamos, NM 87544; [3] University of California Irvine, Irvine, CA 92697; [4] Lawrence Livermore National Laboratory, Livermore, CA 94550; [5] General Atomics, San Diego, CA 92186; [6] SLAC National Accelerator Laboratory, Menlo Park, CA 94025; [7] Sandia National Laboratories, Albuquerque, NM 87185; [8] Laboratory for Laser Energetics, Rochester, NY 14623; [9] University of Michigan, Ann Arbor, MI 48109

E-mail: zylstra@lanl.gov

Abstract. A platform has been developed for generating large and relatively quiescent plasmas in the warm-dense matter (WDM) regime on the OMEGA laser facility. A cylindrical geometry is used to allow charged-particle probing along the axis. The plasma heating is radiative by L-shell emission generated on the outside of the cylinder. The cylinder drive is characterized with x-ray diagnostics. Possibilities for direct characterization of the plasma temperature are discussed. Finally, the unimportance of electromagnetic fields around the target is demonstrated with proton radiography. We expect this platform to be used extensively in future experiments studying charged-particle stopping in this regime.

1. Introduction

The stopping of charged particles has been studied extensively over the last century, in both cold material[1], and in plasmas[2]. Plasma theory is challenging in high-density regimes, where the Coulomb logarithm becomes small, and degeneracy and/or strongly-coupled plasma effects can become important. However, several theories have been developed[3, 4, 5]. The stopping power in this regime is a fundamental test of dense plasma physics, and relevant to the realization of laboratory-scale thermonuclear fusion, where α bootstrap heating occurs via charged-particle stopping. Several recent novel experiments have measured charged-particle stopping in dense plasmas[6, 7, 8]. In this paper we present a platform for generating a large volume of quiescent WDM plasma for use in stopping experiments, as in Ref. [6].

2. Experimental design

The OMEGA laser facility[9] was used to create both a source of protons and the subject plasma described in this work, as shown in Fig. 1. Using exploding pusher targets as a proton source is discussed in the literature[10, 6]. The WDM subject target consists of a tube coated with x-ray generating material, with a cylinder of the desired subject material inserted into the tube. The dimensions and materials depend on the experimental goal. In this first work, a plastic (CH)

Journal of Physics: Conference Series **717** (2016) 012118 doi:10.1088/1742-6596/717/1/012118

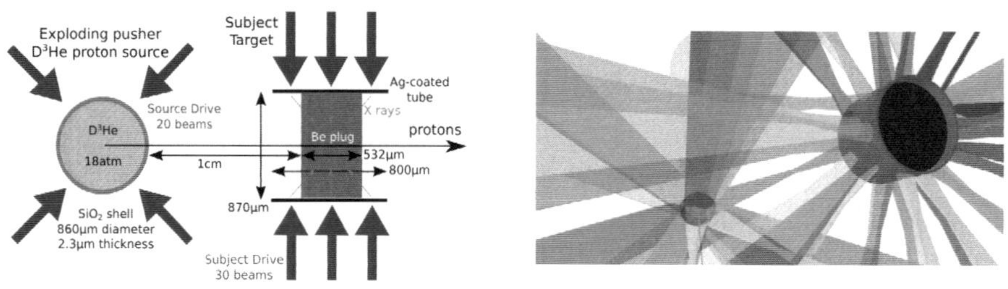

Figure 1. Experimental geometry: cartoon (left) and Visrad CAD model (right). A D^3He proton source, at left, creates protons to probe the subject plasma, at right, which is created by isochorically heating a solid plug with x rays.

tube coated with $1-2$ μm of Ag, with an inner diameter of 870μm, a wall thickness of 24 μm, and 800 μm in length was used with a Be plug; for additional details see Ref. [6].

The arrangement of drive beams on the cylinder target must be chosen to optimally distribute the laser energy in the cylinder's azimuthal and axial directions. In this experiment, 30 OMEGA beams are available to heat the subject target. The chosen design arranges these beams into three rings of 10 beams each, which are pointed to achieve azimuthal symmetry. The three rings are displaced for axial symmetry. The chosen design is shown in Fig. 2, plotting the incident laser intensity versus z and ϕ in the cylinder target's coordinate system.

Each beam is run without DPP and defocused to a $\sim$ 100 μm diameter spot. The three rings are clearly visible at $z\sim-250$, 0, and 250 μm. The axial displacement for the first and last rings were chosen to match the Be plug dimensions (500 μm nominal total thickness). The final laser intensity in the spots is very high ($2-4\times10^{15}$ W/cm^2), which is desirable for generation of the Ag L-shell photons at 3-4keV in the corona surrounding the cylindrical target, which volumetrically heats the Be plug as the attenuation length in solid Be is $300-500$ μm, comparable to the cylinder's dimensions. The total drive energy was 15kJ in a 1ns square pulse.

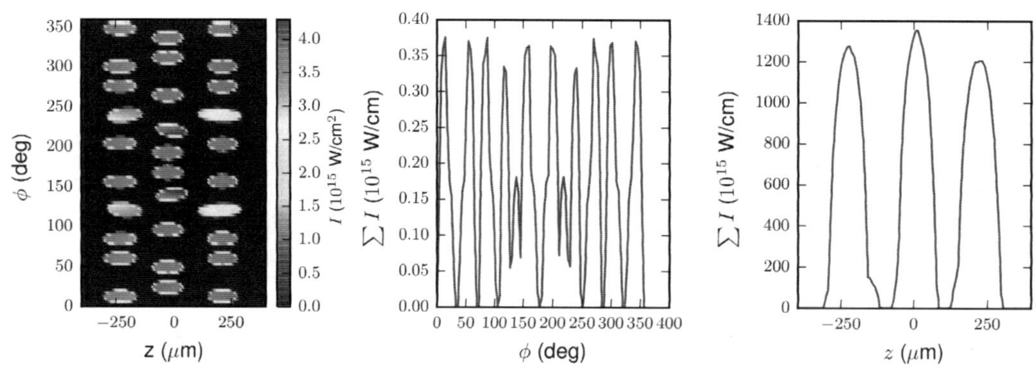

Figure 2. Laser drive used on the subject cylinder: surface intensity map (left), azimuthal energy distribution (center), and axial energy distribution (right).

3. X-ray characterization

Several instruments were used to diagnose the cylinder's x-ray emission. In particular, framing camera (XRFC) data and streaked spectrometer (SSCA) data were taken to qualitatively diagnose the spatial and temporal emission. The XRFC data is shown in Fig. 3. The emission in time clearly lasts approximately 1ns, as expected from the laser drive. Early in time (top strip) the emission is dominated by the laser spots on the cylinder's surface. Later in time (strips 2 and 3) the emission region clearly increases, as the emission becomes more dominated by the hot coronal plasma surrounding the cylinder.

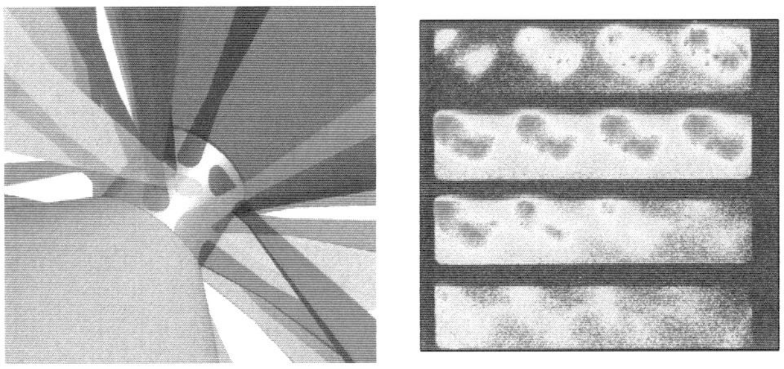

Figure 3. Left: XRFC view from TIM 1. Right: XRFC data as $\log I$ from shot 72016. A four-strip camera was used with 12× magnification. Time increases left to right and top to bottom. The time spanned by a single strip is 200ps, and the timing between strips is 400ps.

The SSCA data are shown in Fig. 4. In the raw data the energy dispersion direction is vertical, while time increases to the right (streak direction). This data can be qualitatively analyzed by looking at the emission profile in time of one of the bright lines, and looking at the spectrum for a single time in the middle of the pulse, as shown in the two plots in Fig. 4. The emission time history clearly shows that most of the Ag L-shell emission occurs early in the laser pulse. 1-D HYADES simulations indicate that the $1 - 2$ μm thick Ag layer burns through at around 0.5ns, corresponding to the rapid decrease in L-shell emission. It was expected that the coronal plasma would be hot enough to continue L-shell line emission, but the HYADES simulations suggest that the corona may be too hot - simulated electron temperatures reach $6 - 10$ keV towards the end of the pulse, which would mean that the Ag will be ionized to a He-like state. Future experiments could use a thicker Ag layer or a shorter laser pulse.

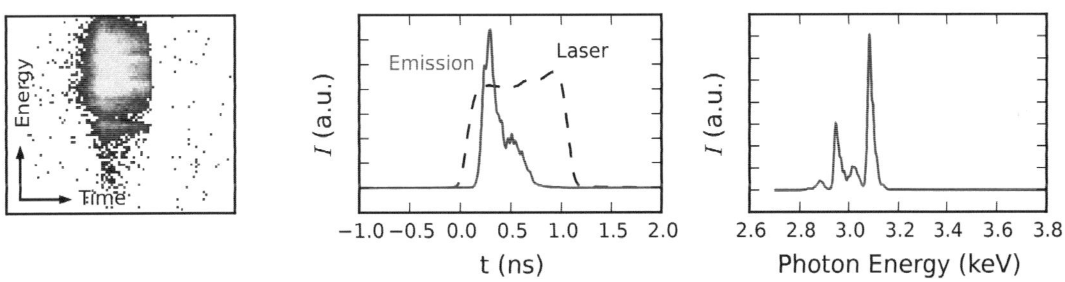

Figure 4. Left: Streaked spectrometer (SSCA) data from shot 72028 as background-subtracted $\log I$. Corresponding temporal emission profile (center) and time-snapshot spectrum (right) are plotted. The time and energy axes are only approximately calibrated, due to a lack of absolute calibration for this instrument.

4. Temperature measurements

Ideally the plasma conditions would be measured directly with a surrogate stopping power target. This was attempted with imaging x-ray Thomson scattering using the IXTS diagnostic[11]. We observed an unexpectedly high background originating from the cylinder heating, i.e. without the Thomson scattering x-ray source. This resulted in a very low signal-to-background in the scattering region, which precludes any interpretation of this data. Future experiments using this platform may be able to obtain high-quality x-ray Thomson scattering data by using better collimating shields. Another route that is being explored is dopant absorption spectroscopy.

5. Proton radiography

The implosion proton source was timed so that the proton probing occurs at 1.4 ns, which prevents target electrostatic charging effects[12] while preserving the temperature. The absence of target charging at the proton probing time was verified using proton radiography[10], as shown in Fig. 5. The expected view (left) is generated from a Visrad model of the experiment. Due to the use of larger implosion targets (see Fig. 1), the spatial resolution is degraded relative to standard D^3He proton backlighting[10]. However, the cylinder target with large ρL generates a central region without proton fluence, since protons transiting the cylinder are ranged out in the detector filtering. The expected cylinder size from geometric magnification is shown by the red dashed circle superimposed on the radiograph. The diameter of the image is consistent with the geometric magnification, demonstrating that field effects are not present. In the corona surrounding the driven cylinder, magnetic fields generated by the Biermann battery ($\nabla n \times \nabla T$) or laser-plasma interaction instabilities are present. While the electric charging of the target decays rapidly after the laser turns off[12], these magnetic fields can persist in the low-density coronal plasma, and show up in the radiograph as quasi-filamentary structures around the periphery of the image. These fields will not affect the stopping power measurement.

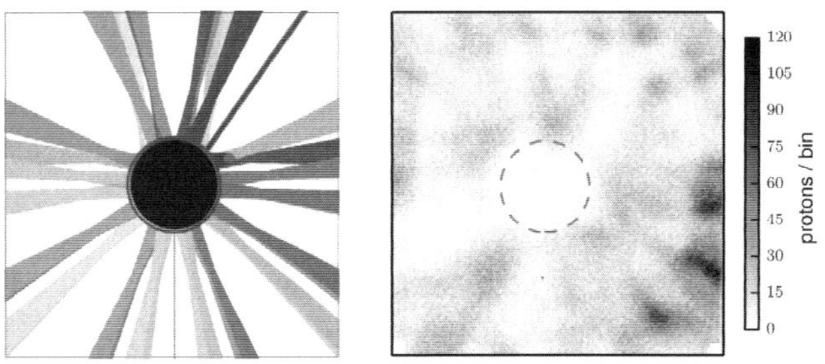

Figure 5. Proton radiography from shot 72028. Left: Radiography field of view from Visrad model, right: measured D^3He-p radiograph. The expected geometric size of the cylinder is shown by the red dashed circle.

6. Conclusion

A platform has been developed to generate a large volume of semi-quiescent WDM plasma, with a primary application being charged-particle stopping power experiments. 30 of the 60 OMEGA laser beams are used to drive a cylinder, which isochorically and indirectly heats the inner sample via high-energy x rays. X-ray diagnostics of the cylinder are presented showing that the spatial, temporal, and spectral characteristics of the drive are as expected. Finally, proton radiography is used to demonstrate that there are no electromagnetic fields impacting stopping-power measurements taken along the axis.

This work was supported by the U.S. DoE, LLNL, the Fusion Science Center at the University of Rochester, the National Laser User's Facility, and the NSF.

References

[1] Ziegler J, Biersack J and Littmark U 1985 (Pergamon, New York)
[2] Trubnikov B 1965 *Particle Interactions in a Fully Ionized Plasma* (Consultant's Bureau)
[3] Li C and Petrasso R 1993 *Phys. Rev. Lett.* **70** 3059–3062
[4] Maynard G and Deutsch C 1985 *J. Physique* **46** 1113–1122
[5] L S Brown, D L Preston, and R L Singleton Jr 2005 *Physics Reports* **410** 237 – 333
[6] Zylstra A B *et al.* 2015 *Phys. Rev. Lett.* **114**(21) 215002
[7] Frenje J A *et al.* 2015 *Phys. Rev. Lett.* **115**(20) 205001
[8] Hayes A C *et al.* 2015 *Physics of Plasmas* **22** 082703
[9] Boehly T *et al.* 1997 *Optics Communications* **133** 495–506
[10] Li C K *et al.* 2006 *Rev. Sci. Instrum.* **77** 10E725
[11] Gamboa E J *et al.* 2012 *Rev. Sci. Instrum.* **83**(10) 10E108
[12] Hicks D G *et al.* 2000 *Phys. Plasmas* **7** 5106–5117

A pathway to laser fusion energy in Japan

Hiroshi AZECHI
Institute of Laser Engineering, Osaka University, Osaka Japan.
Email: azechi@ile.osaka-u.ac.jp

Abstract. High-density compression of DT to one thousand times its liquid density is the critical path of inertial fusion and was demonstrated in Japan and US in late 1980's. The Osaka group has achieved high-density compression that meets one of the critical requirements for thermonuclear ignition and burn. Although the compression densities were well reproduced by computer simulations, the neutron yields were much lower than the simulation predictions by three orders of magnitudes, suggesting catastrophic collapse of a hot spark, from which thermonuclear reactions are triggered. In order to overcome this difficulty the international ICF community has adopted two approaches: one is to generate a larger hot spark than the mixed layer with MJ-Class lasers, such as NIF and LMJ. The other approach is to externally heat the compressed fuel. The second approach is the fast ignition. After the proof-of-concept experiment in 2002, we started the Fast Ignition Realization Experiment (FIREX) project to complete the world most powerful high-energy peta-watt laser "LFEX" as a heating laser.

1. Introduction

In the past half-century, being stimulated by the laser fusion concept by J. Nuckolls [1], various large-scale programs took off throughout the world aiming at ultimate goal of thermonuclear ignition. Namely, OMEGA (original) [2], GEKKO-XII [3], and NOVA [4] in 1980's, followed by OMEGA (upgrade) [5], National Ignition Facility (NIF) [6], Laser Mega Joule (LMJ) [7], and more recently high energy peta-watt lasers, EP [8], LFEX [9], ARC [10] and PETAL [11]. Chinese Shenguang series [12] have also joined in and contributed to the world community. It appears, however, that start of any next laser fusion program depends on the success of ignition at NIF. In this article, in order to make good use of the next generation programs, I would like to review the key elements of successes and failures in the past and present programs along the laser fusion history in Japan.

2. High-density Compression

In early 1970's, J. Nuckolls released a concept of inertial fusion [1], indicating that scientific breakeven could be achieved with only a kJ of laser energy if the fusion fuel is compressed to ten thousand times liquid density. Meanwhile, pessimists suspected such a rosy view in the earliest times saying that at high laser intensity necessary for fusion, plasma becomes collision-less, resulting in low laser absorption. This pessimism was dispelled by anomalous absorption found by C. Yamanaka [13], K. Nishikawa [14] and many other activities, starting full-scale investigation in the world. (The late study clarified anomalous absorption as to be eliminated, but its importance is taken over by "anomalous absorption conference", one of the most importance conferences.)

From the early to mid 1980's, new laser fusion facilities were built one after another, such as OMEGA at the University of Rochester [2], GEKKO-XII at Osaka University [3], and NOVA at Lawrence Livermore National Laboratory [4]. Development of the GEKKO laser facilities is schematically illustrated in Figure 1. The main target in early experiments was the realization of high temperature required for nuclear fusion, and the resultant high neutron yield attracted public attention as a symbol of nuclear fusion. The University of Rochester set the world record of neutron yield to reach $1x10^{10}$. After that, Osaka University increased the yield to over $1x10^{12}$ by the implosion mode called stagnation free [15]. The result appeared as a cover story of Nature. Furthermore, the University successfully achieved $1x10^{13}$ neutrons with a fuel temperature of 100 million degrees required for nuclear fusion through optimization of the implosion mode [16]. The Lawrence Livermore National Laboratory recorded twice the number of neutrons with the same scheme but larger laser energy. (The record was broken ten years later when OMEGA (upgrade) achieved $1x10^{14}$ neutron yield.)

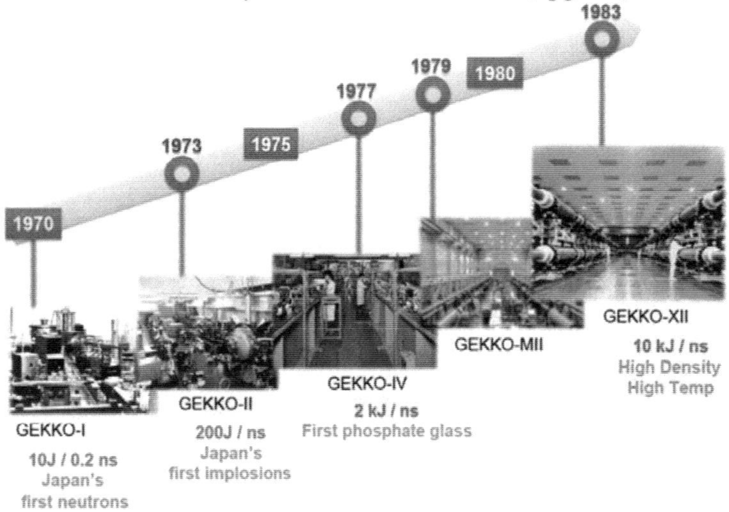

Figure 1. Development of GEKKO laser series.

In the late 1980's, the research focus shifted from high temperature to high-density compression. C. Yamanaka formed three research teams with the same goal but different approach, stimulating intensive race and peer-review research. Among them, three key technologies were developed: uniform CD (deuterated polystyrene) shells made with density-matched emulsion technique developed by M. Takagi [17], random phase plates for uniform laser irradiation by Y. Kato [18], and a wealth of nuclear diagnostics, including secondary fusion reactions [19] for fuel ρR and mix measurements and scintillation streak camera [20] for burn time and duration measurements. Achievements of 100-200 times liquid density (XLD) by R. McCrory [21] and 600 XLD by H. Azechi and N. Miyanaga [22] became epoch-making experiments which verified the formation of a main fuel—one of two requirements of laser fusion. (The record was broken at OMEGA and NIF in 2010's [23, 24].) Although the compression densities were well reproduced by computer simulations, the neutron yields were much lower than the simulation predictions by three orders of magnitudes, suggesting catastrophic collapse of a hot spark, from which thermonuclear reactions are triggered. This collapse occurred even though the Rayleigh-Taylor instability is significantly moderated by material ablation that removes the perturbation away from the unstable surface. The quantitative ablative stabilization was first experimentally verified [25].

So it requires extremely high spherical symmetry to form a hot spark at the centre of the fuel. The direct drive scheme in which a fuel target is directly irradiated by laser has difficulty in forming a hot spark because of the limitation of the number of beams and lack of irradiation uniformity. Stimulated by random phase plates, various technologies of beam smoothing were developed one after another, including the induced spatial incoherence (ISI) of Naval Research Laboratory [26], the smoothing by spectral dispersion (SSD) of University of Rochester [27], and the partially coherent light (PCL) technology of Osaka University [28], lowering residual non-uniformity to only 1% level or less.

On the other hand, the indirect-drive scheme in which laser beams are converted into X-rays driving a fuel capsule to implode is superior in irradiation smoothing over the direct-drive scheme, so the United States had focused its main effort on the indirect-drive scheme. In the early 1990's, they succeeded in hot spark formation, albeit at the expense of energy efficiency injected into a fuel target [29]. The National Academy of Science vigorously reviewed a sequence of results, and issued the famous Koonin Report [30] stating that 11 out of 12 assignments which are conditions for starting the ignition program have been achieved, and because the only failed condition of forming a hot spark under similar conditions to ignition is difficult to realize with the current experiment facility, this failure is not a valid excuse for delaying the program. In response to this, construction of NIF was started and commissioned. The similar LMJ project is in progress in France. Figure 2 illustrates that high temperature and high density demonstrations provided scientific bases of these ignition programs.

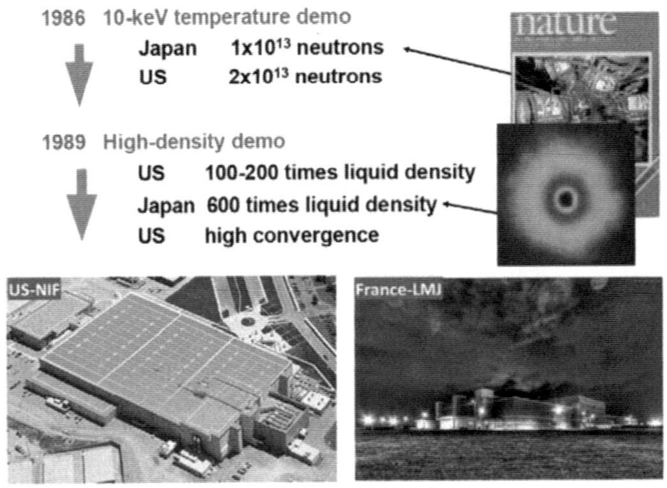

Figure 2. High temperature and high density demonstrations in 1980's provided scientific bases of ignition programs.

The times from the early to mid 1990's when there were discussions on construction of NIF overlapped with the period when science and technology policy was seriously reviewed. The US quit underground testing after Cold War, and in its place, established the Science Based Stockpile Stewardship [31] in which nuclear forces are maintained and managed through virtual simulation experiments. The NIF is regarded as a multi-purpose research facility to promote nuclear fusion energy development and fundamental science, as well as a facility to clarify the equation of state and the opacity for X-rays of material under high-temperature and high-density conditions.

In passing, it was 1980, in the midst of Cold War, when the indirect-drive scheme was proved to be classified. The review itself of "Cannonball Target" proposed by H. Azechi independently from the indirect-drive scheme was refused by Physical Review Letters for security reasons [32]. Subsequently Max Planck Institute for Quantum Optics in Germany and Osaka University published their experimental results of indirect drive [33], making weaker the validity of them being classified, and in the end of 1993, reflecting the end of the Cold War, most of the research results were declassified on the grounds that there was no concern for nuclear proliferation.

3. Birth of the Concept of Fast Ignition

During this period, researchers also sought for a more advanced ignition scheme and proposed a new ignition idea in the early 1990's that a fuel is imploded in high density beforehand, and another laser is injected to the implosion core to heat fuel before hydrodynamic disassembly [34]. This scheme is called "fast ignition". Fuel plasma used for fast ignition has a solid structure, so the scheme can obtain a larger gain with lower laser energy as compared to the central ignition. Russian group also proposed similar idea [35].

Here, let me describe the story of the birth of fast ignition. A brainstorming meeting on a new laser fusion scheme was held in 1983 before operation of GEKKO-XII, and T. Yamanaka proposed a concept of "additional heating" using fast electrons generated by picosecond laser [36]. This proposal was almost the same as the current fast ignition, but the realization of the idea required "future" invention of an ultrashort pulse laser technology. The proposal was not published but only printed as an internal document. At the end of 1980's, concept of additional heating, high density fuel to be heated, and high energy short pulse laser were all in the Laboratory. However, it was an American researcher who combined them and proposed the idea of fast ignition for the first time. This incident caused Japanese researchers to seriously think about the ability to seek for a fundamental solution of a problem by calling upon a great amount of insight from different people.

4. Progress of Fast Ignition Study

In the mid 1990's, with the completion of OMEGA (upgrade) and approval of construction of NIF, the US inertial fusion community shifted from competition to a new era of unification. In light of this situation, K. Mima of Osaka University immediately upon the appointment as a director decided to focus the laboratory on fast ignition and constructed GEKKO-PW peta-watt laser. European research institutes also started basic research of fast ignition. The US, to the contrary, shut down NOVA-PW laser [37] that could provide peta-watt outputs and concentrated on NIF

The first task of the fast ignition study was to make a laser reach the high density fuel neighborhood. Hole boring that creates a laser propagation path by removing corona plasma using light pressure, or relativistic anomalous penetration, were experimentally verified one by one [38]. It was learned that laser propagation in plasma accompanies energy dispersion, and to avoid it, R. Kodama and K. Mima of Osaka University set a perspicacious little hook. They attached a hollow cone to the fuel ball to inject heating laser from the inside of the cone. The fuel temperature reached about 1 keV in an experiment conducted in 2002, resulting in the neutron-yield increased by three orders of magnitude as compared to a case without heating [39].

5. Fast Ignition Realization Experiment Program (FIREX)

Since high-density compression as high as 600 times liquid density and heating to 1 keV temperature have been achieved, the next milestone for the research became realization of fusion ignition and burn. This program is called Fast Ignition Realization Experiment (FIREX).

The FIREX program is separated into two stages: in the first stage, a compressed fuel will be heated with a heating laser (10 kJ/10 ps) to the ignition temperature of about 5 keV. At this stage, alpha particles escape from the fuel without significant heating, causing no ignition. In the subsequent second stage, lasers for compression and heating are both upgraded to 50-100 kJ level and thereby fuel plasma larger than the range of an alpha particle can be generated and fusion ignition and burn is expected. Because inertial confinement time becomes longer according to increase in laser power and fuel size, high fusion gain required for a commercial reactor can be obtained. It is expected that there will be no serious problems in a step from ignition and burn to high fusion gain, so the research on core plasma physics is thought to be completed at the FIREX stage.

At this time, there was a strong need to reorganize fusion activities to invite ITER program into Japan. A fusion working group was established in the Council for Science and Technology with T. Yamanaka, H. Azechi, and K. Mima as members. After two years serious discussion, the Council submitted a report titled "Future Direction of National Fusion Research" saying that FIREX will expand the possibility of realization of fusion energy through a scheme qualitatively different from the magnetic fusion scheme by applying the latest laser technology and science in an extreme state, and the first stage of FIREX will be one of major projects to be prioritized by the nation [40]. The increase of heating laser energy in the past quarter century is illustrated in Figure 3.

Our recent studies identify three scientific challenges to achieve high heating efficiency in the fast ignition scheme with the current GEKKO and LFEX laser systems: (i) control of relativistic electron beam (REB) energy distribution, (ii) guiding and focusing of REB to a fuel core, and (iii) formation of a high areal-density core [41]. The control of the electron energy distribution has been experimentally confirmed by improving the intensity contrast of the LFEX laser up to $>10^9$ and an ultra-high contrast of 10^{11} with plasma mirrors. After the contrast improvement, 50% of the total REB energy is carried by a low energy component of the REB, whose slope temperature is close to the ponderomotive scaling value (~1 MeV). To guide the electron beam, we apply strong external magnetic field to the REB transport region [42-44]. Guiding of the REB by 0.6 kT field in a planar geometry has already been demonstrated at LULI 2000 laser facility, in a collaborative experiment lead by CELIA-Univ. Bordeaux [45]. Considering more realistic fast ignition scenario, we have performed a similar experiment using the kilo-Joule LFEX laser to study the effect of guiding and magnetic mirror on the electron beam. A high density core of a laser-imploded 200 μm-diameter solid CD ball was radiographed with picosecond LFEX-produced K-alpha backlighter. Comparisons of the experimental results and integrated simulations using hydrodynamic and electron transport codes suggest that 10% of the efficiency can be achievable with the current GEKKO and LFEX laser system [46].

Figure 3. Increase of heating laser energy

6. Lessons Learnt from the Previous Programs

To summarize the lessons from the previous programs, we follow three critical elements for success proposed by Mencius, ancient Chinese philosopher, that is, 1) Heavenly Timing, 2) Geographical Advantage, and 3) Harmony among People. The lessons we could learn from GEKKO-XII are 1) Heavenly Timing: Oil crisis in 1973 and 1979 urged political and financial support; 2) Geographical Advantage: good distance from the policy-making capital; and 3) Harmony among People: there was a director's strong leadership. Contrary, at the time of the high neutron yield and high-density compression, the fusion committee in Japan had already selected a large magnetic fusion machine, Large Helical Device (LHD), as the next fusion facility, and as a result, our proposal of the ignition project, KONGOH, finally failed to take off.

The lessons we have learned from FIREX-I are 1) Heavenly Timing: very narrow window opened between LHD completion and ITER cite decision; 2) Geographical Advantage: technical data of fast ignition were just in time for the Committee discussion; and 3) Harmony among People: Data, committee discussion, and new proposal were coherently made, even without charismatic leadership.

7. Next Program

The current situation is characterized as disappearance of Cold War and Oil Crisis that had strongly driven large-scale science. In light of this situation, we have reviewed our next generation program of FIREX-II to be conducted as an international one. In parallel to the ignition program, we set another program using a repetitive multi kJ laser. This program would result in considerably good understanding of fusion science, as well as basic high energy-density science and industrial applications.

References

[1] J. Nuckolls *et al.*, Nature **239**, 139 (1972)
[2] J. Bunkenberg *et al.*, IEEE J. Quantum Electron. **QE-17**, 1620 (1981).
[3] C. Yamanaka *et al.*, Phys. Rev. A **6**, 2335 (1972)..
[4] J.T. Hunt and D.R. Speck: Optical Engineering **28**, 461 (1989).
[5] T. R. Boehly *et al.*, Opt. Commun. **133**, 495 (1997).
[6] G. H. Miller, E. I. Moses, and C. R. Wuest, Nucl. Fusion **44**, 228 (2004).
[7] M.L. Andre, Fusion Engineering and Design **44**, 43 (1999).
[8] J.H. Kelly *et al.*, Journal de Physique IV **133**, 75 (2006).
[9] N. Miyanaga *et al.*, Journal De Physique IV **133**, 81 (2006).
[10] Tomás Díaz, Science and Technology Review **12**, 12 (2011).
[11] A. Casner *et al.*, High Energy Density Physics **17**, 2 (2015).
[12] H. Zhang *et al.*, Phys. Plasmas **21**, 093401 (2014).
[13] C. Yamanaka *et al.*, Phys. Rev. A **6**, 2335 (1972).
[14] K. Nishikawa: J. Phys. Soc. Jpn. **24**, 916 (1967), *ibid* 1152 (1967).
[15] C. Yamanaka and S. Nakai, Nature **319**, 757 (1986).
[16] H. Takabe *et al.*, Phys. Fluids **31**, 2884 (1988).
[17] M. Takagi *et al.*, J. Vac. Sci. Technol. **A9**, 2145 (1991).
[18] Y. Kato *et al.*, Phys. Rev. Lett. **53**, 1057 (1984).
[19] H. Azechi *et al.*, Appl. Phys. Lett. **49**, 555 (1986); H. Azechi *et al.*, Phys. Rev. Lett. **23**, 2635 (1987).
[20] H. Azechi *et al.*, Appl. Phys. Lett. **55**, 945 (1989).
[21] R. McCrory *et al.*, Nature **335**, 225 (1988).
[22] H. Azechi *et al.*, Laser and Part. Beams **9**, 193 (1991).
[23] V. Goncharov *et al.*, Phys. Rev. Lett. **104**, 165001 (2010).
[24] A. J. Mackinnon *et al.*, Phys. Rev. Lett. **108**, 215005 (2012).
[25] H. Azechi *et al.*, Phys. Rev. Lett. **98**, 045002 (2007).
[26] R. H. Lehmberg and J. Goldhar, Fusion Tech. **11**, 532 (1987).
[27] S. Skupsky *et al.*, J. Appl. Phys. **66**, 3456 (1989).
[28] H. Nakano *et al.*, J. Appl. Phys. **73**, 2122 (1993).
[29] M.D. Cable *et al.*, Phys. Rev. Lett. **73** 2316 (1994).
[30] http://www.nap.edu/catalog/5730.html
[31] R. Jeanloz, Physics Today **53**, 44 (2000).
[32] This was finally published in H. Azechi *et al.*, Jpn. J. Appl. Phys. **20**, L477 (1981).
[33] R. Sigel *et al.*, Phys. Rev. A **38**, 5779 (1988); H. Nishimura *et al.*: Phys. Rev. A **43**, 8323 (1991).
[34] M. Tabak *et al.*, Phys. Plasmas **1**, 1626 (1994).
[35] N. Basov, *J. Sov. Laser Res.* **13**, 396 (1992).
[36] T. Yamanaka, *Enhanced heating by psec laser pulse*, Experimental Proposals for Kongoh Project Phase I, Institute of Laser Engineering Internal Report, Osaka University, 5-6 (1983).
[37] M.D. Perry, SPIE Conf. on Generation, Amplification and Measurement of Ultra-short Laser Pulses III, 28-30 Jan. 1996, San Diego.
[38] K.A. Tanaka *et al.*, Phys. Plasmas **7**, 2014 (2000); R. Kodama *et al.*, *ibid*. **8**, 2268 (2001).
[39] R. Kodama *et al.*, Nature **412**, 798 (2001); *ibid*. **418**, 933 (2002).
[40] http://www.mext.go.jp/b_menu/shingi/gijyutu/gijyutu4/toushin/1213875.htm
[41] S. Fujioka *et al.*, Phys. Rev. E **91**, 063102 (2015).
[42] D. Strozzi *et al.*, Phys. Plasmas **19**, 072711 (2012).
[43] S. Fujioka *et al.*, Sci. Rep. **3**, 1170 (2013).
[44] T. Johzaki *et al.*, Nucl. Fusion **55**, 053022 (2015).
[45] M. Bailly-Grandvaux *et al.*, submitted to Nat. Phys.
[46] S. Fujioka *et al.*, Phys. Plasmas **23**, 056308 (2016).